# Fatigue and Durability of Structural Materials

S.S. Manson

G.R. Halford

ASM International®
Materials Park, Ohio 44073-0002
www.asminternational.org

Copyright © 2006
by
ASM International®
All rights reserved

No part of this book may be reproduced, stored in a retrieval system, or transmitted, in any form or by any means, electronic, mechanical, photocopying, recording, or otherwise, without the written permission of the copyright owner.

First printing, March 2006
Second printing, May 2007

Great care is taken in the compilation and production of this book, but it should be made clear that NO WARRANTIES, EXPRESS OR IMPLIED, INCLUDING, WITHOUT LIMITATION, WARRANTIES OF MERCHANTABILITY OR FITNESS FOR A PARTICULAR PURPOSE, ARE GIVEN IN CONNECTION WITH THIS PUBLICATION. Although this information is believed to be accurate by ASM, ASM cannot guarantee that favorable results will be obtained from the use of this publication alone. This publication is intended for use by persons having technical skill, at their sole discretion and risk. Since the conditions of product or material use are outside of ASM's control, ASM assumes no liability or obligation in connection with any use of this information. No claim of any kind, whether as to products or information in this publication, and whether or not based on negligence, shall be greater in amount than the purchase price of this product or publication in respect of which damages are claimed. THE REMEDY HEREBY PROVIDED SHALL BE THE EXCLUSIVE AND SOLE REMEDY OF BUYER, AND IN NO EVENT SHALL EITHER PARTY BE LIABLE FOR SPECIAL, INDIRECT OR CONSEQUENTIAL DAMAGES WHETHER OR NOT CAUSED BY OR RESULTING FROM THE NEGLIGENCE OF SUCH PARTY. As with any material, evaluation of the material under end-use conditions prior to specification is essential. Therefore, specific testing under actual conditions is recommended.

Nothing contained in this book shall be construed as a grant of any right of manufacture, sale, use, or reproduction, in connection with any method, process, apparatus, product, composition, or system, whether or not covered by letters patent, copyright, or trademark, and nothing contained in this book shall be construed as a defense against any alleged infringement of letters patent, copyright, or trademark, or as a defense against liability for such infringement.

Comments, criticisms, and suggestions are invited, and should be forwarded to ASM International.

*Prepared under the direction of the ASM International Technical Books Committee (2004–2005), Yip-Wah Chung, FASM, Chair.*

*ASM International staff who worked on this project include Scott Henry, Senior Product Manager; Bonnie Sanders, Manager of Production; Madrid Tramble, Senior Production Coordinator; and Kathryn Muldoon, Production Assistant.*

Library of Congress Cataloging-in-Publication Data
Manson, S.S.
Fatigue and durability of structural materials / S.S. Manson, G.R. Halford.
p. cm.
ISBN-13: 978-0-87170-825-0
ISBN-10: 0-87170-825-6
1. Building materials—Fatigue. 2. Building materials—Service life. I. Halford, Gary R. II. Title.

TA418.38.M326 2005
624.1′8—dc22        2005054562
SAN: 204-7586

ASM International®
Materials Park, OH 44073-0002
www.asminternational.org

Printed in the United States of America

We dedicate this book to our beloved grandchildren

Shira Lisa Entis
Jonathan Joshua Entis
Chloe Teressa Green
Alexandra Eugena Green
Jennifer Lisa Ames
Jeffrey Manson Ames
Erin Lea Stone
Jeremy Francis Stone
Paul Isaac Manson
Cecily Rose Manson
Joshua Simon Ames

Brian Patrick Packert
Brendan James Packert
Colin Richard Packert
Drew Joseph Halford
Grace Charlotte Halford
Rose Elise Halford

# Contents

Preface by S.S. Manson .................................................................. vii
Preface by G.R. Halford .................................................................. ix
About the Authors ........................................................................... xi
Abbreviations and Symbols ............................................................ xii

Chapter 1  Introduction ..................................................................... 1
Chapter 2  Stress and Strain Cycling ................................................. 9
Chapter 3  Fatigue Life Relations ..................................................... 45
Chapter 4  Mean Stress ..................................................................... 75
Chapter 5  Multiaxial Fatigue .......................................................... 105
Chapter 6  Cumulative Fatigue Damage .......................................... 123
Chapter 7  Bending of Shafts ........................................................... 157
Chapter 8  Notch Effects .................................................................. 179
Chapter 9  Crack Mechanics ............................................................. 201
Chapter 10  Mechanism of Fatigue ................................................... 237
Chapter 11  Avoidance, Control, and Repair of Fatigue Damage ........... 267
Chapter 12  Special Materials: Polymers, Bone, Ceramics,
             and Composites ............................................................... 325

Appendix  Selected Relevant Background Information ..................... 375

Index ................................................................................................. 441

# Preface

by S.S. Manson

---

The past half century has witnessed a virtual revolution in the development of two fields which are the subject of this book: the introduction of advanced materials as structural components in severely loaded machines exposed to high temperatures and temperature gradients, and the development of technology of life computation for such components, of which one of the major failure mechanisms is fatigue. This book is based on the experience of the authors during this period. Although it emphasizes our research both as individuals and as colleagues for half a century, it also includes the work of numerous others who have provided useful results that have moved progress in these fields.

My first report on fatigue appeared in 1953. An intense interest and activity in this rapidly changing field has continued since. Collaboration with Dr. Gary Halford started in 1966 when he joined NASA at its Cleveland center where I served as Chief of the Materials and Structures Division. This cooperation continued after I retired in 1974 to join the faculty of Case Western Reserve University, and even after I retired from CWRU two decades later. We started to write this book well before I left CWRU. Thus, this book has been in the making for a long time, perhaps longer than we care to admit. But to compensate for the slowness of its progress toward publication, it is fair to say that we have been continuously adding content from our own research, and from that developed elsewhere, as warranted.

Initially this book was prepared as a text on fatigue, and its content fashioned after my regular curriculum presentations at Case Western Reserve University, short course presentations at the Pennsylvania State University, and shorter presentations at MIT, The Technion in Israel, and numerous other universities. In later-year presentations it was broadened under the title *Relation of Materials to Design* to include content developed at NASA. Its current context is still largely related to fatigue but includes other subjects representative of the material presented in these courses.

I am grateful to NASA for the support it has rendered me during my employment there, and later in grants provided to continue my activities initiated there. I am also grateful to the Oak Ridge National Laboratory, the Electric Power Research Institute, and the Metals Properties Council for their grants to conduct the research described in this book. My most heartfelt gratitude is for my long-time colleague, toward my co-author, Gary R. Halford. It has been a genuine joy to work with him as a friend, and co-author.

As always, I express my deep appreciation to the Almighty for the gift of life and participation in the developments contained in this book.

S. Manson
December 2005

# Preface

by G.R. Halford

---

This book and a planned second volume dealing with high-temperature durability represent the culmination of many years of collaborative research with my highly respected colleague, S.S. Manson. Few researchers have had the luxury of being able to work together continuously for as long as we have. And few colleagues have been able to work together as amicably as we have. We were fortunate to be involved in numerous advancements to the field through individual and joint publications spread over five decades. Our combined years of experience exceeds a century. This book provides a repository of the most significant of our contributions to the art and science of material and structural durability. Valuable contributions from other researchers are also included as appropriate.

I cannot sufficiently thank NASA for the rare opportunity provided me to have been allowed to work in this field for the duration of my employment. A prime advantage provided by a large government research organization was that we had valuable technical contacts with not only the aerospace industry, but also with many other industries, including electric power generation, off-highway and automotive manufacturing, metals producers, chemical and petroleum producers, and numerous other industries that faced serious material and structural durability issues. We were thus privileged to have exposure to countless durability issues of a diverse nature. From such a vantage point, it was possible to develop generic models having a broad range of applicability.

I would also like to thank the University of Illinois in Urbana-Champaign, its Department of Theoretical and Applied Mechanics, and in particular, Professor JoDean Morrow. I could never have been in a position to participate in this work without their providing me with the appropriate educational background. Finally, my late parents, Herbert C. and Faye S. Halford, brother Donald W. Halford, my wife, Pat M. Halford and our children, Kirk, Gwen, and Shawn must be acknowledged for instilling me with balanced senses of patience, work ethic, responsibility, dedication, and respect—all interspersed with a tinge of humor.

ord<br>
Gary R. 2005<br>
Decer

# About the Authors

**S.S. Manson** is Professor Emeritus, Case Western Reserve University. Professor Manson joined the National Advisory Committee for Aeronautics (the precursor to NASA) at Langley, VA in 1941 and transferred to Cleveland in 1943. There, he performed cutting-edge theoretical and experimental stress analysis and durability research associated with the materials used in piston engines and the newly evolving gas turbine engines. His research interests drew him into the entirely new area of low-cycle fatigue, particularly thermal fatigue. The basic law of low-cycle fatigue that he developed remains in use 50 years later, i.e., the Manson-Coffin law. His research expanded into the study of creep, creep-rupture and time-temperature parameters, for which he created several of great practical value. He has received numerous awards for his work, including the Gold Medal from the Franklin Institute for development of the Manson-Coffin law of low-cycle fatigue, the NASA Exceptional Scientific Achievement Award, and the Nadai Award bestowed by the American Society of Mechanical Engineers. His book *Thermal Stress and Low-Cycle Fatigue* was published in 1966. He remained at NASA until 1974, serving most of the time as Chief of the Materials and Structures Division. At that time, he moved on to become Professor of Mechanical and Aerospace Engineering at Case Western Reserve University. There he continued to teach on the subject of the mechanical behavior of materials and perform research together with his students and colleagues to develop better durability lifing models. He currently lives in California.

**G.R. Halford** is a Distinguished Research Associate, NASA Glenn Research Center, Cleveland, Ohio. Following his education in the Department of Theoretical and Applied Mechanics at the University of Illinois under the guidance of Professor JoDean Morrow, he joined the NASA Center in 1966. Dr. Halford, in conjunction with Professor S.S. Manson, has been actively involved in research and development of advanced life prediction methods for low- and high-temperature fatigue analysis of high-performance mechanical systems. Most notable is the total strain version of the method of strain-range partitioning (SRP). That methodology sees use in several industries. Dr. Halford has been involved with durability issues in virtually every propulsion system of interest to NASA. In the aeronautics arena, he has dealt with subsonic, supersonic, and hypersonic propulsion systems. In space propulsion and power, he has dealt with ion engines, solid propellant rockets, liquid rockets of all sizes and description, as well as solar and nuclear energy conversion and storage systems. The severe durability limitations of these systems have spawned much of the research into advanced life prediction methods that are the subject of the planned second volume of this book. Dr. Halford has authored or co-authored over 200 technical papers, coordinated over 60 grant/contractor reports, edited several technical conference volumes, and delivered over 70 invited technical lectures.

# Abbreviations and Symbols

| | | | |
|---|---|---|---|
| $a$ | Crack length | $n$ | Strain hardening exponent, number of applied cycles |
| $A$ | Cross-sectional area; creep coefficient; ratio of amplitude to mean | $N$ | Number of cycles |
| $b$ | Maximum possible crack length; Basquin exponent; thickness where $t$ is time | $p$ | Pressure |
| | | psi | Pounds per square inch |
| | | $P$ | Force (load) |
| $B$ | Bulk modulus; Bridgman correction factor; number of repetitions of a cyclic loading sequence | $q$ | Notch-sensitivity factor |
| | | $Q$ | Activation energy |
| | | $Q^{-1}$ | Loss coefficient in damping |
| HB | Brinell hardness number | $r$ | Radius |
| $c$ | Half-depth of beam; radius of shaft; plastic strain vs. life (Manson-Coffin) exponent | $R$ | Ratio for cyclic loading (min/max); plastic-zone size; Rockwell hardness |
| $C$ | Fatigue crack growth coefficient | RA | Reduction of area |
| $d$ | Diameter; derivative | $S$ | Nominal, average, or engineering stress |
| $e$ | Engineering strain; base of natural logarithms ($e = 2.718\ldots$) | SWT | Smith-Watson-Topper parameter |
| | | $t$ | Time; thickness |
| $E$ | Modulus of elasticity | $T$ | Temperature; torque |
| $F$ | Force; finite width factor of fracture mechanics | $u$ | Energy per unit volume, displacement |
| $G$ | Shear modulus; strain energy release rate | $U$ | Energy |
| | | $v$ | Displacement |
| $h$ | Height; half-height of cracked bodies | $V$ | Volume |
| | | $w$ | Width, displacement |
| $I$ | Area moment of inertia about an in-plane axis | $W$ | Hysteresis energy; work |
| | | $x, y, z$ | Spatial coordinates |
| $J$ | Polar moment of inertia; $J$-integral; damping coefficient | $X$ | Safety factor |
| $k$ | Stress or strain concentration factor; spring constant | | |
| ksi | Thousand psi | | |
| $K$ | Stress or strain concentration factor; fatigue-strength reduction factor; stress-intensity factor of fracture mechanics; strength coefficient for stress-strain curves | | |
| $L$ | Length | | |
| $m$ | Fatigue limit reduction factor; fatigue crack growth exponent; creep exponent | | |
| $M$ | Bending moment | | |

## Greek Letters

| | |
|---|---|
| $\alpha$ | Coefficient of thermal expansion; angle; relative crack length ($a = a/b$) |
| $\beta$ | Neuber constant |
| $\gamma$ | Shear strain; Walker exponent; surface energy |
| $\delta$ | Slope reduction factor; crack-tip opening displacement; phase angle |
| $\varepsilon$ | Normal strain; true strain |
| $\eta$ | Tensile viscosity |

| | | | |
|---|---|---|---|
| θ | Angle | m | Mean ($\sigma_m$); melting ($T_m$) |
| λ | Stress biaxiality ratio $\varepsilon_2/\varepsilon_1$ | max | Maximum ($\sigma_{max}$) |
| ν | Poisson's ratio | min | Minimum ($\sigma_{min}$) |
| ν | Notch-tip radius | n | Nominal stress ($\sigma_n$) |
| Σ | Summation | oct | Value for octahedral planes ($\tau_{oct}$) |
| σ | Normal stress at a point; true stress | p | Plastic strain ($\varepsilon_p$); proportional limit ($\sigma_p$) |
| τ | Shear stress | r | Residual ($\sigma_r$); rupture ($t_r$) |
| ω | Angular velocity | ss | Steady-state creep ($\varepsilon_{ss}$) |
| | | t | Total ($\varepsilon_t$); theoretical ($K_t$) |
| | | tc | Transient creep ($\varepsilon_{tc}$) |
| | | tr | Tertiary creep ($\varepsilon_{tr}$) |

## Subscripts: Meaning (Example)

| | | | |
|---|---|---|---|
| a | Amplitude ($\sigma_a$); axial ($\varepsilon_a$) | T | Transition life ($N_T$) |
| A, B, C, D | Rockwell Hardness Scales ($R_A$); general coefficients | u | Ultimate ($\sigma_u$) |
| | | x, y, z | Direction ($\sigma_x$); axis ($I_z$) |
| c | Creep ($\varepsilon_c$); critical ($K_c$); value at $y = c$ location ($\sigma_c$) | xy, yz, zx | Plane ($\sigma_{xy}$) |
| | | y | Yield ($\sigma_y$) |
| d | Diametral ($\varepsilon_d$) | 1, 2, 3 | Principal direction ($\sigma_1$) |
| e | Elastic ($\varepsilon_e$); fatigue endurance limit ($\sigma_e$) | ε | Strain ($R_\varepsilon$) |
| | | σ | Stress ($R_\sigma$) |
| eq | Equivalent ($\sigma_{eq}$) | | |
| f | Final ($A_f$); failure ($N_f$); true fracture strength ($\sigma_f$); or fatigue-strength reduction factor ($K_f$) | | |

## Modifiers: Meaning (Example)

| | | | |
|---|---|---|---|
| fp | Fully-plastic value ($M_{fp}$) | δ | Increment or interval ($\delta\sigma$) |
| g | Gross section ($S_g$) | Δσ | Deflection; range of variable ($\Delta\sigma$) |
| i | Initial ($A_i$); summation index, $i$th level of loading | dot | Time rate ($\dot{\varepsilon}$) |
| I | Mode one | prime | Value for cyclic loading ($n'$); other special values |
| II | Mode two | | |
| III | Mode three | x̄ | Overbar used to denote composite material variables and properties |
| Ic | Critical plane strain ($K_{Ic}$) | | |
| j | Summation index where $i$ means *initial* | |x| | Vertical bars denote composited *matrix material* variables and properties |

# CHAPTER 1

# Introduction

The term *metal fatigue* has become very familiar to the lay public. Often, the media dramatize it by reference to causes involved in spectacular mechanical failures resulting in high loss of human life, large financial losses, or severe setbacks in public goals. Indeed, publicized failures of aircraft, power turbines, bridges, or structures of high visibility have convinced the public that metal fatigue can lead to dire consequences. It is also sometimes represented as something very mysterious and not well understood—nor readily predicted—by designers and engineers.

Use of the word *fatigue* was coined by Poncelet in France before the middle of the nineteenth century (Ref 1.1) to imply that the material involved essentially got tired due to repeated loading, and eventually disintegrated. Reference has, in fact (facetiously), been made by Carden (Ref 1.2) to the first incident of acoustic fatigue, in 1200 BC, when Joshua destroyed the walls of Jericho by setting them into vibration (not really *metal* fatigue, but structural fatigue nevertheless). Throughout modern history, particularly since the Industrial Revolution, when machinery became important, there has been an awareness of mechanical failures, identified as "fatigue," which have been costly, inconvenient, and in some cases, fatal. Today there is a special emphasis on mechanical failures, mostly fatigue, but also due to other causes, not only because machines are designed to function to the limit of the capability of the materials of their construction, but because of the media focus on disasters that seemingly occur so frequently.

## Some Statistics

The fact that most mechanical failures are linked to fatigue is illustrated in Table 1.1, which shows the results of an extensive study by the National Bureau of Standards (currently NIST, National Institute of Standards and Technology) of 230 parts for which failure reports were available (Ref 1.3). From the table, it is clear that of the 230 failures, 141 were associated with fatigue. Various aspects of the fatigue mechanism are listed: design deficiencies (24), improper maintenance (52), fabrication defects (31), and so forth. It is clear that the causes of fatigue have many faces, all of which need to be addressed.

How important fracture, due mainly to fatigue, is to our economy was brought out in a 1983 report by Battelle Columbus Laboratories

Table 1.1  Summary of number of types of failures of 230 failed components

| | Type of failure | | | | | | | |
|---|---|---|---|---|---|---|---|---|
| Cause | Fatigue | Overload | Stress corrosion or hydrogen embrittlement | Corrosion | Stress rupture | High-temperature oxidation | Wear or excessive deformation | Totals |
| Improper maintenance | 52 | 18 | 11 | 4 | 0 | 2 | 15 | **102** |
| Fabrication defects | 31 | 6 | 0 | 1 | 0 | 1 | 0 | **39** |
| Design deficiencies | 24 | 5 | 4 | 1 | 1 | 1 | 1 | **37** |
| Defective material | 10 | 3 | 1 | 0 | 0 | 0 | 1 | **15** |
| Abnormal service | 13 | 7 | 1 | 0 | 2 | 0 | 0 | **23** |
| Undetermined | 11 | 2 | 1 | 0 | 0 | 0 | 0 | **14** |
| **Total** | **141** | **41** | **18** | **6** | **3** | **4** | **17** | **230** |

Source: Ref 1.3

in conjunction with the National Bureau of Standards (now NIST) (Ref 1.4). A summary of their results is shown in Table 1.2. Although several types of failures were considered—not only fatigue—many other failure mechanisms were involved.

The total cost to the American economy in 1982 was $119 billion, representing approximately 4% of that year's gross national product. While this amount seems to be large for component breakage alone, it should be noted in item (3) of the table that numerous costs are associated with any failure: personal, medical, and consequential, not to mention legal fees that arise out of accidents.

## Dramatic Examples

Of special importance are the consequential costs that can dwarf the direct costs. The removal from service of a power-generating component of an aircraft, either because of a failure, or because of fear of a possible failure, can disable a whole facility, or an entire fleet of aircraft. Consider, for example, the DC-10 jumbo jet aircraft failure that occurred in 1979. Figure 1.1 (Ref 1.5) is an excerpt from the *Time* magazine description of the accident that occurred when this jumbo jet suddenly plunged back to earth near the Chicago O'Hare airport, killing 275 people. Unexpected hairline cracks were detected to have preceded the final fracture of the pylon of the ill-fated aircraft (the pylon is the outboard structure that carries the engine on the wing). Because it was feared that other DC-10 aircraft might suffer the same fate, the entire U.S. fleet of 138 DC-10s and 132 planes owned by other countries were grounded. The lost use of these aircraft amounted to an estimated loss of $5 million per day. Fortunately, it was learned shortly thereafter that the fatigue cracks were initiated, in part, by improper maintenance procedures and not to inherent design faults or normal service loadings. The airplanes were returned to normal service within several days. However, this was not without consequential daily costs and not without the loss of 275 lives and costs of subsequent litigation. In any case, this disaster is an illustration of the entry in Table 1.1 wherein fatigue is caused not by design or material deficiency, but by improper maintenance.

The case illustrated in Fig. 1.2 was also brought about by poor maintenance practice. It represents the disregard of recognized finiteness of fatigue life in high-performance equipment. The figure includes the headline from *Newsweek's* account of the Aloha Airlines accident of a Boeing 737 in May 1988 (Ref 1.6). As seen in the photograph, the upper half of the fuselage ripped away completely while the plane was in flight. An Aloha flight attendant was blown away during the incident. Fortunately, the pilot was able to land the disabled plane at a nearby airport, saving the lives of the 95 people remaining aboard. The fuselage had failed in fatigue along a line of rivet holes, but only after the failure of the adhesive that also contributed to the bonding of this critical joint. Constant exposure to the salt-laden environment of island hopping had caused adhesive deterioration, and thus the rivets had to carry the entire load of ground-air-ground flight cycles. Although the plane was very near its safe retirement age, it

**Table 1.2 Summary of a Battelle/National Bureau of Standards study on cost of fracture to U.S. economy expressed in 1982 dollars**

**Types of materials studied**

Metals and alloys (including steel, cast iron, and aluminum)
Inorganic materials (such as polymers, concrete, ceramics, and glass)
Composite materials (including laminated structures, glass fibers, and reinforced concrete)

**Types of fracture considered in assessing costs**

Overload
Brittle fracture
Ductile rupture
Fatigue
Creep
Creep rupture
Stress corrosion
Fretting fatigue
Thermal shock
Buckling

**Costs taken into account**

Costs incurred because of actual failures
Cost of pain, injury, death, and medical treatment to victims
Business delays or property damage
Insurance administration
Cargo losses
Environmental cleanups
Prevention through design and inspection

**Sectors of the economy that pay for fracture and prevention**

| | |
|---|---|
| Motor vehicles and parts | 12.5 billion |
| Residential and nonresidential construction | 10.0 billion |
| Aircraft and parts | 6.7 billion |
| Structural metal and other fabricated metal products | 5.5 billion |
| Petroleum refining and industrial chemicals | 4.2 billion |
| Primary metals (iron, steel, aluminum, etc.) | 3.5 billion |
| Food and kindred products | 2.4 billion |
| Tires and inner tubes | 2.4 billion |
| Others | 70.1 billion |

Total cost to U.S. economy = $119 billion/year in 1982 dollars (approximately *$232 billion/year in 2004 dollars*)

Source: Ref 1.4

should have remained flightworthy. Better inspection of the known critical areas might have prevented the failure.

In addition to the extensive studies by Battelle and the National Bureau of Standards, more limited studies have demonstrated that the problem is indeed pervasive. Campbell and Lahey (Ref 1.7) summarized numerous serious aircraft accidents involving fatigue fracture, as shown in Table 1.3.

The U.S. space shuttle is another example of a structure that has encountered numerous fatigue problems over its years of development. The following excerpt from *Aviation Week and Space Technology* (Ref 1.8) describes a problem that occurred in 2003:

> The hydrogen propulsion system cracks that have grounded the space shuttle fleet have been traced to high-cycle metal fatigue.
>
> The cracks can be safely repaired for the near term, but raise questions about whether key orbiter hydrogen lines can sus-

---

**Debacle of the DC-10**

*We are a little uneasy. We have no handle on this one yet. Was it aging metal in a high-time machine? Was it stress? And what kind of stress? Was it quality control of the metal? And if we find out, what kind of fix can we ask to have made? We don't know.*

Those worried—and worrisome—comments came last week from a member of a band of experts who normally know all the answers: the National Transportation Safety Board's "go teams" of plane-crash investigators. Over the years they have been able to pinpoint a "probable cause" in 97% of all U.S. air accidents. Yet even these legendary investigators remained in doubt about the precise cause of the worst U.S. air tragedy in history—the crash of an American Airlines DC-10 jumbo jet near Chicago's O'Hare International Airport on Memorial Day weekend that killed 275. While the experts hunted for both a cause and a cure, 138 DC-10s in the U.S. and 132 more around the world were grounded. As the airlines using DC-10s lost an estimated $5 million a day, the public developed new doubts about the industry's vaunted competence and, equally important, the ability of its federal regulators to protect travelers against disaster.

Source: Excerpted from TIME Magazine, June 18, 1979, p 14 (Ref 1.5)

---

**Fig. 1.1** *Time* magazine excerpt of DC-10 jumbo jet aircraft failure in 1979

**4 / Fatigue and Durability of Structural Materials**

tain 20 more years of shuttle flight operations without a costly retrofit.*

Ground equipment also provides many examples of use limitation by fatigue. One of these examples is mentioned in the following news item from August 2002 (Ref 1.9):

Amtrak says it hopes its premier Acela Express service will gradually resume operations this week. All 18 of the high-speed trains remained out of service Friday as Amtrak, federal railroad officials and representatives of the manufacturer worked out a temporary repair plan to cracks underneath many of the locomotives. Amtrak President David Gunn said that if the repair plan works, "it's a matter of days" before at least some of the trains return to duty in the Boston-New York-Washington corridor, Amtrak's busiest. The experience shows why Amtrak should go with established technology rather than with new designs, such as Acela Express, when it seeks to buy additional high-speed trains, Gunn said. The U.S. government should make it easier for Amtrak to buy trains designed in Europe and already in use there, he said.

---

*As of 2005, NASA plans a maximum of 10 more flights for each of the three remaining space shuttles prior to their retirement in 2010.

**Fig. 1.2** Aloha Airlines Boeing 737 accident in Hawaii. The headline is a recreation of one that appeared in a *Newsweek* article about the accident (Ref 1.6). Photo source: Associated Press Wide World Photos. Reproduced with permission

## Early Recognition of Metal Fatigue as a Problem

It was, in fact, the Industrial Revolution, and specifically, the carriage and railroad industry, that first focused serious attention on the gravity of the fatigue problem with frequent failures of axles and wheels, as well as tracks and bridges. It became obvious that the entire enterprise was at risk if the lifetime and reliability of these components were not substantially increased. First, if the problem were to be solved, it had to be better understood. In the early 1850s, August Wöhler undertook the first systematic study of fatigue (Ref 1.1). What he learned served as a starting point for a technology, and even some science, that would be developed by many investigators in the ensuing century and a half. The authors of this book have devoted a large fraction of their lives to participating in this quest.

## An Overview of this Book

Our work with NASA, the aerospace industry, and nonaerospace industries has involved us with various materials and structural durability problems. Many have involved high temperature; others have involved cryogenics, while some have dealt with both temperature extremes within a given application. Most of the durability areas have been associated with aspects of fatigue, but not entirely.

Initially, our intent was to publish a single book to address all of these topics. Its length, however, as well as consideration of diversity of readership, eventually led to the choice of two separate books. The present book is concerned primarily with temperatures below the "creep range." A planned future book will address temperatures wherein creep, and other thermally activated processes, are important considerations.

This book is devoted almost entirely to fatigue or to peripheral aspects involved in the study of this subject. It starts with material under strain cycling because it is strain cycling that leads to fatigue. Specifically, *plastic* strain is emphasized because it is reversible plastic straining that moves imperfections around and causes them to coalesce into larger and larger internal and surface flaws that lead to failure. Treatment of uniaxial strain reversal of constant amplitude is discussed first, describing how some materials can actually become harder or softer due to cyclic straining. A criterion is developed for predicting, in advance, which materials will harden and which will soften.

Complex loading, involving successive cycling of varying amplitude is treated next. Specific rules for material behavior are first provided without detailed proof, so an analyst can proceed expeditiously without being burdened by too many theoretical considerations. However, more detailed analysis is provided later in the chapter of the underlying reasons for the procedure, introducing several models that lead to the rules. Most materials fall within the scope of the standard model, but some materials, such as gray cast iron, require the addition of special modules, which are discussed. A technique developed in Japan in the 1960s by Professor T. Endo (Ref 1.10), known as the *rainflow* cycle-counting procedure, which simplifies the treatment of highly complex loading, is described and illustrated.

The fatigue life equations are discussed in Chapter 3. Fundamentally, life is dependent on the plastic strain, as developed by the senior author (Ref 1.11), and independently by L.F. Coffin, Jr. (Ref 1.12). This development is generally known either as the Manson-Coffin equation or the Coffin-Manson equation. The chapter describes the background and reasoning used by the author in the development of the equation. Also discussed are a number of approximate equations that are useful for estimating fatigue life when the plastic strain or other characteriz-

Table 1.3 A survey of serious aircraft accidents involving fatigue fracture

| Aircraft category | Number of accidents | Number of aircraft destroyed | Number of fatalities |
|---|---|---|---|
| **Fixed wing** | | | |
| Landing gear | 542 | 9 | 21 |
| Engine or transmission | 408 | 144 | 536 |
| Propeller | 232 | 52 | 162 |
| Wing | 137 | 119 | 400 |
| Flight controls | 48 | 28 | 68 |
| Tail | 37 | 21 | 244 |
| Miscellaneous | 37 | 11 | 285 |
| Fuselage | 25 | 9 | 145 |
| **Fixed wing totals** | **1466** | **165** | **1861** |
| **Rotary wing** | | | |
| Engine or transmission | 136 | 40 | 44 |
| Tail rotor | 100 | 43 | 81 |
| Main-rotor system | 55 | 31 | 167 |
| Miscellaneous | 45 | 18 | 22 |
| Flight controls | 37 | 18 | 34 |
| Airframe | 35 | 13 | 25 |
| Landing gear | 11 | 3 | 6 |
| **Rotor-wing totals** | **419** | **165** | **379** |
| **Overall totals** | **1885** | **558** | **2240** |

Source: Ref 1.7

ing parameters for the material are not available to apply the Manson-Coffin equation. One of these, the method of universal slopes, proposed by the senior author together with M.H. Hirschberg (Ref 1.13), has been used widely in many applications. This method is described together with other approximate equations that have found extensive use with good results in many scientific and industrial applications. This chapter also discusses high-cycle fatigue, which can occur when loading is applied at moderate frequency over a long lifetime of the structure, or at extremely high frequency for shorter-lifetime structures, even if the applied strain is in the nominally elastic range.

Mean stress is discussed in Chapter 4. Numerous formulations for calculation or prediction are described, but the most interesting feature is a discussion of the reason why tensile mean stress reduces fatigue life more than would be expected on the basis of higher maximum stress involved. It turns out that, somehow, the presence of mean stress causes the material to behave as if its stress-strain curve were truncated—that is, cut off at a lower point than it is able to achieve in monotonic loading. The introduction of mean compressive stress is also a highly effective way to improve the fatigue behavior of a specimen or machine component. This chapter also discusses the basis of fatigue life improvement by compressive mean stress.

Chapter 5 treats multiaxial fatigue. The presence of biaxial or triaxial stresses can influence fatigue life substantially, and this subject has been studied by many investigators. Several existing theories are discussed, including a new one introduced by the authors of this book. An interesting finding by the senior author and a graduate student is that atomic structure can be an important factor in governing which theory is best applicable to a given material. Face-centered cubic metals are better served by a different formulation from that used for body-centered cubic metals.

Cumulative fatigue damage—that is, when a specimen or part is subjected to several loading levels throughout its life—is the subject of Chapter 6. There had long been an objective in research circles to be able to treat this problem by what was conceived as a *double linear damage rule,* one applicable before the initiation of a crack, another after a crack has started. For many years the senior author pursued this line of analysis according to a particular formulation he posited. In later years, however, it became clear that this formulation was incorrect. The transition from one behavior to another occurred well before the crack was initiated at a readily observable level. In the 1980s, the authors, working jointly, developed a double linear damage rule that is not related to a physical crack initiation event. In fact, two valuable procedures were derived: one labeled the *double linear damage rule* (DLDR), the other, the *double linear damage curve* (DLDC). Either rule is nearly as easy to apply as the commonly used *Miner's linear damage rule* (LDR). In this chapter, we apply these new developments successfully to results of two-level loading tests and to multiple-level tests in the published literature. Also presented in this chapter are the necessary procedures to deal with cumulative damage in the case of irreversible hardening. Here, the high-strain loading in one part of the history hardens the material so that the low-strain loading is applied to a material that is harder than originally assumed. This harder material is expected to have different fatigue characteristics from those of the virgin material, thus complicating cumulative damage analyses based solely on virgin material fatigue properties. This chapter describes the treatment of tests on type 304 stainless steel that undergoes irreversible hardening.

In Chapter 7 we treat the bending of circular shafts because they are so common in rotating machinery. Our experience has led us to develop a useful closed-form solution for cases that normally require trial-and-error integration. The chapter also discusses the difference in the treatment of a rotating shaft, wherein all surface elements undergo the same stress and strain history, and a nonrotating circular shaft, wherein only some surface elements are subjected to the highest stress and strain while most of the surface elements are subjected to low loading. The discussion centers about the volumetric effect of stress in fatigue because, for the same maximum surface stress, the rotating shaft subjects a larger volume of material to the maximum stress while the stationary shaft subjects only a small volume. Weibull's statistical theory is discussed briefly in this connection.

Stress concentration effects are treated in Chapters 8 and 9. Chapter 8 is restricted to well-defined notches, where the stresses can be analytically determined. Of special interest is a study to determine the development of stresses and strains at various stages of the loading cycle. Beneficial and detrimental effects of residual stress are dominant features of this discussion because residual stress is of such great importance in fatigue life. The crack effects, as discussed in Chapter 9, lead us to consideration of *fracture mechanics*. The field has been particularly useful in explaining brittle fractures in large structures such as naval ships, nuclear reactors, and aerospace vehicles. In these cases, the construction may be of *ductile* materials, but the failures are in a *brittle manner*.

Taking the mystery out of fatigue is treated in Chapter 10. The role of plasticity in agglomerating microscopic imperfections into larger flaws, with associated stress concentrations, is discussed using dislocation theory. Most engineers are familiar with dislocations, but for those who need the basics, or who prefer to refresh their background, a fundamental primer is provided in the Appendix. Surface deterioration, volumetric effects, environmental effects, and the role of foreign particles are discussed in their roles in the development and propagation of cracks.

All of the fundamentals developed in the earlier chapters are brought to bear in Chapter 11 to discuss how to avoid, control, or repair fatigue damage. This chapter is largely a compendium of engineering processes that have been successful in treating the fatigue problem. Decisions made at every stage, from the sizing of parts, choice of materials, mechanical and thermomechanical processing, fabrication, surface preparation, and introduction of favorable residual stresses, provide the designer, assembler, and user an opportunity to strengthen the structure and protect it against the hazards of fatigue failure. Specific examples are provided to illustrate good versus faulty practice.

While most of the content of the book is directed to the application of metals in engineering structures, modern technology has created new materials and used them in special applications. Polymers, ceramics, and composites combining metals with nonmetals have come into common use for applications wherein they are superior in service function, or are less expensive than the metals they can replace. Bone is another special material, evolved by nature for animal or human functioning. Can we carry over the technology developed largely for metals to applications involving these special materials? Chapter 12 describes some experiences of the authors to answer this question. Some tests were conducted in the laboratory by the authors and students, and in some cases the results are drawn from the work of colleagues (e.g., bone studies). The results are very encouraging for polymers. In many respects, the technology developed for metals can be carried over to polymers. Bone, too, is analogous in cycling characteristics to metals that soften with cycling. It lends itself better to analysis by hysteresis energy. Bone can be self-healing to a much greater extent than metals, however. Much more work is needed in the study of bone to bring it to the level of predictability in fatigue that has been achieved for metals. Engineered ceramics, composited ceramics, and ceramic fiber-reinforced metal-matrix composites have been studied intensively, and recent findings are discussed in this chapter.

The last chapter is an Appendix. The information contained within it is equally applicable to the planned second volume; it is, in fact, intended to be repeated in that volume for the convenience of readers who may not have immediate access to the present volume. The Appendix is used for many purposes. First, it provides an opportunity to define many terms and concepts that might otherwise require repetition in multiple locations. Second, it permits brief reviews of subject matter with which many readers are already familiar, but which a limited readership might need in order to follow the text, for example, the theory of dislocations. Similarly the theory of heat treatment and its terminology may be well known to some but not to others. Thus, both of these specialties are given in the Appendix rather than being discussed in multiple places as they occur in the text. Metallurgical aspects and terms of atomic structure are also covered. The types of tests (hardness, tensile, fatigue, etc.) used to characterize materials, or to evaluate them, are discussed, as is the equipment for obtaining the various properties used in analysis. These topics can be given adequate coverage in the Appendix without devoting unnecessary interruptions to clarify discussion each time they are used.

One analytical subject is included in the Appendix—that is, a special technique for inverting an equation in the general form:

$$y = Ax^\alpha + Bx^\beta$$

where $\alpha$ and $\beta$ are nonintegers. When $x$ is the known quantity, $y$ can be easily calculated. However, there is no closed-form solution for calculating $x$ if $y$ is the known variable. This type of equation is encountered in many applications discussed in the book, for example, in the Manson-Coffin-Basquin equation, in the expression for the cyclic stress-strain curve, and in analysis of notches. The Appendix describes how the solution was generated and how it can be applied to other problems discussed in the book.

## REFERENCES

1.1 S.P. Timoshenko, *History of Strength of Materials*, McGraw-Hill Book Co., New York, 1953. See also R.C. Juvinall, *Stress, Strain, and Strength*, McGraw-Hill Book Co., New York, 1967, p 194–195.

1.2 E.A. Carden, Low Cycle Fatigue under Multiaxial Stress Cycling, paper 108, *Proceedings of the Second International Symposium*, Japan Society of Mechanical Engineers, Tokyo, Sept 1967

1.3 R.P. Reed, J.H. Smith, and B.W. Christ, "The Economic Effects of Fracture in the United States," Special Publication 647-1, National Bureau of Standards, U.S. Department of Commerce, Gaithersburg, MD, 1983

1.4 Anon., Battelle Columbus Laboratories, Metals and Ceramics Information Center, Issue 120/121, Feb/March 1983. See also J.J. Duga, W.H. Fisher, R.W. Buxbaum, A.R. Rosenfield, A.R. Buhr, E.J. Honton, and S.C. McMillan, Fracture Costs US $119 Billion a Year, Says Study by Battelle/NBS, *Int. J. Fract.*, Vol 23, 1983, p R81–R83. These references are brief synopses of an extensive report by the same seven Battelle Columbus Laboratories authors titled "The Economic Effects of Fracture in the United States," National Bureau of Standards, U.S. Department of Commerce, Gaithersburg, MD, Special Publication 647-2, 1983.

1.5 Anon., Debacle of the DC-10, *Time Magazine*, June 18, 1979, p 14

1.6 Anon., Whoosh! She was Gone, *Newsweek Magazine*, May 9, 1988, p 18

1.7 G.S. Campbell and R. Lahey, A Survey of Serious Aircraft Accidents Involving Fatigue Fracture, *Int. J. Fatigue*, Vol 6 (No. 1), Jan 1984

1.8 C. Covault, Shuttle Cleared to Fly Again, *Aviat. Week Space Technol.*, Aug 12, 2002, p 27–28

1.9 Anon, Amtrak Repairing High-Speed Trains, *The Plain Dealer*, Cleveland, OH, Aug 19, 2002

1.10 *The Rainflow Method: The Tatsuo Endo Memorial Volume*, Y. Murakami, Ed., Butterworth-Heinemann Ltd., Oxford, UK, 1992

1.11 S.S. Manson, Behavior of Materials under Conditions of Thermal Stress, *Heat Transfer Symposium*, University of Michigan, 1952

1.12 L.F. Coffin, Jr., "A Study of the Effects of Cyclic Thermal Stresses on a Ductile Metal," *Trans. ASME*, Vol 76, American Society of Mechanical Engineering, New York, 1954, p 931–950

# CHAPTER 2

# Stress and Strain Cycling

## Introduction

Metal fatigue has long been associated with variations of stress and strain. In Wöhler's time, about 150 years ago, it was his identification of this fact that led him to the development of the rotating bending machine that has been used so effectively throughout the years to gain an understanding of the fatigue phenomenon and to characterize the fatigue resistance of many materials. In more recent years, as interest has grown in the study of high-strain, low-cycle fatigue, it has been recognized that the rotating bending test imposes limitations in the ability to measure and control stresses and strains, especially when plastic strains are present. Bending also subjects a specimen to a wide variation of stress-strain conditions through its cross section, and even if these conditions can be properly analyzed, only a small volume of material undergoes the maximum stress and strain. By using axially strained specimens, wherein the entire test section volume is under the same conditions, greater volumes of material can be fatigued. Furthermore, using axial straining allows stress and strain to be measured and controlled more accurately. New testing equipment has been highly developed for controlling stress, strain, or any combination of parameters, so as to give special meaning to the control variables. In this chapter, we concentrate our discussion of material behavior under uniaxial stress and strain cycling.

The novice reader is advised to skim the section "Complex Pattern of Strain Cycling Analyzed by Mechanical Models." It is included for the serious student of cyclic stress-strain modeling.

## Simple Strain Cycling

We first consider strain cycling for three major reasons: it is more typical of the conditions that actually develop in many service situations, especially at failure-prone locations such as notches and holes; strain can be either computed or more readily measured; and strain-controlled testing avoids complications that can develop under cyclic loading.

**Initial Straining.** Consider first a material strained monotonically to a given point A along the stress-strain curve of Fig. 2.1. Initially, the strain is elastic and the stress is linearly proportional to the strain, but when the stress exceeds the yield strength, plastic strain occurs in addition to the elastic strain. We assume that elasticity follows a linear relation to stress, and plasticity follows a power-law relation to stress. At point A, the total strain is $\varepsilon_A$ and the stress is $\sigma_A$ such that:

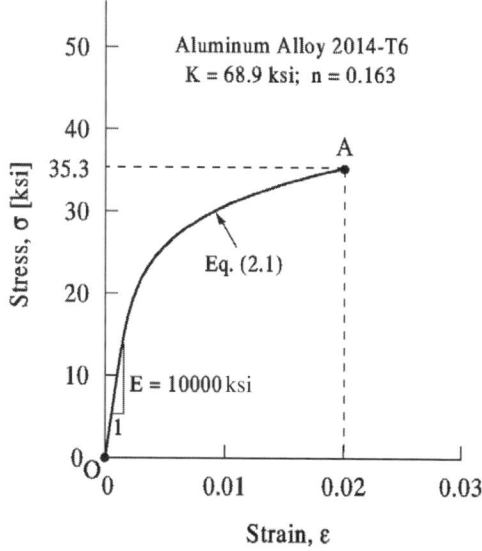

**Fig. 2.1** Typical monotonic tensile stress-strain curve

$\varepsilon_A$ = elastic strain + plastic strain

$$\varepsilon_A = \sigma_A/E + (\sigma_A/K)^{1/n} \qquad (Eq\ 2.1)$$

where $E$ is the modulus of elasticity (slope of the stress-strain curve in the initial linear regime), $n$ is a constant designated the "strain-hardening exponent," and $K$, the "strength coefficient," is a constant relating plastic strain to stress, that is equal to the stress at a plastic strain of 1.0. Equation (2.1) is known as the Ramberg-Osgood equation (Ref 2.1). For quantitative purposes later in the chapter, constants in Eq 2.1 are shown in the figure for a specific material, aluminum alloy 2014-T6. The monotonic stress-strain curve will be used as the starting point for cyclic stress-strain analysis. For this purpose, it will be referred to as the *single-amplitude template* curve.

**Constant Amplitude Cycling of Stable Materials.** If the material for which a monotonic stress-strain curve as shown in Fig. 2.1 is stable, and hence exhibits stress-strain paths that are repeatable from cycle to cycle, these paths define what are known as the cyclic stress-strain curve of the material. Figure 2.2 shows a cyclic stress-strain path for this case. For purposes herein, we concentrate only on *time-independent* stress-strain material behavior*. The initial straining $\varepsilon_A$ to point $A$ follows the ordinary monotonic stress-strain curve. If we now reverse the direction of straining, starting at $A$, and apply a negative amplitude of strain $\varepsilon_B$ ($= -\varepsilon_A$), the path followed is $AB$ (clockwise). This path does not look like $OA$, but, as we shall show, the two paths are related. Reversing again at $B$ to the original strain amplitude $\varepsilon_A$, we follow path $BA$ (clockwise). This path is identical to $AB$ with both axes reversed. Repeating the strains over

---

*In the treatment of cyclic stress-strain behavior of metals and alloys in this volume, we do not consider time-dependent material behavior. We discuss rate and frequency effects in a planned second volume in connection with high-temperature behavior. This is not meant to imply that all materials behave in a time-independent manner when loaded at temperatures below the thermally activated creep regime. For example, low-alloy steels and some other body-centered cubic (bcc) metals are well known to exhibit frequency and straining rate dependencies. This is so because of their interaction with small elemental atoms, such as carbon, that are attracted to the energetically-comfortable interstitial positions within the bcc lattice. The authors have had limited direct experience with the time-dependent behavior of low-alloy, carbon steels at ambient temperatures. The reader is referred to *ASM Handbook* articles on the subject. Titanium alloys also can exhibit time-dependent deformation due to "creep" of a nonthermally activated nature.

and over simply follows $BAB$ (clockwise) for many cycles. Shortly before failure, point $A$ may drop somewhat and a slight cusp may develop in the region of $B$ due to fatigue crack development. For most of the life, however, this loop, designated the *hysteresis loop*, is repetitive and the stress range $\Delta\sigma$ is constant. The equation for the stress-strain paths $AB$ and $BA$ is obtained by doubling both the stress and the strain of the monotonic stress strain path (termed the *single-amplitude template*), and using $A$ and $B$, respectively, *as the origin*. The resultant curve is called the *double-amplitude template* and is given by:

$$2\varepsilon = 2\sigma/E + (2\sigma/K)^{1/n} \qquad (Eq\ 2.2)$$

Alternatively, Eq 2.2 can be written in a more general form in terms of cyclic range of stress range and total strain range:

$$\Delta\varepsilon = \Delta\sigma/E + (\Delta\sigma/K')^{1/n'} \qquad (Eq\ 2.3)$$

where $\Delta$ represents the range of the variable. If the exponent $n'$ is equal to $n$ and $K' = (2)^{(1-n)}K$, the cyclic loop equation is the same as the initial stress-strain curve but with both stress and strain values doubled. Note that the return path $AB$ is identical to $BA$ for cyclically stable materials. Experimentally, the exponent $n'$ may differ somewhat from $n$ and $K'$ may not be related exactly to $K$ according to the aforementioned relation. However, for simplicity, if these properties do not differ appreciably, the monotonic

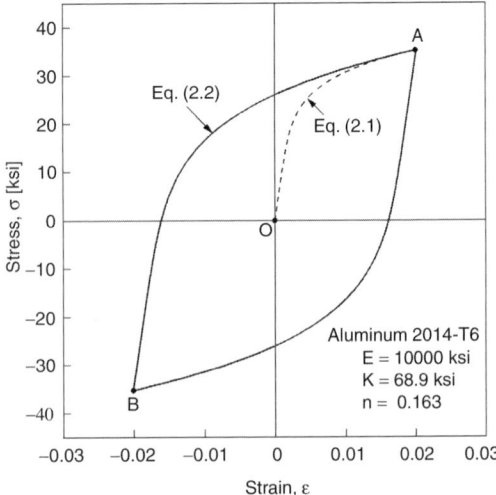

**Fig. 2.2** Cyclic stress-strain behavior for a cyclically stable material whose monotonic curve is as shown in Fig. 2.1

properties can be used to approximate the cyclic values.

While few materials exhibit true cyclic stability (that is, following the behavior described previously), some come close enough to be considered stable for purposes of engineering design. Many materials may, however, either harden or soften during cycling. That is, the peak stresses, or stress ranges, may increase or decrease with progressive cycling, as shown schematically in Fig. 2.3. In an initially soft material, the required stress ranges to maintain a constant strain range may increase according to the path $A'B'C'$. Or, if the material is initially hard, the stress range required for maintaining a constant strain range may decrease according to the path $A''B''C''$ as repetitive cycling progresses.

**Cyclic Hardening Materials.** For initially soft materials, such as annealed metals, the early cycles may show increasingly higher stresses developing as measured at the tip ends of the hysteresis loop. Physically, what is happening is that the back-and-forth flow of dislocations (see Appendix) changes from cycle to cycle making it harder for them to overcome barriers developed during earlier cycles. A substructure of dislocations develops as cycling proceeds, requiring higher stresses to overcome the dislocation tangles. However, a new equilibrium microstructure eventually develops so that the hysteresis loop becomes stable. Or, sometimes fatigue failure may intervene in the later stages of loop stabilization. The schematic behavior of the stress-strain paths developed is shown in Fig. 2.4(a). Initially, the straining to achieve amplitude $\varepsilon_A$ occurs along $OA$ (the ordinary monotonic stress-strain curve). Upon compressive straining to negative amplitude, $\varepsilon_B$, the required compressive stress at $B$ is higher than the tensile stress at $A$. The second time the strain $\varepsilon_A$ is applied, the path followed is $BA'$, and the second time a compressive strain is applied, the path is $A'B'$. Similarly, each successive straining requires a slightly different path as shown in the figure; the end-point stresses gradually increase. The rate of increase is high at first, but diminishes rapidly, approaching a stabilized value, after which very little further hardening occurs. For many materials the stress-strain hysteresis loop remains stable until failure initiates. Near the failure point, the peak tensile strain may drop significantly while the compressive peak strain may drop only slightly as a result of fatigue cracking.

**Cyclic Softening Materials.** Materials that have been previously work hardened will soften from cycle to cycle as they are strained repeatedly. The reason, again, is instability of the internal microstructure. The material may, for example, have previously been hardened by drawing through a die so that it contains a large number of interlocked dislocations that are worked loose by subsequent cyclic plasticity. An orderly substructure of dislocations develops wherein subsequent cycles can generate the required strain with less stress. Or, for previously heat treated materials, the metallurgical structure may change by breaking up dispersed particles to such a small size that they can no longer block dislocation slip. In fact, the particles may become so small that they dissolve into the matrix if the temperature is high enough. In any case, successive cycles of plasticity reduce the stresses required to induce a specified strain. But, as with hardening, the softening diminishes with applied cycles, eventually reaching a stable hysteresis loop before the life of the specimen is exhausted. Figure 2.4(b) shows this case. The path $A, A' A'' \ldots A^N$ corresponds to progressively lower tensile stresses, while $B, B', B'' \ldots B^N$ corresponds to progressively lower compressive stresses.

Figure 2.5 displays the cyclic hardening and softening behavior of two engineering alloys (Ref 2.2): annealed 304 austenitic stainless steel (Fig. 2.5a) and hardened AISI type 4340 aircraft quality steel (Fig. 2.5b). At the far left of each figure are the stable cyclic and the initial monotonic stress-strain curves. The stress and strain values of the monotonic curve have been doubled for direct comparison with the cyclic ranges of stress and strain. When a linear scale is used for cycles, as shown on the right, it becomes

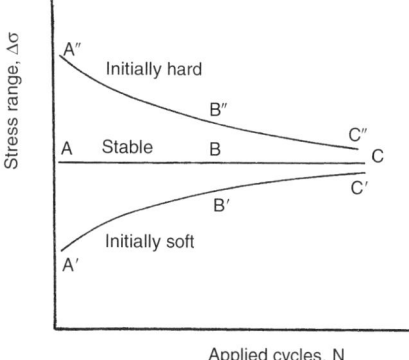

**Fig. 2.3** Patterns of hardening and softening for metals depending on their initial hardness

obvious how quickly the stresses reach their saturation (or steady-state) value. The cyclic stress range variation is rapid at first, approaching a saturation value at as little as 10 to 20% of the life.

**Criteria for Cyclic Hardening or Softening.** Whether a material hardens or softens does not depend so much on its composition but rather on its dislocation state. The same material may either harden or soften depending on its prior processing. The initial internal dislocation structure is directly dependent on the processing. Figure 2.6 shows, for example, some cyclic straining results for oxygen-free high-conductivity (OFHC) copper in three initially different states (Ref 2.3). As an example, a metal hardened by drawing a bar through a die to substantially reduce its diameter will become much stronger. The strength is imparted by the creation of an overabundance of dislocations. Upon being cyclically strained, however, considerable softening occurs as dislocations are mechanically rearranged and annihilation takes place. As an equilibrium level of dislocations forms, the stress-strain response stabilizes and continues to repeat until fatigue failure occurs. This cyclic stress-strain response during initial softening is shown in Fig. 2.6(c). When the same material is fully annealed by holding at a high temperature for a period of time, and slowly cooled, the vast majority of any existing dislocations are liberated and the material becomes soft and easily deformed. When an annealed metal or alloy is plastically strained during cycling, hardening occurs by the process of dislocation generation and tangling. At intermediate or partial levels of annealing, either hardening or softening may occur, depending on whether the initial condition

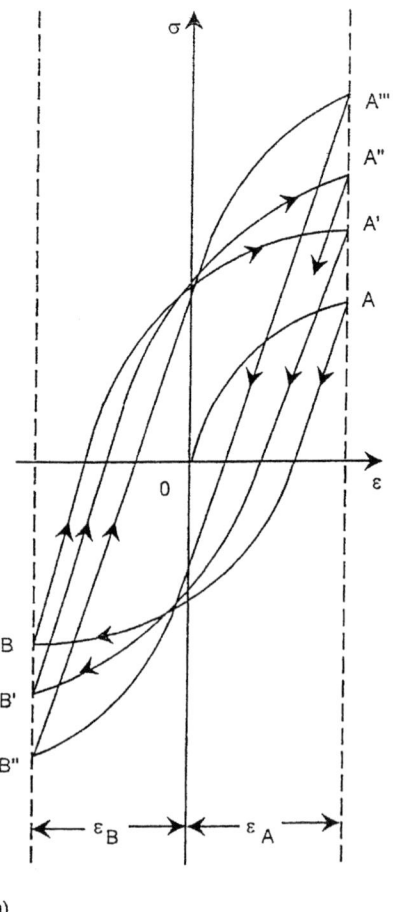

(a)

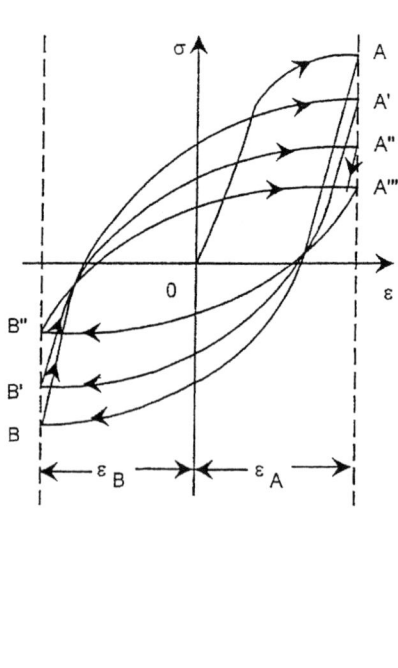

(b)

**Fig. 2.4** Typical pattern of response of materials during strain cycling (schematic). (a) Cyclically hardening material. (b) Cyclically softening material

## Chapter 2: Stress and Strain Cycling / 13

of the material has a higher or lower concentration of dislocations than associated with a given strain range. There is a general tendency for the cyclic stress-strain response to approach a single equilibrium state (Fig. 2.3). Each strain range has its own equilibrium stress level; the higher the strain range, the higher is the stress level.

The aforementioned observations suggest a hardening or softening criterion based on the initial stress-strain curve of the material, using only properties that are usually available, i.e., the ultimate tensile strength, $\sigma_u$, and the yield strength, $\sigma_y$. The yield stress occurs at a low plastic strain (usually 0.002) and the ultimate tensile strength occurs at a high plastic strain (approximately equal to $n$). The ratio of the tensile strength to the yield strength is an indicator of the amount of monotonic strain hardening a material undergoes during a conventional monotonic tensile test. If the monotonic strain hardening exponent $n$ is high (e.g., the ratio of $\sigma_u/\sigma_y$ is high), it can be expected that the material

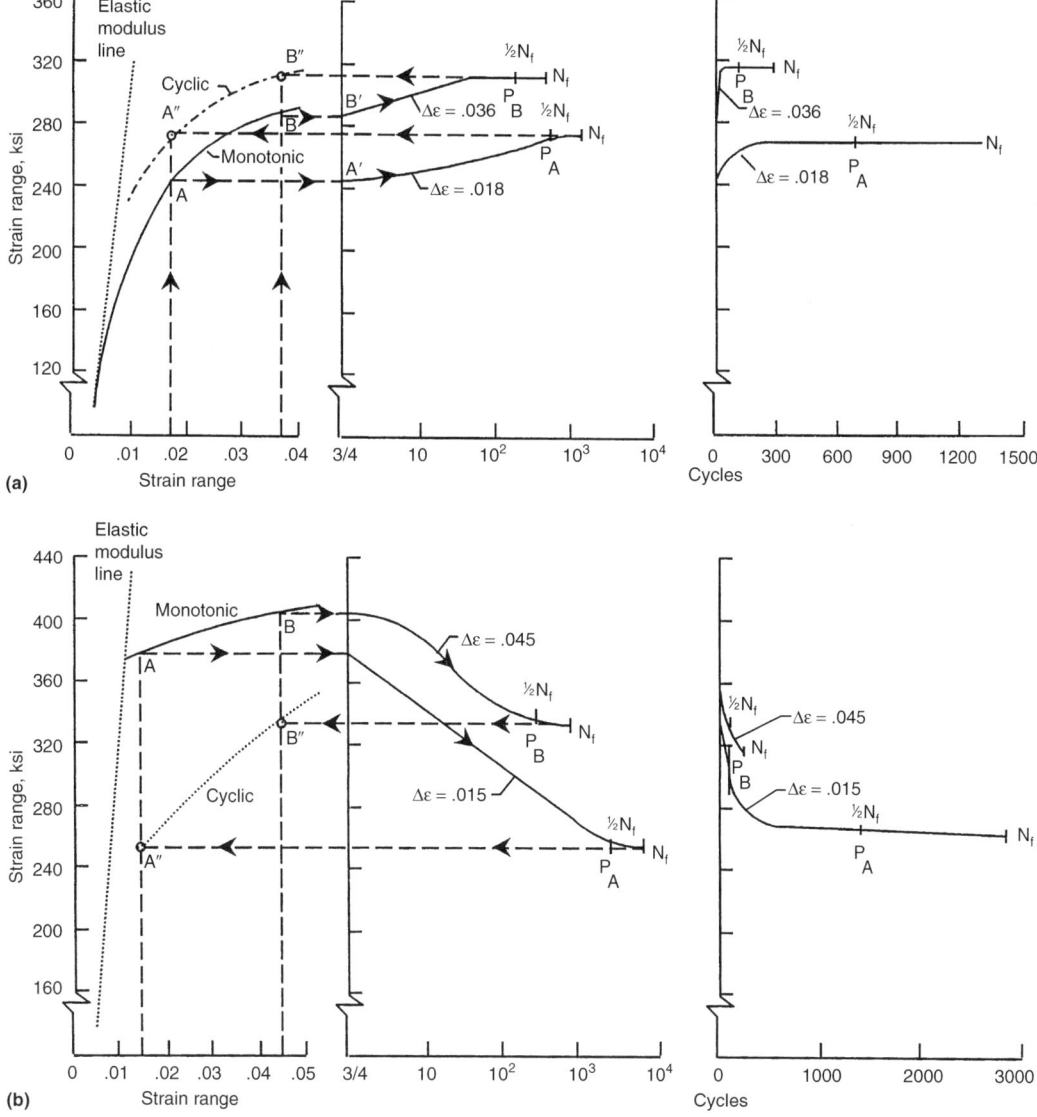

**Fig. 2.5** Response of annealed and hardened steels under cyclic straining. (a) Annealed 304 stainless steel (cyclically hardening). (b) Hardened 4340 steel (cyclically softening). Source: Ref 2.2

also cyclically strain hardens during reversed plastic straining. Conversely, if the material has been cold worked by any of various fabrication operations (e.g., rolling, forging, swaging), the yield strength is increased significantly more than the ultimate tensile strength is increased. Hence, the strength ratio becomes smaller. The fabrication processes introduce excess tangles of dislocations that are trapped in a metastable state. Subsequent *elastic* cycling does not provide sufficient mechanical agitation to disturb the tangled internal state. However, cyclic *plastic* straining necessarily moves dislocations that can trigger dislocation rearrangement and their subsequent release or annihilation. This causes the cyclic yield strength to decrease and cyclic strain softening occurs. Correspondingly, the cyclic strain-hardening exponent $n'$ tends to increase.

The ratio $\sigma_u/\sigma_y$ in a monotonic tensile test might thus be expected to provide a quantitative measure of whether a material is likely to harden or soften during cycling. We have compared this ratio with the experimental observation of a number of materials that harden or soften (Ref 2.4). From these results we have come to the conclusion that if $\sigma_u/\sigma_y > 1.4$, the material is likely to cyclically strain harden; if <1.2, the material is likely to soften. For values between 1.2 and 1.4, the results are inconclusive; some materials hardened, others softened, and a few remained stable.

An alternative gage for indicating cyclic strain hardening or softening is to examine the value of the strain-hardening exponent $n$, if it is known. Again, the concept of the criterion is similar: materials with a high rate of strain hardening, i.e., high strain-hardening exponent, $n$ (>0.14), are likely to cyclically strain harden; those with low $n$ (<0.10) are likely to soften (see Appendix 2A1 at the end of this chapter). A derivation is also presented in this appendix that shows the relationship between the monotonic strain-hardening exponent $n$ and the monotonic strength ratio $\sigma_u/\sigma_y$.

**Concept of a Cyclic Stress-Strain Curve for a Hardening or Softening Material.** In the early stages of constant-amplitude strain cycling of an unstable material, it may either harden or soften substantially. Generally, most materials stabilize rather early in life, and thus the hysteresis loop during the stabilized cycling is usually used in the manner of Fig. 2.2 and Eq 2.2. If, however, cyclic stabilization does not occur early, or even at all, a common practice is to use the hysteresis loop at the cyclic half-life to represent the average cyclic stress-strain curve.

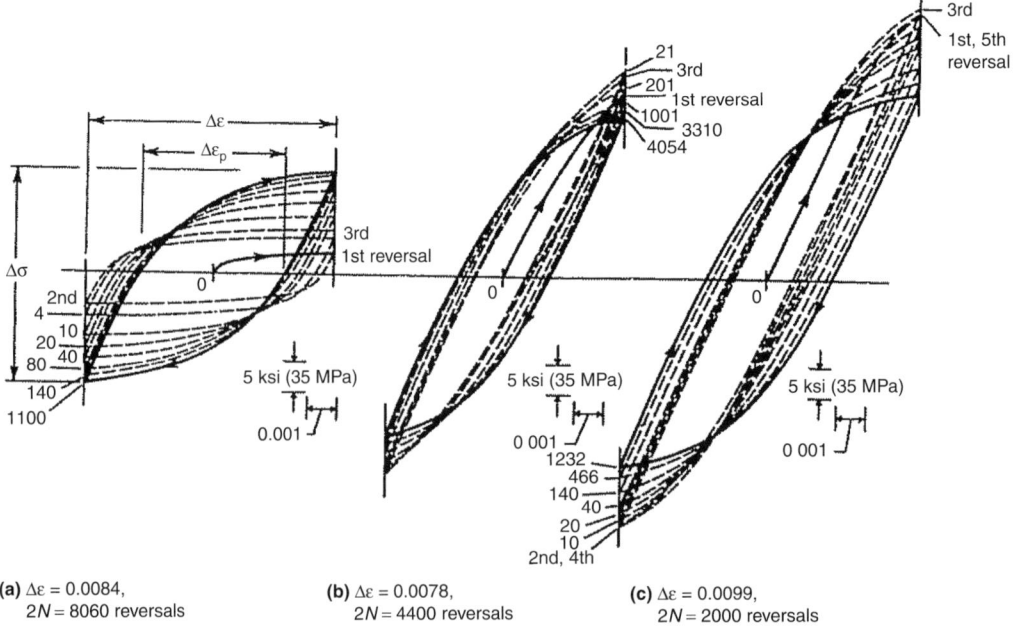

(a) $\Delta\varepsilon = 0.0084$, $2N = 8060$ reversals
(b) $\Delta\varepsilon = 0.0078$, $2N = 4400$ reversals
(c) $\Delta\varepsilon = 0.0099$, $2N = 2000$ reversals

**Fig. 2.6** Cyclic hardening and softening of oxygen-free high-conductivity (OFHC) copper under strain control depends on its initial hardness (Ref 2.3). (a) Fully annealed showing cyclic hardening. (b) Partially annealed showing a small degree of hardening. (c) Extremely cold worked showing cyclic softening. Source: Ref 2.3

Under certain circumstances, as discussed in Appendix 2A2 at the end of this chapter, it may be necessary to track the transition cycles from the first cycle to the point of stability or even to the point of anticipated failure. For such purposes, the hardening or softening transition curves may be approximated by two straight line segments that can adequately represent the nonlinear material behavior.

## Complex Straining Histories

In this section we discuss treatment of complex straining history for a stable material, or for a hardening/softening material using the aforementioned definition of the effective stabilized hysteresis loop of an initially nonstable material. We use the complex straining history shown in Fig. 2.7(a), applied to a material with single-amplitude stress-strain curve of Fig. 2.1. For a stable material, the cyclic stress-strain curve is the double-amplitude curve shown in Fig. 2.2. The resultant complex cyclic stress-strain path is shown in Fig. 2.7(b). The rules used in constructing this path are given subsequently. At the conclusion of this section, the rules are generalized for application to any arbitrary complex straining history.

To assist in determining how to proceed with the analysis, we first examine the stress-strain path of the ordinary monotonic stress-strain curve as in Fig. 2.1. The steps in the construction are shown in Fig. 2.7(b) as curve OA and are listed subsequently. It should be noted that the loops are always traversed in a "clockwise" direction and the coordinates are rotated by 180 degrees at each reversal point:

1. We designate the curve OA as the *single-amplitude template* given by Eq 2.1 for the monotonic stress-strain curve.
2. At A, the first strain reversal will take us along ADB to B. This curve is the *stable cyclic stress strain-curve* given by Eq 2.2 (using stress-strain axes that have been translated from O to A and rotated by 180°). This is known as the *double-amplitude template*.
3. Point B is now used as the origin and the stress-strain axes are again rotated 180°. We trace the double-amplitude template (Eq 2.2) from B through G, and end at the strain at C (same as at point A).
4. Point C is taken as the current origin and the double-amplitude template is used to go to the strain at D.
5. From D, following the double-amplitude template, we strain in the tensile direction to E.
6. At E we again reverse the direction of straining and head toward F. But, before reaching the strain at F, we encounter our first dilemma: At D we would intersect, at a substantial angle, a previously existing double-amplitude template that produced ADB from point A. If we were to continue the double-amplitude curve of origin E, we would have followed EDF, and higher

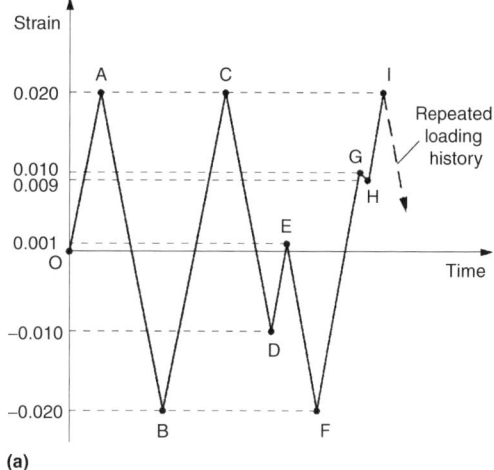

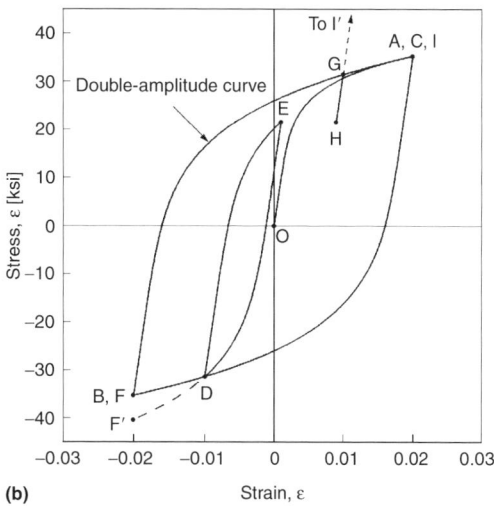

**Fig. 2.7** Analysis of complex straining pattern using templates of the single- and double-amplitude stress-strain curves of a stable material. (a) Strain history. (b) Constructed stress-strain paths

stresses would be achieved without going to higher strains than previously encountered—a situation that is inconsistent with a cyclically stable material. Thus, a rule must be invoked to prohibit this dilemma. This is known as the *nonpermissible rule,* or *memory rule,* which states that hysteresis loops cannot cross each other in such cycling of a stable material. Thus, the construction of the path from strain point $E$ to strain point $F$ must follow the segments $ED$ and $DF$. From $E$ (the new origin), the double-amplitude template can take us only as far as point $D$. At $D$ we must follow along the previous double-amplitude template to point $F$.

7. From $F$, we strain in the tensile direction to point $G$ following the previously established double-amplitude curve.
8. The traverse from point $G$ to point $H$ is again via the double-amplitude template with origin at $G$. Actually, for this problem, only the elastic region of the double-amplitude curve is required to reach $H$.
9. With $H$ as the new origin, we again trace a double-amplitude curve to try to go to the strain associated with $I$. Again, however, we encounter the dilemma that $HG$, if extended, would intersect the double-amplitude curve $FGI$ at $G$ and would want to continue to point $I'$, which is at a much higher stress than at $I$. However, since the path $GI'$ is nonpermissible, we stop at $G'$ and then follow the previously established curve $FGI$. So, in traversing from $H$ to $I$, the stress-strain path must follow the path $HG$, then $GI$ along the existing double-amplitude curve.
10. If the straining history included more reversals, the procedure followed would always be the same: (a) Initial straining from the origin to the maximum strain of the cycle along the initial single-amplitude stress-strain curve. (b) Each reversal follows a path along the double-amplitude curve with the reversal point as origin. (c) Each uses the double-amplitude curve of this origin until it intersects a previously existing hysteresis loop. (d) If additional strain is required, the path to that strain is along the intersected double-amplitude curve, which is drawn from its origin.

After the construction is completed, we examine the closed hysteresis loops that have been generated. In this case, three loops can be seen: the outside loop $ADBC$, the intermediate loop $DED$, and the elastic "loop" $GHG$. We also note the mean stress of each loop. The outside loop has zero mean stress, $DED$ has a small compressive mean stress that would normally be ignored, and $GHG$ has a large tensile mean stress but a small strain range.

Guided by the aforementioned example, we can formulate three rules that enable us to construct the stress-strain paths (hysteresis loops) for any specified repeated stress- or strain-time history.

- *Rule 1:* The starting point is obtained from the ordinary stress-strain curve (single-amplitude template) for a stabilized material by determining the stress at the highest strain of the cyclic history.
- *Rule 2:* Each time a strain-reversal occurs, we construct the next leg by following the double-amplitude stress-strain curve using the point of strain reversal as the origin and tracing to the strain at the point of next reversal.
- *Rule 3:* If the stress-strain path, following a reversal, intersects a previously existing stress-strain curve, continuing farther beyond the intersection follows the double-amplitude template curve, which had its origin at the reversal point from which the intersected hysteresis loop was constructed.

This treatment is for the cycling of either a material that displays no hardening or softening, or for a material after the major amount of hardening or softening has been completed. The analyst who wishes to take account of the early life when hardening or softening may be pronounced may wish to use changing templates that reflect the changes in hardness, which is dictated by his or her judgment and experience. Again, the analyst is referred to Appendix 2A2 at the end of this chapter for a suggested simplified means for treating initial transient stress-strain behavior.

Once the cyclic stress-strain analysis is completed by defining each hysteresis loop created during the complex straining, these loops are used in the subsequent computation of the fatigue life. Each closed hysteresis loop is considered as an event that can be used in further analysis. In the problem illustrated, we have three closed hysteresis loops, two of which have a mean stress. All could contribute to the fatigue life as discussed later in Chapter 4, "Mean

Stress," and Chapter 6, "Cumulative Fatigue Damage."

**Analysis by Rainflow Counting.** For very complex straining histories, it sometimes is not expedient to make a complete analysis of the type described previously in order to establish the exact hysteresis loops and their exact nesting within the system. The analyst may be willing to sacrifice some accuracy in order to expedite the analysis. *Rainflow counting* has been used for this purpose and is also especially useful in manual analysis of complex straining. The procedure has advantages and disadvantages as discussed subsequently.

The method is based on an ingenious analogy conceived by Tatsuo Endo (Ref 2.5) between the generation of hysteresis loops and the flow of raindrops down the roof of a Japanese pagoda. The roof of the pagoda is chosen in relation to the pattern of the strain history to be analyzed. The procedure can best be described in connection with the simple straining pattern of Fig. 2.7(a); however, very complex histories can also be analyzed in exactly the same manner.

To carry out the pagoda analysis, the time history of strain is rotated clockwise by 90°, and the resulting shape is imagined to form a series of rooftops of the pagoda. We start where the strain is at its maximum. Thus, Fig. 2.7(a) becomes the top view in Fig. 2.8. Now, we imagine a raindrop to appear inside the acute angle of each reversal point (apex) and allow it to drop under gravity from one rooftop to the next, being limited by only two restrictions:

- *Rule 1:* A raindrop must stop flowing when it passes an apex of the same sense (pointing in the same direction) but of equal or greater amplitude than the apex from which the raindrop originated.
- *Rule 2:* A raindrop may not flow in any section of a rooftop where a previous raindrop had flowed.

**Illustration of Rainflow Counting.** Starting with point $A$ in Fig. 2.8, the raindrop flows to $B$, drops, and stops at $B'$ as it passes opposite $C$ (rule 1). Next, a drop originates at $B$, flows to $C$, drops, and stops at $C'$ as it passes opposite of $F$ (rule 1). Then, a raindrop is started at $C$ and flows toward $D$ and drops to $D'$ along $EF$ and continues to flow to $F$, where it drops and stops at $F'$ as it passes opposite $I$ (rule 1). To continue, we start a new raindrop at $D$ that flows to $E$ and drops and stops at $E'$ when it passes opposite $F$ (rule 1). From $E$, another new drop flows to $D'$ and is stopped because it encounters previous flow (rule 2). To resume, we originate a raindrop at $F$, which continues to $G$, drops to $G'$, continues to $I$, and then drops. Since $I$ is the last point in the straining history, there is no place for the drop to go, so it remains suspended. But, we bypassed some raindrop paths on the way down, so we must go back and fill these in. Returning to $G$, a drop flows to $H$, and drops to $H'$ and stops as it passes opposite level $I$ (rule 1). Finally, the raindrop flowing from $H$ toward $I$ must stop at $G'$ because rain has already flowed from $G'$ to $I$ (rule 2). This completes the illustration of rainflow counting.

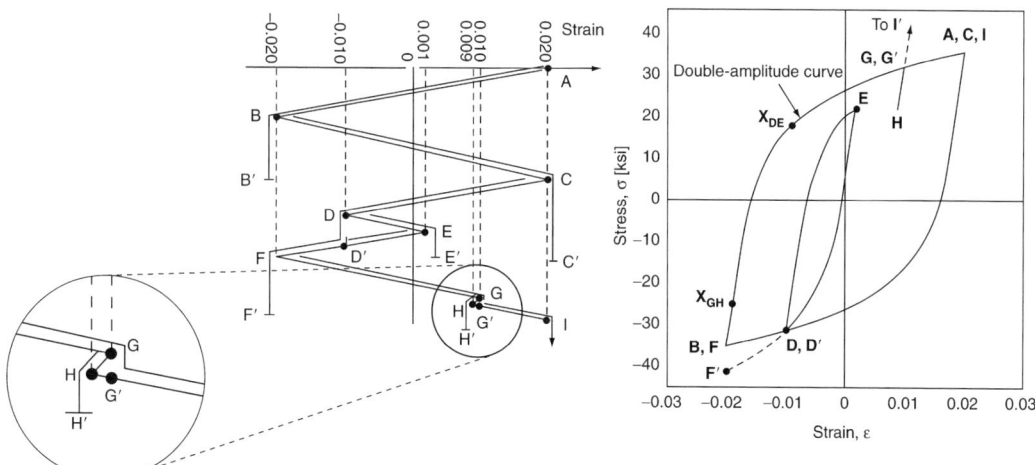

**Fig. 2.8** Combining the double-amplitude cyclic stress-strain curve with rainflow concepts to construct realistic hysteresis loops, which identify mean stresses. Top part is 90° rotation of Fig. 2.7(a). Bottom portion same as Fig 2.7(b)

Each raindrop path corresponds to a strain excursion. There are positive and negative strain excursions, and each corresponds to a respective positive and negative half loop. However, during the actual construction of a rainflow analysis, no attempt is made to match up loop halves as they occur. Matching of halves to form complete loops is done after the entire rainflow analysis has been completed. At that time, positive half loops combine with negative half loops of the same range of strain (and same mean strain) to generate full loops. The strain excursion pairing is shown in Fig. 2.9, and the resultant hysteresis loops appear in the bottom portion of Fig. 2.8.

To generalize, a rainflow analysis of the type just described, even for very complex straining, provides us with a multiplicity of strain excursions and half loops that correspond to the start and stop of the flow of each raindrop. Each half loop matches up with another half loop to form all the hysteresis loops that are generated during the complex loading. Although in the simple example it is clear how the half loops combine to form hysteresis loops, truly complex histories do not always lend themselves to such ready pairing. In hysteresis loop analyses, as described earlier, exact pairing is readily apparent. Knowing the exact pairing is necessary because each of the half loops involved may have a different mean stress. A rainflow cyclic strain analysis, per se, does not reveal mean stresses of the loops anyway; therefore, direct matching is not required if only rainflow analyses are performed. Nevertheless, the procedure is very useful because it can be conducted rapidly, and where mean stress effects are not of overriding importance, serves as a very appropriate substitute for a complete analysis.

**Manual Analysis Combining Rainflow with the Double-Amplitude Stress-Strain Curve.** While the method of Fig. 2.7 results in the determination of all closed hysteresis loops, and their relative locations, so that mean stress for each loop is known, there are conditions for which a full-blown analysis may not be justified. For example, it may be desired to make a manual analysis of several alternative strain histories to establish a first-order approximation of their relative severities, or it may be desired to determine which of several materials, each having a different cyclic stress-strain curve, might be most suitable for an application involving a known strain history. For this purpose, we can use several of the principles already established to make a simple graphical analysis either manually, or, in fact, by a simplified computerized procedure. To illustrate the procedure, we can again make use of the simple strain history of Fig. 2.7(a); more complex straining histories simply produce more complex results, but the principles involved are the same.

Because we have already established that all the loops (beyond the initial straining) consist of segments of the double-amplitude cyclic stress-strain curve, and that the external loop is bounded by two symmetrical segments of this loop, we can draw this loop first. In Fig. 2.8 this is loop $ABC$*. For convenience in carrying out

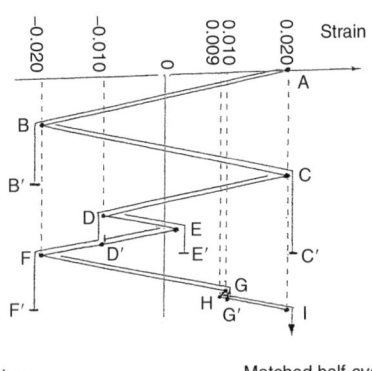

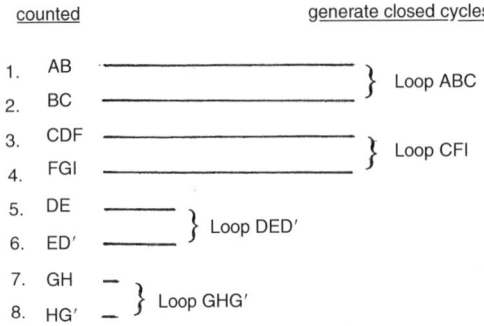

**Fig. 2.9** Rainflow analysis of the strain history shown in Fig. 2.8 (top portion same artwork as Fig 2.8 top)

---

*In this problem we have chosen the maximum and minimum strains of equal amplitude, so the loop is symmetrical with respect to the axes. If the two amplitudes are not equal, for example, if the maximum strain is 2% and the minimum strain is $-1$%, it is still necessary to construct the outside loop symmetrical with respect to the axes. The center value is taken as zero, and the maximum and minimum values adjusted accordingly. In this example, the max and min values would again become $+1.5$% and $-1.5$%. What happens when the two strains are not equal is that the loops do not close after application of the first block, and the whole loop system moves in the first few blocks until they settle at a block system in which the outside loop becomes symmetrical. After the system becomes symmetrical with the axes, the loops will repeat from block to block.

the graphical construction, it is useful to make a template of the curve AB by pasting a plot of the curve on a piece of cardboard (the back support of a paper pad serves well). It then becomes possible to construct rapidly any segment of the curve by tracing along this template. The more sophisticated analyst will use a simple computer program to achieve the same result.

To start the manual analysis, we locate point A on the double-amplitude curve at a strain of 2%, as required. To construct the straining AB, we simply place the origin of the template at point A and trace the double-amplitude curve in the compressive direction, carrying the curve to the required −2% strain at B. Segment BC is also simply traced with point B as the origin, tracing until the strain becomes +2% at point C. Up to this point, the rainflow concept has not been necessary.

The first time we find need for guidance from the rainflow construction is in the tracing of the strain ED. We start at D as the origin at −1% strain and trace the double-amplitude curve to a small tensile strain at point E. From E to F it is again a simple matter to begin the tracing from the origin E. However, as we trace we soon intersect, at a nonzero angle, the existing curve CDF at point D. But we know that hysteresis loops do not cross each other at a nonzero angle, so at D we cannot continue by further tracing of the curve with E as the origin. If we examine the rainflow diagram, however, we note that at D the rainflow of ED stops and that the rain from D to F is really a continuation of the rainflow from C to F. So, by the analogy between strain generations and rainflow, we must continue the curve that had its origin at C. Thus, DF is a continuation of the curve that started at C, not the one that started at E. We trace the piece DF by using the curve with its origin at C. Of course, this portion of the stress-strain curve already existed so that no new construction is really necessary.

Continuing on, we place the origin of the double-amplitude template at F and traverse to a tensile strain of 1% at G along the curve FGI. At G we reverse the direction of straining and go to point H in accordance with the template. Only elastic deformation is involved in this excursion of strain. Again, we call on the template and place its origin at H and strain toward +2%. This is also a case where the rainflow concept is helpful. At point G, we encounter a previous stress-strain path, BC, that cannot be crossed. Thus, we must follow the segment GI instead of GI′.

Thus, we have been able to construct the true hysteresis loops in their proper locations. As before, we note that loop GHG has a high positive mean stress, and DED is a loop that has a small negative mean stress. These mean stresses will later be involved in determining the fatigue life associated with each loop (chapter 4, "Mean Stress").

**Corollary of the Construction of Hysteresis Loops for Complex Straining.** While we are discussing the loops of Fig. 2.8, and the mean stresses of the individual loops, we can make an observation that will later prove very useful in discussing the analytical relations for fatigue life when mean stress is present. The fact, obvious from Fig. 2.8, is that for any one of the hysteresis loops that result from complex straining, the stress range and strain range together follow the relation given by the double-amplitude cyclic stress-strain curve. This observation follows from the very rules of construction, as discussed. We can also see it if we plot the stress range versus the strain range of each loop on the double-amplitude curve. Thus, the segment $BX_{GH}$ corresponds exactly to the same length line segment GH, and the loop segment $BX_{DE}$ corresponds exactly to the loop segment DE. Both segments clearly lie on the curve BGI with all their origins at point B. If more loops were present, they, too, would lie on the double-amplitude cyclic stress-strain curve if the stress range of the loop were plotted versus strain range.

## Experimental Determination of the Cyclic Stress-Strain Curve

For accurate determination of the cyclic stress-strain curve, it is necessary to perform some experiments on the material. Fortunately, it requires very few specimens, sometimes only just one, to obtain an acceptable determination of this curve, although, if many specimens are available, more accurate results can be obtained. Some of the methods based on various features of the curve we have already discussed are:

- *Method A:* Using the fact that the cyclic stress-strain curve represents the stabilized stress values for various strain values, the obvious approach is to test a series of specimens, each at a different strain range, observing the stabilized ranges of stress that develop, and plotting the curve directly. That, in essence, is the concept used to determine the cyclic stress-strain curve shown in Fig. 2.5.

- *Method B:* Recognizing that for many materials, the stabilized value of stress is achieved well before the specimen fails (implying that not much fatigue damage has occurred before near-stabilization of stress, as discussed later) suggests to us that a single specimen can be used to study several strain levels, perhaps to as high a strain as is usually required. Figure 2.10 shows the basic idea behind this method. The smallest strain range $\Delta\varepsilon_1$ of the series is applied first for a number of cycles, and the stress is observed. When the rate of change of stress indicates that saturation (stable value) is near, the strain range is changed to the next higher value $\Delta\varepsilon_2$, and the process is repeated. Similarly, $\Delta\varepsilon_3$ is applied when the stress range associated with $\Delta\varepsilon_2$ is near stabilization, and so on to higher strain ranges. A plot of the results produces the cyclic stress strain curve. If the amplitude of stress is plotted against the amplitude of strain, the curve can be compared directly with the ordinary monotonic stress-strain curve; if ranges of stress and strain are used, the double-amplitude cyclic stress-strain curve results.

- *Method C:* Recognition that after enough blocks of a given complex straining have been applied, the condition of the material stabilizes and leads to a third method shown in Fig. 2.11. Here, modulated strain amplitude is applied in a servocontrolled machine (axial straining) and the required modulated pattern of stress determined. The stabilized values of stress versus the applied strains in the later blocks provide values of stress to plot against strain for determination of a cyclic stress-strain curve. Again, if amplitudes of stress and strain are used, the cyclic stress-strain results; if ranges of strain and stress are used, the double-amplitude stress-strain curve results. This method is of interest because it is easy to perform and the test need not be attended by a technician.

- *Method D:* Perhaps the simplest method of obtaining the cyclic stress-strain curve is

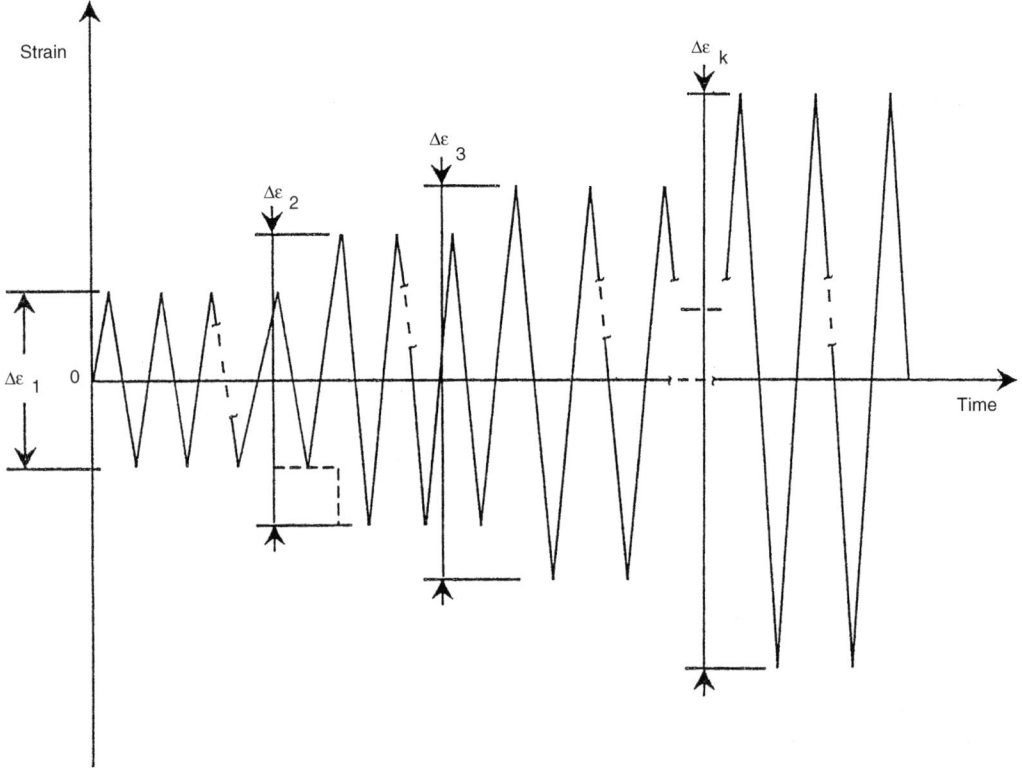

**Fig. 2.10** A typical multiple-step straining history for determining the cyclic stress-strain curve of a material using a single specimen

based on the fact illustrated in Fig. 2.8 that the outside loop of the stress-strain history provides the double-amplitude stress-strain curve. Thus, any arbitrary convenient block of straining, for example, those of Fig. 2.10 and 2.11, or that representing a type of straining of special interest to the investigator, can be used. An analog *X-Y* recorder or a digital computer recording is required so that stress can be plotted against strain, and the outside loop determined as the double-amplitude cyclic stress-strain curve. The cyclic stress-strain amplitude curve is then obtained by dividing both stress range and strain range by 2.

- *Method E:* A final method is now considered. Based on the observation that if a specimen has been cycled many times to the point of stress-strain stability, and then returned incrementally to its initial origin (as done at the end of a block in Fig. 2.11, i.e., stress and strain are both equal to zero), then the subsequent monotonic stress-strain curve will be the same as the cyclic stress-strain amplitude curve. The far right-hand curve in Fig. 2.12 illustrates the cyclic stress-strain curve derived from this procedure. The material is first subjected to enough blocks of increasing/decreasing symmetrical cycles (the incremental step test shown in Fig. 2.11) so as to stabilize its stress-strain behavior, and then "uncycled" by applying symmetrical, progressively decreasing strain amplitudes, as shown, finally returning the material to its original origin at 0. Then, the monotonically conducted stress-strain curve *OAB* becomes the cyclic stress-strain amplitude curve. It is important to recognize, however, that many cycles must be used. Applying only a few applications of strain will produce a static stress-strain curve that is neither the ordinary stress-strain curve (because the cycling has already altered the original static properties) nor the cyclic stress-strain curve (because the cycling did not stabilize the material). Alternatively, if the straining cycles are in coarse increments, the entire process of straining may be repeated a number of times in order to stabilize the cycling behavior.

The advantage of method E is that it can be accomplished even if no equipment is available to measure the cyclic stress, and if no *X-Y* recorder is available to plot hysteresis loops. While the loops are shown in Fig. 2.12, it is not really necessary to display them graphically. The only equipment needed is that which will "cycle" and "uncycle" the material through a progression of increasing and decreasing strains.

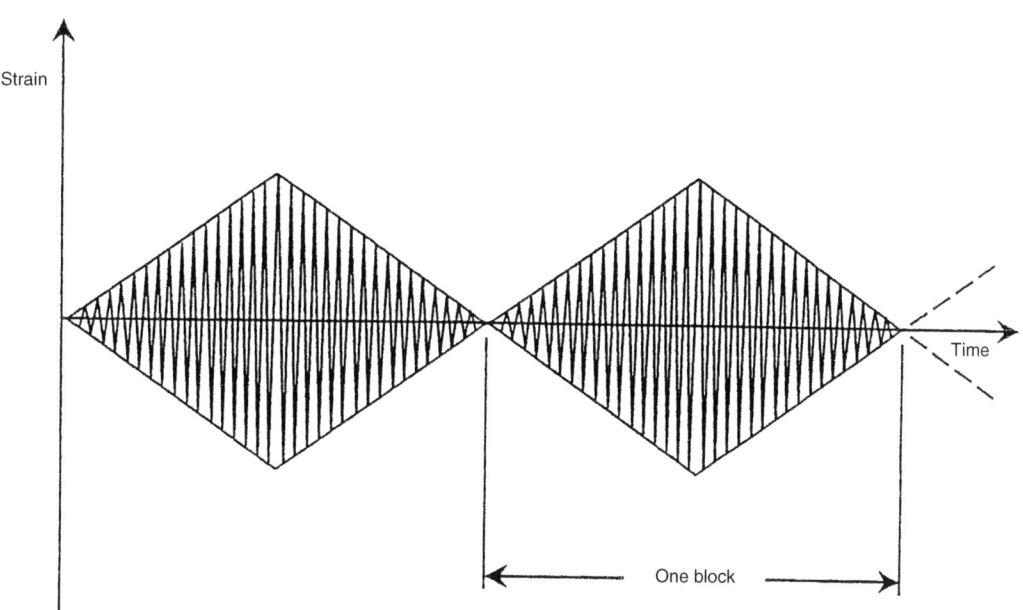

**Fig. 2.11** Blocks of linearly increasing-decreasing straining history for an incremental step test. Source: Ref 2.6

Then, a static stress-strain curve will reveal the cyclic stress-strain properties.

A study by Nachtigall (Ref 2.6) compared several methods of determining the cyclic stress-strain curve. He included methods A, B, and E in his study. His results are shown in Fig. 2.13. The agreement among the three methods is quite good for this alloy steel.

| Starting with a virgin specimen, the cyclic strain is incrementally increased, producing larger and larger hysteresis loops. | When the maximum strain range is reached, the strain range is incrementally decreased until the behavior is nominally elastic. | The strain is gradually increased to the same maximum strain range. Note change in the cyclic stress-strain curve. | The strain is cyclically decreased again. Notice that the cyclic stress-strain curve is nearly stabilized (is the same as before). | Finally, a static tension test is conducted on the same specimen. Note similarity between this curve and the stable cyclic stress-strain curve. |

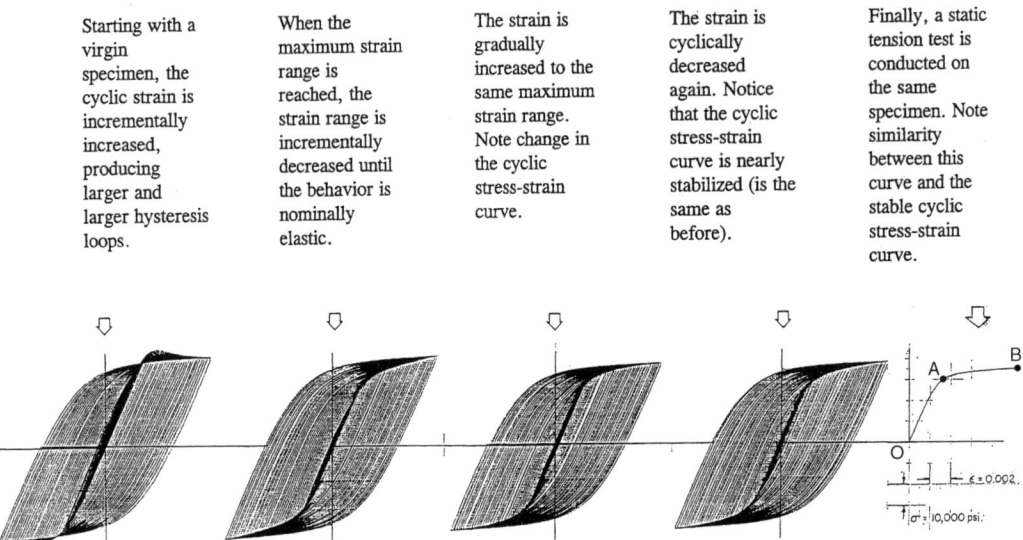

**Fig. 2.12** Incremental step cycling of cold-worked oxygen-free high-conductivity (OFHC) copper followed by monotonic tension. Source: Ref 2.7

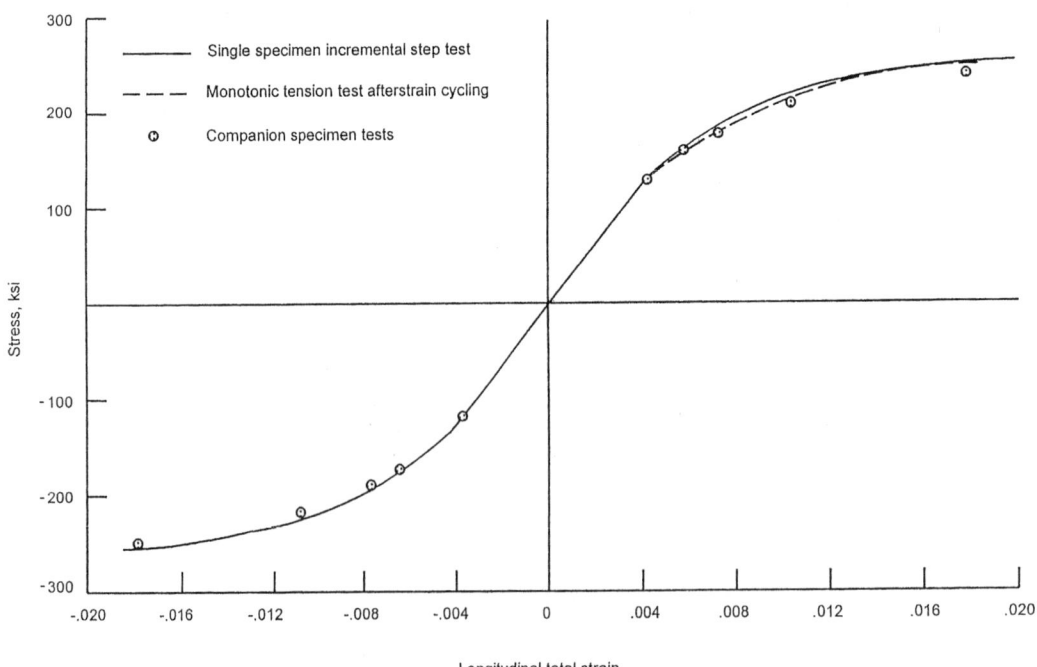

**Fig. 2.13** Comparison of cyclic stress-strain curves obtained by Nachtigall (Ref 2.6) for D6AC steel obtained by three different test methods

## Complex Pattern of Strain Cycling Analyzed by Mechanical Models

It is appropriate to present a mechanical model that provides insight into the reasoning behind the process. We shall, in fact, present two mechanical models: one that has conventionally been used for many years, the other similar in appearance, but with a detail of difference that greatly facilitates the actual processing of the results.

**The Conventional Mechanical Model.** Elastic materials are conventionally modeled as springs to reflect strain proportionality to stress, and strain restoration when stress is removed. To introduce the characteristic of plasticity, an element containing friction, such as a slider on a friction surface, together with a second spring in parallel, is added in series with the first spring. In Fig. 2.14, the second element will not deform for strains less than the slider friction resistance $R_2$, and only $S_1$ deflects. For strains greater than $R_2$, both $S_1$ and $S_2$ deflect, the deflection of $S_1$ being $P/E_1$ (where $E_1$ is the elastic spring constant of $S_1$), and the deflection of $S_2$ being $(P - R_2)/E_2$ since the slider absorbs a strain $R_2$, and the spring $S_2$ absorbs the strain $P - R_2$. The break in the force-deflection curve is analogous to a sharp yield point in a stress-strain curve, and the resistance $R_2$ may be thought of as analogous to resistance to plastic flow. Once started, however, plastic flow occurs freely, the dislocations resisting with a constant force until they meet obstacles that require additional force and strain hardening takes place. The equivalent of strain hardening in the spring and slider model is the slope of the curve AB.

If, after straining to B, the force is reduced, $S_1$ contracts a proportional amount, but $S_2$ cannot contract until the force is reduced to a point where the frictional resistance of $R_2$ in the opposite direction is overcome. The condition for this to occur is that the pull of $S_2$ on $R_2$, together with the external force, must be greater than $R_2$. Thus, for $P < 2R_2$, the return path will be a straight line parallel to OA, and when the force is reduced to zero, a residual deflection OC will be left in the material. If $P - 2R_2 > 0$, nonlinear deflection in compression may occur even before the force reaches zero.

Thus, while the addition of an element such as ($S_2$, $R_2$) imparts a rudimentary characteristic of known stress-strain behavior, it is not sufficient to model realistic behavior of most materials. To become realistic, we must add many different nonlinear elements similar to ($S_2$, $R_2$); however, for purposes of simple discussion, let us assume that two nonlinear elements will suffice. From an analysis of such a model, sufficient behavioral features will develop that will enable us to generalize. Furthermore, let us discuss the problem in connection with specific numerical values assigned to all the parameters in the chosen model so that physical significance becomes apparent.

Consider the combination of springs and sliders shown in Fig. 2.15 and apply to it a force of +4000 lbf. We must consider the forces in increments since different behavior occurs within the increments. However, it is clear that between the increments shown, the information is linear so we need consider only the end points.

It is instructive, especially for the uninitiated reader, to follow the discussion in detail, making reference to the displacement values for each element listed in the table of Fig. 2.15 and the corresponding plot of force versus total deflection in Fig. 2.16.

Up to point A in Fig. 2.16 ($P = 2000$ lbf), only $S_1$ deflects, while elements ($S_2$, $R_2$) and ($S_3$, $R_3$) remain immobilized because the external force is not sufficient to unlock these elements. Between 2000 and 3000 lbf, element $S_1$ continues to deflect and the element ($S_2$, $R_2$) also participates. However, the deflection in this element is obtained by subtracting $R_2$ from the applied force and dividing the result by the spring-constant of $S_2$. However, element ($S_3$, $R_3$) remains immobile. The point B corresponds to 3000 lbf. From 3000 to 4000 lbf (points B to C), all three elements are active. During this interval, elements $S_1$ and $S_2$ continue to increase their deflections. At point B, element $S_3$ contributes its deflection according to the same rationale as described for element $S_2$; i.e., the deflection is given by subtracting $R_3$ from $P$ and dividing by the spring constant of $S_3$.

At point C, let us reverse the force in increments, eventually applying a compressive force of $-4000$ lbf. When we return to 3000 lbf at point D, $S_1$ contracts by 0.5 in., but ($S_2$, $R_2$) and ($S_3$, $R_3$) remain in their immobilized maximum extended positions. At zero force, point F, a residual tensile deflection of 4 in. exists, and element ($R_2$, $S_2$) will finally begin to move in the reverse direction. As before, the amount of motion is determined by the contraction of $S_2$ after account is taken of both the external force and the drag that $R_2$ now applies in the positive direction. The element ($S_3$, $R_3$) is still not ready to start displacing in the negative direction, so the straight line return $FG'$ continues until point $G$, where the external force is $-2000$ lbf, together with the residual force in $S_3$ overcome the resistance $R_3$, and return displacement starts to occur in the element $(S_3, R_3)$. From $G$ to $I$ all three elements contribute to reverse displacement as force goes from $-2000$ to $-4000$ lbf. At $I$ the force-displacement state is symmetrical to point $C$ relative to the origin.

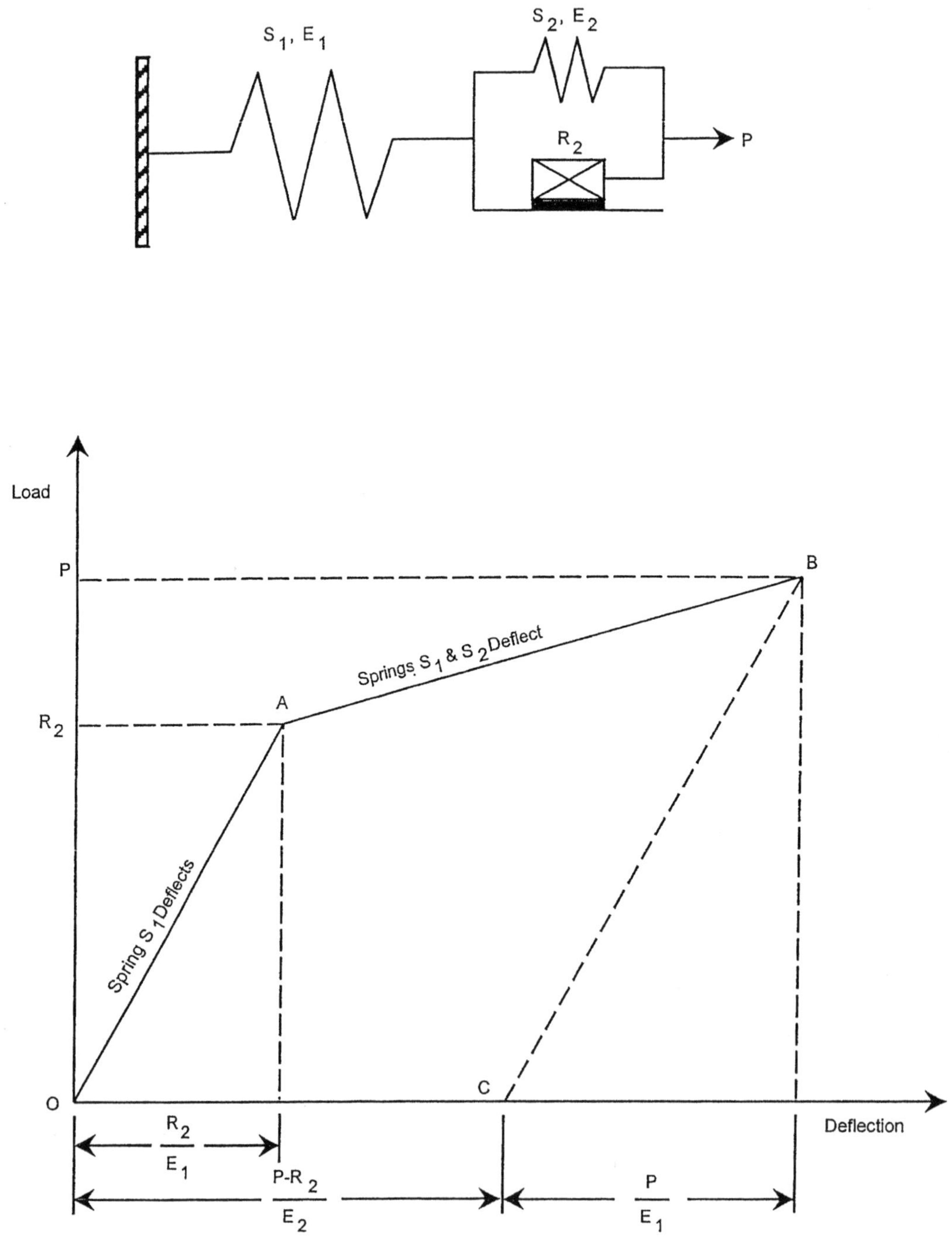

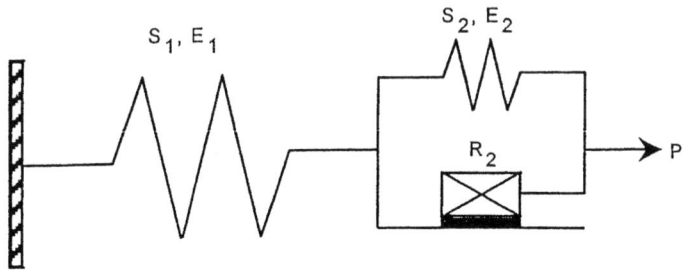

**Fig. 2.14** Conventional model involving springs and slider for representing hysteresis

Reversing the straining follows a path governed by the same principles as the forward straining, and the path becomes *IJKLMNC*. How each of the three elements contributes to the displacement is shown in the table of Fig. 2.15. It is clear that each straining increment produces some displacement in each of the elements, and that careful track must be kept of the force and its direction to determine how much of the external force is absorbed by the slider, so that the remainder may cause spring displacement. In this case, with only three elements, the bookkeeping is fairly simple. However, if there were 30 or more elements involved, the bookkeeping problem could become very complicated. It is also clear from Fig. 2.16 that the double-amplitude curve is twice the single amplitude curve.

There is, however, a simpler way to set up this problem so that the results are the same but the bookkeeping is greatly simplified. This approach is described in the next section.

**The Simplified Mechanical Model Using Displacement-Limited Elements.** To reduce complexity, consider the model shown in Fig. 2.17. Again, for simplicity, we show three elements involving either a spring alone or in combination with a slider. However, "stops" are provided to limit the amount of deformation of each spring. For convenience, we imagine bars attached to the end of each spring, which prevents further displacement once each spring reaches the limit of its allowable deflection. The stops prevent further deflection, additional imposed deflection being transferred to another element.

For simplicity, assume a stop for spring $S_1$ at 2000 lbf, which is the force corresponding to the drag of $(S_2, R_2)$. Therefore, up to a force of 2000 lbf, only $S_1$ deflects. However, if more force is applied the remaining force is absorbed by $(S_2, R_2)$. Placing a "stop" on the deflection of $(S_2, R_2)$ when the total force is $+3000$ lbf (corresponding to the drag force in element $[S_2, R_2]$) will then activate straining in element $(S_3, R_3)$. From this point, $(S_2, R_2)$ no longer contributes to deflection. Thus, we have now replaced the system of Fig. 2.15 by that of Fig. 2.17 in which the "breakpoints" are the same. The spring constants must, however, be changed if the deflections in the two systems are to be the same at all forces. Thus, we make spring $S_2$ have a spring constant, $S_2' = 2000/3$ lbf/in., i.e., $[(1/2000) + (1/1000)]^{-1}$. Similarly, the spring constant of $S_3$ must be replaced by $S_3' = 2000/7$ lbf/in., i.e., $[(1/2000) + (1/1000) + (1/500)]^{-1}$. Once we

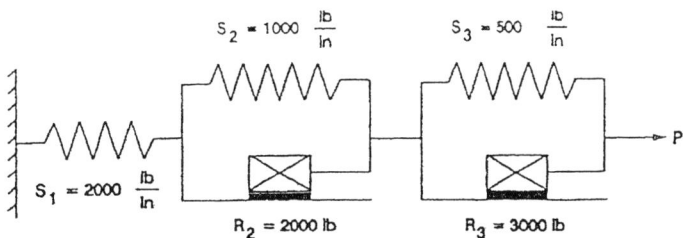

| Point on Fig. 2.16 | Applied load P (lb) | Loads in elements, lb |  |  |  |  | Displacements of elements, in. |  |  | Total displacement (in.) |
|---|---|---|---|---|---|---|---|---|---|---|
|  |  | $S_1$ | $S_2$ | $R_2$ | $S_3$ | $R_3$ | $S_1$ | $S_2 = R_2$ | $S_3 = R_3$ |  |
| 0 | 0 | 0 | 0 | 0 | 0 | 0 | 0 | 0 | 0 | 0 |
| A | +2000 | −2000 | 0 | −2000 | 0 | −2000 | +1.0 | 0 | 0 | +1.0 |
| B | +3000 | −3000 | −1000 | −2000 | 0 | −3000 | +1.5 | +1.0 | 0 | +2.5 |
| C | +4000 | −4000 | −2000 | −2000 | −1000 | −3000 | +2.0 | +2.0 | +2.0 | +6.0 |
| D | +3000 | −3000 | −2000 | −1000 | −1000 | −2000 | +1.5 | +2.0 | +2.0 | +5.5 |
| E | +2000 | −2000 | −2000 | 0 | −1000 | −1000 | +1.0 | +2.0 | +2.0 | +5.0 |
| F | 0 | 0 | −2000 | +2000 | −1000 | −1000 | 0 | +2.0 | +2.0 | +4.0 |
| G | −2000 | +2000 | 0 | +2000 | −1000 | +3000 | −1.0 | 0 | +2.0 | +1.0 |
| H | −3000 | +3000 | +1000 | +2000 | 0 | +3000 | −1.5 | −1.0 | 0 | +2.5 |
| I | −4000 | +4000 | +2000 | +2000 | +1000 | +3000 | −2.0 | −2.0 | −2.0 | −6.0 |
| J | −3000 | +3000 | +2000 | +1000 | +1000 | +2000 | −1.5 | −2.0 | −2.0 | −5.5 |
| K | −2000 | +2000 | +2000 | 0 | +1000 | +1000 | −1.0 | −2.0 | −2.0 | −5.0 |
| L | 0 | 0 | +2000 | −2000 | +1000 | −1000 | 0 | −2.0 | −2.0 | −4.0 |
| M | +2000 | −2000 | 0 | −2000 | +1000 | −3000 | +1.0 | 0 | −2.0 | −1.0 |
| N, B | +3000 | −3000 | −1000 | −2000 | 0 | −3000 | +1.5 | +1.0 | 0 | +2.5 |
| C | −4000 | −2000 | −2000 | −1000 | −1000 | −3000 | +2.0 | +2.0 | +2.0 | +6.0 |

**Fig. 2.15** Three-element model

# 26 / Fatigue and Durability of Structural Materials

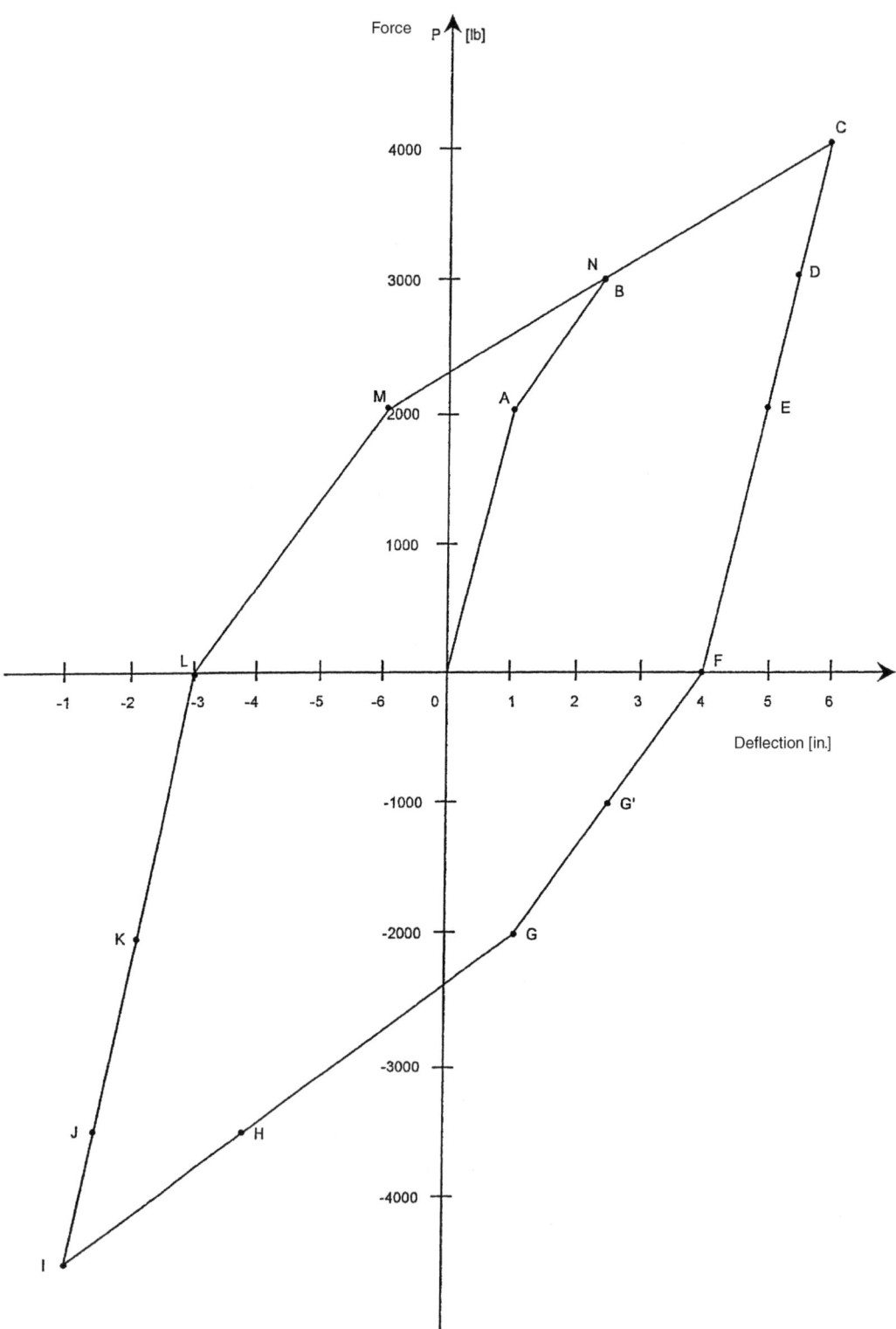

**Fig. 2.16** Force-deflection diagram for the three-element model of Fig. 2.15

have done this, the systems of Fig. 2.17 and 2.15 have exactly the same force-deflection characteristics. The big difference, however, is that whereas in Fig. 2.15 all elements may change deflection during some of the straining, in Fig. 2.17 only one element changes in each range of the strain. In region $R_I$, only $S_1$ deflects; in region $R_{II}$, only $S_2$ deflects (while $S_1$ and $S_3$ remain at constant deflection) and forces in region $R_{III}$ or greater, only $S_3$ deflects. It is evident that if there were many elements, say 30 or more, the scheme of Fig. 2.17 would be much easier to analyze than that of Fig. 2.15.

To make the system of elements have the same compression and tension characteristics, we must also put "stops" at corresponding points to the left of the equilibrium positions of the springs in the same way as there are limits to its tension characteristics.

A further simplification of the bookkeeping can be accomplished by normalizing all displacements (or forces, if preferred) relative to their total allowable deflection in a given direction. Thus, every element is allowed to displace ±1 unit from its equilibrium position. For the simple case of Fig. 2.17, forces from 0 to 2000 lbf will cause unit displacements of 0 to 1 in element $S_1$; from 2000 to 3000 lbf, element $S_1$ has a deflection of 1.0, while the deflection of $(S'_2, R_2)$ varies from 0 to 1.5 in, or if we divide the actual displacement by 1.5, element $(S'_2, R_2)$ will also displace 1 unit, and similarly for element $(S'_3, R_3)$, we divide the actual displacement by 3.5 to get a unit displacement between forces of 2000 to 3000 lbf.

Now we can show how much simplification occurs when we use the scheme of Fig. 2.17 compared with that of Fig. 2.15. The table in

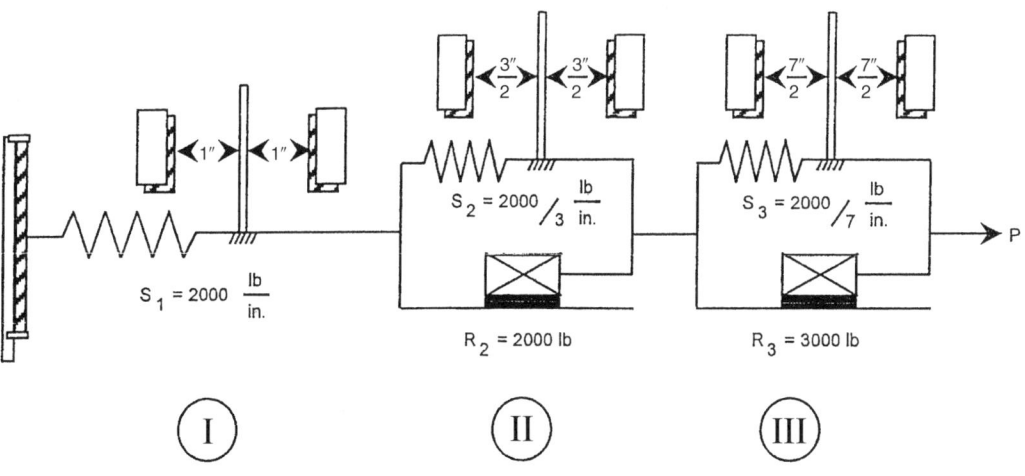

| Point on Fig. 2.16 | Applied load P (lb) | Actual displacement, lb ||||  Fraction of allowable displacements |||
|---|---|---|---|---|---|---|---|---|
| | | Total | I | II | III | I | II | III |
| 0 | 0 | 0 | 0 | 0 | 0 | 0 | 0 | 0 |
| A | +2000 | 1.0 | 1.0 | 0 | 0 | 1.0 | 0 | 0 |
| B | +3000 | 2.5 | 1.0 | 1.5 | 0 | 1.0 | 1.0 | 0 |
| C | +4000 | 6.0 | 1.0 | 1.5 | 3.5 | 1.0 | 1.0 | 1.0 |
| D | +3000 | 5.5 | 0.5 | 1.5 | 3.5 | 0.5 | 1.0 | 1.0 |
| E | +2000 | 5.0 | 0 | 1.0 | 3.5 | 0 | 1.0 | 1.0 |
| F | 0 | 4.0 | −1.0 | 1.5 | 3.5 | −1.0 | 1.0 | 1.0 |
| G | −2000 | 1.0 | −1.0 | −1.5 | 3.5 | −1.0 | −1.0 | 1.0 |
| H | −3000 | −2.5 | −1.0 | −1.5 | 0 | −1.0 | −1.0 | 0 |
| I | −4000 | −6.0 | −1.0 | −1.5 | −3.5 | −1.0 | −1.0 | −1.0 |
| J | −3000 | −5.5 | −0.5 | −1.5 | −3.5 | −0.5 | −1.0 | −1.0 |
| K | −2000 | −5.0 | 0 | −1.5 | −3.5 | 1.0 | −1.0 | −1.0 |
| L | 0 | −4.0 | 1.0 | −1.5 | −3.5 | 1.0 | −1.0 | −1.0 |
| M | +2000 | −1.0 | 1.0 | 1.5 | −3.5 | 1.0 | 1.0 | −1.0 |
| N, B | +3000 | 2.5 | 1.0 | 1.5 | 0 | 1.0 | 1.0 | 0 |
| C | +4000 | 6.0 | 1.0 | 1.5 | 3.5 | 1.0 | 1.0 | 1.0 |

**Fig. 2.17** Analysis of a simple loop for a three-element model with displacement "stops"

Fig. 2.17 shows the essential results. The first three columns show essentially each of the points in the straining history of Fig. 2.16, and the next three columns the individual contributions to the total displacement of each of the three elements. In the final three columns, the displacements are normalized to their total allowable displacement; i.e., because the total displacement of element I is 1.0 in. (in the positive direction), that of element II is 1.5 in., and that of element III is 3.5 in., we divide all displacements in the second three columns by 1.0, 1.5, and 3.5 for elements I, II, and III, respectively.

What is of special interest in the table is that all elements in the last three columns are characterized by displacements that can vary from $-1$ to $+1.0$, and that they are introduced in sequence: first element I, then II, then III. As one element uses up its capacity to deflect, the next one takes over in sequence, and so forth.

It is also evident that once the elements are chosen to provide a selected deflection at each breakpoint, it really is not necessary to deal with individual values of spring constants. We need only know what the deflections are at each breakpoint and we can reference all deflections in an element as functions of the total deflections allowable within the elements, taking into account the algebraic sign of the displacement from its equilibrium position. If we specify force history we can derive strain history, or, if we specify strain history, we can derive force history.

The analogy to the use of a stress-strain curve rather than a force-deflection curve is self-evident. Thus, to determine stress history for a given strain history we need only replace the stress-strain curve by an equivalent curve with a sufficient number of breakpoints along the curve so that straight lines joining the breakpoints are a reasonable representation of the curve, and then operate on the strain history as implied by the model discussed to determine stress history. Conversely, we can specify stress history and determine strain history by an analogous process. The actual procedure is discussed in the next section in the description of Wetzel's method (Ref 2.8). It is interesting to note that while Wetzel did not provide any physical model as the basis for his analytical method, expressing his rules on the basis of empirical observations, the model shown in Fig. 2.17 can in fact be used to provide a physical basis for the method, which he posited without proof.

**Shape of the Straining and Reverse Straining Path.** Before applying Wetzel's method to an illustrative problem, it is appropriate to make an observation regarding the stress-strain path during both the initial straining and in subsequent straining. Note that if the concept of replacing material behavior by a series of springs and sliders, such as those of either Fig. 2.15 or Fig 2.17, correctly represents material behavior, then the initial straining path, for example, OABC in Fig. 2.16, will consist of a series of straight lines. Although we used only three elements in this example, we could, of course, choose as many as we wish, and in this way track any arbitrary stress-strain path we choose. In practice, when the procedure is computerized, we might choose 30 or 40 straight-line elements to conform very closely to the real stress-strain curve. Furthermore, we note that if an analysis is made for a fatigue problem, it is presumed that the straining history will be repeated many times, and therefore, the stabilized cyclic stress-strain curve is the appropriate one to model. If one is interested especially in the first few cycles of strain, the monotonic and early-straining cyclic stress-strain curves may be appropriate ones to use, but the procedures are basically the same except that they are applied to changing stress-strain curves from cycle to cycle.

What is of even greater interest is the shape of the outside hysteresis loop relative to the initial stress-strain curve. Note in Fig. 2.16 that line CDEF is parallel to OA and by nature of the construction is exactly twice its length. Similarly, FG is parallel to AB, and twice its length. Finally, GHI is parallel to BC and twice its length. Thus, the straining path CDEFGHI is the same as the straining path OABC, except that it is twice as large, i.e., the double-amplitude path. Of course, the reversed straining path IJKLMNC is the same as CDEFGHI rotated 180°. Thus, if we know the initial straining path (i.e., the cyclic stress-strain curve), we can determine the shape of the subsequent straining paths as the double-amplitude curve of the cyclic stress-strain curve. We will determine, by applying Wetzel's procedure, which is described next, that all elements of the straining history, no matter how complex, will consist of segments of this double-amplitude curve, although where the segment starts, where it stops, and the location of its origin will not always be obvious. Wetzel's procedure always produces the correct segment. Thus, we now have a physical explanation for why the double-amplitude curve works.

**Wetzel's Method.** Although Wetzel (Ref 2.8) derived his procedure intuitively by observing practical results rather than basing it on the modified model of Fig. 2.17, the results are identical, as already noted. And because his method found widespread acceptance, we shall illustrate an example by reference to his method.

The procedure first replaces the stress-strain curve by a series of straight lines that reasonably define this curve. For our illustrations, we shall use only three straight-line segments, but in practice, 30 or more elements may be used to get close coincidence between model and material behavior. The curve modeled is the cyclic stress-strain curve because the major part of the life for most engineering materials involves the material in a state when equilibrium of hardening or softening has been achieved. The analysis starts when the material is at its maximum strain, which is usually known, and the analysis assumes that the strain history (i.e., strain control) is specified while stress history is to be determined. Of course, stress control can be analyzed by the same procedure. Because the starting point is the maximum strain, Wetzel assigns a value to each of the "availability coefficients" (analogous to our fraction of strain of [1, 1, 1]) in the table of Fig. 2.17, meaning that all the elements are in their most strained position. Wetzel's rules are simple, and their logic follows directly from the illustrations in the table in Fig. 2.17:

1. Apply the elements in succession: first element I, then II, then III, etc., to whatever extent they are required to satisfy the imposed strain history. Keep track of the extent of the use of each element. Availability coefficients may vary from 1 to $-1$; when an element reaches its extreme value, proceed to the next element.
2. At each strain reversal start over again, proceeding from element I to the others in succession until the requirement of that strain reversal is satisfied.

The procedure can best be illustrated by a simple example, which contains most of the features normally found in complex problems.

Figure 2.18(a) shows the cyclic stress-strain amplitude curve for a hypothetical material. For an initially stable material, this is the same as

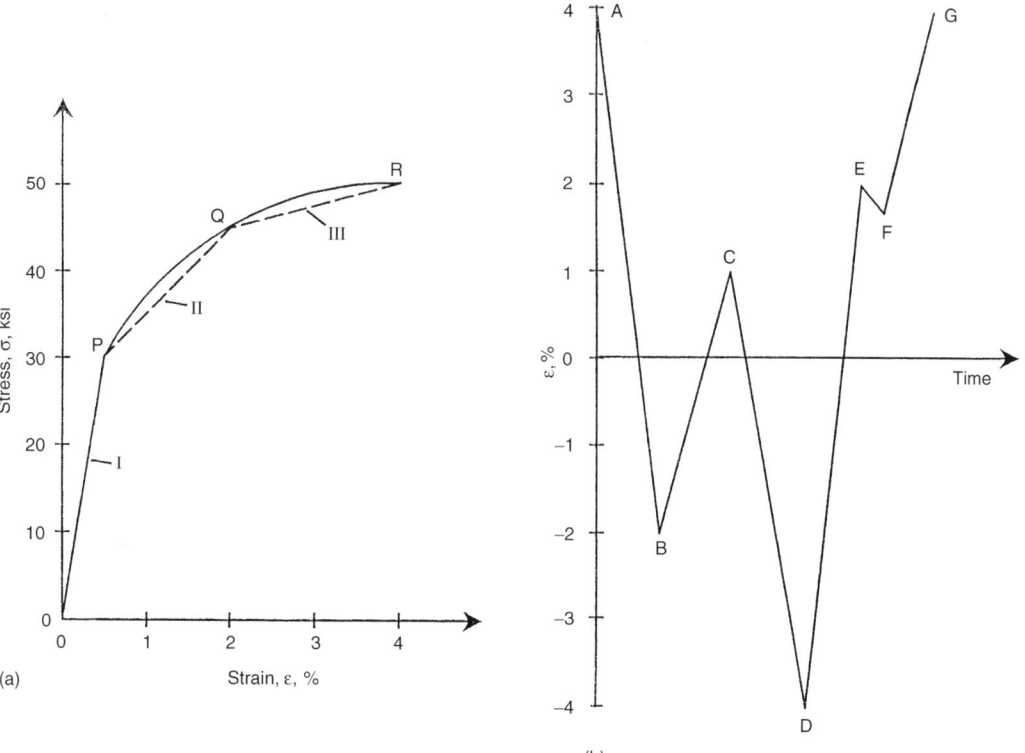

**Fig. 2.18** Cyclic stress-strain curve and strain history for a straining block. (a) Cyclic stress-strain amplitude curve. (b) Strain history

the monotonic stress-strain curve. For illustration we represent the curve by three straight-line segments *OP* (element I), *PQ* (element II), and *QR* (element III). Figure 2.18(b) shows the strain history to which the material is to be subjected. We assume that the history repeats.

To analyze the straining, we use Fig. 2.19 and the inset table. Point *A* is plotted at the maximum strain so its strain is 4% and stress is 345 MPa (50 ksi). The numerical values of the availability coefficients for the three elements are 1, 1, 1 as noted in the table. To implement the first cyclic straining *AB* from 4% to −2%, we first apply element I in the compressive direction. As availability coefficients can go only from 1 to −1, we draw a line parallel to element I, line *AU*, twice (415 MPa, or 60 ksi) the length of element I (210 MPa, or 30 ksi), which brings us to point *U* where the availability coefficient of element I is −1, but since we have not yet used elements II and III, their availability coefficients remain 1 and 1, respectively, as noted in the second column of the table of Fig. 2.19.

Because at point *U* we have not yet reduced the strain to −2%, we proceed by applying element II. So *UV* is parallel to element II and can be drawn as much as twice its length while the availability coefficient goes from 1 to −1 at point *V*. At point *V*, then, the availability coefficients of both elements I and II are −1, but

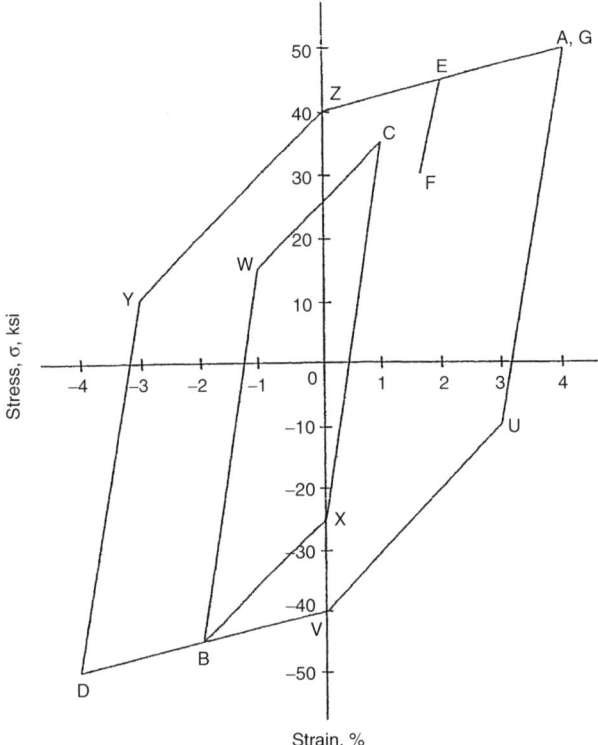

| Point | A | U | V | B | W | C | X | D | Y | Z | E | F | G |
|---|---|---|---|---|---|---|---|---|---|---|---|---|---|
| Element I | 1 | -1 | -1 | -1 | 1 | 1 | -1 | -1 | 1 | 1 | 1 | 1/2 | 1 |
| Element II | 1 | 1 | -1 | -1 | -1 | 1/3 | 1/3 | -1 | -1 | 1 | 1 | 1 | 1 |
| Element III | 1 | 1 | 1 | 0 | 0 | 0 | 0 | -1 | -1 | -1 | 0 | 0 | 1 |
| Strain (%) | 4 | 3 | 0 | -2 | -1 | 1 | 0 | -4 | -3 | 0 | 2 | 1 3/4 | 4 |
| Stress (ksi) | 50 | -10 | -40 | -45 | 15 | 35 | -25 | -50 | 10 | 40 | 45 | 30 | 50 |

**Fig. 2.19** Resulting stress-strain behavior and bookkeeping scheme

that of element III is still +1 because it has not yet been used.

Because the desired control strain of −2% has not yet been reached, we must continue to further reduce the strain; element III must now be used. As seen in Fig. 2.19 and column 4, the control condition is achieved at point $B$ where the availability coefficient of element III turns out to be zero (incidentally to this problem, but not generally).

At $B$ there is a reversal in strain direction, so we must start over again to use elements I, II, and III in order. From $B$ to $C$ the strain must change from −2% to +1%. We start with element I (which from the fourth column of the table has, at $B$, an availability coefficient of −1) and apply it in the positive direction. The maximum amount of use of this element that can be made is twice its length in the direction parallel to it, bringing us to point $W$, where its availability coefficient is now 1. (Note, however, that the availability coefficients of elements II and III at $W$ are the same as at $B$ because no use has been made of the elements in proceeding from $B$ to $W$.) At $W$, however, we have not yet achieved the desired strain of +1%; therefore, we must now introduce element II. Drawing $WC$ parallel to the direction of element II, we find that it intersects the desired 1% strain when the length of $WC$ is 1⅓ the length of element II. Thus, the availability coefficient of element II at $C$ becomes ⅓ (i.e., changing from −1 to 0 for a single length, and from 0 to ⅓ for the other ⅓ length to arrive at $C$). In the column under $C$ in the table, element I has an availability coefficient of 1.0 (the same as at $W$ because element I is not used to proceed from $W$ to $C$), element II has an availability coefficient of ⅓ as just calculated, and element III has an availability coefficient of 0 (same as at $W$ because it, too, was not used to go from $W$ to $C$).

In the same way, all the other strains in the straining history can be tracked. The uninitiated student may find it very instructive to follow through the entire straining history displayed in Fig. 2.18(b) and detailed in the table of Fig. 2.19. In general, the procedure is straightforward, but several features are of special interest, especially since they lead to the generalizations we discuss later. Note, for example, the path from $C$ to $D$; the hysteresis loop actually followed is $CXBD$. Note that at $C$ the availability coefficients are (1, ⅓, 0), so in applying element I we can yet reduce it to −1, bringing us to point $X$, where the availability coefficients are (−1,  ⅓, 0). At this point we can add only 1⅓ the length of element II in the negative direction, bringing us to point $B$ (where we have, in fact, been before) where availability coefficients are (−1, −1, 0). Because the desired strain of −4% has not yet been reached, however, we need only one times the length of element III to get to point $D$, where all the availability coefficients are (−1, −1, −1). Thus, the path from $C$ to $D$ used elements II and III in only truncated pieces governed by their lesser availabilities as described by their availability coefficients. As path $CXB$ intersected $VB$, a preexisting piece of a hysteresis loop, it was necessary to make a turn and follow a continuation of the preexisting hysteresis loop, rather than crossing it acutely. This feature of "memory," the material's recognition that it had previously been brought to a given state by an earlier path, is the valuable feature of the analysis in which the "bookkeeping" of the stress-strain path is retained via the medium of availability coefficients.

The same feature of "memory" is noted in the strain path $FG$. At $F$, the availability coefficients are (½, 1, 0). Thus, in reversing the strain back to the positive direction, only ½ the length of element I can be used, bringing us to point $E$ in Fig. 2.19. When we attempt to utilize element II in the positive direction from $E$, we find that it already has an availability coefficient of 1.0, so that it cannot be used at all. We must then proceed to element III for which the availability coefficient is 0. Thus, as we traverse to $G$, the availability coefficient goes from 0 (at $E$) to 1 (at $G$).

**Generalization of Hysteresis Loop Construction.** While, for simplicity of discussion, we have used only three elements to represent the cyclic stress-strain curve, it is clear that if we had used 30 or even 50 elements, their aggregate would very accurately represent the smooth stress-strain curve, and if we program the process on a computer, we can accurately track any arbitrary straining history according to the basic procedure outlined. Several simple generalities would emerge from such a procedure:

- Instead of requiring fractional parts of an element in order to track a given strain, we can eliminate fractions and use only complete elements, assigning only the availability coefficients 1.0, 0, and −1.0 according to whether a given element is used in the positive strain direction, not used at all, or used in the negative direction. Such a procedure

simplifies the computations considerably. The computer program developed in (Ref 2.8) basically followed this procedure.
- All closed hysteresis loops generated by a straining history are loops bounded by "symmetrical" (reflected about both the stress and strain axes) segments of the double-amplitude cyclic stress-strain curve. Thus, for example, the loop *EF* contains only the elastic section of the double-amplitude curve; the loop *BWCXB* contains the whole elastic segment and a small region of plasticity, while *DYZEAUVBD* contains the elastic segment together with all the plasticity pertinent to the problem. But all loops are generated from the double-amplitude curve. Thus, the stress ranges and strain ranges for all loops satisfy Eq 2.3. This observation is very important, as we discuss later in connection with mean-stress effects on the life relations.
- All closed hysteresis loops are nested within the outside loop, which is defined by the difference between the maximum and minimum strain in the straining history. Closed loops do not cross one another, although they may "blend" into one another.
- When a strain path intersects a preexisting loop, it continues along the double-amplitude stress-strain curve, not from the origin of the last reversal, but from the origin of the loop that it intersects. Thus, for example, in proceeding from *C* to *D*, the path *CXB* intersected *AUVB*; the continuation of the straining is along the double-amplitude curve of the preexisting cyclic stress-strain curve intersected, *AUVB*. Similarly, when *FE* intersects *DYZE*, it continues along this curve to *G*.
- The outer hysteresis loop has a zero mean stress, but the nested loops may have mean stress of positive (e.g., *EF*) or negative (e.g., *BWCXB*) value. These loops may involve different fatigue lives than loops of the same amplitude, but zero mean stress, as we discuss in Chapter 4, "Mean Stress."

## Model Analysis of More Complex Materials

The model consisting of springs and sliders is adequate for representing the behavior of most metals, which is extremely fortunate because it greatly simplifies stress-strain analysis. Unfortunately, some important engineering materials do not lend themselves to this simple treatment. Gray cast iron is one such material. It is relatively cheap, easy to manufacture into complex shapes, has high conductivity, low elastic modulus, and can absorb a considerable amount of heat energy without thermal distortion. It also has good lubricating characteristics. For these reasons, it is commonly used in the transportation and machine-tool industries where these characteristics are highly desirable. In particular, our interest has involved gray cast iron as an automotive disk brake rotor, and the fatigue considerations associated with repeated thermal straining.

**Gray Cast Iron as an Important Engineering Material.** The advantageous characteristics of gray cast iron derive largely from its metallurgical structure. Basically, the material has a steel matrix, which can, by proper thermal treatment, be given many of the desirable properties of steel. But the carbon content is purposely made very high so that part of the carbon takes the form of clusters of free graphite. The graphite forms cabbage-like clusters having relatively sharp-tipped leaves or flakes (Ref 2.9) within a steel matrix. Because these graphite flakes are weak and brittle, they do not participate effectively in carrying tensile stress, but carry compression well. Thus, their monotonic and cyclic stress-strain curves are different in tension and in compression, which makes the use of spring-slider elements alone inadequate for modeling hysteresis loops. Special elements are required to model cast iron. Many different methods have been applied by various investigators to model the behavior of cast iron, but relatively few have been applicable to the study of cyclic behavior. Together with graduate student Kurt Heidmann, we have, however, been quite successful in devising a model, which incorporated the following necessary features:

- Compression behavior, which is different from tensile behavior
- Nonlinear stress-strain behavior, which is due in part to plasticity and in part to the cracking of both the graphite flake and the matrix in its vicinity
- A progressively changing elastic modulus, as inferred from tests involving the application of a force to various levels, and temporary releases of these forces

To accomplish the requirements, three different types of elements were employed as shown in Fig. 2.20:

- A general plasticity element, consisting of a spring and slider, as shown in Fig. 2.20(a)
- A gap element, which provides no constraint against tensile motion, but rigidly prevents compressive motion. This element is also combined with a spring and slider, as shown in Fig. 2.20(b).
- A prestressed gap element that allows tensile motion only after the stress on this element exceeds a preestablished value, $\sigma_1$. This element is combined with a spring, as shown in Fig. 2.20(c).

The stress-strain response of each type of element is shown in Fig. 2.20 adjacent to each schematic model. These elements are intended to model what happens in both tension and compression, as outlined in Fig. 2.21. In tension, the deformation must include crack opening, while no such opening occurs in compression. How the elements combine to produce these displacements is shown in Fig. 2.22. The quantitative choices for each of the elements for the material we studied are shown in Fig. 2.23, these values being determined by observing the tensile and compressive stress-strain response, as well as the response to small stress releases at each of a number of stress levels. Using these parameters in a computer model, a series of hysteresis loops were generated, as shown in Fig. 2.24, together with experimental observations. The agreement between actual and computer-model predictions is quite good. How a specimen would respond to a complex straining history is shown in

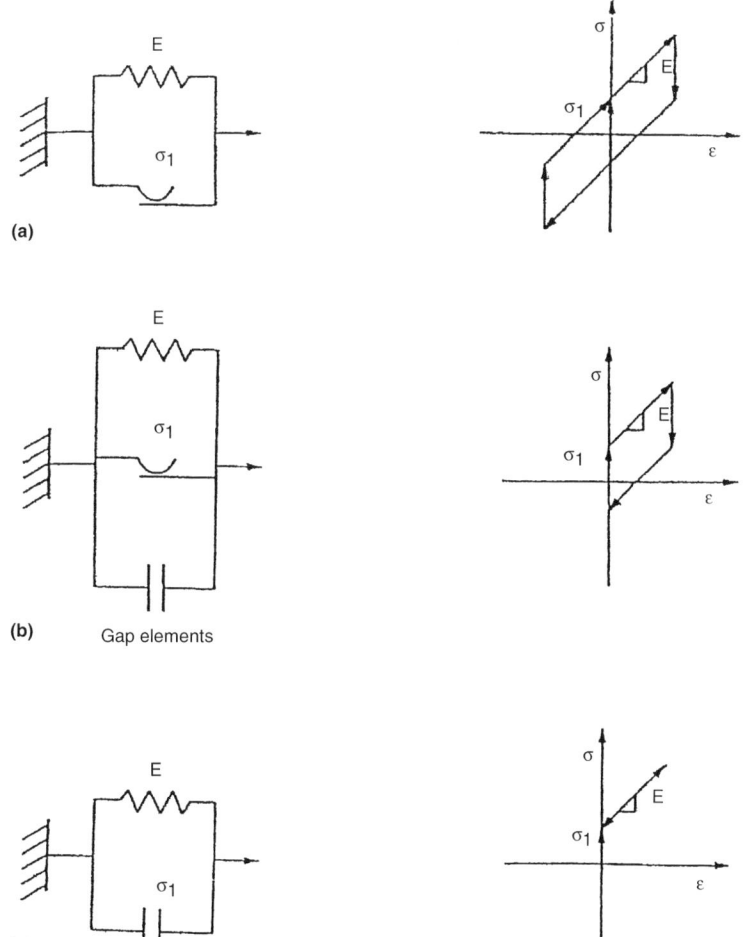

**Fig. 2.20** Compound spring-slider-gap elements. (a) General plasticity (spring-slider). (b) Tensile-only plasticity (spring-slider-gap). (c) Nonlinear plasticity (spring-gap). Source: Ref 2.9

**34 / Fatigue and Durability of Structural Materials**

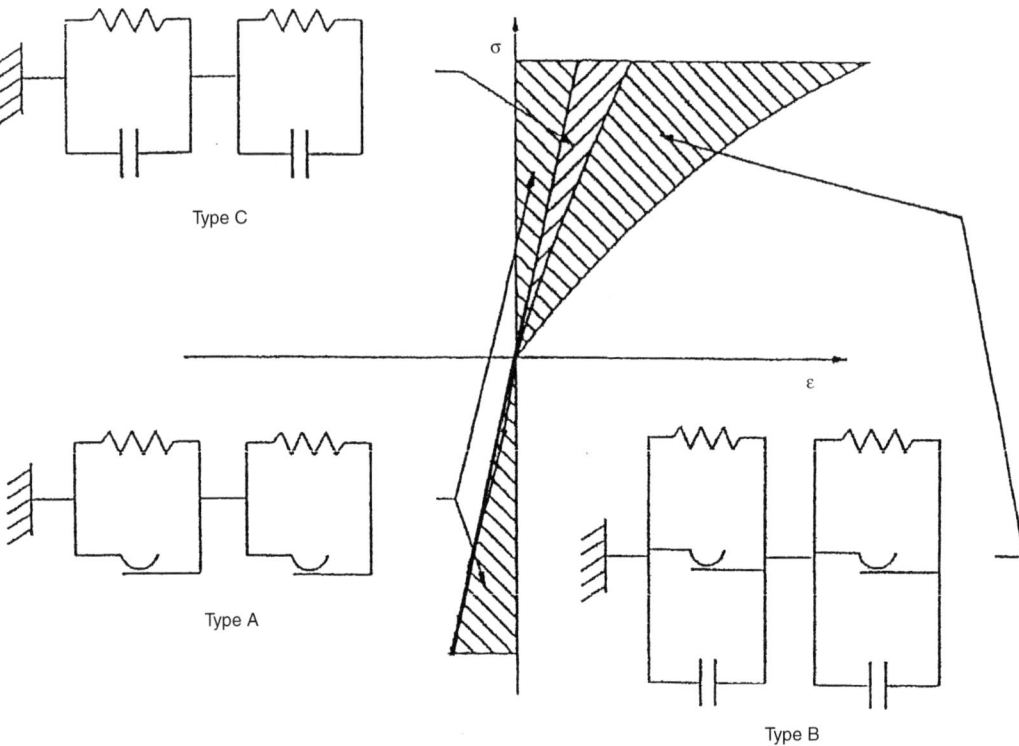

**Fig. 2.21** Deformation models in tension and compression for gray iron. Source: Ref 2.9

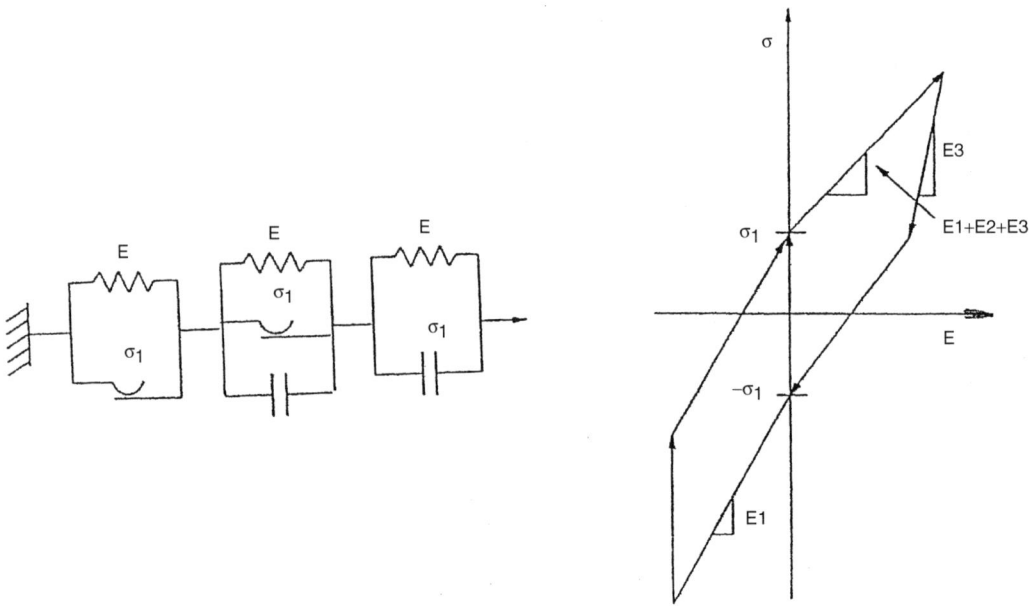

**Fig. 2.22** Combined compound element behavior. Source: Ref 2.9

Fig. 2.25. Interestingly, as with the simple spring-slider models, the result is a series of nested hysteresis loops.

It would be desirable to extend this type of analysis to high-temperature behavior and to temperature-gradient situations. Heidmann's analysis is a good start and can serve as a paradigm for further study. Continuum damage mechanics also has been used to attack this type of problem. Promising results have been achieved by Downing (Ref 2.10) and Fash (Ref 2.11) in their doctoral theses of the 1980s.

## Force versus Stress-Cycling

A distinction should be made between force cycling and true stress cycling. Consider a test conducted under conditions of force cycling wherein the force is simply reversed from a tensile value to its equal, but negative value in compression. Even if the force is supported equally by all elements of the cross-sectional area, the true stress $\sigma$ will not be equal in tension and in compression because of cross-sectional area effects. When tension force is applied, the cross-

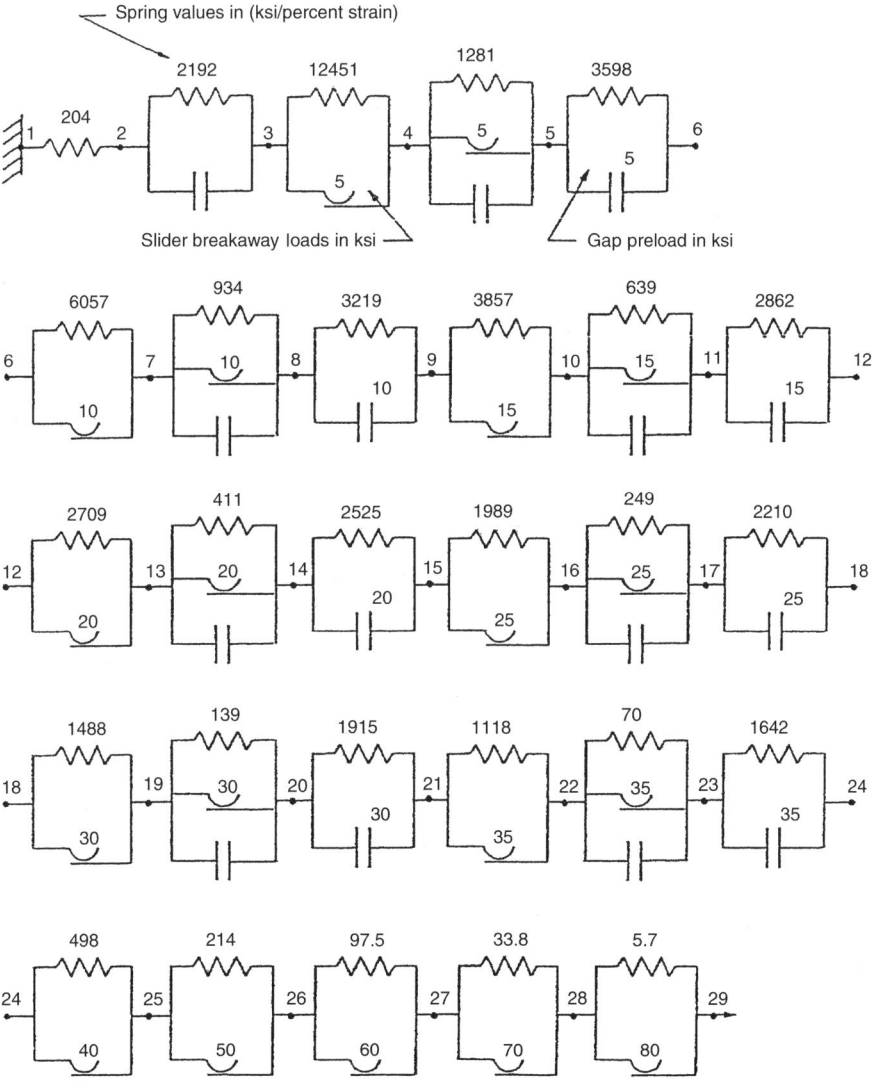

**Fig. 2.23** Spring-slider-gap model for gray iron with spring constants. Source: Ref 2.9

sectional area is reduced according to Poisson's ratio, increasing the true tensile stress $\sigma_t$ over its nominal engineering value $S_t$. In compression, the cross-sectional area is increased, thus reducing the stress. It can be shown that a tensile strain increases the true stress by about the same percentage value as the strain $\varepsilon$ magnitude itself, i.e., $\sigma_t \approx (1 + \varepsilon)S_t$. Thus, a 1% tensile strain gives rise to $\sigma_t \approx 1.01 S_t$. Similarly, a compressive (negative) strain of 1% reduces the true compressive stress $\sigma_c \approx (1 - \varepsilon)S_c = 0.99 S_c$. The effect can be significant if large plastic strains are involved. Obviously, once tensile strains become large enough, necking will commence and tensile rupture will quickly follow.

If the critical region of a uniformly strained cross section is to be truly stress cycled, the force must be adjustable in tension and in compression to maintain true stress of the extremes of the cycle. Tests are rarely conducted in this manner, not only because of the relative experimental difficulty, but also because of complications that arise in connection with both strain-hardening and strain-softening materials. No

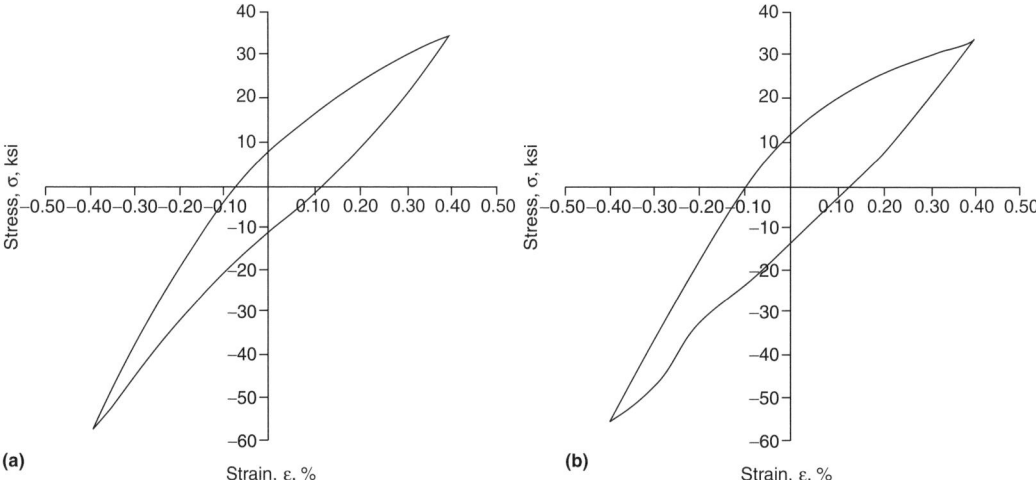

**Fig. 2.24** Computer modeled cyclic stress-strain behavior compared with experimental results for gray cast iron. (a) Experimental. (b) Computed. Source: Ref 2.9

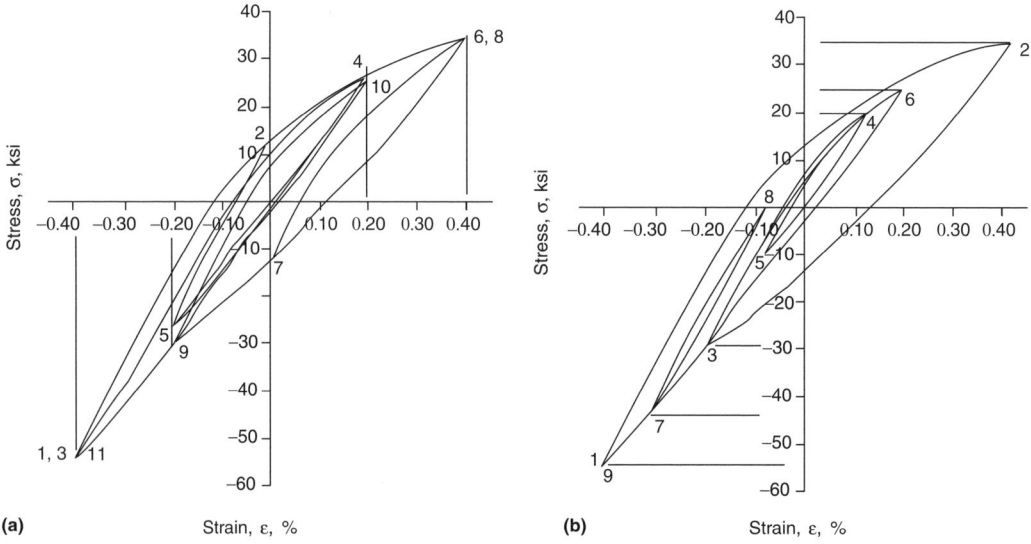

**Fig. 2.25** Computer model behavior under complex straining pattern for gray cast iron. (a) Strain control. (b) Stress control. Source: Ref 2.9

difficulties arise for perfectly stable (i.e., neither softening nor hardening) materials, but very few materials satisfy this exact description. We consider next the special behavior according to whether the material is softening or hardening.

**Softening Materials.** If the material softens, the amount of deformation that occurs for a given force may increase from cycle to cycle simply because, as the material softens, more strain is associated with the same force. This is especially true under force cycling. The material thus ratchets in the tensile direction and the specimen becomes longer with every cycle. This behavior is shown in Fig. 2.26 (Ref 2.12) for a cyclically strain softening 4340 steel. The stresses shown are nominal stress ranges $\Delta S$ for completely-reversed force cycling. It is seen that for nominal stress ranges above $\Delta S = 2185$ MPa (317 ksi), the strain ratchets to quite large tensile strains. In fact, specimens tested under these conditions may not fail by fatigue, but rather by a tensile necking process not unlike that found in a monotonic tension test. Figure 2.27 shows the permanent stretch, i.e., just the plastic strain that remains in the material after each straining. Even for the relatively low stress of $\Delta S = 2185$ MPa (317 ksi), wherein no ratcheting occurs in the early cycles, eventually, ratcheting starts at about 100 cycles, after the material has been sufficiently softened. Figure 2.28 shows in detail the tensile, compressive, and residual strains for the particular loading of $\Delta S = 2500$ MPa (364 ksi).

**Hardening Materials.** In studying hardening materials a decision has to be made at the outset whether the stress is to be applied to the material in its initially soft condition, or whether it should be applied only once the hardened condition has been achieved. To appreciate the need for this decision, we should consider that there are some materials of practical interest that have an ultimate tensile strength in the dead-soft condition that is *less* than the stabilized stress amplitude (after cyclic hardening) that can be achieved at an operational condition in service. This service condition in an actual part cannot be simulated with a laboratory specimen. If simulation were to be attempted under force-controlled testing, the sample would fail on the very first tensile force application. Thus, if large constant-amplitude force cycling is applied to a material in its initial soft condition, the initial plastic strain range would be many times larger than that for the stabilized condition. Such large strains can

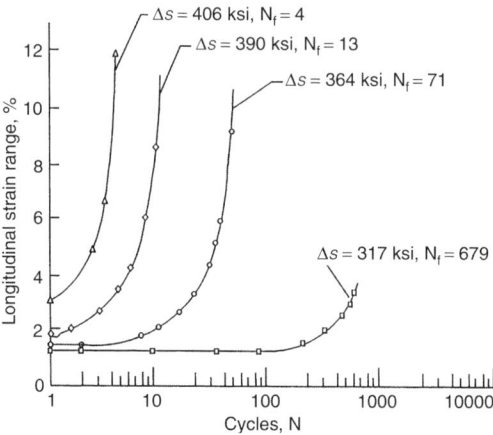

**Fig. 2.26** Strain-range ratcheting under force cycling of cyclically strain softening 4340 steel

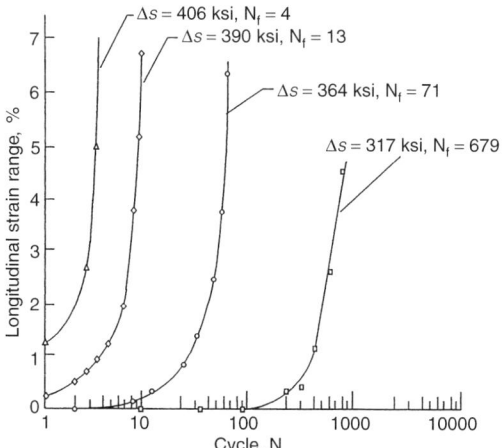

**Fig. 2.27** Permanent stretch under force cycling of cyclically strain softening 4340 steel

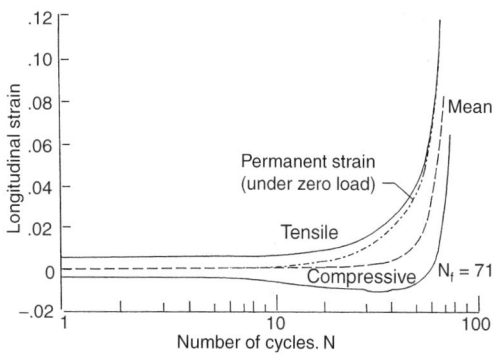

**Fig. 2.28** Strain behavior during force cycling of cyclically strain softening 4340 steel, $\Delta S = 2500$ MPa (364 ksi)

be responsible for plastic strain-induced surface roughness and cycle-dependent compressive buckling. In fact, it may be impossible to generate force-controlled fatigue results in the low- to intermediate-cycle fatigue life regime. Even low stress amplitude force cycling will exhibit large initial plastic strain ranges that decrease with applied cycles to nearly undetectable levels as hardening occurs. Long lives may result that are beyond the range of interest in certain practical applications.

The end result of a force-controlled fatigue testing program conducted with a highly hardening alloy may be of limited value and applicable over only a narrow range of interest. These limitations and those of the extreme ratcheting behavior observed previously for softening materials under force-controlled cycling, particularly in the high-strain, low-cycle fatigue regime, underscores the major reason for developing strain-controlled fatigue testing capability several decades ago.

# APPENDIX 2A1: Alternative Criteria for Predicting Hardening or Softening

There are two parameters that are useful in estimating whether a material is likely to harden or to soften in strain cycling. One is the ratio of ultimate tensile strength to yield strength $S_u/\sigma_y$ as discussed in this chapter. The other is the strain-hardening exponent $n$ in a monotonic tensile test. A derivation of the interrelationship between these two quantities is given in the next section.

## Derivation

For most metallic materials, the true stress-true strain curve for monotonic straining can be represented by the classical Ramberg-Osgood relation (Ref 2.1) as represented by Eq 2.1:

$$\varepsilon = \varepsilon_e + \varepsilon_p = \sigma/E + (\sigma/K)^{1/n}$$

where $E$ is the modulus of elasticity; $K$ is the monotonic strength coefficient, i.e., the true stress at a plastic true strain of 1.0; and $n$ is the monotonic strain-hardening exponent. When two points are known along the stress-versus-plastic strain curve, an expression can be derived relating $n$ to the coordinates of these points. The two most commonly known points on the engineering stress-strain curve are at the yield strength ($S_y \approx \sigma_y$) and the ultimate tensile strength ($S_u$). The coordinates of the yield strength are the simplest to specify, i.e., at yield, $\sigma = \sigma_y \approx S_y$, and $\varepsilon_p = 0.002$. At yield, the strains are small enough that no distinction need be made between engineering and true values of stress and strain. However, this is not true at the point of ultimate tensile strength ($S_u$). Unfortunately, the plastic strain at the ultimate tensile strength is generally not reported, nor is it easy to measure in a standard tensile test. For purposes herein, this strain must be calculated analytically. An approximate engineering equation is derived subsequently.

The key is recognizing that necking commences as the ultimate tensile strength is reached. Necking and the ultimate force, i.e., when $dF = 0$, result from an instability brought on by the instantaneous rate of area reduction ($dA/A$) increasing faster than the rate of strain hardening ($d\sigma/\sigma$). The true stress is $\sigma = F/A$, where $F$ is the axial tensile force carried by the instantaneous cross-sectional area, $A$. Rewriting, $F = \sigma A$. By differentiation, $dF = \sigma dA + Ad\sigma$. At necking, $dF = 0$. Thus, after rearranging terms, $dA/A = -d\sigma/\sigma$. In the large plastic regime near the ultimate tensile strength, deformation takes place under an essentially constant volume condition, i.e., (area) × (length) = $AL$ = constant. Differentiating, $AdL + LdA = 0$, and hence, $dL/L = -dA/A$. But, $dL/L$ is, by definition, the incremental true strain $d\varepsilon$. At large strains, the elastic strain is small relative to the plastic strain, i.e., $d\varepsilon \approx d\varepsilon_p$. Thus, $dA/A = -d\varepsilon_p = -d\sigma/\sigma$. Written more conveniently for the purpose at hand, $d\varepsilon_p/d\sigma = 1/\sigma$. If we now differentiate the Ramberg-Osgood equation:

$$d\varepsilon/d\sigma = 1/\sigma = (1/E) + (1/K)^{1/n}(1/n)(\sigma)^{[(1/n)-1]} \quad \text{(Eq 2A1.1)}$$

Equation 2A1.1 can be simplified to:

$$(\sigma/E) + (\sigma/K)^{1/n}(1/n) = 1 \quad \text{(Eq 2A1.2)}$$

Expressed in terms of the elastic and plastic strains:

$$\varepsilon_e + (\varepsilon_p/n) = 1 \quad \text{(Eq 2A1.3)}$$

or:

$$\varepsilon_p = n(1 - \varepsilon_e) \quad \text{(Eq 2A1.4)}$$

At the ultimate tensile strength, the elastic strain is generally much less than unity, i.e., $\varepsilon_e \ll 1$. Thus, $\varepsilon_p \approx n$ at the point of necking. Because constancy of volume is satisfied during plastic flow, it allows the true stress at this point to be approximated by $\sigma \approx S_u(1 + \varepsilon_p) \approx S_u(1 + n)$. We now have two coordinate points on the true-stress-versus-plastic-true-strain curve. Evaluating the Ramberg-Osgood equation at these two approximated coordinate points:

$$0.002 = (\sigma_y/K)^n \quad \text{(Eq 2A1.5)}$$

$$n = [S_u(1 + n)/K]^{1/n} \quad \text{(Eq 2A1.6)}$$

Solving each equation for $K$ and equating:

$$\sigma_y/(0.002)^n = K = S_u(1 + n)/n \quad \text{(Eq 2A1.7)}$$

Solving for $S_u/\sigma_y$:

$$S_u/\sigma_y = (500n)^n/(1 + n) \quad \text{(Eq 2A1.8)}$$

A plot of Eq 2A1.8 is shown in Fig. 2A1.1. The two end segments along the curve are of interest:

$S_u/\sigma_y < 1.2$ (corresponds to softening)
    implies that $n < 0.07$,
    also corresponds to softening.

$S_u/\sigma_y > 1.4$ (corresponds to hardening)
    implies that $n > 0.11$,
    also corresponds to hardening.

The central segment of the curve represents an indeterminate region where materials may harden, soften, or be initially stable. However, before this potentially new criterion is adopted, it is best to assess how well the derived Eq 2A1.8 represents experimental tensile test results.

Figure 2A1.2 compares Eq 2A1.8 (shown as dashed line) with experimental tensile test results for various steels and aluminum alloys used in the automotive industry (Ref 2.13). Experimental values of $n$ are seen, on average, to be larger than calculated from the derived equation.

The discrepancy may be attributed to the fact that measured values of $n$ were determined in the relatively small strain regime, whereas the calculated values of $n$ are based on plastic strains large enough (maybe an order of magnitude larger) to encompass the onset of necking at the ultimate tensile strength. During a monotonic tensile test, as strain hardening occurs, the capacity to further strain harden decreases. Because $n$ is a measure of the rate of strain hardening, its instantaneous numerical value tends to decrease as monotonic straining continues. This viewpoint could explain the discrepancy observed in Fig. 2A1.2. The implication would be that the Ramberg-Osgood relation doesn't necessarily represent the low plastic strain regime

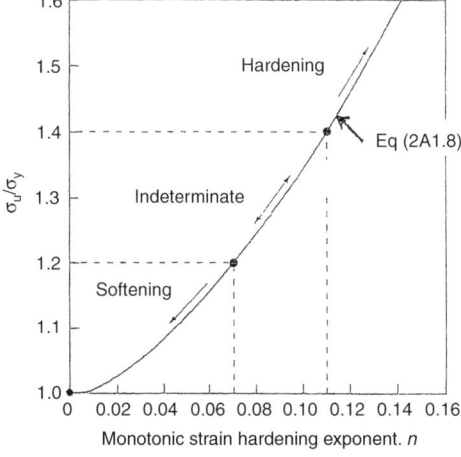

**Fig. 2A1.1** Derived relation between strain-hardening exponent $n$ and ratio of ultimate tensile strength to offset yield strength, $S_u/\sigma_y$, Eq 2A1.8

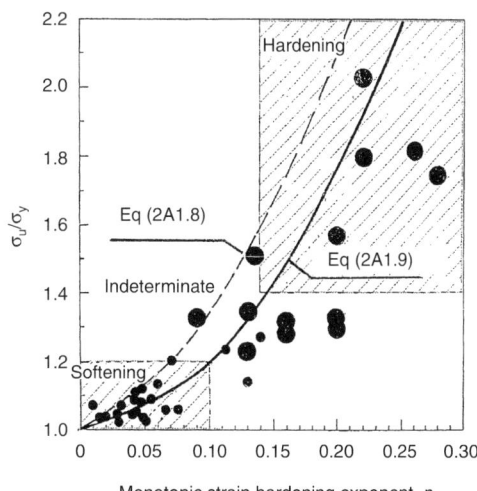

**Fig. 2A1.2** Correlation of experimentally determined values of $n$ and strength ratio $S_u/\sigma_y$ with Eq 2A1.9 showing regions of hardening and softening

and the high plastic strain regime with equal fidelity. Regardless of the explanation, a criterion for judging whether hardening or softening will occur must generally be based on empirical results. Thus, it is a simple matter to modify Eq 2A1.8 to agree with the experimental results by introducing the multiplier $(1 - n)$ on the right hand side of the equation. This modification retains a one-to-one relation between $S_u/\sigma_y$ and $n$:

$$S_u/\sigma_y = (500n)^n[(1 - n)/(1 + n)] \quad \text{(Eq 2A1.9)}$$

This equation is superimposed in Fig. 2A1.2 as a solid line, where it is seen to represent the central trend of the data.

The tensile data source also included information as to whether each material cyclically hardened or softened. Data representing cyclic softening are shown by the small solid symbols, whereas data representing cyclic hardening are shown by larger symbols. The following values of $n$ can be used as reasonable criteria for hardening or softening, at least for alloys of steel and aluminum that have seen common use in the automotive field:

$n > 0.14$ hardening occurs

$n < 0.10$ softening occurs

We now have two interrelated empirical criteria, based on monotonic stress-strain properties, $n$ or $S_u/\sigma_y$, for assessing the likelihood of cyclic strain hardening or softening. An even more stringent criterion would be to impose both simultaneously. These are represented by the rectangular cross-hatched areas for which:

$n > 0.14$ and $S_u/\sigma_y > 1.4$ hardening occurs

$n < 0.10$ and $S_u/\sigma_y < 1.2$ softening occurs

Use of this criterion, however, leaves a large region for which the criterion gives no guidance, i.e., either hardening or softening, or even stable behavior could be expected. Perhaps there is a better way to use both the strength ratio and the strain-hardening exponent to form a more inclusive criterion. This is explored in the following section.

**A More Definitive Criterion for Hardening or Softening.** A new criterion is proposed that includes more of the hardening and softening data than those previously discussed. The criterion is developed by placing a pair of parallel diagonal lines that intersect the curves shown in Fig. 2A1.2 at approximately right angles. The lines are given by simple empirical linear equations relating the strength ratio and the strain-hardening exponent. The lines are separated, leaving a space between for which hardening, softening, or stable behavior could occur. Below the lower line, softening is expected, while above the upper line, hardening is expected. The criterion is given by the following equations:

Lower line:

$$S_u/\sigma_y = 1.5 - 2.5(n)$$
softening occurs if below line    (Eq 2A1.10)

Upper line:

$$S_u/\sigma_y = 1.6 - 2.5(n)$$
hardening occurs if above line
(Eq 2A1.11)

Note that $n$ and $S_u/\sigma_y$ are both *measured* values, not ones computed from Eq 2A1.9. Figure 2A1.3, which repeats all the information from Fig. 2A1.2, shows that considerably more of the data are captured by this new criterion than was previously possible. The reader is reminded that the data shown in the figure are experimental results for a variety of steels and aluminum alloys used in the automotive industry. Caution is advised in using the new criterion for other alloy systems until appropriate data have been generated.

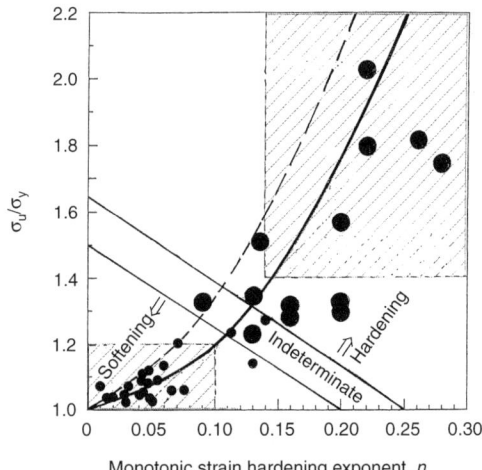

**Fig. 2A1.3** A new and more definitive criterion for predicting cyclic hardening or softening behavior based on $S_u/\sigma_y$ and $n$

# APPENDIX 2A2: Cases for Which Hardening or Softening Requires Special Consideration

Under rare circumstances, certain materials may exhibit hardening or softening characteristics for which the procedures discussed earlier in this chapter are insufficient. They would not adequately represent the cyclic stress-strain behavior over the entire range of fatigue life. An example might be that the hardening or softening process is extremely severe and thus treatment is required to divide the cyclic life into two or more regions to represent the behavior in a simple, tractable manner. Another example could be that hardening or softening never reaches a saturation state before fatigue fracture occurs, and the assumption of a constant cyclic stress-strain curve throughout the majority of life is not accurate enough. A third example is a pervasive situation that has seen little consideration in the field of structural design to resist fatigue failures. This example is not as obvious as the first two. It is discussed first because it is more general and has greater impact on more designs than the first two examples, and the end solution to all three is similar. Concern in this example arises when design safety factors are high, implying a much lower fatigue life than observed in specimen fatigue tests. When no compensation is made in the cyclic rate of hardening or softening, consistency is lost between cyclic stress-strain response and fraction of life expended. For accurate analyses, the representation of the cyclic hardening and softening behavior should take into account the number of applied cycles relative to the design life, not the average life behavior of laboratory samples.

## Cases Involving Use of Large Factors of Safety on Fatigue Life

Significant amounts of scatter in fatigue lives occur in carefully prepared fatigue test samples as well as in structural components made of these materials. Although scatter in fatigue life is greatest in the high-cycle fatigue regime, it can be significant even in low- to intermediate-cycle fatigue wherein cyclic hardening and softening could be encountered. Safety factors on life to ensure low probabilities of failure begin at a minimum of a factor of two and may go as high as a factor of ten or more for crucial equipment wherein human safety is involved. These factors are usually, but not always, taken from the average fatigue life curve generated from laboratory samples. The scatter in life primarily results from the *highly localized* fatigue failure mechanism that is probabilistic in nature (See Chapter 10, "Mechanism of Fatigue"). But, the cyclic stress-strain response is governed by cyclic plasticity mechanisms that, unlike the failure mechanism, are of a bulk nature. Thus, there is considerably less scatter in cyclic stress-strain response than in failure behavior. Consequently, when a laboratory sample is cyclically strained, its hardening or softening characteristics depend almost exclusively on the magnitude of the cyclic strain range and is independent of what the scatter in fatigue life may be. When observations are made about how rapidly a laboratory fatigue sample hardens or softens to its stable saturation level, those observations are made relative to *average* behavior, not the extreme lower-bound life behavior that is captured by large factors of safety. Figure 2A2.1 shows, as examples, the behavior of two materials tested in the author's laboratory (Ref 2.4); one hardening, the other soft-

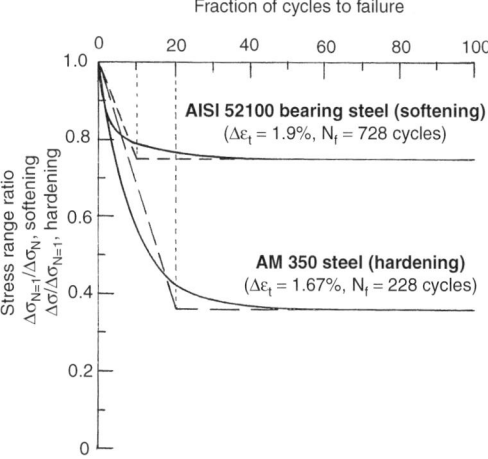

**Fig. 2A2.1** Cyclic strain hardening/softening behavior of two steels. (a) AM 350 alloy. (b) 52100 bearing steel (Ref 2.4)

ening. The horizontal scale is the life fraction $N/N_f$ that ranges from zero to 1.0 at failure of the laboratory specimen. The vertical scale is the ratio of the stress range on the $N$th cycle to the stress range on the first cycle. Cyclic hardening materials start at a stress range ratio of 1.0 and increase until saturation occurs, whereas cyclic softening materials start at a strain range ratio of 1.0 and decrease until saturation occurs. The nonlinear hardening curves for these materials are shown as the solid lines, and bilinear approximations are given by the dashed lines. For these example materials, the annealed AM 350 alloy reached the majority of its hardening by about 20% of its life, while the 52100 bearing steel had completed the bulk of its softening by about 10% of its life.

If an analyst were to take into account the hardening or softening that occurs at these strain ranges, the simplest approach would be to assume that both materials were at their stabilized conditions for the entire life. Adding a level of sophistication, the analysis could then include the linearized hardening or softening transition period during the early portion of the cyclic life and follow the horizontal saturation line for the balance.

Introducing a large factor of safety on life of, say, five, the first and simplest approach above would have been grossly in error for the annealed AM 350 alloy. This is because the design life of the component part would have been reached at the life ratio of 0.2 in the figure. Hence, the material would have been hardening during the entire duration of the design life, and in fact still would not have truly reached saturation at that point.

Applying the same life factor of five to the 52100 bearing steel, we observe that the design life is also reached at an average specimen life fraction of 0.2. In this case, the initial bilinear dashed curves would have indicated saturation being reached at 0.1, or half the design life. However, the initial straight line approximation was based on the entire curve from 0 to 1.0. Bilinear approximations to the curve over only the range of 0 to 0.2 would have to be somewhat different to give a better representation of the softening behavior, and hence a more accurate design analysis.

Smaller factors of safety on fatigue life would not create as large a problem in representing the hardening/softening behavior, but nevertheless would influence any approximations to the transition region.

## Excessive Hardening or Softening or Never Reaching Saturation

Dealing with the first two circumstances discussed at the beginning of this appendix (excessive hardening or softening or never reaching saturation) in essence have been addressed by the preceding discussion involving large safety factors. The three circumstances involve having to follow the hardening or softening that takes place within the lifetime of interest, whether it is an average specimen life or a shorter design life. Approximations of any nonlinear transient hardening/softening behavior by simpler line segments must consider the degree of change in the cyclic stress-strain response as well as the cyclic life used in design, not the average fatigue behavior of laboratory specimens. Despite the attention given previously, it should be emphasized that it is rare in structural analyses to have to account for such details in order to achieve sufficiently fatigue resistant structures.

## REFERENCES

2.1 W. Ramberg and W.R. Osgood, "Description of Stress-Strain Curves by Three Parameters," NACA TN-902, National Advisory Committee for Aeronautics, Washington, DC, 1943

2.2 S.S. Manson, Fatigue: A Complex Subject, *Exp. Mech.*, Vol 5 (No. 7), 1965, p 193–226

2.3 F.R. Tuler, "Cyclic Stress-Strain Behavior of OFHC Copper," master's thesis, Department of Theoretical and Applied Mechanics, University of Illinois, Urbana-Champaign, 1962 (See also Ref 2.6)

2.4 R.W. Smith, M.H. Hirschberg, and S.S. Manson, "Fatigue Behavior of Materials under Strain Cycling in Low and Intermediate Life Range," NASA TN D-1574, National Aeronautics and Space Administration, Washington, DC, April 1963

2.5 Anon., *The Rainflow Method in Fatigue (The Tatsuo Endo Memorial Volume)*, Y. Murakami, Ed., Butterworth-Heinemann Ltd., Oxford, 1992

2.6 A.J. Nachtigall, Cyclic Stress-Strain Determination for D6AC Steel by Three Methods, NASA TM-73815, National Aeronautics and Space Administration, Washington, DC, 1977

2.7 F.R. Tuler and J. Morrow, Cycle-Dependent Stress-Strain Behavior of Metals.

T. & A. M. Report 239, Department of Theoretical and Applied Mechanics, University of Illinois, Urbana-Champaign, March 1963. See also: J. Morrow, "Cyclic Plastic Strain Energy and Fatigue of Metals," *Internal Friction, Damping, and Cyclic Plasticity,* ASTM STP 378, American Society for Testing and Materials, Philadelphia, 1965, p 45–84

2.8 R.M. Wetzel, "A Method of Fatigue Damage Analysis," Ph.D. thesis, Department of Civil Engineering, University of Waterloo, Waterloo, Ontario, Sept 1971

2.9 K.R. Heidmann, "Technology for Predicting the Fatigue Life of Gray Cast Iron," Ph.D. thesis, Department of Mechanical and Aerospace Engineering, Prof. S.S. Manson, Advisor, Case Western Reserve University, Cleveland, OH, May 22, 1985

2.10 S.D. Downing, "Modeling Cyclic Deformation and Fatigue Behavior of Cast Iron under Uniaxial Loading," Ph.D. dissertation, Prof. D.F. Socie, Advisor, Department of Mechanical Engineering, University of Illinois, Urbana, 1983

2.11 J.W. Fash, "An Evaluation of Damage Development during Multiaxial Fatigue of Smooth and Notched Specimens," Ph.D. dissertation, Prof. D.F. Socie, Advisor, Department of Mechanical Engineering, University of Illinois, Urbana, circa 1986

2.12 S.S. Manson, *Thermal Stress and Low-Cycle Fatigue,* McGraw-Hill, 1965, p 187

2.13 R.W. Landgraf, M.R. Mitchell, and N.R. LaPointe, *Cyclic Deformation Behavior of Engineering Alloys,* Ford Motor Company, scientific research staff, Special Technical Publication, Dearborn, MI, June 1972

# CHAPTER 3

# Fatigue Life Relations

## Introduction

**The Traditional S-N Curve.** In attempting to introduce some engineering discipline into the mystery of the systematic failures encountered in the Prussian railway industry, August Wöhler (Ref 3.1) (See also Ref 1.1, 1.2, and 1.4 in this volume) undertook in the mid 1800s to demonstrate that the cause of the failures was repeated stressing. Why a repeated stress was so damaging compared with a steady stress of even higher magnitude was not then understood, but Wöhler proved that this was so by experiment. He conceived a testing machine (Fig. 3.1) (Ref 3.2), simulating a railcar axle, in which a sample could be subjected to completely reversed bending without actually changing the direction of the loading. By rotating the shaft that included the specimen, a static force (supplied by the springs, P in Fig. 3.1) imposed completely reversed bending stresses in all the surface fibers of the specimen. Changing the applied load varied the bending stress. Wöhler's new testing machine duplicated in miniature the loadings of railcars on their axles.

Wöhler's focus was to determine a stress level below which an "indefinite" number of reversals could be sustained without specimen fracture. Actually, he was seeking to determine what was later defined as an "endurance limit," but to approach this stress from above, he conducted tests for which the fatigue life was finite. Although he actually never plotted his data, the four reports that he published led engineers later to construct the S-N (stress versus log [number of cycles to failure]) curve that is attributed to him (Fig. 3.2). Basically, the stress is plotted on the vertical axis and cycles to failure on the horizontal axis. The scale for cyclic life is usually made logarithmic in order to plot conveniently the large variations of life as stress amplitude is changed, and in order to accommodate the large scatter in life found at stresses near the "endurance limit."

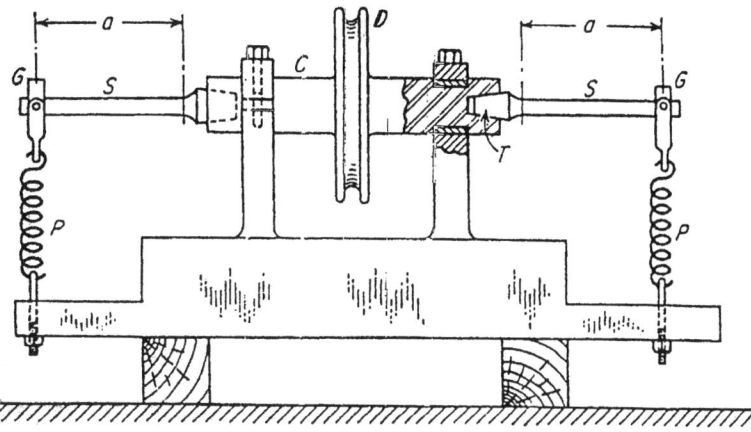

**Fig. 3.1** Wöhler's rotating-cantilever, bending fatigue-testing machine. D, drive pulley; C, arbor; T, tapered specimen butt; S, specimen; a, moment arm; G, loading bearing; P, loading spring. Source: Ref 3.2

It is interesting that Wöhler also studied the effect of notches as stress-raisers, and noted that if a notch is present, the *nominal* stress in the vicinity of the notch that could be endured was considerably reduced, as noted in Fig. 3.2.

The concept of an *S-N* curve as the tool needed for fatigue design served the engineering professions for nearly 100 years, and it is to Wöhler's great credit that he demonstrated that stress *reversal* was the principal reason for metal fatigue, as compared with the polemics that were expressed by the many engineers of his time to explain the then-mysterious phenomenon.

**The Need for a New Framework.** Wöhler's interest in long-life fatigue—an absolute economic requirement for the locomotive equipment of his time—meant that most of the stress levels used in his tests were in the nominally elastic range of the materials. Since the middle of the twentieth century, however, many new applications have arisen wherein it is virtually impossible to restrict the working stresses to the elastic range. Economic design now dictates that some plasticity be allowed, and finite life accepted, at least in the critical regions of many engineering components. While the *S-N* curve lends itself to some treatment in the plastic range, it is not the ideal framework for treating such problems.

Consider, for example, an idealized material for which the stress-strain curve is shown in Fig. 3.3(a). Suppose a material is strained to point *A*, and then completely reversed. The cyclic stress-strain path (hysteresis loop) is *OAOA'O* in Fig. 3.3(b). Alternatively, the material could be strained to point *B*, the reversal giving loop *BB'B"B'"* in Fig. 3.3(c). Or, it can be strained to *C* with attendant loop *CC'C"C'"* in Fig. 3.3(d). If we assume that the material shows no strain hardening along *ABC* in Fig. 3.3(a), then the stress ranges are the same for the three loops shown. But surely, one might expect the fatigue lives to be different. As we shall see later, it is the amount of cyclic *plastic* strain that governs life, and we can be quite certain that the fatigue lives will not, indeed, all be equal. But an *S-N* curve for un-notched material, basing life only on stress range, would surely not tell us how the fatigue lives would differ. It was this consideration and others that led to the adoption of a *strain-based fatigue approach* now used most commonly.

## The Strain-Based Approach to Fatigue Analysis

**Background.** In 1952, when Manson (Ref 3.3) first began to consider fatigue life associated with plastic strain for application to the cyclic fatigue of gas turbine disks, there was no prior information available on this subject. However, the late 1940s work of Sachs and coworkers (Ref 3.4) was useful to him to gain a first approximation of how 24S-T aluminum behaved. Sachs was concerned with the effect of successive reversed straining of the material on its remaining ductility because his main interest was metal formability, but a limited amount of very low cyclic life fatigue data could be extracted from his results, as shown in Fig. 3.4(a).

Here, the graph shows how much ductility remained in this aluminum alloy after each time it was subjected to a complete reversal of a 12% range of strain. Each data point represents a different test coupon. When the remaining ductility was less than 12%, it would, of course, fail before the next reversal was completed. Thus, it was possible to extract the fatigue life for 12%, which as seen from the figure is 7 cycles. Other tests, involving different strain-range reversals, as shown in Fig. 3.4(b), were similarly studied to estimate fatigue lives over a range of strain reversals. Figure 3.4(c) shows the logarithmically linear relation between plastic strain range

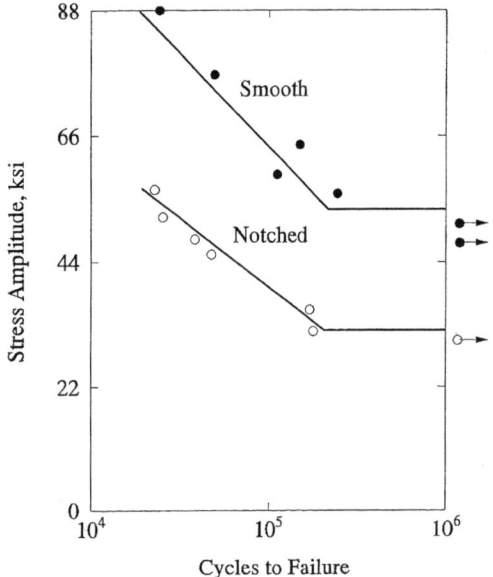

**Fig. 3.2** Re-plot of some of Wöhler's data as *S-N* curves for smooth and notched steel specimens

and fatigue life that could be derived from these data. Plots of Sachs's data lead to the relation (Ref 3.3):

$$N = C/(\varepsilon_p)^c \qquad (Eq\ 3.1)$$

where $N$ is the number of cycles of plastic strain $\pm \varepsilon_p$, $\varepsilon_p$ is plastic strain, $c$ is the exponent,* and $C$ is the proportionality constant.

---

*Through numerical error, Manson expressed $c$ as being in the neighborhood of 3. Actually, Sachs's data closely fitted $c = 2$, as was later also found by Coffin. But it was also recognized that other materials would likely involve different values of $c$, which later turned out to be true. Thus, $c$ was expressed as a material constant rather than a universal constant.

Later we were able to obtain much better data (Ref 3.5) to verify the near-perfect linearity between plastic strain range and fatigue life (Fig 3.4d). Coffin (Ref 3.6) also arrived independently at the conclusion that cyclic life bore an inverse power-law relation to plastic strain range. At first, he maintained that the exponent $c$ in Eq (3.1) was 2 for all materials, but later accepted that it is better to represent each material according to its own experimentally determined exponent.

The power-law equation between cyclic life and plastic strain range has come to be known as the Manson-Coffin equation, or the Coffin-Manson equation, according to the preference of the analyst. This is appropriate because both in-

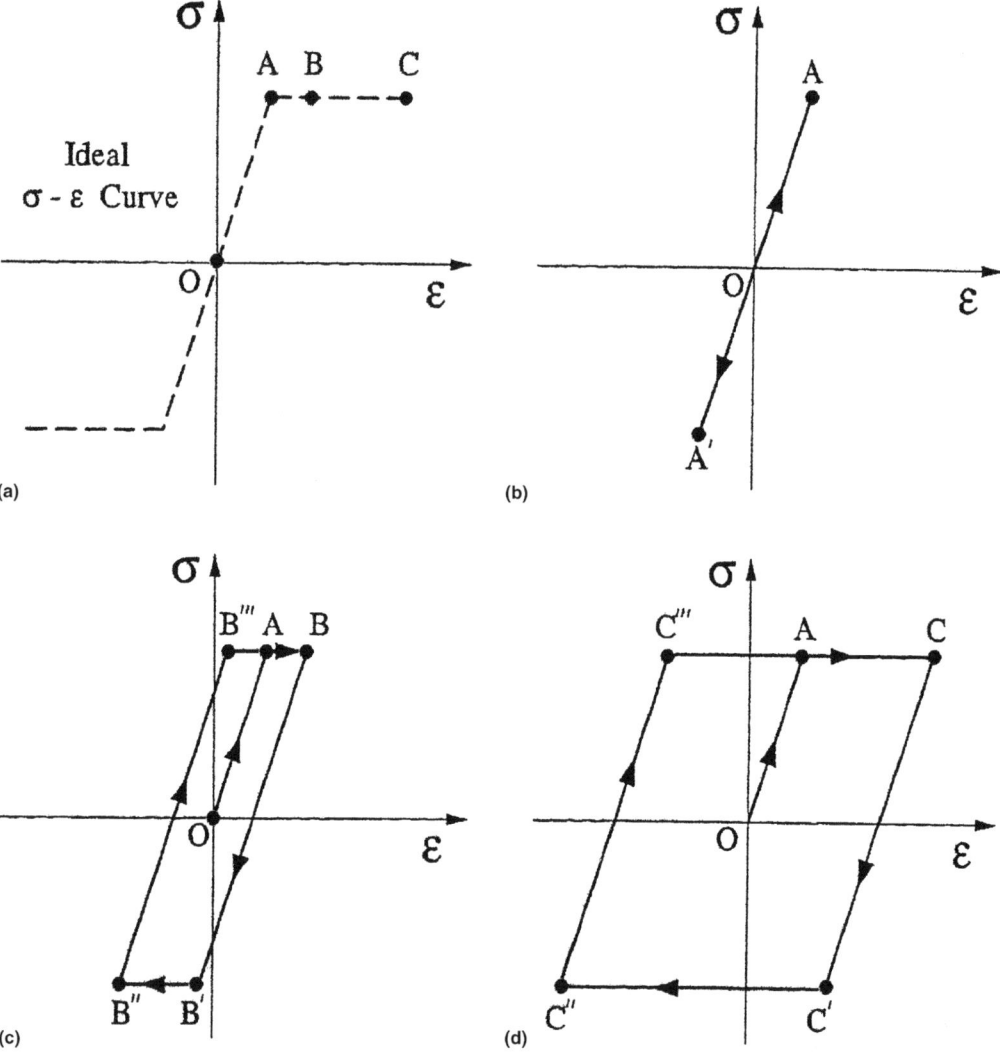

**Fig. 3.3** Stress and strain reversals for a material following an idealized stress-strain curve

vestigators independently arrived at the same equation at about the same time. In this text, we shall refer to it as the Manson-Coffin equation.

**Extension of the Equations to Include Elastic Strain Range.** While the Manson-Coffin equation provided the fundamental insight that fatigue is the result of plastic strain reversal, its direct use for engineering applications is difficult because it is difficult to measure plastic strain directly. However, total strain range is relatively easy to measure using extensometers, resistance strain gages, or numerous other methods. The second goal, therefore, was to express life in terms of total strain range consisting of both plastic and elastic components. Examining the problem in retrospect, the transfer should have been easy. Since it was well known that a power-law relation exists between stress and plastic strain, it should easily have been concluded that a power-law relation must also exist between elastic strain range and cyclic life. Thus, an expression relating total strain range to the sum of the two power-laws of cyclic life could have been recognized immediately. Unfortunately, this insight came neither easily nor directly. A more circuitous path was followed to arrive at this somewhat trivial conclusion.

One of the problems was the considerable stress amplitude variation due to cyclic strain hardening or softening that occurred during cycling either at constant plastic strain range or at constant total strain range. Which stress to use, and in what way, was an obstacle to the smooth transition of the concept. Coffin and Tavernelli (Ref 3.7) studied the fatigue of copper, noting the complex changes in stress during cycling of highly hardening or softening materials, such as shown in Fig. 3.5. They surmised that it would

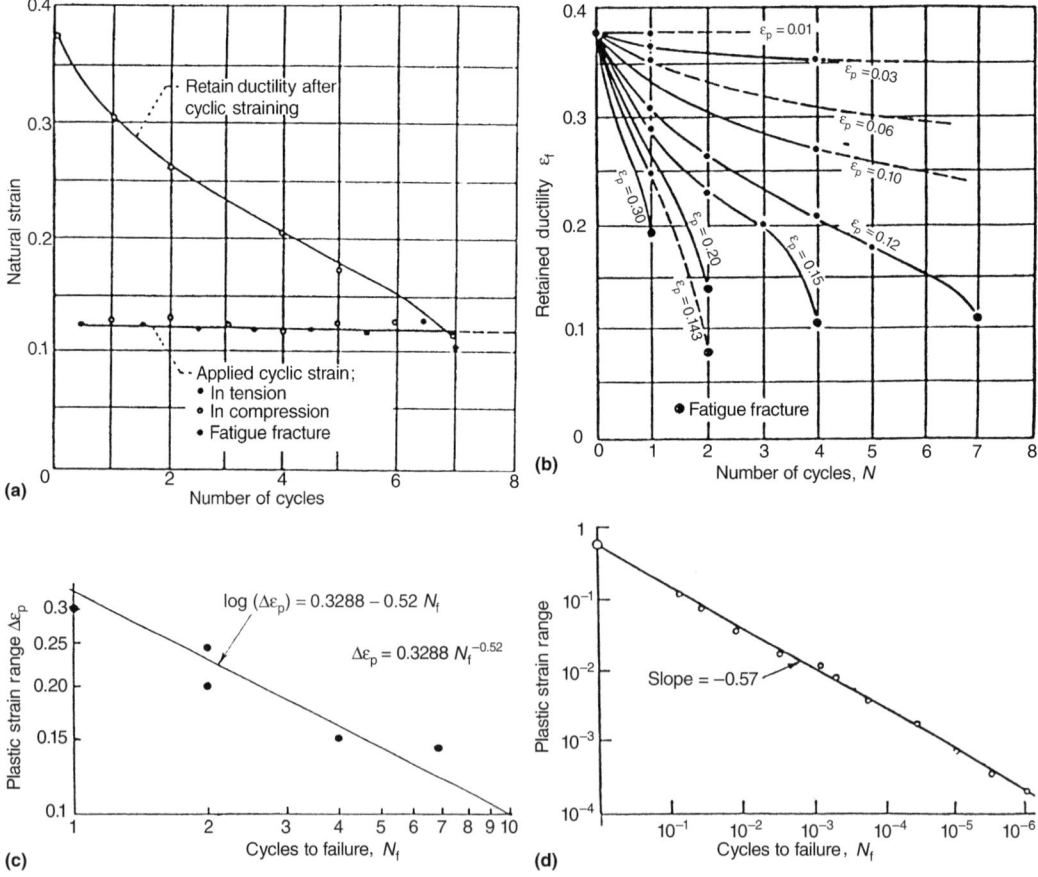

**Fig. 3.4** Strain-based approach. (a) Retained ductility of 24ST aluminum alloy. Source: Ref 3.4. (b) Fracture characteristics of 24ST aluminum alloy as a function of the number of cycles of cyclic straining. Source: Ref 3.4. (c) Reconstruction of Sach's data to suggest a power-law relation between fatigue life and plastic strain range. (d) Subsequent NASA data (unpublished on A-286, 34% cold-rolled and aged) showing linearity of the Manson-Coffin line

be futile to attempt to involve stress in the life relation.

When pursuing an association of fatigue life with both elastic and plastic strains (Ref 3.8), it was noted, however, that reasonably straight lines could be drawn through the failure points (the end points of all the curves in Fig. 3.5), as re-plotted in Fig. 3.6 on double logarithmic co-ordinates. In fact, as seen in Fig. 3.6, these straight lines were the same for as-received, annealed, or cold-worked oxygen-free high-conductivity (OFHC) copper. Similar behavior is also shown in the figure for annealed and prestrained 2S aluminum (Ref 3.7). Thus, although some of these materials either hardened or softened considerably, their end points were very nearly the same. That is, they tended toward the same metallurgically stable condition. Since the scale in Fig. 3.5 is logarithmic in cycles, a very large fraction of the total cycles operate very near the stress at the end point, and therefore a valid life relation for fatigue could be expressed in terms of the end-point stresses.

Later, it was suggested that in cases of drastic stress variations during cycling, the stress used in the life relations be the stress at the half-life.* However, for many materials the stress stabilizes well before the half-life, and the two stresses are nearly the same.

Thus, as early as 1960 (Ref 3.8), the model had evolved that total strain range could be expressed as the sum of two power laws of cyclic life, one associated with the plastic strain, the other associated with stress at half-life or in some cases, the end point.

Later, Morrow (Ref 3.9) was reminded that Basquin (Ref 3.10) had expressed cyclic life as a power law of stress range. Actually, Basquin did not study strain cycling, and if he had applied load cycling to cyclically softening materials, he would have been surprised by the results and would not have been able to incorporate them within his equation. Thus, although Basquin did express the power-law equation between stress and cyclic life, it was not within the historical framework of the modern strain cycling approach. Nor were his results instrumental in the development of the modern theory. However, because of his chronologically early proposal of the stress-life equation, it is common to call the elastic line in the fatigue life equation the Basquin line and the plastic line the Manson-Coffin line. Perhaps it would be more historically correct and appropriate to designate the strain-based life approach as the Manson-Coffin-Basquin model.

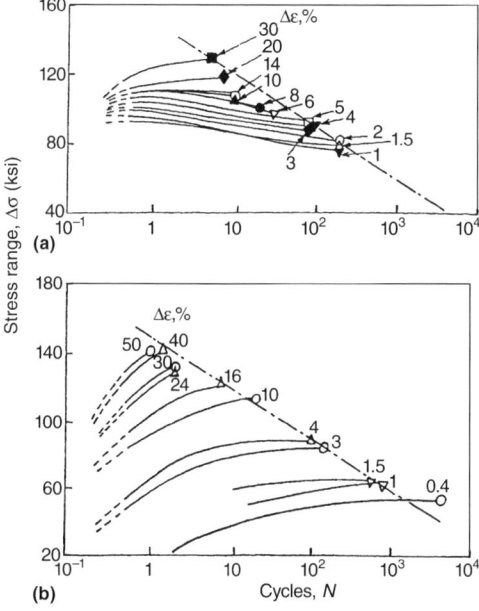

**Fig. 3.5** Analysis by Manson (Ref 3.8) using data of Coffin and Tavernelli (Ref 3.7) to demonstrate the validity of an elastic line even for highly hardening or softening material. (a) Cold-worked copper. (b) Annealed copper

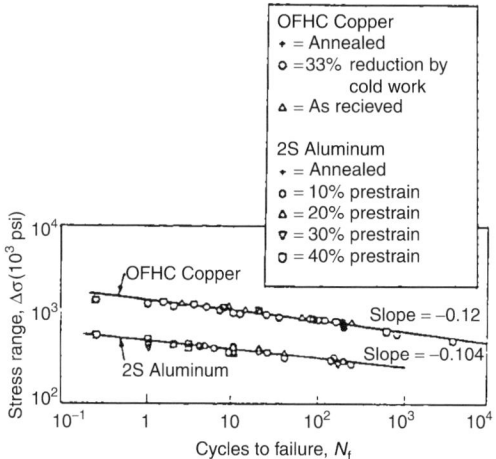

**Fig. 3.6** Relation between stress range at fracture and number of cycles to failure for oxygen-free high-conductivity copper and 2S aluminum alloy in various conditions of prestrain. Source: Ref 3.8

---

*The first known reference to half-life was given by Coffin (Ref 3.6) in his classic paper of 1953. Because of variations in the plastic strain range during his simulated thermal fatigue tests, he selected the half-life as being a reasonable point in the life to measure the plastic strain range to represent the average throughout the test. This designation is now in widespread use.

**The Basic Equation.** As originally proposed in Ref 3.8, the basic relation between total strain range and cyclic life consists of two power-law terms, one for the plastic strain, and the other for elastic strain. Thus,

$$\Delta\varepsilon = \text{total strain range} = \text{elastic strain range} + \text{plastic strain range} = X(N_f)^x + Y(N_f)^y \quad \text{(Eq 3.2a)}$$

where $\Delta\varepsilon$ is the total strain range; $X$, $x$ are the coefficient and exponent terms, respectively, relating to elastic strain range; $Y$, $y$ are the coefficient and exponent terms, respectively, relating to plastic strain range; and $N_f$ represents cycles to failure. It is presumed that the coefficients and exponents would be experimentally determined for each material of interest. This is the format that we prefer, and we use it in many applications described in this chapter. It was, in fact, the only format used between 1952 and 1964 after which an alternate notation was introduced by Morrow (Ref 3.9).

**Alternate Notation of Morrow.** Although in principle there is no difference from the equations developed in Ref 3.8, the notation introduced by Morrow is also commonly used. The literature contributions of Morrow in which he used this notation are themselves quite prolific; furthermore, the many excellent students that he trained have issued an extensive literature, so it is important to be familiar with this notation. The Morrow notation is represented by:

$$\frac{\Delta\varepsilon}{2} = \left(\frac{\sigma'_f}{E}\right)(2N_f)^b + \varepsilon'_f(2N_f)^c \quad \text{(Eq 3.2b)}$$

Here, $\Delta\varepsilon/2$ is the strain amplitude (i.e., half the total strain range), and $2N_f$ is the number of reversals (two reversals for each of the $N_f$ cycles of failure). Morrow reasoned that in complex loading histories, reversals of loading would be more recognizable than cycles, and therefore he expressed the life relation in terms of reversals instead of cycles. The terms $\varepsilon'_f$ and $\sigma'_f/E$ are strain intercepts at $2N_f = 1.0$, and $c$ and $b$ are slopes of the plastic and elastic lines, respectively, as shown in Fig. 3.7. The intersection of the two straight lines is known as the "transition point," and its coordinates are designated $\Delta\varepsilon_T/2$ and $2N_T$, respectively. The total strain amplitude, $\Delta\varepsilon/2$, at the transition fatigue life, is thus $\Delta\varepsilon_T/2$. Many materials have been characterized according to the Manson-Coffin-Basquin model. Figure 3.8 shows typical results for two engineering alloys, AISI 4130 steel and Man-Ten steel, at room temperature. A more complete listing for many engineering alloys is shown in Table 3.1.

These materials were experimentally characterized by many laboratories and summarized by Landgraf, Mitchell, and LaPointe (Ref 3.11). In addition to $\sigma'_f$, $\varepsilon'_f$, $b$, and $c$ listed in the table, there are also listed a number of static properties for each material; these properties have been useful for numerous correlations permitting approximate prediction of fatigue properties in cases wherein cyclic data are either completely absent or available in limited quantities. We discuss some of these correlations later in this chapter.

Although the Morrow notation is common in the literature, and many materials are characterized according to this notation, it can be confusing to some engineers because the digit "2" serves different purposes on each side of the equation. Our preference is for the simple notation of Eq 3.2a. To remain consistent with a preponderance of literature, however, we will use either Eq 3.2a or 3.2b—whichever is the more appropriate for the application discussed.

**Anomalous Behavior.** While most materials are characterized by very good linear log, log life relationship, several types of anomalies have been observed. Figure 3.9 shows a schematic plastic line that has a distinct discontinuity. Above the discontinuity the deformation is homogeneous, as in most materials. Below the discontinuity the deformation occurs in distinct, highly localized regions (as shown schematically in the figure by dark bands). Such behavior

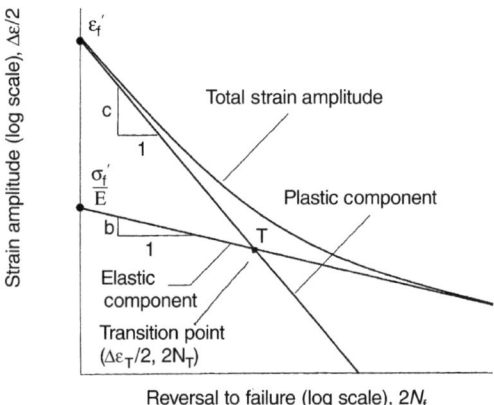

**Fig. 3.7** Morrow's notation for use in the Manson-Coffin-Basquin model for fatigue in strain cycling

has been observed (Ref 3.12) in an aluminum alloy containing lithium and magnesium.

Figure 3.10(a) shows another type of breakline behavior, in this case for 304 stainless steel (Ref 3.13). For this material, as well as other austenitic materials, cyclic straining induces a shear strain-based transformation of some of the austenite to martensite, a harder and stronger phase. The amount of transformation depends on the magnitude of the applied plastic strain. At the higher strain ranges, the transformation occurs early in the life, and a large part of the life involves cycling of the hardened material. At the lower strain ranges, the transformation does not occur until much of the life has been exhausted. Basically, at each plastic strain, the test really involves cumulative fatigue on a material of progressively changing properties, and the Manson-Coffin line should not be straight at all. From an engineering point of view, however, two straight-line segments, as shown by lines $AB$ and $BC$, can represent the behavior.

Compare the representation of Fig. 3.10(a) with that of Fig. 3.10(b), which was the representation chosen in our original testing of this material (Ref 3.5), when we were unaware of the transformation phenomenon, and sought to represent the plastic line as straight; we simply drew the best-fit line through the data by least-squares procedures. However, we did recognize that the troubling behavior was associated with a peculiar metallurgical phenomenon that must have been occurring. We noted that as the cycling progressed in this material, it became more and more magnetic, and we made note of this fact in our 1964 report. Of course, what was really happening was that the initially nonmagnetic austenite was rendered magnetic by the martensitic transformation.

Analysis of fatigue for an unstable material must be treated very carefully, especially in cumulative fatigue problems wherein several strain levels are involved. We discuss this subject in Chapter 6, "Cumulative Fatigue Damage."

Figure 3.11 shows another type of nonlinearity resulting from surface degradation. The plastic line for 1113 steel (Ref 3.14) is quite linear up to $10^5$ cycles for testing in air. In the higher life range, considerable test time is required. Thus, there is greater opportunity for the surface to be attacked by oxygen or other chemicals in contact with it. That linearity of the plastic line can persist up to $10^6$ cycles of life and beyond is also seen in the figure for tests conducted in a high vacuum wherein oxidation is precluded. Coffin (Ref 3.15) has discussed this subject extensively.

Another aspect of anomalous behavior that merits discussion is the break in the elastic line that is observed for some materials, commonly associated with the classic "endurance limit." Figure 3.12 shows the way $S$-$N$ curves are usually drawn for materials displaying a distinct endurance limit. The usual representation is a sloping curve and a horizontal line beyond that life. The intersection is commonly in the vicinity of $10^6$ to $10^7$ cycles to failure. This representation is, of course, inconsistent with the modern concept of continuing the elastic line along a downward slope indefinitely.

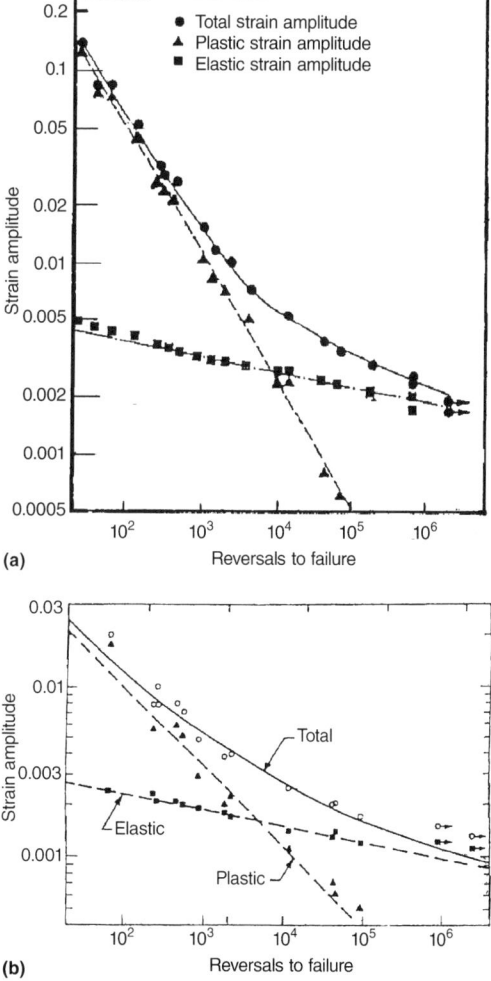

**Fig. 3.8** Typical results showing linearity of elastic and plastic lines of the Manson-Coffin-Basquin model: (a) 4130 steel (annealed, 258 HB). Source: Ref 3.5. (b) Man-Ten steel (hot rolled, 150 HB). Source: Ref 3.11.

An *endurance limit* may exist, but only for certain metals and only for certain conditions of cyclic loading. One class of low-alloy carbon steels is the most common to exhibit this behavior. For these steels, carbon atoms seek the plentiful interstitial sites within the body-centered cubic (bcc) iron lattice and thereby block dislocations as they attempt to glide through the iron. Thus, there appears to be a stress that, if not exceeded, will not cause dislocation motion, i.e., plastic flow. And, it has been known for over a century that plastic flow is a necessary requisite for fatigue damage. If, therefore, that critical endurance limit stress is never exceeded, a specimen may last indefinitely under cycling. Thus, the specimen may exhibit what appears to be an endurance limit. However, if that stress is exceeded in testing or in service and the dislocations are forced to break away and glide past their carbon obstructions, the material will not

**Table 3.1   Room temperature fatigue properties of selected steels, aluminum, and titanium**

| No. | Material | HB | $\sigma'_f$, MPa (ksi) | $\varepsilon'_f$ | b | c | $n' = b/c$ | E, MPa ($10^6$ psi) | $\Delta\varepsilon_T/2$ | $2N_T$ cycles | $\sigma_f$, MPa (ksi) | $S_u$, MPa (ksi) | RA, % |
|---|---|---|---|---|---|---|---|---|---|---|---|---|
| 1 | SAE 1005-1009 | HRLC | 538 (78) | 0.11 | −0.073 | −0.41 | 0.18 | 200.0 (29.0) | 0.00121 | 60600 | 841 (122) | 414 (60) | 64 |
| 2 | Gainex | HRLC | 807 (117) | 0.86 | −0.071 | −0.65 | 0.11 | 201.3 (29.2) | 0.00208 | 10644 | 814 (118) | 521 (75.5) | 61 |
| 3 | Man-Ten | 150 | 972 (141) | 0.85 | −0.110 | −0.59 | 0.19 | 206.9 (30.0) | 0.00143 | 50438 | 1000 (145) | 565 (82) | 69 |
| 4 | SAE 1045 | 225 | 1227 (178) | 1.00 | −0.095 | −0.66 | 0.14 | 200.0 (29.0) | 0.00261 | 8222 | 1227 (178) | 724 (105) | 65 |
| 5 | SAE 1045 | 390 | 1586 (230) | 0.45 | −0.074 | −0.68 | 0.11 | 206.9 (30.0) | 0.00467 | 828 | 1862 (270) | 1344 (195) | 59 |
| 6 | SAE 1045 | 410 | 1862 (270) | 0.60 | −0.073 | −0.70 | 0.10 | 200.0 (29.0) | 0.00573 | 768 | 1862 (270) | 1448 (210) | 51 |
| 7 | SAE 1045 | 450 | 1793 (260) | 0.35 | −0.070 | −0.69 | 0.10 | 206.9 (30.0) | 0.00571 | 390 | 2103 (305) | 1586 (230) | 55 |
| 8 | SAE 1045 | 595 | 2723 (395) | 0.07 | −0.081 | −0.60 | 0.14 | 206.9 (30.0) | 0.0102 | 25 | 2723 (395) | 2241 (325) | 41 |
| 9 | 10B62 | 430 | 1779 (258) | 0.32 | −0.067 | −0.56 | 0.12 | 193.1 (28.0) | 0.0057 | 1334 | 1779 (258) | 1641 (238) | 38 |
| 10 | AISI 4130 | 365 | 1696 (246) | 0.89 | −0.081 | −0.69 | 0.12 | 200.0 (29.0) | 0.00457 | 2082 | 1820 (264) | 1427 (207) | 55 |
| 11 | SAE 4142 | 475 | 2172 (315) | 0.09 | −0.081 | −0.61 | 0.13 | 206.9 (30.0) | 0.00756 | 58 | 2172 (315) | 1931 (280) | 35 |
| 12 | SAE 4142 | 560 | 2654 (385) | 0.07 | −0.089 | −0.76 | 0.12 | 206.9 (30.0) | 0.0103 | 12 | 2654 (385) | 2241 (325) | 27 |
| 13 | Ti-6Al-4V | ... | 3806 (552) | 1.05 | −0.105 | −0.69 | 0.15 | 117.2 (17.0) | 0.0174 | 382 | 1717 (249) | 1234 (179) | 41 |
| 14 | SAE 4142 | 400 | 1896 (275) | 0.50 | −0.090 | −0.75 | 0.12 | 200.0 (29.0) | 0.0055 | 406 | 1896 (275) | 1551 (225) | 47 |
| 15 | SAE 4142 | 450 | 1999 (290) | 0.40 | −0.080 | −0.79 | 0.10 | 206.9 (30.0) | 0.0061 | 308 | 1999 (290) | 1758 (255) | 42 |
| 16 | SAE 4142 | 475 | 2068 (300) | 0.20 | −0.082 | −0.77 | 0.11 | 200.0 (29.0) | 0.0073 | 74 | 2068 (300) | 2034 (295) | 20 |
| 17 | AISI 4340 | 243 | 1200 (174) | 0.45 | −0.095 | −0.54 | 0.18 | 193.1 (28.0) | 0.00249 | 15114 | 1089 (158) | 827 (120) | 43 |
| 18 | AISI 4340 | 409 | 1999 (290) | 0.48 | −0.091 | −0.60 | 0.15 | 200.0 (29.0) | 0.0050 | 2010 | 1558 (226) | 1469 (213) | 38 |
| 19 | SAE 5160 | 430 | 1931 (280) | 0.40 | −0.071 | −0.57 | 0.12 | 193.1 (28.0) | 0.0059 | 1624 | 1931 (280) | 1669 (242) | 42 |
| 20 | SAE 9262 | 260 | 1041 (151) | 0.155 | −0.071 | −0.47 | 0.15 | 206.9 (30.0) | 0.00274 | 5376 | 1041 (151) | 924 (134) | 14 |
| 21 | SAE 9262 | 410 | 1855 (269) | 0.38 | −0.057 | −0.65 | 0.09 | 200.0 (29.0) | 0.0065 | 524 | 1855 (269) | 1565 (227) | 32 |
| 22 | AISI 304 | 160 | 2413 (350) | 1.02 | −0.150 | −0.77 | 0.19 | 186.2 (27.0) | 0.00451 | 1142 | 1572 (228) | 745 (108) | 74 |
| 23 | AISI 310 | 145 | 1655 (240) | 0.60 | −0.150 | −0.57 | 0.26 | 193.1 (28.0) | 0.00188 | 24722 | 1158 (168) | 641 (93) | 64 |
| 24 | AM 350 | ... | 2799 (406) | 0.33 | −0.140 | −0.84 | 0.17 | 193.1 (28.0) | 0.0080 | 86 | 2055 (298) | 1317 (191) | 52 |
| 25 | 18Ni maraging | 460 | 2137 (310) | 0.80 | −0.071 | −0.79 | 0.09 | 186.2 (27.0) | 0.0075 | 366 | 2137 (310) | 1862 (270) | 56 |
| 26 | 18Ni maraging | 480 | 2241 (325) | 0.60 | −0.070 | −0.75 | 0.09 | 179.3 (26.0) | 0.0085 | 296 | 2241 (325) | 1999 (290) | 55 |
| 30 | 2014-T6 Al | ... | 848 (123) | 0.42 | −0.106 | −0.65 | 0.16 | 68.95 (10.0) | 0.0062 | 658 | 600 (87) | 510 (74) | 25 |
| 31 | 2014-T4 Al | ... | 1014 (147) | 0.21 | −0.110 | −0.52 | 0.21 | 70.3 (10.2) | 0.00703 | 688 | 634 (92) | 476 (69) | 35 |
| 32 | 5456 Al | ... | 724 (105) | 0.46 | −0.110 | −0.67 | 0.16 | 68.95 (10.0) | 0.0050 | 854 | 524 (76) | 400 (58) | 35 |
| 33 | SAE 1015 | 80 | 827 (120) | 0.95 | −0.110 | −0.64 | 0.17 | 206.9 (30.0) | 0.00129 | 30366 | 724 (105) | 414 (60) | 68 |
| 34 | SAE 950X | 155 | 627 (91) | 0.35 | −0.075 | −0.54 | 0.14 | 206.9 (30.0) | 0.00141 | 27028 | 752 (109) | 441 (64) | 65 |
| 35 | VAN80 | 225 | 1055 (153) | 0.21 | −0.080 | −0.53 | 0.15 | 194.4 (28.2) | 0.00284 | 3376 | 1220 (177) | 696 (101) | 68 |
| 36 | RQC 100 | 298 | 1014 (147) | 0.60 | −0.076 | −0.67 | 0.11 | 202.7 (29.4) | 0.00271 | 3164 | 1358 (197) | 820 (119) | 68 |
| 37 | SAE 1045 | 500 | 2275 (330) | 0.25 | −0.080 | −0.68 | 0.12 | 206.9 (30.0) | 0.0073 | 182 | 2275 (330) | 1827 (265) | 51 |
| 38 | AISI 4130 | 258 | 1276 (185) | 0.92 | −0.083 | −0.63 | 0.13 | 220.6 (32.0) | 0.00268 | 10596 | 1420 (206) | 896 (130) | 67 |
| 39 | SAE 4142 | 380 | 1827 (265) | 0.45 | −0.080 | −0.75 | 0.11 | 206.9 (30.0) | 0.00553 | 354 | 1827 (265) | 1413 (205) | 48 |
| 40 | SAE 4142 | 450 | 2103 (305) | 0.60 | −0.090 | −0.76 | 0.12 | 200.0 (29.0) | 0.0061 | 418 | 2103 (305) | 1931 (280) | 37 |
| 41 | SAE 4340 | 350 | 1655 (240) | 0.73 | −0.076 | −0.62 | 0.12 | 193.1 (28.0) | 0.00461 | 3534 | 1655 (240) | 1241 (180) | 57 |
| 42 | AISI 52100 | 518 | 2586 (375) | 0.18 | −0.090 | −0.56 | 0.16 | 206.9 (30.0) | 0.0075 | 292 | 2193 (318) | 2013 (292) | 11 |
| 43 | SAE 9262 | 280 | 1220 (177) | 0.41 | −0.073 | −0.60 | 0.12 | 193.1 (28.0) | 0.00355 | 2744 | 1220 (177) | 1000 (145) | 33 |
| 44 | N-11 | 660 | 3172 (460) | 0.08 | −0.077 | −0.74 | 0.10 | 206.9 (30.0) | 0.0127 | 12 | 3172 (460) | 2586 (375) | 33 |
| 45 | AISI 304 | 327 | 2275 (330) | 0.89 | −0.120 | −0.69 | 0.17 | 172.4 (25.0) | 0.0055 | 1616 | 1696 (246) | 951 (138) | 69 |
| 46 | AM 350 | 496 | 2689 (390) | 0.098 | −0.102 | −0.42 | 0.24 | 179.3 (26.0) | 0.0082 | 366 | 2179 (316) | 1903 (276) | 20 |
| 47 | 18Ni maraging | 450 | 1655 (240) | 0.30 | −0.065 | −0.62 | 0.10 | 186.2 (27.0) | 0.0059 | 568 | 1896 (275) | 1517 (220) | 67 |
| 48 | 2024-T351 Al | ... | 1103 (160) | 0.22 | −0.124 | −0.59 | 0.21 | 73.1 (10.6) | 0.0074 | 314 | 558 (81) | 469 (68) | 25 |
| 49 | 7075-T6 Al | ... | 1317 (191) | 0.19 | −0.126 | −0.52 | 0.24 | 71.0 (10.3) | 0.0088 | 368 | 745 (108) | 579 (84) | 33 |
| 50 | SAE 1005 | HRLC | 641 (93) | 0.10 | −0.109 | −0.39 | 0.28 | 200.0 (29.0) | 0.00085 | 200000 | 848 (123) | 345 (50) | 80 |

HRLC, hot rolled, low carbon steel (specific hardness value not provided in source). RA, reduction in area. Source: Ref 3.11

subsequently exhibit an "endurance limit." Dowling (Ref 3.16) has discussed this behavior for an alloy steel as shown in Fig. 3.13. While the S-N curve did seem to flatten out for specimens that had never been subjected to high stress, those with periodic overstrain (overload) continued their downward slope to stresses well below the initial "apparent" endurance limit. It is not unrealistic to assume that structural alloys will be exposed occasionally to stresses that are higher than their endurance limit during the spectrum of their loading history. Hence, it would be wise to ignore an endurance limit. Such a limit may be exhibited in pristine, carefully controlled, constant-amplitude, laboratory fatigue tests of highly polished coupons, but this response cannot be expected in service. Most other metals and alloys do not exhibit an endurance limit under any circumstance.

For conservative design, it is best to assume that the elastic line continues at constant slope down to the lowest stress level in the loading history. However, for some structures that are expected to serve very long lives, up to $10^{10}$ or $10^{11}$ cycles, it turns out to be a handicap to consider the slope to be unchanged from its value in the lower life range. Manufacturers of cyclically loaded equipment prefer to take advantage of the tendency for the S-N curve (or the elastic life line) to level, at least partially, in the very high life ranges. We discuss, in a subsequent section, a reasonable compromise that we have suggested for this purpose.

## Predictive Methods

The most accurate approach to determine the fatigue characteristic of a material is to establish both the complete elastic and plastic lines by actual experiment. However, it frequently is impractical to attempt an extensive experimental program either because facilities are lacking, because the simulated environmental conditions are too difficult or expensive, or because it is desirable to first scan a large number of materials

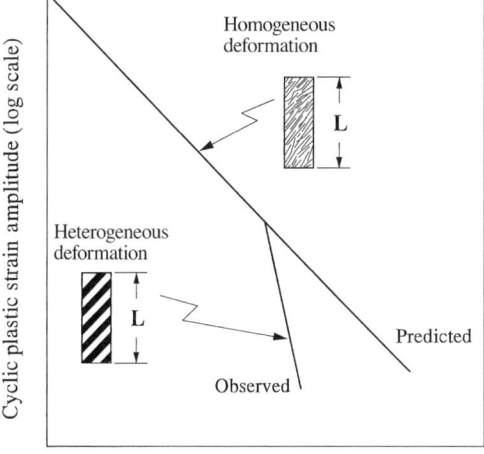

**Fig. 3.9** Discontinuity in Manson-Coffin line in aluminum alloys and correlation of the discontinuity with change in mode of plastic deformation. Source: Ref 3.12

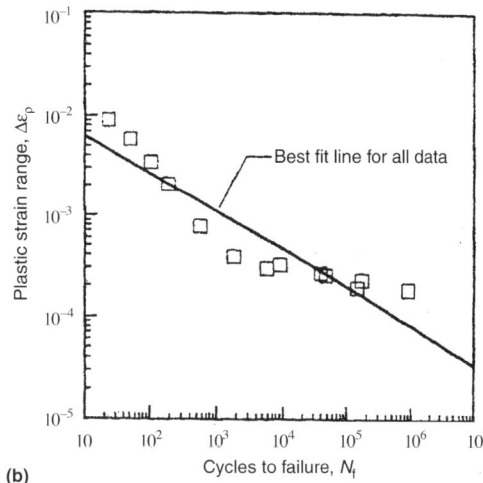

**Fig. 3.10** Manson-Coffin strain-life relation for AISI type 304 stainless steel at room temperature. (a) Realistic representation of the data. (b) Data as originally published. Source: Ref 3.5

in order to find a few promising ones that can later be studied more extensively. It is also desirable to be able to estimate the life in advance of laboratory tests in order to choose the test parameters intelligently. Several methods have been developed for this purpose; some involve no prior knowledge of fatigue behavior, while others make use of a small amount of cycling information.

### Predictions Based Only on Monotonic Tensile Properties

The ensuing section contains details primarily of interest to the fatigue specialist. In 1965 (Ref 3.17) two approaches were proposed (although reported by Manson, the actual study was conducted with the association of Marvin H. Hirschberg, of the NASA Lewis Research Center in the early 1960s) that use only static tensile test properties. The equations relating ultimate tensile strength, ductility, and modulus of elasticity to fatigue life for any prescribed strain range were obtained through correlation of the fatigue data for 29 materials. These equations are the method of universal slopes and the four-point correlation. Two decades later, a more extensive study was conducted on 50 materials for which data had been obtained at NASA Lewis and at industrial and university laboratories, resulting in a method designated as the modified universal slopes equation (Ref 3.18).

**The Original Method of the Universal Slopes Equation.** Figure 3.14 shows the basis for the universal slopes equation. The elastic and plastic lines are taken to have slopes of $-0.12$ and $-0.6$, respectively, for all materials—

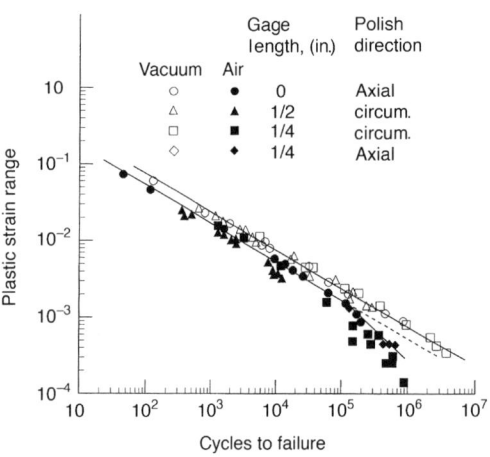

**Fig. 3.11** Effect of environment on Manson-Coffin relation of 1113 steel. Source: Ref 3.14

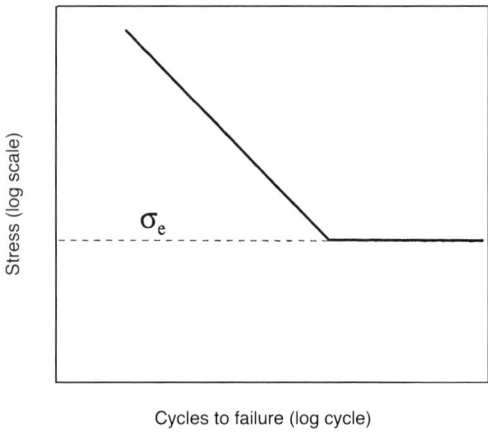

**Fig. 3.12** Traditional way of representing S-N curve of a material with a distinct "endurance limit"

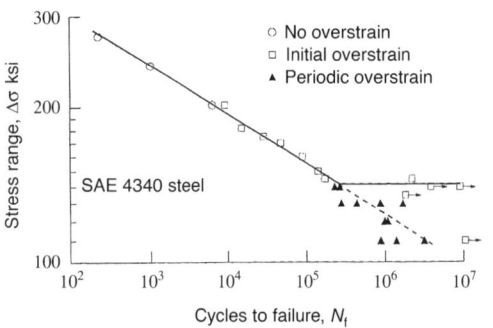

**Fig. 3.13** Removal of an "apparent endurance limit" by occasional overstraining and then testing at stresses lower than the original endurance limit. Fatigue life becomes finite below the endurance limit. Source: Ref 3.16

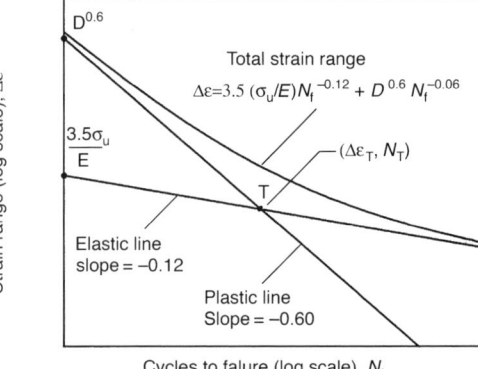

**Fig. 3.14** The original method of universal slopes equation as developed by Manson and Hirschberg

hence, the description "universal slopes." One point on each of these two lines was determined at the intercept on the strain axis at $N_f = 1.0$. These points were based on the average values for the 29 materials, which covered a wide range of strength and ductility. For the elastic line, this intercept point was found to depend only on the parameter $S_u/E$ where $S_u$ is the commonly determined ultimate tensile strength, and $E$ is the elastic modulus. For the plastic line, the intercept point at $N_f = 1.0$ was found to depend only on ductility $D \; (= \; \ln[100/(100 \; - \; \%RA)])$, where RA is the reduction of area in a tensile test). On the basis of these parameters, the Manson-Hirschberg Universal Slopes Equation becomes:

$$\Delta\varepsilon = (3.5 S_u/E)(N_f)^{-0.12} + (D)^{0.6}(N_f)^{-0.60} \quad \text{(Eq 3.3)}$$

Thus, only the tensile properties $S_u$, $D$, and $E$ are required to determine the relations between cyclic life and strain range. We discuss the uses to which the equation has been put in a later section.

**The Modified Method of the Universal Slopes Equation.** By 1986, when the fatigue properties of many more materials had been characterized than the 29 materials used in 1964, a new study was undertaken to reexamine the relation with the intent of improving its accuracy. The study is reported in Ref 3.18. Least squares curve fitting using log-log coordinates of the strain-life data of the 50 materials listed in Table 3.1 resulted in the equation:

$$\Delta\varepsilon = 1.17\left(\frac{S_u}{E}\right)^{0.832}(N_f)^{-0.09} + 0.0266(D)^{0.155}\left(\frac{S_u}{E}\right)^{-0.53}(N_f)^{-0.56} \quad \text{(Eq 3.4)}$$

The first term replaces the elastic life line, $(3.5 S_u/E)(N_f)^{-0.12}$, in the original universal slopes equation. It is seen that the coefficient is still a term in only $S_u/E$, but the slope has become $-0.09$ instead of $-0.12$, which occasionally had been found to be too steep for some materials. There is some compensation for the decreased slope since the intercept is *always* less than that of the method of universal slopes intercept. In general, the two equations cross in the finite-life regime. The lower slope implies a lesser drop in stress when extrapolations are made to life values greater than $10^6$ cycles, which should be satisfying to users interested in extrapolation to the very high cyclic life range.

The second term, representing the plastic line, did bring about a surprise. Although the exponent $-0.56$ differed little from the original value of $-0.6$, the coefficient was far different in appearance. Instead of involving only a ductility term, the new equations contained both a strength term ($\approx \sqrt{S_u/E}$) and a ductility term with a considerably lower exponent. For most materials, both representations may be nearly equally good, because there is, in general, an inverse relation between tensile strength and ductility.

**Comparison between the Two Universal Slopes Equations.** The modified universal slopes equations predicted the average fatigue behavior of the 50 materials slightly better than the original equation. In individual cases, however, one or the other could be better. Figure 3.15 shows some direct comparisons of the two methods. In Fig. 3.15(a) and (b), both equations gave about the same results and agreement with test results is excellent; in Fig. 3.15(c) and (d), the modified equations gave predictions that agreed better with the experimental behavior than the original equation; and in Fig. 3.15(e) and (f), the original equation agreed better with the experimental behavior than the modified equation.

Because of the widespread use of the original universal slopes equation, it would appear appropriate to retain its use, rather than introduce a new equation, until a more definitive reason develops to make the change.

**Special Uses of the Method of Universal Slopes Equation.** Because the universal slopes equation requires no information other than the normally available simple tensile properties, it has found special uses where other approaches would be complicated or more difficult. Here, we discuss four such applications within our experience:

*Cryogenic Fatigue.* Some applications, notably space propulsion equipment (and other earth-oriented equipment), require the use of cryogenic fluids that must be stored in pressure vessels. We encountered this problem at NASA when considering materials for the space program. A cryogenic fatigue testing facility was built at considerable expense at the Cleveland, Ohio, facility, and because the cryogenic liquids imposed hazard, elaborate special remote features were necessary. After testing numerous materials, the experimental results were compared with predictions made by the universal

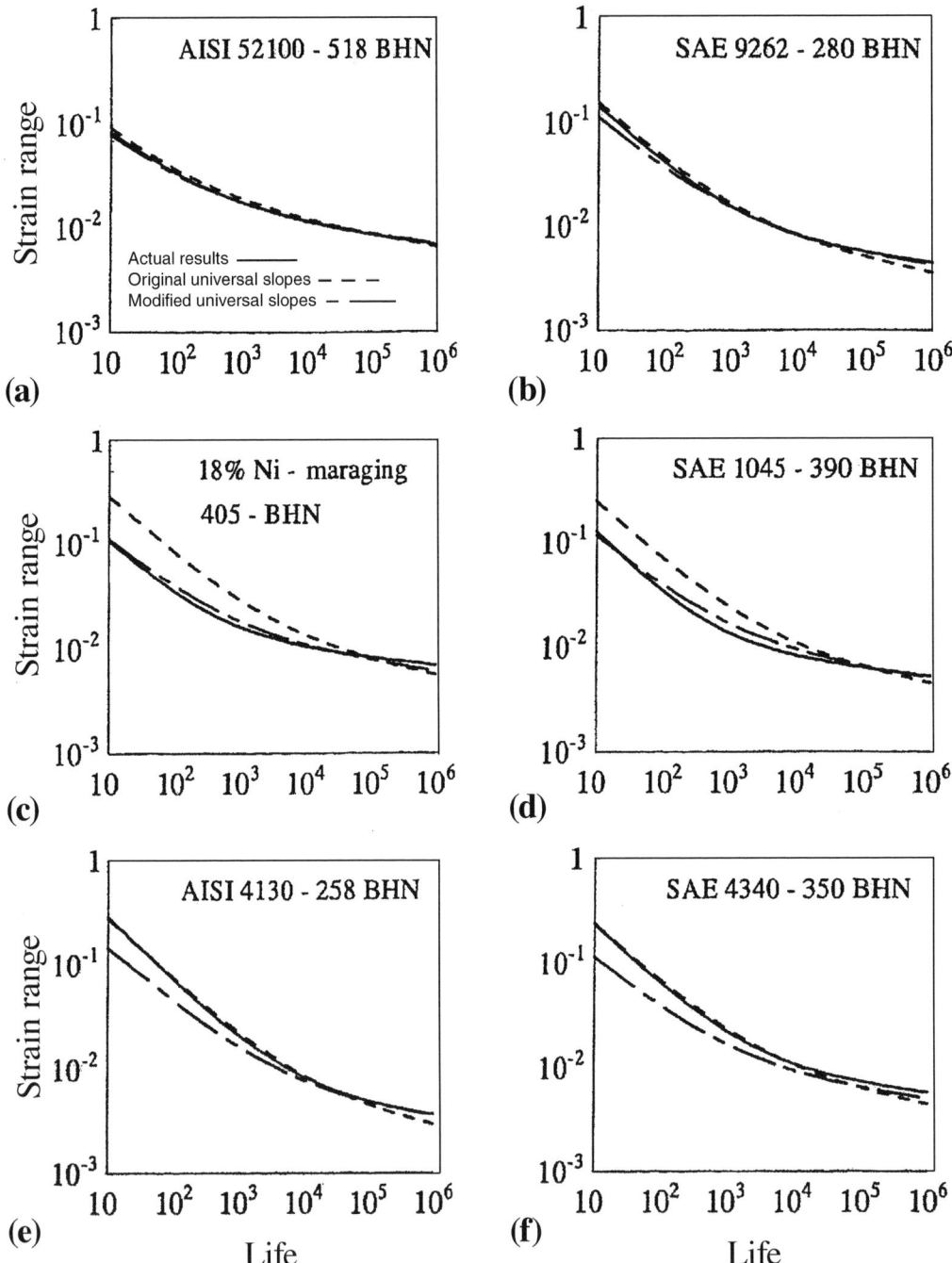

**Fig. 3.15** Comparison of the two universal slopes equations for high-strength steels. (a) and (b) Cases where both equations yield about the same results. (c) and (d) Cases where the modified method of universal slopes equation yields results closer to the experimental values than the original method of universal slopes equation. (e) and (f) Cases where the original method of universal slopes equation yields results closer to the experimental values than the modified method of universal slopes equation

slopes equation. The comparison is shown in Fig. 3.16 (Ref 3.19). In general, reasonably good agreement was observed. Two tables are included in this figure. The first, presented as a baseline for comparison, shows the original degree of agreement between predictions and experiment found for the 29 materials tested at room temperature to develop the correlation. This table shows the fraction of the data that fell within the predictions by several life factors (e.g., 80% of the data were predicted within a life factor of 3.0). The second table shows how these life factors were satisfied by predictions of the Universal Slopes equation for just the cryogenic data. The results are not appreciably different. If new materials were to be considered for future applications under cryogenic conditions, it would seem reasonable to subject these materials first to a universal slopes analysis. Only after they passed this test would an exper-

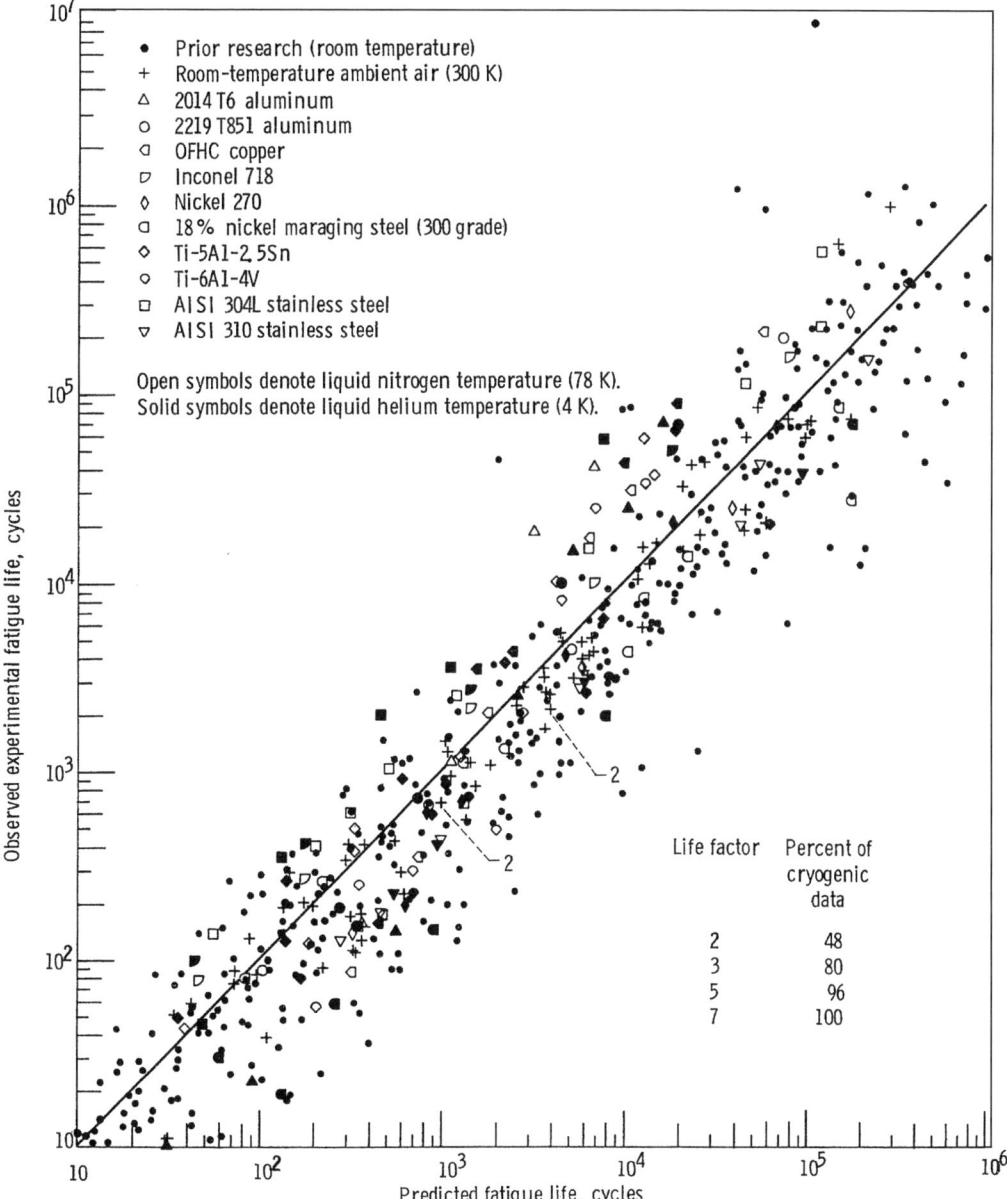

**Fig. 3.16** Predictive capability of the method of universal slopes equation at cryogenic and ambient temperatures. Source: Ref 3.19

imental program be undertaken on the most promising materials, if deemed necessary. Of course, the effect of the cryogenic temperatures on the tensile properties would have to be determined before the method could be applied. It certainly is easier, however, to measure static tensile test properties than fatigue properties.

*Effect of Nuclear Radiation on Fatigue Properties.* In another early application at NASA's Cleveland research laboratory, consideration was given to the development of a nuclear-powered airplane. A special facility was constructed in a remote location that included a reactor for use to evaluate materials in a nuclear radiation environment. Fatigue testing was especially troublesome, but some tests were actually conducted for an austenitic AISI 310 stainless steel. Again, a comparison was made of the experimental fatigue results with predictions by the universal slopes equation, the results of some tests on the AISI 310, being shown in Fig. 3.17. Nonirradiated results are also included in the figure to illustrate the applicability of the universal slopes equation for this alloy. The agreement is excellent with all life predictions being within a factor of two of the observed irradiated fatigue lives. Thus, it was concluded that measurements of only tensile properties in the nuclear environment would be adequate to compare materials for possible fatigue applications. Such tensile test measurements could, of course, be made much more easily and at far less cost and in less time than fatigue measurements. Nuclear irradiation was found, in general, to reduce ductility and increase tensile strength, due to the fact that radiation tends to knock atoms out of their lattice sites, impeding the sliding of dislocations (see Chapter 10, "Mechanism of Fatigue" for discussion of dislocations). These effects on ductility and tensile strength were properly reflected in the fatigue behavior. The nuclear airplane project has long been abandoned in favor of other research, but the results are useful for other radiation applications. Brinkman (Ref 3.21) of the Oak Ridge National Laboratory has also found the method of universal slopes equation useful in prediction of nuclear radiation effects on fatigue properties.

*Materials Selection for the Space Shuttle Main Engines.* When NASA awarded the contract to develop the main engines for the space shuttle, there was little information available on the fatigue properties of the many candidate materials that were considered for the component parts. This was particularly true for the extreme environmental operating conditions within a shuttle main engine. Furthermore, there were insufficient sophisticated testing facilities to generate the required fatigue information within a reasonable period of time. Something had to be done quickly to provide the engine designers with meaningful fatigue characteristics of the many materials and conditions involved. Early in the program, the Rocketdyne Division (currently Boeing) of North American Rockwell developed a complete handbook of approximated fatigue properties based on NASA's method of universal slopes equation. The universal slopes equation was used for several dozens of alloys, each in a multiplicity of conditions for a wide range of temperatures. Note that tensile test properties were needed for each alloy, alloy condition, type of welding, environment (high-pressure hydrogen and oxygen), and temperatures ranging from cryogenic to 465 °C (870 °F). The fatigue lives calculated from the universal slopes equation were reduced by a factor of 3 to reflect the uncertainty in the absolute accuracy of the calculations. Of course, as the program progressed, experimental data were generated on many of the materials selected, but the universal slopes equation remained extraordinarily useful in the process. To this date, the estimated fatigue parameters are still used for certain materials intended for high-pressure hydrogen environments from cryogenic to elevated temperatures. The gratitude of Rocketdyne for this enabling technology was aptly expressed by Newell (Ref 3.22).

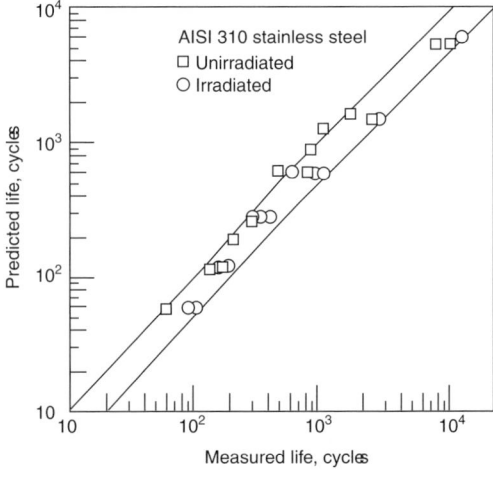

**Fig. 3.17** Method of universal slopes applied to irradiated metals. Source: Ref 3.20

*Estimation of Test Parameters.* When conducting a fatigue program on a material, it is useful to estimate the test parameters in advance so that the test conditions will not result in excessively long or short fatigue lives. The universal slopes equation has been used to advantage for this application. Often, one of the first estimates made is that of the transition strain range and the transition life. If the estimated life is not excessive for test verification, the transition strain range test is the first point to be studied. Further test points can be determined by systematic use of the results of this test and the others that follow.

### The Four-Point Correlation Method

The four-point correlation method is another approach for estimating the elastic and plastic lines from information determined in standard tensile tests, which are usually available. To develop the method, numerous correlations were attempted for points on the elastic and plastic lines, choosing the parameters for correlation on the basis of what was logical to expect. The actual coefficients (two for each life relation) were then determined from behavior of the 29 materials for which data were available at the time (Ref 3.17). The coordinates of the coefficients finally chosen are shown in Fig. 3.18.

For the elastic line, two points were chosen:

- *Point 1:* At ¼ cycle (presumed to represent the first tensile loading), one would expect the elastic strain amplitude to be related to $\sigma_f/E$, where $\sigma_f$ is the true fracture strength. Thus, the range would be of the order of $2\sigma_f/E$. Actually, the correlation showed that a better representation is $2.5\sigma_f/E$. This result is reasonably consistent with the later finding that an appropriate assumption is for the intercept to be $2\sigma_f/E$ at $2N_f = 1.0$.
- *Point 2:* The most consistent results were $10^5$ cycles, where the strain range is $0.9S_u/E$.

For the plastic line, the two points were chosen as follows:

- *Point 3:* At 10 cycles, the plastic strain range is $¼D^{3/4}$, where $D$ is the ductility of the material.
- *Point 4:* At $10^4$ cycles, $\Delta\varepsilon_p$ was determined as $(0.0132 - \Delta\varepsilon_e)/1.91$, and because $\Delta\varepsilon_e$ was already known from the elastic line, $\Delta\varepsilon_p$ can be determined.

The four-point correlation gives approximately the same predictions as the method of universal slopes equation, as shown in Fig. 3.19 (Ref 3.17). It has not been used extensively, however, largely because of its slightly more involved formulation.

### A Critique of Other Correlations in the Literature

In addition to the relations involved in the method of universal slopes equation and four-point correlation method, a number of other relations have been proposed by various investigators that can be useful when attempting to predict the fatigue resistance of materials. Some of these have been summarized by Mitchell (Ref 3.23). Following is a brief discussion of several of these approximations. Where possible, their validity is examined by reference to the actual parameters for the 50 materials of Table 3.1.

**Strength and Ductility Coefficients.** Based on the work of Morrow and his students, approximations have been suggested that $\sigma'_f = \sigma_f$, and $\varepsilon'_f = \varepsilon_f$. The implication is that the end points of both the static and cyclic stress-strain curves are the same. A comparison of these parameters is shown in Fig. 3.20. While it appears that $\sigma'_f = \sigma_f$ is a reasonably good correlation, $\sigma'_f$ is somewhat larger than $\sigma_f$. Furthermore, only 75% of all the results fall within a deviation of less than 25% away from perfect correlation. This observation, considered along with the observation that the nominal fatigue strength exponent is on the order of −0.10, implies a deviation in fatigue life prediction of greater than an order of magnitude for 25% of the data. This is a far poorer correlation than found for the method of universal slopes for which only 4% of the materials had implied predictions of life greater than a factor of 10 at room temperature

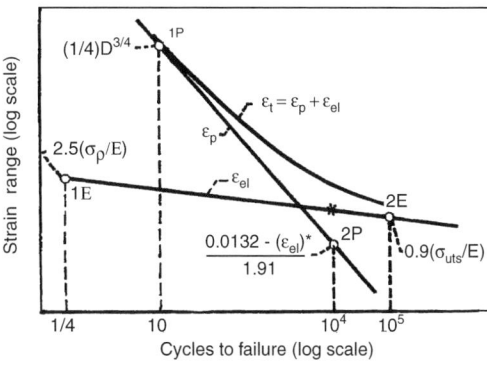

**Fig. 3.18** The four-point correlation method. Source: Ref 3.17

and none at cryogenic temperatures (Fig. 3.16). From Fig. 3.20, it is obvious that the correlation between ductility coefficients $\varepsilon_f'$ and $\varepsilon_f$ is much poorer than for the strength coefficients. However, the implied deviation in projected fatigue life (for a ductility exponent of $-0.6$) is considerably less than that implied for the strength coefficient correlation.

**Strength and Ductility Exponents.** Morrow (Ref 3.24), using energy considerations, has proposed that $b = n'/(1 + 5n')$ and $c = -1/(1 + 5n')$. These equations are, of course, consistent with the relation that $n' = b/c$, but they require that the *cyclic* strain-hardening exponent already be known before $b$ and $c$ can be determined. Actually, it is possible to get $n'$ from just one specimen, whereas $b$ and $c$ require multiple specimens to be determined. Therefore, if accurate, these relations could be very useful. Their degree of correlation is shown in Fig. 3.21.

Another scheme that can be used to estimate $b$ is based on the general observation (at least for one class of materials, carbon steels) that at about $10^6$ cycles to failure, the completely reversed fatigue stress amplitude is approximately one-half the ultimate tensile strength. If it is assumed that at $N_f = \frac{1}{2}$, the stress amplitude is $\sigma_f' = \sigma_f$, then the straight line joining the points $[N_f = \frac{1}{2}, \sigma = \sigma_f']$ with the point $[N_f = 10^6, S = 0.5S_u]$ the slope $b$ becomes $-1/6 \log (2S/S_u)$. Figure 3.22 shows the accuracy of this assumption. The greatest deviations from a perfect correlation are for austenitic stainless steels and aluminum alloys (Table 3.1). These alloys are well known to have fatigue strengths considerably less than one-half the ultimate tensile strength. Some aluminum alloys have been known to have fatigue strengths at $10^6$ cycles to failure of

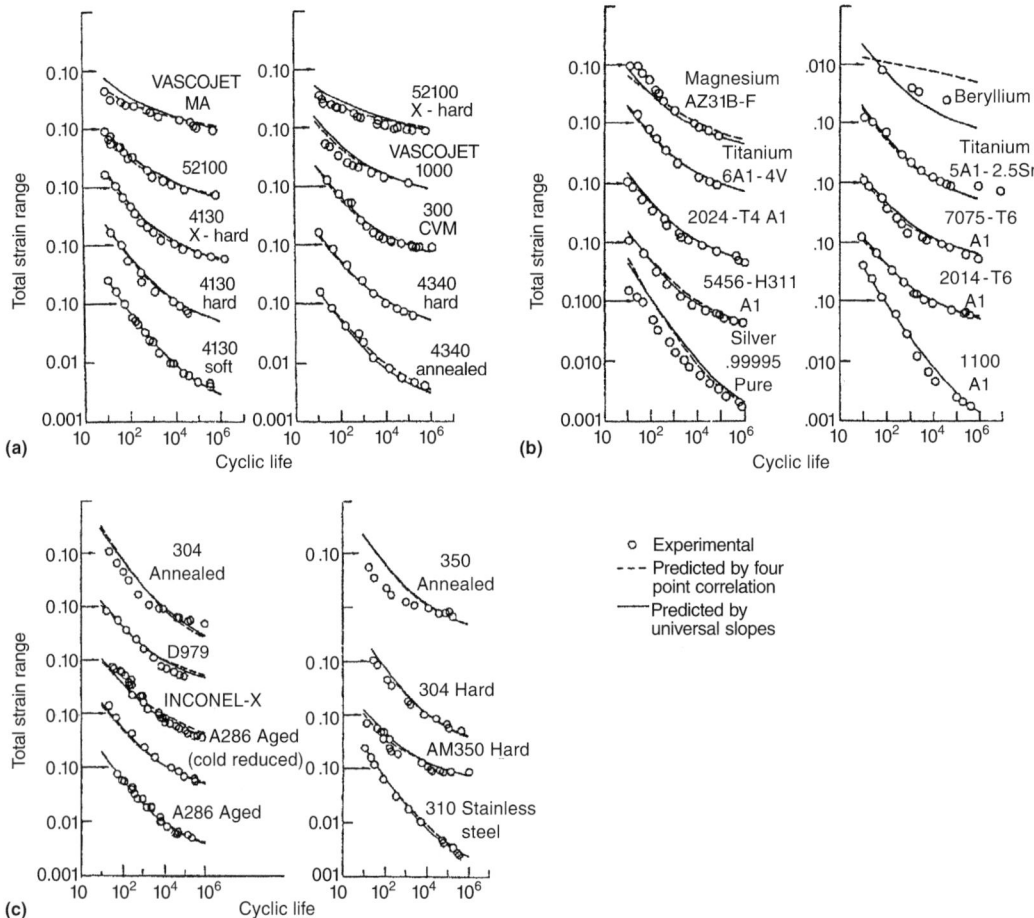

**Fig. 3.19** Comparison of predicted and experimental axial fatigue results using the method of universal slopes equation and the four-point correlation method. (a) Low-alloy and high-strength steels. (b) Nonferrous metals. (c) Stainless steels and high-temperature alloys. Source: Ref 3.17

only one-fourth to one-fifth the ultimate tensile strength.

The corresponding value of $c$ can be obtained from a consideration of the transition point that is described in the following section after estimates for the transition point are discussed.

**Transition Fatigue Life.** Landgraf (Ref 3.25) has made the interesting observation that for steels the transition life depends very closely on the hardness, and he has provided the graphical relation shown in Fig. 3.23. In Ref 3.26, this graph is expressed by the relation:

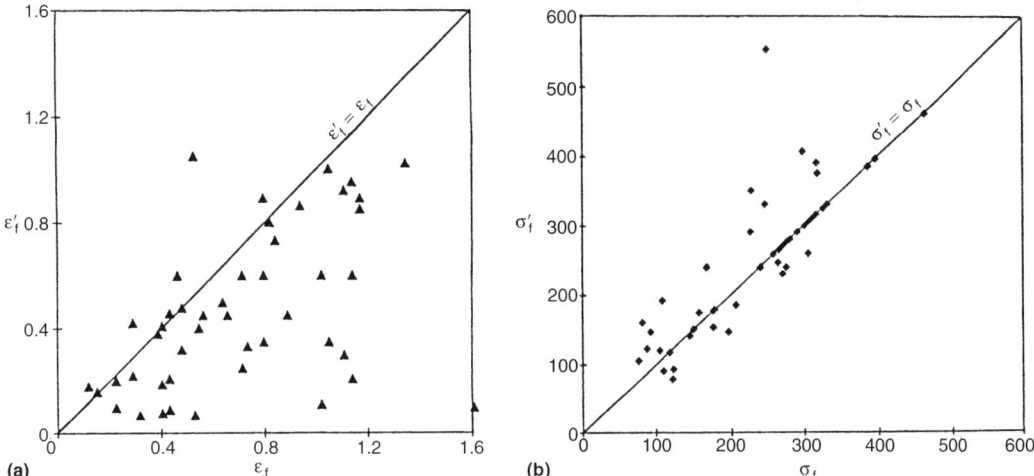

**Fig. 3.20** Validity of Morrow's assumption for approximating (a) $\varepsilon_f'$ and (b) $\sigma_f'$ for the materials listed in Table 3.1

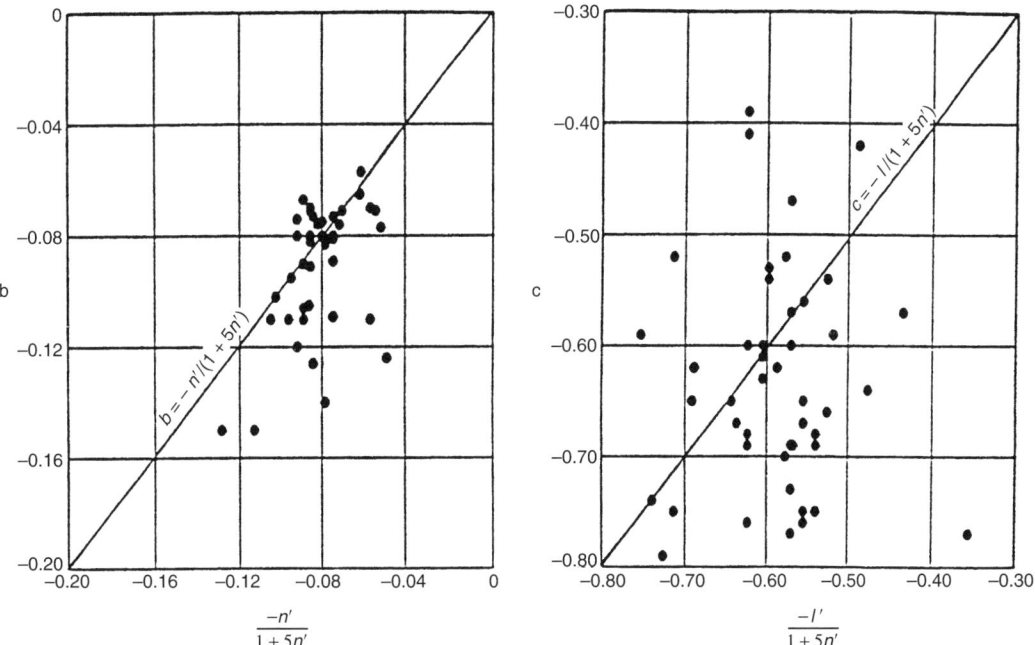

**Fig. 3.21** Validity of Morrow's assumption for approximating (a) $b$ and (b) $c$ from $n'$

$$N_T = 3.436 \times 10^5/(2.555 \times 10^7)^{HB/1000}$$
(Eq 3.5)

Using the relation $S_u$(psi) = 500 HB, we developed the equation:

$$N_T = \frac{53,444}{(20.13)^\gamma}$$
(Eq 3.6)

where for steel $\gamma = S_u/1000$ ($S_u$ in ksi). We further speculated that the same equation could be used for materials other than steel by adjusting for elastic modulus, giving $\gamma = (S_u/100)(30 \times 10^4/E)$, where $E$ = elastic modulus in ksi. Figure 3.24 shows the degree of accuracy that can be expected according to this correlation. Here, $N_T$ is the actual transition point for the materials listed in Table 3.1.

At this point, we can complete the discussion of how to estimate the slope of the plastic line if the estimated transition life according to Eq 3.6 is used in conjunction with the second of the methods discussed previously wherein the elastic line is already established. On the elastic line the transition point may be positioned, and this point also serves as a point on the plastic line. The other point may be taken at $\varepsilon_f' = \varepsilon_f = D$ at $N = \frac{1}{2}$, or the point established at $N = 10$, where $\varepsilon_p = (\frac{1}{2})(D)^{3/4}/4$ and the plastic line may be drawn. The approach is, in fact, analogous to the procedure used in the four-point correlation, except that a known relation is used at $N = N_T$ instead of $N_f = 10^4$ cycles. Because of the many possibilities involved, and because this method is rarely used, we show no correlations.

**Near-Convergence of Total-Strain-versus-Life Curves.** The four-point correlation method made use of the near-convergence of the total-strain-versus-life curves for many materials in the life range of about $10^4$ cycles. Other investigators have also represented this observation of near convergence. Peterson (1962) (Ref 3.27) has expressed it as actual convergence at 2% strain range at $10^3$ cycles, while Manson, in a discussion to Tavernelli and Coffin (1962) (Ref 3.28), has expressed it as 1% strain range at $10^4$ cycles.

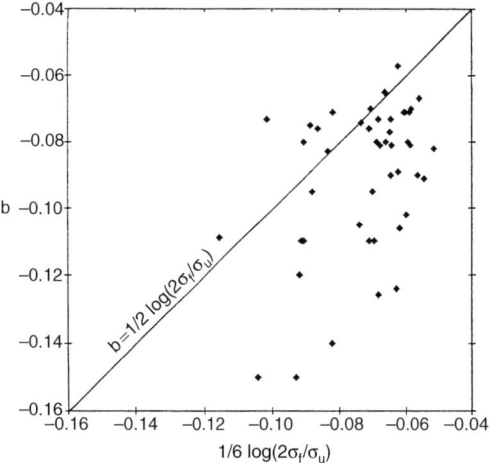

**Fig. 3.22** Estimating $b$ by assuming $\sigma = \sigma_f$ at $2N_f = 1$ and $\sigma = 0.5 S_u$ at $2N_f = 2 \times 10^6$

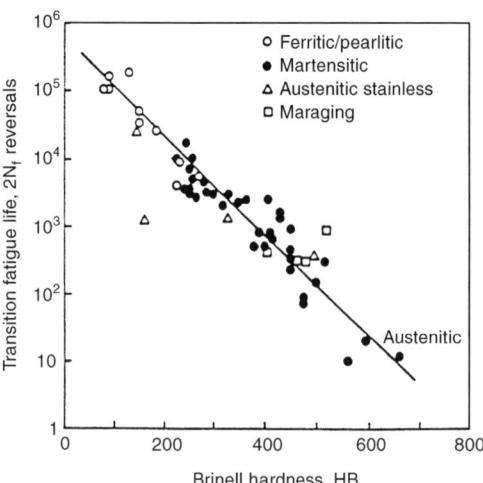

**Fig. 3.23** Graphic relation for transition life vs. Brinell hardness for steels. Source: Ref 3.25

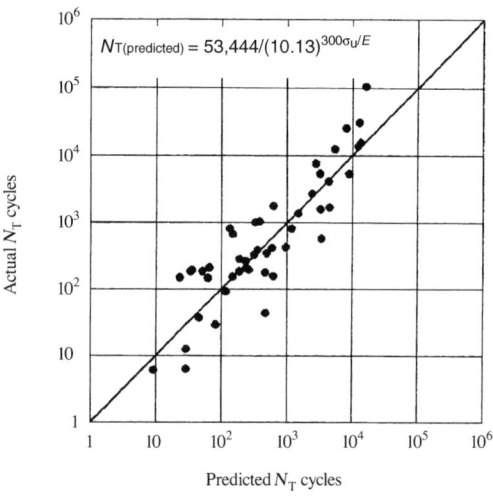

**Fig. 3.24** Correlation of transition fatigue life $N_T$ with a modified form of the proposal by Landgraf (Ref 3.25) for the materials listed in Table 3.1

## Predictive Methods Based on Limited Cyclic Data

In principle, only two fatigue tests are needed to establish both the elastic and plastic lines. In fact, however, many more tests would be needed if realistic data scatter were considered. It would seem that at least three specimens would be required to establish a first approximation of the two life lines. A reasonable approach is to choose strain ranges that would result in cyclic lives of about $10^3$, $10^4$, and $10^5$. Here, the universal slopes equation may prove very useful.

An additional bit of information can be obtained from each of the fatigue tests, relating to the shape of the stabilized hysteresis loop. All loops would depict the *double-amplitude* cyclic stress-strain curve as discussed in Chapter 2. By superimposing all three loops from the same origin, a good representation of the average double-amplitude cyclic stress-strain curve can be obtained, from which the value of $n'$ can be determined. Since $n'$ gives us the ratio of the slopes $b$ and $c$, this information can also be fed into the least-squares solution of the optimum elastic and plastic lines that fit the experimental data.

If only one specimen can be dedicated for the purpose of estimation, it would seem that its most useful purpose would be to establish the value of $n'$ from its cyclic stress-strain curve. One way of doing this is to estimate the transition strain range from the method of universal slopes equation and to conduct a continuous cycling test at twice that strain range. From the hysteresis loop, the value of $n'$ can be determined. In Ref 3.26 we proceeded in the following way.

First, we used the cyclic stress-strain curve to determine the value of $n'$, which gave us the value of $b/c$, as shown in Fig. 3.25. By finding the intercept of the cyclic stress-strain curve with a line of slope $E/2$, we determined the point C, that is, the transition point strain amplitude, because at this point the elastic and plastic strains are equal.

Because the primary purpose of the analysis in Ref 3.26 was to predict high-cycle fatigue, our focus was on the elastic line. We then calculated $b$ and $c$ from the Morrow equations of the section titled "Strength and Ductility Exponents." To place the elastic line, we made use of the approximations used in the four-point correlation method, shown in Fig. 3.18. However, since we were imposing a slope of $b$ on this line, a least-squares analysis was conducted to establish a line of given slope that minimized the distance from two points. The intersection of this line with the horizontal line at $\Delta\varepsilon = \Delta\varepsilon_T$, which was known from Fig. 3.25, gives the result:

$$N_T = 1.3 \times 10^4 \{[1.2E^2(\Delta\varepsilon_T)^2]/[\sigma_f S_u]\}^{3.5/b}$$

(Eq 3.7)

To check the validity of Eq 3.7, we compared, in Fig. 3.26, the true transition life for each of the 50 materials of Table 3.1 to the predictions of this equation. The correlation is good considering that only one test specimen is required for

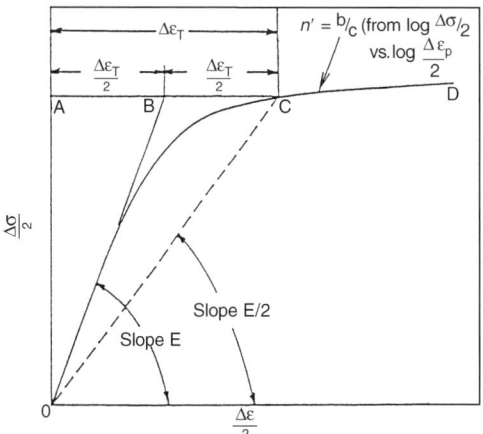

**Fig. 3.25** Use of the cyclic stress-strain curve to obtain the strain-hardening exponent and transition strain range. Source: Ref 3.26

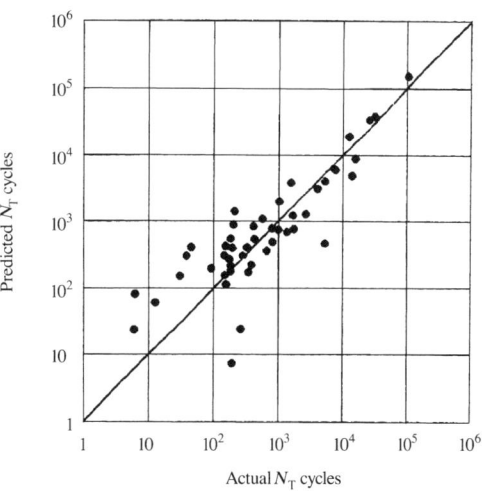

**Fig. 3.26** Correlation of transition life relation with predictions by Eq 3.7

this prediction. (Despite the use of one cyclic test result, the correlation is only slightly improved compared with the method of universal slopes analysis that uses no cyclic data.)

In Ref 3.26, it is shown how well this procedure was able to predict the curves of total strain range versus life in the range from $10^4$ to $10^7$ cycles for all 50 materials listed in Table 3.1. Example results for materials 43–50 are shown in Fig. 3.27. The agreement with the experimental fatigue behavior is good, but there is room for improvement.

## High- versus Low-Cycle Fatigue

It is important to define what we mean by the term *high-cycle fatigue*. The question is "why draw a distinction at all between low-cycle fatigue (LCF) and high-cycle fatigue (HCF)?" The answer to this question depends to some extent on why it is asked.

**Consideration of Absolute Life values.** In recent years there has been a tendency to tie the definition of LCF and HCF to the transition life, as we discuss later. There are valid reasons for such connection associated with applicability of elastic analysis, mean stress retention, and so forth. Thus, for a high-strength, low-ductility material with a transition life of only 10 cycles, lives of the order of 1000 cycles could be considered high-cycle fatigue. On the other hand, there are very weak, ductile materials with transition lives on the order of $10^5$ cycles, so that the HCF region does not start until $10^7$ cycles, based on the transition life criterion discussed later. It may be difficult, however, for the potential user of a material to regard a life of a thousand cycles as having achieved "high-cycle fatigue" while the potential user of the soft material may not be happy about having to run tests out into the $10^7$ cycle range before he can say he has achieved "high cycle fatigue" life.

Thus, if the definition is for the purpose of establishing the type of testing equipment required, testing durations, or durability estimates of machine components, the distinction between LCF and HCF may well have to be made on the basis of actual cyclic life. The cutoff points for each of the regions have to be arbitrary, and subjectively based on the user's requirements. One reasonable subdivision, however, might be somewhere between $10^4$ and $10^5$ cycles.

**Transition Fatigue Life as a Criterion for HCF.** The major reason for distinguishing between LCF and HCF lies in the relative amounts of plastic and elastic strains present. If the plastic strain is small relative to the elastic strain, it may be adequate to base design on elastic analysis.

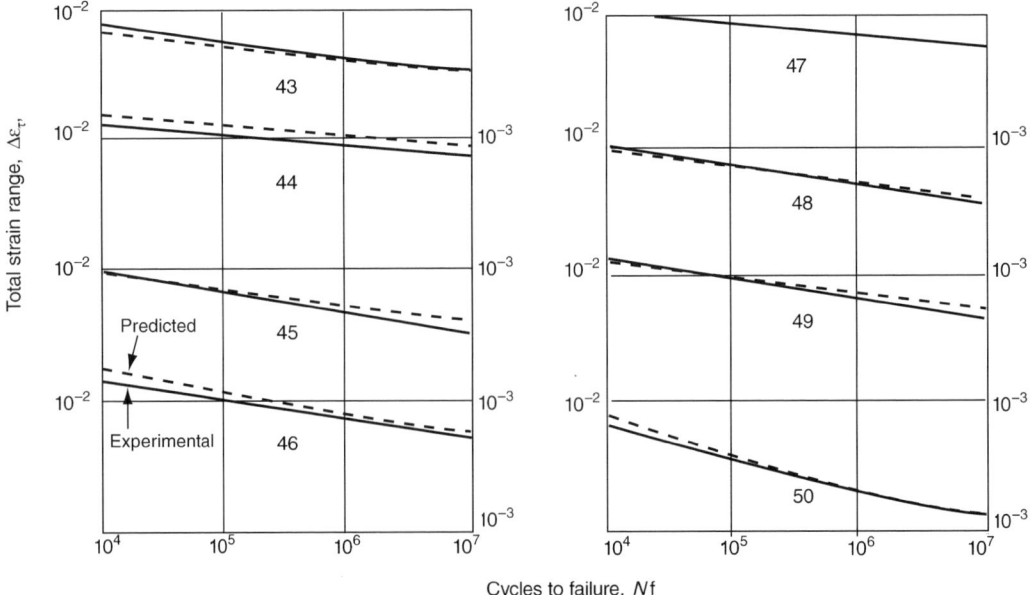

**Fig. 3.27** Comparison of experimental strain range vs. fatigue life curves with predictions by the transition point correlation formula for several representative materials from Table 3.1. Solid line, experimental; dashed line, predicted. Source: Ref 3.26

Elastic analyses are significantly easier and much less costly to perform. Furthermore, the elastic materials properties are most frequently well known, whereas inelastic, or plastic, properties are more obscure. Plasticity brings with it additional behavior that must be accounted for. Examples are cyclic stress redistribution, cyclic relaxation of mean stresses, cyclic plastic strain ratcheting, and alteration of the degree of multiaxiality. In the area of fracture mechanics analysis of cracked members, linear-elastic fracture mechanics can be routine, but elastic-plastic fracture mechanics is tedious and costly. There has been a trend to associate LCF with applications wherein the plastic strain dominates, and HCF with applications wherein the elastic component of the strain dominates. Obviously, this type of association gives important significance to the intersection point of the elastic and plastic lines that is the transition point.

We learned the hard way that there is a physical significance to the location of the transition point. In fact, when we first introduced the terminology, the word *transition* referred to two types of transition: transition in failure location, as well as transition from dominance of plastic strain to dominance of elastic strain. In our program we tested various materials; some were ductile, some brittle. Our purpose (Ref 3.18) was to cover a spectrum of materials in an investigation to validate Eq 3.3 and to determine appropriate material constants. In most cases, the specimens were button headed as shown in Fig. 3.28, but for some materials, including the bearing steel AISI 52100, we did not have bar stock of large-enough diameter for button-head specimens, so threaded ends were used instead. For the LCF tests, we encountered no problems; failure was in the expected minimum cross section of the hourglass specimen. (Hourglass specimens used diametral extensometers. Uniform gage-length specimens and axial extensometry are discussed briefly in the Appendix.) However, for the lower strain ranges, failures occurred in the threaded ends of the specimens, and for some materials in the sharp corner of the button heads as well. This was a very curious result: why should one section of the specimen be weaker at one load level, while another section became the weak link at another? After considerable contemplation, it eventually became clear that it was a question of the dominance of the elastic component of the strain relative to the plastic component, and as the transition of strain occurred, so did transition of failure location. The terminology, which referred both to strain range and life, as seen in (Fig. 3.29), has been adopted for common usage and has become valuable in designating the distinctions between low- and high-cycle fatigue.

**Some Quantitative Relations Involving Transition Fatigue Life and Strain Range.** It is appropriate to review here some of the concepts discussed earlier and to introduce some formulations that have been useful.

Referring to Fig. 3.29, designated by $T$ the intersection point of the elastic and plastic lines, and by $\Delta\varepsilon_T$ and $N_T$, the strain range and cyclic life at the point, it is seen that if we locate point $T_1$ on the life curve, then the total strain range to produce this life is $2\Delta\varepsilon_T$. Above this strain range, the life is dominated by the plastic strain. Between a strain range of $2\Delta\varepsilon_T$ and $\Delta\varepsilon_T$ the dominant component is the elastic strain, but the curve $T_1B$ is still far from the elastic line $DE$. One way of separating the low- and high-cycle fatigue range, therefore, is to use the life $N_B$ at the strain $\Delta\varepsilon_T$. This life is much higher than $N_T$, being of the order of 10 $N_T$ to 100 $N_T$, and be-

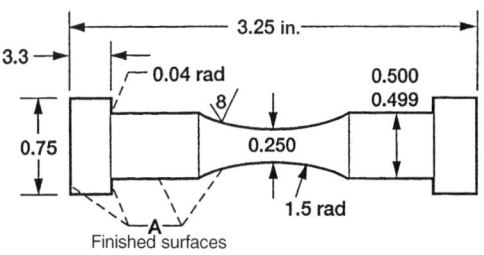

**Fig. 3.28** Button-head test specimen used in fatigue evaluation program (dimensions in in.)

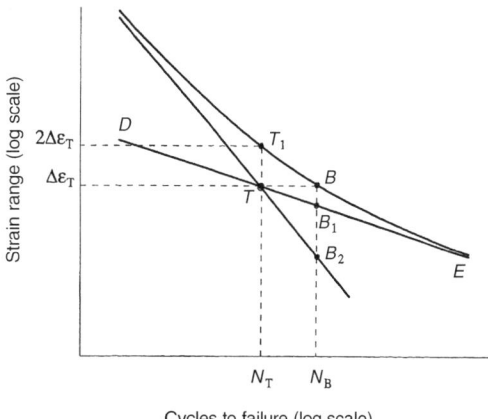

**Fig. 3.29** Definition of transition point

coming larger the smaller the absolute value of $b$ (note that this life is infinity when $b = 0$). At the life $N_B$, the elastic strain range is $\Delta\varepsilon_{B1}$, which is clearly much larger than the plastic strain range $\Delta\varepsilon_{B2}$. While for some materials this difference may be large enough to accomplish the practical purpose of establishing a demarcation between LCF and HCF (such as allowing elastic analysis, using elastic fracture mechanics, avoiding consideration of mean stress relaxation, etc.), it may fall short of these goals for materials with extreme combinations of the exponents $b$ and $c$. It is better, then, to leave some flexibility in the definition.

This subject has been discussed in Ref 3.26. Considering all the materials listed in Table 3.1, a definition can be arrived at as:

- *High-cycle fatigue* is for strain ranges, $\Delta\varepsilon$, less than approximately 75% of the transition strain range, $\Delta\varepsilon_T$.
- *Transition-cycle fatigue* is for strain ranges, $\Delta\varepsilon$, that lie between the transition strain range, $\Delta\varepsilon_T$, and 75% $\Delta\varepsilon_T$.
- *Low-cycle fatigue* is for strain ranges $\Delta\varepsilon$, greater than the transition strain range, $\Delta\varepsilon_T$.

This compromise keeps the plastic strain range at approximately less than 10% of the elastic strain range, which is a reasonable criterion for permitting the use of elastic analysis and neglecting effects of stress relaxation. Further discussion of this criterion appears in Chapter 4, "Mean Stress."

### Concluding Remarks on Strain-Controlled Fatigue Testing

**Plastic Strain Governs Fatigue.** Earlier in the chapter, it is emphasized that reversed plastic straining has been fully accepted as the underlying cause of fatigue—whether in the low-cycle, high-cycle, or the ultra-high-cycle regimes. Thus, it was natural that plastic strain cycling and the magnitude of the range of plastic strain were important in the study of fatigue and, in particular, LCF. Control of the magnitude of the plastic strain range was obviously more treatable by using strain-controlled fatigue testing equipment.

**Constraint at Notches in Structural Components.** Perhaps the most practical justification is that coupon testing, using an extensometer, better represents how a material is typically "loaded" in a structural component at a notch. Structural fatigue failures notoriously initiate at geometric discontinuities in the form of notches (grooves, fillets, holes, and any other form that concentrates stresses and strains in a highly localized zone). In fact, the concentrations of stress and strain can be a factor of three or greater than is experienced by remote portions of the structural component. The vast bulk of the component could be operating at stress-strain levels that would give cyclic lives much greater than required of the component. At the discontinuity, however, the cyclic life to initiate a crack could be below the allowable design life. Hence, the behavior of this region of material is of greatest concern. Close examination of how the material is being cycled in this constrained region reveals that "strain-control" best describes how it is "loaded," even though the component may be undergoing force-regulated "loading." The bulk of the component acts in an elastic manner and does not permanently deform. Consequently, after unloading, the shape of the component is returned elastically to its original geometry, except for the small region of material at the notch. The amount of cyclic deformation applied to this critical region of material is essentially constant, regardless of the amount of plasticity or the degree of hardening or softening experienced. Understanding how this material behaves under strain-controlled cycling is crucial to determining fatigue life. Results of force-controlled testing at the potentially very high stress levels at the notch would not be as representative of the material response behavior. This is so because severe strain ratcheting could occur, and, under cyclic strain softening conditions, excessively large plastic strain ranges result. Thus, a strain-controlled fatigue test of a uniform gage-length specimen is more appropriate than a force-controlled test. This topic is of vital interest to the design of fatigue-resistant structural components. An entire chapter (Chapter 8, "Notch Effects") is devoted to it.

**Thermal Fatigue.** The study of thermal fatigue is another area wherein strain-control fatigue testing is far more appropriate than is force-control. Thermal fatigue in components involves imposed strains that are governed by differential thermal expansions between areas in the component that are heated or cooled to different levels. Thermal fatigue and high-temperature fatigue test are described in the Appendix. A planned second volume is devoted entirely to high-temperature materials behavior.

## Ultra-High-Cycle Fatigue

Just as it is difficult to define HCF without some reference to the material involved, it is even more difficult to define ultra-high-cycle fatigue for materials in general. Some cases are very clear. For example, no one would argue with the designation ultra-high-cycle fatigue as applied to vibration of the rotor wire of an electrical generator that operates at 60 cycles per second and is expected to do so for 30 or 40 years. Many cases arise when the design life is $10^{10}$ to $10^{12}$ cycles; these are obviously ultra-high cycle numbers. For other cases, however, the descriptions may be hazy as discussed in the next section. It should be pointed out that very long cyclic lives are generally accumulated over long periods of time. During that time, interactions with the surrounding atmosphere can significantly alter the cyclic strength of materials.

### Models for Treating Ultra-High-Cycle Fatigue

Several models have been proposed for combining the modern concepts of strain-cycling fatigue with consideration for the ultra-high-cyclic range.

**The Langer Model.** Langer (Ref 3.29) was one of the first investigators to consider this subject. In 1958 he proposed that the Manson-Coffin line be combined with an elastic line that is horizontal (Fig. 3.30). Of course, this approach does not take into account that the elastic line must have finite slope even in the low-cycle range.

**Model Containing Features of the Langer and the Manson-Coffin-Basquin Models.** An improved approach is to combine the sloping elastic and plastic lines with a cutoff strain level (Fig. 3.31). Here there is no violation in the low-cycle and moderately high-cycle range, but an endurance limit is indeed assumed below a limiting strain range.

**Model Based on Elastic Life Line of Progressively Reduced Slope.** Halford (Ref 3.30) suggested a more general description of the ultra-high-cycle elastic fatigue behavior for use in conjunction with predetermined LCF curves. Examining available data, he observed that the features of the Manson-Coffin-Basquin model are generally valid up to lives of about $10^6$ cycles, but beyond this range the slope of the elastic line undergoes characteristic changes. The main interest was to establish conservative design strains for use in nuclear pressure vessel and piping design. As an initial step, Halford assumed the validity of the Manson-Coffin-Basquin model in the life range below $10^6$ cycles. Beyond this life, however, he observed characteristic changes in the slopes of the elastic line according to the pattern shown in Fig. 3.32. Two cases are considered, depending on whether the elastic line slope in the region from $10^5$ to $10^6$ cycles is steeper, or shallower, than $b = -0.12$. For the steeper slope (Fig 3.32a), the slope becomes $-0.10$ in the life region from $10^6$ to $10^7$ cycles, $b = -0.06$ from $10^7$ to $10^8$ cycles, and $b = -0.04$ for all lives beyond. Where the slope in $10^6$ to $10^7$ cycle range is less than $b = -0.12$, he found the transition to the asymptotic slope $-0.04$ should take place in two steps, as shown in Table 3.2.

Using the existing fatigue curve representation of each segment, it is possible to determine the strain range at $10^6$ cycles. Using Fig. 3.32,

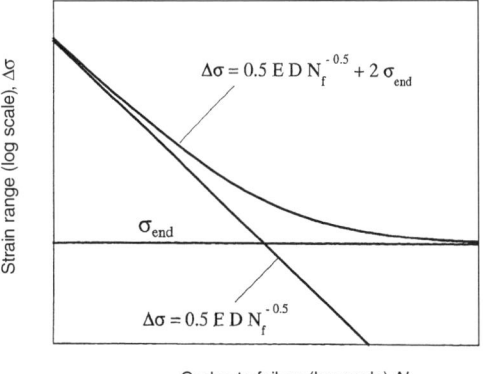

**Fig. 3.30** Langer's model for fatigue in ultra-high-cycle range

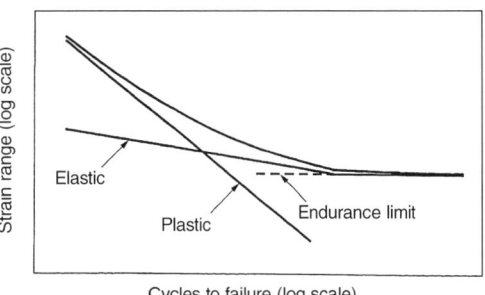

**Fig. 3.31** Model combining features of the Manson-Coffin-Basquin and Langer models

these results are shown in columns 1 and 2 of Table 3.2. Thus, once the strain range to cause $10^6$ cycles to failure is known, the strain range at all other levels are known.

Alternatively, the model of Fig. 3.32 and Table 3.2 has been expressed by an analytical expression:

$$b = -0.12 + \left(\frac{0.082}{90}\right) \tan^{-1} \times [0.403(\log N_f - 5.5)^{2.544}] \quad \text{(Eq 3.8)}$$

We have also generalized the approach to allow for the existence of an arbitrary starting value of slope in the $10^5$ to $10^6$ cycle range, as well as an arbitrary terminal value in the $10^{10}$ to $10^{11}$ cycle range:

$$b = b_o - \left[\frac{(b_o - b_i)}{90}\right] \tan^{-1}[\alpha(\log N_f - 5.5)^\beta] \quad \text{(Eq 3.9)}$$

Here $b_o$ is the slope of the elastic line in the life range from $10^5$ to $10^6$ cycles, i.e., when the average value of $\log N_f = 5.5$. As $\log N_f$ increases, the arc tangent function increases, eventually approaching 90°, so that $b$ approaches $b_o - (b_o - b_1)$. Furthermore, the value of $b$ varies smoothly from $b_o$ to $b_1$, with flexibility in the pattern of transition being provided by the adjustable constants $\alpha$ and $\beta$. Agreement is excellent between Eq 3.9 and the original model with its discrete slopes.

Once the $b$ values are known, the values of $\Delta\varepsilon$ can be determined by integration.

$$b = \frac{d(\log\Delta\varepsilon)}{d(\log N_f)} \quad \text{(Eq 3.10)}$$

Therefore,

$$\log\Delta\varepsilon = \int b\,d(\log N_f)$$

integrated from 6 to $\log N$. We can integrate Eq 3.10 either numerically by Simpson's rule or attempt an approximate closed-form solution. Approximate closed-form integration can be expressed as:

$$\log\Delta\varepsilon = -0.12[\log N - 6] + \left(\frac{0.082}{90}\right)[88\log N + 1.2 \times 10^9(\log N)^{-9} - 624] \quad \text{(Eq 3.11)}$$

Over the range of $10^6$ and $10^{10}$ cycles to failure, the original model, numerical solution, and approximate closed-form solution differ from one another by less than 2% in strain range. Because

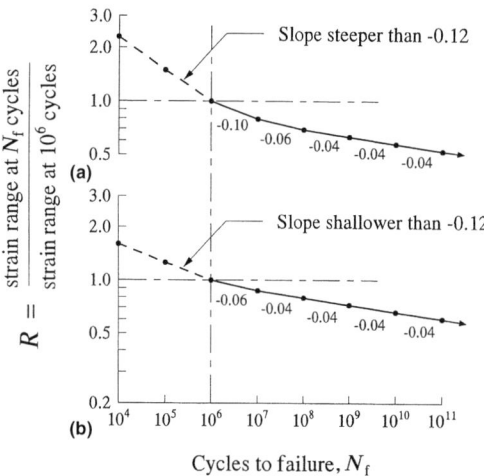

**Fig. 3.32** Model for extrapolating high-cycle fatigue beyond $10^6$ cycles by using elastic line segments of progressively reduced slope. (a) Slope steeper than $-0.12$. (b) Slope shallower than $-0.12$.

**Table 3.2** Implementation of the model using an elastic line consisting of progressively reduced slope

| | Steep slope formulation | | | | Shallow slope formulation | | | |
|---|---|---|---|---|---|---|---|---|
| | Given | | Eq 3.8 | | Given | | Eq 3.8 | |
| R | $N_f$ | $N_f$ | R | | R | $N_f$ | $N_f$ | R |
| 1.00000 | $10^6$ | $1.12 \times 10^6$ | 0.99861 | | 1.00000 | $10^6$ | $0.799 \times 10^6$ | 0.99987 |
| 0.79433 | $10^7$ | $0.860 \times 10^7$ | 0.78720 | | 0.87096 | $10^7$ | $1.11 \times 10^7$ | 0.87305 |
| 0.69183 | $10^8$ | $1.32 \times 10^8$ | 0.69854 | | 0.79433 | $10^8$ | $1.01 \times 10^8$ | 0.79405 |
| 0.63096 | $10^9$ | $1.07 \times 10^9$ | 0.62991 | | 0.72444 | $10^9$ | $0.993 \times 10^9$ | 0.72406 |
| 0.57544 | $10^{10}$ | $0.964 \times 10^{10}$ | 0.57268 | | 0.66069 | $10^{10}$ | $0.988 \times 10^{10}$ | 0.66034 |
| 0.52482 | $10^{11}$ | $0.912 \times 10^{11}$ | 0.52184 | | 0.60256 | $10^{11}$ | $0.987 \times 10^{11}$ | 0.60224 |

R, ratio of the strain range at $N_f$ to the strain range at $10^6$

all three are approximations, one can be considered as numerically accurate as the others.

## Some Ultra-High-Cycle Fatigue Predictions

One of the primary interests of predictive analysis relates to the ultra-high-cycle fatigue range, which we arbitrarily define here as lives in excess of $10^6$ cycles. First, let us examine some of the data available in this range to establish typical materials behavior. Figure 3.33 shows a compilation provided by Soo (Ref 3.31), for a number of materials available at various temperatures.

Note that the ordinate is stress rather than strain, but in the ultra-high-cycle range it can be expected that the two be essentially related by the elastic modulus. Also note that the ordinate is plotted to a linear scale, rather than logarithmic, although for any one material the range of ordinates covered is relatively small, and a replot to a logarithmic scale would not change the shape appreciably. Slopes are, however, different depending on whether linear or log coordinates are used. Therefore, it may well be that the steepness of the Inconel 718 curve at high stress cannot be compared with the shallowness of the aluminum curve at the low-stress level.

Some materials seem to show a tendency toward the development of an endurance limit, while others show a progressive decrease of allowable stress with increasing life. Since data scatter is, however, an important feature of ultra-high-cycle fatigue, the true shapes of these curves are not really known until many duplicate tests are conducted. Soo has not provided predictive analyses for these data, but the discussion of Soo and Chow (Ref 3.32) shows that useful results can be obtained using the universal slopes method.

An interesting study of predictive capability in the ultra-high cyclic life range has been provided by Thiruvengadam and Conn (Ref 3.33). They used an ultrasonic generator to impose high-frequency loadings of the order of 14 KHz so that a billion cycles could be accumulated in only 20 h. They then compared their experimental results with predictions by the universal slopes and four-point correlation methods. However, because of the high strain rates involved in these tests, they considered not only the monotonic (static) tensile properties but also the changes in $\sigma_u$, $D$, and $E$ associated with the dynamic loading. They used a split-Hopkinson bar tensile testing facility to measure properties at very high (impact) strain rates. For 316 stainless steel at room temperature (RT), ultimate tensile strength $S_u$ may, for example, increase about 60%; ductility $D$ falls about 50%; while the modulus of elasticity $E$ remains approximately constant. These comparisons are made to those property values determined at conventional static monotonic tensile loading rates.

Typical results of their study are shown in Fig. 3.34 for 316 stainless steel at RT. The continuous curves show predictions based on universal

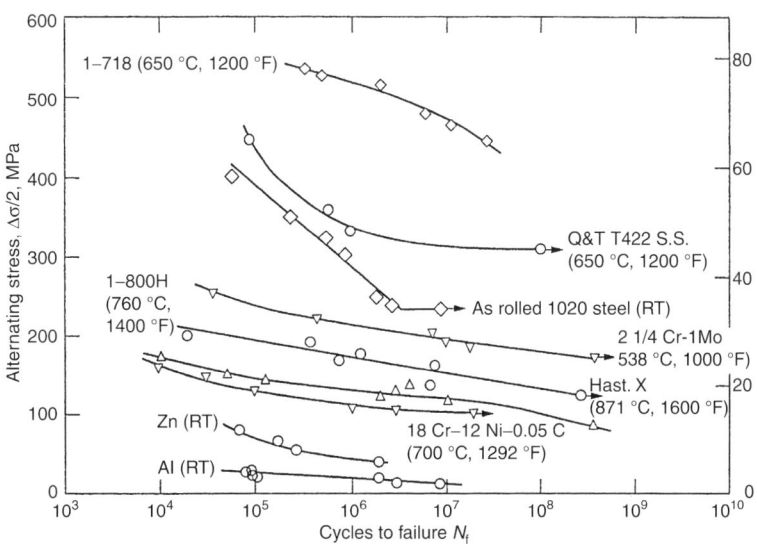

**Fig. 3.33** High-cycle-fatigue data for a range of metals and alloys. Source: Ref 3.31

slopes and four-point correlation using static monotonic tensile properties, while the dotted curves show predictions from the dynamic monotonic properties using the same two methods. It is seen that the curves using the static properties provide a lower bound for the actual data, while the predictions based on dynamic properties are straddled by the data. It is not clear from the figure how much scatter there really is in the data, or whether a true shallow slope exists. However, it can be concluded that there is some predictive utility to both the universal slopes method and the four-point correlation method in the ultra-high-cycle range and that the shallower slope of the experimental curve is in general agreement with Halford's observation that slope decreases in the very high-cycle-fatigue range.

A comparison between various predictions for fatigue in the ultra-high-cyclic life range for 304 stainless steel at 425 °C (800 °F) is shown in Fig. 3.35 taken from Ref 3.33. The life relation in the range up to $10^6$ cycles is shown in the figure; it differs from the standard strain-range-versus-fatigue-life equation in that it expresses strain range in terms of three power laws and a constant, rather than two power laws. However, it is the curve that Westinghouse investigator, Manjoine (Ref 3.34) regarded as the best fit for their data and also suitable for extrapolation into the very high-cyclic life range. It is, in essence, a form derived from the Langer model (Ref 3.29), containing the constant 0.0024, which becomes the asymptotic strain range for curve A when the other terms essentially vanish. Curve B shows the extrapolation according to the Halford model. The other extrapolations are based on various earlier recommendations, as indicated. Halford's recommendation was curve B as a compromise between the conservative extrapolations of curves C, D, and E, and the rather optimistic implication of curve A that a true endurance limit exists.

## General Comments on Ultra-High-Cycle Fatigue Extrapolation

It is not uncommon for equipment operating at high frequency, or at moderate frequency for long periods, to accumulate a large number of cycles. For example, an electric power generating coil subjected to a 60 Hz excitation will accumulate about $10^{11}$ cycles in 30 years, which is not an unreasonable service life for such a component. The question arises as to how to test such materials for so long a life. Frequencies can be increased for testing purposes, but the necessary time is still measured in years. Sometimes ultrasonic frequencies are used. While these tests demonstrated that the universal slopes

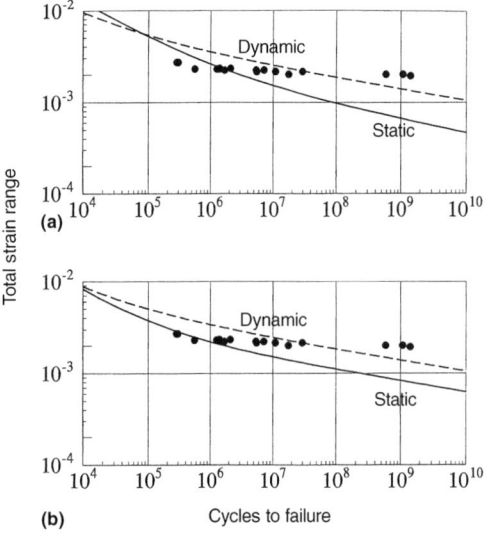

**Fig. 3.34** Comparison of high-frequency data with fatigue-life predictions for annealed 316 stainless steel at room temperature. (a) Four-point correlation method. (b) Method of universal slopes. Source: Ref 3.33

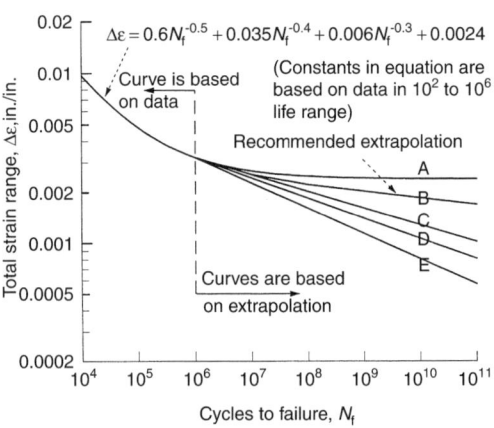

**Fig. 3.35** Comparison of several methods for extrapolation of data for 304 stainless steel at 425 °C (800 °F) to life values in the ultra-high-cycle fatigue range. Curve A, direct extrapolation of equation based on existing data ($\Delta\varepsilon_t = 0.6 (N_f)^{-0.5} + 0.35(N_f)^{-0.4} + 0.06(N_f)^{-0.3} + 0.0024$); curve B, recommend extrapolation based on "steep slope" in $10^5$–$10^6$ life regime (Halford, 1978); curve C, constant slope of $-0.10$ (Halford recommendation to Creep Fatigue Task Force: ASME Boiler and Pressure Vessel Code, 1970); curve D, constant slope of $-0.12$ (Method of Universal Slopes, 1965); curve E, constant slope of $-0.15$ (C. Jaske recommendation to Creep-Fatigue Task Force, ASME Boiler and Pressure Vessel Code, 1970)

equation and the four-point correlation method provide reasonable extrapolations out to very high lives, other tests have shown, however, that ultrasonic excitation can produce spurious failure modes and misleading properties. Wood (Ref 3.35) has, for example, shown that such testing may result in local interior failures. Apparently, local regions are overheated and produce failure of a type that would not occur if the same number of cycles had been applied over a longer period. Soo and Chow (Ref 3.32) show similar discrepancies in the very-high-cyclic life range when mechanical loading tests were compared with ultrasonic excitation tests. The conclusion is to avoid the artificial forms of excitation that do not resemble the service loadings.

What, therefore, is a reasonable predictive approach? The answer would seem to be to run at as high a frequency as still reasonably induces the same mode of deformation. Axial loading or bending at 10,000 to 20,000 cpm is not difficult to achieve. Using air or vapor cooling, if necessary, the temperature of the test specimen can be kept at reasonable values. If these tests are conducted to cyclic values within $1/10$ or $1/20$ of those desired, linear extrapolation of the last log cycle may be acceptable, or a mathematical formula based on behavior of other materials can be used. Such approaches are probably preferable to using artificial loading and deformation modes and carrying the tests out to the entire desired cyclic life. Even at service frequencies, 1.5 to 3 years can be used to get 30 year estimates. With increased frequencies the tests can usually be accomplished in less than a year, which is not an unreasonable test time.

## REFERENCES

3.1 A. Wöhler, Über die Festigkeit-Versuche mit Eisen und Stahl. *Z. Bauwesen (J. Engineering Construction),* Vol 8, 1858 (in German). See also Vol 10, 13, 16, and 20, 1870.

3.2 E.C. Hartmann and F.M. Howell, Laboratory Fatigue Testing of Materials, *Metal Fatigue,* G. Sines and J.L. Waisman, Ed., McGraw-Hill, New York, 1959, p 89–111

3.2A J.A. Ewing and J.C.W. Humfry, The Fracture of Metals under Repeated Alternations of Stress, *Philos. Trans. R. Soc. (London),* Vol A197, 1903, p 241–249 (plus 4 pages of plates)

3.3 S.S. Manson, Behavior of Materials under Conditions of Thermal Stress, lecture presented at *Symposium on Heat Transfer,* University of Michigan, June 27–28, 1952. See also NACA TN 2933, 1953.

3.4 S.I. Liu, J.J. Lynch, E.J. Ripling, and G. Sachs, Low Cycle Fatigue of the Aluminum Alloy 24ST in Direct Stress, *Trans., Am. Inst. Mining and Metall. Eng.,* Vol 175, 1948, p 469–490 (discussion, p 490–495)

3.5 R.W. Smith, M.H. Hirschberg, and S.S. Manson, "Fatigue Behavior of Materials under Strain Cycling in Low and Intermediate Life Range," NASA TN D-1574, April 1963. See also *Proceedings, Tenth Sagamore Army Materials Research Conference,* J.J. Burke, N.L. Read, and V. Weiss, Ed., Syracuse University Press, 1964, p 133–173, and Ref 3.17.

3.6 L.F. Coffin, Jr., "A Study of the Effects of Cyclic Thermal Stresses on a Ductile Metal," presented at a joint session of the Metals Engineering and Applied Mechanics Divisions at the Annual Meeting of the American Society of Mechanical Engineers, New York, Nov 29–Dec 4, 1953. See also *Trans. ASME,* Vol 76, 1954, p 931–950.

3.7 L.F. Coffin, Jr. and J.F. Tavernelli, The Cyclic Straining and Fatigue of Metals, *Trans. Am. Inst. Mining and Metall. Eng.,* Vol 215, 1959, p 794–807

3.8 S.S. Manson, Thermal Stresses in Design, Part 19: Cyclic Life of Ductile Materials, *Mach. Des.,* Vol 21, July 7, 1960, p 139–144. See also S.S. Manson, *Thermal Stresses and Low-Cycle Fatigue,* McGraw-Hill, Inc., New York, 1966, p 147.

3.9 J. Morrow, Fatigue Properties of Metals, presented at a meeting of Division 4 of the SAE Iron and Steel Technical Committee, Nov 4, 1964

3.10 O.H. Basquin, The Exponential Law of Endurance Tests, *Proc., American Society for Testing and Materials,* Vol 10, Part II, 1910, p 625–630

3.11 R.W. Landgraf, M.R. Mitchell, and N.R. LaPointe, *Cyclic Deformation Behavior of Engineering Alloys,* Ford Motor Co., Scientific Research Staff, Special Technical Publication, Dearborn, MI, June 1972

3.12 T.H. Sanders, Jr., D.A. Mauney, and J.T. Staley, Strain Control Fatigue as Tool to Interpret Fatigue Initiation of Aluminum

Alloys, *Fundamental Aspects of Structural Alloy Design,* Plenum Press, New York, 1977, p 487–518

3.13 S.S. Manson and S. Hailu, Maverick Behavior of the Irreversibly Hardening 304 Stainless Steel in Sequential Loading, *Material Durability/Life Prediction Modeling—Materials for the 21st Century,* PVP-Vol 290, S.Y. Zamrik and G.R. Halford, Ed., ASME, New York, 1994, p 1–9

3.14 D.E. Martin, Plastic Strain Fatigue in Air and Vacuum, *J. Basic Engineering,* Vol 87 (No. 4), Dec 1965, p 850–856

3.15 L.F. Coffin, Jr., "Fatigue at High Temperature," *Fatigue at Elevated Temperatures,* ASTM STP 520, American Society for Testing and Materials, 1973, p 5–34

3.16 N.E. Dowling, Fatigue Life and Inelastic Strain Response under Complex Histories for an Alloy Steel, *J. Test. Eval.,* Vol 1 (No. 4), July 1973, p 271–287

3.17 S.S. Manson, Fatigue—A Complex Subject, *Exp. Mech.,* Vol 5 (No. 7), 1965, p 193–226

3.18 U. Muralidharan and S.S. Manson, "A Modified Universal Slopes Equation for Estimation of Fatigue Characteristics of Metals," presented at the Winter Annual Meeting of the American Society of Mechanical Engineers (Anaheim, CA), Dec 1986, preprint 86-WA/Mats-17, 1986

3.19 A.J. Nachtigall, "Strain-Cycling Fatigue Behavior of Ten Structural Metals Tested in Liquid Helium, Liquid Nitrogen, and Ambient Air," *Properties of Materials for Liquefied Natural Gas Tankage,* ASTM STP 579, American Society for Testing and Materials, 1975, p 378–396

3.20 C.E. DeBogdan, Tensile and Fatigue Data for Irradiated and Unirradiated AISI 310 Stainless Steel and Titanium-5 percent Aluminum-2.5 percent Tin, *Application of the Method of Universal Slopes,* NASA TM-X-2883, Sept 1973

3.21 C.R. Brinkman and G.E. Korth, "Fatigue and Creep-Fatigue Behavior of Irradiated Stainless Steel—Available Data, Simple Correlations, and Recommendations for Additional Work in Support of LMFBR Design," Aerojet Nuclear Company Report, ANCR-1096, Feb 1983

3.22 J.F. Newell, A Note of Appreciation for the MUS, *Material Durability/Life Prediction Modeling, Materials for the 21st Century,* PVP-Vol 290, S.Y. Zamrik and G.R. Halford, Ed., American Society of Mechanical Engineers, New York, 1994, p 57–58

3.23 M.R. Mitchell, Fundamentals of Modern Fatigue Analysis for Design, *Fatigue and Fracture,* Vol 19, *ASM Handbook,* ASM International, 1996, p 227–249

3.24 J. Morrow, "Cyclic Plastic Strain Energy and Fatigue of Metals," *Internal Friction, Damping, and Cyclic Plasticity,* ASTM STP 378, American Society for Testing and Materials, 1965, p 45–84

3.25 R.W. Landgraf, Control of Fatigue Resistance through Microstructure—Ferrous Alloys, *Fatigue and Microstructure,* American Society for Metals, 1978, p 439–466

3.26 S.S. Manson, Predictive Analysis of Metal Fatigue in the High Cycle Life Range, *Methods for Predicting Material Life in Fatigue,* American Society of Mechanical Engineers, New York, 1979, p 145–182

3.27 R.E. Peterson, Fatigue of Metals—Engineering and Design Aspects, The 1962 Edgar Marburg Lecture, *Mater. Res. Standards,* Vol 3 (No. 2), Feb 1963, p 122–139

3.28 J.F. Tavernelli and L.F. Coffin, Jr., Experimental Support for Generalized Equation Predicting Low Cycle Fatigue, *Trans. Am. Soc. Mech. Eng. J. Basic Engineering,* Series D, Vol 84 (No. 4), Dec 1962, p 533–541 (See discussion by Manson.)

3.29 B.F. Langer, "Design Values for Thermal Stresses in Ductile Metal," ASME preprint 58-MET-1, 1958

3.30 G.R. Halford, A Recommended Procedure for Extrapolating Strain Fatigue Curves beyond One Million Cycles to Failure, prepared for the Working Group on Creep-Fatigue of the ASME, March 1978

3.31 P. Soo, High Cycle Fatigue, *A Decade of Progress,* American Society of Mechanical Engineers, S.Y. Zamrik and D. Dietrich, Ed., 1982, p 499–506

3.32 P. Soo and J.G.Y. Chow, "Correlation of High- and Low-Cycle Fatigue Data for Incoloy-800H," Brookhaven National Laboratory Report BNL-NURE-50574, Oct 1976

3.33 A. Thiruvengadam and A.R. Conn, High-Frequency Fatigue and Dynamic Properties at Elevated Temperature, *Proc. Soc. Experimental Stress Analysis,* Vol XXVIII (No. 2), July 1971, p 315–320

3.34 M. Manjoine, personal communication, 1978

3.35 W.A. Wood, "Four Types of Metal Fatigue," George Washington University Report AD-746-121, 1972

# CHAPTER 4

# Mean Stress

## Introduction

Mean stress effects can have an extremely detrimental influence on fatigue, sometimes mysterious because their presence is hidden, for example, as tensile residual stress. Correspondingly, one of the most potent allies in the prevention of fatigue is favorable compressive mean stress. Mean stress is defined as the algebraic average of the maximum and minimum stresses in a complete, closed cycle, i.e., $\sigma_m = (1/2)(\sigma_{max} + \sigma_{min})$. Correspondingly, the alternating stress is half the range of stress, or, $\sigma_a = (1/2)(\sigma_{max} - \sigma_{min})$. It is important to understand how mean stress becomes induced in a component. In that way their presence can be anticipated and properly treated in design.

## Origin of Mean Stress

Basically, there are two generic ways to generate mean stress during fatigue testing: through force control or through strain control.

### Force-Induced Mean Stress

The most common situation for inducing mean stress during testing is by an externally applied cyclic force with a mean value. Consider the three cases of axial loading of a specimen as shown in Fig. 4.1. If loading results in linear elastic response, we get the case of Fig. 4.1(a) wherein the stress-strain path is 0A0. The mean stress is the average of 0 and A. If we now consider loading into the nonlinear plastic regime to point B in Fig. 4.1(b), but then cycle with linear elastic response between B and C, the mean stress is the average of B and C. Figure 4.1(c) shows unloading from tensile point B to point D in compression followed by reloading to B, thus forming a hysteresis loop. Plasticity is encountered in both directions of loading and a small tensile mean stress develops as shown. Again, the mean stress is the algebraic average of the extreme stresses between points B and D. In force-controlled tests of smooth axial specimens, wherein force and stress are linearly related, the externally applied mean force creates and enforces the internal mean stress. Of course, cycle-dependent strain ratcheting may occur under force control when cyclic plasticity is involved. This behavior is discussed subsequently.

### Strain-Induced Mean Stress

Mean stress may also be induced by strain control (Fig. 4.2). Here, the strain is repeatedly enforced between minimum and maximum values. In Fig. 4.2(a) the strain is small, well within the elastic regime, and no plasticity occurs between 0 and A, resulting in the mean stress shown. This case is identical to that of Fig. 4.1(a). In Fig. 4.2(b) plasticity occurs on the initial straining 0AB, but the subsequently applied strain BD is moderate. Again, little or no plasticity occurs. Finally, in Fig. 4.2(c) the strain is large enough to produce plasticity on the first loading and on subsequent reversals, and a closed hysteresis loop BDB develops with the mean stress shown. Cyclic relaxation of mean stress will generally be expected in the presence of detectable cyclic plasticity. This behavior is illustrated and discussed in the following section.

### Ratcheting and Cyclic Mean Stress Relaxation

Force control, and strain control under certain limited conditions, invariably produces strain ratcheting when cyclic plasticity is present. Strain ratcheting produces a mean strain (defined in the same manner as mean stress). The

# 76 / Fatigue and Durability of Structural Materials

amount and rate of the ratchet depends on three factors: the amount of cyclic plasticity induced, whether the material is cyclically hardening or softening, and the magnitude of the mean stress involved. Schematic representations of the response of cyclically hardening and softening

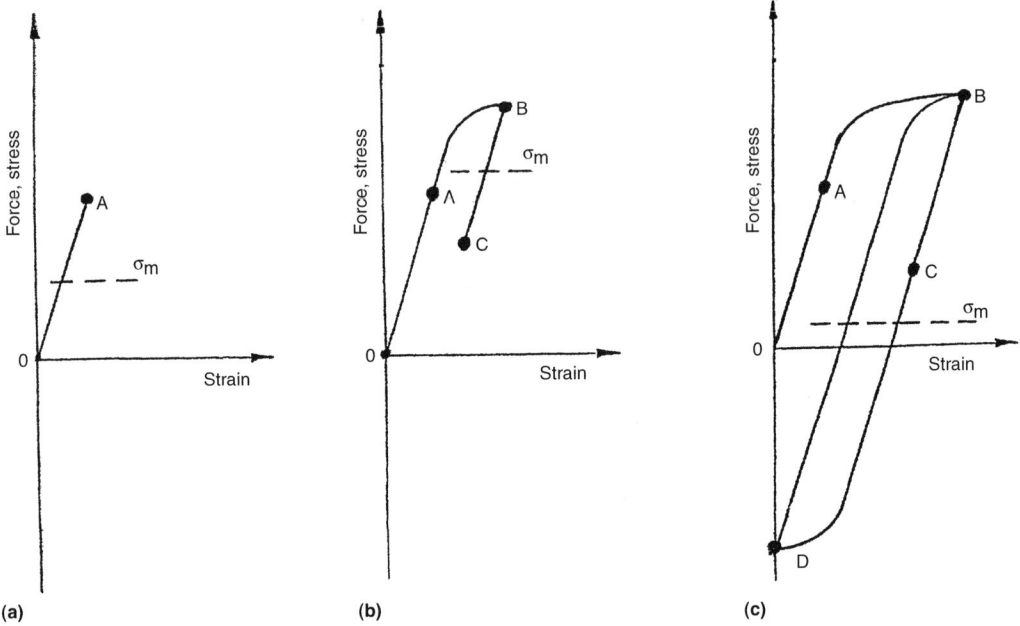

**Fig. 4.1** Generation of mean stress under force-controlled limits

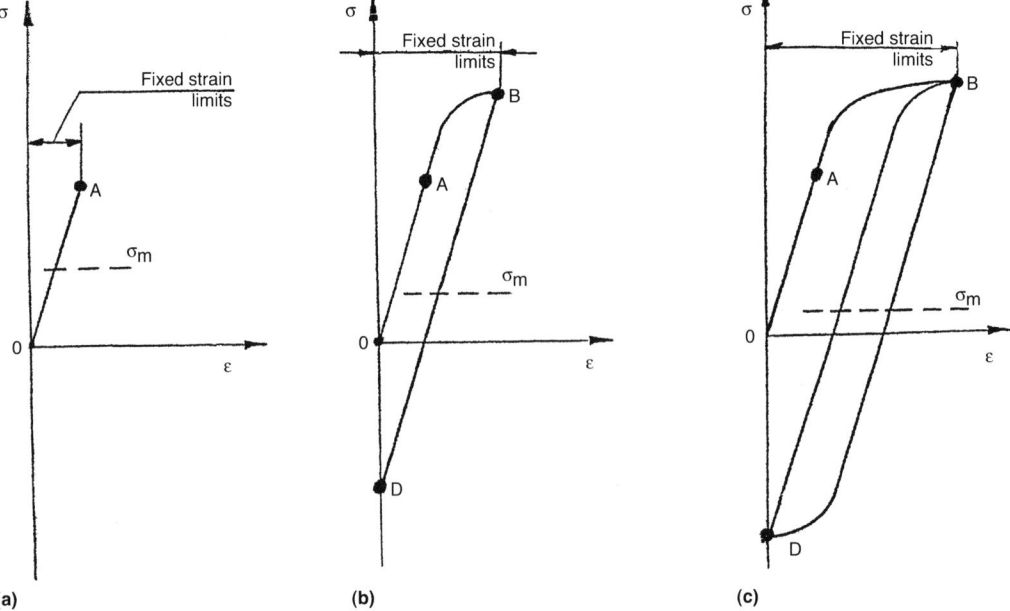

**Fig. 4.2** Mean stresses induced by cycling between fixed strain limits

materials under zero-to-max force control and zero-to-max strain-controlled cycling are illustrated in Fig. 4.3. The upper half of the figure shows material response under force control while the lower half shows strain control. The illustrations show the first eight or more hysteresis loops and, where possible, the stabilized loop, $O'B'$. Plastic strain ratcheting, $\varepsilon_{p,r}$, occurs parallel to the strain axis and is given by the distance $OO'$. For ease of illustration, the first segment $OAB$ of the four initial stress-strain curves are taken to be identical.

**Force-Control.** For cyclically hardening materials, the plastic ratchet strain per cycle steadily decreases on each subsequent cycle and tends to approach zero as stabilization occurs reaching nearly elastic response (Fig. 4.3a, $O'B'O'$). Clearly, hardening has occurred. The initial tensile forcing produced noticeable plasticity, but at stabilization, the same amount of stress produces virtually zero plasticity.

Conversely, for cyclically softening materials, ratcheting can become a runaway process if the maximum stress is high enough. The ratchet strain per cycle steadily increases from cycle to cycle (Fig. 4.3b), until localized necking and rupture occur prior to classical fatigue crack initiation and propagation. The failure mode is similar to that of a monotonic tension test. It becomes obvious that load-controlled axial fatigue testing at high stresses is inappropriate for determining fatigue characteristics of a softening material.

**Strain-Control.** Under zero-to-maximum strain control shown in the lower half of the figure, two responses occur. First, the initial mean stress $\sigma_{m,i}$ tends to cyclically relax toward zero, and secondly, any cyclic hardening or softening causes the hysteresis loop to become narrower or wider, respectively. A cyclically hardening material may reach a nearly linear elastic response before the initial mean stress reaches zero (Fig. 4.3c), whereas the cyclic softening material will invariably develop a hysteresis loop with sufficient inelasticity to cause complete relaxation of the mean stress (Fig. 4.3d). As a result of this relaxation, there is a small net plastic ratchet strain that is equal to the horizontal distance $OO''$.

Because the conditions at the root of a notch in a structural component are closer to strain control than they are to force control, an initial mean stress will likely decrease due to cyclic relaxation if the cyclic strain is great enough. This response of the material can be beneficial to fatigue resistance if the initial mean stress is tensile, or it can be detrimental if the initial mean stress is compressive. Compressive mean stresses (in the form of residual stresses) are frequently introduced intentionally to fatigue-prone areas of a component so as to prolong fatigue life. Unfortunately, the benefit may not be fully realized if the local cyclic strain is too large and the residual stresses cyclically relax to zero. As demonstrated in Fig. 4.3, the loss would be most severe for an initially cold-worked material because of its high propensity for cyclic strain softening. This subject is covered in greater detail in Chapter 11, "Avoidance, Control, and Repair of Fatigue Damage."

**Mean Strain.** Mean strain ($\varepsilon_m$) may also occur during stress and strain cycling. For elastic cycling, the mean strain is directly related to the mean stress by the modulus of elasticity, i.e., $\varepsilon_m = \sigma_m/E$. However, when plasticity is involved, this relation is no longer valid. For example, during plastic strain ratcheting under force control, mean strain constantly increases, although the mean stress remains constant. From the standpoint of fatigue life, there is little concern for mean strain provided its magnitude remains a small fraction of the tensile ductility of the material. This conclusion will be better understood following the discussion later in this chapter in the section "Insight into Cause of Life Reduction due to Tensile Mean Stress." It is of historical interest to point out that the cyclic straining test data of Liu, Lynch, Ripling, and Sachs (Ref 3.4), discussed in Chapter 3, "Fatigue Life Relations," were generated to determine the retained ductility of 24ST aluminum after various cycles of large plastic straining. It was this data set that attracted Manson's attention and led him to develop the plastic strain law of low-cycle fatigue that relates ductility to fatigue life. It is thus reasonable to presume that imposition of a large mean tensile strain would reduce subsequent ductility and hence fatigue life.

## Mean Stress Development in Complex Loading History

We have already demonstrated, in the section "Complex Pattern of Strain Cycling Analyzed by Mechanical Models," in Chapter 2, "Stress and Strain Cycling," how a complex loading pattern can result in the generation of a sequence of hysteresis loops of varying size. Most will likely have a mean stress. In the earlier discus-

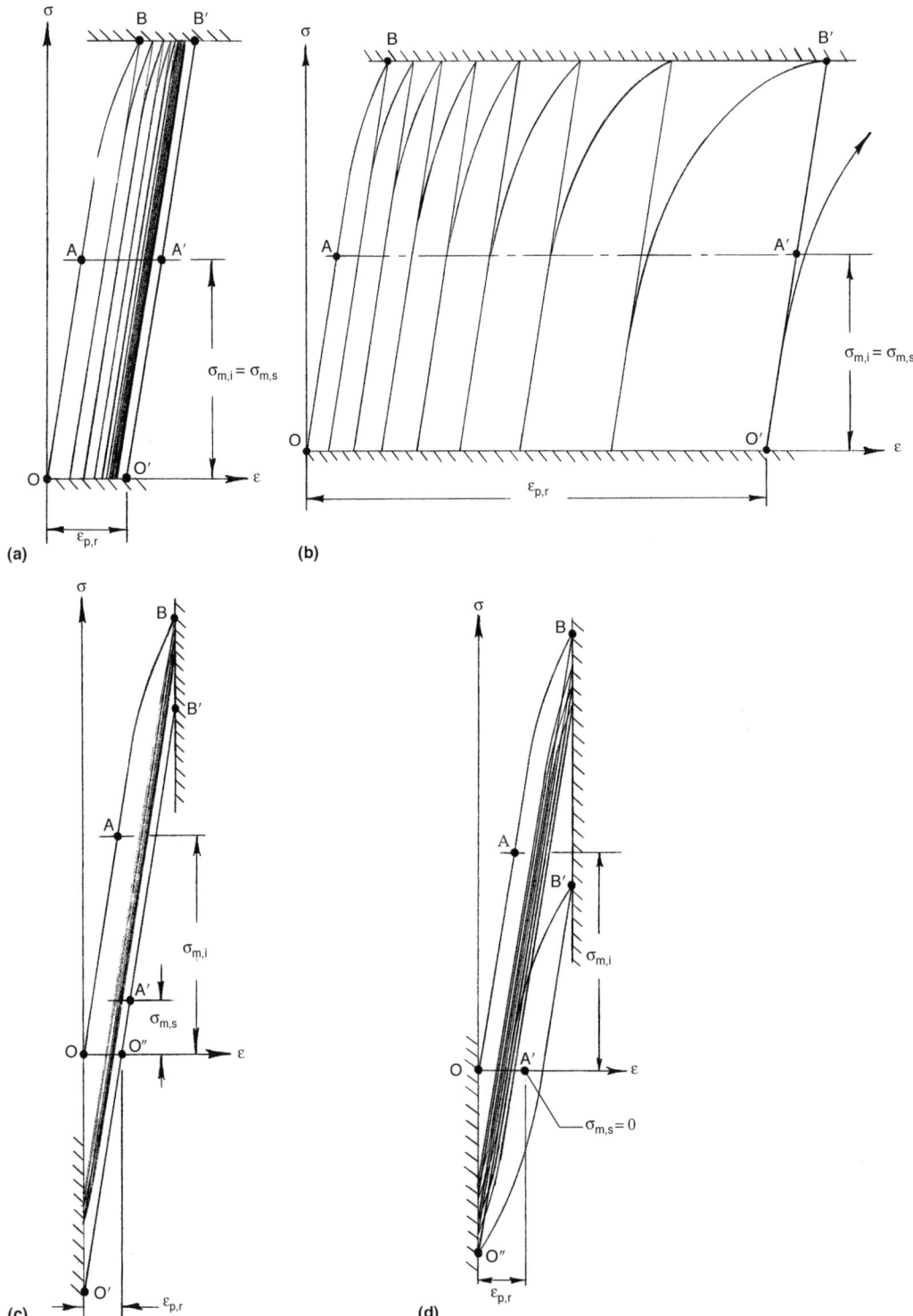

**Fig. 4.3** Schematic patterns of strain ratcheting under force control for (a) cycling hardening and (b) cyclic softening, and patterns of cyclic stress relaxation under strain control for (c) cyclic hardening and (d) cyclic softening

sion we purposely limited the loading histories to very simple cases so that the principles could be displayed with a minimum of complexity. In practice, the loops may be manyfold and quite complex. Figure 4.4 shows a realistic sequence of loops encountered at the root of a mission-loaded notched specimen used in the Society of Automotive Engineers cumulative fatigue damage analysis program (Ref 4.1).

It is important to take into account the fact that cyclic relaxation of mean stress has little opportunity to occur fully for individual hysteresis loops contained within a repeating complex block of loops involving greater plasticity. In each block, the larger encompassing loop (usually with negligible mean stress) regenerates the mean stress within the smaller loops. Thus, under these circumstances, it becomes necessary to account for the damaging effect of mean stress in these smaller hysteresis loops, many of which may have observable plasticity. If these smaller encapsulated loops were allowed to repeat many times, their mean stress would assuredly relax toward zero, thereby mitigating life-influencing mean stress effects. However, the repetitive nature of the complex block of loading interrupts the relaxation process and, in fact, regenerates the mean stresses in each smaller loop. Hence, an assumption of their mean stress relaxing to zero is misleading to a damage analysis.

An interesting example of how mean stress can be regenerated by subsequent loading, especially if large plastic strains are involved, is shown in Fig. 4.5. This figure shows the unpublished results of tests we had conducted on 316 stainless steel at 480 °C (900 °F). After subjecting the specimen to the large strain loop $ABCDA$ for a sufficient number of cycles to stabilize the loop, several cycles of oscillation between strain values of $C$ and $D$ were applied. Cyclic stress relaxation occurred, as shown in the figure by $C'D'$ and $C''D''$. When the mean stress relaxed by 20%, the complete large strain range $AC$ was again applied. It took approximately three cycles to reestablish the stabilized loop $ABCDA$, and resumption of cycling between strains $C$ and $D$ repeated the pattern of initial high stress, followed by cyclic relaxation. Thus, it must be emphasized that cyclic stress relaxation does not always result in total disappearance of a mean stress. It has an effect while it is present, and it can be regenerated by the subsequent loading history. Hence, more damage is imparted than would be the case for uninterrupted relaxation. In the illustration of Fig. 4.5 we show the case of a mean tensile stress, but the same result would have been obtained if the cycling had been imposed near the compressive apex $A$, resulting in cycling with a progressively relaxing compressive mean stress. In this case, fatigue life is prolonged by not allowing uninterrupted relaxation of the compressive mean stress.

## Stress Relaxation
## Pattern in Strain-Control Loading

The mean stress associated with an initially imposed mean strain tends to relax as cycles of alternating load are applied. Consider, for example, the results of Fig. 4.6 (Ref 4.2). Axially loaded specimens of A-36 steel were subjected to an initial strain $\varepsilon_m$ of $+0.005$ followed by various alternating strain amplitudes of $\Delta\varepsilon/2$ ranging from $\pm0.0010$ to $\pm0.0050$. The normalized mean stress, i.e., ratio of the mean stress at cycle $2N$ to the initial value

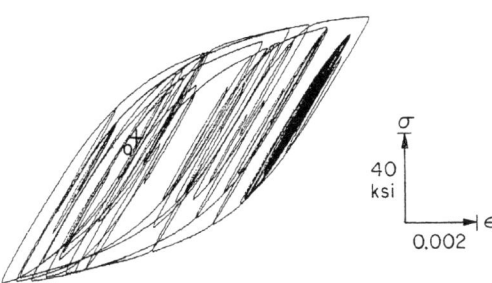

**Fig. 4.4** Development of hysteresis loops for man-ten steel under complex load history. Smooth specimen simulation of notch root stress-strain response of a notched specimen used in the SAE cumulative fatigue damage program. Source: Ref 4.1

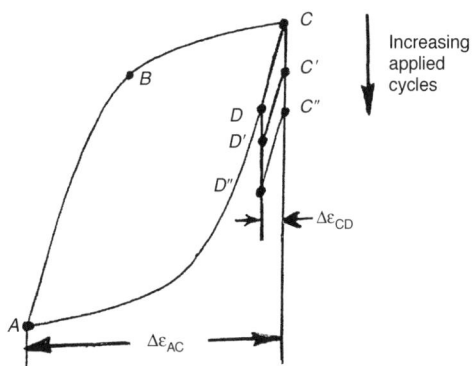

**Fig. 4.5** Cycling at $\Delta\varepsilon_{CD}$ causes stress relaxation, but mean stress can be regenerated by a subsequent cycle of $\Delta\varepsilon_{AC}$.

(at $2N - 1 = 1$), i.e., $\sigma_{m,1}$ is plotted against $2N - 1$. With progressive strain cycling, the mean stress gradually diminishes. The larger the strain amplitude, the faster the mean stress approaches zero. The implication is, of course, that if a mean stress is built into a part by imposing an initial strain, the mean stress will not necessarily be retained indefinitely.

Another way of displaying relaxation of mean stress due to strain cycling is shown in Fig. 4.7 from Conway and Sjodahl (Ref 4.3). The material is AF2-1DA, an experimental nickel-base superalloy, tested isothermally at 760 °C (1400 °F). Cycling was similar to that shown in Fig. 4.3; that is, straining was imposed between zero and selected positive values. Rather than showing how the mean stress relaxed with cycling, however, the figure shows the mean stress that remained at the half-life of the cycling, when the cycling was continued to failure. Thus, for example, up to strains associated with point B ($\approx 0.2\%$ strain), both the mean and alternating stresses are equal because of the linear elastic behavior. At higher strains, the alternating stress increases, while the mean stress decreases. From B to C the mean stress decreases almost linearly, indicating that partial relaxation of mean stress occurred, while beyond strains of C ($\approx 0.6\%$) the mean stress relaxes completely before half the fatigue life is consumed. The figure does not provide detailed information on how fast the mean stress relaxes, but as shown in Fig. 4.6, the relaxation could be quite rapid (note the logarithmic scales).

## Mean Stress Effects on Fatigue Life

As noted previously, mean stress is very potent both in reducing fatigue life and in increasing it. An informative example from Stadnick and Morrow (Ref 4.4) is seen in Fig. 4.8. Here, mean stress was induced by residual effects in a notched specimen.

We discuss in Chapter 8, "Notch Effects," how to calculate stress in a notched part, and how small nuances of loading can introduce either a tensile or compressive stress in the hysteresis loops generated. It is sufficient now to note that, depending on how the transition takes place from the high load or strain to the low load or strain cycling, in Fig. 4.8(b) the transition is such that the hysteresis loop for the small strain range has a tensile mean stress, while in Fig. 4.8(c) the small hysteresis loop has a compressive mean stress. As seen in Fig. 4.8(a), the fatigue lives for the small loops are much higher for the case of compressive mean stress (load history B) than for the case in which the mean stress is tensile (load history A). Depending on the strain range of the small loop, the fatigue lives can differ by as much as 50:1. Because the two loadings are the same, except for the mean stress, it is apparent that mean stress is very important in affecting life.

## Various Ratios Describing Relation of Mean to Alternating Stress

Before discussing the effect on fatigue life of mean stress, it is appropriate to define several

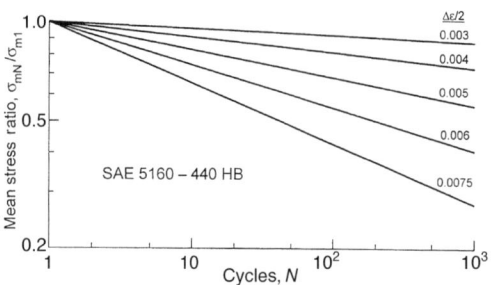

**Fig. 4.6** Cyclic mean stress relaxation of SAE 5160 steel (440 HB) at constant +0.005 mean strain. Source: Ref 4.2

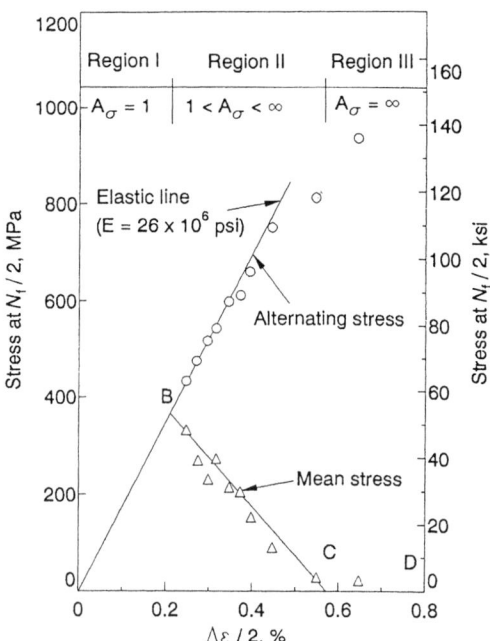

**Fig. 4.7** Alternating/mean stress at half-life in strain-cycling from zero to max. Source: Ref 4.3

Chapter 4: Mean Stress / 81

alternative ways mean stress is described in the literature.

**A-Ratio.** The first commonly used descriptor for mean stress representation was the A-ratio. It is the ratio of the stress amplitude ($\sigma_a$) of the fatigue cycle divided by the mean ($\sigma_m$):

$$A = \frac{\sigma_a}{\sigma_m} \quad \text{(Eq 4.1)}$$

For rotating bending fatigue (the type of fatigue tests conducted by Wöhler (Ref 4.5), the loading and the stresses are completely reversed,

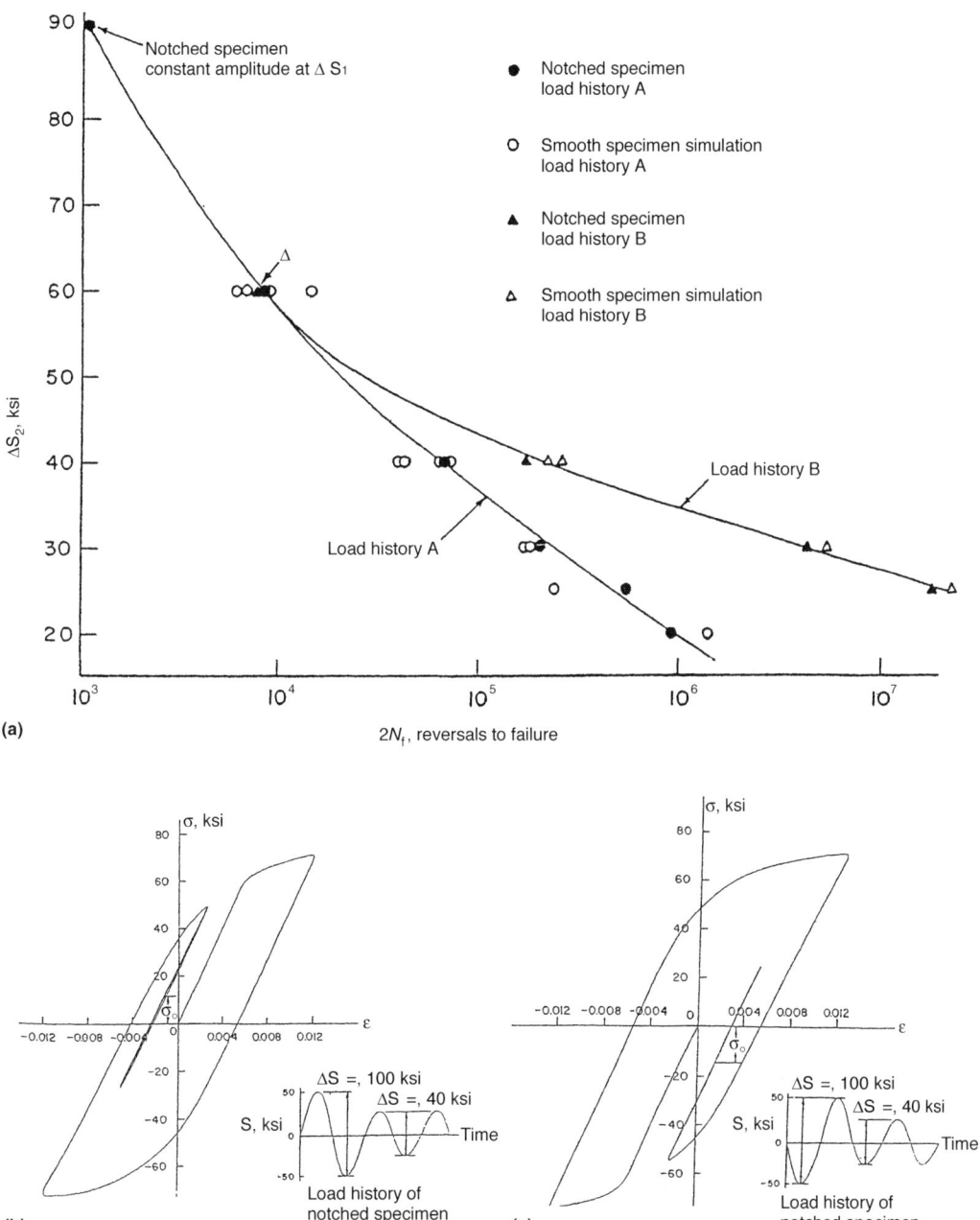

**Fig. 4.8** Effect of mean stress on fatigue life. (a) Notched specimen fatigue lives resulting from two types of initial overload; also lives of un-notched specimens subjected to strain histories estimated to occur at notch. (b) Details of load history A. (c) Details of load history B. Source: Ref 4.4

so the mean stress is zero and $A = \infty$. As mean stress approaches zero from a tensile value, $A \to +\infty$, and as it approaches zero from compression, $A \to -\infty$. The A-ratio becomes zero only when the amplitude goes to zero and there is no fatigue loading whatsoever. Because amplitude is in the numerator, this descriptor tends to emphasize the amplitude relative to the mean rather than the more meaningful other way around. The A-ratio turns out to be a numerically awkward descriptor, even for use with those test results that formed the initial baseline of most fatigue testing.

Usage of the A-ratio has become less common since the advent of the study of low-cycle fatigue. However, because some mean stress fatigue results are still quoted in the literature in terms of the A-ratio, its description is included herein. Rarely is the A-ratio used to describe mean stress during low-cycle fatigue.

**R-Ratio.** Currently, the R-ratio is the predominately used mean stress descriptor. It is the ratio of the algebraic minimum stress ($\sigma_{min}$) to the maximum algebraic stress ($\sigma_{max}$) in a cycle:

$$R = \frac{\sigma_{min}}{\sigma_{max}} \qquad \text{(Eq 4.2)}$$

Perhaps the initial reason for adopting this definition was the fact that $R = 0$ for zero-to-max tensile loading, a common loading condition for many structural components. Thus, $R = 0$ was an advantageous baseline for axial force-controlled testing in the nominally elastic, high-cycle fatigue regime. This advantage, however, was soon lost once low-cycle fatigue testing came into its own with the development of reliable extensometers, closed-loop servocontrol systems, bending-free alignment fixtures, and backlash-free grips that permitted reliable axial loading into compression as well as tension. Once loading levels increased in the low-cycle regime, plasticity occurred at stress concentrations where the local stresses could eventually approach completely reversed cycling. Even for smooth specimens with zero-to-max *strain* control ($R_\varepsilon = 0$), the *stress* would quickly become completely reversed ($R_\sigma = -1$) in the presence of sufficiently high strains to produce cyclic plasticity. Yet, under the same test control at low strain levels, the initial zero-to-max mean stress doesn't relax, resulting in a total strain-range-versus-life curve that had no mean stress at the low-life end, but tensile mean stress at the high-life end. In other words, the "baseline" curve would represent a variable mean stress condition that a user might be unaware of, and hence would be ill-prepared to judge when it was necessary to perform mean stress corrections. Primarily for this reason, strain-controlled, low-cycle fatigue testing has most frequently been conducted using completely reversed *straining* ($R_\varepsilon = -1$) so that the *stress* response starts out, and remains, completely reversed ($R_\sigma = -1$) throughout the entire test. But, the use of $R = -1$ (or $A = \infty$, as discussed previously) is a cumbersome numerical representative for a "zero baseline" testing condition. Unfortunately, neither A nor R conjures an immediate and clear image of the degree of mean stress with respect to the alternating component. Considerable familiarity with their use is required before one gains an instant recognition of the mean stress condition as defined by R and A, i.e., its sign and its relative magnitude compared with the alternating stress. Neither $R = -1$ nor $A = \infty$ gives the novice the impression that the mean stress is zero. This shortcoming is rectified by defining the V-ratio*.

**V-Ratio.** The V-ratio is simply the inverse of the A-ratio:

$$V = \frac{1}{A} = \frac{\sigma_m}{\sigma_a} \qquad \text{(Eq 4.3)}$$

Halford and Nachtigall (Ref 4.6) suggested the letter V because of its physical resemblance to an inverted A. Most importantly, the ratio was proposed because of its self-evident attributes. When V is zero, the mean stress is zero. When V is positive, the mean stress is positive. When V is negative, the mean stress is negative, and finally, when V is infinite, the mean stress is infinite relative to its amplitude. A quick glance at the value and sign of the V-ratio gives an immediate image as to the sign and relative magnitude of mean to alternating stress. The ratio V is simply related to A and R by:

$$V = \frac{1}{A} = \frac{1 + R}{1 - R} \qquad \text{(Eq 4.4)}$$

---

*Independently proposed and used by Halford and Nachtigall in 1979 (Ref 4.6), it was recently learned that as early as 1970, Smith, Watson, and Topper (Ref 4.7) had defined the ratio of mean to amplitude as C in a footnote listing the numerous alternative ways of writing their stress-strain function for mean stress effects, the Smith-Watson-Topper parameter.

And, vice-versa, the ratios $R$ and $A$ are interrelated by the expressions:

$$R = \left(\frac{1-A}{1+A}\right) = \left(\frac{V-1}{V+1}\right) \quad \text{(Eq 4.5)}$$

$$A = \left(\frac{1-R}{1+R}\right) = \left(\frac{1}{V}\right) \quad \text{(Eq 4.6)}$$

While the $V$-ratio is not currently in widespread use in the literature, its merits have been utilized extensively by Halford and his coworkers at NASA Glenn. Figure 4.9 illustrates the interrelationships between the three mean stress descriptors.

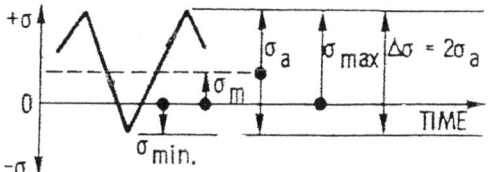

$$R = \frac{\sigma_{min.}}{\sigma_{max}} = \frac{1-A}{1+A} = \frac{V-1}{V+1}$$

$$A = \frac{\sigma_a}{\sigma_m} = \frac{1-R}{1+R} = \frac{1}{V}$$

$$V = \frac{\sigma_m}{\sigma_a} = \frac{1+R}{1-R} = \frac{1}{A}$$

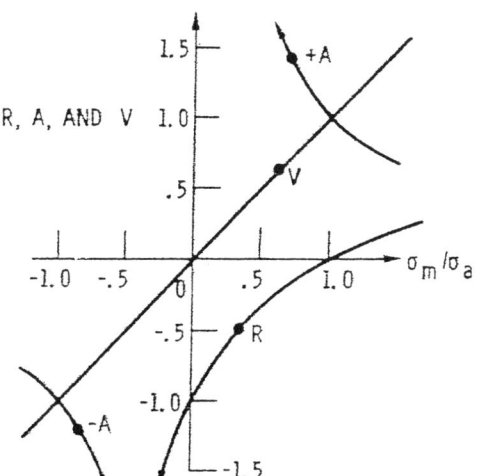

**Fig. 4.9** Relations between mean stress descriptors. Source: Ref 4.6

## Early Studies of Mean Stress Effects

More than a century ago, Gerber, in 1874 (Ref 4.8), and Goodman, in 1899 (Ref 4.9), studied the pattern of loss of fatigue life due to tensile mean stress. They concerned themselves with high-cycle fatigue, and stresses within the elastic range. Their respective schemes for representing the effects of tensile mean stress with high-cycle fatigue life are shown in Fig. 4.10(a). Plotted is the allowable alternating stress to produce lives of, say, $N_1 = 10^5$ and $N_2 = 10^7$ cycles, as a function of tensile mean stress. The allowable alternating stress falls as mean tensile stress increases, approaching zero as the mean stress approaches the ultimate tensile strength. Gerber fitted his data with a parabola, while Goodman more conservatively estimated the trend with a straight line. The points $A$ and $A'$ for each case where mean stress is zero are fatigue strengths from the ordinary fatigue characterization for completely reversed stress. A generalization is shown in Fig. 4.10(b). Here, the vertical axis is replaced by the ratio of actual alternating stress to the completely reversed alternating stress that will cause failure at any chosen life level, and the horizontal coordinate is the ratio of the mean stress to the ultimate tensile strength. Curves for all life levels were presumed to collapse to a single curve—a straight line for the Goodman formulation, and a parabola for the Gerber formulation. The analytical representation is:

$$\frac{\sigma_a}{\sigma_r} + \frac{\sigma_m}{\sigma_u} = 1 \quad \text{(Goodman)} \quad \text{(Eq 4.7)}$$

$$\frac{\sigma_a}{\sigma_r} + \left(\frac{\sigma_m}{\sigma_u}\right)^2 = 1 \quad \text{(Gerber)} \quad \text{(Eq 4.8)}$$

Because Goodman did not present his equation in the form shown in Fig. 4.10, the figure is commonly referred to as the *modified* Goodman diagram*. Goodman's original presentation of his formulation was by straight lines that showed both the maximum and minimum stresses reached during the cycling. This is shown graphically in Fig. 4.11. The advantage of his display form is that it emphasizes that

---

*According to Maxwell and Nicholas (Ref 4.10), based on the historical research of Sendeckyj (Ref 4.11), Haigh was more responsible for this form of presentation than Goodman was and hence it is historically more appropriate to refer to it as the *Haigh diagram*. In fact, the term Haigh diagram is preferred in the European Union.

84 / Fatigue and Durability of Structural Materials

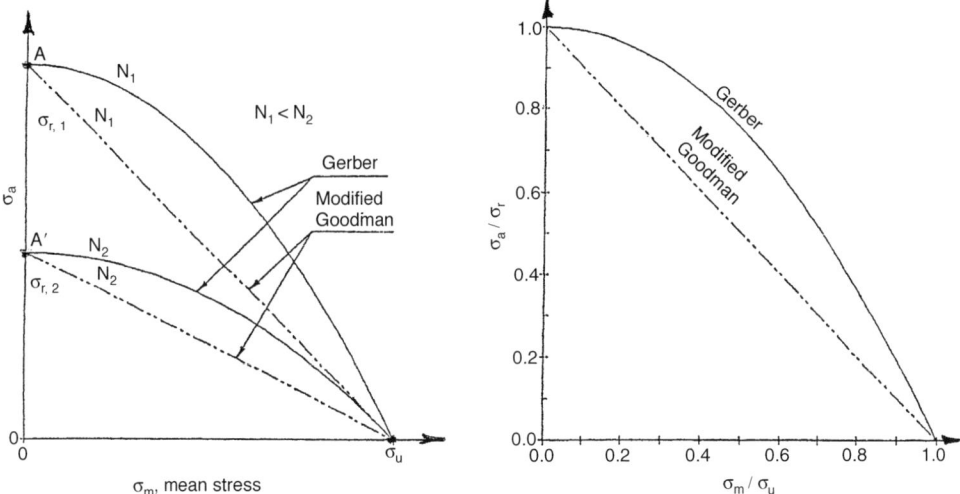

**Fig. 4.10** Early mean stress theories. (a) Specific life levels. (b) Generalization for all life levels

higher stresses are reached during the cycling than that reached during completely reversed loading. Obviously, this is part of the reason for the reduction in fatigue life due to the mean stress, but not the entire reason, as discussed later.

An extensive review of the mean stress literature up to about 1980 has been given by Conway and Sjodhal (Ref 4.3).

### More Recent Approaches

Because the subject of mean stress effects continues to receive a great deal of attention, we shall not attempt to be comprehensive in describing all the nuances of the methodology and shall briefly outline only two methods that are commonly used. We shall then outline a new method, which we have recently developed that embraces, in a general form, many of the features of common current methods.

**Morrow's Method.** Although not explicitly expressed, this development was intended primarily for the long lives associated with stresses in the elastic range. Morrow (Ref 4.12) arrived at his method through consideration of both a preponderance of nominally elastic, high-cycle fatigue data that had been published in Ref 4.13 and Basquin's power law equation relating fatigue strength to fatigue life. This equation is directly related to the elastic strain range line of the Manson-Coffin-Basquin diagram previously discussed. No consideration was given to the plastic strain line because all of the mean stress data were in the nominally elastic, high-cycle fatigue regime wherein the degree of plasticity was of no consequence to an engineering analysis of fatigue life.

Without the presence of mean stress, the elastic line intersects the vertical axis ($2N_f = 1$) at the point $\sigma = \sigma_f'$ (Fig. 4.12). As previously discussed, the vertical axis represents the ordinary tensile test if one regards this test as being half of a cycle. Thus, if a tensile test were conducted on a specimen already subjected to an initial mean tensile stress of $\sigma_m$, its remaining true fracture strength would appear to be only $\sigma_f' - \sigma_m$. This mean stress-altered true fracture strength would then represent the intercept of the elastic line with a mean stress of $\sigma_m$. Morrow also im-

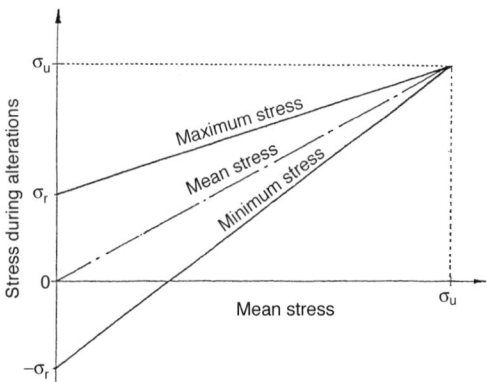

**Fig. 4.11** Goodman's original representation of mean stress effects by displaying both the maximum stress and the minimum stress during alternations

plicitly assumed that the elastic line for the material with mean stress would be parallel to the elastic line with zero mean stress, thus, in Fig. 4.12 the line $A'B'$ is parallel to $AB$, and has an intercept at $2N_f = 1$ of only $\sigma'_f - \sigma_m$.

Now, if we consider in Fig. 4.12 any life level $2N_{f,1}$ where, *without* mean stress, the allowable strength is $\sigma_{a,1}$ (point $B$). The allowable strength *with* mean stress is $\sigma'_{a,1}$ (point $B'$). Because the ratio of $B'$ to $B$ has the same value as the ratio of the stresses at $2N_f = 1$, (points $A'$ and $A$), we can write:

$$\left(\frac{\sigma'_{a,1}}{\sigma_{a,1}}\right) = \left(\frac{\sigma'_f - \sigma_m}{\sigma'_f}\right) = 1 - \left(\frac{\sigma_m}{\sigma'_f}\right) \quad \text{(Eq 4.9)}$$

If this equation is plotted on the same coordinates as used for the modified Goodman diagram, the result is as shown in Fig. 4.13. The straight line intersects the vertical axis at $\sigma_{f,1}$, the stress required to produce a life of $2N_{f,1}$ when there is no mean stress, and proceeds toward the point where $\sigma_m = \sigma_f$ on the horizontal axis. Invariably, $\sigma_f > S_u$ is required because $\sigma_f$ can be estimated (Manson [Ref 4.14] gave credit to Jack O'Brien for this useful relationship) by $\sigma_f = S_u(1 + D)$, where $D$ is the true ductility in a tensile test and $S_u$ is the ultimate tensile strength.

More generally, if we normalize the line to include all life levels, as we did in Fig. 4.10(b), we get the result shown in Fig. 4.14. The Morrow approach is seen to be a compromise between the Gerber and the modified Goodman diagrams in the low and moderate mean stress range, but it lies above both for mean stresses close to the ultimate tensile strength. However, such high mean stresses are rarely of practical interest due to plasticity that causes cyclic mean stress relaxation and strain ratcheting.

The Morrow mean stress equation (Ref 4.12) was originally related to the number of reversals to failure by the expression:

$$\sigma_a = (\sigma'_f - \sigma_m)(2N_f)^b \quad \text{(Eq 4.10)}$$

In terms of the associated elastic strain amplitude, the life relation becomes (Fig. 4.15):

$$\frac{\Delta\varepsilon_e}{2} \left(\frac{\sigma'_f - \sigma_m}{E}\right)(2N_f)^b \quad \text{(Eq 4.11)}$$

At the time Eq 4.11 was proposed, there was no concern for circumstances of mean stress in the presence of measurable cyclic plastic strains.

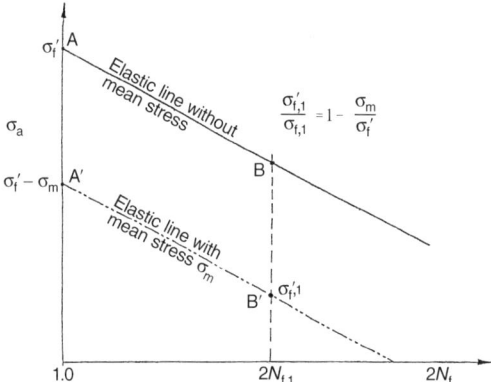

**Fig. 4.12** Morrow's concept of the effect of mean stress as a displacement of the elastic line to retain consistency of results in a simple tensile test

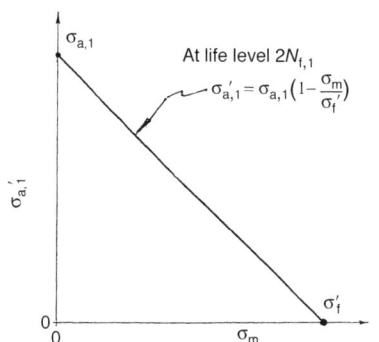

**Fig. 4.13** Morrow's concept of the displacement of the elastic line when shown in terms of the Haigh (modified Goodman) diagram

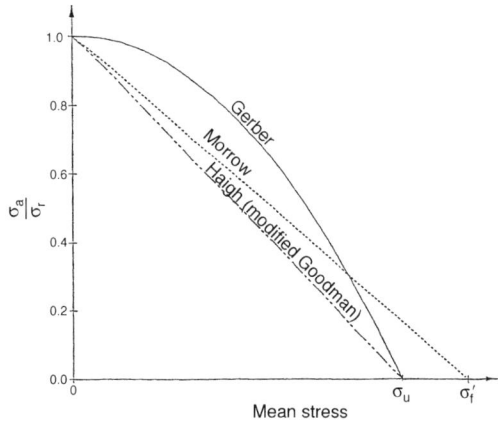

**Fig. 4.14** Comparison of concepts of Haigh (modified Goodman), Gerber, and Morrow for treating mean stress effects

As a result, there was no reason to consider any changes to the Manson-Coffin-Basquin fatigue relation other than to shift the elastic line. The plastic line remained fixed as the miniscule plastic strain involved in the mean stress studies of the time was known to be negligible. By adding the unaltered plastic line to the mean stress-modified elastic line:

$$\frac{\Delta \varepsilon_t}{2} = \left(\frac{\sigma'_f - \sigma_m}{E}\right)(2N_f)^b + \varepsilon'_f(2N_f)^c \quad \text{(Eq 4.12)}$$

As suggested in Ref 4.15, Eq 4.11 can be made more general by the addition of a multiplier $k$ to the mean stress term:

$$\frac{\Delta \varepsilon_e}{2} = \left(\frac{\sigma'_f - k\sigma_m}{E}\right)(2N_f)^b \quad \text{(Eq 4.13)}$$

This is equivalent to applying a material-dependant variable scale to the mean stress axis. In this way, mean stress results for a given material and operating condition can be more accurately fit by an empirical evaluation of $k$. Selecting a value of $k = \sigma_u/\sigma_f$, for example, reduces the Morrow equation to the modified Goodman equation. If mean stress data were unavailable, $k$ would naturally be selected as 1.

It was pointed out at about the same time (Ref 4.6) and (Ref 4.16) that if the Morrow mean stress-modified elastic strain range term were to be added directly to the completely reversed plastic strain amplitude versus life relation, a potential inconsistency would develop. The implied cyclic stress-strain curve would become strongly dependent on the magnitude and sign of the mean stress. Figure 4.15 aids in understanding the dilemma. The completely reversed plastic, $AB$, and elastic, $CD$, strain range life lines are shown as solid lines for zero mean stress. For a tensile mean stress, according to the Morrow model, the elastic line is lowered by the amount of the elastic strain associated with the mean stress $(\sigma_m/E)$, i.e., dashed line $C'D'$. Similarly, the elastic line is raised to $C''D''$ (also shown as a dashed line) for a compressive mean stress.

For a given life under zero mean stress, consider points $P$ and $Q$ on the completely reversed plastic and elastic lines. The corresponding hysteresis loop for this straining condition is shown schematically in Fig. 4.16(a). When the elastic line is moved down due to a tensile mean stress, but the plastic strain range at point $P$ remains fixed, the elastic strain range (now at point $Q'$) becomes significantly smaller, and hence the stress range of the hysteresis loop for $PQ'$ appears as in Fig. 4.16(b) (shown for an arbitrary mean stress ratio, $V_\sigma = +1.0$). The hysteresis loop for the same magnitude mean stress ratio, except in compression ($V_\sigma = -1.0$), is shown in Fig. 4.16(c) with its implied large stress range. All three hysteresis loops have the same plastic strain range, but the stress ranges are calculated to vary by an absurdly large and physically unrealistic factor of 4:1.

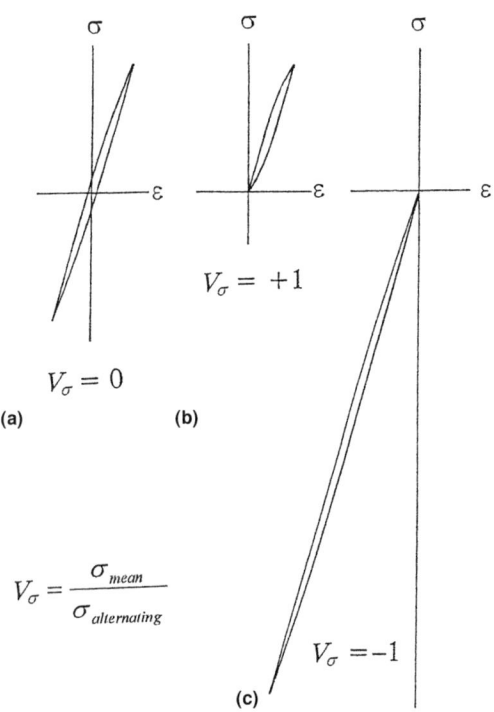

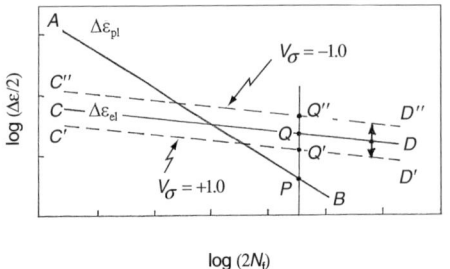

**Fig. 4.15** Illustration of Morrow's initial mean stress approach of shifting the elastic line vertically while leaving the plastic line fixed

**Fig. 4.16** Hysteresis loops implied by moving only the elastic line due to mean stress. (a) Zero mean stress ($V_\sigma = 0$). (b) Tensile mean ($V_\sigma = +1$). (c) Compressive mean ($V_\sigma = -1$)

If the analysis were repeated for the same tensile and compressive mean stress ratio over a range of lives and strain ranges, one would arrive at three parallel, but drastically different, cyclic stress-strain curves as shown in Fig. 4.17. These curves do not represent a physically viable situation. To substantiate this claim, a series of three stress-strain cycles was imposed in tests by the authors and the resultant hysteresis loops were recorded. The stress range in these experiments was controlled constant at 297.9 MPa (43.2 ksi), and the plastic strain ranges were measured as the maximum widths of the resultant hysteresis loops. Figure 4.18 shows the measured hysteresis loops to have the same plastic strain ranges of approximately $1 \times 10^{-4}$, regardless of the mean stress ($V_\sigma = -1$, 0, or $+1$). This result is in sharp contrast to the implications of Fig. 4.17. Clearly, the cyclic stress-strain curve must remain nominally independent of the magnitude of the mean stress. Consequently, if the elastic line is shifted in life due to mean stress, the plastic line must also be shifted by an amount that retains a unique cyclic stress-strain relation independent of mean stress. The only physically acceptable shift is to move both the elastic and plastic lines the same amount parallel to the life axis. This topic is explored further in a later section because it is a contributor to a more comprehensive method for dealing with mean stress effects.

**Halford-Nachtigall Modified Morrow Model.** It is appropriate to mention briefly a modified version of the Morrow model first derived in 1979 by Halford and Nachtigall and published in Ref 4.6. It is based on translating both the elastic and plastic lines the same amount as suggested above. The amount of shift in life is expressed in terms of the mean stress descriptor $V_\sigma$. The importance of the derivation is that it results in an exceptionally simple form that permits rapid calculations of the effect of mean stress on fatigue life. Knowledge of only the mean stress ratio $V_\sigma$ is needed to assess the change in fatigue life.

We start with a simple modification to the original Morrow proposal wherein we substitute cycles to failure $N_f$ for reversals to failure $2N_f$ (an example of the error introduced by this substitution is given later in this section). Thus, the stress amplitude $\sigma_a$ is related to the fatigue life $N_{fm}$ (with mean stress), the mean stress $\sigma_m$, and the fatigue strength coefficient $\sigma'_f$ by the expression:

$$\sigma_a = (\sigma'_f - \sigma_m)(N_{fm})^{-b} \quad \text{(Eq 4.14a)}$$

For zero mean stress, the relation reduces to:

$$\sigma_a = \sigma'_f(N_{fm})^{-b} \quad \text{(Eq 4.14b)}$$

Solving for $\sigma'_f$ from Eq 4.14(b), substituting into Eq 4.14(a), and further recognizing that, by definition, $\sigma_m = V_\sigma \sigma_a$:

$$\sigma_a = (\sigma_a N_{fb} - V_\sigma \sigma_a)(N_{fm})^{-b} \quad \text{(Eq 4.14c)}$$

Canceling $\sigma_a$ and rearranging terms, we arrive at the exceptionally simple relation:

$$(N_{fm})^{-b} = (N_f)^{-b} - V_\sigma \quad \text{(Eq 4.14d)}$$

where $N_{fm}$ is fatigue life with a mean stress ratio $V_\sigma = \sigma_m/\sigma_a$; $N_f$ is fatigue life with zero mean stress, $V_\sigma = 0$; and $b$ is the Basquin exponent ($\approx -0.05$ to $-0.15$).

Thus, knowing only the fatigue property $b$ and a value of the mean stress ratio $V_\sigma$, it is

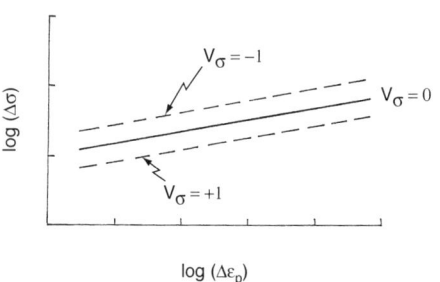

**Fig. 4.17** Erroneous cyclic stress-strain dependence on mean stress implied by shifting only the elastic line while leaving the plastic line fixed

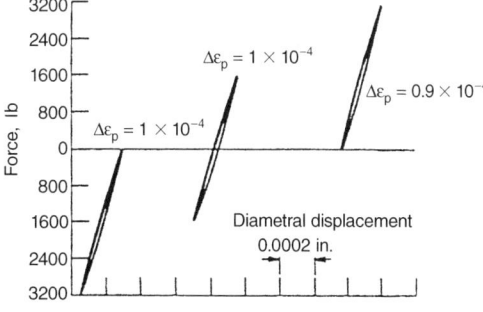

**Fig. 4.18** Hysteresis loops with positive, negative, and zero mean stress, showing that measured cyclic stress-strain response is independent of mean stress. Material, 316 stainless steel at room temperature. $\Delta\sigma$, 43.2 ksi. Source: Ref 4.16

possible to estimate rapidly the shift in fatigue life due to a mean stress. This ability to scope out the potential effects of mean stress is a valuable tool in the arsenal of equations for estimating fatigue response of materials. For example, with $N_f = 10^5$, $b = -0.10$, and $V_\sigma = +0.5$, the life $N_{fm} = 1.8 \times 10^4$. This represents a factor of 5.6× loss in fatigue life. As a check on the magnitude of the error introduced by replacing reversals ($2N_f$) with cycles ($N_f$) in creating Eq 4.14(a), we take the original life of ($2N_f$) = $2 \times 10^5$ and calculate the value of ($2N_{fm}$) to be $4 \times 10^4$, i.e., a factor of 5.0 loss in fatigue life. This life is 10% less conservative than that computed directly from Eq 4.14(d).

Equation 4.14(d) utilizes the fatigue property $b$, but otherwise doesn't specify use of any correlating variable for representing the fatigue life relation. The correlating variable that goes with fatigue life $N_f$ could be stress, elastic strain, plastic strain, total strain, strain energy, or even the Smith-Watson-Topper parameter that is discussed in the next section. The only requirement is that the correlating variable must be held constant while using the equation to assess the effects of mean stresses. This is equivalent to saying that the working fatigue curve for a given mean stress ratio $V_\sigma$ is shifted horizontally (parallel to the fatigue life coordinate) in accordance with Eq 4.14(d). Another interesting and useful aspect of the expression is that it relates the change in fatigue life directly to the change in mean stress ratio; knowledge of the magnitudes of the stresses is unnecessary. Extensive use of Eq 4.14(d) has been made by Halford at the NASA Glenn Research Center.

**Stress-Strain Function of Smith-Watson-Topper (SWT).** In 1970 Smith, Watson, and Topper (Ref 4.7) published the concept that life is governed by a parameter proportional to the product, ($\sigma_{max} \Delta \varepsilon/2$). Their intent was to be able to apply the parameter to cases that involved plasticity as well as those limited to the elastic range. For completely reversed loading, $\sigma_{max}$ and $\Delta \varepsilon/2$ have a unique one-to-one relation so that life can be expressed more simply in terms of either the alternating strain range or the maximum stress. However, when a mean stress is present, this direct relation is altered, but the efficacy of the parameter remains the same. Voluminous test results have been generated to validate the SWT hypothesis. To implement the method it is merely necessary to know the ordinary material characterization according to the Manson-Coffin-Basquin model for completely reversed loading. Because such characterization provides information on the maximum stress (elastic line) and total strain range (elastic plus plastic lines), the relation between fatigue lives and the product strain range and maximum stress can be extracted without any mean stress testing. Once the relation has been established for the case of completely reversed loading, it can be applied to other cases involving mean stress if the maximum stress and the alternating stress are known. Figure 4.19 shows the application of the method to mean stress results for 4340 steel. The parameter in this case is expressed as $(\sigma_{max} E \Delta \varepsilon/2)^{0.5}$ where the elastic modulus is used to multiply their parameter, and the square root extracted of the product in order to maintain units of stress.

Smith, Watson, and Topper did not present any analyses to compare the results of their approach with the Goodman, Gerber, or the Morrow analyses, and did not indicate whether these other methods are inappropriate for cases wherein their stress-strain function serves well. Nor did they indicate under what conditions it is easier to determine the maximum stress and strain range than it is to determine the mean stress and stress range for a given application. Either parameter can be determined with equal ease or difficulty for any given application.

An analysis comparing the SWT concept to the Morrow concept is shown in Fig. 4.20 for 4130 steel. In Fig. 4.20(a), the SWT parameter was taken to be applicable, and the Morrow-type curves were computed for five life levels between $10^3$ and $10^7$ cycles.

If the Morrow-type analysis were correct, all these individual curved lines would become a

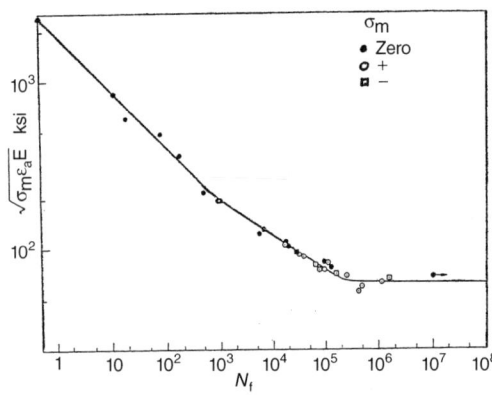

**Fig. 4.19** Smith-Watson-Topper (SWT) analysis applied to fatigue results of 4340 steel. Source: Ref 4.7

single straight line from coordinates (0, 1) to (1, 0), which is obviously not the case. Similarly, in Fig. 4.20(b), we have assumed that the Morrow-type analysis is correct and calculated how the stress-strain function would plot against life for the same material at values of mean stress from 0.1 to 0.7 of $\sigma_f$. Instead of obtaining a single curve, a family of distinct curves resulted. Obviously, the two methods for representing mean stress effects on fatigue life will always give different predictions of life. It would seem to be the burden of the analysts favoring the SWT procedure to reconcile the discrepancy since the Morrow procedure, being more or less consistent with the Goodman and Gerber approaches, which have received extensive study for nearly a century, would be more credible. The SWT procedure evolved from fatigue results in the intermediate-to high-cycle fatigue regime, whereas the Goodman and Gerber approaches, and subsequently, the Morrow approach, were derived from very high-cycle fatigue data. Unfortunately, the otherwise useful SWT model is ill-suited for large compressive mean stresses, especially when the maximum tensile stress drops appreciably below the stress amplitude for completely reversed high-cycle fatigue. In the extreme of a very tiny maximum tensile stress or even a negative maximum stress, the parameter loses its engineering utility and would erroneously predict infinite life.

## A New Comprehensive Model (Manson)

We have evolved a comprehensive model for treating mean stress effects that would appear to have validity in both the elastic and plastic range because it is a logical extension of the Manson-Coffin-Basquin diagram, which, of course, embraces both elastic and plastic strains. The basis of the model is discussed next.

### Basis of the Model

As a starting point, the comprehensive model uses the Manson-Coffin-Basquin (MCB) fatigue diagram for displaying the cyclic straining level versus cyclic life. The specific mean stress model could be any of a number of proven models; Goodman, Haigh, Gerber, Soderberg, Morrow, and so forth, as discussed in an earlier section. All such models capture the influence of tensile mean stress as detrimental to fatigue life while compressive mean stress is beneficial. In all cases, the net effect, for example, of a tensile mean stress, is a displacement of the elastic line of the MCB fatigue diagram in Fig. 4.21 to lower strains, i.e., $B$ to $B'$ (same as $T$ to $T''$) or lower lives, $T$ to $T'$. The extent of decreased elastic strain resistance may vary from model to model. However, whether the elastic line is considered to be moved down in strain range for a given life, or moved to lower lives for a given strain range, is not of importance because the net effect is exactly the same.

A significant contribution of the comprehensive model is the previously discussed recognition that in the presence of a mean stress, the plastic line cannot remain unaffected. Otherwise, a physically unrealistic cyclic stress-strain curve results. Therefore, a rule must be invoked that keeps the elastic and plastic lines in a fixed relationship one to the other so that the cyclic stress strain curve is independent of mean stress. This is accomplished by maintaining the slopes for the elastic ($b$) and plastic ($c$) lines in the presence of mean stress. This ensures that the cyclic strain-hardening exponent remains fixed, be-

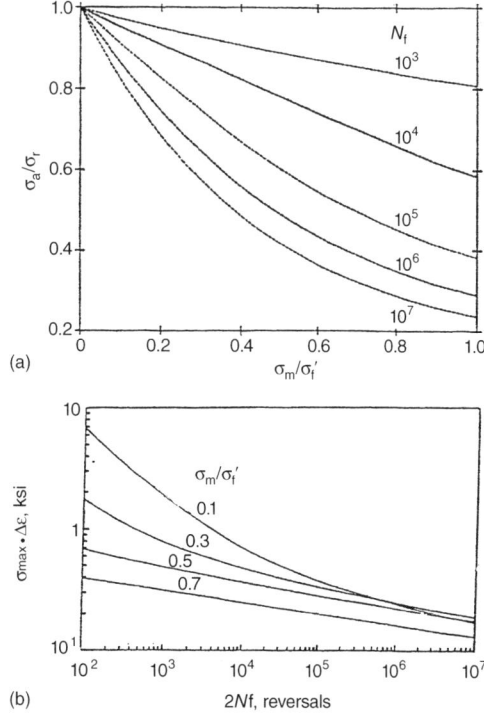

**Fig. 4.20** Comparison of concepts of Smith-Watson-Topper (SWT) and Morrow applied to fatigue results of 4130 steel. (a) Isolife curves that result when the SWT parameter is taken to be applicable. (b) SWT type of plot obtained when the Morrow type analysis is assumed correct

cause $n' = b/c$. With $n'$ fixed, we need to fix only one point $P$ along the stress-strain curve, shown in the insert in Fig. 4.21, to guarantee that the cyclic stress-strain curve remains constant. A convenient point to pick is where the elastic and plastic strain ranges are equal. In terms of the life relations, these strain ranges are known as the *transition strain range*, $\Delta\varepsilon_T$. The corresponding life is the *transition fatigue life*, $N_T$. The net effect of the presence of a mean stress is to change the value of the transition life $N_T$ to $N_{T'}$ in Fig. 4.21, while keeping the transition strain range constant. Hence, the entire MCB fatigue curve is translated parallel to the fatigue life axis.

In summary, the locations of the shifted elastic and plastic lines are established by the following conditions:

- At $N_f = 1$ the intercept of the elastic line, $B'$, must be $(B - 2k\sigma_m/E)$, where $k$ is a constant depending on the basic mean stress model of interest. For example, for Morrow's model, $k = 1$, and for the modified Goodman model, $k = S_u/\sigma_f$.
- The cyclic stress-strain curve of the material with mean stress must be the same as the cyclic stress-strain curve for completely reversed loading.
- The slopes of the elastic and plastic lines must be the same for the case with mean stress as for completely reversed stress. This is a natural consequence of the cyclic stress-strain curve being independent of mean stress, because at any given life, the elastic and plastic strains must always fall on the cyclic stress-strain curve.
- In keeping with the cyclic stress strain curve being independent of mean stress, the plastic line must be moved parallel to the life axis exactly the same amount as the elastic line.

A simple way of expressing this condition is to state that the transition point $T$ must move horizontally to point $T'$. Only if the transition point moves horizontally can the relation between the elastic strain and plastic strain remain unaffected, and hence the cyclic stress-strain curve also remains unaffected by mean stress.

It is illustrative to show these rules applied to the generalized mean stress model depicted in Fig. 4.21. It can be shown that the amount of displacement along the life axis between $T$ and $T'$ is:

$$\frac{N_{T'}}{N_T} = \left[1.0 - \left(\frac{2k\sigma_m}{BE}\right)\right]^{1/b} \quad \text{(Eq 4.15)}$$

where $b$ is the slope of the elastic line, $E$ is the modulus of elasticity, and $B$ is the intercept of the elastic line under zero mean stress. Thus, the intercept $C'$ of the plastic line at $N_f = 1$ with a mean stress $\sigma_m$ is:

$$C' = (2)^c \left[(0.5) - \left(\frac{k\sigma_m}{BE}\right)\right]^{c/b} C \quad \text{(Eq 4.16)}$$

and the total strain range versus $N_f$ equation becomes:

$$\varepsilon = (2)^b \left[(0.5) - \left(\frac{k\sigma_m}{BE}\right)\right] B(N_f)^b$$
$$+ (2)^c \left[(0.5) - \left(\frac{k\sigma_m}{\sigma'_f}\right)\right]^{c/b} C(N_f)^c \quad \text{(Eq 4.17)}$$

Because both the elastic and plastic lines are translated in life by the same amount, an alternate way of expressing the effect of mean stress is to build the displacement into the fatigue life scale. In Fig. 4.22 the material behavior with mean stress is characterized by a single Manson-Coffin-Basquin set of elastic and plastic lines, which is identical to the completely reversed set. Thus, the fatigue life scale, normally $N_f$, is replaced by the product $N_f[1 - (2k\sigma_m/BE)]^{1/b}$.

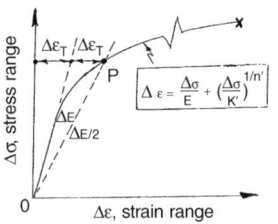

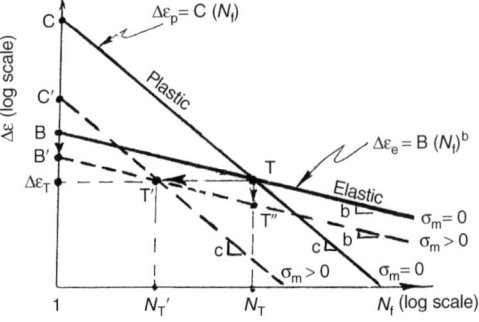

**Fig. 4.21** Basis for comprehensive model for treating mean stress effects

Obviously, when $\sigma_m = 0$, the scale reverts back to $N_f$.

It cannot be over emphasized that the underlying principle permitting Eq 4.17 to be written is the recognition that:

**The cyclic stress-strain behavior of a material cannot be a function of the magnitude or sign of a mean stress.**

This statement is a virtual theorem for material stress-strain response. It becomes self-evident after considering that the path taken by a stress-strain response is predestined by the state of the material *prior* to having completed a cycle of loading, whereas a mean stress can be defined only *after* a completed cycle. Obviously, the instantaneous path cannot depend on something that must yet happen in the future. This is a powerful realization that is helpful in the analysis of cyclic loading wherein small amounts of plasticity and mean stresses are involved. As discussed in the next section, invoking use of the cyclic stress-strain curve also provides insight into why mean stress can have such a profound effect on fatigue life.

Any fatigue life model that accounts for influences on basic fatigue life due to mean stress, multiaxiality, cumulative damage, and so forth should not violate the cyclic stress-strain response behavior of a material. In fact, conformity to the aforementioned theorem should be a high priority for any model proposed for dealing with bulk property effects that influence fatigue life.

## Illustration of the Comprehensive Model Using the Morrow Notation

Because of the widespread use of the Morrow notation of strain amplitudes, $\Delta\varepsilon/2$, and reversals, $2N_f$, it is informative to express the comprehensive model in these terms. It can be shown readily that the amount of displacement of the transition life is:

$$\left[1.0 - \left(\frac{k\sigma_m}{\sigma_f'}\right)\right]^{1/b} \qquad \text{(Eq 4.18)}$$

Thus, the intercept of the plastic line at $2N_f = 1$ is:

$$\varepsilon_f'\left(1.0 - \frac{k\sigma_m}{\sigma_f'}\right)^{c/b} \qquad \text{(Eq 4.19)}$$

and the total strain amplitude versus $2N_f$ equation becomes:

$$\frac{\Delta\varepsilon}{2} = [1.0 - k(\sigma_m/\sigma_f')](\sigma_f'/E)(2N_f)^b$$
$$+ [(1.0 - (k\sigma_m/\sigma_f')]^{c/b}\varepsilon_f'(2N_f)^c \qquad \text{(Eq 4.20)}$$

Repeating Fig. 4.22 for the Morrow notation with $k = 1.0$, we have Fig. 4.23. The life scale, normally $2N_f$, is replaced by the product $2N_f[1.0 - (\sigma_m/\sigma_f')]^{1/b}$. Obviously, when $\sigma_m = 0$, the scale reverts back to $2N_f$.

## Insight into Cause of Life Reduction due to Tensile Mean Stress

One of the interesting and valuable uses of the comprehensive model for mean stress effects is that it provides insight as to why mean stresses have a seemingly disproportionate effect on life when it is well understood that cyclic plasticity is the root cause of metal fatigue.

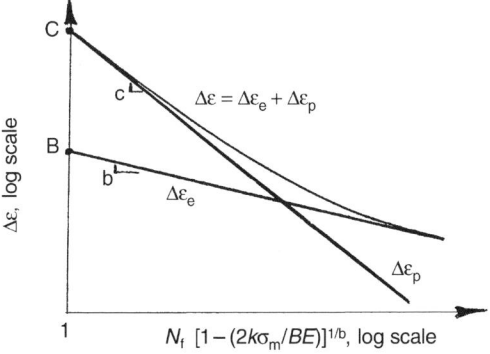

**Fig. 4.22** Method of incorporating mean stress effects in the comprehensive model by retaining the elastic and plastic lines of reversed loading and changing the life scale to include a term containing mean stress

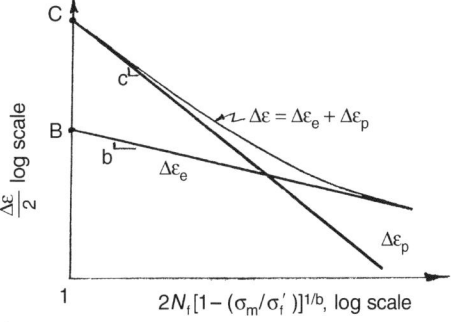

**Fig. 4.23** Casting Fig. 4.22 in terms of Morrow's notation

**Tensile Mean Stress Reduces Ductility.** One of the first obvious explanations for tensile mean stress life reduction is that, for a given stress range, the maximum stress reached is greater if there is a tensile mean stress than if the stress is completely reversed. The Goodman representation shown in Fig. 4.10 makes this point clearly. One might attempt to explain the loss in fatigue life on the basis that the higher maximum stress carries the operation to a higher point on the stress-strain curve, using up more of the ductility of the material. Thus, cycling is imposed on a material of lower remaining ductility, and therefore the life is lower. While this explanation is correct qualitatively, it doesn't quantitatively explain the large effect.

Consider the following illustrative examples based on the material whose cyclic stress-strain curve is shown in Fig. 4.24. The strain-life relation of the material under zero mean stress conditions is taken as:

$$\frac{\Delta\varepsilon}{2} = 0.0123(2N_f)^{-0.106} + 0.42(2N_f)^{-0.65}$$

(Eq 4.21)

Assuming a completely reversed loading with a total strain range of $\Delta\varepsilon = 0.008$, the expected life is 39,200 cycles. Now, assume a tensile mean stress of 30 ksi is imposed prior to imposing the same cyclic strain. Consideration of the mean stress would cause the material to reach point $P'$ (maximum stress = 67.2 ksi, maximum strain = 0.0169) at the maximum loading, while without mean stress, the maximum stress (37.2 ksi) and strain (0.004) would have been at $P$ so that yielding to $P'$ uses up 0.0102 or 2.4% of the material's available ductility of 0.42. If only 2.4% of the ductility were used up, however, 97.6% of the ductility would still remain. The resultant cyclic stress-strain curve of the material loaded to $P'$, and then released to zero, might be approximated by $O'P'Q$ having ductility 0.4098 and exactly the same fracture stress of 123 ksi. The change imparted to the termination point of the cyclic stress-strain curve is quite small. Applying the total strain range of 0.008 would only cause the life to drop to 38,800 cycles to failure. This example calculation does not offer a viable explanation as to why a tensile mean stress can have such a large effect on fatigue life. It is certainly not a matter of the maximum stress having reached a higher value and hence having used up a significant percentage of the ductility of the material.

However, let us now examine the more realistic situation that is captured by the comprehensive mean stress model. By retaining the cyclic stress-stain curve, the model requires a significant reduction in the intercept of the plastic line (as well as of the elastic line) as a tensile mean stress is applied. This significant reduction in the *apparent* ductility of the material is the key to understanding why tensile mean stress can cause such a large reduction in fatigue life. We select the same materials properties as used in the aforementioned example. It is assumed that the total strain range is 0.008 in./in. and that the mean stress is +30 ksi. Evaluating the numerical coefficients and exponents in the appropriately-modified Eq 4.21, we have

$$\frac{\Delta\varepsilon}{2} = 0.0093(2N_f)^{-0.106} + 0.0756(2N_f)^{-0.65}$$

(Eq 4.22)

Thus, for the current situation, the *apparent* stress-strain curve has a cyclic fracture strength value of only 93 ksi and a cyclic fracture ductility of only 0.0756. These values are required to be on the plastic portion of the cyclic stress-strain curve, $\varepsilon_p = (\sigma/142)^{6.135}$ and, in fact, are as shown in Fig. 4.25. Presuming there is no relaxation of the mean stress while cycling, the predicted fatigue life from Eq 4.22 is only 2,810 cycles. This is a loss of nearly 93% of the 39,200 cycle life for completely reversed loading. Note that the apparent ductility in Eq 4.22 is now only 0.0756; a correspondingly large reduction of 82% from the initial ductility of 0.42. Because fatigue is so intimately associated with cyclic

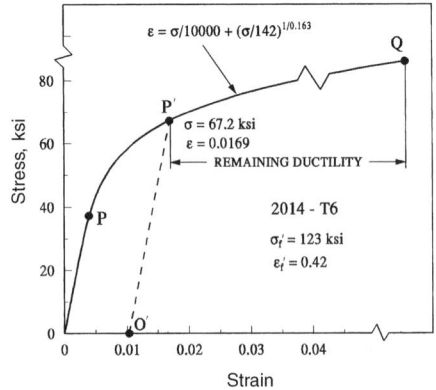

**Fig. 4.24** Analysis of tensile mean stress effects on the basis of loss of ductility caused by maximum stress in cycle

plasticity regardless of life level, it is insightful to realize that the large fatigue life reductions due to tensile mean stresses are indeed associated with large reductions in ability to flow plastically, i.e., ductility.

Figure 4.26 helps in further visualizing the close connection between ductility loss and fatigue life loss due to tensile mean stress. Here, the total strain amplitude versus $2N_f$ curve is shown for both completely reversed loading (Eq 4.21), and for +30 ksi mean stress (Eq 4.22). We follow the procedure outlined earlier in the section "Basis of the Model" for accounting for the influence of mean stress on fatigue life while preserving the cyclic stress-strain curve. The total strain amplitude curve as well as the linear elastic and plastic lines are translated to the left equally by an amount to produce the lowering from $\sigma'_f/E$ to $\sigma''_f/E$ of the elastic line intercept and the simultaneous lowering of the plastic line intercept from $\varepsilon'_f$ to $\varepsilon''_f$. Regardless of the specific mean stress model, the initial intercepts $\sigma'_f/E$ and $\varepsilon'_f$ are always reduced for a tensile mean stress: the greater the mean stress, the greater the reduction. Because of the differing slopes of the elastic and plastic lines, the reduction in ductility is always substantially greater than the reduction in the elastic intercept.

**Compressive Mean Stress Enhances Ductility.** The previous example of the dramatic influence of mean stress on the available ductility and hence upon fatigue life has been shown for the case of tensile mean stress. It is well established that *compressive* mean stresses can be

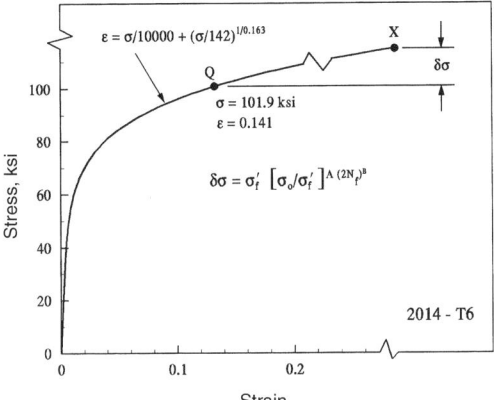

**Fig. 4.25** Analysis of tensile mean stress effects on fatigue life on the basis of loss of ductility caused by necessity to maintain the cyclic stress-strain curve exactly the same fracture stress of 848 MPa (123 ksi)

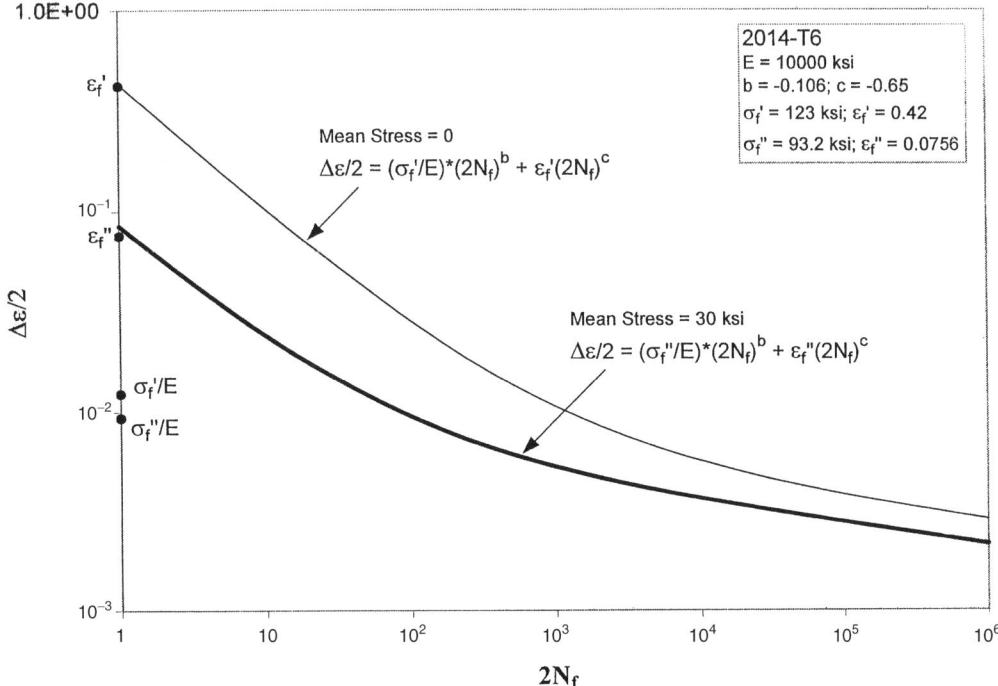

**Fig. 4.26** Shifting of Manson-Coffin-Basquin fatigue lines to account for tensile mean stress effects. Note large reduction in plastic line intercept required to maintain cyclic stress-strain curve.

used to enhance fatigue life through use of surface treatments that impart beneficial compressive residual stresses (see Chapter 11, "Avoidance, Control, and Repair of Fatigue Damage"). Analyses similar to those used previously help to explain the fatigue life benefit by showing that greater *apparent* ductility is available under these conditions. For example, had the mean stress in the previous example been at −30 ksi, the Manson-Coffin-Basquin fatigue equation would have been:

$$\frac{\Delta\varepsilon}{2} = 0.0153(2N_f)^{-0.106} + 1.66(2N_f)^{-0.65}$$

(Eq 4.23)

The new intercepts reflect an increase in the terminal point of the stress-strain curve to 153 ksi and a ductility of 1.66. Once again, as required, these terminal points fall on an extension of the cyclic stress-strain curve, $\varepsilon_p = (\sigma/142)^{6.135}$. At a strain range of 0.008, the computed life is 310,000 cycles to failure—an increase in life of a factor nearly 8×. The corresponding increase in ductility is 4×, thus further linking the increase in life caused by compressive mean stress to an enhancement of the apparent ductility of the material as required by enforcing the cyclic stress-strain curve to be independent of mean stress. It is important to emphasize that the significance of this requirement is independent of any particular mean stress model.

From these examples we have gained valuable insight into the important reason for the large influence of mean stress on fatigue life. Tensile mean stress truncates the usable part of the stress-strain curve of a material, reducing its effective fracture strength and drastically reducing its ductility, so that, for a given stress (or strain) range, the life is reduced much more than commensurate with the loss of ductility associated with only the maximum stress reached in the cycle. Similarly, compressive mean stress effectively extends the usable portion of the stress-strain curve, increasing its fracture strength and significantly increasing the ductility.

Of course, there are other factors, such as would be pointed out by fracture mechanics-oriented analysts that higher surface stress intensity is associated with the higher maximum stress (see Chapter 9, "Crack Mechanics"). However, this argument has only limited applicability because in most mean stress applications, the study is made on highly polished smooth specimens, and the action of interest occurs long before an appreciable-sized crack is initiated.

Another factor of potential significance to the explanation of why mean stresses have such a large influence on fatigue life is the basic nature of the fatigue mechanism. As is emphasized in Chapter 10, "Mechanism of Fatigue," regions of cyclic plasticity within the crystallographic grains of metals is an unquestionable site for metal fatigue to occur. In the longer life regime, the plasticity is highly localized, and crystallographic slip may be constrained to only a few slip bands. One can imagine under high tensile mean stresses, these tiny regions undergoing extremely localized ratcheting that breaks atomic bonds without affording the opportunity for rebonding, localized cracking could occur much sooner than without the mean stress. There would be a virtually imperceptible overall ratcheting of the sample because so few sites would be contributing. For example, if 10 slip bands each contributed 100 atomic distances of ratcheting, the overall contribution would be 1000 atomic units of about $3.5 \times 10^{-10}$ m each. Total displacement is thus only $3.5 \times 10^{-7}$ m, or 14 μin. Such a small extension, over millions of cycles would likely go undetected using commercially available fatigue extensometers. Mechanical extensometers cannot be used during most high-cycle fatigue testing because the cycling frequency is usually too high for their slower dynamic response. Noncontacting extensometry techniques would have to be employed in conjunction with metallographic techniques. This would be an exceptionally fertile area for basic fatigue mechanism research.

### Further Generalization of the Model

We can further generalize the new comprehensive model to accommodate Gerber behavior and curves of other exponents by assuming that the fractional drop in stress intercept (at one cycle to failure) is $[1.0 - (k\sigma_m/\sigma'_f)^\alpha]$ instead of accepting that it is $[1.0 - (k\sigma_m/\sigma'_f)]$ according to the notation of Morrow. Furthermore, since we have observed that some materials display a different curvature of the Gerber type at different life levels, we have found that a more complete generalization can be achieved by choosing:

$$\alpha = Q(2N_f)^P \quad \text{(Eq 4.24)}$$

Thus, for given values of $Q$ and $P$, $\alpha$ can be less than, or greater than, unity, according to the value of $N_f$. When $\alpha < 1.0$, the Gerber-type curve is convex upward; when $\alpha > 1$, the curve is convex downward. The value of $Q$ is always positive, while $P$ can be positive or negative, so that $\alpha$ is always positive. The generalized model is shown in Fig. 4.27(a). Here it is seen that the characterization of the material is still according to the elastic and plastic lines of the original Manson-Coffin-Basquin diagram. To take tensile mean stress into account, only the horizontal life scale is changed from the original $2N_f$ to $2N_f[1.0 - (k\sigma_m/\sigma'_f)]^d$, where $\alpha = Q(2N_f)^P$. For current purposes, $k$ is taken as unity. Compressive mean stress is discussed in a following section. If the material constants $Q$ and $P$ are known, the diagram permits determination of mean stress for any set of operating parameters, as discussed later. Figure 4.27(b) shows that for the proper choice of $Q$ and $P$ the generalized Gerber-type relations can be forced to change progressively from convex to concave as fatigue life is changed. Some materials display such a characteristic, as discussed in the following section.

### Some Applications to Material Behavior

We have analyzed several materials according to the generalized theory for which extensive mean-stress data were available (Ref 4.17, 4.18). Our results are shown in Fig. 4.28. The smooth curves represent the calculations according to the theory, using the constants $Q$ and $P$ tabulated in the figure. The symbols are values derived from the smoothed experimental curves. The agreement is seen to be quite good. Of special note is the 9Ni-4Co-0.45C steel (Ref 4.17), shown in Fig. 4.28(d). Here, there is a distinct change from a concave downward curve at $10^2$ cycles of fatigue life, to concave upward at $10^5$ cycles of fatigue life. The generalized model was able to embrace realistic behavior over the entire life range for which data were available.

### Accounting for Compressive Mean Stress

In the discussion thus far we have, for simplicity, considered only tensile mean stress. When the mean stress is compressive, life is increased rather than decreased, for a given alternating stress range. To account for this behavior we must use a negative value for $\sigma_m$. However, we cannot use a negative number in the expression modifying the horizontal life scale when $\alpha$ is a noninteger, or is fractional. To overcome this problem, we resort to:

$$2N_f = 2N_{fm}[1.0 \pm (\sigma_m/\sigma'_f)^\alpha]^{1/b}$$
$$+ \text{ if } \sigma_m \leq 0$$
$$- \text{ if } \sigma_m \geq 0 \quad \text{(Eq 4.25)}$$

where $\alpha = Q(2N_f)^P$; $2N_f$ is life at a given stress range $\Delta\sigma$ with zero mean stress; $2N_{fm}$ is life at the same stress range $\Delta\sigma$ in the presence of a mean stress; $\sigma_m$ is applied mean stress; $\sigma'_f$ is the fatigue strength coefficient; $b$ is the fatigue strength exponent; and $Q$ and $P$ are material constants (for example, Fig. 4.28).

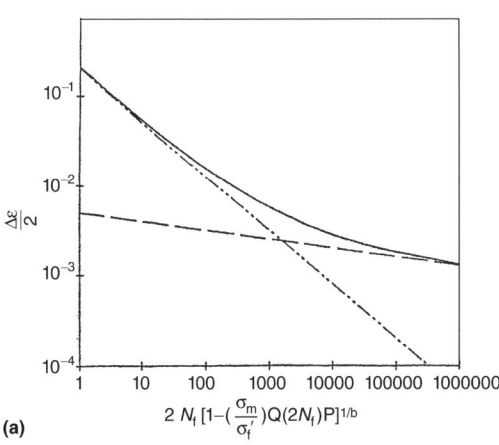
(a)

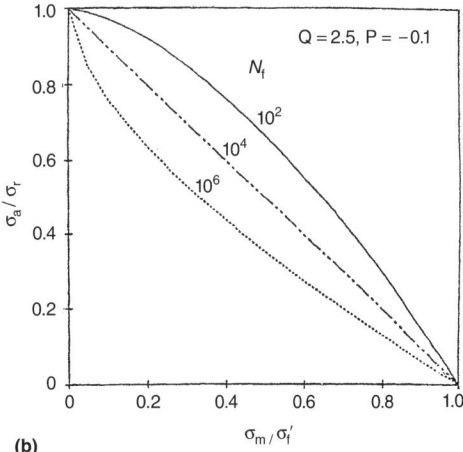
(b)

**Fig. 4.27** Framework for generalized model for the mean stress effects. (a) The model using the basic Manson-Coffin-Basquin characteristic of the material and changing only the life scale to account for mean stress. (b) A typical example of the application of the new model using $Q$ and $P$ constants that cause progressive changes in curvature from convex to concave as life is changed

**96 / Fatigue and Durability of Structural Materials**

In the parenthetical expression, which is raised to the exponent $\alpha$, the stress used is the absolute value of the mean stress. However, the algebraic sign in front of the parenthesis is positive when $\sigma_m$ is positive, negative when mean stress is negative. How this assigned behavior affects the life in the range of compressive mean stress is discussed subsequently in connection with Fig. 4.29.

### Implied Effects of Compressive Mean Stress

While well-documented experimental data on compressive mean stress are rather scarce, the aforementioned unifying formulation can indicate what we can expect. In Fig. 4.29, we have extended into the compressive mean range the data from Fig. 4.28(b) for 2024-T4 wrought aluminum. The progression is seen to proceed smoothly from tensile to compressive mean stresses. Figure 4.29(b) shows the cross plot of the data to demonstrate the powerfulness of mean stress in affecting fatigue life at various selected values of alternating stress ranges. Higher lives at a given alternating stress range as the mean stress becomes compressive. Few experiments are available to verify the predicted behavior, but we know that compressive mean stresses improve fatigue life, as illustrated in Fig. 4.8, and as evidenced by the fact that one of the most powerful ways of improving fatigue life is to introduce intentionally compressive mean stresses in the form of residual stresses. This subject is discussed further in Chapter 11, "Avoidance, Control, and Repair of Fatigue Damage."

### Characterizing a Material According to the Comprehensive Model

**General.** As shown in Fig. 4.27, straight lines result on log-log coordinates for both the elastic

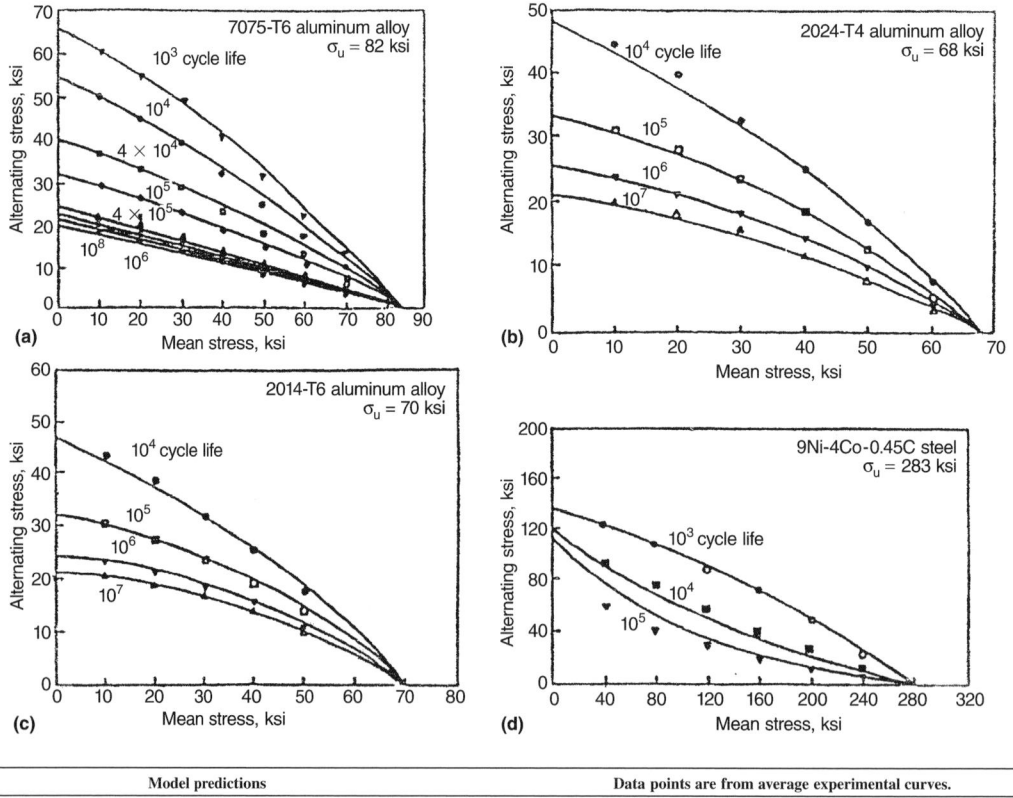

**Fig. 4.28** Comparisons between observed and calculated axial fatigue strengths

and plastic strain-range components when strain amplitude is plotted against:

$$2N_f\left[1 - \left(\frac{\sigma_m}{\sigma_f'}\right)^{Q(2N_f)^P}\right]^{1/b}$$

For most applications involving mean stress, the elastic strain is dominant in the life ranges of interest. While we illustrate procedures using the elastic line, it is evident how the procedure can be extended to the plastic line, and even the total strain-range curve.

Denoting the expression $Q(2N_f)^P$ by $\alpha$, the equation for the elastic line in terms of stress becomes:

$$\frac{\Delta\sigma}{2E} = \frac{\sigma_f'}{E}\left\{2N_f\left[1 - \left(\frac{\sigma_m}{\sigma_f'}\right)^{\alpha}\right]^{1/b}\right\}^b \quad \text{(Eq 4.26)}$$

This simplifies to:

$$\frac{\Delta\sigma}{2} = \sigma_f'\left[1 - \left(\frac{\sigma_m}{\sigma_f'}\right)^{\alpha}\right](2N_f)^b \quad \text{(Eq 4.27)}$$

Letting $\sigma_r$ be the stress amplitude for which the life is $N_f$ when the mean stress is zero:

$$\sigma_r = \sigma_f'(2N_f)^b \quad \text{(Eq 4.28)}$$

and denoting $\Delta\sigma/2$ by $\sigma_a$ we can express $\sigma_a$ as:

$$\sigma_a = \sigma_r\left(1 - \frac{\sigma_m}{\sigma_f'}\right)^{\alpha} \quad \text{(Eq 4.29)}$$

Solving for $\alpha$:

$$\alpha = \frac{\log\left(1 - \frac{\sigma_a}{\sigma_r}\right)}{\log\left(1 - \frac{\sigma_m}{\sigma_f'}\right)} \quad \text{(Eq 4.30)}$$

or

$$Q(2N_f)^P = \frac{\log(1 - (\sigma_a/\sigma_r))}{\log(1 - (\sigma_m/\sigma_f'))} \quad \text{(Eq 4.31)}$$

Taking the logarithm of both sides:

$$\log Q + P\log(2N_f) = \log\left[\frac{\log(1 - \sigma_a/\sigma_r)}{\log(\sigma_m/\sigma_f')}\right] = Z \quad \text{(Eq 4.32)}$$

This equation is in the form $y = mx + b$, so a straight line should result when $Z$ is plotted against $\log(2N_f)$. Because any experimental data point involves a known life for a known mean and alternating stress, a plot of $Z$ versus $\log(2N_f)$ should yield a straight line of slope $P$ and intercept $\log Q$, as shown in the sketches of Fig. 4.30.

**Applications.** We now consider two cases: when a material has already been characterized by another model or graphical representation, but we wish to recharacterize it by the comprehensive model, and when we wish to conduct an efficient test program on a new material that is to be characterized by the comprehensive model.

*Recharacterizing a Material.* Figure 4.31 shows the mean stress characteristics of the low-alloy steel 300M (copied from *Mil Handbook-5C*, Ref 4.19). As is typical in the *Mil Handbook* figures, two sets of scales are used, one with minimum and maximum stress coordinates and the other using mean stress and alternating stress as coordinates. The axes are, of course, related.

Figure 4.32 shows the stress amplitude-based elastic line for this material as deduced by determining the intercepts of the eight given life lines on the axis of zero mean stress. Normally, the basic uniaxial, completely reversed constants would be known for a material to be characterized for mean stress effects, but here we needed

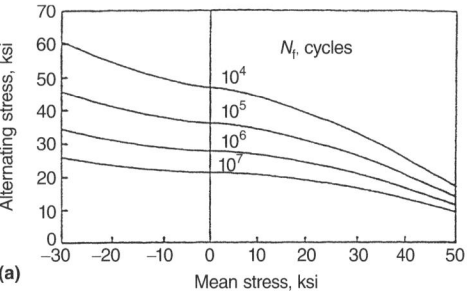

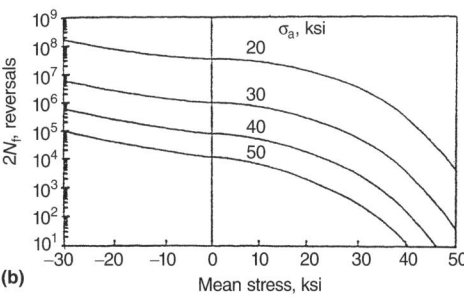

**Fig. 4.29** Extension of data for 2024-T4 to include compressive mean stresses (original data from Ref 4.17). (a) Fatigue life as a parameter. (b) Stress amplitude as a parameter

98 / Fatigue and Durability of Structural Materials

to deduce them from the provided fatigue data, from which it was determined:

$$\sigma'_f = 476 \text{ ksi}$$

$$b = -0.110$$

These are the constants used in the ensuing analysis.

The first step is to estimate mean and alternating stress values for a number of points on each line of constant life as shown in Fig. 4.31. Points are chosen at mean stresses of 0, 40, 80,

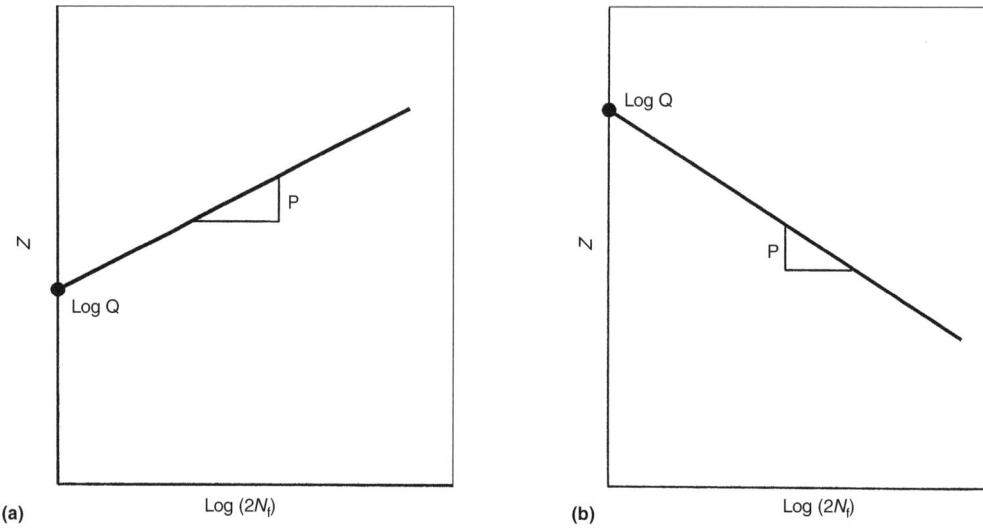

**Fig. 4.30** Sketch showing nature of Eq 4.32. (a) Positive slope $P$. (b) Negative slope $P$

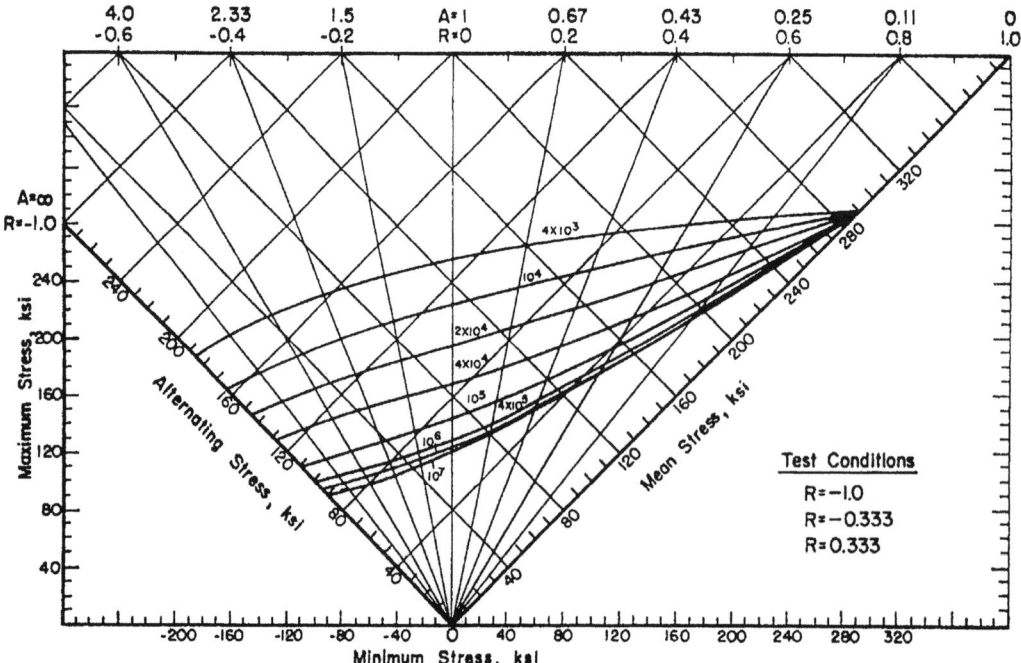

**Fig. 4.31** Typical constant-life fatigue diagram for low-alloy steel 300M at room temperature. Source: Ref 4.10

120, 160, and 200 ksi, and alternating stresses estimated for each life line wherever intersections occurred. For these points, plots are made of Z versus $\log(2N_f)$ according to Eq 4.32. This plot is shown in Fig. 4.33, determining the constants $Q = 4.83$ and $P = -1.39$. As is seen in Fig. 4.33, there is considerable scatter in the "data" points plotted from each life level. This scatter is due partly to the estimates required in reading the curves, and perhaps from the fact that the faired curves chosen by the handbook analyst did not always agree well with the material behavior. However, a replot of the calculation for the comprehensive model representation with $Q = 4.83$ and $P = -1.39$ coincided over the entire range with the curves shown in Fig. 4.31. An enlarged set of curves for $N = 4 \times 10^3$, $2 \times 10^4$, and $10^6$ is shown in Fig. 4.34. Here, the continuous curves represent the comprehensive model, while the "data" symbols represent the curves for these life levels as shown in Fig. 4.31. Again, what is of special interest is that the comprehensive model can represent both upward and downward curvature as required by the experimental data trends.

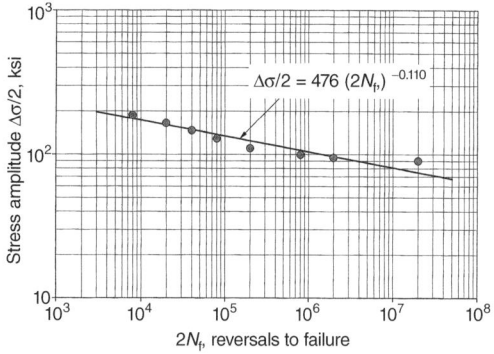

**Fig. 4.32** Deduced completely reversed stress amplitude vs. fatigue life relationship for 300M

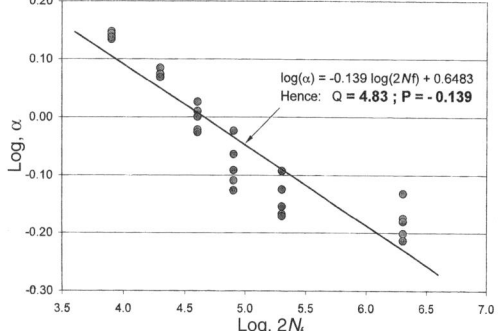

**Fig. 4.33** Determination of constants $Q$ and $P$ for 300M steel

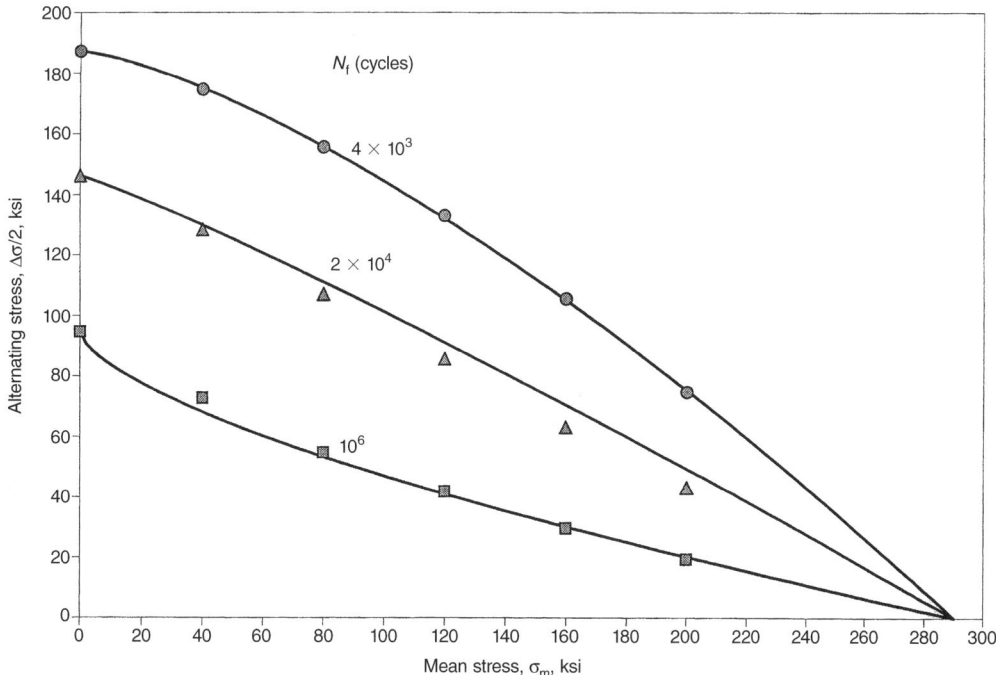

**Fig. 4.34** Comprehensive model for mean stress representation of fatigue results for 300M steel. $Q = 4.83$; $P = -0.139$

**Table 4.1  Six tests to determine constants Q and P in the comprehensive model**

| Test No. | Mean stress, MPa (ksi) | Alternating stress, MPa (ksi) | Target life cycles | Actual life cycles |
|---|---|---|---|---|
| 1 | 415 (60)  | 876 (127) | 10,000    | 15,937  |
| 2 | 415 (60)  | 683 (99)  | 100,000   | 71,792  |
| 3 | 415 (60)  | 531 (77)  | 1,000,000 | 297,636 |
| 4 | 830 (120) | 648 (94)  | 10,000    | 18,420  |
| 5 | 830 (120) | 503 (73)  | 100,000   | 67,132  |
| 6 | 830 (120) | 393 (57)  | 1,000,000 | 227,312 |

*Experimental Programs to Establish the Constants for a New Material.* We now consider one way an experimentalist could proceed to run a program to obtain the comprehensive model constants $P$ and $Q$ for a new material. It is assumed that the completely reversed stress characterization is already available so $\sigma'_f$ and $b$ are known. For illustration, we pretend that the material to be characterized is the 300M material already discussed and assume that the true behavior of the material is as shown in Fig. 4.31, or as briefed in Fig. 4.34 by the three continuous curves. But we do *not* know the values of $Q$ and $P$, which are to be determined by analysis and experiment.

The questions to be decided relate to the number of tests to be conducted and the test parameters. Strictly, only two tests are required to determine the straight line that defines $Q$ and $P$, provided there is no scatter. However, to allow realistically for some redundancy and scatter, we shall assume that six tests will be allocated. For practical reasons, the mean and alternating stresses are selected as shown in Table 4.1. Three tests are to be conducted at each of the two mean stresses (60 and 120 ksi), which correspond to about 20 and 40% of the ultimate tensile strength. These tests are to target lives in the vicinity of $10^4$, $10^5$, and $10^6$ cycles to failure. The alternating stresses are estimated using a simple model such as the modified Goodman. These parameters are not critical; they were approximated so that the experiments will not run too long.

Shown in the last column are the deduced lives if the material truly conformed to the model established for this material in Fig. 4.31 and 4.34.

Let us now assume the tests are conducted and the material indeed yields the lives shown in the last column. Using the analysis already established, the results would be the perfect straight line fit of log $\alpha$ versus log($2N_f$) as seen in Fig. 4.35(a) (since we have basically forced this development by assuming conformity with the determined model constants). The constants determined naturally would be $Q = 4.83$ and $P = -1.39$. The resultant analytic mean stress diagram is given by the curves in Fig. 4.35(b) with the "data" points representing the originally selected values of stress from Fig. 4.31.

However, it is well known there will be scatter in the data, so it is reasonable to attempt to determine how seriously such scatter would distort the results. To start the analysis, scatter in the fatigue life data is applied via a computerized statistical program that randomly assigns a ± deviation of up to 1.5× to the mean life. Then, a new table of experimental results is hypothesized and similarly analyzed as shown in Fig. 4.36, where each data point falls in the analysis

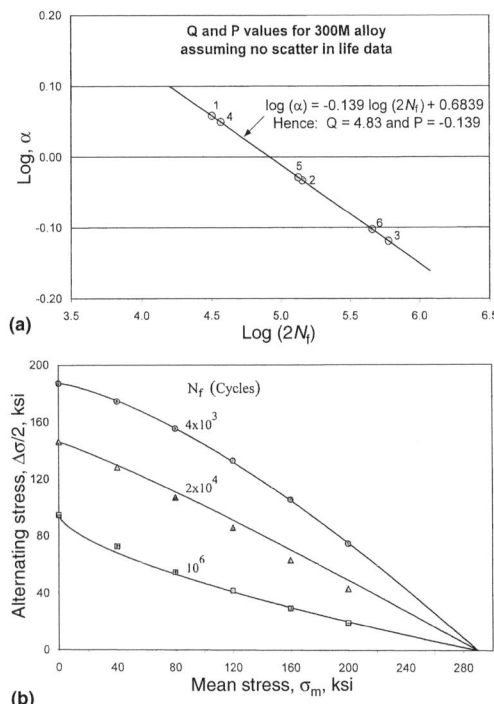

**Fig. 4.35** Results of applying the comprehensive model for mean stress effects for fatigue behavior of 300M steel assuming zero scatter: (a) Log-log plot of $\alpha$ vs. $2N_f$. The number by each data point is the test number in Table 4.1. (b) Mean stress vs. alternating stress diagram. $Q = 4.83$; $P = -0.139$

is identified by the number adjacent to the data point. It is seen in Fig. 4.36 that there is relatively little effect on the log ($\alpha$) versus log ($2N_f$) line and that the life predictions by the comprehensive model are very close to the hypothesized behavior.

Similar analysis for $\pm 2\times$ scatter in the life data is shown in Fig. 4.37. Of course, these calculations are just hypothetical, but they suggest the concept is applicable and practical and should be pursued in an experimental program to test the aforementioned results.

### Summary Remarks on Comprehensive Mean Stress Modeling

Several important and useful new aspects of mean stress modeling have been brought out in this chapter. It is appropriate to summarize these before moving on to other factors influencing fatigue life of structures and materials.

A comprehensive model for mean stress effects on fatigue life has been documented. Its generality allows it to encompass classic mean stress models as well as contemporary models associated with use of the basic Manson-Coffin-Basquin equation for representing fatigue behavior for completely reversed straining. Principal features include:

- Model consistency with a unique cyclic stress-strain curve that is independent of the sign and magnitude of the applied mean stress
- The model provides insight into the large influence of mean stress on fatigue life because it connects the mean stress to cyclic plasticity through the cyclic stress-strain curve. Large tensile mean stress corresponds to a large reduction in usable ductility and hence a lowered capacity for cyclic plastic straining. Correspondingly, a compressive mean stress enhances ductility and provides a higher capacity to absorb cyclic plasticity.
- The fatigue failure curve for an applied mean stress can be predicted directly from the zero

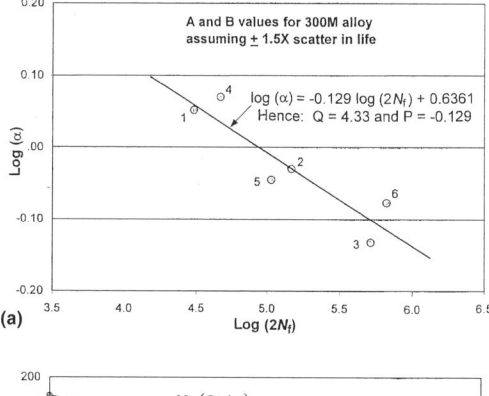

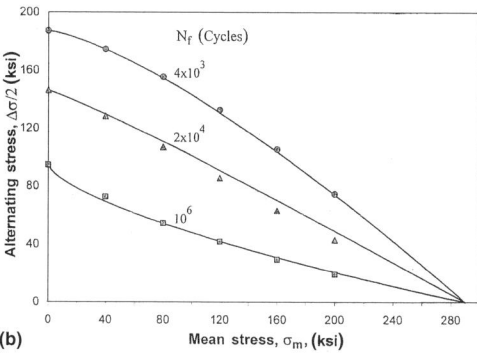

**Fig. 4.36** Results of applying the comprehensive model for mean stress effects for fatigue behavior of 300M steel assuming $\pm 1.5\times$ scatter in life. (a) Log-log plot of $\alpha$ versus $2N_f$. The number by each data point is the test number in Table 4.1. (b) Mean stress vs. alternating stress diagram. $Q = 4.33$ and $P = -0.129$

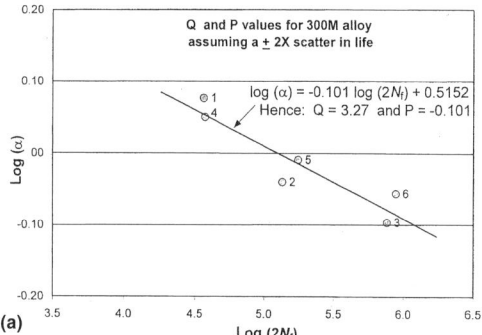

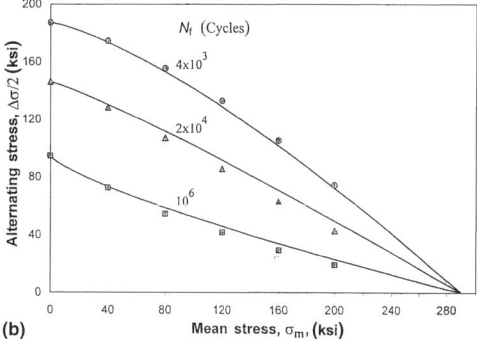

**Fig. 4.37** Results of applying the comprehensive model for mean stress effects for fatigue behavior of 300M steel assuming $\pm 2\times$ scatter in life. (a) Log-log plot of $\alpha$ versus $2N_f$. The number by each data point is the test number in Table 4.1. (b) Mean stress vs. alternating stress diagram. $Q = 3.27$; $P = -0.101$

mean stress curve by shifting the curve parallel to the life axis by an amount directly relatable to the magnitude and sign of the mean stress. If the mean stress relaxes, the amount of the shift in life also changes and could be computationally followed throughout the life if necessary.

- Importantly, the comprehensive mean stress model serves as a guide for conducting a testing program to evaluate the constants in the model. If the completely reversed fatigue constants (intercept $B$ or $\sigma'_f$ and slope $b$) are already known, the only remaining constants to be determined are $k$ in Eq 4.15 and $Q$ and $P$ in Eq 4.24. For this purpose, the only restriction is that the loading condition, which includes mean stress, must be maintained constant throughout the duration.

The new model can address directly the life-reducing effects of mean stress, even when appreciable cyclic plasticity is involved. Cycles of this nature certainly can occur during cumulative fatigue loading conditions wherein mean stresses can be reinitiated repeatedly by the history of loading. Such mean stress conditions would not normally be captured by, nor used by, other contemporary models. Those models, when applied to cycles with appreciable plasticity, would tacitly assume that any initial mean stress would simply relax to zero and be of no consequence to life. This is because the calibration of these models (e.g., Smith-Watson-Topper) is accomplished by conducting tests with an initial mean stress that may tend to relax to zero after relatively few cycles. At the half-life, wherein the stress and strain parameters are evaluated, the mean stress could be anywhere between its initial value and zero and that parameter correlated with the observed life. The comprehensive model can be used to assess the severity of intermittently repeated mean stress cycles due to occasional large strain cycles within a spectrum. Other currently constituted models cannot do so without modification.

## REFERENCES

4.1 R.W. Landgraf, F.D. Richards, and N.P. LaPointe, Fatigue Life Prediction for a Notched Member under Complex Load Histories, *Fatigue under Complex Loading: Analyses and Experiments,* Society of Automotive Engineers, Warrendale, PA, 1977, p 95–106

4.2 R.W. Landgraf and R.C. Francis, Material and Processing Effects on Fatigue Performance of Leaf Springs, *Trans. Society of Automotive Engineers,* Vol 88, Section 2, Society of Automotive Engineers, Warrendale, PA, 1979, p 1485–1494

4.3 J.B. Conway and L.H. Sjodahl, *Analysis and Representation of Fatigue Data,* ASM International, 1991, p 154

4.4 S.J. Stadnick and J. Morrow, "Techniques for Smooth Specimen Simulation of the Fatigue Behavior of Notched Members," *Testing for Prediction of Material Performance in Structures and Components,* ASTM STP 515, American Society for Testing and Materials, 1972, p 229–252

4.5 A. Wöhler, Versuche über die Festigkeit der Eisenbahnwagen-Achsen, *Z. Bauwesen (J. Engineering Construction),* 1860 (in German)

4.6 G.R. Halford and A.J. Nachtigall, "The Strainrange Partitioning Behavior of an Advanced Gas Turbine Disk Alloy, AF2-1DA," NASA TM-79179, 1979. See also *J. Aircr.,* Vol 17 (No. 8), 1980, p 598–604

4.7 K.N. Smith, P. Watson, and T.H. Topper, A Stress-Strain Function for the Fatigue of Metals, *J. Mater.,* Vol 5 (No. 4a), American Society for Testing and Materials, 1970, p 767–778

4.8 W.Z. Gerber, Investigation of the Allowable Stresses in Iron Construction, *Bayer. Arch. Ing. Ver. (Bavarian Archives of Engineering Associates),* Vol 6, 1874, p 101 (in German)

4.9 J. Goodman, *Mechanics Applied to Engineering,* Longman, Green & Company, London, 1899. See also Vol 1, 9th ed., 1930.

4.10 D.C. Maxwell and T. Nicholas, "A Rapid Method for Generation of a Haigh Diagram for High Cycle Fatigue," *Fatigue and Fracture Mechanics: 29th Volume,* ASTM STP 1332, T.L. Panontin and S.D. Sheppard, Ed., American Society for Testing and Materials, 1999, p 626–641

4.11 G.P. Sendeckyj, History of Constant Life Diagrams, *High Cycle Fatigue of Structural Materials,* T.S. Srivatsan and W.O. Soboyejp, Ed., The Metallurgical Society, 1998, p 95–107

4.12 J. Morrow, Fatigue Properties of Metals, *Fatigue Design Handbook,* Sec. 3.2, SAE Advances in Engineering, Vol 4, 1968, p 21–29 (draft submitted April, 1964)

4.13 H.J. Grover, S.A. Gordon, and L.R. Jackson, *Fatigue of Metals and Structures,* U.S. Government Printing Office, Washington, DC, revised June 1, 1960, Navweps 00-25-534

4.14 S.S. Manson, Fatigue: A Complex Subject—Some Simple Approximations, *Exp. Mech.,* 1965, p 193–226

4.15 J. Walcher, D. Gray, and S.S. Manson, Aspects of Cumulative Fatigue Damage Analysis for Cold End Rotating Structures, preprint 79-1190, *AIAA/SAE/ASME 15th Joint Propulsion Conference,* (Las Vegas, NV), June 18–20, 1979

4.16 S.S. Manson and G.R. Halford, Correction to: Practical Implementation of the Double Linear Damage Rule and Damage Curve Approach for Treating Cumulative Fatigue Damage, International Journal of Fracture, Vol 17 (No. 2), 1981, p 169–192. *Int. J. Fract.,* Vol 17 (No. 4), 1981, p R35–R42. See also NASA TM-81517, 1980.

4.17 R.C. Juvinall, *Engineering Consideration of Stress, Strain, and Strength,* McGraw-Hill, New York, 1967, p 272–274

4.18 D.F. Bulloch, T.W. Eichenberger, and J.T. Guthrie, "Evaluation of the Mechanical Properties of 9Ni-4Co Steel Forgings," AFML-TR-68-57, Wright Patterson Air Force Base, OH, March 1968. Originally ITAR restricted; restriction removed March 1972.

4.19 *Military Standardization Handbook, Metallic Materials and Elements for Aerospace Vehicle Structures,* Vol 1, MIL-HDBK-5C, United States Department of Defense, Sept 15, 1976

# CHAPTER 5

# Multiaxial Fatigue

## Introduction

Few subjects in fatigue analysis have attracted as much attention in recent years as the effect of multiaxiality. Frequently one or more national or international conferences are held annually that deal with the effect of multiaxiality. It is an attractive subject to engineers because of its presence in so many designs. Analysts enjoy the elegance of the mathematics that lends itself to ingenious manipulation, and experimentalists, who find challenging ways to verify hypotheses, have provided a wealth of literature on which to base methodologies. In actual applications, however, engineers tend to use simple concepts in their calculations, compensating for inaccuracies by overdesign. Despite the extensive literature that exists on this subject, there is still no consensus on the "best way" to approach design. What is needed is a methodology that recognizes the complexity but uses realistic simplifications to incorporate its essence without overburdening the designer to have to perform either complex analyses or specialized experiments from which to develop input data. Rarely is the designer in a position to conduct experiments while the design is in progress.

In this chapter we review some of the common theories and, more importantly, large amounts of data that have been generated to check the validity of these theories. Our goal is to identify an approach that captures the essence of valid theories, but yet is simple enough to be incorporated in practical design. So far as is possible, we seek an approach that uses only uniaxial fatigue properties.

## Some Important Factors That Govern Fatigue Behavior

Before discussing several of the approaches that have been studied in recent years to account for multiaxiality, it is useful to clarify a few concepts and practices used by various investigators to interpret their data. We can thus avoid disturbing the continuity of discussion in later analysis. Some of the factors discussed subsequently are mentioned elsewhere in this book (primarily in the Appendix, "Selected Background Information"), but it is useful to review them briefly here in the specific context that they arise in the subject at hand.

### Crystalline Structure

Although the investigators in many of the studies described were well aware of the crystalline structure of their materials, this factor was usually not considered in interpreting their results. Specifically, as we describe later, body-centered cubic (bcc) metals seem to follow different fatigue criteria than do face-centered cubic (fcc) metals. There may in fact be a difference in behavior for hexagonal close-packed (hcp) metals, but no data are available at this time to study the possibility.

Whether the difference in behavior between bcc and fcc materials is due simply to atomic structure, or to other factors implied by the atomic packing, is not clear at this time. In the Appendix, we discuss the differences of these two structures. The fcc packing involves four atoms per unit cell, a half atom at the center of each face, and one-eighth atom at each corner, while the bcc structure has two atoms in each unit cell, with one central atom and one-eighth of an atom at each corner. Another significant difference exists between the number of slip-systems available in each packing arrangement; fcc has only 12, whereas, bcc possesses 48. If the number of slip systems were the only feature that governed multiaxial behavior, it would be reasonable to expect that bcc metals would be less sensitive to applied multiaxiality. A greater number of potential slip systems should imply that more systems are available, regardless of the orientations of multiple loading axes. However, plastic flow under multiaxial loading may differ from plastic flow under uniaxial loading in each of these two systems due to other factors not considered here. In any case, we have found that attempting to collapse data for the two structures using the same theory is not as successful as treating them separately. Later in the chapter we provide separate relationships, each exclusively applicable to one or the other crystalline structure.

## Testing Methods

It is difficult to test a material under multiaxiality to precisely known stresses or strains. Over the years various types of specimens and loading methods have been developed which have been proven to be useful. Some of these specimens are discussed here. Among these are:

**Simple Cruciform Specimen.** The simple cruciform type of specimen of Fig. 5.1 might appear to be ideally suited to applying biaxial loading of arbitrary magnitude and direction. However, in practice there are complications that have caused most experimentalists to choose other methods. The test area of interest is, of course, at the center of the specimen, marked $A$. Typically, this section is machined to be thinner than the remainder of the specimen, thus amplifying the stress and inducing failures within the test section area. However, a less-than-ideal specimen design can result in failure at the stress-strain concentration points $B$ in the fillet. Furthermore, the stress distribution in the center of the specimen may be nonuniform, especially when large plastic deformations are involved. Thus, few of the recent testing programs have used this type of specimen.

**Refined Cruciform Specimen.** A refined cruciform specimen such as shown in Fig. 5.2 (Ref 5.1) is much better suited for biaxial fatigue testing. The slots surrounding the four loaded sides are an attempt to reduce the constraint imposed on the central test section caused by the bidirectional loading and the associated clamping forces. Considerable research has gone into the refinement of this style of biaxial test specimen (see for example, Ref 5.2). The greater the complexity of the slot arrangement, the more costly is the specimen and the less likely it could be used for multiple fatigue tests.

**Circular Solid and Tubular Specimens.** The most common type of biaxial fatigue specimen in use is in the form of a solid or thin-walled tubular shaft as shown in Fig. 5.3. Twisting of the shaft produces one of the simplest forms of biaxiality, *simple shear,* wherein the maximum

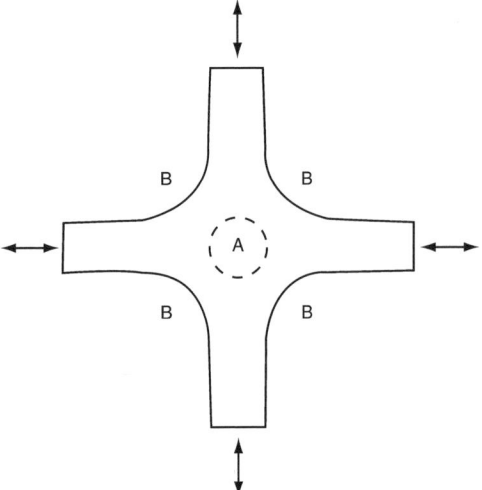

**Fig. 5.1** Simple cruciform specimen

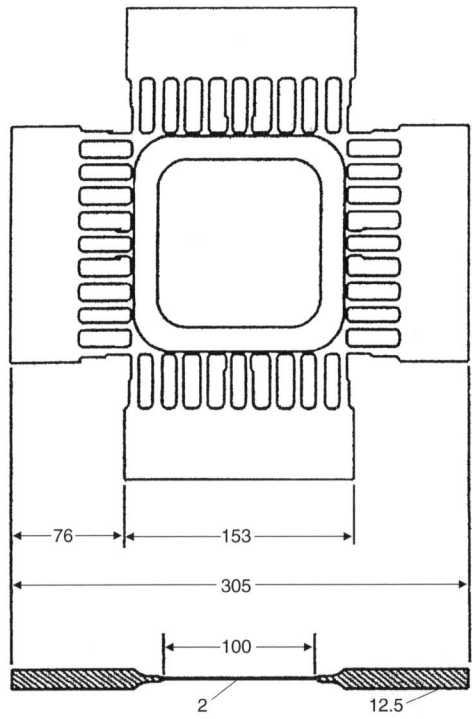

**Fig. 5.2** Refined cruciform specimen (dimensions in mm). Source: Ref 5.1

principal stress is equal in magnitude but opposite in sign to the minimum principal stress. Of course, the shaft could also be subjected to an axial force along its length to create more complex states of biaxial stress and strain. Furthermore, if the shaft is solid it may have a notch that can provide a stress concentration within a biaxial field. A solid shaft will, however, have a linear strain gradient and, hence, under nominally elastic loading conditions, a linear stress gradient under torsion. As plasticity occurs at higher forces, the stress gradient becomes nonlinear to the same extent as the stress-strain curve is nonlinear. This makes it difficult to measure, directly, the shear stress with any good degree of accuracy. However, a hollow shaft of relatively thin wall can provide a shell of reasonably uniform stress both in torsion and in tension loading. A tube also provides a hollow interior that can be pressurized, thus adding to loading flexibility and permitting a truly controlled *triaxial* loading condition. With fluctuating internal (and external) pressurization, axial loading, rotational torque, and even bending moments, a rather large range of fatigue multiaxiality loading ratios can be achieved. Any one of several loading machines can be immersed in a hydrostatic compression chamber so that whatever loading is imposed on the immersed specimen will also have a hydrostatic component. We refer later to a valuable study that permitted extension of the variables to assist in constructing a more generalized theory.

**Plates** can be loaded in a variety of ways to produce biaxial stress-strain states (Fig. 5.4). A plate can be loaded at its corners, producing a saddlelike deformation, known as anticlastic bending (see for example, Ref 5.4). By shaping the plate as diamonds with various aspect ratios, the biaxiality ratio may be varied. Or, circular plates may be bolted to a pressure chamber and fluid pressure may be alternated on one or both sides. By making the openings of the pressure chamber elliptical with various aspect ratios, a range of stress ratios again can be achieved. Such a scheme is more valid for biaxial stress ratios than for strain ratios, unless a special system is provided to measure strains and controlling them through servomechanisms. These methods were once very popular at The Pennsylvania State University (Ref 5.5).

**Ball and Roller Bearings.** Rolling contact bearings can be forced against each other or against a flat plate to produce a complex triaxial subsurface stress state, e.g., Fig. 5.5. Again, we shall use test data from specimens of this type to extend the range of the variables and to generalize theory.

## Stress-Based versus Strain-Based Theories

Some theories express the operating parameters in terms of stress, others in terms of strains. While strain is the variable most commonly used in modern fatigue analysis, primarily because of its ease of measurement, stresses are also very important. We have certainly favored the development of strain as the operating parameter, yet the approach we have formulated for treating multiaxiality involves stress terms. It should be emphasized, however, that whichever way an approach is formulated, whether in stress or strain, the corresponding treatment in terms of the other parameter is implicit through the stress-strain relations. This fact is clear today but has not always been so because today we deal in more carefully defined stress-strain curves. In the early days of fatigue theory development it was not, for example, uncommon to represent a relatively flat stress-strain $OAB$, as shown in Fig. 5.6, by two segments, one of which was a horizontal line $A'B'$. Thus, many strain levels could be associated with the yield stress, $\sigma_y$. It was, in fact, because of this dilemma that Manson (Ref 5.6) resorted to a fatigue life representation in terms of plastic strains that became one of the origins of the Manson-Coffin equation for low-cycle fatigue. Current usage, however, favors the representation of the actual stress-strain relation ($OAB$) by the Ramberg-Osgood stress-strain relation. This relation is consistent with the Manson-Coffin-Basquin equation when the life, $N_f$, is eliminated. Thus, in discussing several of the

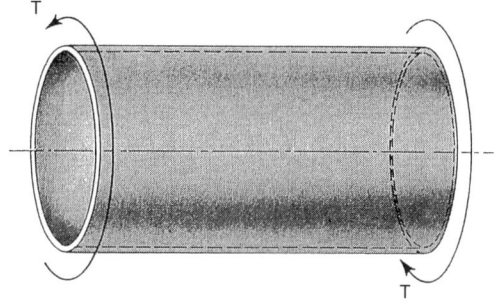

**Fig. 5.3** Thin-walled tubular specimen. Source: Ref 5.3

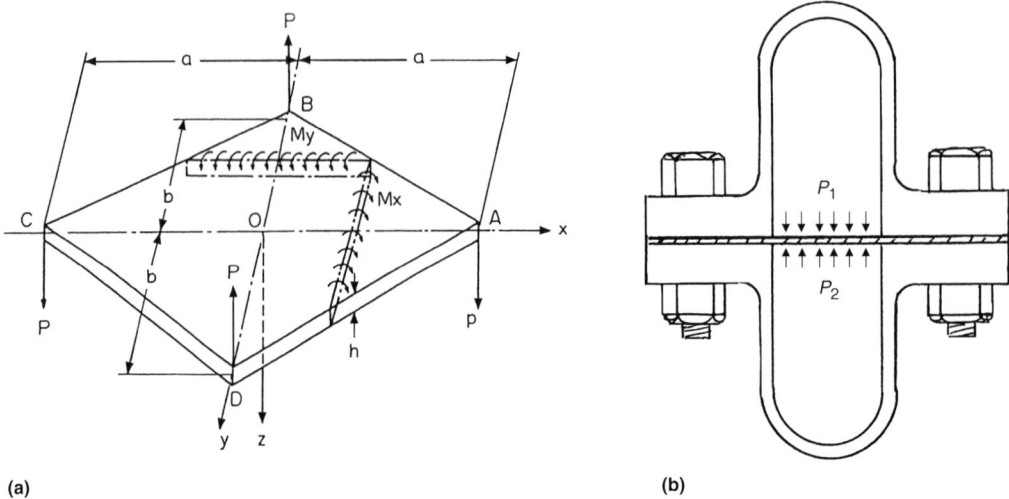

**Fig. 5.4** Plate specimen. (a) Anticlastic bending. $M_y$ and $M_x$ are bending moments about the $y$ and $x$ axes, respectively. Source: Ref. 5.4. (b) Differential pressure loading Source: Ref. 5.5

currently used theories, we do not emphasize whether they are stress- or strain-based, assuming that whatever form used by the originators for convenience or simplicity, the implementation could be accomplished by the use of the other parameter as well.

## The Stress-Strain Relationships

For convenient reference, we repeat here the stress-strain relations frequently encountered in treating problems in the plastic regime:

$$\varepsilon_{x,t} = \varepsilon_{x,e} + \varepsilon_{x,p} = \sigma/E + (\sigma/K)^{1/n} \quad \text{(Ramberg-Osgood)} \quad \text{(Eq 2.1)}$$

$$\varepsilon_x = [\sigma_x - \mu(\sigma_y + \sigma_z)]/E + (\varepsilon_{ep}/\sigma_{eq})[\sigma_x - (1/2)(\sigma_y + \sigma_z)] \quad \text{(Eq 5.1)}$$

where

$$\varepsilon_{eq,p} = \frac{\sqrt{2}}{3}\sqrt{(\varepsilon_{x,p} - \varepsilon_{y,p})^2 + (\varepsilon_{y,p} - \varepsilon_{z,p})^2 + (\varepsilon_{z,p} - \varepsilon_{x,p})^2} \quad \text{(Eq 5.2)}$$

and

$$\sigma_{eq} = \frac{1}{\sqrt{2}}\sqrt{(\sigma_x - \sigma_y)^2 + (\sigma_y - \sigma_z)^2 + (\sigma_z - \sigma_x)^2} \quad \text{(Eq 5.3)}$$

where $\sigma_x$, $\sigma_y$, $\sigma_z$ are the principal stresses. Similar expressions for $\varepsilon_y$ and $\varepsilon_z$ can be readily obtained in terms of the appropriate stresses by systematic rotation of the indices $x$, $y$, and $z$.

The relation between $\sigma_{eq}$ and $\varepsilon_{eq,p}$ is that of the uniaxial stress-strain curve, so that if the equivalent stress $\sigma_{eq}$ is known, the equivalent plastic strain can be determined from the power law Ramberg-Osgood relation, Eq 2.1:

$$\varepsilon_{eq,p} = (\sigma_{eq}/K)^{1/n} \quad \text{(Eq 5.4)}$$

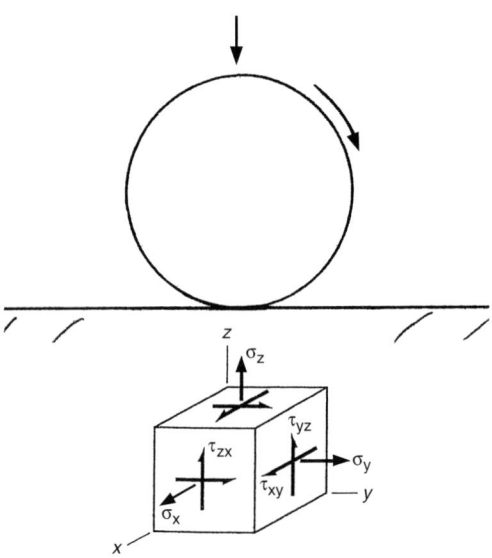

**Fig. 5.5** Ball or roller bearing on flat surface showing general 3-D state of stress below the surface

Thus, it is readily apparent that if the three principal stress components are known, the calculation of all strains is possible. The problem is much more complicated, however, if the strains $\varepsilon_x$, $\varepsilon_y$, and $\varepsilon_z$ are specified and it is necessary to calculate the stresses. We cannot start with Eq 5.1 because we do not yet know how much of each strain component is plastic; therefore, a trial-and-error procedure would normally be required. However, such a procedure is not simple because three unknowns must be operated upon simultaneously. A much simpler procedure was presented by Manson in Ref 5.7 and for convenience this method is outlined here.

Manson and Mendelson (Ref 5.8) have shown that an equivalent total strain $\varepsilon_{eq,t}$ can be represented by using Eq 5.2 but using total strains for the three coordinated directions, i.e.:

$$\varepsilon_{eq,t} = \frac{\sqrt{2}}{3}\sqrt{(\varepsilon_x - \varepsilon_y)^2 + (\varepsilon_y - \varepsilon_z)^2 + (\varepsilon_z - \varepsilon_x)^2} \quad \text{(Eq 5.5)}$$

Once the equivalent total strain has been computed, the equivalent plastic strain can be determined from the following readily derivable (see Ref 5.8) equation:

$$\varepsilon_{eq,t} = \varepsilon_{eq,p} + \frac{2(1 + \mu)}{3}\frac{\sigma_{eq}}{E} \quad \text{(Eq 5.6)}$$

where $\sigma_{eq}$ is the equivalent stress corresponding to the total-strain state. Thus, formulating the equivalent total strain from the known three components of total strain gives us an additional relation between $\varepsilon_{eq,p}$ and $\sigma_{eq}$, that together with the stress-strain relation already expressed in Eq 5.4, permits the calculation of both $\varepsilon_{eq,p}$ and $\sigma_{eq}$. Thus, substituting Eq 5.4 into Eq 5.6 produces the result:

$$\varepsilon_{eq,t} = (\sigma_{eq}/K)^{1/n} + (2/3)(1 + \mu)(\sigma_{eq}/E) \quad \text{(Eq 5.7)}$$

Equation 5.7 is a transcendental equation in $\sigma_{eq}$, similar to the general class of such equations discussed in the Appendix, "Selected Background Information," namely, double-term, power-law equations of the form $Y = AX^d + BX^\beta$. Alternatively, Eq 5.7 can be solved simply by trial and error. In the current situation, the latter solution is the quickest.

Once $\sigma_{eq}$ is determined, $\varepsilon_{eq,p}$ is calculated from Eq 5.4. Then the values of $\sigma_{eq}$ and $\varepsilon_{eq,p}$ are substituted into Eq 5.1 for each of the three components, leading to three simultaneous linear equations involving $\varepsilon_x$, $\varepsilon_y$, and $\varepsilon_z$ that can readily be solved. We illustrate the procedure for a specific numerical example in the next section.

## A Numerical Example for Determining Stresses When Strains Are Specified

Consider a material for which we have the following tensile properties. In the elastic regime, $E = 3 \times 10^7$ psi and $\mu = 0.3$, and in the plastic range, $\varepsilon_{eq,p} = (\sigma_{eq}/K)^{1/n}$, where $K = 2 \times 10^5$ psi and $1/n = 10$. Note that the plastic Poisson's ratio is 0.5 to reflect constancy of volume. Now, assume the three principal strains are in the $x$, $y$, and $z$ directions and are given by 0.007, 0.003, and $-0.006$. Thus, $\varepsilon_{eq,t}$ can be determined from Eq 5.5. Substituting into Eq 5.7:

$$\varepsilon_{e,t} = (\sigma_{eq}/2 \times 10^5)^{10} + (2/3)(1 + 0.3)(\sigma_{eq}/3 \times 10^7) = 0.00769$$

Rewriting this equation:

$$\sigma_{eq} = 2.66 \times 10^5 - 3.38 \times 10^{-46}(\sigma_{eq})^{10}$$

Solving by trial and error, $\sigma_{eq} = 1.16 \times 10^5$ psi. From Eq 5.4, $\varepsilon_{eq,p} = 0.00433$, and from Eq 5.7, $\varepsilon_{eq,e} = 0.00336$.

From here, we can solve the three simultaneous equations (one of which is represented by Eq 5.1) relating the known total strain components to the unknown stress components. For example, $\sigma_x = 1.57 \times 10^5$ psi, $\sigma_y = 1.17 \times 10^5$ psi, and $\sigma_z = 0.262 \times 10^5$ psi. The reverse problem of computing the three component strains from three given stresses is straightforward. In fact, the aforementioned numerical example provides a source of calibration for such computations.

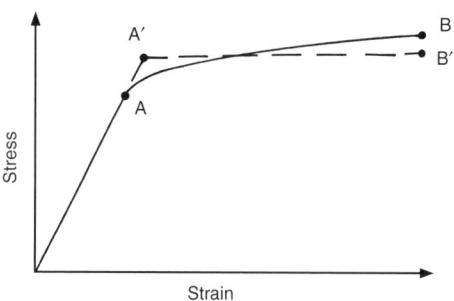

**Fig. 5.6** Flat-top stress-strain curve

## Critical Planes

To understand the importance of multiaxiality in fatigue, we need to recognize the two stages of fatigue: crack initiation and crack propagation. It is best to start with a consideration of uniaxial loading and then progress to biaxiality and then triaxiality.

The principles can best be described by starting with the uniaxial case in Fig. 5.7(a). If the applied axial stress is $\sigma_y$, it is readily proved by consideration of equilibrium that on a 45° plane the shear stress is $\sigma_y/2$, and the normal stress is also $\sigma_y/2$. With cycling, a shear crack will eventually develop along $AB$, as discussed in Chapter 10, "Mechanism of Fatigue." Once it starts, propagation will generally proceed along $BC$ because the stress normal to this plane is $\sigma_y$, whereas the stress normal to the plane $AB$ is only half this value. Thus, it is not uncommon in uniaxial fatigue to observe crack initiation in the 45° direction, but propagation to continue normal to the applied force. Thus, the crack may follow the path $ABC$.

Next we consider the biaxial case, shown in Fig. 5.7(b). The alternating shear stress is $\frac{1}{2}(\sigma_y - \sigma_x)$, and the stress normal to the 45° plane is $\frac{1}{2}(\sigma_y + \sigma_x)$. It is assumed that $\sigma_y$ is the maximum principal stress. If the crack starts in the shear plane $AB$, it might progress (as it did in Fig. 5.7a) along $BC$ once it gets started, because presumably the stress normal to this plane is $\sigma_y$, which is greater than $\frac{1}{2}(\sigma_y + \sigma_x)$. Depending on the geometry of the crack, however, crack propagation may develop in the plane normal to $BC'$, which is the plane normal to the plane of maximum shear. Thus, there are two parameters that govern cracking in biaxial fatigue, the maximum shear and normal stresses. There are also at least three planes involved in the initiation and propagation processes.

Finally, let us consider the three-dimensional case where all three principal stresses, $\sigma_1$, $\sigma_2$, and $\sigma_3$ exist. Treating this case in detail, bringing in all important shear planes and planes along which crack growth is likely, could make the analysis complicated and cumbersome. Fortunately, the history of stress analysis has focused on a methodology that can greatly simplify the treatment, while retaining enough reality to be useful. This is the concept of replacing the three-dimensional stress system by an equivalent system involving the octahedral planes. Other aspects of the octahedral plane system are discussed in the Appendix, and an excellent discussion can be found in Ref 5.9. Here, we review only those details useful for our further analyses.

The octahedral planes are planes that make equal angles with the three principal planes. As seen in Fig. 5.8, the general system of three principal stresses is replaced by the octahedron formed by eight octahedral planes that make

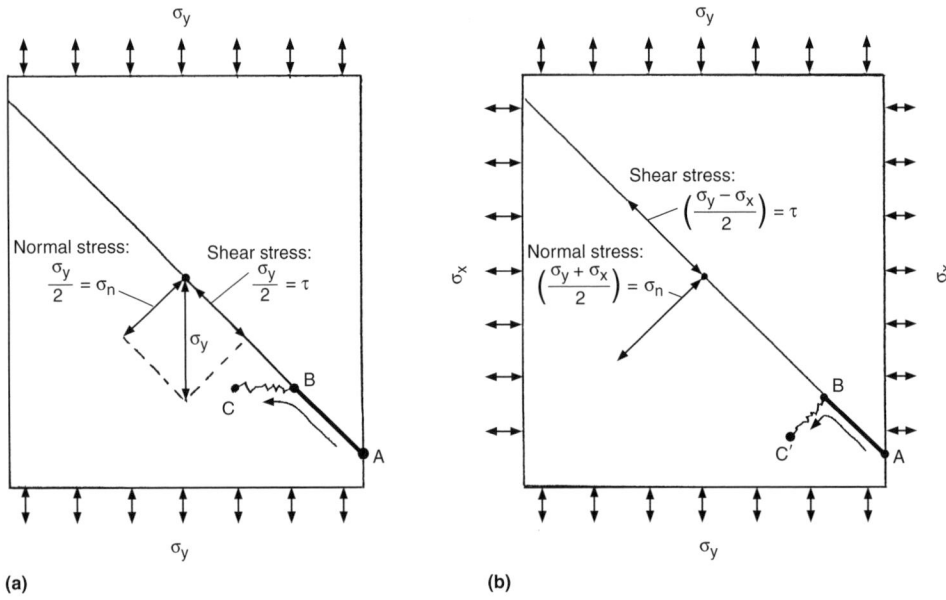

**Fig. 5.7** Crack initiation and propagation under different fatigue loading. (a) Uniaxial. (b) Equibiaxial

equal angles with the three principal directions. The four planes above the $\sigma_2 - \sigma_3$ plane, and the four below comprise the elemental planes of the octahedron. The interesting fact about this equivalent system is that the shear and normal stresses on all of the eight planes are the same. Thus, by rigorous stress analysis it can be shown that:

$$\sigma_{oct} = \frac{\sigma_1 + \sigma_2 + \sigma_3}{3} \quad \text{(Eq 5.8)}$$

$$\tau_{oct} = \frac{1}{3}\sqrt{(\sigma_1 - \sigma_2)^2 + (\sigma_2 - \varepsilon_3)^2 + (\sigma_3 - \sigma_1)^2} \quad \text{(Eq 5.9)}$$

Because the octahedral planes are all acted upon by the same normal stress, $\sigma_{oct}$, we can regard $\sigma_{oct}$ as an effective hydrostatic stress, and $\tau_{oct}$ as a single equivalent shear stress for the system. Of course, it must be pointed out that while the general state of stress can be regarded as the equivalent of the octahedral system, the reverse is not strictly true. Given the octahedral shear and normal stresses, we cannot back calculate the $\sigma_1$, $\sigma_2$, and $\sigma_3$ stresses deterministically. Many combinations are possible, but the assumption of the theories that follow is that all will produce the same mechanical history.

It needs also to be emphasized that the mechanical behavior is not necessarily limited to slip in the octahedral planes, and fracture normal to them. The important slip planes may still be the planes of maximum shear, and crack growth directions may still be the normal planes of the general stress state system (Fig. 5.8a), but the equivalent system (Fig. 5.8b and c) serves as a satisfactory proxy for establishing when important stress magnitudes are reached for a process to occur, but not necessarily defining the details of how the process will occur. Thus, an analysis based only on octahedral shear and normal stresses may require additional input to make it more determinate. It is especially interesting, however, that the octahedral plane analysis takes into account all three stresses simultaneously in a simple mathematical formulation, while other approaches involving analyses of one plane at a time usually neglect the effect of the intermediate stress. Thus, while not fully rigorous, the octahedral plane theory is a reasonable proxy for an accurate representation of true behavior.

As we discuss various theories, we have the opportunity to specify in greater detail how the parameters involved in the octahedral plane treatment can be usefully deployed.

## Some Common Multiaxial Fatigue Theories

In this chapter, we later describe the method we favor for general treatment of multiaxial fatigue. However, before we do so it is appropriate to mention other methodologies that have been proposed in recent years, some of which are currently used in numerous engineering applications. For brevity, this discussion does not attempt to be complete; rather, it serves only to indicate some of the difficulties involved in the problem and how various methods have attempted to overcome these difficulties.

**Methods Based on Maximum Normal Stress or on Shear Stress Considerations Alone.** The Rankine theory, based on failure where the maximum principal stress reached a critical value, was at first considered. It became clear that it could not be adequate when attention was directed to the case of hydrostatic tension. Although difficult to achieve experimentally, ideally such a condition would involve no shear

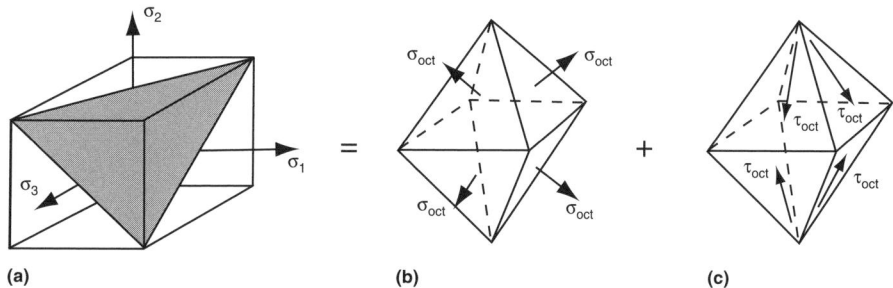

**Fig. 5.8** The general state of stress and its equivalent system based on shear and normal stresses on the octahedral formed by the eight octahedral planes. (a) Principal planes with one octahedral plane shown by the shaded area. (b) Normal stresses on the eight octahedral planes. (c) Shear stresses on the eight octahedral planes. Source: Ref 5.14

stresses, so no crack could start by alternating plastic flow. Brittle rupture, of course, could occur ideally when the hydrostatic tension became sufficiently high (but this is not fatigue). Nor would the stress be equal to the maximum stress reached in the case of axial fatigue, so the theory would not account in the same manner for two important conditions of stress state.

Postulating a failure condition based on maximum shear stress (Tresca condition) would also violate one of the desirable features described earlier in the section, "Critical Planes," namely, that the crack growth characteristic would be left unspecified. It would be easy to increase both $\sigma_1$ and $\sigma_2$ in the Tresca relation involving ($\sigma_1 - \sigma_2$) without affecting the shear stress, but affecting the maximum stress that is well known to govern crack growth. Similarly, the von Mises criterion (Eq 5.1) is not by itself a sufficient fatigue condition (but including the octahedral normal stress would be a step in the right direction, as discussed later). This is so especially since the octahedral shear stress (numerically is proportional to the von Mises condition) is only a proxy for maximum shear stress, and not the actual *maximum* shear. Similarly, using only the octahedral normal stress (Eq 5.8) would not be adequate because it is but a generalization of the Rankine hypothesis and does not include a shear component for crack initiation.

**Methods That Include Both Shear and Normal Stresses.** The next step is to include both maximum shear and normal stresses. Sines (Ref 5.10) developed in 1955 a theory based on the octahedral plane stresses, in the form:

$$(\tau_{oct})_{alt} = c - k(\sigma_{oct})_{mean} \quad \text{(Eq 5.10)}$$

where $c$ and $k$ are material constants.

In 1959 Davis and Connelly (Ref 5.11) proposed an alternate combination of the octahedral shear and normal stress, in the form of what was called the triaxiality factor ($TF$):

$$TF = \frac{\sigma_{oct}}{\tau_{oct}}$$

$$= \frac{\sigma_1 + \sigma_2 + \sigma_3}{\sqrt{(\sigma_1 - \sigma_2)^2 + (\sigma_2 - \sigma_3)^2 + (\sigma_3 - \sigma_1)^2}} \quad \text{(Eq 5.11)}$$

They did not propose this relation for use with cyclic deformation because their area of interest was not fatigue. Rather, their interest was thick plates and cylinders, and their proposition was that the $TF$ governed the ductility of the material and therefore the static rupture strength for the plates and cylinders.

Manson and Halford (Ref 5.12) in 1977, being sensitive to the importance of ductility in fatigue, further developed Davis and Connelly's concept of the triaxiality factor's influence on creep rupture ductility, and applied the $TF$ to multiaxial fatigue. However, because Davis and Connelly did not specify any quantitative relation between ductility and $TF$, Manson and Halford introduced what they called a multiaxiality factor ($MF$). As a baseline for expressing this relation, Manson and Halford built on the observation of Zamrik and Blass (Ref 5.13) that 304 stainless steel could support twice the von Mises plastic strain range in completely reversed torsion for the same fatigue life as in completely reversed axial loading. Almost immediately it was recognized that for other materials this relation was not necessarily true; completely reversed torsion and completely reversed axial loading required the same octahedral shear plastic strain for such materials. Thus, a cloud was cast on the general validity of the Manson-Halford proposed relation. Manson and Jung later clarified (Ref 5.14) this discrepancy and generalized the results on the basis of material crystallography, as discussed later.

Another approach that includes both shear and normal stress considerations is that of Miller and his associates. In 1973, Brown and Miller (Ref 5.15) presented a theory based on crack initiation on one plane and crack growth in a different plane. They used constant life contours, known as $\Gamma$-plots that basically expressed ($\varepsilon_1 - \varepsilon_3$) as a function of ($\varepsilon_1 + \varepsilon_3$). In effect, the ($\varepsilon_1 - \varepsilon_3$) term is a measure of the shear strain that initiates the crack, and the $\varepsilon_1 + \varepsilon_3$ term is a measure of the normal strain that propagates the crack. In fact, they considered two types of crack growth, one along the surface, and the other inward from the surface, but in any case, their theory involved both initiation and propagation of cracks. Later, in 1985, Kandil, Miller, and Brown (Ref 5.16) found that the explicit function relating ($\varepsilon_1 - \varepsilon_3$) and ($\varepsilon_1 + \varepsilon_3$) is:

$$(\varepsilon_1 - \varepsilon_3)/2 + S(\varepsilon_1 + \varepsilon_3)/2 = \text{constant} \quad \text{(Eq 5.12)}$$

where the constant $S$ is evaluated by fitting the fully reversed torsional data to the fully reversed axial results. Thus, their method allowed for torsional fatigue to be independent of axial fatigue, and the generalized relation for any multiaxial state to make use of both axial and torsion fatigue, requiring both types of tests to be conducted before the general relation could be expressed. Good results can be obtained by their theory, but a sizeable database is required.

**A More Detailed Discussion of the Multiaxiality Factor.** Because the method based on the multiaxiality factor has been developed to a point of very practical use, as we describe subsequently, it is appropriate to devote some discussion to its origin and later development.

When, in 1977, we introduced the multiaxiality factor ($MF$) (Ref 5.12), it was intended to be a multiplier of ductility so that when the material was cycled at a $TF$ other than 1.0 (wherein its ductility is $D$), it would behave as if its ductility were ($MF \times D$). At $TF = 1.0$, of course, $M$ must be 1.0. The main condition we attempted to impose was that $MF = \frac{1}{2}$ at $TF = 0$, because in torsion, $TF = 0$ (i.e., $\sigma_1 + \sigma_2 + \sigma_3 = 0$, which is true because $\sigma_2 = -\sigma_1$, and $\sigma_3 = 0$), and we were trying to express the fact that for reversed torsion the strain supported for a given life was twice the strain supported in reversed axial fatigue. Of course, there are numerous mathematical relations between $MF$ and $TF$ that will satisfy this requirement. After considering several alternatives we chose the relations:

$$MF = TF \quad \text{for } TF \geq 1.0 \quad \text{(Eq 5.13)}$$

and

$$MF = 2/(2 - TF) \quad \text{for } TF < 1.0 \quad \text{(Eq 5.14)}$$

While relations satisfy the required conditions, they display an added logic; above $TF = 1.0$, Eq 5.13 is the simplest expression that satisfies the relation we would expect for $MF$ to increase as $TF$ increases; below $TF = 1.0$, Eq 5.14 expresses a variation that is logically symmetrical to the relation above $TF = 1.0$ on a semilog plot. For example, if $TF$ is three units above 1.0, then $MF = 4.0$. However, if $TF$ is three units below 1.0, (that is, if $TF = -2$), Eq 5.14 states that $MF = 1/[2 - (-2)] = \frac{1}{4}$, so that if ductility is decreased by a given amount for a given number of $TF$ units above 1.0, it will also result in the same fractional increase in ductility for the same number of units of $TF$ below $TF = 1.0$. The analytical relation of these equations is graphically expressed in Fig. 5.9. The symmetry about the point (1,1) can be seen on this semilog plot that expresses the numerical features expressed previously in the more general sense.

Also shown in Fig. 5.9 is another relation expressed by Manjoine (Ref 5.17). Manjoine was not concerned with fatigue, but rather high-temperature creep, and we were aware of it through personal communication rather than prior publication (contrary to our referencing a report by him in our 1977 paper), and we considered it as an alternate relation. We felt, however, that the expression implied too severe a reduction in fatigue at the higher triaxiality factors. We therefore limited our consideration to Eq 5.13 and 5.14. Both the formulations of $MF$ versus $TF$, of

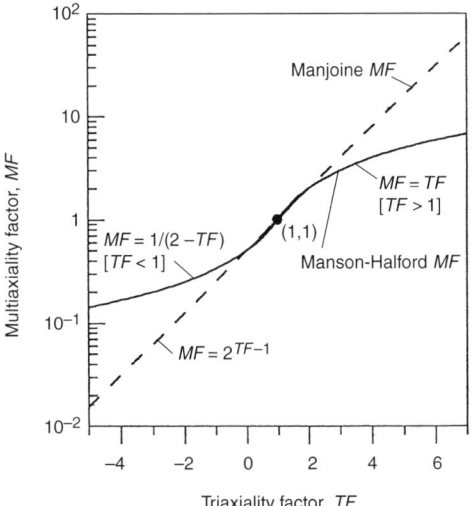

**Fig. 5.9** Multiaxiality factor relations. Source: Ref 5.14

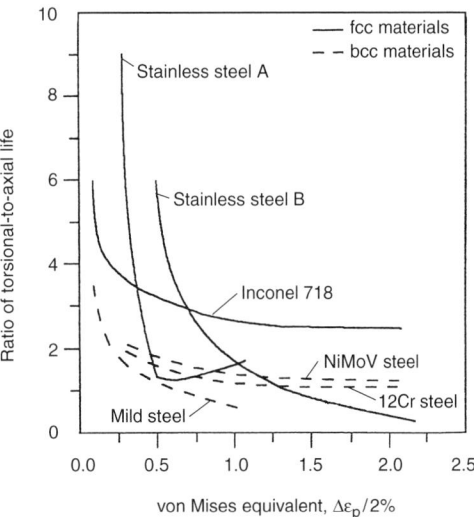

**Fig. 5.10** Survey by Doquet and Pineau (Ref 5.20) of several face-centered cubic and body-centered cubic materials

course, imply that in shear ($TF = 0$), the supportable strain in torsion is twice that in axial loading. But, as noted later, this behavior was true for only some materials. In other materials, torsion and axial loading strain ranges were the same for the same fatigue life.

In 1980 Marloff and Johnson (Ref 5.18) also investigated the relation $MF = 2^{TF-1}$, and they were reasonably satisfied with the results. In the same year Mowbray (Ref 5.19) proposed a variation of the Manson-Halford equation:

$$MF = \frac{1}{1 - mTF} \quad \text{for} \quad 0 \le m \le 0.5 \quad \text{(Eq 5.15)}$$

This equation produces a value of $MF = 1.0$ for $TF = 0$ and would be useful for those materials for which torsion and axial loading give the same fatigue life. Thus, it can be seen that from the very beginning there has been some ambiguity in the results comparing pure torsion with axial loading.

In 1994, some insight was gained into the problem. In a paper, Manson and Jung (Ref 5.14) presented some results of a study by Doquet and Pineau (Ref 5.20) that are shown in Fig. 5.10. The figure displays the ratio of the life in torsion to that in reversed axial loading for given values of von Mises equivalent plastic strain.

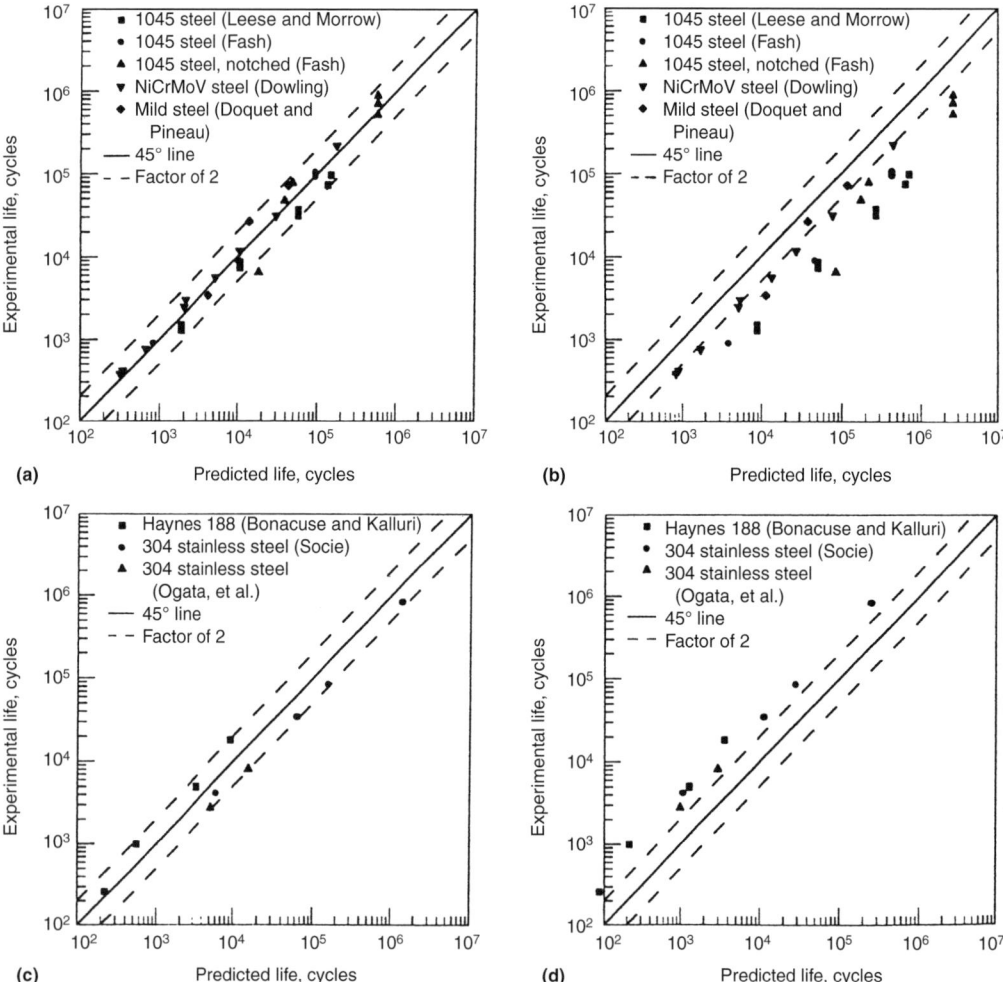

**Fig. 5.11** Suitability of multiaxiality factors with body-centered cubic and face-centered cubic materials. (a) bcc materials under torsion ($MF = 1.0$). (b) bcc materials under torsion ($MF = 0.5$). (c) fcc materials under torsion ($MF = 0.5$). (d) fcc materials under torsion ($MF = 1.0$). Source: Ref 5.14

While some of the behavior shown in this figure is not easy to explain, one striking feature is evident: for two of the bcc materials, the NiMoV steel and the 12Cr steel, the curves level out to about 1.0 in the large strain region, while for the fcc material, Inconel 718, the ratio levels out to about 2.5. Because in the high strain region where plasticity dominates, this ratio depends directly on *MF*, it appears that at the very least it is necessary to establish different *MF* versus *TF* relations for bcc materials and for fcc materials.

This conclusion is further emphasized in Fig. 5.11(a) and (b). Here is plotted the life correlations in torsion for a number of bcc materials using *MF* = 1.0 and 0.5. It is clear that 1.0 is the far better value. For the fcc materials shown in Fig. 5.11(c), the value of *MF* = 0.5 appears very good, whereas Fig. 5.11(d) shows that the *MF* = 1 predictions are poor.

It became clear, therefore, that no multiaxiality theory could be developed using the same relations between *MF* and *TF* for all materials. A way out was to provide separate relations for bcc materials and for fcc materials. By then, a considerable amount of experimental data had become available in the literature to justify a pragmatic approach: determine what the *MF-TF* relation is in order to correlate the experimental data for each of the two crystallographic systems using data for as large a range as possible for triaxiality factor *TF*. In examining the data from various sources, it became clear that in many cases the strains were low, largely elastic, and therefore it would be necessary to perform the analyses on the basis of how multiaxiality affected the elastic life line as well as how it affected the plastic line. We must therefore treat this problem before we can proceed.

**Addition of an Elastic Component to the Life Relation.** Let us assume that in the normal Manson-Coffin-Basquin plot for uniaxial loading the line *AB* in Fig. 5.12 represents the plastic line and *CD* the elastic line. The addition of a multiaxiality factor displaces the plastic line to *A'B'*. The elastic line must, however, also be displaced. How much displacement is required for the elastic line is determined by how much the transition point *T* is displaced. This transition point must be displaced horizontally because only under this condition will the cyclic stress-strain curve be unaltered, as we would expect. Thus, through point *T* a horizontal line is drawn to determine *T'* at its intersection with line *A'B'*. Through *T'* the line *C'D'* is drawn parallel to *CD*. We have used this concept in Chapter 4, "Mean Stress," in reestablishing the fatigue effect of mean stress on the elastic and plastic lines; here it is the displacement of the elastic line that is known, but the plastic line is also displaced in order to preserve the cyclic stress-strain curve. Using this approach, the equation for the effect of multiaxiality becomes:

$$MF(\Delta\varepsilon_{eq}) = MF^{(1-b/c)}B(N_f)^b + C(N_f)^c \quad \text{(Eq 5.16)}$$

where *b* and *c* are the exponents in the basic uniaxial life relation, *B* and *C* are the coefficients in the elastic and plastic components, and $\Delta\varepsilon_{eq}$ is the total equivalent strain range under the multiaxial condition. This derivation was first carried out by Bonacuse and Kalluri (Ref 5.21) and further details can be found in their report.

**Combining Multiaxiality with Mean-Stress Effects—A Speculative Extension.** How can we combine what we have learned about multiaxiality with what we know about the effects of mean stresses? The answer has to be totally speculative because, to our knowledge, no experimental programs have ever been undertaken on this specific subject. However, if we extend logically what we have discussed individually for mean stress effects and for multiaxiality effects, we can speculate on what the result might be.

The concepts we discussed earlier can be combined into a single diagram as shown in Fig. 5.13. The key features of the diagram are:

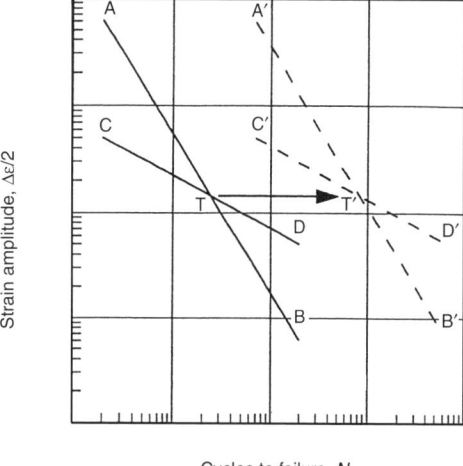

**Fig. 5.12** Extending the life range for applicability of multiaxiality factor

- On the horizontal axis, we show a life scale, modified by the mean stress, considering the possibility of mean stress in all three directions. Thus, the $\sigma_m$ for a uniaxial case has been replaced by $\sigma_{m,1} + \sigma_{m,2} + \sigma_{m,3}$.
- The vertical axis is concerned only with amplitudes of cycling for each of the three directions. Thus, if the strain amplitude in the 1-direction is $\Delta\varepsilon_1$, and in the 2- and 3-directions, $\Delta\varepsilon_1$ and $\Delta\varepsilon_1$, we have combined these three ranges in an analogous manner to that for individual strains:
  a. The elastic line is displaced downward by a factor $2(1 + \mu)/3$.
  b. The plastic line is displaced downward by the $MF$ factor (or upward if $MF < 1$).
  c. The total strain range line is drawn as the sum of the displaced elastic and plastic lines. It is seen that to use the diagram, a considerable amount of information is needed.

## Application of Multiaxiality Concepts to a Large Range of Published Information

A large amount of published information from various sources was studied to determine the pattern between $MF$ and $TF$. Table 5.1 provides a summary of the materials and test conditions included in this study, and Table 5.2, an indication of the geometries involved. Also noted is the definition of failure used by the author in each study. This factor provided quite a complication in the analyses because of the wide differences in the definition. Some authors relied on a certain percentage drop in force; others on the development of a defined length of crack. Because of the scatter that results from analyzing tests from different laboratories and failure definitions, and because of inherent scatter in fatigue data, we followed a consistent practice of eliminating some of the data that contributed to confusing scatter. Where there were duplicate data, points that deviated from the mean by more than one standard deviation were omitted.

As noted, we considered the bcc and the fcc materials separately. The larger amount of data was on bcc materials. For this class:

| | |
|---|---|
| $TF = 1.0$ | Covered by axial reversed strain data |
| $TF = 0$ | Covered by pure torsion |
| $0 < TF < 1.0$ | Covered by torsion and axial loading or torsion and bending |
| $TF > 1.0$ | Covered by torsion and bending |
| $TF < 0$ | Covered by two types of tests: by loading a specimen inside a vessel subjected to high hydrostatic pressure (Ref 5.25), and by using rolling contact fatigue data. |

The classic work of Moyar (Ref 5.26) to demonstrate why rolling contact fatigue appears to

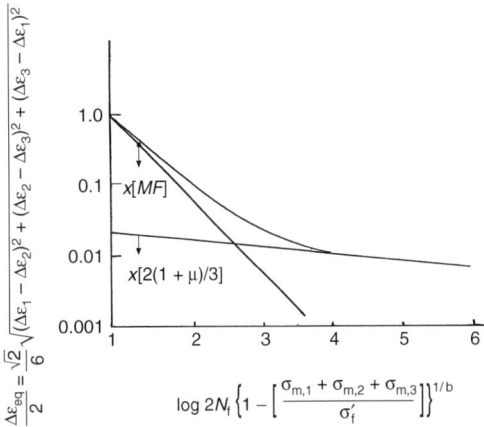

**Fig. 5.13** Mean stress effects under multiaxial loading

Table 5.1  Summary of experimental programs included in the multiaxiality study

| Author(s) (reference) | Material | Type of test(s) | Definition of failure |
|---|---|---|---|
| **Body-centered cubic materials** | | | |
| Leese & Morrow (Ref 5.22) | 1045 steel | Torsion | 10–20% drop |
| Fash (Ref 5.23) | 1045 steel | Torsion, axial-torsion, bending, bending-torsion | 10% drop |
| Doquet & Pineau (Ref 5.20) | Mild steel | Torsion, axial-torsion | 15% drop |
| Dowling (Ref 5.24) | NiCrMoV steel | Torsion | N/A |
| Crossland (Ref 5.25) | Vibrac V30 steel | Hydrostatic-torsion | Final separation |
| Moyar (Ref 5.26) | 52100 steel | Rolling contact | Spalling |
| **Face-centered cubic materials** | | | |
| Bonacuse & Kalluri (Ref 5.21) | Haynes 188 | Torsion | 10% drop |
| Kalluri & Bonacuse (Ref 5.27) | Haynes 188 | Axial-torsion | 10% drop |
| Socie (Ref 5.28) | 304 stainless steel | Torsion, axial-torsion | Final separation |
| Ogata et al. (Ref 5.29) | 304 stainless steel | Torsion, axial-torsion | 25% drop |

be so superior to axial fatigue was very useful. His data required considerable analysis to be useful in this study.

The results of this study are shown in Fig. 5.14 for bcc materials. Obviously, there is some scatter, and in some regions there is reasonable uncertainty as to how best to draw the line to represent the data. In the region $TF < 0$, the relation for *MF* is approximately:

$$MF = (3)^{TF} \qquad \text{(Eq 5.17a)}$$

and in the region of $0 < TF < 1.0$, we can reasonably write:

$$MF = 1.0 \qquad \text{(Eq 5.17b)}$$

and in the region $TF > 1.0$, a reasonable representation is:

**Table 5.2  Types of specimens included in the multiaxiality study**

| Geometry | Type of loading(s) | Author(s) (reference) | Material(s) |
|---|---|---|---|
| Thin cylinder | Torsion | Leese & Morrow (Ref 5.22) | 1045 steel |
| | | Fash (Ref 5.23) | 1045 steel |
| | | Doquet & Pineau (Ref 5.20) | Mild steel |
| | | Dowling (Ref 5.24) | NiCrMoV steel |
| | | Bonacuse & Kalluri (Ref 5.21) | Haynes 188 |
| | | Socie (Ref 5.28) | 304 stainless steel |
| | | Ogata et al. (Ref 5.29) | 304 stainless steel |
| | | Fash (Ref 5.23) | 1045 steel |
| | | Dquet & Pineau (Ref 5.20) | Mild Steel |
| | Torsion and axial | Kalluri & Bonacuse (Ref 5.27) | Haynes 188 |
| | | Socie (Ref 5.28) | 304 stainless steel |
| | | Ogata et al. (Ref 5.29) | 304 stainless steel |
| Solid cylinder | Torsion and hydrostatic | Crossland (Ref 5.25) | Vibrac V30 steel |
| Notched shaft | Torsion | Fach (Ref 5.23) | 1045 steel |
| | Bending | Fash (Ref 5.23) | 1045 steel |
| | Torsion and bending | Fash (Ref 5.23) | 1045 steel |
| Roller bearings | Rolling contact | Moyar [Ref 5.26] | 52100 steel |

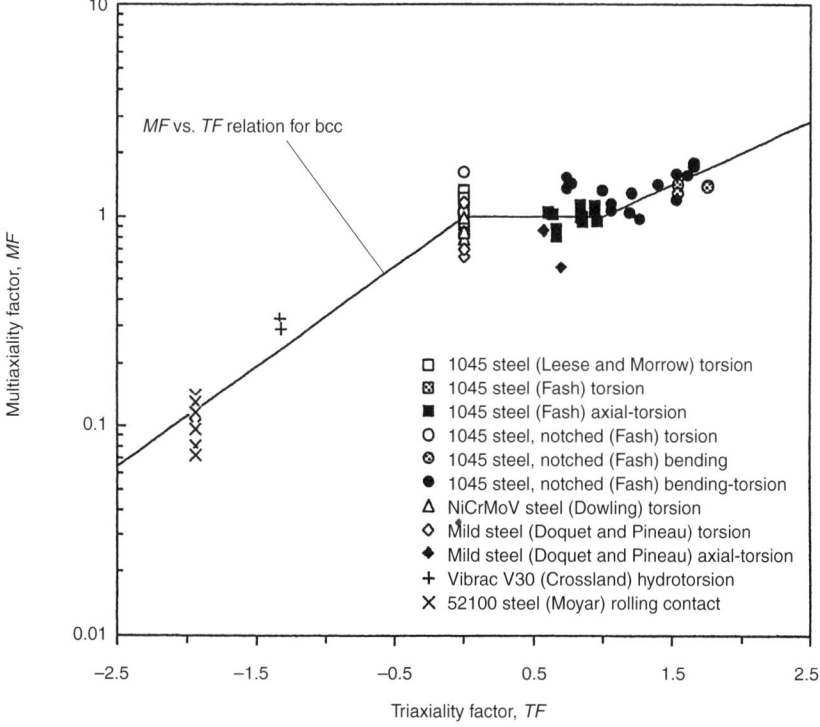

**Fig. 5.14**  Tentative consolidation of *MF* versus *TF* over complete range of available body-centered cubic data. Source: Ref 5.14

$$MF = (2)^{TF=1} \quad \text{(Eq 5.17c)}$$

For the limited amount of fcc material data, there is a shorter range of $TF$ to consider. In the range $0 < TF < 1.0$, the relation for $MF$ is best represented by:

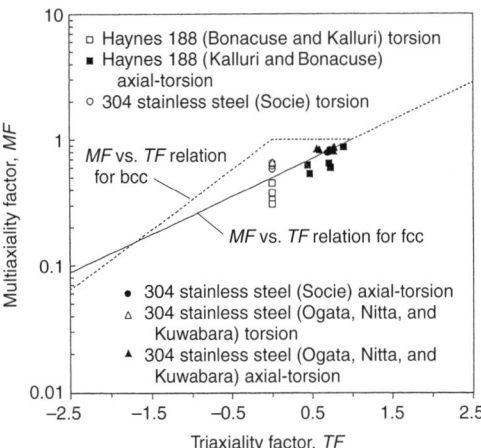

**Fig. 5.15** Tentative consolidation of $MF$ versus $TF$ over complete range of available face-centered cubic data. Source: Ref 5.14

$$MF = (2)^{TF=1} \quad \text{(Eq 5.18)}$$

The relation beyond this range of $TF$ is unclear due to the lack of data. Figure 5.15 shows the results of the analysis for fcc materials. It is still not clear as to which representation, $MF = (2)^{TF}$ or $MF = TF$, is best. Figure 5.15 also shows, in phantom, the representation of bcc data for reference.

### Discussion

The ability of Eq 5.17 and 5.18 to represent all the data is shown in Fig. 5.16 and 5.17. It is clear that this representation of fatigue life is quite good, considering the large range of materials, types of loading, and life ranges. This progress is quite encouraging. What is also clear from the figures is that it requires recognition of crystallographic structure before the prediction can be made accurately. For example, if Eq 5.17(c) is used in an attempt to predict fcc materials, the results are as shown in Fig. 5.18, which are not very good. It is not feasible to

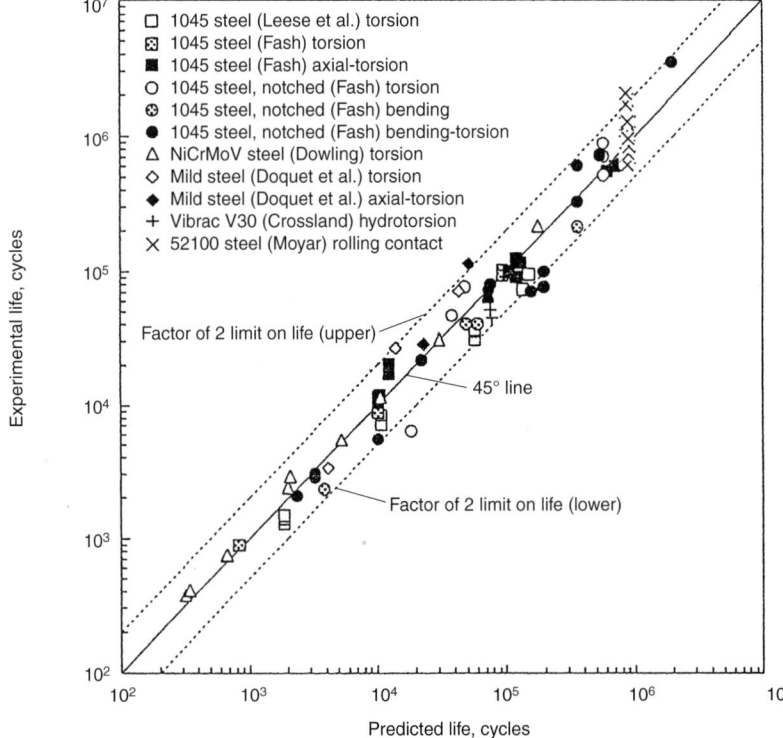

**Fig. 5.16** Body-centered cubic materials for all types of available tests. Source: Ref 5.14

display the predictions for the bcc materials on the basis of the fcc representation because the fcc predictions do not cover the range of actual bcc tests; however, it is clear that the results would not be good.

Other investigators, e.g., Brown and Miller (Ref 5.15), Fatemi and Socie (Ref 5.30), and Fatemi and Kurath (Ref 5.31), have required that both axial and torsional fatigue data be separately determined and have built the results into their theories. This approach would automatically distinguish between bcc and fcc materials but would require additional experimentation. Because the crystalline structure of a material is usually known, the methodology presented herein may be more convenient to implement, as it requires only the fundamental axial fatigue properties.

Reference 5.32 provides a comprehensive review of how well a large number of multiaxial fatigue life models predict the substantial axial and torsional fatigue data available on the titanium alloy Ti-6Al-4V. The alloy was heat treated to produce a structure comprising approximately 60% alpha that has the hcp crystalline structure. The axial fatigue data were generated by participants in the U.S. Air Force's high cycle fatigue program, whereas the torsional data were generated at the University of Illinois, Urbana-Champaign.

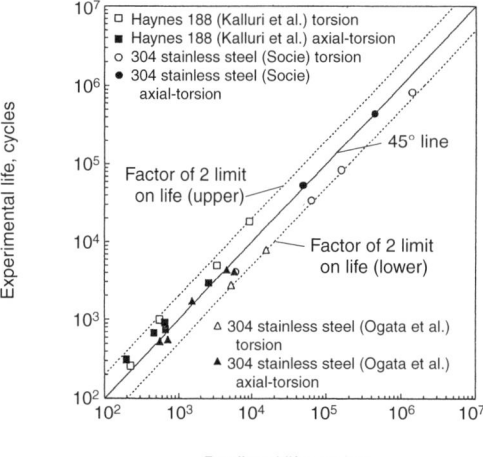

**Fig. 5.17** Face-centered cubic materials for all types of available tests. Source: Ref 5.14

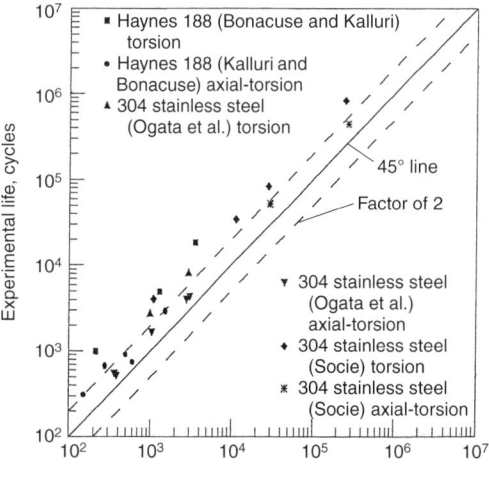

**Fig. 5.18** Use of Eq 5.17(c) to predict face-centered cubic materials. Source: Ref 5.14

## REFERENCES

5.1 D.L. Krause and P.A. Bartolotta, "An In-Plane Biaxial Contact Extensometer," *Multiaxial Fatigue and Deformation: Testing and Prediction,* ASTM STP 1387, S. Kalluri and P.J. Bonacuse, Eds., American Society for Testing and Materials, 2000, p 369–381

5.2 J.R. Ellis and A. Abdul-Aziz, "Specimen Designs for Testing Advanced Aeropropulsion Materials under In-Plane Biaxial Loading," NASA/TM-2003-212090, National Aeronautics and Space Administration, January 2003

5.3 C.E. Richards, *Engineering Materials Science,* Wadsworth Publishing Company, Inc., San Francisco, CA, 1961, p 194

5.4 S.Y. Zamrik and P.Y. Tang, Aspect of Cumulative Fatigue Damage under Multiaxial Strain Cycling, *Mechanical Behavior of Materials, Proceedings of the 1971 International Conference on Mechanical Behavior of Materials,* Vol II, 1972, p 381–390

5.5 K. Ohji and J. Marin, "Review on Uniaxial and Biaxial Low Cycle Fatigue," NASA CR-68645, Pennsylvania State University, 1965

5.6 S.S. Manson, "Behavior of Material under Conditions of Thermal Stress," NACA TN-2933, National Advisory Committee for Aeronautics, 1953

5.7 S.S. Manson, *Thermal Stress and Low-Cycle Fatigue,* original ed., McGraw-Hill,

New York, 1966; reprinted ed., Robert E. Krieger Publishing Co., Inc., 1981, p 86–92

5.8 A. Mendelson and S.S. Manson, "Practical Solution of Plastic Deformation Problems in Elastic-Plastic Range," NACA TN-4088, National Advisory Committee for Aeronautics, 1957

5.9 R.C. Juvinall, *Engineering Considerations of Stress, Strain and Strength*, McGraw-Hill, 1967, p 85–89

5.10 G. Sines, "Failure of Materials under Combined Repeated Stresses with Superimposed Static Stresses," NACA TN-3495, National Advisory Committee for Aeronautics, 1955

5.11 E.A. Davis and F.M. Connelly, Stress Distribution and Plastic Deformation in Rotating Cylinders of Strain Hardening Material, *J. Appl. Mech. (Trans. ASME)*, Vol 81, 1959, p 25–30

5.12 S.S. Manson and G.R. Halford, Discussion to paper by J.J. Blass and S.Y. Zamrik: Multiaxial Low-Cycle Fatigue of Type 304 Stainless Steel, *1976 ASME-MPC Symposium on Creep-Fatigue Interaction*, ASME, 1976, p 129–159; *J. Eng. Mater. Technol.*, Vol 99, July 1977, p 283–286

5.13 S.Y. Zamrik and J.J. Blass, Multiaxial Low-Cycle Fatigue of Type 304 Stainless Steel, *1976 ASME-MPC Symposium on Creep-Fatigue Interaction*, ASME, 1976, p 129–159

5.14 S.S. Manson and K. Jung, Progress in the Development of a Three-Dimensional Fatigue Theory Based on the Multiaxiality Factor, *Material Durability/Life Prediction Modeling*, PVP, Vol 290, S.Y. Zamrik and G.R. Halford, Ed., American Society of Mechanical Engineers, New York, 1994, p 85–93

5.15 M.W. Brown and K.J. Miller, A Theory for Fatigue Failure under Multiaxial Stress-Strain Conditions, *Proc. Inst. Mech. Eng.*, Vol 187 (No. 65), 1973, p 745–755

5.16 F.A. Kandil, K.J. Miller, and M.W. Brown, "Creep and Ageing Interactions in Biaxial Fatigue of Type 316 Stainless Steel," *Multiaxial Fatigue*, ASTM STP 853, K.J. Miller and M.W. Brown, Ed., American Society for Testing and Materials, 1985, p 651–668

5.17 M.J. Manjoine, personal communication, 1978

5.18 R.H. Marloff and R.L. Johnson, The Influence of Multiaxial Stress on Low Cycle Fatigue of Cr-Mo-V Steel at 1000°F, *Weld. Res. Counc. Bull.*, No. 264, Dec 1980, p 1–21

5.19 D.F. Mowbray, *J. Test. Eval.*, Vol 8 (No. 1), 1980, p 3–8

5.20 V. Doquet and A. Pineau, Multiaxial Low-Cycle Fatigue Behavior of a Mild Steel, *Fatigue under Biaxial and Multiaxial Loading*, ESIS 10, K. Kussmaul, D. McDiarmid, and D. Socie, Ed., Mechanical Engineering Publications, London, UK, 1991, p 81–101

5.21 P.J. Bonacuse and S. Kalluri, Elevated Temperature Axial and Torsional Fatigue Behavior of Haynes 188, *Journal of Engineering Materials and Technology*, Vol. 117, April 1995, p 191–199. (Also see NASA TM 105396, AVSCOM TR-91-C-045, June 1992.)

5.22 G.E. Leese and J. Morrow, "Low Cycle Fatigue Properties of a 1045 Steel in Torsion," *Multiaxial Fatigue*, ASTM STP 853, K.J. Miller and M.W. Brown, Ed., American Society for Testing and Materials, 1985, p 482–496

5.23 J.W. Fash, "An Evaluation of Damage Development during Multiaxial Fatigue of Smooth and Notched Specimens," Ph.D. thesis, Department of Theoretical and Applied Mechanics, University of Illinois, Urbana, IL, 1985

5.24 N. Dowling, "Torsional Fatigue Life of Power Plant Equipment Rotating Shafts," DOE/RA/29353-1, U.S. Department of Energy, Washington, DC, 1982

5.25 B. Crossland, Effect of Large Hydrostatic Pressures on the Torsional Fatigue Strength of an Alloy Steel, *Proc. Inst. Mech. Eng.: International Conference on the Fatigue of Metals*, 1956, p 138–149

5.26 G.J. Moyar, "A Mechanics Analysis of Rolling Element Failures," Ph.D. thesis, Department of Theoretical and Applied Mechanics, University of Illinois, Urbana, IL, 1960

5.27 S. Kalluri and P.J. Bonacuse, In-Phase and Out-of-Phase Axial-Torsional Fatigue Behavior of Haynes 188 Superalloy at 760 °C, *Advances in Multiaxial Fatigue*, ASTM STP 1191, D.L. McDowell and J.R. Ellis, Eds., American Society for Testing and Materials, Philadelphia, 1993, p 133–150. (Also see AVSCOM TR-91-C-046, NASA TM 105765, October 1991.)

5.28 D. Socie, Multiaxial Fatigue Damage

Models, *J. Eng. Mater. Technol.,* Vol 109, 1987, p 293–298

5.29 T. Ogata, A. Nitta, and K. Kuwabara, Biaxial Low-Cycle Fatigue Failure of Type 304 Stainless Steel under In-Phase and Out-of-Phase Straining Conditions, *Fatigue under Biaxial and Multiaxial Loading,* K. Kussamaul, D. McDiarmid, and D. Socie, Ed., Mech. Engineering Publications, London, UK, 1991, p 377–392

5.30 A. Fatemi and D.F. Socie, A Critical Plane Approach to Multiaxial Fatigue Damage Including Out-of-Phase Loading, *Fatigue Fract. Eng. Mater. Struct.,* Vol 11 (No. 3), 1988, p 149–165

5.31 A. Fatemi and P. Kurath, Multiaxial Fatigue Life Predictions under the Influence of Mean Stresses, *J. Eng. Mater. Technol. (Trans. ASME),* Vol 110, 1988, p 380–388

5.32 A.R. Kallmeyer, A. Krgo, and P. Kurath, Evaluation of Multiaxial Fatigue Life Prediction Methodologies for Ti-6Al-4V, *J. Eng. Mater. Technol.,* Vol 124, April 2002, p 220–237

# CHAPTER 6

# Cumulative Fatigue Damage

## Introduction

When loading varies during the life of a structure, multiple hysteresis loops of varied sizes may develop. Determining the life of the structure under this condition has come to be known in the vernacular of fatigue terminology as "cumulative fatigue damage analysis." Although it is recognized that fatigue "accumulates" progressively within the loaded volume even if the same hysteresis loop is continuously repeated, the term "cumulative fatigue damage" has traditionally been reserved for variable loading histories.

There are several reasons why variable loading history may be expected to influence the damage in a manner different from what might be expected on a basis of *linear* summation of the individual events occurring within the life history. First, there is the normal loading order effect that has been observed for many materials, and by many investigators, whereby it has been clearly established that a high load (low life level) has a more than commensurate effect on the damage accumulation at a subsequent cycling at a lower load (high life) condition. Conversely, using up a large fraction of the life at a lower loading level has been observed to have a much lower effect than might be expected on subsequent higher load level cycling. It has been proposed that this loading order effect can be explained on the basis of a *double-linear* damage rule—one for *crack initiation* and one for *crack propagation*. This possibility is explored in this chapter and shown to be questionable. However, the observed phenomenon is a valid one, and we propose alternate procedures for explaining and dealing analytically with the effect.

A second reason for uncommon cumulative damage effects relates to synergism and that occurs only in some materials. It is especially applicable to high-temperature loadings where metallurgical effects can occur at one temperature and strain range, and the altered material behaves differently at another temperature than it would have had the metallurgical effect not occurred. Thus, for example, if a metallurgical precipitation can occur at 815 °C (1500 °F), but not at 650 °C (1200 °F), prior service at 815 °C (1500 °F) will cause the precipitation to take place, and later service at 650 °C (1200 °F) will reflect the properties of the altered material rather than its initial properties at 650 °C (1200 °F). If performance prediction is based on the initial properties at 650 °C (1200 °F), considerable error may occur. Other examples of synergism may be involved where slip planes become activated at a prior strain range, and these remain active at a later applied strain range for which these slip planes would not have been active if the prior strain range had not occurred, or, if a higher strain range initially applied can break up precipitates that would normally prevent slip at a lower strain range. In all these cases, basing predictions for multiple loadings on data obtained for singly applied loadings may produce considerable error by overlooking the synergism factor. The effects may be beneficial or detrimental, but in any case, performance will not agree with expectations.

In this chapter we focus most of our attention on the loading order effect that is most pervasive among engineering materials, and wherein considerable progress has been made in recent years. We also briefly mention known cases of synergistic effect, but because these effects are uncommon, and relate only to some materials, this discussion is abbreviated.

**Counting Events for Cumulative Damage Analysis.** In the early days of study of cumulative damage, there was no rational system for counting the events that contributed to the damage. Several arbitrary systems were used—among them, level crossings and range pairing. In recent years it has become recognized that an "event" can best be defined as a closed hysteresis loop. Thus, the "event" identifies both the strain range and the mean stress that are the fundamental parameters required to establish the cyclic life of the event if repeated to failure. In this chapter, we regard the closure of a hysteresis loop as an "event." A brief summary of these methods has been provided by the Society of Automotive Engineers (Ref 6.1).

A small problem develops when considering the actual closure of a hysteresis loop as the establishment of the event because the loop may be traced in stages while another event or partial events intervene. For example, in Fig. 6.1(a) we see a simple loading history *ABCDEFA* that traces out the loops *ADA*, *BCB*, and *EFE* shown in Fig. 6.1(b) according to the principles described in Chapter 2, "Stress and Strain Cycling." However, while loops *BCB* and *EFE* can be clearly identified in time sequence, loop *ADA* is traced in stages. First, the segment *AB* is traced, at which point it is interrupted by the introduction of three loops of *BCB*. Next, the segment *BDE* is traced, but again interrupted by two loops *EFE*. Finally, the segment *EA* completes the hysteresis loop *ADA*. Strictly speaking, then, loop *ADA* is the last to be completed; however, it was the first to be started. In establishing the loading order, we could consider partial loops introduced in the exact time sequence of loading, or we could compromise by considering the whole loop introduced at once somewhere in the loading sequence. Our preference would be to regard loop *ADA* to be introduced first because this would regard the greatest loading first, thus producing the most conservative life calculation; however, the choice may be made in another way by the analyst. We shall seek methods of analysis that permit flexibility of procedure.

**Loading History to be Used in Analysis.** In order to be able to focus quantitatively on the results discussed in this chapter, it is desirable to select a particular history that will be used uniformly throughout the discussion. We shall consider a case involving four loadings with life levels of $10^3$, $10^4$, $10^5$, and $10^6$ cycles. The loadings will be applied in blocks, each block consisting of 1% of the cycles to failure of each loading, applied in the order of increasing life. Thus, we will consider that 10 cycles of the $10^3$ life will be applied first, followed by $10^2$ cycles of the $10^4$ life, and so forth. For simplicity, we can regard the loadings as shown in Fig. 6.2(a). The maximum strain is kept constant, and appropriate cycles are applied at each maximum strain level to carry out the stated requirement. The hysteresis loops are shown in Fig. 6.2(b). Of course, the mean stress of each loop has to be considered in order to determine the cyclic life associated with each loop. Before we apply various models to this idealized loading history in subsequent sections, it is first necessary to introduce and explain the various models. We start with the classical linear damage rule.

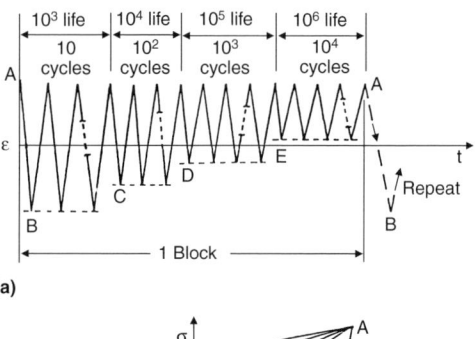

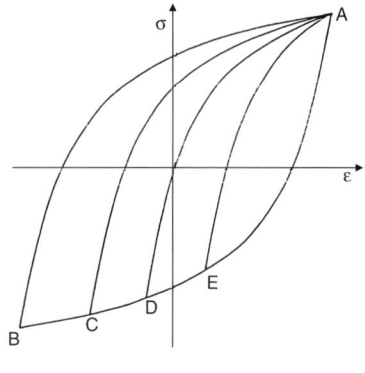

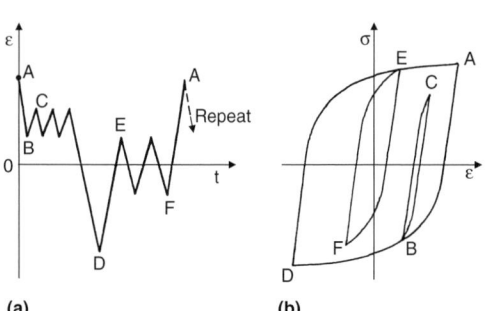

**Fig. 6.1** (a) Example straining history. (b) Corresponding hysteresis loops

**Fig. 6.2** Loading history selected to be used in analysis. (a) Straining history. (b) Corresponding hysteresis loops

## Linear Damage Rule (LDR) and Its Limitations

One of the earliest methods proposed for treating cumulative damage analysis is the linear damage rule, first proposed by Palmgren (Ref 6.2) in 1924 for application to rolling element fatigue of ball bearings made of high-carbon steels. Subsequently, and apparently independently, it was proposed by Langer (Ref 6.3) in 1937 for pressure vessel and piping design using relatively low-carbon steels. Being unaware of Palmgren's work, Miner (Ref 6.4) in 1945 reintroduced the linear rule with proper credit to Langer. He applied it to aircraft structures made of aluminum alloys. Because Miner published considerable data showing how well the rule worked for his test conditions, he garnered the popular credit for establishing the rule. Most know it as Miner's linear damage rule (LDR). It is, in fact, still in use by most analysts today because of its seeming logic and because it is so easy to apply. A critical assessment of Miner's database has been presented (Ref 6.5) wherein it was shown that the database virtually guaranteed that the LDR would adequately correlate his results. While the experimental fatigue testing conditions selected were representative of certain fatigue loadings experienced by aircraft, they were ill-suited for bringing out the shortcomings of the LDR. The various reasons for considering a nonlinear damage rule simply were not present in either the mission loadings considered or in Miner's published database. As a consequence, the LDR gained widespread acceptance early on. It was simple, and there was published evidence that it worked well. The logic behind the LDR is that if a given fraction of life is used up at one loading level, the remaining fraction (1 minus the applied fraction) is available to be used at any other loading level. Thus, for example, if 1% of the life is used up at one level, then 99% is available for use at the same or any other loading level. Each time a cycle fraction is used up at any level, that cycle fraction must be subtracted from the cycle fraction available for use at all subsequent loading levels. Mathematically stated, it is:

$$\sum \frac{n_i}{N_i} = 1 \qquad (6.1)$$

Almost from the beginning of the study of fatigue it became evident that this assumption is not strictly correct. For over half a century, many review reports have been published (see for example, Ref 6.6–6.10), regarding the limitations of the LDR. Alternative methods have been proposed to overcome the limitations. But use of the linear damage rule persists to this day for two main reasons:

- It is extremely easy to use.
- Nearly every alternative method developed requires one or more constants that must be experimentally determined.

The latter feature itself virtually rules out the use of alternative methods for most design purposes. If designers need experimental constants to evaluate the applicability of each candidate material they are considering, they will be disinclined to choose an alternative to the LDR.

The application of the linear damage rule has thus been commonplace, but corrections have been applied to account for expected deviations. Thus, for example, one procedure is to let the summations in Eq 6.1 be less than 1.0, introducing some level of conservatism. For example, the ASME Boiler and Pressure Vessel Code Case 1592 (for design of nuclear power plant components at high operating temperatures) (Ref 6.11) uses a summation that varies from 1.0 down to 0.4 for creep-fatigue interaction, depending on the proportion of each type of fractional damage present. The 0.4 limit corresponds to 0.2 fatigue and 0.2 creep damage fractions for austenitic stainless steels. The value of the constant used may vary among applications depending on experience and desired conservatism. However, the designer never really knows the resulting level of conservatism, or whether there is any conservatism at all.

Over the years the authors have conducted many tests on numerous materials to determine the applicability of the LDR. Figure 6.3 shows a summary of some of these results. These tests were of the two-level type. On the horizontal axis we plot the cycle ratio of the first life level applied and on the vertical axis, the remaining cycle ratio $n_2/N_2$, the cycle ratio at the second level that was carried to failure. If the LDR were applicable, all the points would lie on the line $AB$ for which $n_1/N_1 + n_2/N_2 = 1.0$. Obviously, the experimental points do not conform to this expectation. Most of the points fall below the line, indicating that the remaining cycle ratio was not as high as would be expected according to the LDR. In these cases, the first loading was applied at the lower life level. In a few cases, the experimental points lie above *AB*; here, the first loading was at a lower life level than the second. These results verify the common notion of loading-order effect. It is clear that it would be desirable to replace the LDR concept by one that accounted for the loading order effect; however, it is also important that the replacement concept be simple to apply and not require many experimental constants that would discourage its use in the design stage.

One of the earliest concepts proposed for explaining the loading-order effect was that of a double linear damage rule (DLDR) based on crack initiation and crack propagation. The reasoning was that the fraction of the total life used in initiation is different for the different life level; it being lower for the lower life level. Thus, mixing up the cycle ratios by basing them on the total life level can produce the observed loading order effect.

## Treatment of Loading-Order Effects by a Double Linear Concept Predicated on Crack "Initiation" and "Propagation"

### Background

**Example.** Consider a case of cyclic loading that would give a 1000 cycle life that is followed by loading at a level resulting in a 100,000 cycle life. Let us assume that at the 1000 cycle life level initiation occurs at 20% of life (200 cycles), leaving a propagation life of 800 cycles, while at the 100,000 cycle life level, the initiation life is 80,000 cycles (leaving only 20,000 cycles for propagation). Now let us assume that 100 cycles are applied at the 1000 life level, and the loading is then changed to the 100,000 level and run to failure. The 100 cycles use 50% of the initiation life, so 50% remains. When the change is made to the second level, the first 40,000 cycles (i.e., 0.5 × 80,000) completes the initiation, and the next 20,000 cycles completes the propagation. Thus, the remaining number of cycles at the second level is 60,000 cycles, whereas it would be 90,000 cycles if the LDR applied (i.e., 100/1000 + 90,000/100,000 = 1.0). This simple example explains why the concept of a DLDR applied to "crack initiation" and to "crack propagation" could appear attractive to early research workers in the field. It did explain how first loading at the high load (low life) level could cause a larger truncation at the subsequent lower load (high life) than would be expected on the basis of the LDR applied to

the complete life level of each loading. Similarly, it could also be shown that if the 10% of first loading were applied at the 100,000 cycle level, more than 90% would be left at the 1000 cycle level. Thus, the concept showed promise for accounting for the loading-order effect of summing cumulative fatigue damage.

The first suggestion of a separation of fatigue damage accumulation into two linear damage rules, applied in succession, was made in 1960 by Horace Grover (Ref 6.12). However, he did not provide a quantitative criterion for separating the life into what can be regarded as "initiation" and what part is "propagation." In 1965 Manson et al. (Ref 6.13) made an attempt at quantification. Based on limited data he proposed that if the complete life was $N_f$, the "propagation" period $N_p$ was $14(N_f)^{0.6}$ and the "initiation" was $N_f - 14(N_f)^{0.6}$. This assumption worked well for the set of data analyzed but was later found to be improper for other data. In fact, as we show subsequently, there is no possible choice of "initiation" and "propagation" that can truly account for the loading-order effect over a large range of conditions. However, to set the stage for this conclusion, we apply the recommended quantification to our illustrative problem.

**Application of the Manson Proposal to Illustrative Problem.** Using the criterion that

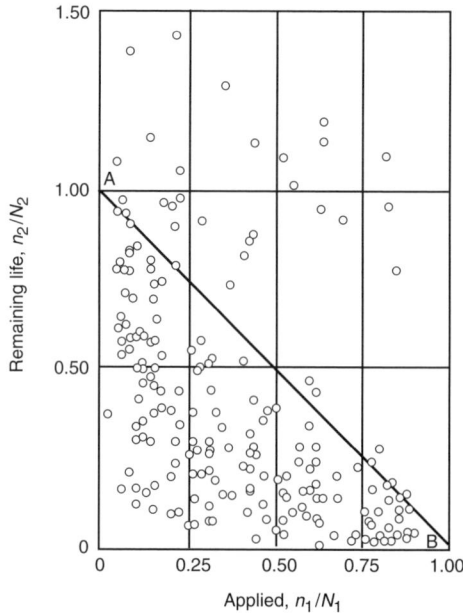

**Fig. 6.3** Two-level test results showing the inadequacy of the linear damage rule (LDR)

$N_p = 14N_p^{0.6}$ or $N_p/N_f = 14(N_f)^{-0.4}$, and therefore that $N_i/N_f = 1 - 14N_f^{-0.4}$, we then estimate the initiation and propagation lives as shown in Table 6.1.

Now let us apply the numbers in Table 6.1 to the problem at hand (Fig. 6.2). In the first block, we apply 1% of each of the loading levels, i.e., 10 cycles of $N_1$, 100 cycles of $N_2$, 1000 cycles of $N_3$, and 10,000 cycles of $N_4$. Thus, the sum of the fractions of the initiation lives used up in the first block is:

$$\frac{10}{120} + \frac{100}{6500} + \frac{1000}{86,000} + \frac{10,000}{954,000} = 0.1208$$

(Eq 6.2)

Because the hypotheses is that "crack initiation" occurs when the summation of the cycle ratios based on the "crack initiation" lives becomes unity, it will take 1/0.1208 (i.e., 8.28) blocks before crack initiation occurs. (We retain fractional blocks for convenience of discussion.) Further blocks then carry out the propagation. In the same way, one block of crack propagation consumes:

$$\frac{10}{880} + \frac{100}{3500} + \frac{1000}{14,000} + \frac{10,000}{56,000} = 0.290$$

(Eq 6.3)

That is, the number of blocks to "propagate" the crack is 1/0.290 = 3.45 blocks.

The total number of blocks to failure is the sum of the initiation blocks and the propagation blocks, or, 8.28 + 3.45 = 11.73 blocks. Because the LDR would predict 25 blocks, we see that this calculation predicts less than half of the cycles predicted by LDR, and therefore may be reasonable considering the known loading-order effects. In fact, we discuss later what we regard to be a more accurate approach and show that the expected life is about 12 blocks so that we have, in effect, obtained a fairly good answer by this method for the life levels considered.

But there are several difficulties with the approach, both conceptual and possibly numerical, in some cases. These difficulties are discussed in the next section.

## Difficulties with the Double Linear Concept Based on Crack "Initiation" and "Propagation"

In order to bring out the limitations of the concept, we consider what we would expect if each of the loadings were applied together with just one other loading, rather than all four together in blocks—for example, if we apply a certain number of loadings of $N_1$, and then follow them by additional loadings of $N_2$ until failure occurs. Let us apply 0.05 (5%, or 50 cycles) of $N_1$ and determine how much of $N_2$ would be left. The 50 cycles use up 50/120 = 42% of the crack initiation life, so by the rule 58% of initiation is left when the 10,000 cycle loading is applied. Therefore, the first 3770 cycles of $N_2$ (i.e., 0.58 × 6500) is needed to complete the initiation. An additional 3500 cycles carries out the complete propagation, and the total number of sustainable cycles at $N_2$ is 3770 + 3500 = 7270. Again, this sounds reasonable because the sum of the cycle ratio based on total life is 0.050 + 0.727 = 0.77, which is lower than unity, as would be expected on the basis of experience that the higher load level (lower life) is applied first. In Fig. 6.4(a) we plot this as point A.

Next, we consider what to expect if we use up the entire initiation in the first loading, i.e., apply 120 cycles at $N_1$ and then go to $N_2$. What is left is only the entire propagation period at $N_2$, 3500 cycles, or a cycle ratio of 0.35, plotted as point B in Fig. 6.4(a). Again, the summation of the cycle ratios based on total life is 120/1000 + 3500/10,000 = 0.47, lower than the 1.0 expected on the basis of LDR.

Finally, we consider a case where more than the "initiation" life is used up at the first loading, for example, applying 500 cycles at $N_1$. Then the whole initiation life of 120 cycles applies at $N_1$, plus the 380 (or 380/880 = 0.432). What is left at $N_2$ is 0.568 (i.e., 1 − 0.432) of the propagation life of 3500 cycles, or 1988 cycles. The point $n_1/N_1 = 0.5$ and $n_2/N_2 = 1988/10000 = 0.1988$ is plotted as point C in Fig. 6.4(a). It is clear that O, A, and B lie on a straight line, and B, C, and J lie on a straight line. Any other choice of initial loading at $N_1$ will result in a remaining cycle ratio at $N_2$, which lies on line OAB or BCJ, as would be expected from the consideration of the DLDR.

Applying similar reasoning to combined loadings between $N_1$ and $N_3$ results in lines ODE and

Table 6.1 "Initiation" and "propagation" lives (cycles) for illustrative problem

| $N_f$ | $N_{Initiation}$ | $N_{Propagation}$ |
| --- | --- | --- |
| $10^3$ ($N_1$) | 120 | 880 |
| $10^4$ ($N_2$) | 6,500 | 3,500 |
| $10^5$ ($N_3$) | 86,000 | 14,000 |
| $10^6$ ($N_4$) | 954,000 | 56,000 |

*EFJ*, and between $N_1$ and $N_4$ ($N_4 > N_3$) results in *OGH* and *HIJ*. Likewise, other combinations of these load levels result in the other plots shown in Fig. 6.4(b), (c), and (d). Note the position of the lines above the LDR 45° line. These correspond to conditions of higher-cycle fatigue loading followed by lower-cycle fatigue loading. In such cases, the life fraction sum would be greater than 1.0. The important feature of all of these plots is that all show double linear representation. The "knee," or "breakpoint," is where the horizontal coordinate is the fraction of life when the first life level reaches its initiation value. The vertical coordinate is the fraction of the life at the second level that constitutes its propagation period. This is true for all the cases shown in Fig. 6.4.

It is the fact that the horizontal coordinate of the kneepoint depends only on the first loading level, and the vertical coordinate depends only on the second loading level, that militates against the concept of a DLDR based on the initiation and propagation phases. To check the evidence, consider Fig. 6.5, which is a summary of

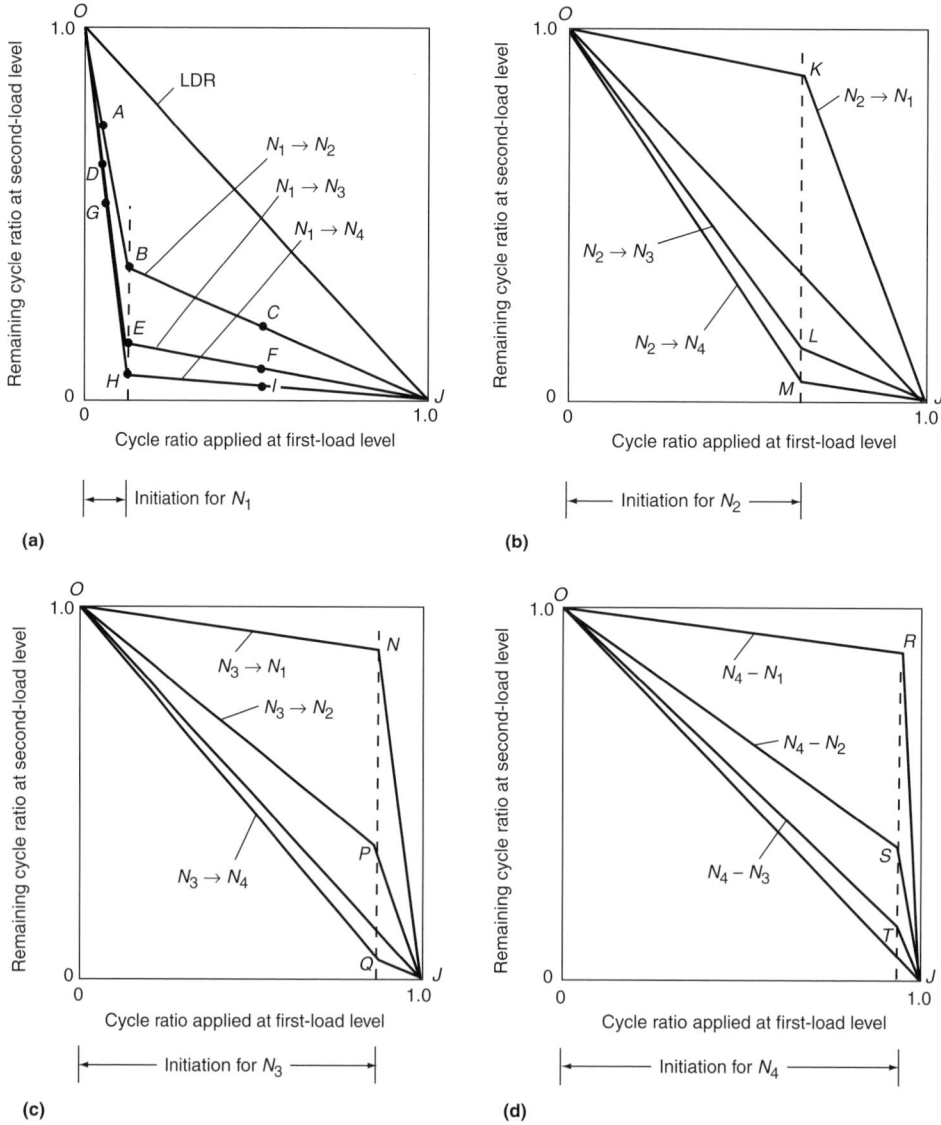

**Fig. 6.4** Concept of the original double linear damage approach based on physical "initiation" and "propagation" of cracks applied to illustrative problem. (a) Initiation at $N_1$. (b) Initiation at $N_2$. (c) Initiation at $N_3$. (d) Initiation at $N_4$

Chapter 6: Cumulative Fatigue Damage / 129

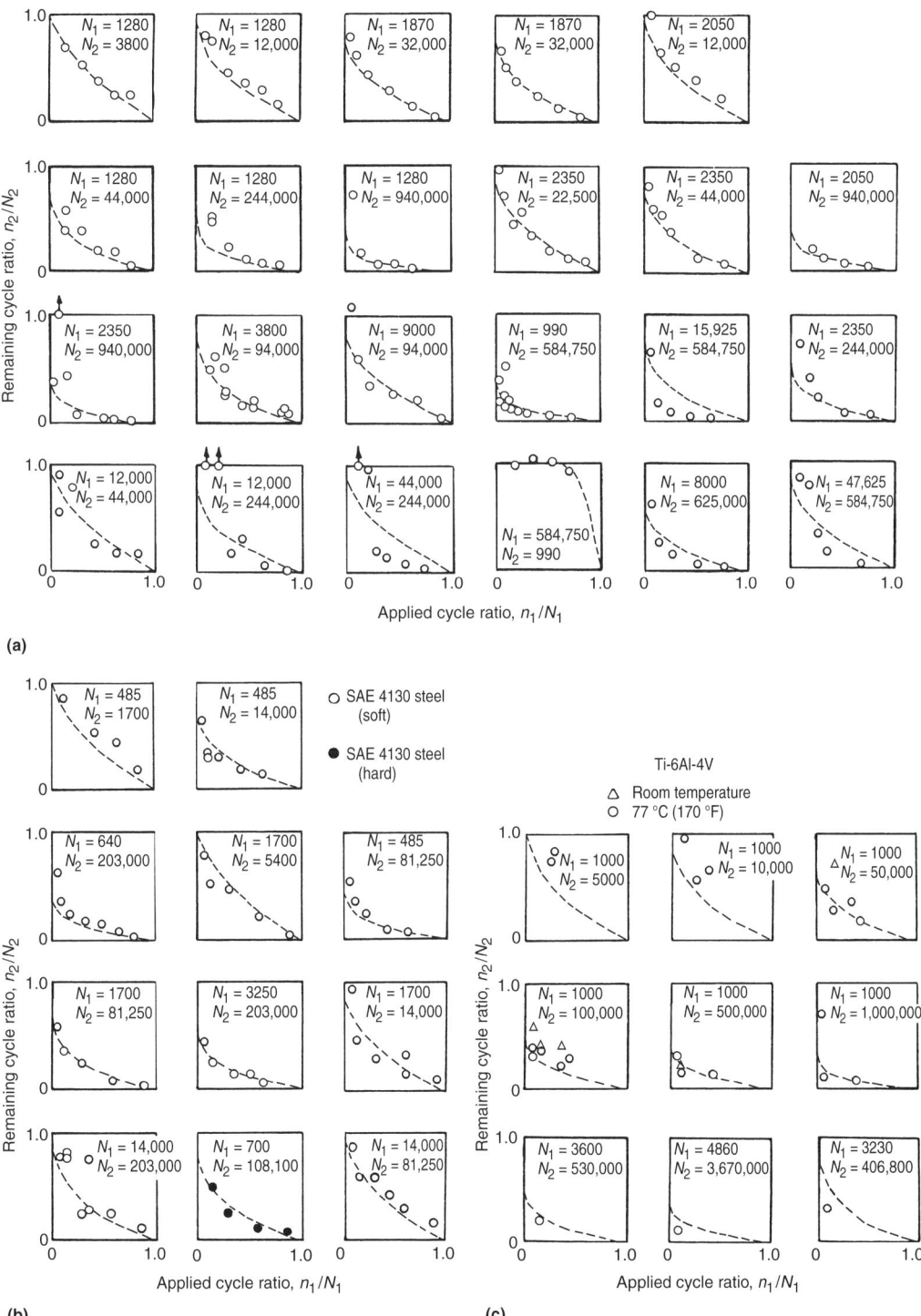

**Fig. 6.5** Two-load-level test results for three engineering alloys. Data from Ref 6.14. (a) 300 CVM steel (500 HB). (b) SAE 4130 steel (soft, 260 HB and hard, 367 HB). (c) Ti-6Al-4V (STA, 377 HB)

numerous two-load-level tests conducted at NASA (Ref 6.14). Three different classes of materials were studied: 300 CVM maraging steel (500 HB), SAE 4130 steel hard (367 HB) and soft (260 HB), and Ti-6Al-4V alloy (solution treated and aged). Considering data scatter, it is not always clear how to draw two straight lines through the data to obtain a well-defined kneepoint. Using our best judgment, we obtained approximate kneepoints and plotted the results shown in Fig. 6.6. If the concept were correct, there would be a one-to-one relation between the vertical coordinate $n_1/N_{f,1}$ and the horizontal coordinate $N_{f,1}$. That is clearly not true! When $N_{f,1}$ is $10^3$, for example, the kneepoint can be as low as 0.05 or as high as 0.45, depending on the second life level. Note that these results do not necessarily require the Manson hypothesis that $N_p = 14(N_f)^{0.6}$. No other assumption could be appropriate because any notation defining the initiation or the propagation life would still also require a single one-to-one relation between the first applied load and the horizontal coordinate of the kneepoint. The coordinates of the breakpoint are calculated for the damage line to be used next, and the decision is made as to which line to use according to whether the damage on that curve is greater than the breakpoint value either before or after the cycle ratio increment is applied.

Another question that can be leveled at the DLDR based on initiation and propagation is that of "crack size" that constitutes initiation. In Fig. 6.7, we show some results (Ref 6.15) for crack initiation on smooth specimens subjected to completely reversed strain cycling. Here, a "crack" was defined as being approximately 0.076 mm (0.003 in.) deep, or 0.254 mm (0.010 in.) along the surface. Such a crack is clearly visible with the 20 power microscope used. To apply these results to our example shown in Fig. 6.4, the "kneepoint" for the cases where $N_1 = 1000$ cycles should be at about $n_1/N_1 = 0.12$. Based on Fig. 6.7, the first "visible" crack when $N_1 = 1000$ is ($n_1/N_1 = 0.75$). Thus, if a crack does develop at $n_1/N_1 = 0.12$, it must indeed be a tiny crack. We made several attempts to detect a crack at the kneepoint but never observed a visible crack using a 30× stereo optical microscope.

## Potential for a Double Linear Rule

While we have demonstrated that a double linear damage concept based on crack initiation and propagation is not viable in a strict sense, there is still potential for the concept if the basis is changed. Later in this chapter we describe an approach that still has the features of a DLDR with all its advantages of simplicity, but in which the breakpoints are flexibly tailored to the problem being studied rather than on the rigid physical concepts.

### Cumulative Damage Analysis Based on Damage Curves

**Introduction.** An alternate approach to treating cumulative fatigue damage is through dam-

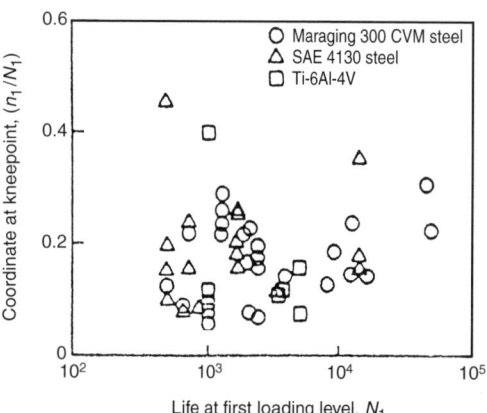

**Fig. 6.6** Lack of correlation between kneepoint coordinates and the concept of a discrete crack initiation event. Source: Ref 6.10

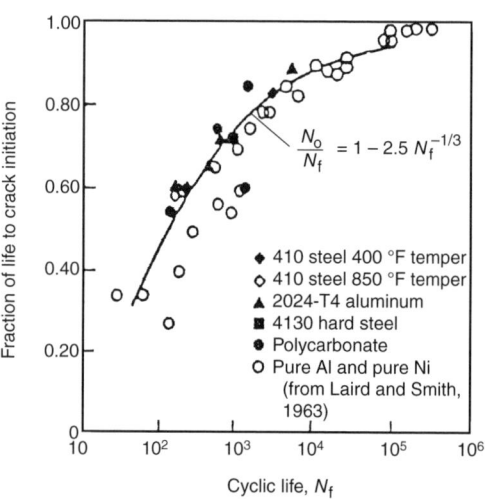

**Fig. 6.7** Relation between percent of life to initiate approximately 0.076 mm (0.003 in.) deep crack and fatigue life for smooth specimens under uniaxial cycling. Source: Ref 6.15

age curves. Suppose, for example, damage curves such as shown in Fig. 6.8 are applicable. For this purpose it is not really necessary to define the term "damage" $D$ quantitatively. It is only necessary to postulate that when $D$ accumulates to 1.0, the metal has reached the end of its useful life, and that in changing from one loading level to the next we can proceed along a horizontal line, implying that no new damage is introduced during the changeover. The amount of damage introduced at each loading level is then tracked according to the cycle ratio applied at each load level. For example, in Fig. 6.8 the cycle ratio $n_1/N_1 = 0.3$ is applied at the $10^3$ life level and the damage is $OA$, then $n_2/N_2 = 0.1$ is applied at the $10^5$ life level, introducing the damage $BC$. At that point, $n_3/N_3$ is applied at the $10^6$ life level, introducing the damage $DE$ with a cycle ratio of only 0.02. It is clear that in this case the summation of the cycle ratios to failure is less than unity. However, if the loading order were reversed, the summation of the cycle ratios would be greater than unity. Thus, the damage curve concept, if the curves line up in the order shown in Fig. 6.8, lends itself to predicting a loading-order effect that is consistent with experimental evidence.

The damage curve concept was first proposed in 1948 by Richart and Newmark (Ref 6.16); however, they did not provide quantitative equations. Others have also explored this approach, including us. In 1981 (Ref 6.17) we proposed a relation in the form:

$$D = \left(\frac{1}{0.18}\right)\left[\alpha_o + (0.18 - \alpha_o)\left(\frac{n}{N}\right)^{(2/3)N^{0.4}}\right] \quad \text{(Eq 6.4)}$$

This equation was based on an analogy to an equation we had developed for early crack growth. However, after some experience we came to the realization that although Eq 6.4 provided a definite set of damage curves, exactly the same answers could be obtained by choosing the damage curve equation in the form:

$$D = \left(\frac{n}{N}\right)^{(N/N_{ref})^{0.4}} \quad \text{(Eq 6.5)}$$

where $N_{ref}$ was an arbitrary reference life. Thus, the other constants in Eq 6.4 were really superfluous and we could limit our attention to Eq 6.5. Furthermore, in Eq 6.5, $N_{ref}$ was arbitrary, so by choosing $N = N_{ref}$ we could cause one of the damage curves to be a straight line. Thus, we could arbitrarily linearize any one of the damage curves, and the others would correspondingly be arranged to produce the same results as if we had linearized any other one. Fig. 6.9 shows this feature of the damage curves. In Fig. 6.9(a) the damage line for $N_{ref} = 10^4$ is made straight. All the other damage lines take their place as shown when determined by Eq 6.5. In Fig. 6.9(b) the value of $N_{ref}$ is taken to be $10^3$, so the damage line for $N_{ref} = 10^3$ becomes the straight line, but all the other damage curves are identical to those of Fig. 6.9(a), except shifted by one decade. Any cumulative damage problem could be solved according to either set of damage curves, and the results would be identical. For example, the illustrative problem of Fig. 6.2 yields the answer of 10.8 blocks when solved according to either set of damage curves, and would also yield the same answer if, for example, either of the damage curves for $N = 10^5$ or $N = 10^6$ were linearized. The LDR predicts 25 blocks in all cases.

The realization that it is not necessary to assign a unique damage for each life level opened a whole new perspective to consideration of cumulative damage analyses by the damage curve concept. We really do not need to define damage according to any particular physical manifestation. Any family of damage curves could obtain the same answer, so long as the damage curves are chosen to bear the proper relation to each other. In Ref 6.15 we referred to the approach as a damage curve analysis. Since that time we have used the terminology damage curve approach (DCA) (Ref 6.5, 6.10).

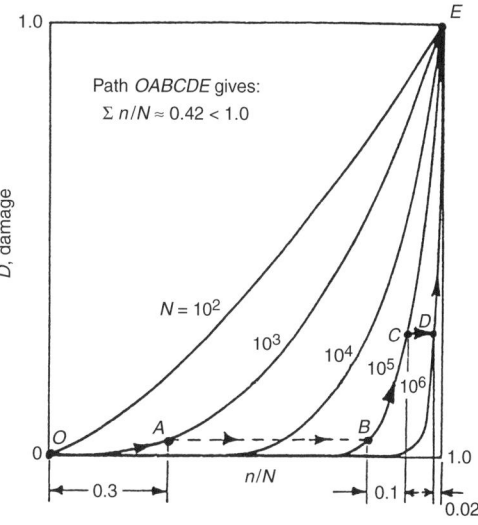

**Fig. 6.8** The concept of damage curves and their implementation to sum damage due to successive loadings

## Extension of the Damage Curve Concept to a Set of Double Linear Damage Lines

At this point, it became clear as to how a new set of damage curves might be developed based on the extensive two level tests shown in Fig. 6.5. First, however, it was necessary to reinterpret these data in terms of the damage curve concept. Figure 6.10(a) shows schematically how any one of the remaining life curves of Fig. 6.5 might be interpreted by passing a straight line through the data for low, and another straight line for high, $n_1/N_1$. (Although we have demonstrated that the crack initiation/crack propagation distinction is improper for representing the data, we still retain the double linear feature of this curve, partly because when adequate data are available, the trend for such representation seems justified, and partly because a very simple conclusion can be deduced by this representation.) To obtain Fig. 6.10(b) we simply rotate Fig. 6.10(a) 90° counterclockwise. From the geometry of the figure it is clear that if $n_2/N_2$ is the remaining life fraction at $N_2$ when $n_1/N_1$ is first applied, exactly the same value of $n_2/N_2$ would be obtained by a damage curve analysis in which we linearized the damage curve for $N_1$ and chose the double linear damage line ($A'C'E'$) for $N_2$.

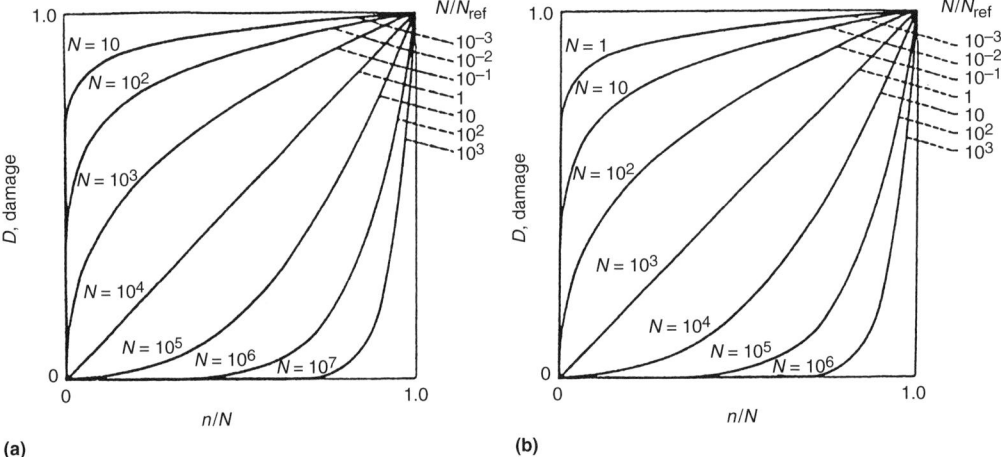

**Fig. 6.9** Damage curves. (a) Reference life $N_{ref} = 10^4$. (b) Reference life of $N_{ref} = 10^3$

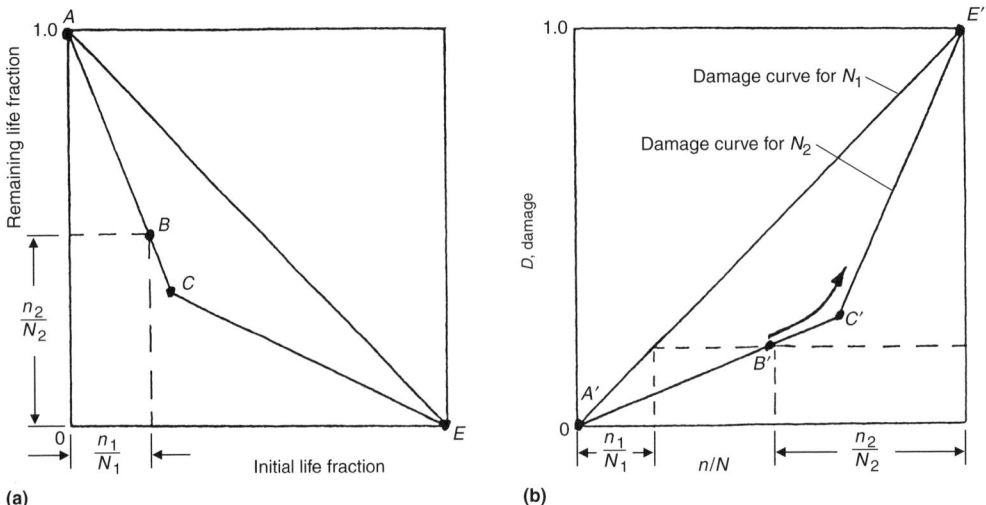

**Fig. 6.10** Remaining life curve (a) when $n_1/N_1$ cycle ratio is applied first and $n_2/N_2$ is determined experimentally, and (b) when reinterpreted as a pair of damage curves in which the damage line for $N_1$ is linearized

In other words, each of the remaining life curves of Fig. 6.5 provides us with data that we can use to establish appropriate double linear damage curves (DLDCs) for general use. The important difference we learned, however, from the damage curves of Eq 6.4 or 6.5 is that we must establish all features of the damage curves in terms of the ratio $N_1/N_2$, rather than $N_1$ and $N_2$ separately. Thus, the kneepoints should also be determined in terms of $N_1/N_2$.

In Fig. 6.11 we show two plots that were constructed using the data of Fig. 6.5, using the coordinates of the kneepoint. On the basis of this analysis we can now construct the family of damage curves shown in Fig. 6.12. These damage curves are completely general and can be

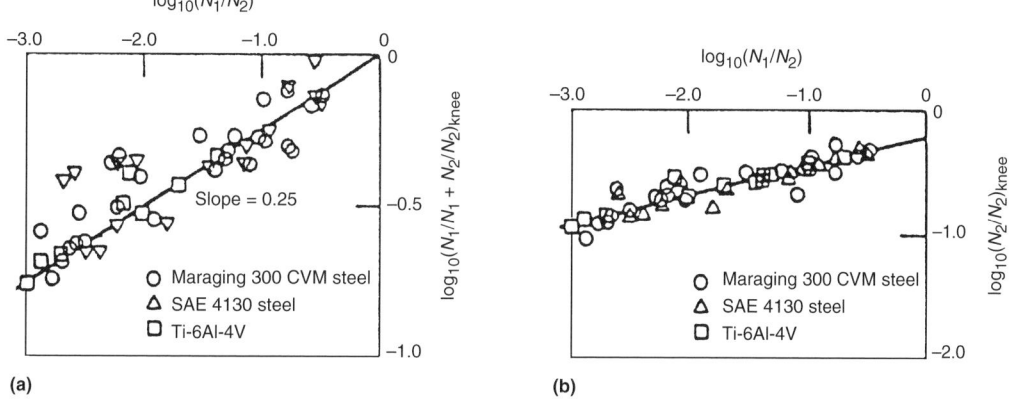

**Fig. 6.11** Determination of exponent and coefficient used to calculate the kneepoint for the double linear damage curve. (a) Exponent. (b) Coefficient. Source: Ref 6.10

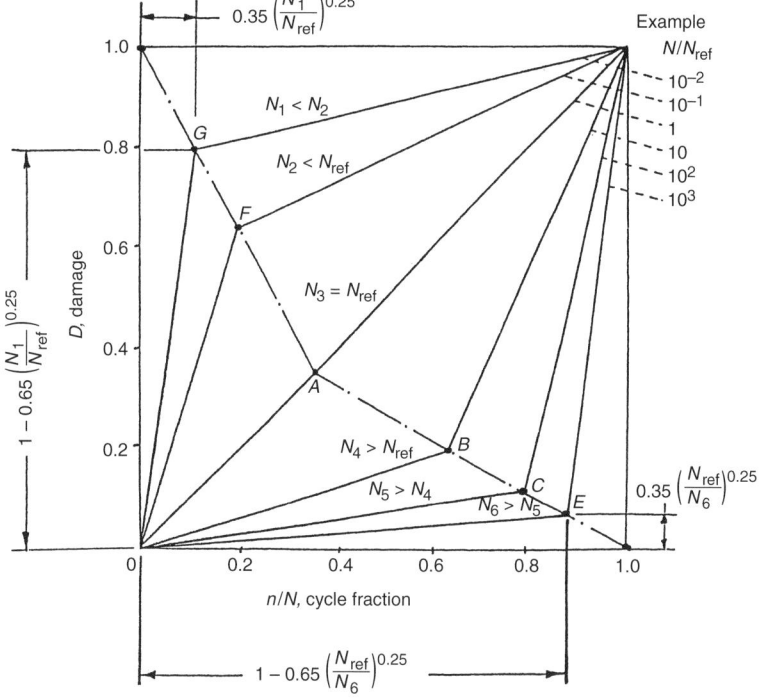

**Fig. 6.12** Construction of double linear damage curve

used by linearizing any one of the curves for one of the events appearing in the loading sequence for a particular problem. Once the life level to be linearized has been chosen, the damage curves for all events of known life can be constructed using the coordinate scheme shown in Fig. 6.12. As seen in the figure, if $N_{ref}$ is the reference life level, the damage curve for $N_{ref}$ is a straight 45° line, and the damage curve for any other life level consists of two straight lines with breakpoint. For $N < N_{ref}$, the breakpoint coordinates are:

$$\text{Horizontal coordinate} = 1 - 0.65\left(\frac{N_{ref}}{N}\right)^{0.25}$$

(Eq 6.6a)

$$\text{Vertical coordinate} = 0.35\left(\frac{N_{ref}}{N}\right)^{0.25}$$

(Eq 6.6b)

For $N > N_{ref}$, the breakpoint coordinates:

$$\text{Horizontal coordinate} = 0.35\left(\frac{N_{ref}}{N}\right)^{0.25}$$

(Eq 6.6c)

$$\text{Vertical coordinate} = 1 - 0.65\left(\frac{N_{ref}}{N}\right)^{0.25}$$

(Eq 6.6d)

### Application of the Double Linear Damage Curve (DLDC) to our Illustrative Problem

Although calculations using the DLDC are basically very simple, and require no information beyond that normally required for the ordinary LDR, practical problems are best handled by a computer program.

The analysis for a sample problem is shown in Fig. 6.13. Figure 6.13(a) shows the overview of the problem with the break points computed from Eq 6.6, while in Fig. 6.13(b) to (d), enlarged views are shown of the first, sixth, and last (11.73) blocks. Figure 6.13(c) is included to show how the addition of a given cycle ratio can cause the damage to move across a break point. In the computer program the selection of which of the two lines the final point occurs is determined by a calculation, which is always made at the beginning of the analysis and stored in the computer.

It is clear from Fig. 6.13 that the calculated life for our illustrative problem is 11.73 blocks compared with the 25 blocks predicted by the LDR. This difference is quite substantial and, if correct, emphasizes the importance of considering a nonlinear damage summation process. We shall now show some evidence for the correctness of the concept.

**Some Applications of the Double Linear Damage Curves (DLDC).** One of the first checks of the DLDC concept is to apply it to the two-level loading tests shown in Fig. 6.5 that already have been shown in Fig. 6.3 not to comply with the LDR. The reanalysis by the DLDC is shown in Fig. 6.14, with more favorable results. Of course, there is still a scatterband around the data, as would be expected. It may be argued that because the constant in the DLDC structure was derived from these tests, their concurrence with predictions by the method is ensured. However, there are a great many data points for many materials, and such good conformity would not likely be expected if the thesis of the method were in error.

We have also conducted additional tests and analyses to check the validity of the method. The first set of tests involving three load levels has been conducted on Haynes 188 alloy, a very important cobalt-base superalloy used in the original space shuttle main engines. (The test program described here was conducted at the NASA Lewis Research Center, Cleveland, OH, by Case Western Reserve University graduate student, Sissay Hailu, as part of his master's degree research in 1991 [Ref 6.18].) Figure 6.15 shows the results. Figure 6.15(a) illustrates the strain cycling fatigue test specimen used for the Haynes 188 alloy. Three strain levels were chosen to involve three different cyclic life regimes (150 to 500), (4500 to 7900), and (66,000 to 224,000). The regimes were selected as far apart as possible, yet in the regime in which tests could be conducted in a practical time frame. Various tests were conducted involving the sequencing shown in Fig. 6.15(b). In these tests it was not always possible to achieve the target strain range exactly, but the strain values were measured and the life values determined from Fig. 6.15(c). The results of the analyses of the data are shown in Fig. 6.15(d) and (e). It is clear that the DLDC predictions agree better with the experimental data than do the LDR predictions. The LDR predictions substantially overestimate life in all cases.

Another application of the DLDC is shown in Fig. 6.16. Here, SAE data (Ref 6.19) were analyzed (by S. Hailu while a Ph.D. candidate at Case Western Reserve University, Cleveland, OH, in the 1990s) both by the LDR and DLDC. Again, it is clear from Fig. 6.16(c) that the LDR normally overestimates life, while the DLDC more accurately predicts cyclic lifetimes.

A final example is provided in Fig. 6.17. The data shown were generated by Webber and Levy (Ref 6.20) long before the development of DLDC. They used three two-level tests, with components in the very high life range (mean fatigue lives of 59,000, 414,000, and 13,800,000 cycles) combined. Complete data were provided in the original report. We have superimposed the expected results using DLDC, which agree remarkably well with the experimental data. What is of special interest here is that the same type of results are obtained when both life levels are quite high, thus providing further evidence against a DLDR based on crack initiation and crack propagation concepts. An examination of Fig. 6.7 would suggest that at all of the life levels involved, the total life is almost entirely crack initiation. Thus, if the LDR concept were correct, we would expect it to apply to these data. Yet the results are the same as would be expected from the DLDC based only on the ratios involved. This result is damaging to the LDR concept as well as to the original DLDR concept based on physical crack initiation and crack propagation. They are very favorable toward the DLDC concept based on life ratios.

**Resurrection of the DLDR by Changing the Physical Basis for the Break Point.** While the DLDR, based on the concept of crack initiation and crack propagation had already long been

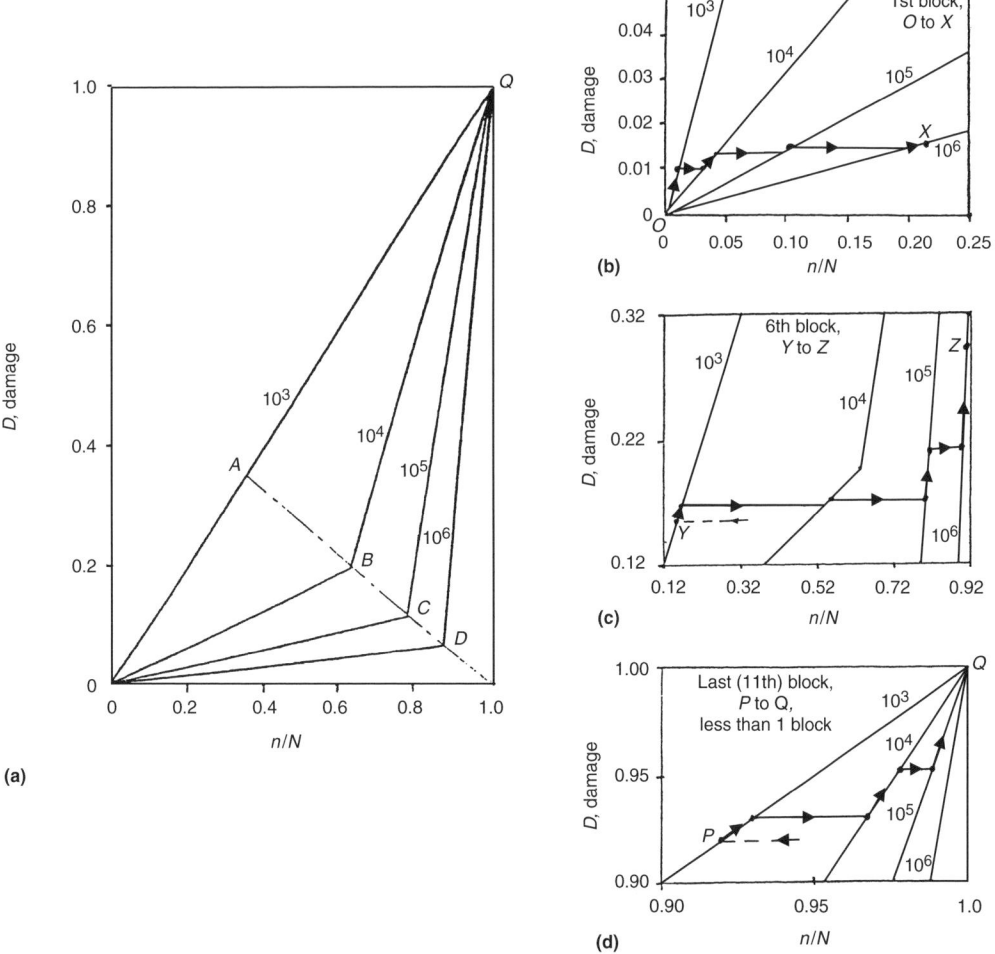

**Fig. 6.13** Application of the double linear damage curve. To illustrative problem (a) Overview. (b) First block. (c) Sixth block. (d) Last (11th) block

136 / Fatigue and Durability of Structural Materials

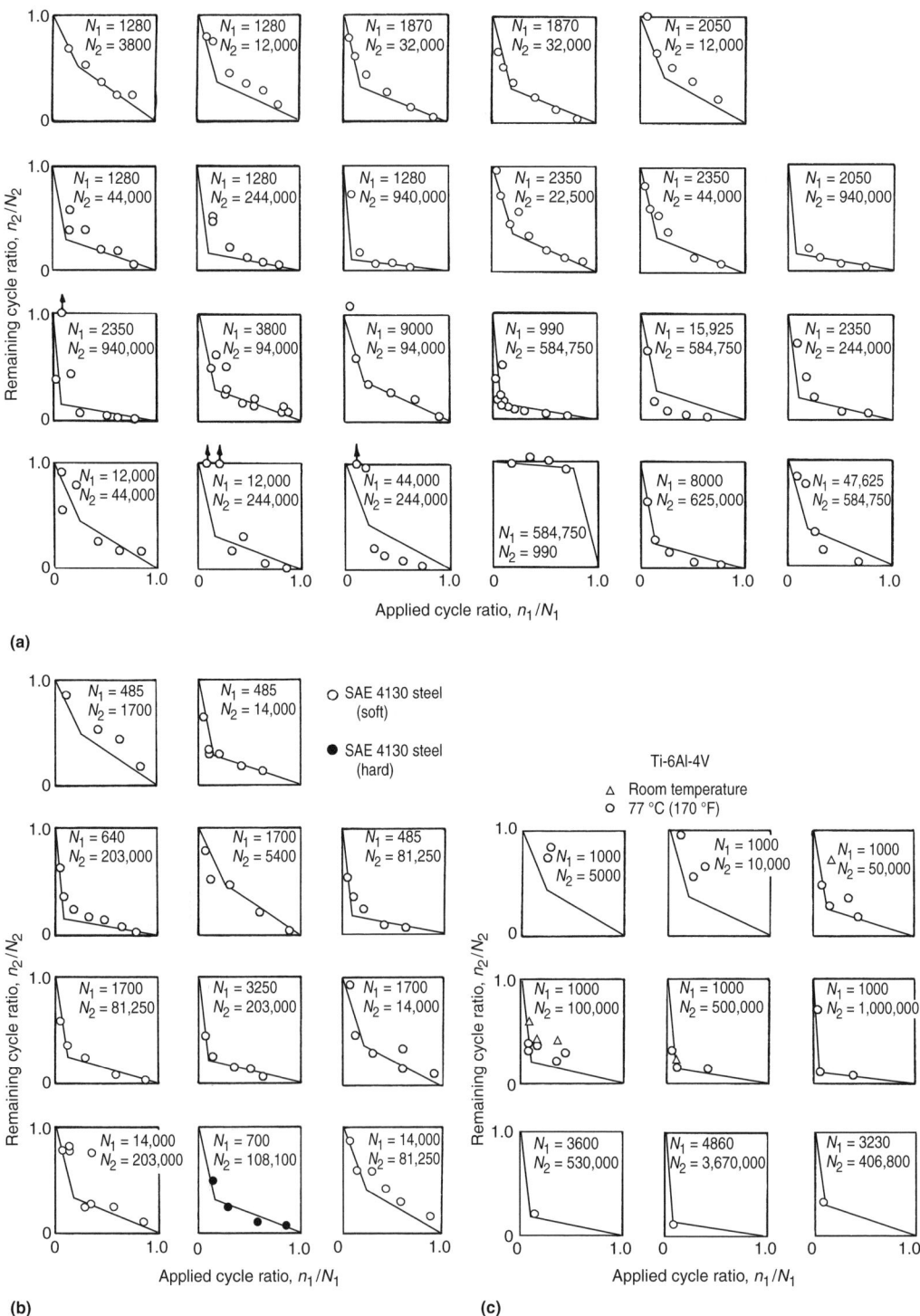

**Fig. 6.14** Application of double linear damage curve to data of Fig. 6.5. (Note that this model was originally referred to as the double linear damage rule in the earliest publications). (a) 300 CVM steel. (b) SAE 4130 steel (both hard and soft). (c) Ti-6Al-4V

Chapter 6: Cumulative Fatigue Damage / 137

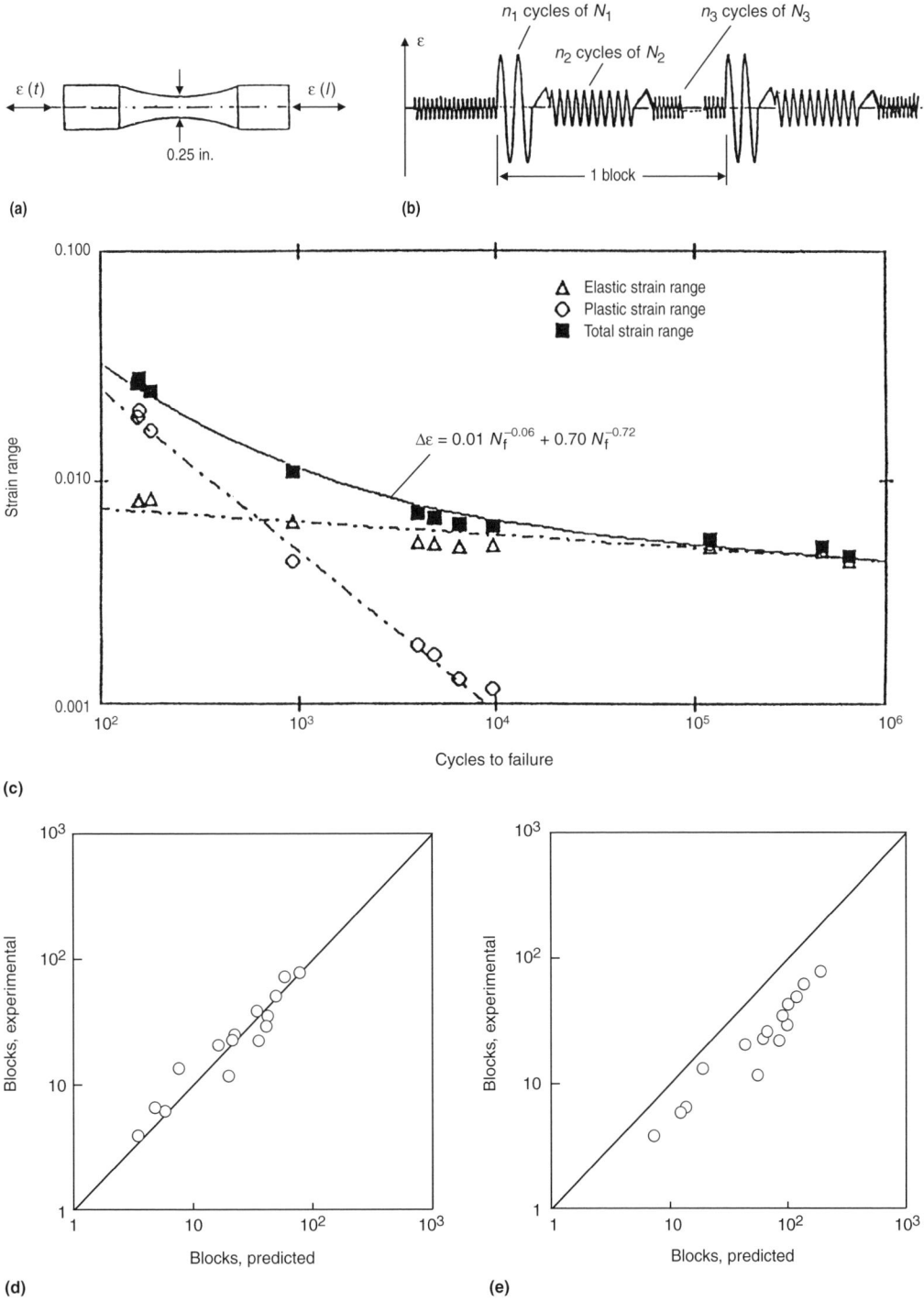

**Fig. 6.15** Experimental program to check the linear damage rule and double linear damage curve for blocks of three loading levels using Haynes 188 alloy. (a) Diametral specimen. (b) Loading history at 760 °C (1400 °F). (c) Strain-life curve, Haynes 188 at 760 °C (1400 °F). (d) Double linear damage curve predictions. (e) Linear damage rule predictions. Source: Ref 6.18

138 / Fatigue and Durability of Structural Materials

abandoned, the success of the DLDCs introduced a new, hopeful view for identifying a DLDR based not on crack initiation and propagation, but rather by recasting the DLDC so as to force the break points along a horizontal line of constant damage. If an alternate simple set of damage lines can be constructed with this characteristic, then the tremendous simplifications offered by a DLDR can be capitalized on.

Consider, for example, the problem analyzed in Fig. 6.13. Here, we have four damage curves; one is a straight line, the others have break points at various levels of damage. Suppose we seek to replace them by a set of damage curves, all of which have their break points at the same ordinate. We shall consider where to choose the ordinate of the break point in detail later, but for the present, let it be assumed that we arbitrarily choose it to be at the same ordinate as that for the damage curve for the $10^5$ life, point $C$. Thus, the damage curves for a life $10^3$ $OAQ$ can still be retained as it was when analyzed earlier because any point along the straight line $OAQ$ can be regarded as a break point. Similarly, the dam-

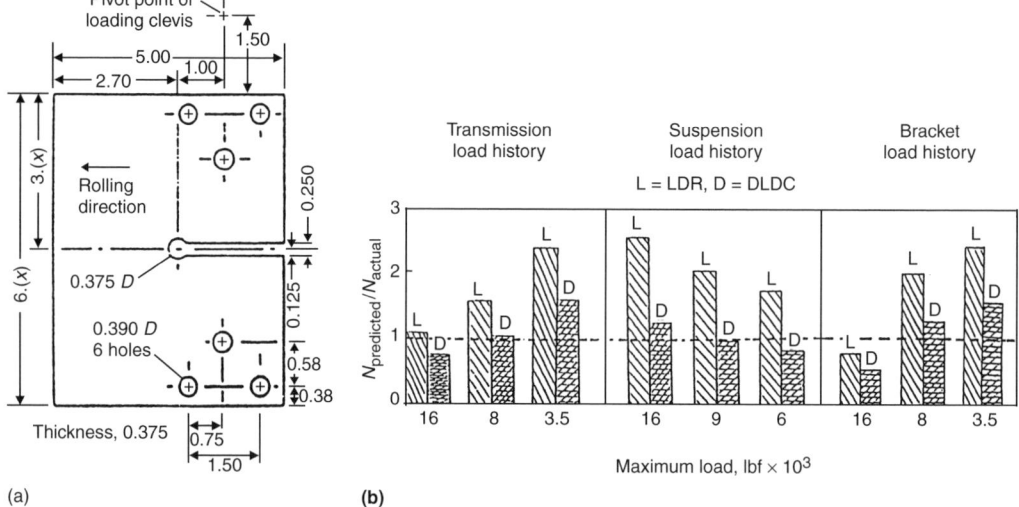

**Fig. 6.16** Comparison between linear damage rule and double linear damage curves for SAE data on man-ten steel for three loading histories: suspension load, transmission load, and bracket vibration. (a) SAE specimen (Ref 6.20). (b) Linear damage rule compared with double linear damage curve

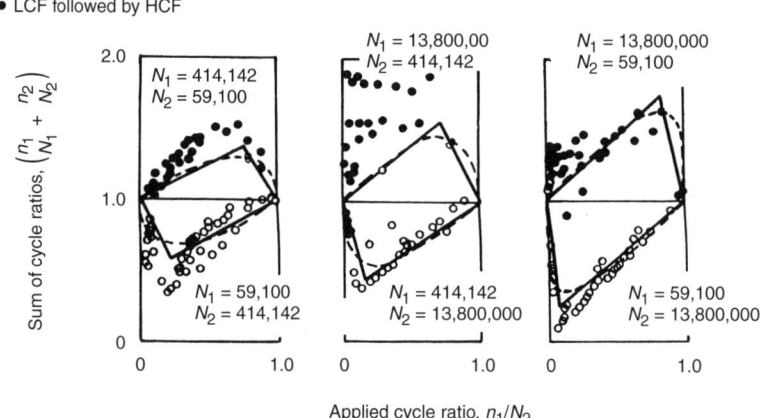

**Fig. 6.17** Two-load-level test results and double linear damage curve predictions for D.T.D. 623 aluminum alloy, data from Ref 6.20

age curve $OCQ$ can be retained without change because it has its break point at the chosen level. The damage curves $OBQ$ and $ODQ$ of Fig. 6.13 must, however, be compromised. In Fig. 6.18, the compromised damage curves are shown as $OB'Q$ and $OD'Q$. The choice of the horizontal coordinates $B'$ and $D'$ can be made on the basis of several possible self-evident suggestions. For example, to get $B'$ we could average where $OB$ and the downward extension of $BQ$ intersect the horizontal line through $C$. Such an approach of averaging intersection values might, however, introduce difficulty when considering the damage line at a much higher life. In such a case, one of the intersection points could be at a value of $N_I/N_f$ greater than 1.0 which is impossible, depending on the value of the life for which the compromised damage curve is to be determined:

$$\frac{N_I}{N_f} = \exp Z(N_f)^\Theta \quad \text{(Eq 6.7)}$$

where $N_I$ is the life of phase I for the total life $N_f$, and $Z$ and $\Theta$ are constants to be determined from the fact that the kneepoints for two life levels are known.

In this equation, if $\Theta$ is negative it cannot produce the anomalous behavior of $N_I/N_f$ to be greater than 1.0.

Recognizing that points $A'$ and $C'$ in Fig. 6.18 must fit Eq 6.7, we arrive at the general development.

Thus, in essence, we consider the fatigue process to consist of two phases analogous to, but not equivalent to, crack initiation and crack propagation. To distinguish these phases we designate them as phase I and phase II. Note, however, that these phases are not identified with a definable physical process. They depend, instead, on the ratio of the life level considered to be the reference life level, and change as different life levels become involved. The life levels $N_1$ and $N_2$ are those which, for the particular problem being analyzed, are regarded to be identified with the two most damaging events in the history being analyzed. An alternate approach, while not discussed, gives comparable overall results and defines $N_1$ and $N_2$ as the lowest and highest lives that exhibit life fraction $n/N$ of some minimum amount on the order of 0.001 to 0.010 (Ref 6.5).

After several trials, we settled on the following procedure to be acceptable:

1. Use as the reference life, which results in the 45° damage line, the lowest life level in the blocks as was done in Fig. 6.13(a). Usually, the lowest life level causes a large amount of the damage, so it is desirable not to compromise the damage curve for this event. This feature is ensured by choosing it as the 45° line. This choice also leads to simplification of the ensuing construction.

2. Let the assumed constant value of the transition level of damage, below which we call phase I, and above which we call phase II, be $D_{I-II}$, as shown in Fig. 6.19(a). Let $N^*$ be the particular life level that would have its break point at an ordinate of $D_{I-II}$. This life level may actually be associated with an event in the block, or it may be an intermediate level. To obtain its value, we take the averages of all the kneepoints involved in the history. As a first approximation, the weighting process is based on the relative importance of each of $k$ events in a block using the relative cycle ratios:

$$(n/N)_i / [(n/N)_1 + (n/N)_2 + \ldots + (n/N)_k]$$
$$= (n/N)_i \Sigma[(n/N)_i]$$

for each $i$th event. Usually this approach will be satisfactory, but the process can be iterated on the basis of damage calculated from the DLDR after it is once determined. Thus, if each relative cycle ratio is multiplied by the damage $D_{knee,i}$ at each kneepoint for life $N_i$, the average value of $D_{knee,i}$ ($= D_{I-II}$) is given by:

$$D_{I-II} = \frac{\sum_{i=2}^{k} \left(\frac{n}{N}\right)_i D_{knee,i}}{\sum \left(\frac{n}{N}\right)_i} \quad \text{(Eq 6.8)}$$

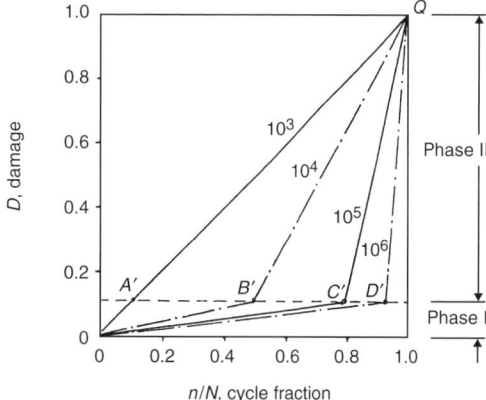

**Fig. 6.18** Reconstructed family of damage curves for our standard illustrative problem

But, from Eq 6.6c:

$$D_{knee,i} = 0.35\left(\frac{N_1}{N_i}\right)^{0.25} \quad \text{(Eq 6.9)}$$

Therefore:

$$D_{I-II} = 0.35 \sum_{i=2}^{k}\left[\left(\frac{n}{N}\right)_i\left(\frac{N_1}{N_i}\right)^{0.25}\right] \quad \text{(Eq 6.10)}$$

3. We now replace all the damage curves by a set of damage curves all of which have their break point at an ordinate of $D_{I-II}$, as shown in Fig. 6.19(b) using the assumption that:

$$\frac{N_{1,i}}{N_i} = 1 - P\left(\frac{N_1}{N_i}\right)^Q \quad \text{(Eq 6.11)}$$

Two known values of $N_I$ are available at $N_1$ and $N^*$. For the reference 45° line the value of $N_{I,1}/N_1 D_{I-II}$, and for $N^*$, $N_I^*/N^* = (1 - 0.65)(N_1/N^*)^{0.25}$. The first condition leads to $P = 1 - D_{I-II}$. From the second condition, and using the $N^*$ defined by the equation:

$$D_{I-II} = 0.35\left(\frac{N_1}{N^*}\right)^{0.25} \quad \text{(Eq 6.12)}$$

it follows that:

$$Q = 0.25 + \frac{\ln\left[\frac{0.65}{1 - D_{I-II}}\right]}{4\ln\left[\frac{D_{I-II}}{0.35}\right]} \quad \text{(Eq 6.13)}$$

4. Thus, the life levels in a block can be broken into two parts, $N_{I,i}$ and $N_{II,i}$, such that:

$$N_{I,i} = N_i\left[1 - P\left(\frac{N_1}{N_i}\right)^Q\right] \quad \text{(Eq 6.14)}$$

$$N_{II,i} = N_f - N_{I,i} \quad \text{(Eq 6.15)}$$

where $P$ and $Q$ are given by Eq 6.13 and 6.11.

### Summary of DLDR

The double linear damage rule (DLDR) is a scheme for dividing each loading level in a sequence (block) into two parts, one to reach a given level of damage, the other to carry the damage to failure. The transition damage, designated $D_{I-II}$, does not have any physical significance (such as the demarcation between crack

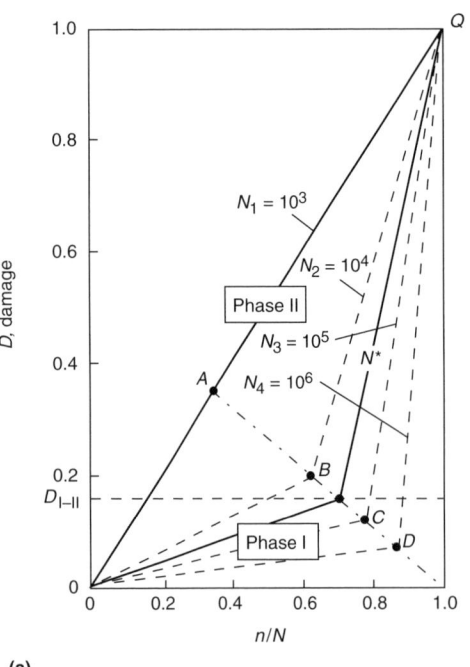

(a)

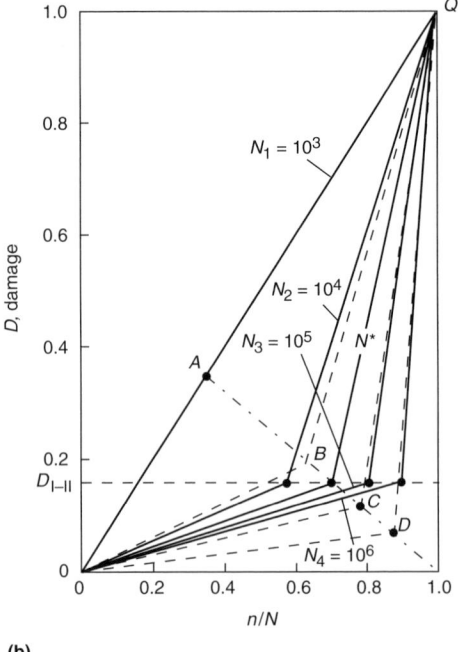

(b)

**Fig. 6.19** The double linear damage rule construction to determine the level of damage separating phases I and II. (a) Identification of $N^*$. (b) Establishment of $D_{I-II}$

initiation and crack propagation), and it depends on the content of loadings in the block. Changing the number of loading cycles for any of the components in a block will change the transition damage. The overriding advantage of using the DLDR compared with the double linear damage curve (DLDC) procedure is the reduced amount of computation needed in the DLDR compared with the DLDC approach, while obtaining answers that do not differ by large factors. The breakdown of each life level is accomplished by use of Eq 6.14 and 6.15. Note that the total life consisting of the two phases for a given loading level is constant. Within each of the two phases, the summation of the cycle ratios is taken as unity, just as in the LDR. However, each time a loading of any life level is applied, the cycle ratio is obtained by dividing unity by the life of phase I, to obtain the damage increment that is summed to determine the completion of phase I. After phase I cycle ratios sum to unity, regardless of order applied, the process changes to adding cycle ratios based on phase II, until they, too, add up to unity, after which it is assumed that "failure" takes place. The damage done by each loading level is the sum of its phase I cycle ratio and its phase II cycle ratio, (i.e., $n_1/N_1 + n_2/N_2$) where $n_1$ is the number of cycles of a given loading during phase I, and $n_2$ is the number of cycles of that load level during phase II. Thus, the damage differs when treated by the DLDR than when treated by LDR. While the LDR determination of relative damage is a good starting point to estimation of $D_{I-II}$ in the first calculation, refinement can be made by iteration using in each case the damage determined by DLDR calculation of the damage in the previous iteration. For the calculations we have conducted, the preponderant number of cases required no more than the initial LDR assumption of damage to get a satisfactory answer.

A number of versions of nonlinear or double linear damage accumulation schemes the authors have developed over the years have been discussed in detail. To aid in keeping these straight, Table 6.2 and Fig. 6.20 provide a concise summary.

## Application of DLDR to Several Problems

It is interesting to review several applications of the DLDR to determine how it compares with the linear damage rule (LDR) and to the double linear damage curve (DLDC) procedures.

**Two-Level Tests.** When only two loading levels are involved, the DLDR method may yield results considerably different from the LDR, but they will always be exactly the same as those of the DLDC method. How much difference develops between calculations based on LDR and DLDR (or DLDC) depends on the relative content of each of the two loading levels. Figure 6.21 shows the essential importance of the variables. Here, we are considering loading blocks wherein, according to the LDR, 1% of the life is imposed in each block. According to the LDR, therefore, the material should sustain 100 blocks before failure. According to DLDR (or DLDC), however, the number of blocks that can be sustained will depend on both of the life levels of the two component events, and the relative

Table 6.2 Summary of linear, nonlinear, and double-linear cumulative fatigue damage concepts discussed by authors

| Cumulative damage concept | Basis of model | Ref. No. (year) |
| --- | --- | --- |
| LDR, Miner's(a) (linear damage rule) | Damage accumulates identically at all life levels. Cycle fractions sum to 1.0 at failure at all life levels. See Eq 6.1. | Ref 6.4 (1945) |
| DLDR, Manson et al. (double linear damage rule) | Physical size crack (function of life) constitutes "initiation." Failure occurs after sequential linear summation of LDR applied to "initiation" life, then "propagation" life; each reaches 1.0. See Eq 6.7 to 6.15. | Ref 6.13 (1965) Ref 6.14 (1967) |
| DCA, Manson and Halford's (damage curve approach) | Nonlinear damage accumulates along power-law damage curves that are well-defined functions of life level. | Ref 6.17 (1981) |
| DLDC, Manson and Halford's (double damage curve approach) | Each DCA power-law damage curve is approximated by a pair of linear lines with a prescribed break point that is a function of life and damage levels. | Ref 6.17 (1981) |
| DLDR, Manson and Halford's (double linear damage rule) | Simplifies DLDC concept with break points forced along a line of constant damage separating phase I and II damage. Each phase has a fatigue curve that adds to become the total fatigue life curve. Failure occurs after sequential linear summation of LDR applied to "phase I" life, then "phase II" life, each reaches 1.0. | Ref 6.17 (1981) |

(a) Palmgren (Ref 6.2) in 1924 and Langer (Ref 6.3) in 1937 preceded Miner.

142 / Fatigue and Durability of Structural Materials

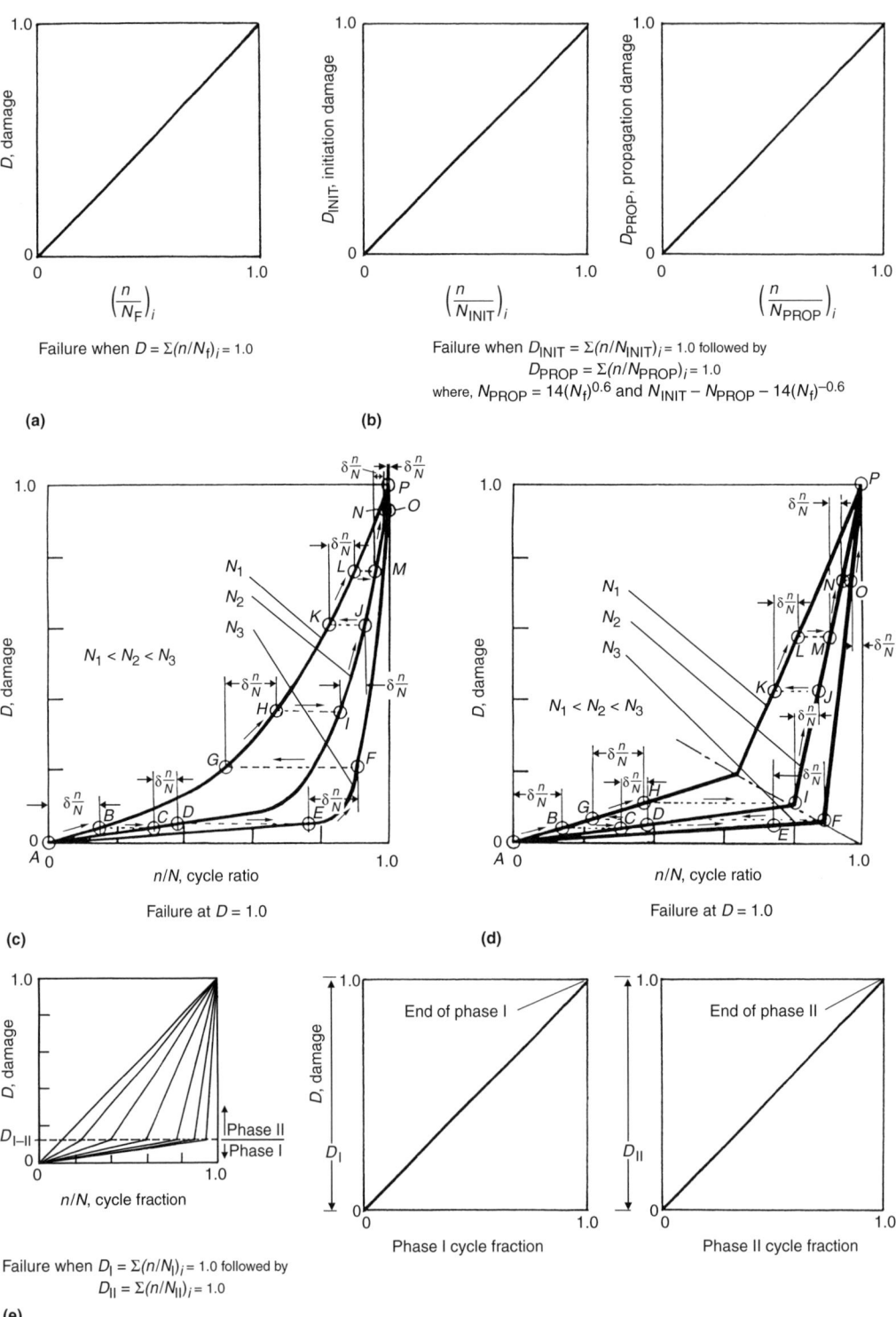

**Fig. 6.20** Schematic summary of cumulative fatigue damage concepts discussed in this chapter. (a) Linear damage rule. (b) Double linear damage rule, physically based crack "initiation"/"propagation." (c) Damage curve approach. (d) Double linear damage curve. (e) Double linear damage rule, based on general phase I and II concept

amount of each loading level included within a block. The smaller the content of one event or the other, the less will be its effect in making the DLDR different from LDR. If the second event disappears altogether, all calculations give the same result because only one loading is present. If both loadings are present, however, the number of sustainable blocks always is lower according to DLDR than LDR. For a given ratio of the life levels, the most detrimental effect occurs when the cycle ratios of each event within the block are approximately equal, as seen in Fig. 6.21. (Actually the peak of each curve is at an $n_2/N_2$ slightly greater than $n_1/N_1$, where $N_2$ is the higher of the two life levels. This is so because the effect of a high life cycle following a low life cycle is more damaging than the benefit derived from a low life cycle following a high life cycle.) The difference between the two types of calculations also depends strongly on the ratio of the lives of the two events. The larger the ratio of life levels, the greater is the loss of life. This behavior is regarded as an LCF interaction with HCF. But note that the effect depends only on the ratio of $N_2$ to $N_1$, not on their absolute magnitudes. Thus, the same loss of life occurs when $N_1$ is 100 cycles and $N_2$ is 10,000 cycles (both in the relatively low-cycle range) as when $N_1$ is 100,000 cycles and $N_2 = 10,000,000$ cycles (both in the relatively high-cycle life range).

**Multiple-Level Tests.** Our standardized example of Fig. 6.2 involving four levels of equally damaging (1% life fraction) loading is a good case for numerically comparing the DLDR with the other rules (DLDC, DCA, and LDR). Calculations are shown in Fig. 6.22. Here we again take the 45° line as the $10^3$ cycle life, and use the kneepoints shown in Fig. 6.13(a). Three conditions are examined in which $N_1$ is always $10^3$, but $N_2$ takes on values of $10^6$, $10^5$, and $10^4$. This range of combinations of $N_1$ and $N_2$ was chosen because no two of the four life levels were contributing the maximum life fractional damage as called for by the DLDC and DLDR approaches. The transition kneepoints $D_{I-II}$ were computed to be 0.062, 0.11, and 0.197, respectively. Application of an LDR to phase I resulted in 4.68, 7.47, and 11.34 blocks. Subsequent computation by linear damage accumulation in phase II indicated 6.79, 4.56, and 2.44 blocks to give failure in 11.47, 12.03, and 13.78 total blocks, respectively. As expected, the combi-

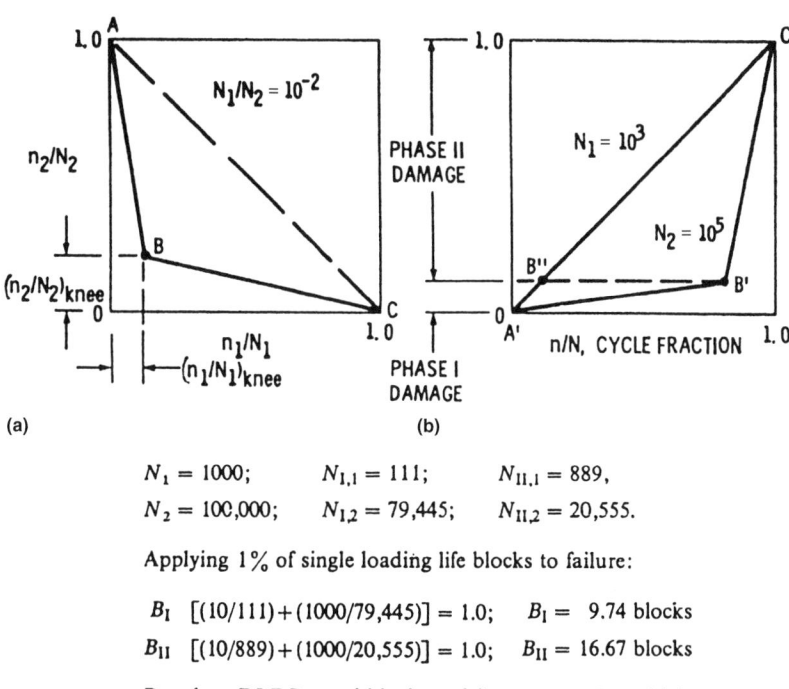

$N_1 = 1000;$  $N_{I,1} = 111;$  $N_{II,1} = 889,$
$N_2 = 100,000;$  $N_{I,2} = 79,445;$  $N_{II,2} = 20,555.$

Applying 1% of single loading life blocks to failure:

$B_I$ $[(10/111)+(1000/79,445)] = 1.0;$  $B_I = 9.74$ blocks
$B_{II}$ $[(10/889)+(1000/20,555)] = 1.0;$  $B_{II} = 16.67$ blocks

Based on DLDR, total blocks at failure $= B_I + B_{II} = 26.41.$
Based on Miner linear damage rule, total blocks at failure $= 50.$

**Fig. 6.21** Double linear damage rule applied to block loading involving two loading levels. (a) Conventional interaction lines. (b) Reiteration as damage lines. Source: Ref 6.10

144 / Fatigue and Durability of Structural Materials

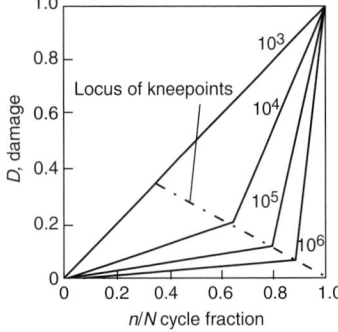

(a)

At the kneepoints:
$D_{knee} = 0.35(N_1/N_2)^{0.25}$
$(n/N)_{knee} = 1 - 065(N_1/N_2)^{0.25}$
where:
$N_1 = 10^3$ cycles to failure
$N_2 = 10^4, 10^5,$ or $10^6$

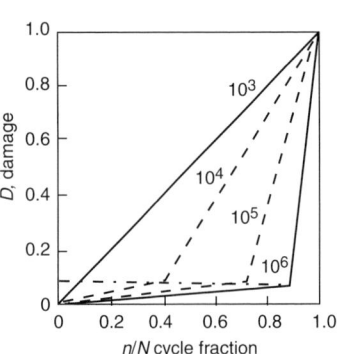

(b)

| Total life, $N$ | Phase I life, $N_I$ | Phase II life, $N_{II}$ |
|---|---|---|
| $10^3$ | 62 | 938 |
| $10^4$ | 3,745 | 6,255 |
| $10^5$ | 70,658 | 29,342 |
| $10^6$ | 884,412 | 115,588 |

Calculation of number of blocks if apply 1% life fraction at each life level shown:
For Phase I:
$(10/62) + (10^2/3745) + (10^3/70,658) + (10^4/884,412) = 0.2135$
Thus, 4.68 blocks for Phase I.

For Phase II:
$(10/938) + (10^2/6255) + (10^3/29,342) + (10^4/115,588) = 0.1472$
Thus, 6.79 blocks for Phase II.
Total blocks = 11.47

(c)

| Total life, $N$ | Phase I life, $N_I$ | Phase II life, $N_{II}$ |
|---|---|---|
| $10^3$ | 111 | 889 |
| $10^4$ | 4,908 | 5,092 |
| $10^5$ | 79,445 | 20,555 |
| $10^6$ | 928,300 | 71,700 |

Calculation of number of blocks if apply 1% life fraction at each life level shown:
For Phase I:
$(10/111) + (10^2/4908) + (10^3/79,445) + (10^4/928,300) = 0.1338$
Thus, 7.47 blocks for Phase I.

For Phase II:
$(10/889) + (10^2/5092) + (10^3/20,555) + (10^4/71,700) = 0.2191$
Thus, 4.56 blocks for Phase II.
Total blocks = 12.03

(d)

| Total life, $N$ | Phase I life, $N_I$ | Phase II life, $N_{II}$ |
|---|---|---|
| $10^3$ | 197 | 803 |
| $10^4$ | 6,345 | 3,655 |
| $10^5$ | 88,044 | 11,956 |
| $10^6$ | 964,987 | 35,013 |

Calculation of number of blocks if apply 1% life fraction at each life level shown:
For Phase I:
$(10/197) + (10^2/6345) + (10^3/88,044) + (10^4/964,987) = 0.0882$
Thus, 11.34 blocks for Phase I.

For Phase II:
$(10/803) + (10^2/3655) + (10^3/11,956) + (10^4/35,013) = 0.4091$
Thus, 2.44 blocks for Phase II.
Total blocks = 13.78

**Fig. 6.22** Double linear damage rule applied to block loading involving four loading levels. (a) Representation by damage lines. (b) Application for $N_1 = 10^3$ and $N_2 = 10^6$. (c) Application for $N_1 = 10^3$ and $N_2 = 10^5$. (d) Application for $N_1 = 10^3$ and $N_2 = 10^4$. Source: Ref 6.10

nation of $N_1 = 10^3$ and $N_2 = 10^6$ gives the most conservative estimate of the number of sustainable blocks (11.47).

By comparison, the DLDC approach predicted 11.73 blocks, whereas DCA and LDR estimated 10.8 and 25.0. For repeated mission loading blocks that include multiple loading levels, the new approaches just discussed will always predict reduced mission durability compared with the LDR, and hence are always more conservative than LDR predictions. For the sample problem, the reduction is on the order of a factor of two. Obviously, the reduction depends on the mission mix of numbers of applied cycles and the range of life levels involved. A few low-cycle fatigue cycles interspersed between many very high-cycle fatigue loadings can result in dramatic losses in mission lifetime. An order of magnitude loss is well within the realm of possibility for structural elements subjected to low-amplitude, high-frequency excitation and occasional high strain amplitude low-cycle fatigue loading. Losses on the order of a factor of up to 10 in life have been demonstrated in the laboratory using smooth specimen fatigue testing at room temperature (Ref 6.10, 6.13, 6.15, and 6.20).

**Other Comparisons.** Double linear damage rule calculations were also made for the cases shown in Fig. 6.15 and Fig. 6.16. No substantial differences were found between the DLDR results and the DLDC results shown in these figures. In no case was the difference more than 9%. This was accomplished with a significant reduction calculational procedure. Our experience suggests that in all cases studied to date the DLDR produced comparable, although slightly more conservative, results to the DLDC procedure with considerably less labor.

**Application to Complex History of a Compressor Disk of a Gas-Turbine Engine.** In Ref 6.21 we treated a problem of a compressor disk originally studied for a Teledyne company application. This problem involved 14 cyclic events per block of service. The parameters for the loading are shown in Table 6.3. Calculations were made for both LDR and DLDR methods, but only a relatively small difference was observed; by LDR the number of sustainable blocks was 307, whereas for DLDR it was 276, only 10% lower. Examining the parameters of the problem, it becomes clear why so little difference appeared in the calculations. The main event was No. 8, with a basic fatigue life of 2500 cycles. In fact, if only this event were present, the number of sustainable blocks would have been 1250. The other events not only were of relatively smaller significance, but their life levels were also fairly close to that of event No. 8. The highest life event, No. 6, had a life level of 64,000, only about $25\times$ the major one. For significant impact, the extreme life levels must be at least $100\times$ (Ref 6.5). For this problem, therefore, we should not expect much deviation between LDR and DLDC computations.

Suppose, however, a large number of high-frequency vibrations were introduced into the preceding mission history. Even if the strain amplitude were small, its effect would be seen to be much more damaging when analyzed by

Table 6.3 Application of double linear damage curves (DLDC) to a complex duty cycle (Table 1, p 12, Walcher, Gray, and Manson (Ref 6.21)

| | Loading conditions | | No. cycles per mission, $n$ | Cycles to failure, $N_f$ | Cycles to end of Phase I, $N_I$ | Cycles to complete Phase II, $N_{II}$ | $n/N_I$ | $n/N_{II}$ |
|---|---|---|---|---|---|---|---|---|
| Event | $\Delta\varepsilon$ | $\sigma_0$, MPa (ksi) | | | | | | |
| 1 | 0.00254 | 695 (101) | 4 | 37,180 | 23,643 | 13,537 | $1.692 \times 10^{-4}$ | $2.955 \times 10^{-4}$ |
| 2 | 0.00791 | 394 (57) | 2 | 7,200 | 2,474 | 4,726 | 8.084 | 4.232 |
| 3 | 0.00735 | 359 (52) | 1 | 13,650 | 6,354 | 7,296 | 1.574 | 1.371 |
| 4 | 0.01017 | 268 (39) | 6 | 5,550 | 1,632 | 3,918 | 36.770 | 15.310 |
| 5 | 0.00396 | 616 (89) | 3 | 17,400 | 8,873 | 8,527 | 3.381 | 3.518 |
| 6 | 0.00198 | 727 (105) | 2 | 64,000 | 45,518 | 18,482 | 0.439 | 1.082 |
| 7 | 0.00848 | 172 (25) | 1 | 33,000 | 20,382 | 12,618 | 0.491 | 0.793 |
| 8 | 0.01564 | 62 (9) | 2 | 2,500 | 390 | 2,110 | 51.280 | 9.479 |
| 9 | 0.01045 | 3 (0.4) | 1 | 31,325 | 19,092 | 12,233 | 0.524 | 0.817 |
| 10 | 0.00932 | 66 (10) | 1 | 42,540 | 27,897 | 14,643 | 0.359 | 0.683 |
| 11 | 0.01074 | 145 (21) | 1 | 9,390 | 3,706 | 5,684 | 2.698 | 1.759 |
| 12 | 0.01271 | 127 (18) | 1 | 4,440 | 1,122 | 3,318 | 8.913 | 3.014 |
| 13 | 0.01158 | 188 (27) | 1 | 4,900 | 1,327 | 3,573 | 7.536 | 2.799 |
| 14 | 0.00452 | 557 (81) | 2 | 20,605 | 11,124 | 9,481 | 1.798 | 2.110 |
| Totals | | | | | | | $125.539 \times 10^{-4}$ | $49.922 \times 10^{-4}$ |

Missions to complete Phase I damage = 1/.0126 = 79; missions to complete Phase II damage = 1/.005 = 200; total missions to failure = 79 + 200 = 279

DLDR than by LDR. A small study is shown in Table 6.4. Here we assume that the vibration has a small strain amplitude so that its basic life is $10^7$ cycles. The table shows how the lives would depend on the number of vibration cycles per block that are present. As this number increases, the effect according to DLDR is much greater than according to LDR. For example, if $10^4$ cycles occurred per block, only 1/10 of 1% of its life, the number of blocks to failure according to the LDR would be reduced from 307 blocks to 235 blocks, only 23%. However, by the DLDR, the number of blocks would be reduced from 279 blocks to 92, a reduction of 3× from that predicted by LDR. In this case, of course, the highest loading level life is 4000 times the lowest loading level life, so the effect of high cycle fatigue-low cycle fatigue (HCF-LCF) interaction can be substantial. This procedure shows the importance of using a DLDR calculation instead of an LDR calculation while when considering HCF-LCF interaction when the range of life level loadings is high. Of course, the damage content of the widely diverse loadings must be large enough to cause a significant effect. Note, for example, that if only $10^3$ cycles of the $10^7$ loading is present per cycle (0.01%), the difference between the DLDR and LDR calculations is much smaller. Note, also, that as the block becomes totally dominated by the HCF loading ($10^7$ cycles/block), the number of blocks drops to only one, and there is nothing to interact with, so the DLDR and LDR give the same trivial answer.

While the DLDR calculation captures the effect obtained by DLDC, the results are not identical. The ratio DLDR/DLDC can vary from unity to 1.2× while LDR/DLDR can range from unity to 2.5×.

## Anomalous Behavior: Synergistic Effects

While our experience with two-level tests and limited multilevel tests have demonstrated the validity of the DLDC analysis (and correspondingly, the equivalent DLDR analysis) to yield quite accurate predictions, it is clear that cases can arise in which the method may break down. Several reasons are evident as the possible causes of the breakdown:

- The prior loading may alter the material such that the follow-on loading may be on an essentially new material. Omitting the effects of physical or metallurgical material alteration that are phenomena independent of damage per se might lead to significant error in the predictions.
- The prior loading may introduce synergistic effects on the follow-on loading so that the latter are not what they would seem to be based on the initial virgin material.
- New phenomena may be introduced by the combined loading that would not be present if each of the loading were singly applied. Crack tip blunting due to overloading is an example.

### Irreversible Hardening

We shall consider a number of situations that we have observed that fall into one (or perhaps a combination of more than one) of these categories. It is possible that numerous other situations exist that have not yet been experienced by us.

**Example Applied to Annealed 304 Stainless Steel.** While we have successfully applied DLDC and DLDR to a number of materials and loadings, we encountered considerable difficulty when we attempted to apply them to annealed 304 stainless steel. To illustrate the problem, consider a simple case. When a specimen was cycled at a strain range of $\Delta\varepsilon_1 = 4.25\%$, failure occurred at $N_1 = 100$ cycles. When strained at $\Delta\varepsilon_2 = 0.84\%$, without any other loadings, the life obtained was $N_2 = 15,000$ cycles. We then conducted a test in which $n_1 = 46$ cycles were applied at $\Delta\varepsilon_1$ level, and then the strain range was changed to $\Delta\varepsilon_2$. The expected remaining life could be calculated as shown in Fig. 6.23. We first construct line *ABC* by locating the knee-

Table 6.4 Effect of adding high-cycle fatigue ($N_f = 10^7$) loading to mission of Table 6.3

| No. of cycles/block at $N_f = 10^7$ | No. blocks to failure LDR | DLDR | DLDR/LDR |
|---|---|---|---|
| $10^0$ | 307 | 279 | 0.91 |
| $10^1$ | 307 | 278 | 0.91 |
| $10^2$ | 306 | 262 | 0.86 |
| $10^3$ | 298 | 179 | . . . |
| $10^4$ | 235 | 92 | 0.39 min |
| $10^5$ | 75 | 46 | 0.61 |
| $10^6$ | 10 | 9 | 0.90 |
| $10^7$ | 1 | 1 | 1.00 |

LDR, linear damage rule; DLDR, double linear damage rule

point B at $(n_1/N_1)_{knee} = 0.1000$ and $(n_2/N_2)_{knee} = 0.1857$. We then spot point P on this line at $n_1/N_1 = 0.46$, and determine the remaining life ratio $n_2/N_2 - 0.1114$. So, for $N_2 = 15,000$ cycles, the expected remaining life $n_2 = 0.1114$ $N_2 = 1671$ cycles. The experimentally observed remaining life, however, was 7120 cycles, which is about 4.3 times as much as the predicted life.

To understand what has gone awry, we must consider what is special about annealed 304 stainless steel. Figure 6.24 shows how the hardening develops both for single-level cycling and for two-level cycling. For the 4.25% strain range test, the path followed is AEB. In the first cycle, it requires a stress range of about 1138 MPa (165 ksi) to produce a strain range of 4.25%, but the stress increases progressively until it is about 2413 MPa (350 ksi) when the specimen fails. In fact, the material is still hardening at B; it never quite reaches saturation, as is the case for most materials. For the 0.84% strain range test, it requires a stress range of only 883 MPa (128 ksi) in the first cycle, and the material softens slightly in the early loading, following the path CD. Saturation occurs very early in life, and the stress remains almost constant until failure occurs at D.

Figure 6.24 also shows the path traced AEE'F during a two-level test involving the application of 35 cycles at $\Delta\varepsilon_1 = 4.25\%$ level followed by cycling at $\Delta\varepsilon_2 = 0.84\%$ level until failure occurred. Of course, AE is expected to lie along the single-level hardening curve AB, but when the strain range is lowered, the path followed E'F is at a stress much higher (about 60% higher) than that required for single-level cycling at $\Delta\varepsilon_2$ level. As we discuss in the next section, this behavior is not common for most materials.

As an adjunct to the preceding observation, we show in Fig. 6.25 the Manson-Coffin plot of plastic strain range versus life for this material obtained many years ago (Ref 6.23). Figure 6.25(a) shows a least-squares fit of a straight line through all the data. It was troubling even then that the data did not fit the common perception of what a Manson-Coffin line should look like, but that was early in the development of the LCF model, and the attempt to understand the reason for this anomalous behavior was deferred for later study. Figure 6.25(b) clearly shows that there are two regions for the behavior of this material, probably because different phenomena govern the behavior in the two regions. Since then, several other investigators have observed this double linear behavior of the Manson-Coffin line on other materials, and some have indeed attempted to represent each segment by a different mathematical equation. But whether such a representation serves a useful purpose depends on how it is to be used. For our purpose, this procedure would not help resolve the dilemma discussed in connection with Fig. 6.23.

Much research has been conducted on 304 stainless steel in the intervening years (Chanani and Antolovich, Ref 6.24; Baudry and Pineau, Ref 6.25; Bayerlein et al., Ref 6.26). It is beyond

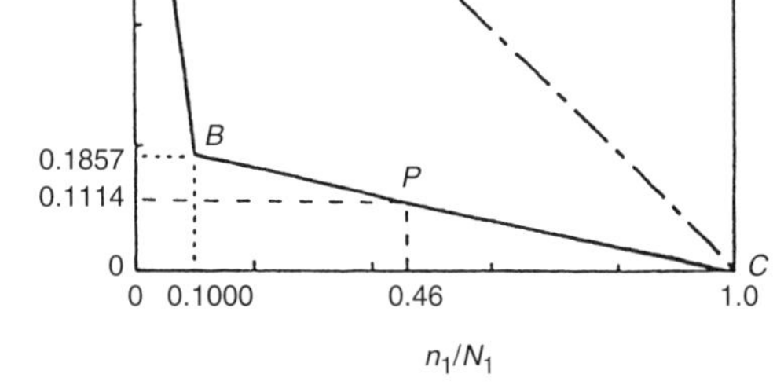

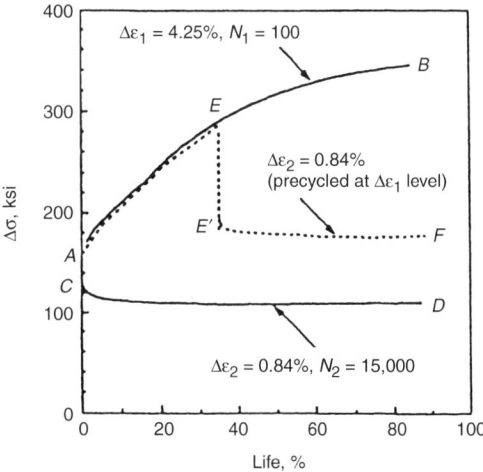

**Fig. 6.23** Application of double linear damage curve to special problem for 304 stainless steel. Source: Ref 6.22

**Fig. 6.24** Room temperature hardening behavior of 304 stainless steel under single- and two-level strain cycling. Source: Ref 6.22

our scope to review the interesting results of these references. Basically, however, it was found that when 304 stainless steel (and other austenitic materials as well) are cyclically strained above a critical level, a martensitic phase is transformed from some of the austenite matrix of the alloy. It is easy to determine this fact by metallurgical study, but it may also be observed by surrounding the test specimen with a magnetometer. Initially the material, which is austenite, is not magnetic, but as the martensitic transformation occurs, it becomes more and more magnetic. The hardening for this material is thus due, at least partially, to the transformation into a hard phase, and perhaps also to dislocations that pile up during the plastic straining, as is common for many other cyclically strain-hardening materials.

**Concept of Reversible and Irreversible Hardening.** Many materials harden when strained cyclically. But the hardening is often due to the pile up of dislocations that require higher and higher stresses in subsequent cycles in order to overcome the stress field developed by the dislocation pileup. It is also common, when the hardening is due to dislocation pileup, to reach saturation fairly early in the life at a given strain level. Once the dislocations have reached a stable level, cycling annihilates as many dislocations as are formed, and a given plastic strain can be sustained over the main part of the life at very close to the same stress. If the plastic strain changes, the dislocations very quickly tend to seek the equilibrium condition associated with the new strain level.

So, ideally, if we change strain levels, the type of behavior we would expect is indicated in Fig. 6.26(a). If hardening at the first strain level $\Delta\varepsilon_1$ occurs along $AE$, and strain range is changed to a lower level $\Delta\varepsilon_2$ at $E$, the stress rapidly drops to $E'$. After a few cycles at $\Delta\varepsilon_2$ (to give the dislocations a chance to gravitate to a new equilibrium configuration), the curve $EE'E''$ fairs into the hardening curve of the second strain level. Figure 6.26(b) shows the actual behavior of an aluminum alloy in two-step loading. This figure was contained in Ref 6.16. Comparing Fig. 6.26(a) with (b), we notice that while the actual behavior of the aluminum alloy is not quite the same as the idealized behavior, it certainly approximates it. We will refer to the behavior shown in Fig. 6.26(b) as *reversible hardening*.

Now, reconsider the behavior of 304 stainless steel shown in Fig. 6.24. Once the material has been hardened at the higher strain level, changing strain range *does not* reverse the behavior to that observed at the lower strain range. A solid-state transformation has irreversibly hardened the material, and it acts, not as the initial material, but rather as one containing the harder phase as well. Hence, in the second loading, we must consider that the strain is imposed on a *new material,* and not on the original material. Thus, the question resolves itself into how the material modified by the irreversible hardening behaves.

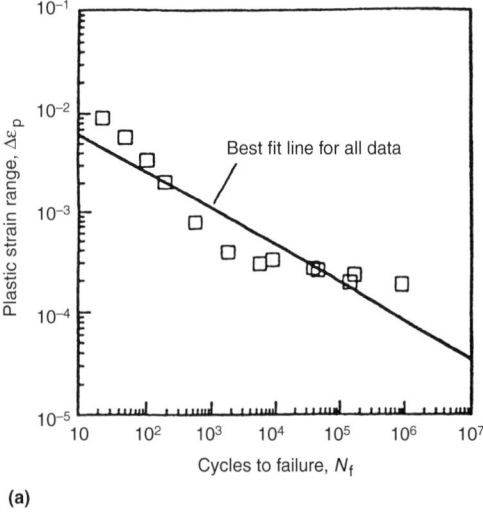
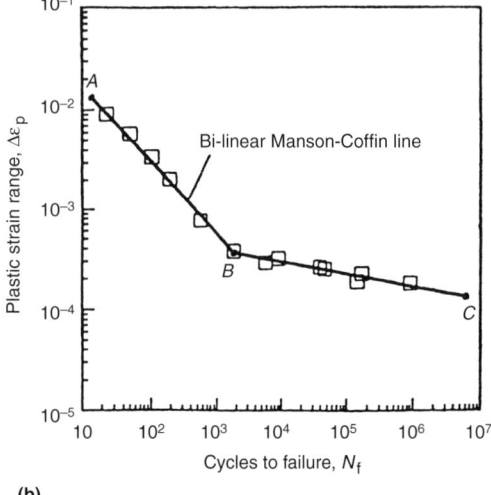

(a)  (b)

**Fig. 6.25** Manson-Coffin strain-life relation for annealed 304 stainless steel at room temperature. (a) As originally shown in Ref 6.23. (b) A more realistic representation of the data (Ref 6.22)

Chapter 6: Cumulative Fatigue Damage / 149

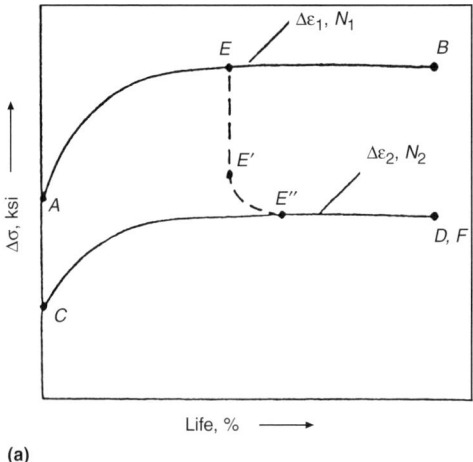
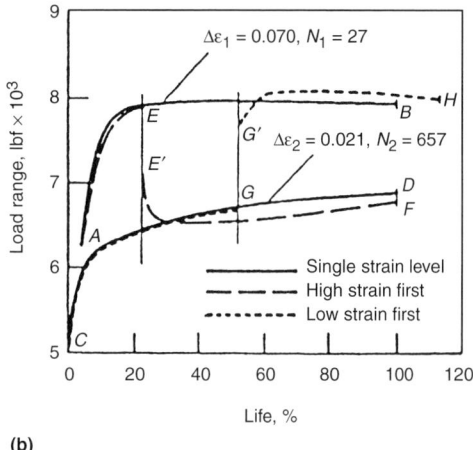

**Fig. 6.26** Characteristic behavior in two-level test for a reversible hardening material. (a) Ideal behavior. (b) Actual behavior of 2024-T-4 aluminum. Source: Ref 6.16

While several approaches are possible, we describe in the next section one of these materials.

**A Simple Procedure for Treating Irreversible Hardening.** Consider again the illustrative example. Cycling at $\Delta\varepsilon_1$ level causes a martensitic phase to form. When the strain range is lowered to $\Delta\varepsilon_2$, the cycling is actually performed on a "new" material, at a plastic strain lower than that observed under constant amplitude cycling at $\Delta\varepsilon_2$ level. Hence, the basic life at which the specimen is being cycled under the second condition is not the same as the 15,000 cycle life associated with $\Delta\varepsilon_2$ in the virgin material. Determining this life becomes the problem. If the same amount of precipitate could be induced in some other way—say, by a suitable heating schedule—then the life at $\Delta\varepsilon_2$ level could be measured directly. However, this would be a difficult way to solve the problem. So, if we try to measure directly the underlying life of the material hardened by cycling, we have to take into account the substantial damage that has been imposed by the prior cycling. Instead, we can use the very two-level test that we have conducted to back-calculate the life using DLDC. The sacrifice is that we must use the experimental results of the test instead of predicting it. The prize, however, is that once the life relation for the material containing the precipitate is established using information obtained from this one test, we can analyze other cumulative fatigue situations without requiring additional information.

The schematic of the underlying procedure is shown in Fig. 6.27. Here we seek to establish the elastic and plastic lines for the material containing the martensite precipitated by prior cycling. We assume that the material now behaves in the conventional way, having a single linear elastic and plastic line over the entire life range, rather than the broken curves of Fig. 6.25. One point on each line, A and A', is established from a constant amplitude fatigue test conducted at the high strain level $\Delta\varepsilon_1$ because martensite will form when cycling at this strain level. In order to establish the coordinates of a second point B and B' of Fig. 6.27, we measure the remaining number of cycles $n_2$, and the elastic $\Delta\varepsilon_{e2}$ and plastic $\Delta\varepsilon_{p2}$ strain ranges developed at the sec-

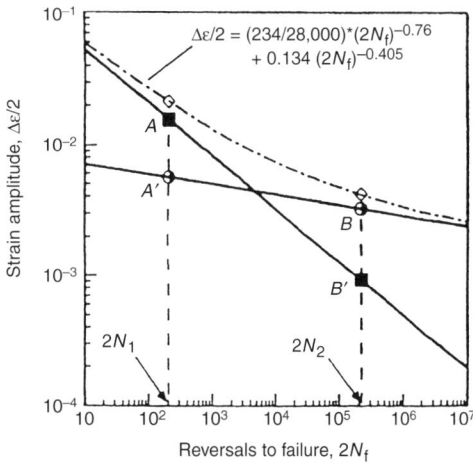

**Fig. 6.27** Basic life relations for 304 stainless steel with martensitic precipitate. Source: Ref 6.22

ond condition in a two-level test. The life level $N_2$ at which these elastic and plastic strain ranges are applicable is not, however, known. But, it can be calculated from the fact that DLDC applies to the test conducted.

Given $N_1$, $n_1$, and $n_2$, it can be shown that the value of $N_2$ that will satisfy the DLDC condition can be computed from one of the following equations numerically (in general, the equations will each have two solutions, but only one of the solutions, $N_2 > n_2$, is valid.):

$$N_2 = \frac{7N_1}{7N_1 + 13n_1}\left[n_2 + \frac{20n_1}{7}\left(\frac{N_2}{N_1}\right)^{1.25}\right] \quad \text{(Eq 6.16)}$$

$$N_2 = \frac{7N_1 n_2}{13(N_1 - n_1)}\left[\frac{20}{7}\left(\frac{N_2}{N_1}\right)^{0.25} - 1\right] \quad \text{(Eq 6.17)}$$

Depending on the location of $n_1/N_1$ relative to the kneepoint of the double linear damage line, only one of these equations will have valid solutions for a given set of $N_1$, $n_1$, and $n_2$.

For our illustrative example, the constant amplitude test at $\Delta\varepsilon_1 = 4.25\%$ produced $\Delta\varepsilon_{e1} = 1.12\%$, $\Delta\varepsilon_{p1} = 3.13\%$, and $N_1 = 100$ cycles, thus setting the location of points $A$ and $A'$ of Fig. 6.27. The results obtained from the two-level test were: $n_1 = 46$ cycles, $n_2 = 7120$ cycles, $\Delta\varepsilon_{e2} = 0.657\%$, and $\Delta\varepsilon_{p2} = 0.184\%$. Substituting these data into Eq 6.16 results in no solution. Applying Eq 6.17, however, yields $N_2 = 109{,}620$ cycles. We can now establish the location of points $B$ and $B'$ of Fig. 6.27. The resulting strain-life relation for the material is shown in the figure. The implication of this construction is that it is this life relation that should be used in cumulative fatigue analysis.

First, let us apply this concept to the other of the nine two-level tests conducted in this program. The results are shown in Table 6.5 and Fig. 6.28. Table 6.5 gives the results of nine tests conducted. Figure 6.28 shows how the results of all these tests would have been predicted using DLDC in conjunction with the life relation of Fig. 6.27 derived from test no. 5. The results are quite good, and all predictions lie within the scatterband of a factor of 1.5, which is at least as good as can be expected in the repeatability of replicate tests at a single condition.

Next, we examine the basic strain-life relations that would have been determined if any one of the other tests had been used as the baseline instead of test 5. These are shown in Fig. 6.29. They are consistent with the cyclic hardening behavior of the material. If the $n_1/N_1$ value in a baseline test is high, the amount of hardening that takes place will also be high, and the plastic strain component at the second loading level will be relatively small. As a result, a life relation derived from such a test will have a relatively steep plastic line and a flat elastic line. The converse is true for tests involving low $n_1/N_1$ values. This fact can be seen in Fig. 6.29 by comparing life curves obtained from tests with $n_1/N_1 = 0.10$ and $n_1/N_1 = 0.70$.

If life relations derived from a baseline test involving a high $n_1/N_1$ are used to predict the remaining life for a test with a low $n_1/N_1$, the predictions will likely be unconservative (the converse also being true) because the material used to establish the life relations would essentially be "different" from that for which life pre-

**Table 6.5  Two-level step tests conducted for stainless steel at room temperature**

$\Delta\varepsilon_1 = 4.25\%$; $N_1 = 100$

| Test No. | $n_1$ (cycles) | $\Delta\varepsilon_2$, % | $n_2$ (cycles) |
|---|---|---|---|
| 1 | 10 | 0.84 | 9,461 |
| 2 | 22 | 0.84 | 7,812 |
| 3 | 25 | 0.84 | 14,440 |
| 4 | 35 | 0.84 | 9,645 |
| 5 | 46 | 0.84 | 7,120 |
| 6 | 70 | 0.87 | 5,481 |
| 7 | 10 | 0.99 | 5,438 |
| 8 | 35 | 1.17 | 1,899 |
| 9 | 70 | 1.40 | 576 |

Source: Ref 6.22

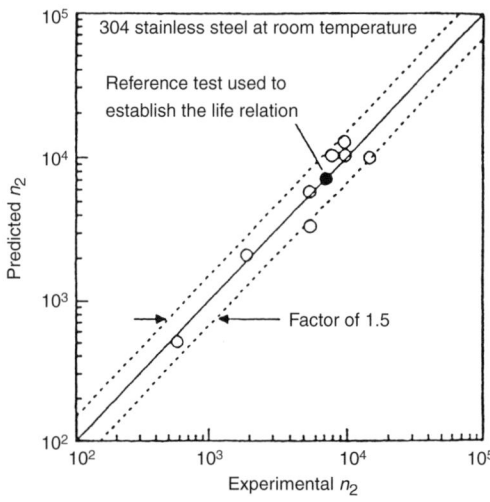

**Fig. 6.28** Comparison of predicted vs. observed experimental lives for two-level step tests for 304 stainless steel at room temperature. Source: Ref 6.22

dictions are being sought. Because the goal is to have a life relation that can be used to analyze any loading sequence, it would be beneficial to choose a baseline test, which involves an intermediate $n_1/N_1$ value. Such a choice will have either low or high $n_1/N_1$ values. Figure 6.30 shows how this approach would have worked for the tests conducted in this study. Here, three of the tests with $n_1/N_1$ in the range of 0.3 to 0.5, taken one at a time, were used to establish the strain-life relations, and the lives of the remaining eight tests were then predicted. In Fig. 6.30(a), the strain-life curves obtained are shown, and it can be seen that all three tests produced nearly the same results. Figure 6.30(b) shows a plot of predicted versus experimental remaining lives for all three cases. The results obtained for all three cases are almost identical, and the agreement between the predicted and experimentally observed lives is excellent. Hence, it is suggested that for baseline tests, an intermediate $n_1/N_1$ value (typically about 0.4) should be used.

The satisfaction that can be derived from the excellent results described thus far is marred by consideration of how the classical LDR analysis would have predicted the results. Figure 6.31 shows a comparison of the predicted and experimental remaining lives for all nine tests conducted. To make the calculations, the strain-life

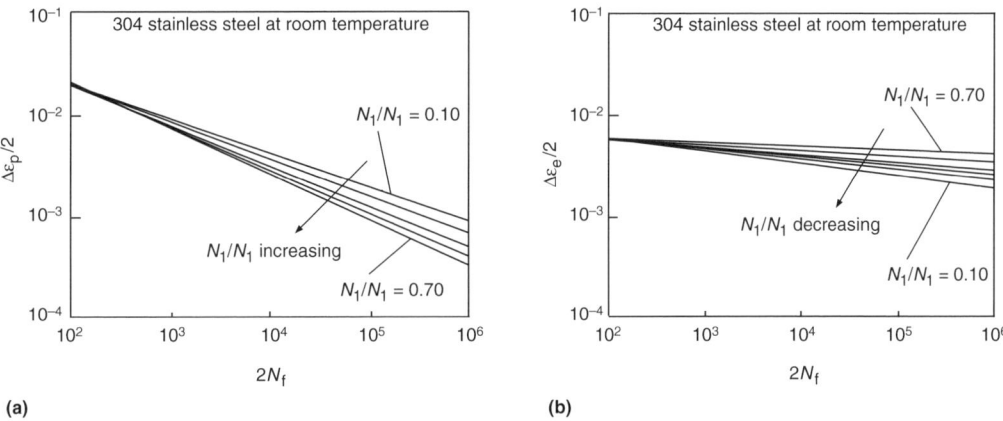

**Fig. 6.29** Dependence of strain-life relations on applied cycle ratio $n_1/N_1$. (a) Plastic line. (b) Elastic line. Source: Ref 6.22

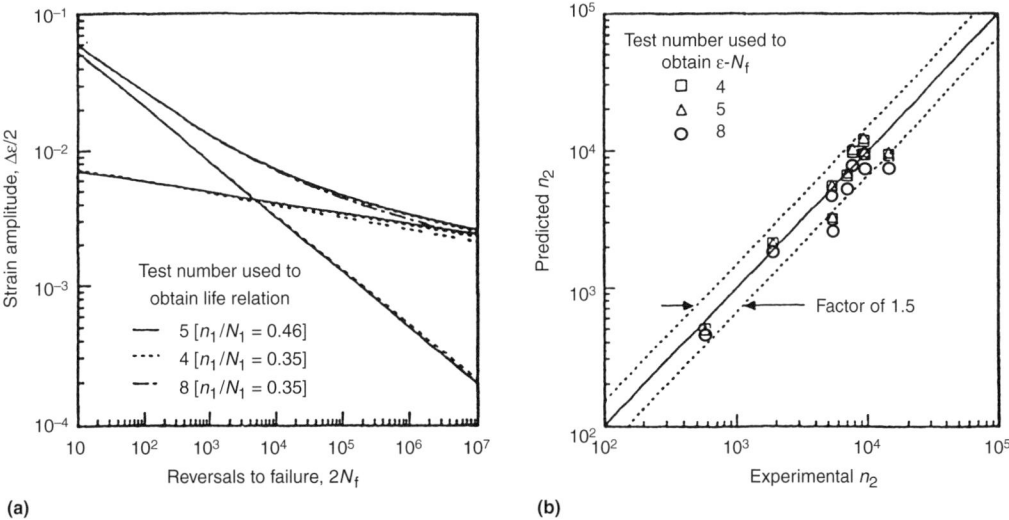

**Fig. 6.30** Strain-life relations and predicted lives obtained using baseline tests for which $0.3 < n_1/N_1 < 0.5$. (a) Life relations. (b) Predicted vs. experimental lives. Source: Ref 6.22

relations obtained from constant amplitude tests, and the usual LDR, was applied. Surprisingly, contrary to how most materials would have been expected to behave, LDR gave good predictions. Many investigators, in fact, have found that for 304 stainless steel the ordinary LDR works quite well.

Looking at it from the viewpoint of the DLDC method, the apparent discrepancy can be explained on the basis of two compensating errors. Consider, for example, test no. 1. An $n_1/N_1$ of 0.10 should give a remaining life of 0.118 of $N_2$ = 80,480, or 9600 cycles. When the LDR is used, and $n_1/N_1$ = 0.10, the remaining damage should be 0.90. In using LDR, however, the life of the second loading level is 15,000 cycles. Thus, the remaining life is 0.90 × 15,000 = 13,500 cycles, which is essentially the same. Therefore, the lower implied damage predicted by LDR is compensated for by using the lower second life level.

An alternate way of examining the LDR is to proceed by exactly the same procedure we described in connection with the DLDC method in Fig. 6.27, Fig. 6.28, and Table 6.5. For example, if we take test no. 5 and establish a life relation consistent with LDR, we get Fig. 6.32, analogous to Fig. 6.30(b). The predictions of LDR are quite good. This concept also has potential but should be studied further to determine advantages and limitations.

It should be noted that a material may behave both as a reversible and an irreversible hardening/softening material depending on test conditions. For example, the austenite-martensite transformation observed in 304 stainless steel occurs only at or near room temperature. Furthermore, even at room temperature, the transformation takes place only if the applied strain range is greater than some critical value, about 1%. Thus, if tests are carried out at high temperature and/or if all applied strain ranges are smaller than the critical value, then the annealed 304 stainless steel may not show the irreversible hardening behavior.

**Inadvertent Mean Stress Effects.** While DLDC and DLDR require the characterization of any component event by its fatigue life value, which involves both strain range and mean stress, often the mean stress is inadvertently overlooked. This is so because circumstantial details may not be fully recognized. Consider, for example, the study of cumulative damage in two-level loading of annealed 304 stainless steel discussed in the previous section. In test B, the first strain range is 4.25% and the second is 0.84%. Suppose that after 40 cycles of 4.25% the changeover to the 0.84% loading is introduced, as shown in Fig. 6.24. Two cases will now be considered: one in which the change takes place when the initial strain is at the bottom of the loop, the other when it is at the top.

As seen in Fig. 6.33, changing over at the bottom of the large hysteresis loop produces a small loop with a tensile mean stress and inherently lower life, while changing over at the top pro-

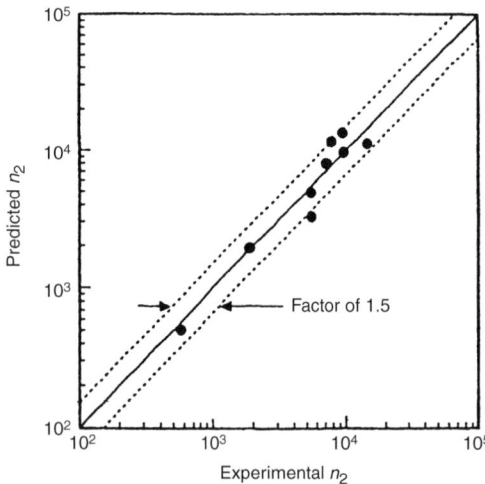

**Fig. 6.31** Application of an ordinary linear damage rule to two-level tests where the basic life relations determined from constant amplitude tests are used. Source: Ref 6.22

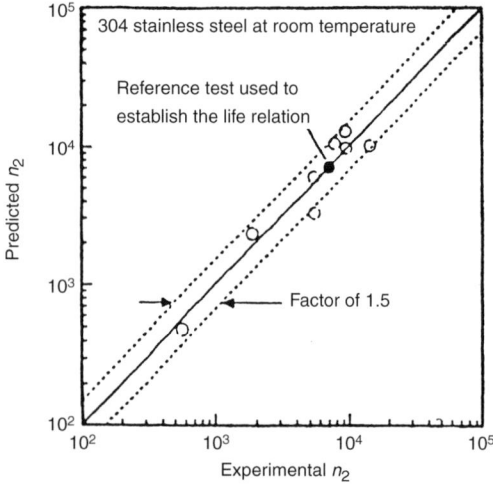

**Fig. 6.32** Life correlation obtained using a two-level test in conjunction with linear damage rule to determine a new effective life relation. Source: Ref 6.22

duces a life-enhancing compressive mean stress. For the mean stresses shown, the mean stress ratios are +0.286 and −0.286, respectively. Applying the modified-Morrow mean stress model, Eq 4.14(b) from Chapter 4, with $b = -0.076$ (from Fig. 6.27, for example), the life associated with the tensile mean stress would be 2100, while the compressive mean stress would increase the life to 82,000 cycles to failure. The life for completely reversed straining (zero mean stress) is 15,000 cycles. It is assumed that the mean stress would be maintained at the initial level without cyclic relaxation for as many of the small strain cycles as exist in the block of loading considered. If multiple blocks of loading are applied, the large strain cycle would reestablish the mean stress each time, countering the possibility of cyclic relaxation. It should be obvious that significantly different mission lives would result, depending on how the transition from the large to the small strain cycles occurs. Knowing the details of the cyclic stress-strain response following each transition is thus very important in assessing the cumulative fatigue damage done under nonuniform loading histories.

The preceding example considered a straining situation wherein gross cyclic plasticity was involved in both straining levels. If the unloading from the large loops was less drastic, the unloading path for the smaller strain range could be nominally elastic. For the present circumstances, that would impose mean stresses of opposite sign to the ones just discussed. Such a series of experiments would be revealing as to the influences of mean stress magnitude, sign, and magnitudes of the levels of cyclic straining on cumulative fatigue damage assessment.

**Metallurgical Instabilities.** Stress and strain together with exposure to high temperatures can produce metallurgical changes that alter the mechanical response characteristics of a material. Examples are alloy softening due to thermal annealing, oxidation of surfaces exposed to high-temperature oxygen, and precipitation hardening through a process called *strain aging*. The latter two examples are used in the ensuing discussions.

Strain aging is the phenomenon wherein an alloy has an element, such as carbon, in a metastable state of solid solution that will precipitate if it can find energetically acceptable sites. The increased dislocation density produced by plasticity provides sites for the precipitation process. The overall internal energy level is decreased by the element dropping out of solid solution and becoming attached to the dislocation. The element helps to pin the dislocation into its existing position, thus resisting further motion and acting as a barrier preventing ready motion of other newly generated dislocations. The net result is an increased rate of strain hardening and a higher strength response (accompanied by lower ductility) than exhibited by virgin material unexposed to a combination of plasticity and high temperature. Thus, loadings that induce such precipitations during one portion of the loading history will reveal a behavior in another portion that by itself would be different because of no metallurgical precipitation. The phenomenon of strain aging thus could create considerable synergistic impact on cumulative fatigue damage assessment.

While investigating creep-fatigue interaction in the cobalt-base superalloy L-605, at 760 °C (1400 °F), we observed a strong strain-aging effect (Ref 6.27). Carbon, initially in solution with the matrix, precipitated, forming an $M_{23}C_6$-type carbide along the generated dislocations. As the carbides increased in number and size during cycling, the cyclic stress-strain response of the alloy changed dramatically. The alloy was thus increasing its cyclic flow resistance and altering its resistance to fatigue failure. The amount of change depended on several factors: amount and duration of deformation and temperature. The implication to cumulative damage analysis is that the fatigue life at a prescribed load or strain level is not a unique quantity dictated by the magnitude of the loading. Instead, details of prior loadings can alter the fatigue life relation. For example, consider two high-temperature-loading levels. One under high strain range gives

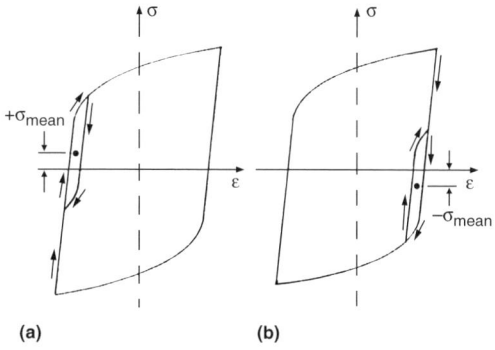

**Fig. 6.33** Inadvertent mean stresses at low $\Delta\varepsilon$ following high $\Delta\varepsilon$. (a) Tensile mean after unloading from compression. (b) Compressive mean after unloading from tension

rise to dislocation generation and carbide precipitation, producing a life, $N_1$. The other is at such a low strain level that no precipitation nuclei are formed, the material does not harden, and the ensuing life is $N_2 > N_1$. If a few high strain cycles are imposed on another sample, carbides will form, the material hardens, and its fatigue resistance will be altered. When the partially fatigued sample is subsequently loaded at the smaller strain level, its fatigue resistance should no longer be associated with $N_2$. Competing processes are occurring simultaneously: fatigue crack nucleation and growth (damage) and material hardening. The difficulty to the analyst is in how to separate these factors to accurately predict remaining fatigue life. Further research is required as to how best to resolve such complex problems.

Another example of altered fatigue resistance due to high-temperature exposure is one we encountered while studying strain-range partitioning (SRP) for creep-fatigue analysis (Ref 6.28). When 316 stainless steel was tested with hold periods at peak compressive stress, an oxide surface layer formed, which cracked during the tensile portion of each cycle. After a few cycles of loading, the surface was riddled with short cracks, most of which had not as yet penetrated the substrate. For small cyclic strains, oxides still formed, but would not crack during the tensile excursions. Hence, a sample loaded at large strains for a fraction of its life and then fatigued to failure at a small strain range would experience an additional damaging phenomenon (a cracked oxide surface layer) not accounted for in the original fatigue curve of the alloy at the low strain level. Determining the correct life level for cumulative damage analysis could become a tedious task. Our approach does not specifically address this issue. Instead, the approach is more general and deals only with life levels and not with how the life level is attained.

The phasing of temperature and strain cycling during thermal fatigue can produce additional complications to the problems of cumulative damage assessment. In a study (Ref 6.29) of thermomechanical fatigue behavior of the nickel-base superalloy MAR M 200, we observed significant differences in life depending on whether in-phase or out-of-phase cycling for which compressive stresses close cracks at the high temperature, retarding oxidation within the cracks. Out-of-phase cycling was considerably less damaging than in-phase. Furthermore, the two phasings result in different modes of cracking at the ubiquitous internal carbides. In-phase cycling produced carbide-matrix interfacial cracking, while out-of-phase cycling caused the carbides themselves to fracture. The implications of these findings to cumulative damage analysis are of concern. For example, consider two straining levels, $\Delta\varepsilon_1$ and $\Delta\varepsilon_2$, giving rise to the same cyclic life, $N_f$. One condition involves in-phase cycling, $\Delta\varepsilon_1$ and the other, out-of-phase, $\Delta\varepsilon_2 > \Delta\varepsilon_1$. If cycling of a sample is started at $\Delta\varepsilon_1$ and is discontinued after a life fraction $n_1/N_f$ and is resumed at $\Delta\varepsilon_2$ until failure occurs, the remaining life fraction according to the current rules (and the LDR as well) would simply be $n_2/N_f = 1 - n_1N_f$, i.e., $n_1 + n_2 = N_f$. Because both conditions produce the same life, any combination would be predicted also to produce the same life. However, it is unlikely that such a simple result would be borne out by experiment, due to different damage accumulation mechanisms existing for the two loading conditions. With different mechanisms of damage, the damage curves of $D$ versus cycle fraction $n/N_f$ would not coincide for the two conditions, and hence damage summations should differ from 1.0.

The lesson to be learned from these examples is that the cumulative damage rules developed in this chapter that are based on a single basic fatigue crack initiation and propagation mechanism, will require refinement for application to unique high-temperature conditions wherein additional damage mechanisms can come into play.

## Concluding Remarks

In this chapter we describe a philosophy of analysis that has evolved in our laboratories over many years. Our emphasis is simplicity of engineering application, and minimizing the baseline information required for implementation. While much of the methodology is discussed in earlier publications, this presentation takes the evolution process a few steps further. The format of the damage curve analysis has been altered and the procedure clarified to be consistent with the implied damage curves used in the double linear damage rule (DLDR). We call the new formulation the double linear damage curve (DLDC) approach. We have gained insight as to how to iterate the choice of these substitute damage curves to minimize error while still retaining simplicity in analysis. Because our basic damage curves are now consistent in both the DLDC and DLDR analyses, the usefulness of the damage curve analysis procedure has been increased. By making the same calculations through the DLDC we can establish how much error is introduced by the compromises brought about to alter them for DLDR analysis. On the other hand, it also becomes clear from such calculations how much simplicity is gained by the DLDR procedure compared with retention of damage curves with attendant computational complexity.

An important feature of our methods is the characterization of an event only by its life, not the parameters that enter into determining the life. Thus, a strain range and associated mean stress that leads to a given life is treated in identical manner as a smaller strain range with associated larger tensile mean stress that also leads to the same life. In this way, improvements in life calculation, per se, do not alter the equations involved in the DLDC or DLDR analyses. We have, in fact, discussed what we regard to be improvement in accounting for mean stress. Thus, while the life values that enter into a damage calculation may be altered by using the new mean stress relations, the equations operating on these life values do not change. Experience with the application of the method to a spectrum of complex loading types is currently limited, but we hope to make detailed computations in generic cases to evaluate the effects of various parameters. Sample computations are included in this report. Finally, we urge caution in the use of the method described, or indeed any other method, to ensure that some unexpected phenomenon, not inherently contained in the framework of the method, is introduced inadvertently. Among these are stress multiaxiality, unidentified deformation and fracture mechanisms, unknown residual stresses (especially at notches and crack tips), oxidation, and metallurgical instabilities. Some of these are illustrated in this chapter.

**REFERENCES**

6.1 H. Jackel, Service History Determination, Chapter 5, *Fatigue Design Handbook,* Society of Automotive Engineers, Vol AE-10, 1988, p 99–123
6.2 A. Palmgren, Die Lebensdauer von Kugellagern, *Z. Vereinesdeutscher Ingenierure,* Vol 68 (No. 14), April 1924, p 339–341 (in German). See also *The Service Life of Ball Bearings,* NASA technical translation of German text, NASA TT 1-13460, 1971.
6.3 B.F. Langer, Fatigue Failure from Stress Cycles of Varying Amplitude, *J. Appl. Mech.,* Vol 59, 1937, p A160–A162
6.4 M.A. Miner, Cumulative Damage in Fatigue, *J. Appl. Mech.,* Vol 67, 1945, p A159–A164
6.5 G.R. Halford, Cumulative Fatigue Damage Modeling: Crack Nucleation and Early Growth, *Int. J. Fatigue,* Vol 19, supplement, (No. 1), 1997, p S253–S260
6.6 N.M. Newmark, A Review of Cumulative Damage in Fatigue, *Fatigue and Fracture of Metals,* W.M. Murray, Ed., John Wiley and Sons, New York, 1952, p 197–228
6.7 L. Kaechele, "Review and Analysis of Cumulative-Fatigue-Damage Theories," RM-3650-PR, The Rand Corporation, Santa Monica, CA, 1963
6.8 M.J. O'Neill, A Review of Some Cumulative Damage Theories, ARL/SM-Report-326, Aeronautical Research Laboratories, Melbourne, Australia, 1970
6.9 J. Schijve, "The Accumulation of Fatigue Damage in Aircraft Materials and Structures," AGARD-AG-157, Advisory Group for Aerospace Research and Development, Paris, 1972
6.10 S.S. Manson and G.R. Halford, Re-Examination of Cumulative Fatigue Damage Analysis—An Engineering Perspective, *Mechanics of Damage and Fatigue,* International Union of Theoretical and Applied Mechanics (IUTAM) Symposium on Mechanics of Damage and Fatigue, July 1985 (Haifa and Tel Aviv, Israel), S.R. Bodner and Z. Hashin, Ed., Pergamon Press, New York, 1985, p 539–571. Volume also published in *Engineering Fracture Mechanics,* Vol 25 (No. 5/6), 1986, p 539–571. See also NASA TM-87325, 1985.
6.11 Anon., ASME Code Case N-47. Boiler and Pressure Vessel Committee, American Society of Mechanical Engineers, 1987
6.12 H.J. Grover, "An Observation Concerning the Cycle Ratio in Cumulative Damage," *Fatigue of Aircraft Structures,* ASTM STP 274, 1960, p 120–124
6.13 S.S. Manson, A.J. Nachtigall, C.R. Ensign, and J.C. Freche, Further Investigation of a Relation for Cumulative Fatigue Damage in Bending, *J. Eng. Ind. (Trans. ASME),* Vol 87, 1965, p 25–35

6.14 S.S. Manson, J.C. Freche, and C.R. Ensign, "Application of a Double Linear Damage Rule to Cumulative Fatigue," *Fatigue Crack Propagation*, ASTM STP 415, 1967, p 384–412

6.15 S.S. Manson, Fatigue: A Complex Subject—Some Simple Approximations, *Exp. Mech.*, Vol 5 (No. 7), 1965, p 193–226

6.16 F.E. Richart and N.M. Newmark, An Hypothesis for Determination of Cumulative Damage in Fatigue, *Proceedings, American Society for Testing and Materials*, Vol 48, 1948, p 767–800

6.17 S.S. Manson and G.R. Halford, Practical Implementation of the Double Linear Damage Rule and Damage Curve Approach for Treating Cumulative Fatigue Damage, *Int. J. Fract.*, Vol 17 (No. 2), 1981, p 169–192. See also Vol 17 (No. 4), 1981, p R35–R42. See also NASA TM-81517, 1980.

6.18 S. Hailu, Fatigue Life Prediction for Haynes 188 Alloy under Variable Amplitude Loading Using Nonlinear Models, master of science thesis, Department of Mechanical and Aerospace Engineering, Case Western Reserve University, Cleveland, OH, S.S. Manson, advisor, May 1991

6.19 L. Tucker and S. Bussa, The SAE Cumulative Fatigue Damage Test Program, *Fatigue under Complex Loading: Analyses and Experiments*, Advances in Engineering, Vol 6, 1977, p 1–53

6.20 D. Webber and J.C. Levy, Cumulative Damage in Fatigue with Reference to Scatter of Results, *Scientific and Technical Memorandum 15/58*, Ministry of Supply, United Kingdom, 1958

6.21 J. Walcher, D. Gray, and S.S. Manson, *Aspects of Cumulative Fatigue Damage Analysis of Cold End Rotating Structures*, American Institute of Aeronautics and Astronautics Preprint 79–1190, 1979

6.22 S.S. Manson and S. Hailu, Maverick Behavior of the Irreversibly Hardening 304 Stainless Steel in Sequential Loading, *Material Durability/Life Prediction Modeling: Materials for the 21st Century*, PVP-Vol 290, S.Y. Zamrik and G.R. Halford, Ed., ASME, 1994, p 49–56

6.23 R.W. Smith, M.H. Hirschberg, and S.S. Manson, "Fatigue Behavior of Materials under Strain Cycling in Low and Intermediate Life Range," NASA TN D-1574, April 1963. See also *Proceedings, Tenth Sagamore Army Materials Research Conference*, J.J. Burke, N.L. Read, and V. Weiss, Ed., Syracuse University Press, New York, 1964, p 133–173

6.24 G.R. Chanani and S.D. Antolovich, Low Cycle Fatigue of a High Strength Metastable Austenitic Steel, *Metall. Trans.*, Vol 5, 1974, p 217–229

6.25 G. Baudry and A. Pineau, Influence of Strain-Induced Martensite Transformation on the Low-Cycle Fatigue Behavior of a Stainless Steel, *Mater. Sci. Eng.*, Vol 28, 1977, p 229–242

6.26 M. Bayerlein, H.-J. Christ, and H. Mughrabi, Plasticity-Induced Martensite Transformation during Cyclic Deformation of AISI 304L Stainless Steel, *Mater. Sci. Eng.*, Vol A114, 1989, L11–L16

6.27 S.S. Manson, G.R. Halford, and D.A. Spera, The Role of Creep in High-Temperature Low-Cycle Fatigue, Chapter 12 in *Advances in Creep Design*, A.I. Smith and A.M. Nicolson, Ed., Applied Science Publishers Ltd., London, 1971, p 229–249

6.28 S.S. Manson, G.R. Halford, and R.E. Oldrieve, "Relation of Cyclic Loading Pattern to Microstructural Fracture in Creep Fatigue," NASA TM-83473, 1983

6.29 R.C. Bill, M.J. Verrilli, M.A. McGaw, and G.R. Halford, "A Preliminary Study of the Thermomechanical Fatigue of Polycrystalline MAR M-200," NASA TP-2280 (AVSCOM TR 83-C-6), Feb 1984

# CHAPTER 7

# Bending of Shafts

## Introduction

Application of the methodology thus far covered to the study of fatigue of circular shafts is valuable for two reasons. First, the circular shaft is one of the most commonly used components in machine technology. Transmission of motion and power is, of course, the logical domain of the rotating shaft. Shear stresses developed during torque transmission and bending stresses generated by the external loading are commonly induced in such shafts. Frequent starts and stops, together with changes in loading, result in common failure by the fatigue mechanism. While the subject has been studied for years in applications involving long life, the finite life application is of relatively recent origin. Here, both elastic and inelastic strain components are of importance, and the solution of the elastic-plastic equations in simple form presents a challenge.

Another reason for interest in this problem is that we are dealing with a geometry in which the stresses and strains are not uniform over the entire volume of the material, contrary to the commonly developed specimens for basic fatigue evaluation in which care is exercised to ensure stress and strain uniformity. Thus, we have the opportunity to discuss how nonuniformity affects fatigue behavior of a part simulating realistic engineering components.

In this chapter, we deal with bending of solid and hollow shafts.* Complications arise in respect to many elements of the problem:

- How to solve the equations for stress and strain when both elastic and inelastic strains are significant
- Whether to consider early loading when hardening or softening may be significant
- How to distinguish between loadings applied to a stationary shaft and to those applied to a rotating shaft
- How to account for the volumetric effect due to the fact that only a small portion of the material is subjected to the high stress, the origin of the failure, while much of the material is at low stress. It is also important to recognize that the development of a failure crack could significantly change the stress distribution.

## Basic Approach to Treatment of Low-Cycle Fatigue in Bending

It is common to express the bending fatigue life of a material in terms of the nominal elastic stress or strain calculated by assuming stress proportional to strain. Yet, in the low-cycle fatigue range and, in fact, over a wide range of life of engineering interest, plastic flow is involved, and stress is not proportional to strain. Thus, considerable discrepancy exists between bending fatigue and axial fatigue strength when the latter is expressed in terms of actual stresses and strains. When plasticity is involved, the elastically calculated stresses in a beam can be significantly higher than the actual stresses. When axially loaded samples are tested, the actual stresses can be measured and reported directly. Because of this major difference in computed stresses, bending fatigue strengths typically appear to be larger than axial fatigue strengths.

Reference 7.1 demonstrated that flexural fatigue is considerably different from axial fatigue but that the two could be brought into coincidence when based on true surface stresses. Figure 7.1 shows some of the results.

The solid curve and circular data points show the axial fatigue characteristic of a 4130 steel.

---
*Only elastic and plastic strains will be considered, as shafts usually are not used at temperatures high enough for creep deformation to occur.

In this case the stresses are the actual values because in axial loading, it is simple to determine the stress from knowledge of load and cross-sectional area. The dashed curve and square data points refer to the bending characteristics. Here, the nominal stress is calculated from the equation $S = Mc/I$, based on linear elastic assumptions, so that the stresses are fictitious values determined on the basis that no plastic flow occurs. It is seen that for this material the nominal bending stress can be 50% higher at the 100 cycle life level, and that significant difference persists between the two curves until well above $10^5$ cycles to failure. Obviously, then, it is important to reconcile differences over a wide range of life levels of practical engineering interest. It was also demonstrated in Ref 7.1 that if the stress calculated in the case of the bending specimen had accounted for plasticity, the life comparison with axial data is much more favorable. However, because of the difficulty of performing the integrations involved in the bending case, general closed-form solutions weren't presented. Each material and geometry had been treated numerically.

In Ref 7.2, the analysis was extended to obtain closed-form generalized relationships. The approach was first to study rectangular rather than circular sections. Not only are rectangular sections of great importance in engineering applications, but they result in relations that can be integrated in closed form. The resulting form of the solutions then provided a model for the case of circular sections. This was accomplished by assuming an analogous model, involving several undetermined constants, and then determining the constants so as to fit a range of known solutions obtained by numerical integration. It then became possible to obtain very accurate, though not exact, closed-form solutions for the case of the circular sections. The procedure and results are described next. To carry out the analysis, use is made of the life relation discussed in Chapter 3, "Fatigue Life Relations":

$$\Delta\varepsilon/\Delta\varepsilon_T = (N_f/N_T)^c + (N_f/N_T)^b \quad \text{(Eq 7.1)}$$

This equation can then be inverted to solve for $N_f/N_T$ by the approach discussed in the Appendix, "Selected Background Information." The result is:

$$N_f/N_T = [(R_\varepsilon)^{z/c} + (R_\varepsilon)^{z/b}]^{1/z} \quad \text{(Eq 7.2a)}$$

where:

$$R_\varepsilon = \Delta\varepsilon/\Delta\varepsilon_T \quad \text{(Eq 7.2b)}$$

$$Z = -0.889c(c/b)^{-0.36}\exp[P(\ln R_\varepsilon)^2 + Q\ln(R_\varepsilon)] \quad \text{(Eq 7.2c)}$$

$$P = -0.001277(c/b)^2 + 0.03893(c/b) - 0.0927 \quad \text{(Eq 7.2d)}$$

$$Q = +0.004176(c/b)^2 - 0.135(c/b) + 0.2309 \quad \text{(Eq 7.2e)}$$

Thus, if the material constants $b$, $c$, $N_T$, and $\Delta\varepsilon_T$ are known, it is possible to express the fatigue life ratio $N_f/N_T$ in terms of those constants and strain-range ratio, $R_\varepsilon$.

Similarly, we can determine the stress at any point where the strain ratio is $R_\varepsilon$ by:

$$\Delta\sigma = E\Delta\varepsilon_T[(R_\varepsilon)^{z/c} + (R_\varepsilon)^{z/b}]^{b/z} \quad \text{(Eq 7.3)}$$

associated with the strain $\Delta\varepsilon$ at any point. These equations will now be used to treat rectangular and circular sections.

## Flexural Bending of a Solid Rectangular Cross Section

We first treat the bending of rectangular cross sections because the resulting equations can be integrated in closed form. As discussed in Ref 7.3, it is more convenient to solve this type of problem in an inverse fashion compared with conventional treatment. Rather than starting with a known bending moment, and determining the resulting surface strain, which would involve the solution of nonlinear equations, it has been found to be better to select a surface strain and calculate the bending moment required to produce this strain. From the selected surface strain

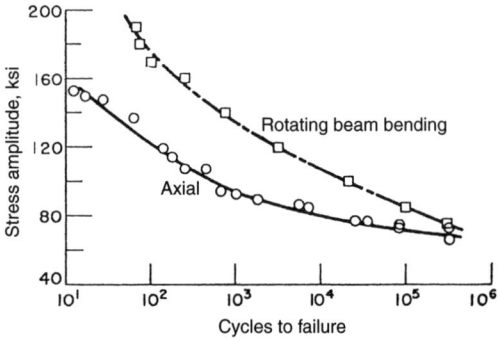

**Fig. 7.1** Fatigue data under axial loading and rotating bending for 4130 steel. Source: Ref 7.1

we establish the fatigue life, and from the required bending moment, the nominal elastic stress. Thus, we can establish the relation between nominal elastic stress and fatigue life.

Figure 7.2 shows the notation used in the analysis. We assume a surface strain $\varepsilon_s$ corresponding to a life $N_S$ when the full bending moment $M$ is applied. Because it is common to assume that plane sections remain plane in bending, strain varies linearly across the thickness so that the strain $\varepsilon_y$ at a distance $y$ from the neutral axis (center of the bar) is $\varepsilon_y = (y/h)\varepsilon_s$. Substituting for $\varepsilon_s$ in terms of life $N_S$, associated with the surface strain, from Eq 7.1:

$$\varepsilon_y = \varepsilon_T(y/h)[(N_S/N_T)^b + (N_S/N_T)^c] \quad \text{(Eq 7.4)}$$

Now, if $N_y$ is the life associated with the strain at $y$, then by Eq 7.1:

$$\varepsilon_y/\varepsilon_T = (N_y/N_T)^b + (N_y/N_T)^c \quad \text{(Eq 7.5)}$$

and the stress associated with the strain $\varepsilon_y$ is as derived from the equation for the total strain-range-versus-fatigue-life equation:

$$\Delta\varepsilon/2 = (\sigma_f'/E)(2N_f)^b + \varepsilon_f'(2N_f)^c \quad \text{(Eq 7.6)}$$

and the expressions for the transition life $N_T$ and the transition strain range $\Delta\varepsilon_T$:

$$N_T = 0.5(E\varepsilon_f'/\sigma_f')^{1/(b-c)} \quad \text{(Eq 7.7)}$$

$$\Delta\varepsilon_T = 2(\varepsilon_f')^{b/(b-c)}(\sigma_f'/E)^{c/(c-b)}$$
$$= 2(\varepsilon_f')^{n/(n-1)}(\sigma_f'/E)^{1/(1-n)} \quad \text{(Eq 7.8)}$$

where $n = b/c$. Thus, the bending stress at location $y$ is:

$$\sigma_y = \varepsilon_T E(N_y/N_T)^b \quad \text{(Eq 7.9)}$$

The increment of bending moment contributed by a strip $wdy$ at the distance $y$ from the neutral axis becomes:

$$dM = y\sigma_y wdy = \varepsilon_T E(N_y/N_T)^b wydy \quad \text{(Eq 7.10)}$$

and the total bending moment, taking into account the symmetrical contribution to the bending moment of the section below the neutral axis is:

$$M = \int dM = 2Ew\varepsilon_T \int_0^{y=h} (N_y/N_T)^b ydy \quad \text{(Eq 7.11)}$$

Here $N_y$ is a function of $y$ according to Eq 7.4 and 7.5. In order to carry out the integration, it is best to establish $dy$ in terms of $N_y/N_T$, rather than expressing $N_y/N_T$ in terms of $y$, which would result in a more difficult integration. Thus:

$$y = h\{[(N_y/N_T)^b + (N_y/N_T)^c]/[(N_S/N_T)^b + (N_S/N_T)^c]\} \quad \text{(Eq 7.12)}$$

and

$$dy = \{h/[(N_S/N_T)^b + (N_S/N_T)^c]\}$$
$$\{b(N_y/N_T)^{b-1} + c(N_y/N_T)^{c-1}\}d(N_y/N_T) \quad \text{(Eq 7.13)}$$

From Eq 7.11, and letting $N_y/N_T = x$, and $(N_S/N_T) = X$, therefore:

$$M = [2Ew\,h^2\varepsilon_T]/[(X)^b + (X)^c]^2$$
$$\int_{-\infty}^{x} [b(x)^{(2b-1)}c(x)^{(b+c-1)}][(x)^b + (x)^c]dx \quad \text{(Eq 7.14)}$$

$$M = [2Ew\,h^2\varepsilon_T]/[(X)^b + (X)^c]^2 \int_{-\infty}^{x} [b(x)^{(3b-1)}$$
$$+ b(x)^{(2b+c-1)} + c(x)^{(2b+c-1)}$$
$$+ c(x)^{(b+2c-1)}]dx \quad \text{(Eq 7.15)}$$

$$M = [2Ew\,h^2\varepsilon_T]/[(X)^b + (X)^c]^2$$
$$\int_{-\infty}^{x} \{(b/3b(x)^{3b} + [(b+c)/(2b+c)]$$
$$(x)^{(2b+c)} + [c/(b+2c)](x)^{(b+2c)}\}dx \quad \text{(Eq 7.16)}$$

In Eq 7.14 to 7.16, the lower limit for $N_y/N_T(=x)$ is taken at $h = 0$, where the strain is

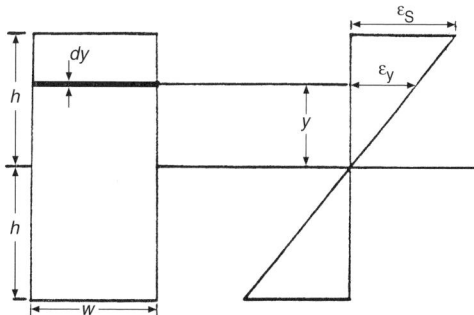

**Fig. 7.2** Rectangular cross-section bar in flexural bending. Source: Ref 7.2

zero and, therefore, the life is infinite. Because all the exponents of $N_y/N_t$ in the integrated expression are negative, the value of all terms is zero at the lower limit, and $M$ is evaluated as:

$$M = 2Ewh^2\varepsilon_T\{[1/3](X)^{3b} + [(b - c)/(2b + c)] \\ (X)^{2b+c} + [c/(b + 2c)](X)^{b+2c}\}/ \\ \{(X)^{2b} + 2(X)^{b+c} + (X)^{2c}\} \quad \text{(Eq 7.17)}$$

The nominal surface stress $S_{\text{bending}}$ is $Mh/I = 12Mh/w(2h)^3 = 12/8wh^2$ because the total height of the rectangle is $2h$ and the moment of inertia of a rectangular section is $(1/12)(\text{base})(\text{height})^3$. Thus, substituting for $M$ from Eq 7.17 and dividing both numerator and denominator by $(X)^{2b}$, the resulting equation for $S_{\text{bending}}$ is obtained as:

$$S_{\text{bending}} = E\varepsilon_T(X)^b\{1 + [3(b + c)/(2b + c)] \\ (X)^{c-b} + [3c/(b + 2c)](X)^{2(c-b)}\}/ \\ \{1 + 2(X)^{c-b} + (X)^{2(c-b)}\} \quad \text{(Eq 7.18)}$$

$$S_{\text{axial}} = E\varepsilon_T(X)^b \quad \text{(Eq 7.19)}$$

$$\frac{S_{\text{bending}}}{S_{\text{axial}}} = \text{correction factor} = \delta \quad \text{(Eq 7.20)}$$

where:

$$\delta = [1 + f_1(X)^{c-b} + f_2(X)^{2(c-b)}]/ \\ [1 + 2(X)^{c-b} + (X)^{2(c-b)}] \quad \text{(Eq 7.21)}$$

where (with $n = b/c$):

$$f_1 = 3(b + c)/(2b + c) = 3(n + 1)/(2n + 1) \quad \text{(Eq 7.22)}$$

$$f_2 = 3c/(b + 2c) = 3n/(1 + 2n) \quad \text{(Eq 7.23)}$$

Fatigue data have not been generated for the same material using rectangular cross-sectional specimens in bending and round cross-sectional specimens in axial loading to show a comparison between prediction and experiment. However, it would be expected that a similar degree of agreement would be achieved as was demonstrated for computation of fatigue life of circular cross sections in rotating bending from experimental axial fatigue results.

## Flexural Bending of a Circular Cross Section

We can now treat circular cross sections by an analogous procedure. We calculate the bending moment required to produce a selected surface strain. From the selected surface strain, a fatigue life can be established and similarly nominal stresses from the required bending moment. Thus, we establish the relation between nominal elastic stresses and fatigue life.

Figure 7.3 shows the notation used in the analysis. We assume a surface strain $\varepsilon_s$ corresponding to a life $N_S$ when the full bending moment $M$ is applied. Strain varies linearly across the section so that the strain $\varepsilon_y$ at a distance $y$ from the neutral axis is:

$$\varepsilon_y = (y/R)\varepsilon_S \quad \text{(Eq 7.24)}$$

$$y = R\sin\theta, \text{ and } dy = R\cos\theta\, d\theta \quad \text{(Eq 7.25)}$$

$$\varepsilon_y = \varepsilon_S \sin\theta \quad \text{(Eq 7.26)}$$

and the stress associated with the strain $\varepsilon_y$ can be found using Eq 7.3:

$$\sigma_y = E\varepsilon_S \sin\theta \left[1 + \left(\frac{\varepsilon_S \sin\theta}{\varepsilon_T}\right)^{z[(1/c + 1/b)]}\right]^{b/z} \quad \text{(Eq 7.27)}$$

where $z$ is given by Eq 7.2b to 7.2e.

The increment of bending moment contributed by a strip at the distance $y$ from the neutral axis becomes $dM = y\sigma_w dy$. Substituting for $y$ and $dy$ from Eq 7.25 and $\sigma_y$ from Eq 7.27:

$$dM = R\sin\theta \cdot E\varepsilon_S \sin\theta \left[1 + \left(\frac{\varepsilon_S \sin\theta}{\varepsilon_T}\right)^{z[(1/c) + (1/b)]}\right]^{b/z} \\ 2R\cos\theta R\cos\theta d\theta \quad \text{(Eq 7.28)}$$

The total bending moment, taking into account the symmetrical contribution to bending moment of the section below the neutral axis, is:

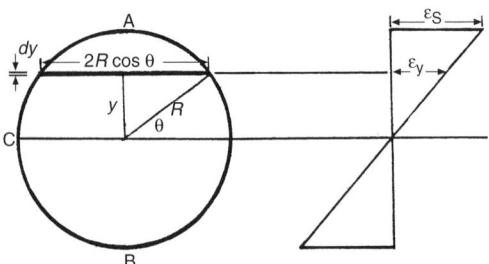

**Fig. 7.3** Circular cross-sectional bar in flexural bending. Source: Ref 7.2

$$M = 2 \int_{y=0}^{y=R} dM = 4R^3 E \varepsilon_S \int_{\theta=0}^{\theta=(\pi/2)}$$
$$\left[ 1 + \left( \frac{\varepsilon_S \sin\theta}{\varepsilon_T} \right)^{z[(1/c + 1/b)]} \right]^{b/z}$$
$$\sin^2\theta \cos^2\theta \cdot d\theta \quad \text{(Eq 7.29)}$$

From Eq 7.1:

$$\frac{\varepsilon_S}{\varepsilon_T} = \left( \frac{N_S}{N_T} \right)^b + \left( \frac{N_S}{N_T} \right)^c \quad \text{(Eq 7.30)}$$

where $z$ is a function of $c/b$ and $\varepsilon_S/\varepsilon_T$ given by Eq 7.2b to 7.2e.

It is clear from Eq 7.29 that the integral does not lend itself to a simple closed-form solution. However, for any given material and a given surface strain, the integral can be computed numerically and the moment can be calculated. The nominal surface stress, $S_{\text{bending}}$, is expressed as:

$$S_{\text{bending}} = \frac{MR}{I} = \frac{64MR}{\pi R^4} = \frac{64M}{\pi R^3} \quad \text{(Eq 7.31)}$$

The ratio of nominal surface stress $S$ to the stress that would be required to produce a life $N_S$ in axial fatigue can be written analogous to Eq 7.21 by assuming the denominator and the exponent $(c - b)$ remain the same for both rectangular and circular cross sections:

$$\delta = \frac{1 + f_1 \left( \frac{N_S}{N_T} \right)^{(c-b)} + f_2 \left( \frac{N_S}{N_T} \right)^{2(c-b)}}{1 + 2 \left( \frac{N_S}{N_T} \right)^{(c-b)} + \left( \frac{N_S}{N_T} \right)^{2(c-b)}} \quad \text{(Eq 7.32)}$$

where $f_1$ and $f_2$ are for the circular cross section. The ratio $\delta$ is calculated numerically, using Eq 7.30 and 7.29, for different values and combinations of parameters involved as follows:

$$\frac{N_S}{N_T} = 10^{-3}, 10^{-2}, \ldots, 10^2, 10^3$$

$$c = -0.3, -0.4, \ldots, -0.9, -1.0$$

$$b = -0.05, -0.06, \ldots, -0.19, -0.20$$

Equation 7.32 can be written as:

$$\left[ \frac{\delta \left[ 1 + 2 \left( \frac{N_S}{N_T} \right)^{(c-b)} + \left( \frac{N_S}{N_T} \right)^{2(c-b)} \right]}{\left( \frac{N_S}{N_T} \right)^{(c-b)}} \right]$$
$$= f_1 + f_2 \left( \frac{N_S}{N_T} \right)^{(c-b)} \quad \text{(Eq 7.33)}$$

Define functions $G$ and $H$ as:

$$G\left( \frac{N_S}{N_T}, c, b, \delta \right) = f_1 + f_2 H\left( \frac{N_S}{N_T}, c, b \right) \quad \text{(Eq 7.34)}$$

If the function $G$ is plotted against the function $H$ for a given $b/c$ ($= n$), for all combinations of $c$, $b$, and $N_S/N_T$, a straight line results. Thus, the slope $f_2$ and intercept $f_1$ are functions of only $n$. The plot of $G$ versus $H$ for different values of $n$ is shown in Fig. 7.4. For clarity, the numerical values of functions $G$ and $H$ for different combinations of $c$, $b$, and $N_S/N_T$ are shown in the figure for only the case of $n = 0.2$. The preceding procedure is repeated for several materials of various values of $n$. The parameters $f_1$ and $f_2$ are calculated for each material with a specific $n$. Because the parameters $f_1$ and $f_2$ are found to be functions of only $n$, an attempt was made to determine these functions whose form is analogous to ones shown for rectangular cross sections in Eq 7.22 and 7.23. The functions are determined by least-squares analysis to be:

$$f_1 = \frac{0.8432 + n}{0.255 + 0.6849n} \quad \text{(Eq 7.35)}$$

$$f_2 = \frac{2.6188}{1.5411 + n} \quad \text{(Eq 7.36)}$$

The curves represented by the preceding functions are shown by solid lines in Fig. 7.5.

For any given material the functions $f_1$ and $f_2$ can be calculated using Eq 7.35 and 7.36. The correction factor $\delta$ and axial stress $S_{\text{axial}}$ can be calculated for any given life using Eq 7.32 and 7.19, respectively. Thus, we can plot nominal bending stress $S_{\text{bending}}$ against life. The plots for several representative materials of engineering interest are shown in Fig. 7.6. The properties of these materials, obtained from Ref 7.4, are listed in Table 3.1 in Chapter 3, "Fatigue Life Relations." Unfortunately, experimental bending data for materials in the same hardness condition are unavailable to compare with the predictions shown by the dashed lines in the figure.

If the fatigue properties of the material are not known or cannot be obtained easily, the method of universal slopes (MUS) solutions can be used with the properties obtained from a tensile test. Transition strain $\Delta\varepsilon_T$ and transition life $N_T$ can be determined from Eq 7.7 and 7.8, respectively.

## Flexural Bending of a Hollow Rectangular Cross Section

In the previous two sections, the equations are derived relating axial stress and nominal bending stress for solid cross sections. However, in many practical applications, the beams are made hollow to reduce weight and save material, still retaining much of the bending (and torsional) strength. Hence, the flexural bending life prediction of beams with hollow sections is addressed in this section. It is noted again that only the effect of plastic flow is considered in reconciling the difference between the axial stress and nominal bending stress for the same life. As discussed earlier, it is instructive to work first with hollow rectangular cross sections that can serve as guidance in deriving equations for hollow circular sections.

Much of the work of this section is derived from the work of Muralidharan Ref 7.5 and 7.6.

**Derivation of Equation for $\delta$ by Integration.** Figure 7.7 shows the variables used in the derivation. We assume a surface strain $\varepsilon_s$ corresponding to a life $N_S$ when the full bending moment $M$ is applied. It is assumed that plane sections remain plane, i.e., strain varies linearly across the thickness and that $\varepsilon_y = (y/h)\varepsilon_s$ is valid. The area is divided into elemental strips

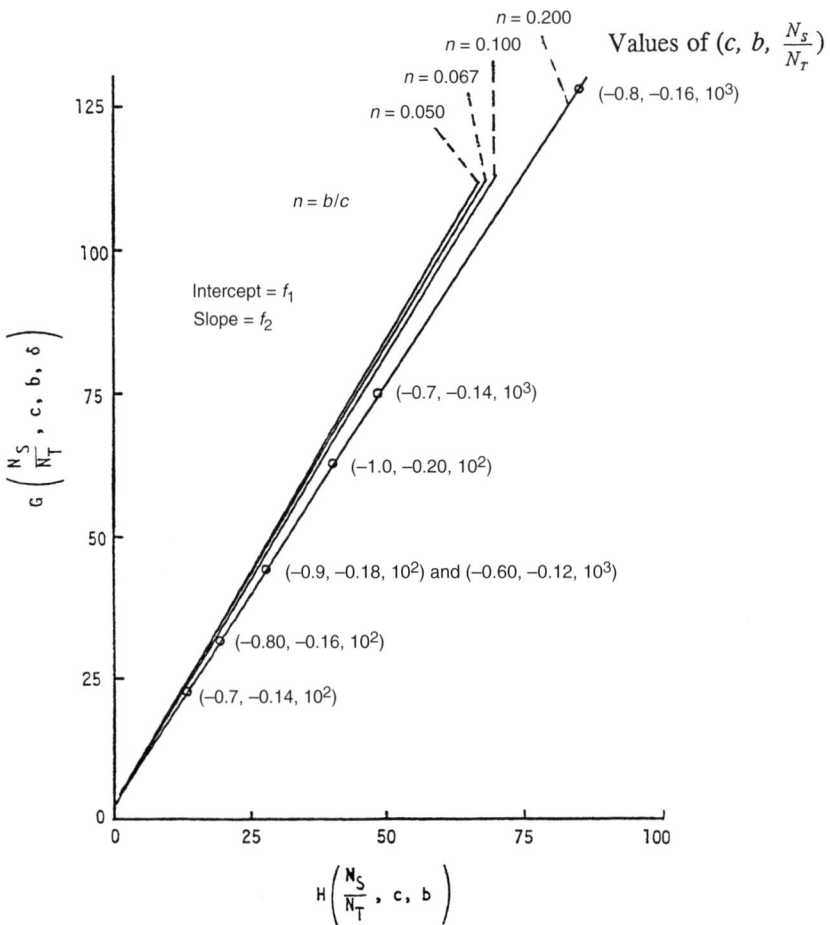

**Fig. 7.4** Plot of $G$ vs. $H$ for a given $c/b$ and for all combinations of $c$, $b$, and $(N_S/N_T)$ showing $f_1$ and $f_2$ are functions of only $n(= b/c)$. Source: Ref 7.2

$dA$ parallel to the neutral axis, and the elemental bending moment $dM$ is computed from knowing the bending stress $\sigma_y$ at a distance $y$ from the neutral axis. Stress is determined from the strain using the cyclic stress-strain curve. The incremental moments are integrated across the entire cross section to give the total bending moment:

$$dM = y\sigma_y\, dA \quad \text{(Eq 7.37)}$$

Substituting for the stress at location $y$ ($\sigma_y$), from Eq 7.9:

$$M = 2E\varepsilon_T w\left[(1-k)\int_0^{kh}\left(\frac{N_y}{N_T}\right)^b y\, dy + \int_{kh}^h\left(\frac{N_y}{N_T}\right)y\, dy\right] \quad \text{(Eq 7.38)}$$

The value of $y$ is substituted in terms of life from Eq 7.12 and Eq 7.38 is integrated. This value of $M$ is substituted into the following equation to compute the nominal bending stress:

$$S_{bending} = 3M/2wh^2(1 - k^4) \quad \text{(Eq 7.39)}$$

$$S_{bending} = \frac{E\varepsilon_T\left(\frac{N_S}{N_T}\right)^b}{(1-k^4)}[\text{term}(1) - \text{term}(2)] \quad \text{(Eq 7.40)}$$

where:

$$\text{Term 1} = \frac{1 + f_1\left(\frac{N_S}{N_T}\right)^{c-b} + f_2\left(\frac{N_S}{N_T}\right)^{2(c-b)}}{1 + 2\left(\frac{N_S}{N_T}\right)^{c-b} + \left(\frac{N_S}{N_T}\right)^{2(c-b)}} \quad \text{(Eq 7.41)}$$

where $f_1$ and $f_2$ are given by Eq 7.22 and:

$$\text{Term}(2) = \frac{k}{\left(\frac{N_S}{N_T}\right)^b}\frac{1}{\left[\left(\frac{N_S}{N_T}\right)^b + \left(\frac{N_S}{N_T}\right)^c\right]^2}[\text{term}(3)] \quad \text{(Eq 7.42)}$$

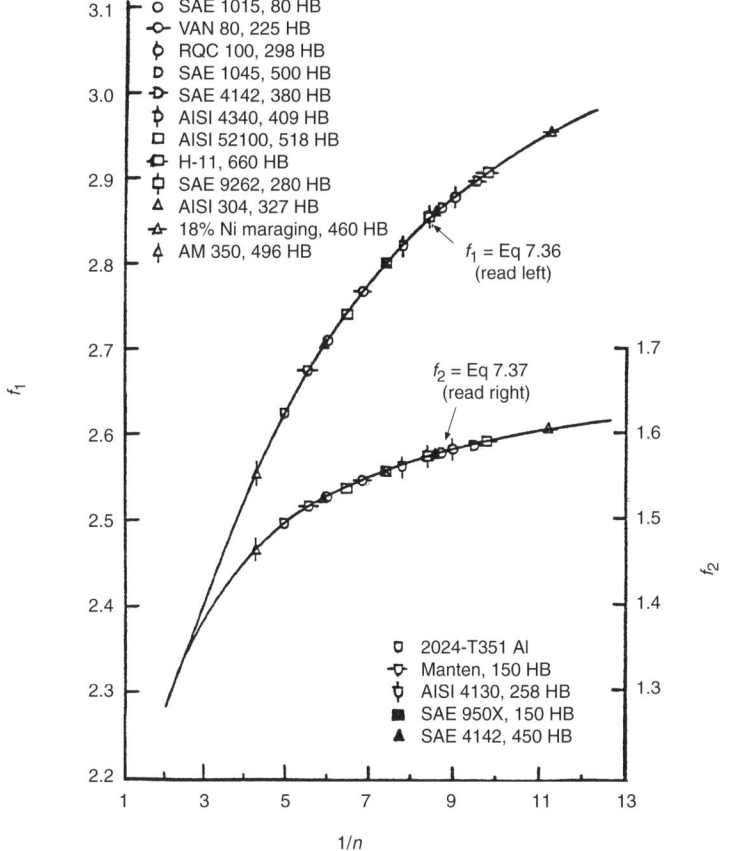

**Fig. 7.5** The parameters $f_1$ and $f_2$ are shown for materials with different $n$ values. The solid lines are curve fits given by Eq 7.36 and 7.37, respectively. Source: Ref 7.2

where:

$$\text{Term}(3) = \left(\frac{N_k}{N_T}\right)^b + f_1\left(\frac{N_k}{N_T}\right)^c + f_2\left(\frac{N_k}{N_T}\right)^{2c-b} \quad \text{(Eq 7.43)}$$

Term 2 of Eq 7.40 can be simplified further by expressing $N_S$ in terms of $N_k$:

$$\varepsilon_k = k\varepsilon_S \quad \text{(Eq 7.44)}$$

$$\frac{\varepsilon_k}{\varepsilon_T} = \left(\frac{N_k}{N_T}\right)^b + \left(\frac{N_k}{N_T}\right)^c \quad \text{(Eq 7.45)}$$

$$\frac{\varepsilon_S}{\varepsilon_T} = \left(\frac{N_S}{N_T}\right)^b + \left(\frac{N_S}{N_T}\right)^c \quad \text{(Eq 7.46)}$$

Substituting these into Eq 7.44:

$$\left(\frac{N_S}{N_T}\right)^b + \left(\frac{N_S}{N_T}\right)^{2c} = \frac{1}{k}\left[\left(\frac{N_k}{N_T}\right)^b + \left(\frac{N_k}{N_T}\right)^c\right] \quad \text{(Eq 7.47)}$$

Substituting Eq 7.47 into Eq 7.42:

$$\text{Term}(2) = k^3\left(\frac{N_k}{N_T}\right)^b \frac{\left[1 + f_1\left(\frac{N_k}{N_T}\right)^{c-b} + f_2\left(\frac{N_k}{N_T}\right)^{2(c-b)}\right]}{1 + \left(\frac{N_k}{N_T}\right)^{c-b} + \left(\frac{N_k}{N_T}\right)^{2(c-b)}} \quad \text{(Eq 7.48)}$$

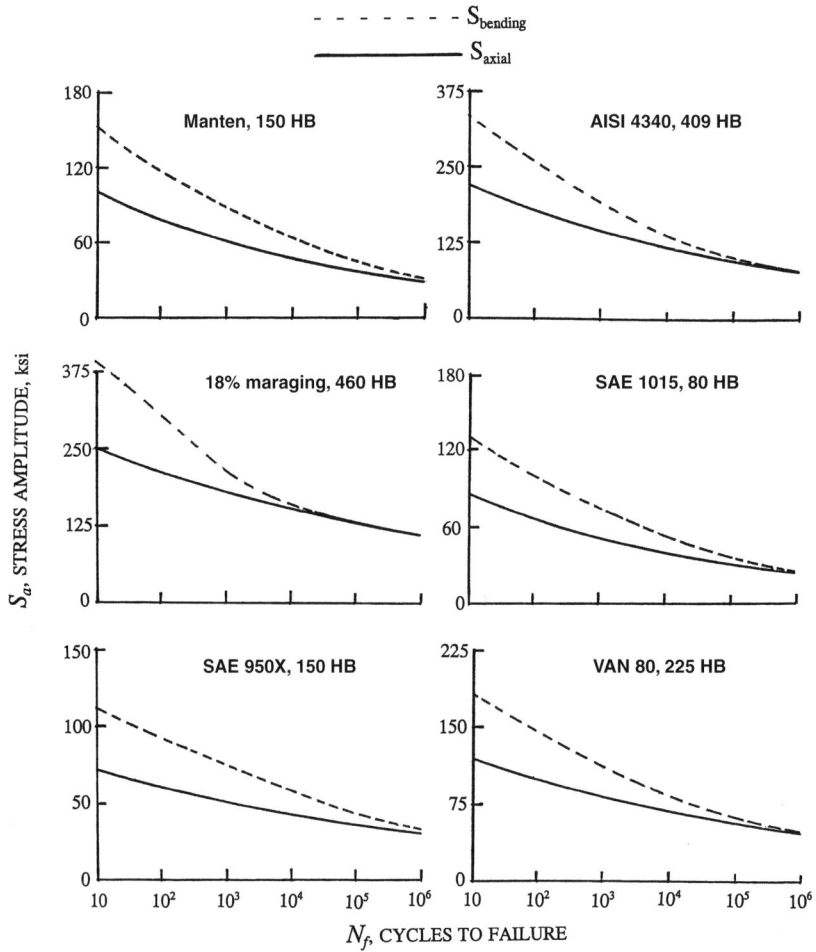

**Fig. 7.6** Axial stress and nominal bending stress are plotted against life for various materials using equations summarized in Table 3.1, Chapter 3. Source: Ref 7.2

by denoting the function:

$$F(x) = \frac{1 + f_1(x)^{c-b} + f_2(x)^{2(c-b)}}{1 + 2(x)^{c-b} + (x)^{2(c-b)}} \quad \text{(Eq 7.49)}$$

from Eq 7.41:

$$\text{Term(1)} = F\left(\frac{N_S}{N_T}\right) \quad \text{(Eq 7.50)}$$

from Eq 7.48:

$$\text{Term(2)} = k^3\left(\frac{N_k}{N_T}\right)F\left(\frac{N_k}{N_T}\right) \quad \text{(Eq 7.51)}$$

from Eq 7.19:

$$S_{\text{axial}} = E\varepsilon_T\left(\frac{N_S}{N_T}\right)^b \quad \text{(Eq 7.52)}$$

hence:

$$\delta = \frac{S_{\text{bending}}}{S_{\text{axial}}} = \frac{1}{1 - k^4}\left[F\left(\frac{N_S}{N_T}\right) - k^3\left(\frac{N_k}{N_S}\right)^b F\left(\frac{N_k}{N_T}\right)\right] \quad \text{(Eq 7.53)}$$

For a given surface life $N_S$, the surface strain $\varepsilon_S$ is known from Eq 7.46. For a known ratio of inner to outer dimensions, $\varepsilon_k$ is found from Eq 7.44, thus $N_K$ is found from the inverted form of Eq 7.45 by using Eq 7.2(a). The function $F$ is known from the equation derived for $\delta$ for solid rectangular cross sections (Eq 7.21). All the terms in Eq 7.53 are known for a given hollow rectangular section and $\delta$ can be computed. It can be seen that when $k$ is zero, solid rectangular section Eq 7.53 reduces to Eq 7.21, as it should.

**Simple Approach to Derivation of $\delta$.** The method just described is straightforward, but the approach becomes more complex when applied to hollow circular sections. The relationship between the equations of hollow and solid sections is not seen as clearly. Hence, a simplified approach is devised that can be applied to hollow cross sections of any shape if the section is symmetric. This concept is shown and checked on rectangular cross sections and later applied to circular cross sections.

The approach is based on the fact that in the derivation of bending moment for a given surface strain, the incremental bending moments are linearly summed. Hence, the moment required to cause a given surface strain on the hollow rectangular section can be computed by computing the moment, $M_1$ for the solid section of outer dimensions as shown in Fig. 7.7(b) and the moment $M_2$ for the solid section of inner dimensions as shown in Fig. 7.7(c), and then subtracting $M_2$ from $M_1$. The only important precaution that must be observed is that at any distance from the neutral axis, the strain in the outer rectangle must be the same as the strain in the inner rectangle. We start by writing the equation for the bending stress $S_{\text{bending}}$ in the hollow rectangular section in terms of the bending moment $M$:

$$S_{\text{bending}} = \frac{3M}{2wh^2(1 - k^4)} \quad \text{(Eq 7.54)}$$

where $M = M_1 - M_2$ and:

$$M_1 = S_{\text{bending},1}\left(\frac{2wh^2}{3}\right) = \delta_1 S_{\text{axial},2}\left(\frac{2}{3wh^2}\right)$$

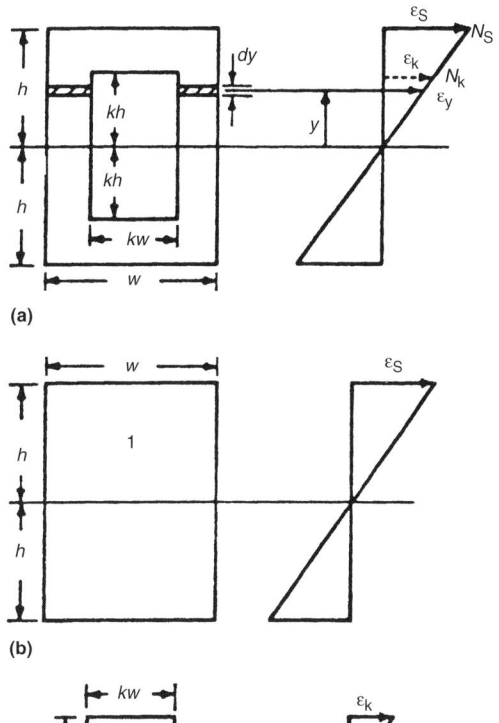

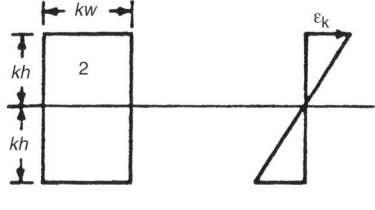

**Fig. 7.7** Schematic showing the variables used in the derivation of Eq 7.54 for $\delta$ for hollow rectangular cross sections. (a) Hollow rectangular cross section. (b) Outermost cross section. (c) Hollow portion of cross section. Source: Ref 7.6

Using Eq 7.21 and 7.19:

$$M_1 = F\left(\frac{N_S}{N_T}\right)E\varepsilon_T\left(\frac{N_S}{N_T}\right)^b \frac{2wh^2}{3} \quad \text{(Eq 7.55a)}$$

$$M_2 = S_{bending,2}\left[\frac{2}{3}(kw)(kh)^2\right] = \delta_1 S_{axial,2}\left[\frac{2}{3}wh^2k^3\right]$$

$$= F\left(\frac{N_k}{N_T}\right)\left[E\varepsilon_T\left(\frac{N_k}{N_T}\right)^b\left(\frac{2}{3}wh^2k^3\right)\right]$$

$$= F\left(\frac{N_k}{N_T}\right)\left[E\varepsilon_T\left(\frac{N_S}{N_T}\right)^b \frac{2}{3}wh^2\right]k^3\left(\frac{N_k}{N_S}\right)^b \quad \text{(Eq 7.55b)}$$

Thus, the ratio of nominal bending to axial stress for a hollow rectangular section can be written as:

$$\delta = \frac{S_{bending}}{S_{axial}} = \frac{1}{E\varepsilon_T\left(\frac{N_S}{N_T}\right)^b}\left[\frac{(M_1 - M_2)}{\frac{2}{3}wh^2(1 - k^4)}\right] \quad (7.56)$$

Substituting Eq 7.55(a) and (b) for $M_1$ and $M_2$, respectively:

$$\delta = \frac{1}{1 - k^4}\left[F\left(\frac{N_S}{N_T}\right) - k^3\left(\frac{N_k}{N_S}\right)^b F\left(\frac{N_k}{N_T}\right)\right] \quad (7.57)$$

This is the same as Eq 7.53 obtained by direct integration previously. This technique is very simple and easy to apply to the cross section of any shape.

## Flexural Bending of a Hollow Circular Cross Section

Figure 7.8 shows the variables used in the derivation. The approach presented in the earlier section "Derivation of Equation for δ by Integration" does not lend itself to closed-form solution as in the case of solid circular sections. Thus, the simpler approach given in the section "Simple Approach to the Derivation of δ" is used to obtain an equation for δ and then verified by numerical integration. They are presented in this section.

**Simple Derivation of δ.** The moment required to cause a given surface strain on the hollow cross section is computed by computing the moment, $M_1$, for the solid circular section of radius $r_0$ as shown in Fig. 7.8(b) and the moment, $M_2$, for the solid circular section of radius $r_i$ as shown in Fig. 7.8(b) and subtracting $M_2$ from $M_1$:

$$S_{bending} = \frac{4M}{\pi r_o^3(1 - k^4)} \quad \text{(Eq 7.58)}$$

where $M = M_1 - M_2$ and

$$M_1 = S_{bending,1}\left(\frac{\pi r_o^3}{4}\right)$$

Using Eq 7.32 and 7.19:

$$M_1 = F\left(\frac{N_S}{N_T}\right)E\varepsilon_T\left(\frac{N_S}{N_T}\right)^b \frac{\pi r_o^3}{4} \quad (7.59)$$

Similarly:

$$M_2 = F\left(\frac{N_k}{N_T}\right)E\varepsilon_T\left(\frac{N_S}{N_T}\right)^b \frac{\pi r_i^3}{4} \quad (7.60)$$

$$= F\left(\frac{N_k}{N_T}\right)E\varepsilon_T\left(\frac{N_S}{N_T}\right)^b \frac{\pi r_o^3}{4} k\left(\frac{N_k}{N_T}\right)^b \quad (7.61)$$

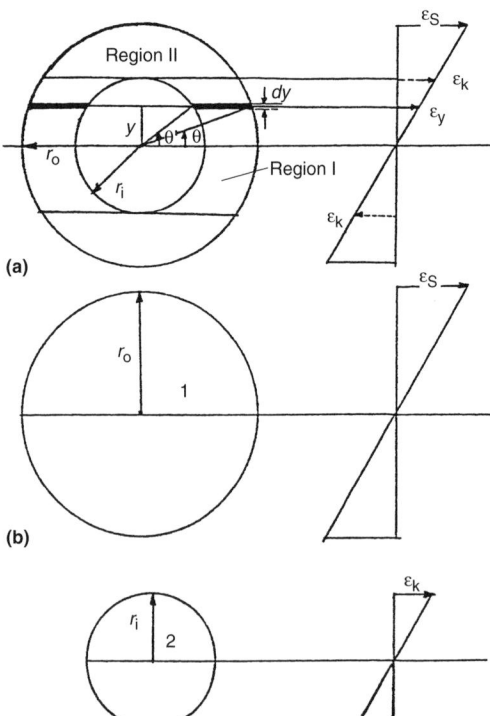

**Fig. 7.8** Schematic showing the variables used in the derivation of the equation for δ for hollow circular cross sections. (a) Hollow circular cross section. (b) Solid outer radius. (c) Solid inner radius. Source: Ref 7.5

$$\delta = \frac{1}{E\varepsilon_T\left(\dfrac{N_S}{N_T}\right)} \cdot \frac{4(M_1 - M_2)}{\pi r_o^3(1 - k^4)} \quad (7.62)$$

Substituting for $M_1$ and $M_2$:

$$\varepsilon = \frac{1}{1 - k^4}\left[F\left(\frac{N_S}{N_T}\right) - k^3\left(\frac{N_k}{N_S}\right)^b F\left(\frac{N_k}{N_T}\right)\right] \quad (7.63)$$

It is important to note that Eq 7.63 and 7.57 are identical except the functions $f_1$ and $f_2$ in $F$ are the ones derived in the section on solid circular sections and given by Eq 7.35 and 7.36.

**Verification of the Equation for δ by Numerical Integration.** The equations for the integration of elemental bending moments across the hollow circular section are developed by the procedure similar to the one given in the section "Flexural Bending of a Circular Cross Section." Refer to Fig. 7.8(a) for the following derivation:

$$k = \frac{r_i}{r_o} \quad (\text{Eq } 7.64)$$

$$dM = y\sigma_y w dy \quad (\text{Eq } 7.65)$$

where $y = r_o \sin\theta$ and $dy = r_o \cos\theta\, d\theta$.

The width of the strip $w$ is different in the two regions shown in the figure.

In region I, $\theta \leq \sin^{-1}(k)$:

$$\text{width, } w = 2r_o \cos\theta - 2r_i \cos\theta'$$

because $r_o \sin\theta = r_i \sin\theta'$

$$\text{width, } w = 2r_o \cos\theta - 2r_i[1 - (r_o/r_i)^2 \sin^2\theta]^{1/2}$$
$$= 2r_o[\cos\theta - (k^2 - \sin^2\theta)^{1/2}]$$

In region II, $\theta \geq \sin^{-1}(k)$:

$$\text{width, } w = 2r_o \cos\theta$$

Substituting these expressions and $\sigma_y$ in terms of $\varepsilon_S$ from Eq 7.27 into Eq 7.65 and integrating:

$$M = 4E\varepsilon_S(r_o)^2 \int_0^{\pi/2} \sin^2\theta \cos\theta$$
$$\cdot [\text{term}(4)] \cdot [\text{term}(5)] \cdot d\theta \quad (\text{Eq } 7.66a)$$

where:

$$\text{Term}(4) = \cos\theta - (k^2 - \sin^2\theta)^{1/2} \text{ for } \theta \leq \sin^{-1}(k)$$
$$= \cos\theta \text{ for } \theta \geq \sin^{-1}(k) \quad (\text{Eq } 7.66b)$$

$$\text{Term}(5) = \left[1 + \left(\frac{\varepsilon_S \sin\theta}{\varepsilon_T}\right)^{z[1/c - 1/b]}\right]^{b/z} \quad (\text{Eq } 7.66c)$$

where $z$ is given by Eq 7.2(c) to (e):

$$S_{\text{bending}} = \frac{4M}{\pi r_o^3(1 - k^4)}$$

$$S_{\text{axial}} = E\varepsilon_T\left(\frac{N_S}{N_T}\right)^b$$

Substituting these expressions into the equation for δ and substituting the value of $M$ from Eq 7.66(a):

$$\delta = \frac{16}{\pi(1 - k^4)}\left(\frac{\varepsilon_S}{\varepsilon_T}\right)\frac{1}{(N_S/N_T)^b}$$
$$\int_0^{\pi/2} \sin^2\theta \cos\theta[\text{term}(4)][\text{term}(5)]d\theta \quad (\text{Eq } 7.67)$$

Terms 4 and 5 are given by Eq 7.66(a) and (b), respectively.

The validity of this equation has been verified by Muralidharan (Ref 7.5) by comparing the results of a large number of cases wherein numerical integration was applied directly to specific geometries to those results obtained by using Eq 7.67. In all cases, there was very close agreement (within 2%) between the two methods.

## Summary and Illustrations for Bending of Rectangular and Circular Shafts

We can now summarize the detailed derivations for simple applications as follows: The fatigue life equation is taken as:

$$\frac{\varepsilon}{\varepsilon_T} = \left[\left(\frac{N_f}{N_T}\right)^b + \left(\frac{N_f}{N_T}\right)^c\right] \quad (\text{Eq } 7.68)$$

This is equivalent to the Manson-Coffin-Basquin formulation expressed in terms of Morrow's notation:

$$\varepsilon = \sigma_f'(2N_f)^b + \varepsilon_f'(2N_f)^c \quad (\text{Eq } 7.69)$$

and the transition strain range:

$$\varepsilon_T = 2(\varepsilon_f')^{(b/b - c)}\left(\frac{\sigma_f'}{E}\right)^{(c/c - b)} \quad (\text{Eq } 7.70)$$

and the transition fatigue life:

$$N_T = \frac{1}{2}\left(\frac{E\varepsilon'_f}{\sigma'_f}\right)^{1/(b-c)} \quad \text{(Eq 7.71)}$$

It is assumed that fatigue life is governed by the surface stress (or strain), and that strain varies linearly across the section from zero at the neutral axis to a maximum at the surface. The stress is not linear across the section but conforms to the strain according to the true stress versus true strain curve that follows a Ramberg-Osgood form (See Chapter 2, "Stress and Strain Cycling"). Thus, if any of the strains in the section pass into the plastic range, the stresses must follow accordingly. The key step is to determine the stress distribution that will support an imposed bending moment. Once this bending moment has been calculated, we can calculate the nominal stress that would occur if the same bending moment were carried by the section while the material remained elastic. The nominal stress would, of course, be higher than the true stress. In fact, by the calculations:

$$\delta = \frac{\text{nominal stress for applied bending moment}}{\text{true stress developed for applied bending moment}} \quad \text{(Eq 7.72)}$$

where:

$$\delta = \frac{1 + f_1(N_S/N_T)^{cb} + f_2(N_f/N_T)^{2cb}}{1 + 2(N_f/N_T)^{cb} + (N_f/N_T)^{2cb}} \quad \text{(Eq 7.73)}$$

where, for a solid rectangular cross section:

$$f_1 = \frac{1+n}{2/3 + (1/3)n} \quad \text{and}$$
$$f_2 = \frac{3n}{1+2n} \quad \text{(Eq 7.74)}$$

and for a solid circular cross section:

$$f_1 = \frac{1 + 0.8432n}{0.6849 + 0.255n} \quad \text{and}$$
$$f_2 = \frac{2.6188n}{1 + 1.5411n} \quad \text{(Eq 7.75)}$$

For a hollow cross section we first define the function $F(x)$, analogous to the expression for $s$ except that the variable $x$ replaces the term $N_S/N_T$:

$$F(x) = \frac{1 + f_1(x)^{cb} + f_2(x)^{2cb}}{1 + 2(x)^{cb} + x^{2cb}} \quad \text{(Eq 7.76)}$$

We then find that for the hollow section, the ratio $\delta$ changes to the form:

$$\delta = \frac{1}{1-k^4}\left[F\left(\frac{N_S}{N_T}\right) - k^3\left(\frac{N_k}{N_S}\right)^b F\left(\frac{N_k}{N_T}\right)\right] \quad \text{(Eq 7.77)}$$

Here, $k$ is the ratio of the inner dimension of the hollow section to the outer dimension, as seen, for example, in Fig. 7.7 for rectangular sections and $r_i/r_o$ for circular sections (Fig. 7.8). The value of $N_k$ is the life corresponding to the strain on the inner surface when the life on the outer surface is $N_S$. Thus, the same equations already derived for the solid sections, rectangular or circular, can now be used for a hollow section by applying Eq 7.77.

For illustration, consider material No. 25, 18% maraging steel with hardness of 460 HB (Table 3.1, Chapter 3, "Fatigue Life Relations"), shown in Fig. 7.9, and let us determine the coordinates at the points $U$ and $V$ at a life level of 100 cycles to failure (log $N_f = 2$). Point $U$ is for a solid circular cylinder and $V$ is for a hollow circular cylinder with inner diameter 75% of outer diameter ($k = 0.75$). For this material:

$$\sigma'_f = 2137.9 \text{ MPa}; \quad \varepsilon'_f = 0.80,$$
$$c = -0.79, \quad b = -0.071, \text{ and } E = 186.2 \text{ GPa}$$

Application of Eq 7.8 and 7.7 gives $\Delta\varepsilon_T = 0.015$ and $N_T = 183$ cycles. From which the basic axial fatigue life equation (Eq 7.1) is:

$$\frac{\Delta\varepsilon}{0.015} = \left[\left(\frac{N}{183}\right)^{-0.79} + \left(\frac{N}{183}\right)^{-0.071}\right] \quad \text{(Eq 7.78)}$$

and the stress equation versus life (Eq 7.9) is:

$$\sigma = E\varepsilon_T\left(\frac{N}{183}\right)^{-0.071} \quad \text{(Eq 7.79)}$$

Five additional examples appear in Ref 7.6 for materials 3, 34, 36, 44, and 46 (Chapter 3, "Fatigue Life Relations," Table 3.1).

It should be noted also that the bending analyses presented previously are valid only for flexural, not rotating, bending. During rotating bending, the cross section deflects about an axis that makes an angle with the loading axis (Ref 7.1). However, once that angle has been determined, the strain distribution about the resultant axis is essentially the same as that for flexural bending. Furthermore, experimental results for rotating bending agreed well with predictions based on the derived flexural bending equations (Ref 7.1). A discussion of the relation of rotating to flexural bending is presented in the following section.

## Relation between Flexural Bending and Rotating Bending

**Fundamental Difference.** Rotating bending is a common form of bending encountered in axles and shafts. Rotating bending fatigue specimens were first used in the 1800s by Wöhler (Ref 4.1 in Chapter 4) to investigate the fatigue resistance of steels, the bill of material for rail car axles. Rotating bending fatigue machines are still in use as a new century begins, despite the considerable differences in fatigue results between rotating bending and axially loading. Initial analyses of these differences have been presented in Ref 7.1. Rotating bending analyses are considerably more complex than for flexural bending. This complexity is illustrated with the aid of Fig. 7.10 and 7.11.

Consider the two elements A and B shown in Fig. 7.10(b). If the cross section were simply being bent back and forth as in flexural bending, these two elements, being at the same distance from the bending axis αα would have exactly the same stress. In rotating bending, however, a problem develops because of a phase relation, with the result that A and B are not at the same stress level unless the entire cross section is elastic. If we were to analyze this problem in the same manner as we did for flexural bending, assuming a strain at the surface of $\varepsilon_o$, the stress at points A and B would be equal and would be obtained from knowledge of the strain, as shown in Fig. 7.10(c). If the cross section were not rotating, A and B would always be at the same distance from the neutral axis, and no question would arise regarding the value of stress in these elements. When we consider rotating bending, however, it is clear that points A and B will ultimately reach a strain level $\varepsilon_m$ when they achieve the maximum distance from the strain axis, but at the instant considered in the figure, point B has already passed its maximum strain, while point A is approaching its maximum. The stresses at each of these points must be determined from the hysteresis loop (Fig. 7.10d) that, for simplicity, makes use of an idealized representation of the stress-strain curve. In this figure, the stress corresponding to the maximum strain $\varepsilon_m$ to be reached by both these points is shown as $\sigma_m$, obtained from Fig. 7.10(c) for the maximum strain $\varepsilon_m$ experienced by both points A and B. To obtain the current stresses, however, we must determine the stress when the strain is $\varepsilon$.

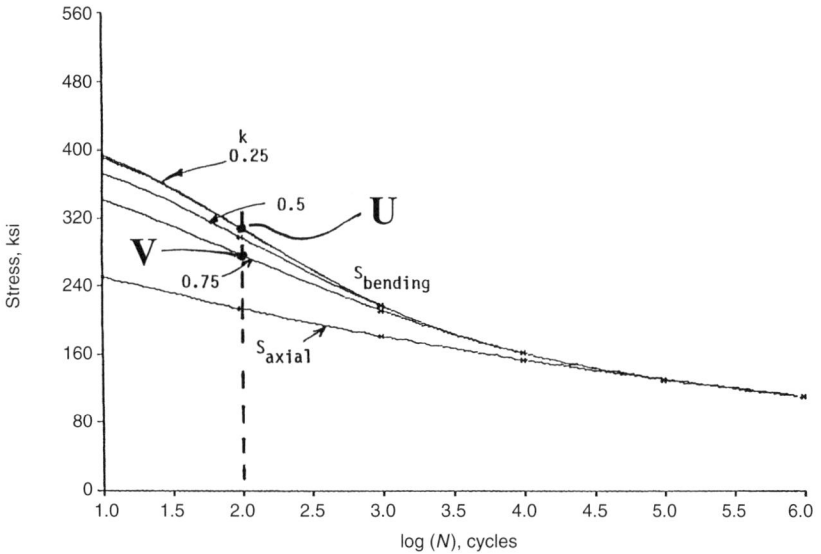

**Fig. 7.9** Nominal bending stress for various values of k and axial stress are plotted vs. logarithmic cyclic life for an example material No. 25 (Table 3.1 in Chapter 3), 18% Ni-maraging steel, 460 HB. Calculated curves are based on Eq 7.78, 7.76, and 7.80 for circular cross sections. Source: Ref 7.5

Therefore, referring again to Fig. 7.10(d), we can move out a distance to $\varepsilon$ from the vertical axis locating the points identified as $\sigma_A$ and $\sigma_B$. These are the stresses at the two elements $A$ and $B$. Thus, even though $A$ and $B$ are at the same distance from the strain axis, their stresses are,

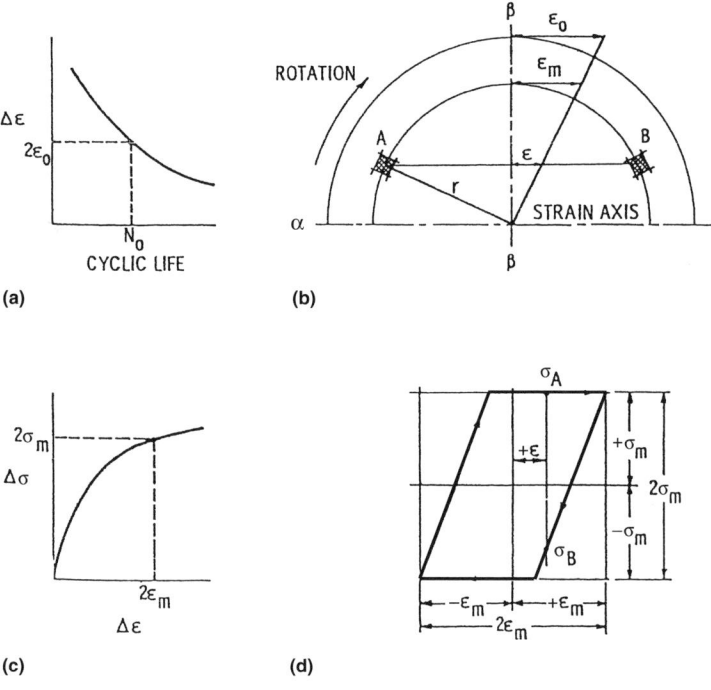

**Fig. 7.10** Determination of stress distribution in a rotating beam from axial strain-cycling fatigue data. (a) Total strain range-life relation for material. (b) Strain distribution in cross section of rotating beam. (c) Cyclic stress range-strain range relation for material. (d) Hysteresis loop for points at radius $r$ of rotating beam. Source: Ref 7.1

$$M_{\alpha\alpha} = \Sigma\sigma \cdot \beta \cdot \Delta A = \Sigma\sigma \cdot r \cdot \sin\theta \cdot \Delta A$$
$$M_{\beta\beta} = \Sigma\sigma \cdot \alpha \cdot \Delta A = \Sigma\sigma \cdot r \cdot \cos\theta \cdot \Delta A$$
$$M_{yy} = \Sigma\sigma \cdot x \cdot \Delta A = \Sigma\sigma \cdot r \cdot \cos(\theta + \varphi) \cdot \Delta A$$
$$= \Sigma\sigma \cdot r \cdot [\cos\theta \cdot \cos\varphi - \sin\theta \cdot \sin\varphi] \cdot \Delta A$$
$$= \cos\varphi \cdot M_{\beta\beta} - \sin\varphi \cdot M_{\alpha\alpha}$$

but

$$M_{yy} \equiv 0$$

Hysteresis angle

$$\varphi = \tan^{-1}\left(\frac{M_{\beta\beta}}{M_{\alpha\alpha}}\right)$$

$$M_{xx} = \Sigma\sigma \cdot y \cdot \Delta A = \Sigma\sigma \cdot r \cdot \sin(\theta + \varphi) \cdot \Delta A$$
$$= \Sigma\sigma \cdot r \cdot [\sin\theta \cdot \cos\varphi + \cos\theta \cdot \sin\varphi] \cdot \Delta A$$
$$= \cos\varphi \cdot M_{\alpha\alpha} + \sin\varphi \cdot M_{\beta\beta}$$

Nominal

$$\sigma_{max} = M_{xx} \cdot \frac{R}{I} = \frac{4M_{xx}}{\pi R^3}$$

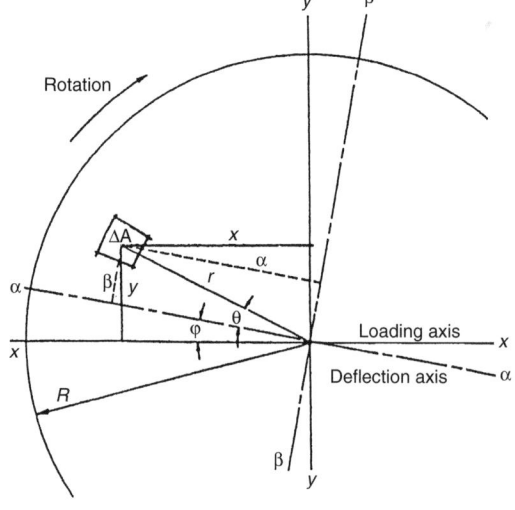

**Fig. 7.11** Determination of bending moment and hysteresis angle in a rotating beam. Source: Ref 7.1

in fact, quite different at the instant considered and, in obtaining the bending moment, allowance must be made for the actual stress present in each element.

Further reflection regarding this problem indicates that a convenient approach for analysis is that indicated in Fig. 7.11. The loading axis xx is the axis about which the bending moment is applied. If one considers the strains at the individual locations in the element, it can be seen that the cross section bends about a different axis αα, which makes an angle φ with the xx axis. By writing an expression for the bending moment contributed by all the elements about an arbitrary axis relative to the displacement axis, and expressing the fact that the bending moment about the axis normal to the loading axis 44 must be zero (because the applied bending moment has no component about this normal axis), we can determine the angle φ between the deflection axis and the loading axis, as shown in the figure. Once the angle φ between the two axes has been determined, the complete stress and strain distribution can be computed, from which the bending moment for an assumed value of maximum surface strain can be established. From the assumed value of surface strain, life can be determined from Fig. 7.10(a). An experimental investigation comparing rotating and flexural bending with this analysis has been performed (Ref 7.1). Some of these comparative results are shown in Fig. 7.12 for four common structural alloys.

The approach presented up to this point in the chapter takes account only the effect of plastic flow in explaining the difference between axial and bending fatigue resistances. According to the analyses, the bending and axial fatigue curves become coincident for most engineering materials at lives greater than approximately $10^6$ cycles to failure because the strains are elastic at those life levels and the classical elastic $Mc/I$ assumption is valid. However, bending fatigue resistance could still be greater than axial for the same nominal stress due to the volumetric effect that has not been taken into account in the preceding analysis. Volumetric effects are addressed in the next section.

**Difference Related to Volumetric Effects**
Another important difference between alternating bending and rotating bending relates to the volume of material that is fatigued at the maximum stress range for each of the two cases. In alternating bending only the very outer fibers that remain at the maximum distance from the center of the shaft perceive the maximum stress range. In rotating bending, however, every fiber of the periphery of the specimen eventually arrives to the position of maximum stress. Thus, more volume of material is subjected to fatigue in the rotating bending specimen than in the flexural bending specimen. As is amplified in the next section, volumetric effects can be considerable, depending especially on the sensitivity of the material to microscopic flaws, and thus especially for the harder materials. Data illus-

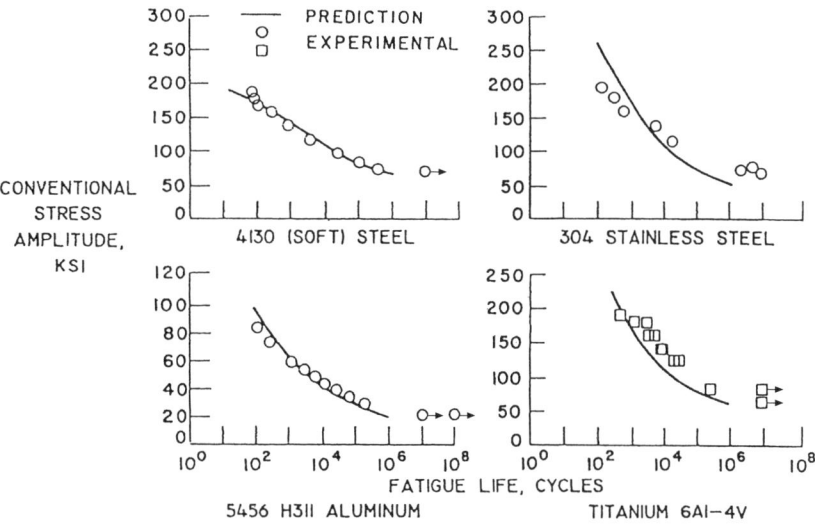

**Fig. 7.12** Prediction of rotating-bending fatigue from reversed strain-cycling behavior for four materials. Source: Ref 7.1

trating this effect are limited. One interesting study by Esin (Ref 7.7) is shown in Fig. 7.13 wherein comparison is made directly for flexural bending and rotating bending for the same material. It is clear that the flexural bending specimens provide higher fatigue results than rotating bending. This figure also compares the bending results to axial push-pull loading. In both cases (Fig. 7.13b and c), the push-pull fatigue results appear inferior to the bending cases. While several factors can be involved in this comparison (for example, whether the stresses shown for bending properly account for stress variation across the sections, as discussed earlier) the push-pull specimens always yield lower fatigue resistance, presumably because of volumetric effects as well.

## Consideration of Volumetric Effects

### Background

An interesting observation can be made by comparing the results of the theoretical analysis of flexural fatigue and axial fatigue with the experimental results of Esin in Fig. 7.13. Whereas all the curves from Fig. 7.6 show that axial stress and bending stress become equal at the very long lives, Esin's experiments do not verify this observation. Instead, the tolerable stresses for axial fatigue at a given life are always lower than those for either flexural or rotating bending. As already noted, this is due to the fact that in axial fatigue the entire cross section is subjected to the applied stress, whereas in bending, only a very small volume of material perceives the high stress while most of the volume is at lower stress. In rotating bending, more material eventually perceives the highest stress, but still a large volume is always at a lower stress. Esin, in explaining his results, argued that even at the high cyclic lives where the stresses are nominally elastic, highly localized plastic pockets are developed, surrounded by a strong elastic matrix. Such development is, however, a highly probabilistic phenomenon, the occurrence and extent of which is dependent on such factors as microstructure, microinhomogeneity, anisotropy, and stress concentration. Plastic hysteresis energy is dissipated each cycle by a macroelement subjected to stress amplitudes above the

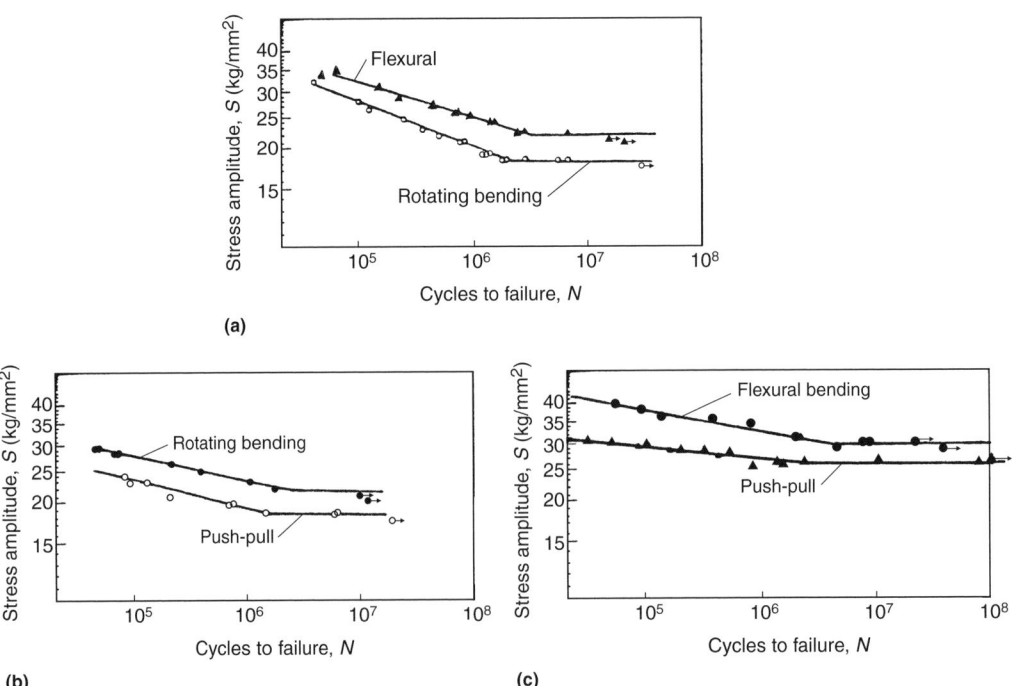

**Fig. 7.13** Experimental results for carbon steels from different types of fatigue testing (Ref 7.7). (a) C1030 (similar to AISI/SAE 1030) in rotating and flexural bending. (b) C10 (DIN) (similar to AISI/SAE 1010) in push-pull and rotating bending (Ref 7.5). (c) C1020 (similar to AISI/SAE 1020) in push-pull and flexural bending

true elastic limit, which is lower than the "endurance limit" of the material. When the plastic hysteresis energy accumulates to the true fracture energy of the material, fatigue failure occurs. Thus, the higher the stressed volume, the greater is the probability that a failure could be induced by a lower average stress. While qualitatively valid, Esin did not develop a quantitative approach for general treatment of the problem.

Kuguel (Ref 7.8) also treated problems of this type by considering the highly stressed volume as a governing parameter in the fatigue of a structure. He defined the "highly stressed volume" as the layer of material subjected to a stress 95% or more of the maximum stress, and he treated several problems successfully by this approach. Peterson (Ref 7.9) and others (see Ref 1 to 16 in Ref 7.10) also recognized the factor of volume under stress. In Chapter 8, "Notch Effects," where Peterson's analysis of the fatigue stress concentration factor is discussed, a term relating to the radius of the notch is involved in addition to the geometric stress concentration factor. Thus, for the same geometric stress concentration factor, the element that had the smaller dimension (characterized by the radius) was less susceptible to fatigue because a smaller volume was subjected to the highest stress (even though the maximum stress was the same for the two geometries). Raske (Ref 7.10) conducted a comprehensive investigation of the subject. He examined four prominent theories from the literature that had evolved to explain why the fatigue resistance of notched fatigue specimens was reduced, to a lesser extent than would have been estimated based on the theoretical elastic stress concentration factor, $K_t$. By combining the salient features of these theories, he was able to arrive at an equation that was demonstrated to predict, more accurately, the experimental fatigue lives of notched fatigue coupons than by any of the individual theories. His equation was written in terms of parameters of the specimen and material and how they reduced the theoretical stress concentration factor to the fatigue notch factor, $K_f$:

$$K_f = K_t[1 - (c\delta/2r)](V_u/V_\delta)^k \quad \text{(Eq 7.80)}$$

where $c$ is the dimensionless stress or strain gradient variable, $\delta$ is the depth of the zone that governs the fatigue process when the local stresses at the root of the notch are elastic, $r$ is the notch radius, $(V_u/V_\delta)$ is the ratio of the highly stressed volumes of the unnotched and notched specimens, and $k$ is the material constant in Kuguel's size effect equation.

Figure 7.14 shows a comparison of Raske's prediction, Eq 7.80, of the experimental notched fatigue curves for 2024-T3 aluminum plates with two different widths (data Ref 7.11).

Of all the methods that can be applied to the treatment of volumetric effects, the Weibull statistical theory of strength (see, for example, Ref 7.12 and 13) lends itself most generally to problems of this type. We have used it in the past to treat thermal shock problems in brittle materials (Ref 7.14) and later extended it, together with U. Muralidharan, for the study of bending fatigue. A brief review of the Weibull concept is given in the Appendix "Selected Background Information." In the next section, we show some of the results obtained by Muralidharan (Ref 7.5).

## Volumetric Effect in the High Cyclic Life Range

In the high cyclic life regime, the strains are elastic, hence, the stress is proportional to strain. The ratio of flexural or rotating bending stress to axial fatigue stress is obtained for a given life, i.e., for the same risk of rupture.

**Flexural Bending of a Rectangular Cross Section.** Consider two bars of the same rectangular cross section and length, one subjected to alternating flexural loading and the other to alternating axial loading. The bar subjected to

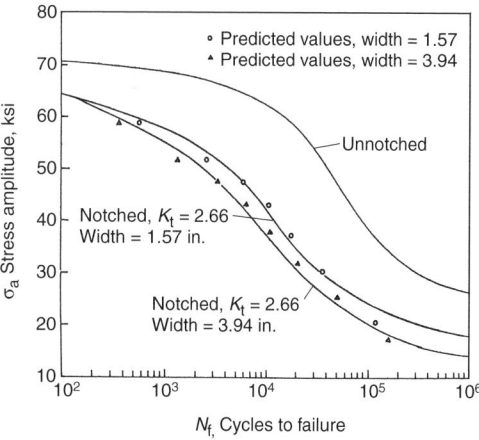

**Fig. 7.14** Predicted fatigue response (○ width = 39.9 mm, or 1.57 in., △ width = 100.1 mm, or 3.94 in.) of notched specimens of 2024-T3 aluminum plate based on predictions by Eq 7.81 (Ref 7.10). Smooth and notched fatigue results (solid curves) from Ref 7.11

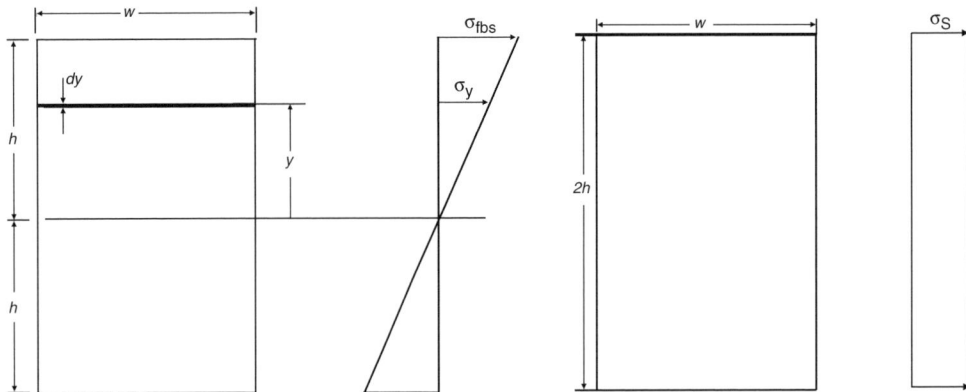

**Fig. 7.15** Rectangular cross-sectional bars subjected to flexural alternating bending and axial cyclic loading. Source: Ref 7.5

flexural bending is expected to sustain longer life than the bar cycled axially for the same maximum stress. This difference is due primarily to the effect of volume since the plastic flow was assumed absent in this regime. The ratio of maximum stress in flexural bending to pure axial loading can be obtained for a given life by equating the risks of rupture in these two cases (Fig. 7.15). The risk of rupture in flexural bending = $R_{fb}$:

$$R_{fb} = 2 \int_0^h \left(\frac{y\sigma_{fbs}}{h\sigma'}\right)^m w\,dy \quad \text{(Eq 7.81)}$$

$$R_{fb} = \frac{2hw}{m+1}\left(\frac{\sigma_{fbs}}{\sigma'}\right)^m \quad \text{(Eq 7.82)}$$

Risk of rupture in axial loading = $R_a$:

$$R_a = \left(\frac{\sigma_s}{\sigma'}\right)^m 2hw \quad \text{(Eq 7.83)}$$

Equating Eq 7.82 and 7.83:

$$\sigma_{fbs} = (m+1)^{1/m}\sigma_s \quad \text{(Eq 7.84)}$$

where $m$ is the material homogeneity factor.

For example, let $m = 20$; $\sigma_{fbs} = 1.16\sigma_s$; and for $m = 2$, we have $\sigma_{fbs} = 1.73\sigma_s$.

**Rotating Bending of a Circular Cross Section.** Consider two bars of the same circular cross section and length, one subjected to rotating bending and the other to axial fatigue loading (Fig. 7.16). The ratio of the maximum stress in rotating bending to pure axial loading can be obtained for a given life by equating the risks of rupture in these two cases. Risk of rupture in rotating bending = $R_{rb}$:

$$R_{rb} = \int_0^R \left(\frac{\sigma_{rbs}}{\sigma'}\frac{r}{R}\right)^m 2\pi r\,dr \quad \text{(Eq 7.85)}$$

$$R_{rb} = \frac{2\pi R^2}{m+2}\left(\frac{\sigma_{rbs}}{\sigma'}\right)^m \quad \text{(Eq 7.86)}$$

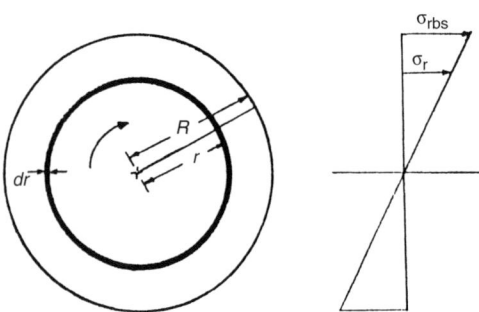

**Fig. 7.16** Circular bar subjected to rotating bending fatigue loading. Source: Ref 7.5

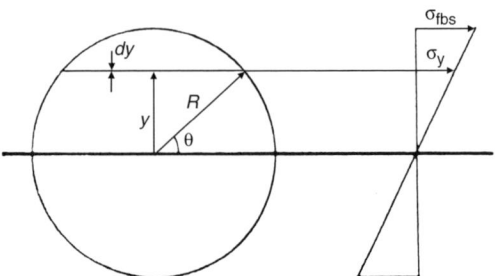

**Fig. 7.17** Flexural alternating bending of circular cross-sectional bar. Source: Ref 7.5

Chapter 7: Bending of Shafts / 175

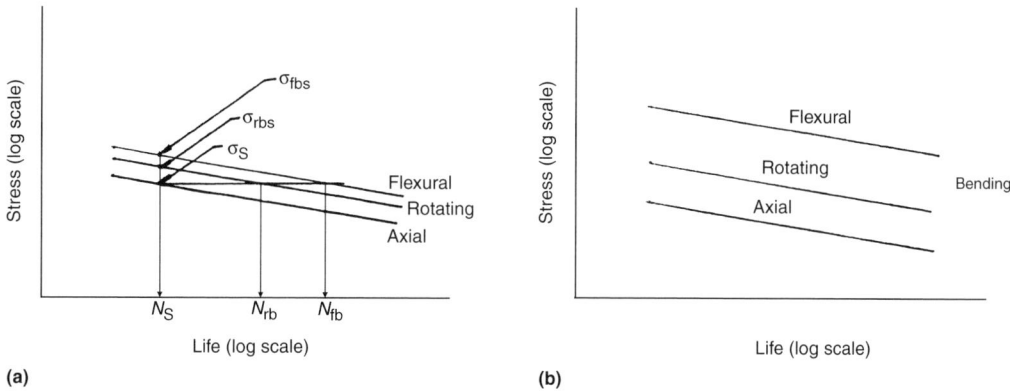

**Fig. 7.18** Comparison of lives under axial, rotating bending, and flexural bending fatigue loading. (a) Constant material homogeneity factor = $m_1$ (b) Constant material homogeneity factor = $m_2$. Note $m_2 < m_1$. Source: Ref 7.5

Equating the risks of rupture in the two cases for the same life:

$$\sigma_{rbs} = \left(1 + \frac{m}{2}\right)^{1/m} \sigma_s \quad \text{(Eq 7.87)}$$

For $m = 20$, $\sigma_{rbs} = 1.13\sigma_s$, and for $m = 2$, $\sigma_{rbs} = 1.41\sigma_s$. As we would expect, in rotating bending, the material can withstand higher surface stress (up to 41% higher) than the axial fatigue stress for the same life. These are compared with the results for flexural bending of circular cross section in the following section.

**Flexural Bending of a Circular Cross Section.** Analysis is made similar to the ones shown in the previous discussion, referring to Fig. 7.17. Risk of rupture in flexural bending = $R_{fb}$:

$$R_{fb} = 4R^2 \left(\frac{\sigma_{fbs}}{\sigma'}\right)^m$$

$$\cdot \int_0^{\pi/2} \cos^2\theta \sin^m\theta \, d\theta \quad \text{(Eq 7.88)}$$

Equating $R_a$ and $R_{fb}$ from Eq 7.83 and 7.88, respectively:

$$\sigma_{fbs} = \left[\left(\frac{4}{\pi}\right) \cdot \int_0^{\pi/2} \cos^2\theta \sin^m\theta \, d\theta\right]^{1/m} \cdot \sigma_s \quad \text{(Eq 7.89)}$$

The multiplicative constant of $\sigma_s$ in Eq 7.89 can be obtained for any given $m$ by numerical integration. For $m = 20$, $\sigma_{fbs} = 1.23\sigma_s$, and for $m = 2$, $\sigma_{fbs} = 2.0\sigma_s$.

**Comparison.** The rotating bending and flexural bending fatigue lives can be compared by the use of Eq 7.86 and 7.89. It is clear that for any value of $m$, the life in rotating bending is higher than the life under axial loading for the same maximum stress. The life in flexural bending is even higher than that obtained in rotating bending. This is so because more volume is subjected to high stress in rotating bending than in flexural bending as seen in Fig. 7.18 and 7.19.

For any given $m$, $N_s < N_{rb} < N_{fb}$, as shown in Fig. 7.18(a). This effect will be much higher

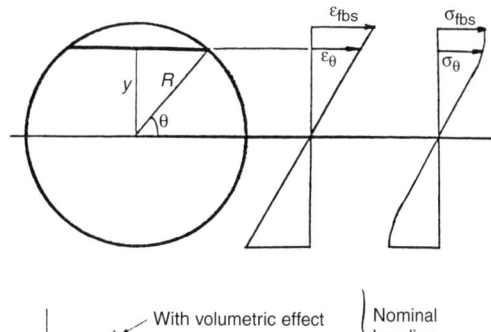

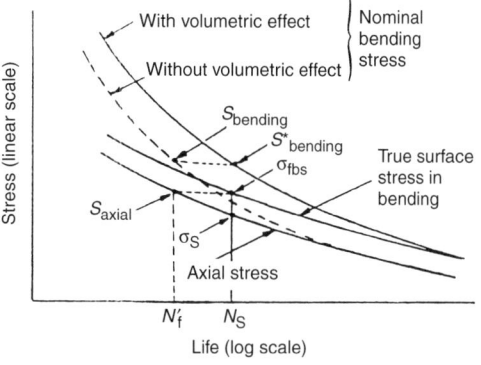

**Fig. 7.19** Schematic for determination of nominal bending stress including the effect of volume. Source: Ref 7.5

for lower values of *m*, i.e., for material with more microinhomogeneities, as illustrated in Fig. 7.18(b). The effect of volume under flexural bending at a given value of *m* for rectangular cross sections is less than that for circular cross sections.

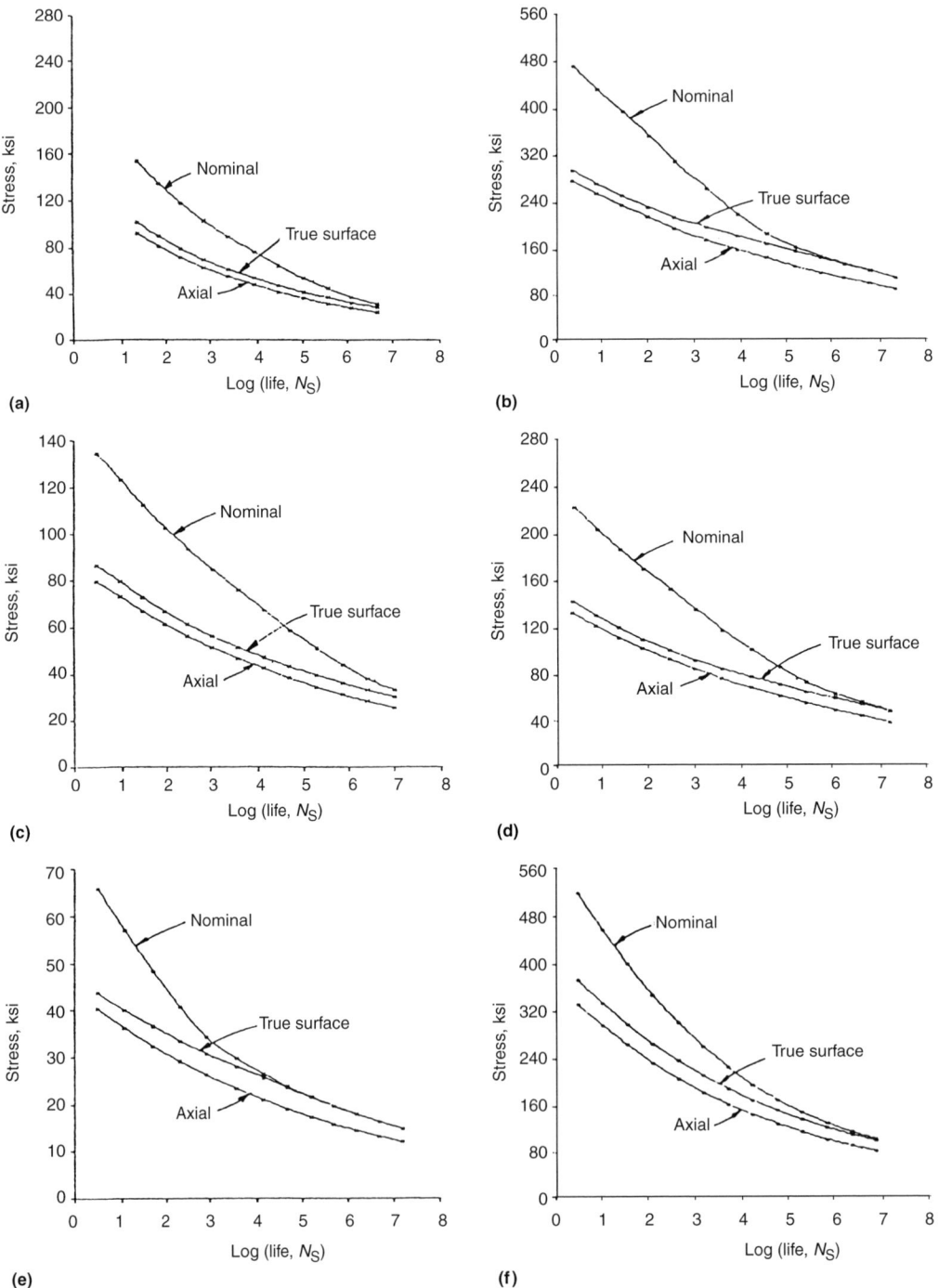

**Fig. 7.20**  Nominal stress including the effect of volume, true surface stress in flexural bending of circular cross sections, and axial stress shown as a function of life for six typical materials for constant *m* value of 20. (a) No. 3, man-ten steel, 150 HB. (b) No. 25, nickel-maraging steel, 460 HB. (c) No. 34, SAE 950X steel, 150 HB. (d) No. 36, RQC100 steel, 298 HB. (e) No. 44, H-11 steel, 660 HB. (f) No. 46, AM350 steel, 496 HB. Source: Ref 7.5

## Extension to Consideration of the Low-Cycle Fatigue Regime

When the material is subjected to flexural/rotating bending fatigue at high strain levels, plastic flow is involved. Hence, this effect, as well as the volumetric effect, needs to be taken into account in computing the nominal stress in bending. An approach is presented here using Weibull's concept (see Appendix) for the case of flexural bending of circular cross sections. The approach is applicable, however, in treating flexural bending of rectangular cross sections and rotating bending of circular cross sections.

When the effect of volume is not considered, the axial fatigue stress and the true surface bending stress for a given life are identical through the entire life range. Because plastic flow is involved in low-cycle fatigue, the nominal bending stress computed by using the $Mc/I$ assumption is higher than the axial stress for the same life. This nominal stress can be obtained from the axial stress by using only axial fatigue properties of the material and two functions that depend only on the geometry of the cross section as previously explained and given by Eq 7.32, 7.35, and 7.36. When the effect of volume is considered, greater true surface stress can be sustained in bending than in axial fatigue for the same life, because lesser volume is subjected to high stress in bending fatigue as compared with axial fatigue. To understand how the volume-modified theory would alter the results if applied to the plasticity-modified stress distribution, we proceed as follows:

If the true axial stress is assumed, the computation of true stress in flexural bending $\sigma_{fbs}$ of a circular cross section for the same life/risk of failure would involve trial and error solutions. However, if we start with the true surface stress in flexural bending $\sigma_{fbs}$, the true axial stress $\sigma_s$ can be determined for the same life by equating the risks of failure in the two cases. From an assumed true surface bending stress $\sigma_{fbs}$, true surface bending strain $\varepsilon_{fbs}$ can be obtained using the cyclic stress-strain curve. Plane sections are assumed to remain plane, thus strains vary linearly across the section from zero at the neutral axis to $\varepsilon_{fbs}$ at the surface. Thus, the strain at the elemental area at a distance $y$ from the neutral axis is:

$$\varepsilon_\theta = \varepsilon_{fbs} \sin\theta \quad \text{(Eq 7.90)}$$

The stress on this element is determined also from the cyclic stress-strain curve:

$$\sigma_\theta = E\varepsilon_\theta \left[1 + \left(\frac{\varepsilon_\theta}{\varepsilon_T}\right)^{z[(1/c - 1/b)]^{b/c}} \right]^{b/z} \quad \text{(Eq 7.91)}$$

where $\varepsilon_T = (\Delta\varepsilon_T/2)$ and $\Delta\varepsilon_T$ is given by Eq 7.70. Risk of failure for the entire volume in flexural bending is $R_{fb}$, where:

$$R_{fb} = 4R^2 \left(\frac{1}{\sigma'}\right)^m \cdot \int_0^{\pi/2} \cos^2\theta (\sigma_\theta)^m d\theta \quad \text{(Eq 7.92)}$$

and $\sigma'$ and $m$ are material constants. Risk of failure in axial loading is given by Eq 7.86. Equating Eq 7.92 and 7.86 for the same risk of failure, we obtain the axial stress:

$$\sigma_s = \left(\frac{4}{\pi} \cdot \int_0^{\pi/2} \cos^2\theta \sigma_\theta^m d\theta\right)^{1/m} \quad \text{(Eq 7.93)}$$

The strain, $\varepsilon_s$, corresponding to the stress $\sigma_s$ can be obtained directly from the cyclic stress-strain curve of the material. The life $N_s$ corresponding to this strain can be obtained using the basic strain-life fatigue curve for the material. Thus, true surface stress in flexural bending, $\sigma_{fbs}$, and axial stress, $\sigma_s$, are known for the same life $N_s$. To determine the nominal bending stress for the same life $N_s$, the analytical relations obtained in the section "Flexural Bending of a Circular Cross Section" can be used. It is important to note that these relations can be used for any given axial stress. If we choose the $S_{axial}$ as $S_{fbs}$, as shown in Fig. 7.19, the life $N'_f$ can be obtained from the known $S_{fbs}$.

Thus, $N'_f/N_T$ is known for this axial stress. The functions $f_1$ and $f_2$ are known from Eq 7.35 and 7.36. All the other material constants are known. Thus, $\delta$ can be computed from Eq 7.32. Hence, the nominal bending stress is known, which is equal to $\delta$ times the $S_{axial}$, i.e., $S_{fbs}$. However, this nominal bending stress $S_{bending}$ corresponds to the life of $N_S$, which is greater than $N'_f$ because of the effect of volume considered previously. Results for six materials are shown in Fig. 7.20.

## REFERENCES

7.1 S.S. Manson, Fatigue: A Complex Subject—Some Simple Approximations, *Exp. Mech.*, Vol 5 (No. 2), 1965, p 193–226

7.2 S.S. Manson and U. Muralidharan, Fatigue Life Prediction in Bending from Axial Fatigue Information, *Fatigue Fract. Eng. Mater. Struct.,* Vol 9 (No. 5), 1987, p 357–372. See also NASA CR-165563, Feb 1982.

7.3 S.S. Manson and U. Muralidharan, A Single Expression Formula for Inverting Strain-Life and Stress-Strain Relationships, *Fatigue Fract. Eng. Mater. Struct.,* Vol 9 (No. 5), 1987, p 343–356

7.4 R.W. Landgraf, M.R. Mitchell, and N.R. LaPointe, "Monotonic and Cyclic Properties of Engineering Materials," Ford Motor Company, 1972. See also U. Muralidharan and S.S. Manson, "A Modified Universal Slopes Equation for Estimation of Fatigue Characteristics of Metals," presented at the Winter Annual Meeting of the American Society of Mechanical Engineers, Dec 1986 (Anaheim, CA), Preprint 86-WA/Mats-17, 1986.

7.5 U. Muralidharan, "Advances in Room Temperature Low Cyclic Fatigue Life Prediction Technology," doctoral thesis, Department of Mechanical and Aerospace Engineering, Case Western Reserve University, Cleveland, OH, 1985

7.6 U. Muralidharan, Fatigue Life Prediction in Flexural Bending of Hollow Cross-Sectional Bars from Axial Fatigue Information, *J. Test. Eval.,* American Society for Testing and Materials, Vol 16 (No. 2), March 1988, p 146–152

7.7 A. Esin, A Method for Correlating Different Types of Fatigue Curves, *Int. J. Fatigue,* Vol 2 (No. 4), Oct 1980, p 153–158

7.8 R. Kuguel, "The Highly Stressed Volume of Material as a Fundamental Parameter in the Fatigue Strength of Metal Members," Department of Theoretical and Applied Mechanics Report 169, University of Illinois, Urbana, IL, 1960

7.9 R.E. Peterson, Discussion of D. Morkovin and H.F. Moore, Third Progress Report on the Effect of Size of Specimen on Fatigue Strength of Three Types of Steel, *Proceedings of the American Society for Testing and Materials,* Vol 44, 1944, p 137

7.10 D.T. Raske, Fatigue Failure Predictions for Plates with Holes and Edge Notches, *J. Test. Eval.,* Vol 1 (No. 5), American Society for Testing and Materials, 1973, p 394–404

7.11 J. Schijve and F.A. Jacobs, "Fatigue Crack Propagation in Unnotched and Notched Aluminum Alloy Specimens," Report NLR-TRM.2128, National Aero- and Astronautical Research Institute, Amsterdam, May 1964

7.12 W. Weibull, A Statistical Theory of the Strength of Materials, *Proc. Royal Swedish Institute for Engineering Research,* No. 151, 1939

7.13 W. Weibull, A Statistical Distribution of Wide Applicability, *J. Appl. Mech.,* Vol 18, 1951, p 293–297

7.14 S.S. Manson, *Thermal Stresses and Low-Cycle Fatigue,* McGraw-Hill, Inc., New York, 1966

# CHAPTER 8

# Notch Effects

## Introduction

Of all the factors that contribute most devastatingly to the occurrence of fatigue, notches undoubtedly are at the top of the list. In his very first studies of notch fatigue, Wöhler, more than 150 years ago, recognized the detrimental effects of notches, and he demonstrated that when a notch is present, the fatigue load-carrying capacity of a part is drastically reduced compared with a corresponding part with no notch. Although Wöhler may not have understood fully the fundamentals of stress and strain within the notch that leads to fatigue, his early studies inspired numerous investigations that have led to the very detailed knowledge currently available on this subject. In this chapter we review the state of the art as it has been developed in recent years.

## Stress Concentrations in the Elastic Range

We first discuss the elastic range because a wealth of work exists for this aspect of the subject, and because a general method exists for converting results in the elastic range to corresponding results in the plastic range, as discussed.

**Fundamental Reasons for Higher Stresses and Strains in the Vicinity of Notch.** A notch is basically a change in cross section such that adjacent regions impose deformation constraints on each other. It is the enforcement of deformation that introduces local strains, sometimes of large magnitude, that has led to the description "strain concentration." These large strains are accompanied by corresponding stresses. The concentrated stresses and strains, if reversed many times, would eventually lead to fatigue cracking and failure.

Consider the simple, yet severe, case shown in Fig. 8.1. A shaft abruptly changes in diameter at the step. The cross-sectional area of the small shaft is $A$, and that of the large shaft is $B$. If an axial load $P$ is applied, the axial stresses, axial elastic strains, and transverse elastic strains become $P/A$, $P/B$, $P/AE$, $P/BE$, $\mu P/AE$, and $\mu P/BE$, respectively, where $E$ is the modulus of elasticity and $\mu$ is the elastic Poisson's ratio. At the step there must be reconciliation between these stresses and strains while maintaining continuity of the material in this localized region. Stresses and strains within the small, localized circular ring of transitional material that blends the smaller to the larger diameter become even larger than in the smaller diameter section a short distance removed from the transition. The ratio of the elevated axial stress to the nominal axial stress (far removed from the discontinuity) in the small shaft is referred to as the theoretical stress concentration factor, $K_t$. For elastic loading, this is the elastic stress concentration factor and it is numerically equal to the elastic *strain* concentration factor because of the linearity of the elastic stress-strain relation for metals.

The value of a stress concentration factor is dictated primarily by the geometric details of the notch. A notch with a small root radius will have a $K_t$ greater than a notch with a large root radius. The nature of the loading may also have an effect, but it is usually of secondary influ-

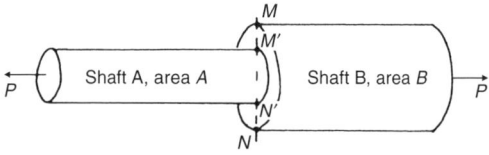

**Fig. 8.1** Stress concentration caused by abrupt change in cross-sectional area

ence. (A notable exception is a circular hole in an infinite plate. Under axial loading, $K_t = 3.0$, under equibiaxial loading it is 2.0, and for shear loading it is 4.0.) For the geometry in Fig. 8.1, axial, bending, and torsion will have different values of $K_t$ for exactly the same geometric detail. Examples are shown in Fig. 8.2. Numerous categories of geometries develop stress concentrations, all of which depend on the type of loading and exact geometry. Numerous collections of formulas for stress concentrations are available, among them Ref 8.1 to 8.4.

**Determination of Stress Concentration Factors.** Most of the tabulated stress-concentration factors refer to the elastic range. Of course, therefore, the strain-concentration factors are equal to the stress concentration factors. These quantities are designated, respectively, $K_S$ and $K_e$. The subscripts S and e refer to nominal engineering stress and strain. For linear elastic behavior, $K_s = K_e = K_t$.

Many different methods can be used to determine the elastic stress and strain concentration factors. Among them is the analytical procedure of solving the classic equations of elasticity. It is rare, however, that exact solutions can be obtained. Mostly, they are approximate solutions derived by simplifying the geometry and mathematics. Classic among them is the monumental work of Heinz Neuber in Germany. In 1937 he published (translated into English in 1946) (Ref 8.5) the results of an extremely extensive study in which he cleverly solved the equations for many geometries by approximations that were not valid in general but were sufficiently accurate at the surface in the critical area where the maximum stresses exist. His work is still used extensively, but many additional studies have since been published.

In recent years the finite-element method has played a dominant role in providing adequately accurate solutions for specific geometries. The method is well known today, and many computer programs exist for implementing the calculations, even by those who may not be thoroughly familiar with the fundamentals. Basically, the body to be analyzed is divided into a number of finite-sized elements, the size depending on the stress gradient in the region. Thus, the elements are very fine where the stress gradient is high and can be coarse where little change in stress occurs. By forcing adjacent sides of the elements to displace together, the compatibility conditions can be satisfied.

The equilibrium conditions for each element are then formulated, although it is common to use an equivalent energy formulation, to express a sufficient number of equations to solve for the

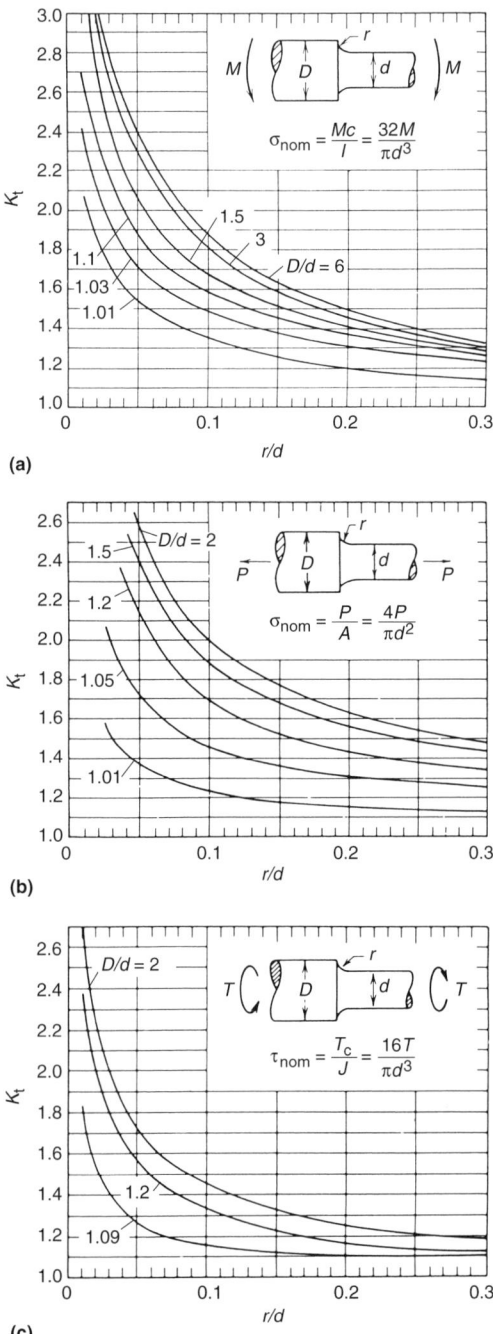

**Fig. 8.2** Shaft with fillet. (a) Bending. (b) Axial load. (c) Torsion. Source: Ref 8.6

interelement stresses, and the displacement of the nodal points (common points of adjacent elements). Using state-of-the-art computers, in many cases relatively compact personal computers, the large system of equations generated can be solved to a high degree of accuracy, and the highest local stresses and strains are determined. The ratio of these stresses to those far away from the geometrical discontinuity is the stress concentration, and because the material is presumed to be elastic, the strain concentration is equal to the stress concentration. Most frequently, $K_t$ isn't computed per se, and the magnitude of the maximum stresses and strains at their location are used directly in fatigue analysis.

Another approach that was used extensively prior to the widespread use of computers is experimental. Numerous experimental techniques have evolved over the years, among them strain gages (wire resistance and foil), Stresscoat (Stresscoat, Inc., Upland, CA) (a brittle, paint-like coating that cracks perpendicular to the maximum principal strain), and photoelasticity with its many variants. Basically, a physical model is necessary, and it is loaded while local strains are measured. The model can be of any size needed to make the proper measurement, although it must possess geometric similitude with the part of interest in all the details that influence the stress distribution. Because elastic stresses scale according to size, a large model may be used and stresses determined for a smaller scale model for which the experimental equipment, strain gages, for example, would not otherwise be suitable.

Fatigue tests can also be used to determine the effective stress concentration of a geometry that might otherwise be hard to analyze. If there is a region of high stress for a given loading, repetition of the loading will eventually start a crack in this critical region. A comparison of the external load required to produce a crack in a given number of cycles to the load required to produce a crack at the same life in a part without the stress concentration can lead to the determination of the effective stress concentration. It must be ensured, however, that the stresses involved are all in the elastic range. Usually, this is the case if the fatigue life is $10^6$ cycles or greater. It should be noted, however, that the fatigue notch factor $K_f$ is invariably lower than the theoretical stress concentration factor $K_t$, as discussed subsequently.

## The Classic use of Stress Concentration Data in the Elastic Range

The elastic stress concentration is a starting point for fatigue analysis whether the condition at the root of the notch is elastic or plastic. Progress has been made both in calculating the plastic condition at the root of a notch and in fatigue life formulations for strains including plasticity.

Even if the strains are largely elastic, at least three factors must be considered in addition to the theoretical stress concentration before fatigue can be properly analyzed:

- The material sensitivity to local strain gradient
- The volume of material subjected to the elevated stress associated with the stress concentration
- The local surface condition (as affected by the machining process used to develop the geometry)

These factors are, in many respects, interrelated. But, for simplicity they can be treated as independent—at least that is how the technology has developed. In the ensuing discussion, we follow the descriptions given by Juvinall (Ref 8.6).

**Neuber's Fatigue Correction.** As already noted, Neuber concentrated his early studies on methods of analytically determining theoretical elastic stress concentration factors. However, he recognized that the fatigue characteristics also involved a materials factor, a geometrical factor, and a size factor. He developed the concept of a "technical stress concentration factor," $K_N$, given by:

$$K_N = 1 + \frac{K_t - 1}{1 + \frac{\pi}{\pi - \omega}\sqrt{\alpha/r}} \quad \text{(Eq 8.1)}$$

The term $\alpha$ is an empirical material constant. It represents a fictitious "building block" or "equivalent grain size" of the material. The concept is that the larger the number of "equivalent grains" present in the high-stress volume, the more effective the stress concentrations will be in causing fatigue. Thus, as $\alpha \to 0$, $K_N \to K_t$. For values of $\alpha$ larger than zero, $K_N$ is always less then $K_t$. Thus, the actual fatigue notch stress concentration is invariably less than the theoretical elastic stress concentration factor.

The term $r$ refers to the tip radius of the notch. The larger the radius, the closer $K_N$ approaches

$K_t$, as would be expected because $r$ is a measure of how large a volume is under the high stress. The volumetric effect is discussed more completely in the Appendix, "Selected Background Information," in connection with the Weibull analysis. Here it is seen that it was Neuber's way of taking size effect into account.

The flank angle ω is defined in Fig. 8.3. Of course, this flank angle is already included in determining the elastic stress concentration $K_t$; however, it is given a dual role in establishing $K_N$. The larger the flank angle, the less effective is the theoretical elastic stress concentration in producing fatigue.

**Kuhn's Application of the Neuber Equation.** Kuhn and his coworkers (Ref 8.7, 8.8) modified Neuber's equation as:

$$K_N = 1 + \frac{K_t - 1}{1 + \frac{\pi}{\pi - (\omega/2)}\sqrt{\alpha/r}} \quad \text{(Eq 8.2)}$$

This form gives the flank angle lesser significance. They also developed empirical values of α for steel and aluminum as shown in Fig. 8.4, taken from Juvinall (Ref 8.6).

**Peterson's Notch Analysis.** Peterson (Ref 8.9) spent a great deal of effort not only to promote Neuber's analysis but also to determine functional forms for the geometric and material parameters in the relationship. In his final equation, he introduced a parameter $q$ described as the notch sensitivity:

$$q = \frac{K_f - 1}{K_t - 1} \quad \text{(Eq 8.3)}$$

or

$$K_f = 1 + (K_t - 1)q \quad \text{(Eq 8.4)}$$

Figure 8.5 shows the notch-sensitivity curves extracted by Peterson from numerous tests on steels of various hardness levels, and a limited amount of data on aluminum. The axial loading data is expressed in terms of the ultimate tensile strength, $S_u$, or in terms of Brinell hardness (essentially using the relation $HB = 2S_u$, where $S_u$ is expressed in ksi). The same curves are used for axial loading and bending (although a small correction might be expected in the elastic range for the volumetric difference between axial loading and bending). However, a stress gradient exists in both cases when a stress concentration is involved. This explains why Peterson found no need to distinguish the application of the same formulation for both cases. For torsion, however, each of the labeled curves refers to a material of 20 ksi less than that for bending, suggesting that in torsion materials are more notch sensitive than they are in bending. This should not be too surprising because in torsion, the theoretical stress concentration factor is 4.0 for a circular hole and therefore increases the notch sensitivity. However, it is noted that Peterson labeled these designations for torsion "tentative." These results presumably apply to the elastic range, not when plasticity is widespread. The matter needs further clarification.

It is clear from Fig. 8.5 that for small notch radii the actual fatigue notch factor is much lower than the theoretical elastic stress concentration factor; for soft steels and aluminum, $q$ can be less than 0.5. For large radii, however, the notch sensitivity is 0.75 or more. This result is a reflection of the size effect wherein the larger

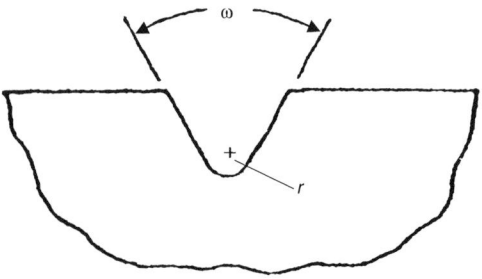

**Fig. 8.3** Parameters in Neuber's equation (ω = 0 for holes and for notches with parallel sides). Source: Ref 8.6

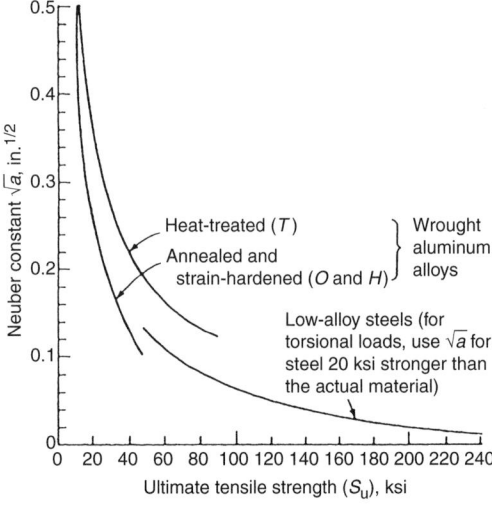

**Fig. 8.4** Neuber constants for steel and aluminum, figure from Ref 8.6, results from Ref 8.7

part has a larger volume of material in the vicinity of the calculated maximum surface stress.

Figure 8.5 also clearly displays the "unforgiveness" of hard materials in fatigue, compared with the softer materials, which have higher ductility, allowing some localized plastic flow that redistributes the very high local surface stress. It might have been more revealing if the parameter for separating the curves were ductility rather than tensile strength, but, of course, we are dealing here with relatively low stress where bulk plasticity is absent.

**Heywood's Representation of Fatigue Notch Factor.** Heywood's (Ref 8.10) representation of extensive fatigue is somewhat different from Neuber's or Peterson's. It is in the form:

$$K_f = K_t/\{1 + 2[(K_t - 1)/K_t]\sqrt{\alpha'/r}\} \quad \text{(Eq 8.5)}$$

The advantage of using his form is that he has provided values of the constant $\alpha'$ for other materials than those studied by Neuber or Peterson. In addition to studying steels and aluminum, which can be treated equally well by the methods already discussed, he has also provided values of $\alpha'$ for cast iron and magnesium:

- Cast iron, flake graphite, $\sqrt{\alpha'} = 0.12$
- Cast iron, spheroidal graphite, $\sqrt{\alpha'} = 5/S_u$, where $S_u$ is in ksi
- Magnesium alloys, $\sqrt{\alpha'} = 0.015$

These values cannot be used directly in the Neuber or Peterson equations; rather, they are useful only in Heywood's Eq 8.5.

**Juvinall's Inclusion of Surface Finish Effects.** While all the methods discussed involve no consideration of external surface finish, Juvinall (Ref 8.6) has included this factor. His examination of extensive data on steel has led to the summary of Fig. 8.6. Here, the endurance limit at $10^6$ cycles is plotted against tensile strength for the mirror-polished surface normally used in R.R. Moore fatigue characterization. The averaged data show the validity of the usual assumption that the endurance limit is half the ultimate tensile strength. However, for various types of surface processing, the endurance limit can be only a fraction of that determined on a polished surface. In the as-forged condition, for example, the endurance limit may be as low as 10 to 20% of the mirror-polished material for the very hard materials, and not much more than half for even the lowest-strength most-forgiving materials. One reason for the loss of fatigue strength for as-forged or hot-rolled surfaces is the surface roughness for steels processed in this way. Also important, however, is the decarburization that is present. Because of the lower carbon content of the surface, it is also essentially of lower hardness (lower strength, higher localized plasticity) relative to the bulk material. Thus, the overall measurement reflects both factors.

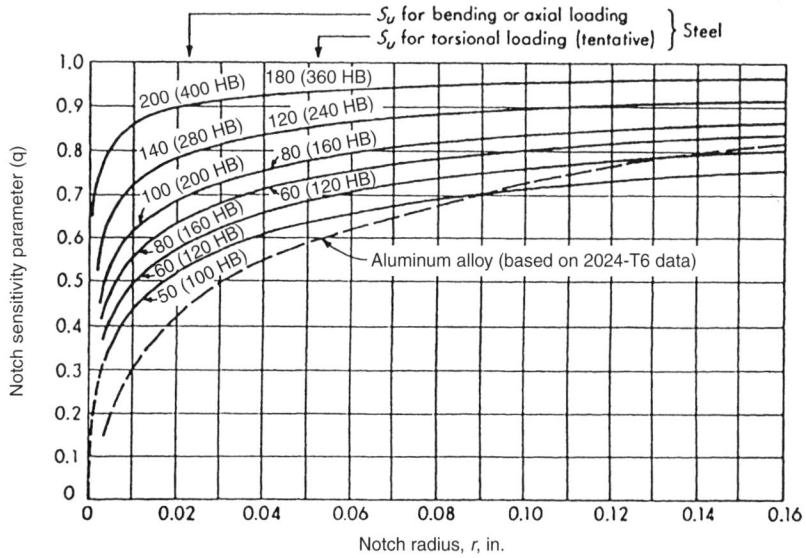

**Fig. 8.5** Notch-sensitivity curves for use with theoretical factors $K_t$, figure from Ref 8.6, results and analysis from Ref 8.9

The results of Fig. 8.6 are redrawn in Fig. 8.7. Here, the fraction of the mirror-polished endurance limit is plotted versus $S_u$ for the same surface treatments shown in Fig. 8.6, and the effect of immersion in tap water and saltwater is also shown. Juvinall designates this factor as $C_s$, a correction factor for surface finish that is properly used to determine the endurance limit of a standard specimen without stress concentration.

For use on parts with stress concentration, Juvinall does not recommend multiplying $C_s$ times the stress concentration factor unless extreme conservatism is desired. Rather, he prefers to multiply the notch sensitivity $q$ by $C_s$. Thus, he recommends that $K_f$ be computed by the equation

$$\frac{K_f - 1}{K_t - 1} = qC_s \quad \text{(Eq 8.6)}$$

so that $C_s$ operates only on the *excesses* of $K_f$ and $K_t$ that are above 1.0. This reduces the severity of amplification of the surface stress. His view is that in this way, $C_s$ compensates for the influence of surface irregularities in the same way as $q$ compensates for the influence of material factors.

Once the value of $K_f$ has been determined from Eq 8.6, its use is to determine an endurance limit of the notched part. This is accomplished by multiplying $K_f$ by the basic endurance limit for the material.

Furthermore, Juvinall does not recommend using the $C_s$ values of Fig. 8.7 for the hot-rolled or as-forged materials because he believes the purpose of introducing $C_s$ is to allow for surface irregularities, not decarburization, which is also included in the $C_s$ factors for material so processed. He, therefore, recommends that for treating hot-rolled or as-forged material, the values of $C_s$ for the machined surface at the same $S_u$ level be used.

**Summary for Determination of Fatigue Notch Factor.** The preceding discussion indicates that there is a wealth of background for determining the "effective" stress concentration for a given geometry, material, specific metallurgy, and surface condition. In general, the first step is to determine the fatigue stress concentration factor in the nominally elastic regime using analytic calculation, numerical analyses (finite element, boundary elements, etc.) or experiment. The result is then modified to account for a material of current interest by applying experimental input or by calculation using the material described previously.

## Extension to the Plastic Range

When plasticity develops at the root of a notch, the stress or strain can no longer be cal-

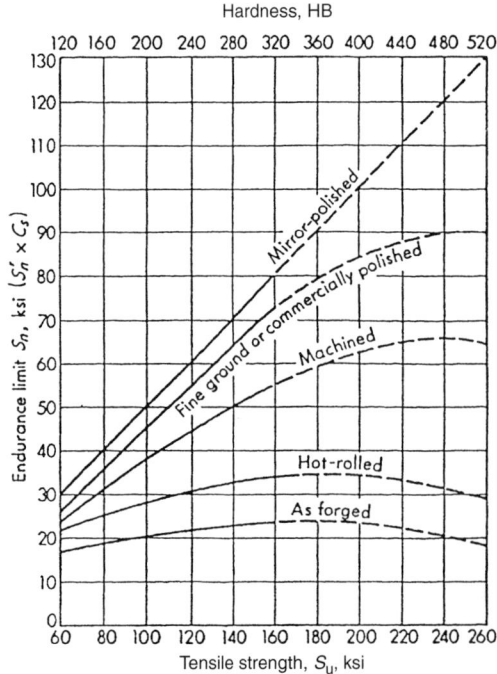

**Fig. 8.6** Effect of surface finish on endurance limit of steel. Source: Ref 8.6

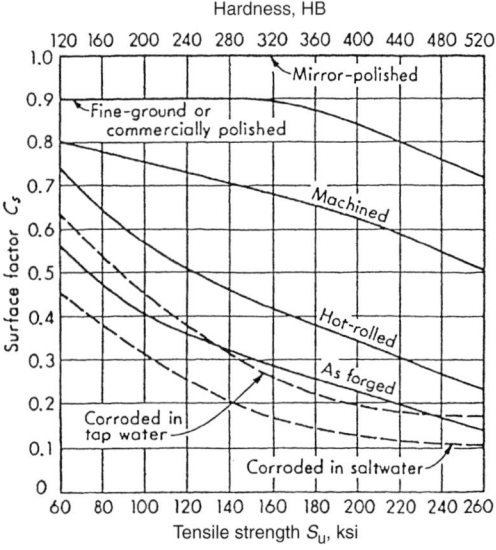

**Fig. 8.7** Reduction of endurance strength due to surface finish for steel. Source: Ref 8.6

culated by multiplying the nominal values by the elastic stress concentration factor $K_t$. If this were done, the stress/strain condition at the root would no longer lie on the stress-strain curve. In Chapter 4, "Mean Stress," it is emphasized that the stress-strain curve must not be violated, otherwise errors will occur. It becomes necessary to develop an approach that takes into account both the notch effect in altering the stress/strain distribution and maintains the condition that the stress/strain relation is the appropriate stress/strain curve of the material. While the early loadings of a cyclic history may involve some hardening or softening, it is common in engineering calculations to neglect these effects, and then to make appropriate corrections if warranted. Thus, the stress/strain relation for the material at the root of the notch is taken as the stabilized cyclic stress/strain curve. The Neuber equation has been commonly used as a means of accounting for the effect of the notch in distorting the stress/strain conditions within the notch. The approach is described subsequently in its simplest form, but later we present an approach that has been developed to increase its accuracy.

**Neuber Type Analysis.** Figure 8.8 illustrates the simplified Neuber condition for the case of a small circular hole in a large plate subjected to uniaxial tensile loading. For purposes of discussion, let us accept that in the elastic range the stress and strain concentration is 3.0. The most important point is point $A$ because this is where the strain range is the highest and where the crack will start. Point $B$, on the other hand, is remote from the region of the notch effect (and hence is unaffected by the presence of the hole), and it is assumed that at this location the stress range $\Delta S$ and strain range $\Delta e$ for the loading cycle can be calculated. The stress range and strain range at the critical point $A$, on the other hand, are as yet unknown, but are designated $\Delta\sigma$ and $\Delta\varepsilon$. Thus, the actual stress and strain concentration factors are:

- Stress concentration factor = $\Delta\sigma/\Delta S = K_\sigma$
- Strain concentration factor = $\Delta\varepsilon/\Delta e = K_\varepsilon$

Neuber's *hypothesis* is that the product of the stress concentration factor and strain concentration factor for all levels of loading remains constant. Because for low loading, in the elastic range, the stress concentration factor and strain concentration factor are equal to $K_t$, according to Neuber's rule:

$$K_\sigma K_\varepsilon = K_t^2 \qquad \text{(Eq 8.7)}$$

**Elastic Loading in Region Remote from Notch.** Consider first the case when the loading starts from zero, and $\Delta S$ and $\Delta e$ are in the elastic range. Then $\Delta S = E\Delta e$, where $E$ is the elastic modulus. Here:

$$\frac{\Delta\sigma}{\Delta S} \cdot \frac{\Delta\varepsilon}{\Delta e} = \frac{\Delta\sigma}{\Delta S} \cdot \frac{\Delta\varepsilon}{\Delta S/E} = K_t^2 \qquad \text{(Eq 8.8)}$$

$$\Delta\sigma \cdot \Delta\varepsilon = \frac{K_t \Delta S}{E} \qquad \text{(Eq 8.9)}$$

Because $K_t$, $\Delta S$, and $E$ are known, and because the starting stress and strain are both zero, Eq 8.9 states that the product of the stress and strain at the notch is equal to a known constant $(K_t\Delta S)^2/E$, which we can designate as $C$. But the point $(\sigma,\varepsilon)$ must also lie on the stress-strain curve. The Neuber relation $\Delta\sigma\Delta\varepsilon = C$ is of the form $xy = \text{const.}$, and is, of course, a hyperbola, so the Neuber relation is usually referred to as the Neuber hyperbola. The actual stress and strain that develop during the initial loading are the coordinates of the intersection of the cyclic stress-strain curve and the Neuber hyperbola.

In the preceding discussion, it is assumed that the condition remote from the notch is elastic. However, the same procedure can be followed even if the remote condition is plastic. The condition simply becomes $\Delta\sigma\Delta\varepsilon = K_t^2\Delta S\Delta e = C$, where $\Delta S$ and $\Delta e$ are the remote stress and strain, even if they are in the plastic range.

The numerical solution for a particular problem can be obtained graphically by drawing the two appropriate curves as shown in Fig. 8.9, or it can be determined numerically by using the equations of the stress-strain curve and the Neuber hyperbola. Thus, if the equation between cyclic strain and stress ranges is:

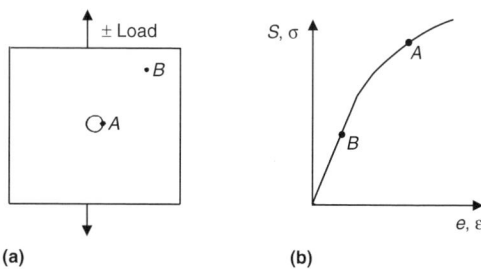

**Fig. 8.8** Neuber condition for small circular hole. (a) Hole in plate. (b) Stress-strain response at hole, $A$, and remote location, $B$

$$\Delta\varepsilon = (\Delta\sigma/E) + (\Delta\sigma/K')^{1/n'} \quad \text{(Eq 8.10)}$$

and the Neuber equation is:

$$\Delta\varepsilon\Delta\sigma = C \quad \text{(Eq 8.11)}$$

the relation becomes:

$$\Delta\varepsilon\Delta\sigma = (\Delta\sigma)^2/E + (\Delta\sigma)(\Delta\sigma/K')^{1/n'} = C \quad \text{(Eq 8.12)}$$

or

$$\frac{\Delta\sigma^2}{E} + \frac{1}{(K')^{1/n'}} \Delta\sigma^{1+1/n'} = C \quad \text{(Eq 8.13)}$$

This equation is in the generalized form $y = AX^a + BX^b$ that can be solved in closed form, as discussed in the Appendix, "Selected Background Information." Of course it can also be solved by successive approximations using numerical techniques. We shall apply the closed-form solution approach in a later numerical example.

For cyclic loading, a procedure for treating load reversal is necessary. The procedure is the same as for initial loading with two exceptions:

- The origin of the construction is transferred to the stress-strain reversal point (corresponding to the load reversal point); i.e., the origin moves to point $A$ in Fig. 8.10.
- The reversal path is the double-amplitude cyclic stress-strain curve.

Thus, while $OA$ in Fig. 8.10 was the single amplitude cyclic stress-strain curve, $ABCD$ is the double-amplitude cyclic stress-strain curve. This curve is drawn relative to the origin of the axes $\varepsilon'$ and $\sigma'$ drawn through the point $A$, and has the equation:

$$2\Delta\varepsilon = (2\Delta\sigma)/E + (2\Delta\sigma/K')^{1/n'} \quad \text{(Eq 8.14)}$$

The reader may find it useful to turn the figure upside-down and view $ABCD$ as the double-amplitude cyclic stress-strain curve relative to the reversed axes drawn through the reversal point $A$. The Neuber hyperbola $N_2$ is also drawn relative to this coordinate system and is constructed in the same way as was $N_1$. The hyperbola constant has the value $(K_t\Delta S)^2/E$ if the remote load during reversal is in the elastic range, or by $K_t^2 \Delta S \Delta e$ if the remote region is in the plastic range with load $\Delta S$ and nominal strain $\Delta e$.

The intersection point $C$ of the hyperbola $N_2$ and the reversal path $ABCD$ is the stress/strain condition of the critical element upon completion of the reversal. This point may be obtained graphically, or by the solution of Eq 8.13 using point $A$ as the origin with stress axis $\sigma'$ pointing down and strain axis $\varepsilon'$ pointing to the left as illustrated in Fig. 8.10.

The solution provides, of course, the change in stress between origin $A$ and point $C$ but can be expressed relative to the initial origin 0 by the conventional rules for translation of coordinate axes. The solution of this equation can be by successive approximations or by the closed form of the generalized equation $y = AX^a + BX^b$.

The case shown in Fig. 8.10 determines the return point $C$ at a negative stress and positive strain. Physically what has happened is that, during the forward loading, tensile plastic strain has been induced at the root of the notch, but during load removal, the compressive plasticity does not completely reverse the tensile plasticity, therefore, as seen in the figure, the residual stress upon load reversal is compressive. This compressive stress can have a significant effect on the fatigue resistance in subsequent loading. If properly used, compressive mean stress can produce highly beneficial effects. We shall illustrate this point in a planned second volume.

In the case illustrated in Fig. 8.10, the hyperbola $N_2$ passes through the origin 0 because the load was completely reversed in this example. However, if the reversal is to another load level, the hyperbola $N_2$ can move to a different loca-

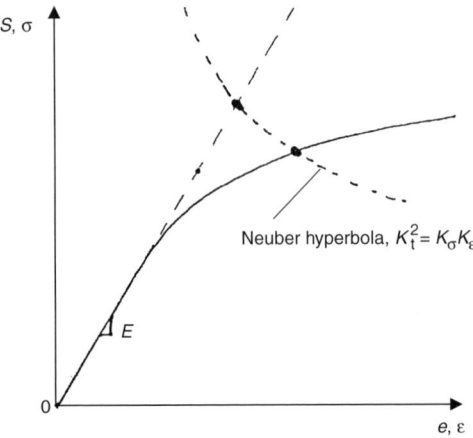

**Fig. 8.9** Intersection of Neuber hyperbola with stress-strain curve for monotonic deformation

tion. This general problem has been studied by numerous groups in the 1960s and early 1970s (Ref 8.11–8.15).

**Crews' Study of Notched Plates.** An extremely interesting and instructive study was the one conducted by John Crews in 1971 (Ref 8.15) to test the validity of the Neuber analysis and several other aspects of notch fatigue. Because his investigation can be used to demonstrate a number of important points emphasized in this chapter, we present sufficient detail to enable us to draw valuable conclusions.

The tests were conducted on large sheets of 2024-T3-aluminum alloy characteristic of the type of structure of interest to the investigator who, as an employee of NASA's Langley Research Center, was concerned with the fatigue of airplane fuselages and wing surfaces. The thin sheets, approximately 305 × 889 mm (12 × 35 in.) size had a 51 mm (2 in.) diam central hole, so as a test specimen it was large enough to accommodate wire-resistance strain gages, which provided accurate information on strains developed in the critical region of the notch. Figure 8.11 shows the test specimen plus an auxiliary unnotched specimen that was used to determine stresses. Through a fairly complex servohydraulic system, the strain that occurred in the notch section was simultaneously enforced on the unnotched (companion) specimen that was installed in an independent loading machine. Because the companion specimen was subjected to the same strain history as the notch region, its measured stress history (directly from the load history) was presumed to be the same. Thus, the stress-strain history at the root of the notch could be determined at all times.

**Stress-Strain History in Completely Reversed Loading.** Figure 8.12 shows the Neuber construction for the first loading in a history consisting of completely reversed loading. The nominal load is in the elastic range, so $N_1$ is constructed according to Eq 8.9. Note that at the point of complete unloading, $B$, the hyperbola $N_2$ produces a residual compressive stress, but tensile strain. In this test, however, the loading is continued into compression and is carried to point $C$. Thus, it is necessary to construct another Neuber hyperbola, $N_3$, to define the location of point $C$. The axes for this hyperbola are at origin $A$, where the last load reversal occurred, not $B$, because no reversal occurred there. Similarly, at $C$ there is a reversal, so this point serves as the origin (point of new axes construction) for the hyperbola, $N_4$, that intersects the double-amplitude stress-strain curve at point $D$. The residual stress at point $D$ is tensile. With reversed loading, therefore, the residual stress upon returning to zero load can be either tensile or compressive, depending on where in the loading history the zero load is reached. This point is very important, and it was used in the Crews' study to show how fatigue life can be either substantially increased or decreased depending on what at first appears to be a minor detail in the loading history.

If the reversed loading were successively repeated, the hysteresis loop would be $ABCDA$, if it were the stabilized cyclic stress-strain curve that governed the behavior. But, in the early

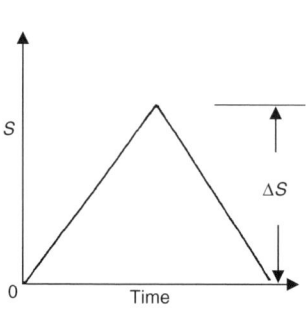

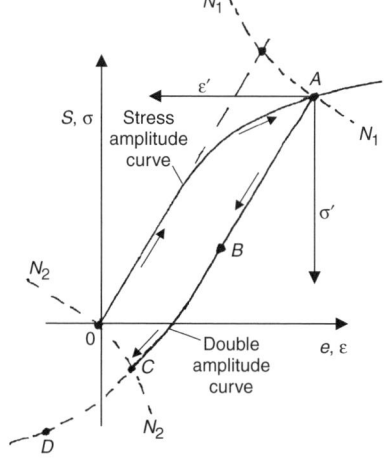

**Fig. 8.10**  Treatment of load reversal

loadings a material may harden (or soften), so it requires a number of completed reversals before complete repetition of stress and strain history occurs. If the actual cyclic stress-strain curves are available, the Neuber analysis can be made using successive cycle calculations. Figure 8.13 shows the calculations made by Crews based on Neuber hyperbolas, indicating how the behavior should be expected to change during the early cycles. Figure 8.14 shows typical recordings from the companion specimen procedure for both the nominal stress and strain (in the region remote from the notch) and from the region at the root of the notch. The general appearance is much like that of Fig. 8.13, but no direct comparisons are made because these are shown in later figures.

**Combined Loadings: Beneficial and Detrimental Effects.** In addition to checking the validity of Neuber-type calculations, Crews's program also checked the concept of introducing beneficial or detrimental residual stresses by loading notched members to appropriate points on the hysteresis loops. Figure 8.15 shows the sequences of loading investigated. In Fig. 8.15(a), a 0 to +20 ksi (nominal stress, $S$) loading is applied to a notched specimen with a $K_t$ of 2.57. The local stress-strain response shown to the right of the figure was nominally elastic, i.e., 0 to 51.4 ksi. The mean stress is +25.7 ksi. In Fig. 8.15(b), a ±40 ksi loading was applied for 10 cycles, and then the loading was changed to 0 to +20 ksi. Carefully note the sequence by which the change was made from the higher to the lower loading, as this is directly responsible for the outcome. From the highest load $B$, the loading was reduced to zero at $C$, and then the 0 to +20 ksi loading ($C$ to $D$) was started.

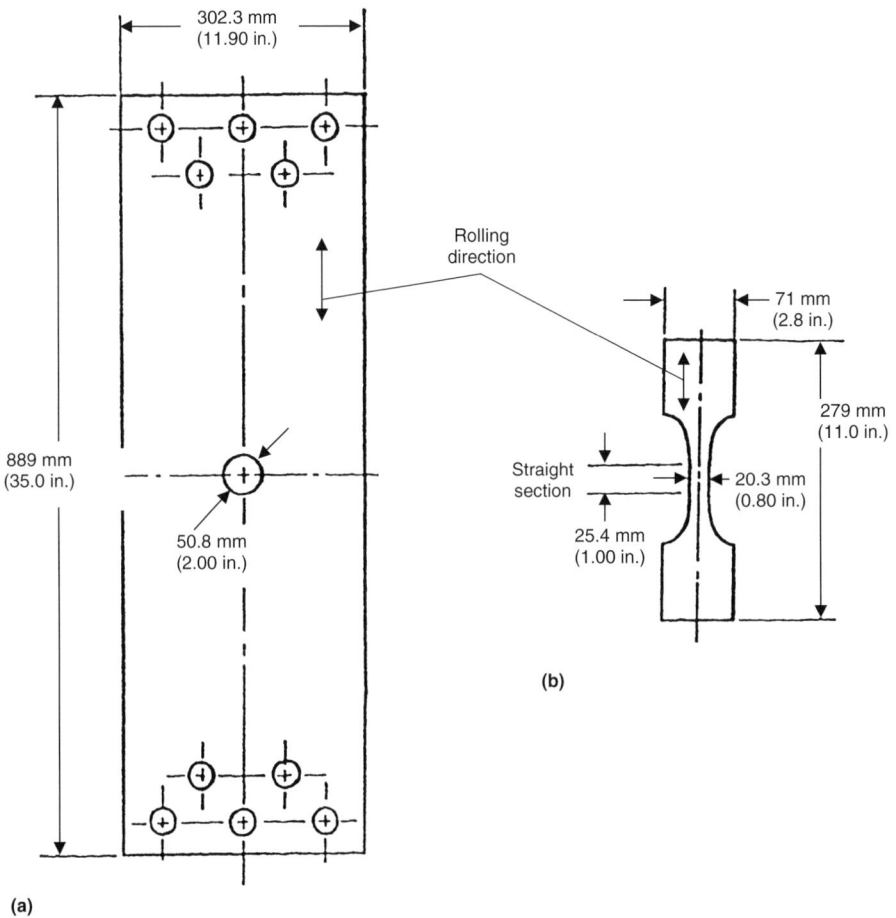

**Fig. 8.11** Specimen configurations and dimensions (thickness = 3.96 mm, or 0.156 in.) (Ref 8.15). (a) Notched specimen. (b) Unnotched companion specimen

Again, the local stress-strain response to this lower loading was nominally elastic with a range of 51.4 ksi. The unloading to C, however, produced a residual compressive stress at the notch root of −7.5 ksi, thus earning the designation "beneficial initial loading." On the other hand, the third type of loading in Fig. 8.15(c) is designated "detrimental initial loading," even

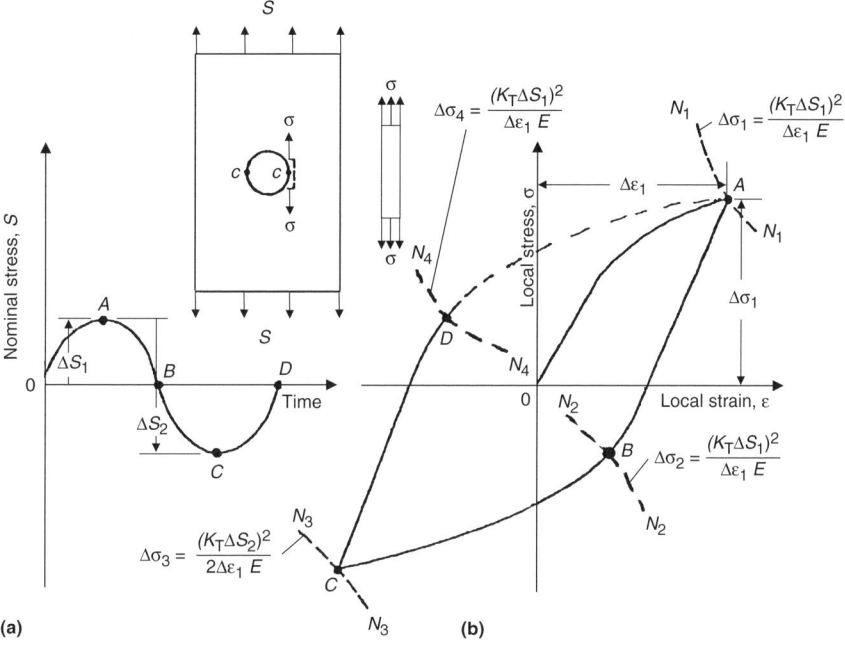

**Fig. 8.12** Nominal stress cycle and corresponding local stress-strain curve by Neuber procedure. (a) Nominal stress cycle. (b) Simulated local stress-strain curve. Source: Ref 8.15

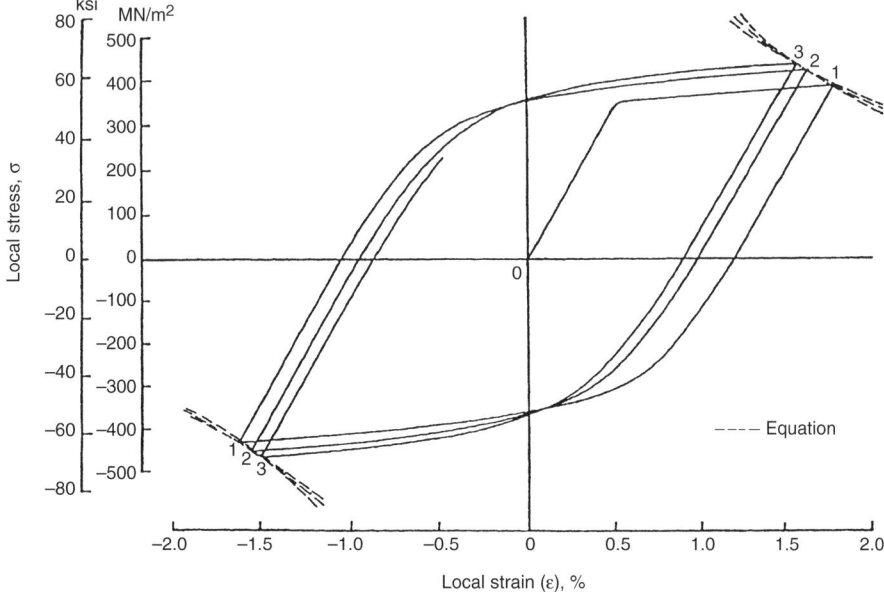

**Fig. 8.13** Typical calculations by Neuber procedure. Source: Ref 8.15

though the basic loadings are the same as in Fig. 8.15(b). The only difference is in the detail of changeover from the high to the low loading. In this case, the last point of the high loading is at *E* and the unloading is to *F* (analogous to and equal in magnitude to the unloading from *B* to *C* in Fig. 8.15b). In this unloading, we are left with a residual tensile stress that is even larger than the one for Fig. 8.15(a). The block of 0 to +20 ksi loading (local stress range remains at 54.1 ksi due to the elastic response) is thus superimposed on a tensile mean stress of 30.5 ksi, implying a detrimental effect on fatigue because it is larger than the 25.7 ksi mean stress for this loading with no prior higher-stress loading. While the difference in unloading may be subtle to the novice, the effects on fatigue life are far from subtle.

The actual comparison between stress-response predictions by Neuber calculations and companion-specimen measurements for the "beneficial initial loading" is shown in Fig. 8.16. Figure 8.16(a) shows exceptionally good agreement between the predicted and measured stresses, while Fig. 8.16(b) shows that the agreement was not quite so good for the strains, but acceptable. The strain ranges agree well, but the mean strains during the 0 to +20 ksi loading

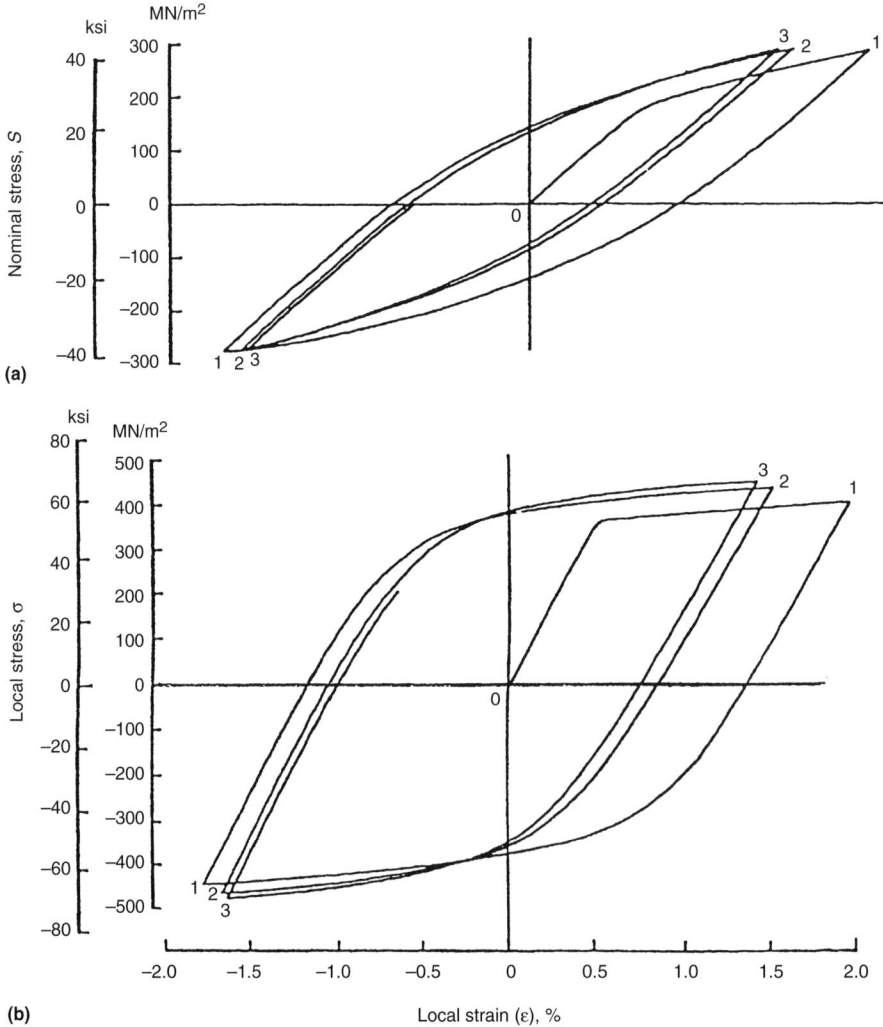

**Fig. 8.14** Typical recording from companion specimen procedure. (a) Remote region. (b) Notch root. Source: Ref 8.15

did not. Of course, fatigue life is more critically related to strain range and mean stress and not so dependent on mean strain.

The comparisons for the "detrimental initial loading" are shown in Fig. 8.17. Again, the agreement is very good for the stresses in Fig. 8.17(a). Note that the mean stress is tensile, as expected. As was the case for the local mean strains in "beneficial prior loading," the local mean strains in the "detrimental prior loading" also show a discrepancy between Neuber predictions and companion specimen measurements (Fig. 8.17b). The strain ranges and mean stresses agree well, however, and they, primarily, govern fatigue life.

Another series of tests were conducted by Crews to determine fatigue life of notched specimens as well as smooth specimens on which were imposed strains as measured at the notch root of cyclically loaded notched specimens. Table 8.1 shows the resultant cyclic lives for the notched specimens fatigued under the three applied conditions:

- Only 0 to +20 ksi loading notched specimen, with neither beneficial nor detrimental initial loadings
- 0 to +20 ksi loading notched specimen applied immediately following 10 cycles of "beneficial initial loading"
- 0 to +20 ksi loading notched specimen applied, immediately following 10 cycles of "detrimental initial loadings"

In each case, four tests were conducted. The geometric mean of the crack initiation cycles for the 0 to +20 ksi loadings alone was 115,600. With beneficial initial loading, the life increased by a factor of four to 458,900 cycles, and with detrimental initial loading it was reduced by a factor of two to 62,900 cycles. Thus, it is clear that there indeed were beneficial and detrimental effects, as predicted by the notch analyses.

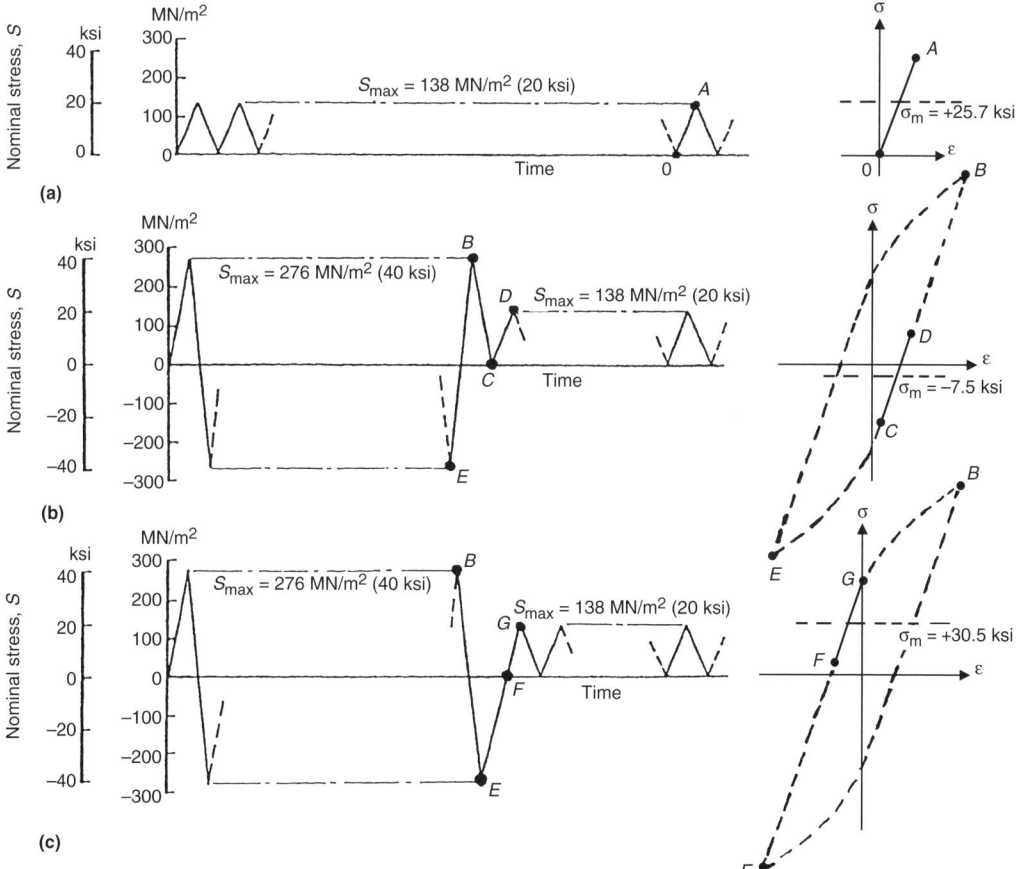

**Fig. 8.15** Loading sequences for notched specimens (Ref 8.15). (a) No initial high loading. (b) Beneficial initial loading. (c) Detrimental initial loading

In Table 8.2 the stresses as calculated for the notched specimen for the three types of loading were simulated on smooth specimens to determine whether the lives would agree with those determined experimentally for the notched specimens. It is clear that qualitatively the results are similar, but quantitatively there is a factor of 2 or 3 difference in the corresponding cycles to failure. In all three cases, the smooth specimen results are lower than the notched specimen results. It is not clear from the report exactly how crack initiation was measured, but, of course, the stress gradients are different for the case of notched versus smooth specimen. Thus, for the smooth specimen, the stress is high across the entire specimen, whereas for the notched spec-

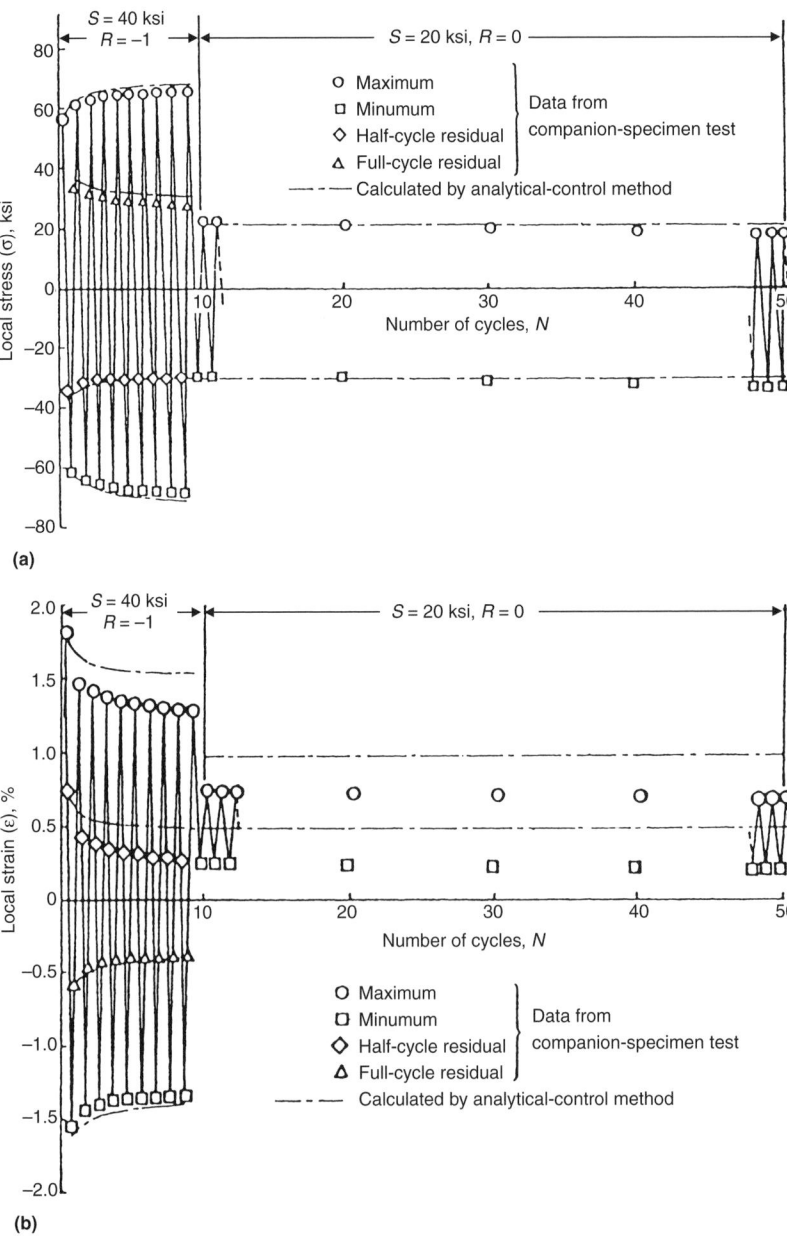

**Fig. 8.16** Local stress-strain history for beneficial prior loading (material, aluminum alloy 2024-T3) (Ref 8.15). (a) Stress history. (b) Strain history

imen the stress is high at the surface of the notch but decreases rapidly with depth below the notch. Because only the surface stresses were maintained equal, it is clear that the notched specimens should last longer, which they did in all three cases.

Finally, in Table 8.3 results are shown for tests conducted on smooth specimens applying only the stresses (with their respective mean values) calculated for the root of the notched specimen at the low-level loading (0 to +20 ksi) immediately following either detrimental or beneficial

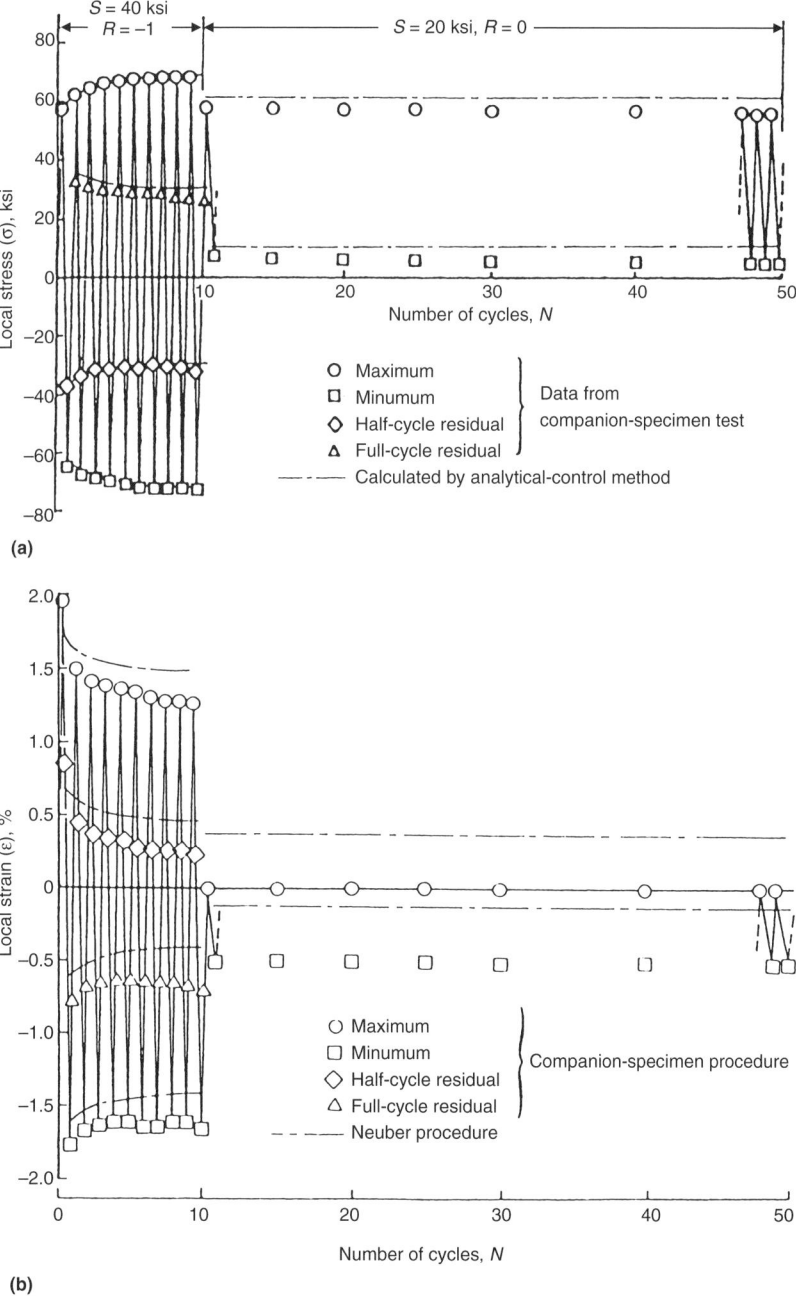

**Fig. 8.17** Local stress-strain history for detrimental prior loading (material, aluminum alloy 2024-T3) (Ref 8.15). (a) Stress history. (b) Strain history

large initial amplitude loadings. Again, qualitatively the results are correct: the case for beneficial loading results in significantly higher life than does the case for detrimental loading. But quantitatively, it is clear that the lives are about 50% higher than for the corresponding cases wherein the prior higher level loading was also applied. Of course, data scatter could introduce some of the difference. However, if the results are viewed in the light of the double-linear damage rule (Chapter 6, "Cumulative Fatigue Damage"), it is clear that the early loadings could easily reduce subsequent life substantially at the lower loading level.

**Discussion.** The results of Crews and numerous other investigators (Ref 8.16–8.19) have demonstrated that reasonably good predictions can be obtained by the use of the Neuber analysis. For many applications the results are adequate. However, two factors stand out: corrections can be made to the Neuber analysis to improve accuracy, and care must be exercised in interpreting the results to recognize that stresses and strains computed apply only to the surface of the notch; once the crack starts to penetrate the surface, the stress and strain gradient must be recognized. We treat the question of possible improvements to the Neuber analysis in the next sections. The issue of separating the crack initiation from its propagation is discussed in Chapter 9, "Crack Mechanics."

## Some Attempts to Improve the Neuber Analysis

In 1979 we conducted a program to investigate the accuracy of the Neuber approach. This study was conducted with colleagues J.C. Walcher and D.F. Gray of the Teledyne CAE Company and was reported in Ref 8.20. The specimen used was a plate with a central hole as shown in Fig. 8.18, made of 6Al-4V titanium, an alloy then being used in a compressor stage of a Teledyne CAE engine. The elastic stress concentration was 2.37. In the plastic regime, the stress and strain concentration factors were established from strain-gage measurements at the notch root. Strains were entered into the cyclic stress-strain curve to determine stresses. These experimental concentration factors were then compared with those obtained from a Neu-

**Table 8.1  Experimentally determined crack-initiation periods for notched specimens**

| Test condition | Crack-initiation periods, $N_o$, cycles | Geometric mean of crack-initiation periods, cycles |
|---|---|---|
| No initial high loading | 90,700 | 115,600 |
|  | 113,600 |  |
|  | 123,100 |  |
|  | 140,800 |  |
| Beneficial initial loading | 233,000 | 458,900 |
|  | 269,100 |  |
|  | 696,400 |  |
|  | 1,016,200 |  |
| Detrimental initial loading | 61,800 | 62,900 |
|  | 62,000 |  |
|  | 62,800 |  |
|  | 65,400 |  |

**Table 8.2  Crack-initiation periods from simulation of sequence effects by using unnotched specimens**

| Simulated test condition | Fatigue lives(a), cycles | Geometric mean of fatigue lives, cycles |
|---|---|---|
| No initial high loading | 19,200 | 35,100 |
|  | 22,700 |  |
|  | 43,500 |  |
|  | 48,100 |  |
|  | 58,400 |  |
| Beneficial initial loading | 140,000 | 190,600 |
|  | 181,000 |  |
|  | 273,300 |  |
| Detrimental initial loading | 17,300 | 22,600 |
|  | 23,100 |  |
|  | 28,800 |  |

(a) Used as estimates for crack-initiation periods for notched specs

**Table 8.3  Crack-initiation periods from simulation of residual stress effects by using unnotched specimens**

| Test condition | Maximum stress MN/m² | Maximum stress ksi | Minimum stress MN/m² | Minimum stress ksi | Fatigue lives(a), cycles | Geometric mean of fatigue lives, cycles |
|---|---|---|---|---|---|---|
| Beneficial initial loading | 147 | 21.3 | −208 | −30.1 | 270,900 | 300,500 |
|  |  |  |  |  | 276,900 |  |
|  |  |  |  |  | 361,700 |  |
| Detrimental initial loading | 424 | 61.5 | 69.6 | 10.1 | 28,300 | 31,700 |
|  |  |  |  |  | 32,400 |  |
|  |  |  |  |  | 34,600 |  |

(a) Used as estimates for crack-initiation periods for notched specs

ber analysis. The concentration results are shown in Fig. 8.18 as a function of nominal stress, $S$. For nominal stresses below 62 ksi, the notch remained elastic, so in the region $OA$, the Neuber and the experimental stress and strain concentration factors agree at 2.37. But, as the nominal stress was increased, the strain concentration factor increased and the stress concentration factor decreased, as would be expected. While the experimental stress concentration factors agreed with the Neuber calculation, there was considerable discrepancy in strain concentration factor. In Fig. 8.18, $AB$ is the strain concentration calculated by the Neuber method and $OB'$ is the experimentally determined factor. The discrepancy between the two factors is about 10% at a nominal stress of 80 ksi, but beyond this stress the two factors start to get closer together. A close examination of Fig. 8.18 suggests that because the slope of $OB'$ is getting steep beyond nominal stresses of 80 ksi, the two curves $AB$ and $OB'$ might in fact intersect again at some high value of nominal stress. Thus, while the Neuber calculation overstates the strain in some ranges of nominal stress, it might, in fact, understate it for loadings beyond some critical value.

In order to find an improved method of analyses we attempted a relation in the form:

$$K_\sigma K_\varepsilon (K_\varepsilon/K_\sigma)^\alpha = K_t^2 \qquad \text{(Eq 8.15)}$$

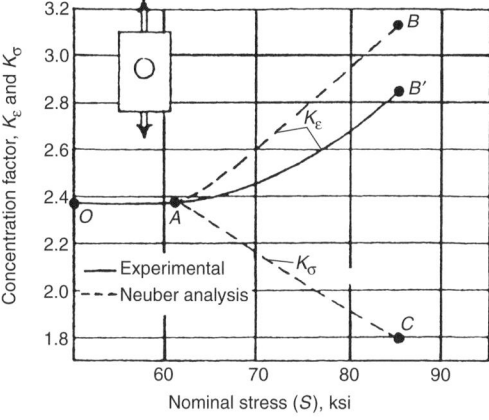

**Fig. 8.18** Comparison of experimental results with Neuber calculations. Source: Ref 8.20

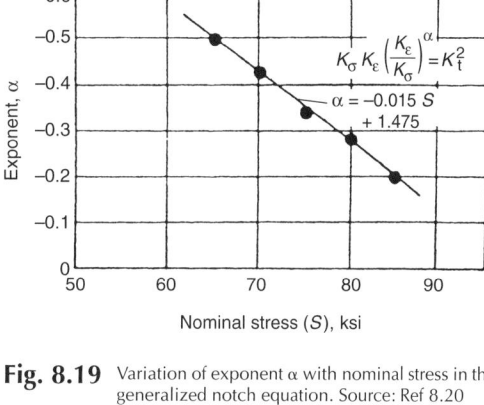

**Fig. 8.19** Variation of exponent α with nominal stress in the generalized notch equation. Source: Ref 8.20

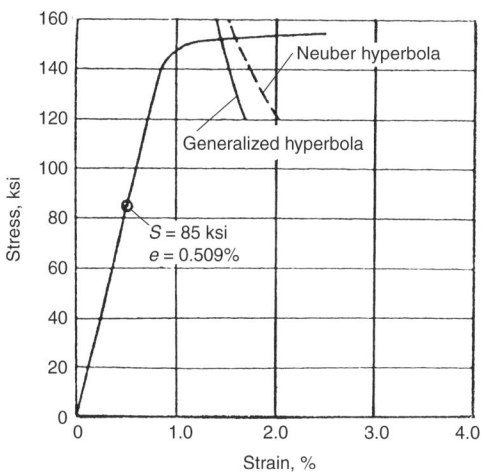

**Fig. 8.20** Comparison of the generalized notch equation and the Neuber equation for a nominal stress of 586 MPa (85 ksi). Source: Ref 8.20

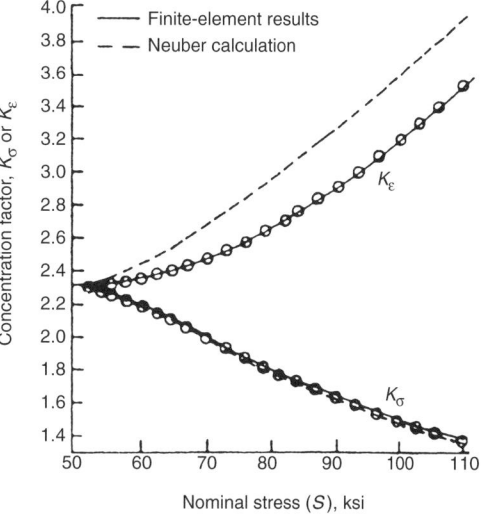

**Fig. 8.21** Comparison of results from finite-element calculations and Neuber analysis. Source: Ref 8.21

Without the factor $(K_\varepsilon/K_\sigma)^\alpha$, the equation is, of course, the Neuber equation. However, it is important to recognize that the equation with the correction factor is valid in the elastic range because in this range, the correction is simply unity. But, in the plastic range, the modified equation can produce different results from the Neuber equation because, in essence, it becomes $K_\sigma^p K_\varepsilon^q = K_t^2$, depending on the value chosen for $\alpha$.

Analyzing the data by choosing $\alpha$ to force a fit with the experimental results produced Fig. 8.19. An excellent straight-line relationship resulted between $\alpha$ and nominal stress. Thus, for example, at a nominal stress of 85 ksi, the exponent is 0.2, and the corresponding modified Neuber equation is:

$$K_\sigma K_\varepsilon (K_\varepsilon/K_\sigma)^{0.2} \equiv K_t^{1.2} K_\sigma^{0.8} = K_t^2$$
$$= (2.37)^2 = 5.62 \quad (\text{Eq 8.16})$$

Thus, when this equation is represented as a generalized hyperbola, the result is as shown in Fig. 8.20, while the Neuber equation is $K_\varepsilon K_\sigma = 5.62$. Each represents a hyperbola, and its intersection with the cyclic stress-strain curve determines the stress and strain at the critical section of the central hole of the plate. It is clear from Fig. 8.20 the stress and strain as determined from the generalized hyperbola are lower than those determined by the Neuber hyperbola (although there is relatively little effect on stress because of the flatness of the stress-strain curve).

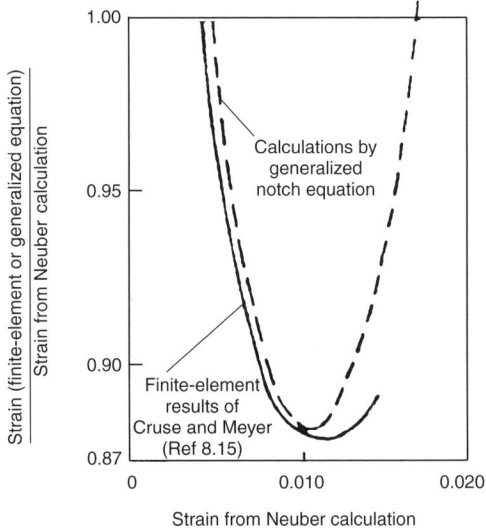

**Fig. 8.23** Comparison of strains from elastoplastic finite-element analysis and strains from the generalized notch equation. Source: Ref 8.20

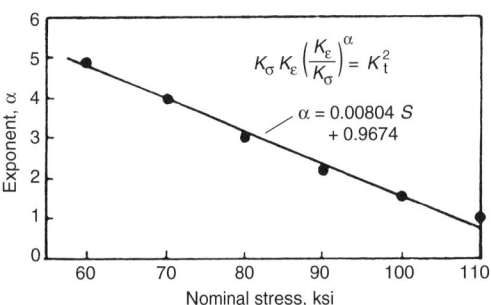

**Fig. 8.22** Variation of exponent $\alpha$ with nominal stress from elastoplastic finite-element results. Source: Ref 8.20

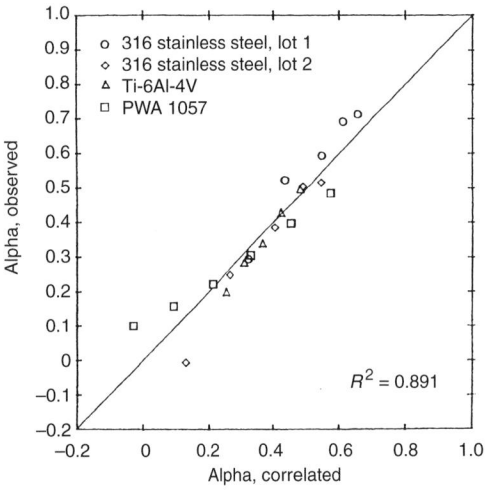

**Fig. 8.24** Correlation of $\alpha = 1.12 + 0.00084(E/S_y) - 1.69(S/S_y)$

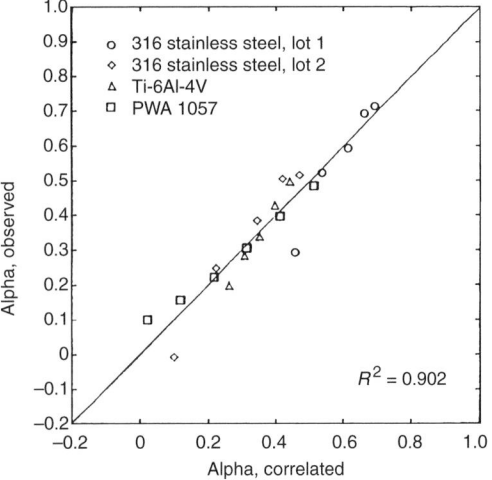

**Fig. 8.25** Correlation of $\alpha = 0.916 + 0.00087(E/S_y) - 1.14(SS_u/S_y^2)$

**Table 8.4  Data used in the analysis**

| Material | Temperature, °F | $K_t$ | $E$, MPa (ksi) | $S_y$, MPa (ksi) | $S_u$, MPa (ksi) |
|---|---|---|---|---|---|
| 316SS, Lot 1 | RT | 2.8 | 196,500 (28,500) | 524 (76) | 552 (80) |
| 316SS, Lot 2 | RT | 2.1 | 196,500 (28,500) | 421 (61) | 552 (80) |
| Ti-6Al-4V | RT | 2.37 | 112,178 (16,270) | 1034 (150) | 1200 (174) |
| PWA 1057 | 900 | 2.32 | 199,948 (29,000) | 972 (141) | 1172 (170) |

RT, room temperature

**A Corroboratory Study.** To determine whether the same basic approach could be applied to another set of results of interest in gas turbine application, a similar analysis was made of the data of Cruse and Meyer (Ref 8.21). They examined a turbine disk nickel-base superalloy (PWA 1057) at 900 °F. The test section of their plate specimen was a 3.9 mm (0.150 in.) thick by 55.9 mm (2.2 in.) wide with an 11.1 mm (0.438 in.) diam central hole. The results are shown in Fig. 8.21 to 8.23. Here, the theoretical stress concentration is 2.32, and the Neuber analysis was compared with elastoplastic finite-element calculations as shown in Fig. 8.21. Again, the Neuber calculations are seen to show higher local notch strains than the more accurate finite-element results. Figure 8.22 shows the calculated values of $\alpha$, similar to Fig. 8.22, from the data of Fig. 8.21. The value of $\alpha$ is again linear with $S$. Figure 8.23 shows an interesting comparison of the relation between the strain at the root of the notch as determined from the finite-element method as a fraction of the strain determined from the Neuber analysis. It is seen that at relatively low nominal stresses, the ratio of the two strains progressively decreases as nominal stress is increased. However, a minimum is reached when the strain as calculated by the Neuber equation is about 0.011 and the nominal stress is 95 ksi. Reference 8.21 carries the calculations only slightly beyond this minimum; however, if the calculations are made according to the generalized notch analysis, using the value of $\alpha$ from Fig. 8.22, the predictions agree well with the calculations of Cruse and Meyer. The curve in Fig. 8.23 turns upward as observed, continuing up as load increases (i.e., the Neuber strain increases), and eventually reaching a point at which the Neuber calculations become unconservative.

**Attempts to Generalize.** The proposal of the Neuber correction used in the Teledyne study was based on an ongoing study by Manson and several of his graduate students. The results, although not published, are reviewed here. Standard circular fatigue specimens were used and small holes were drilled transversely across the minimum section. Miniature strain gages were used to measure strain. Although it was intended initially to do the entire program on a single batch of material, two lots (1 and 2) of 316 stainless steel were inadvertently used in the program. One lot had a much lower yield strength than the other. Table 8.4 shows the properties of all four materials finally used in the analysis, including the Ti-6Al-4V of the Teledyne program and the PWA 1057 discussed earlier.

To analyze all the data in a more general way we assumed that the generalized Neuber equation expressed by Eq 8.15 has the exponent $\alpha$ as a function of nominal applied stress:

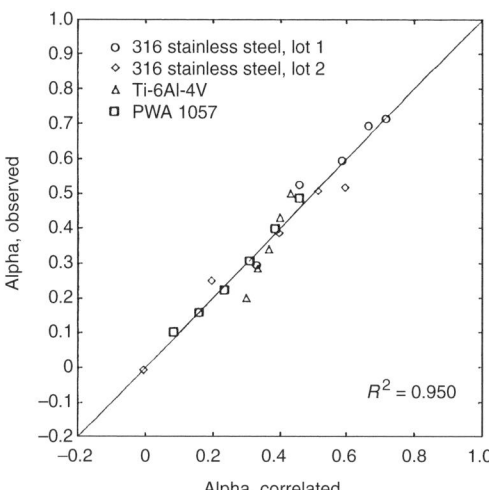

**Fig. 8.26** Correlation of $\alpha = 0.0774 + 117/S_y - 149S/(S_y^2)$

**Table 8.5  Comparison of calculation by Neuber and generalized Neuber equations**

| | At $(\varepsilon_{G.N.}/\varepsilon_{NEUBER})$min | | At $\varepsilon_{G.N.} = \varepsilon_{NEUBER}$ |
|---|---|---|---|
| Material | $\varepsilon_{G.N.}/\varepsilon_{Neuber}$ | $S_{NOM}/S_Y$ | $S_{NOM}/S_Y$ |
| 316SS, 1 | 0.86 | 0.65 | 0.8 |
| 316, 2 | 0.76 | 0.6 | 0.78 |
| Ti-6Al-4V | 0.89 | 0.6 | 0.78 |
| PWA 1057 | 0.90 | 0.6 | 0.78 |

$$\alpha = A + B\sigma_{nom} \quad \text{(Eq 8.17)}$$

Constant $B$ cannot be nondimensional because $\alpha$ would thus have dimensions of stress. Several forms that would make $B$ nondimensional were therefore chosen, and regression analyses were made to fit the available data. Only the most promising results are discussed in the following, in connection with the assumed forms:

$$\alpha = A + \frac{BS}{S_y}$$

or  (Eq 8.18)

$$\alpha = A + B\frac{S_u}{S_y^2}S$$

The correlations are shown in Fig. 8.24 and 8.25. The agreements are not perfect, showing coefficients of correlation of 0.89 and 0.90, respectively, but are acceptable as a starting point. Actually, the best correlation was obtained by using the form shown in Fig. 8.26, but the form is not nondimensional, so its use is questionable.

## Discussion and Concluding Remarks

It appears conclusively that the Neuber equation is an approximation that can be improved. If the equations shown in Fig. 8.24 and 8.25 are assumed to be applicable (or if they are corroborated or replaced by improved relations), the

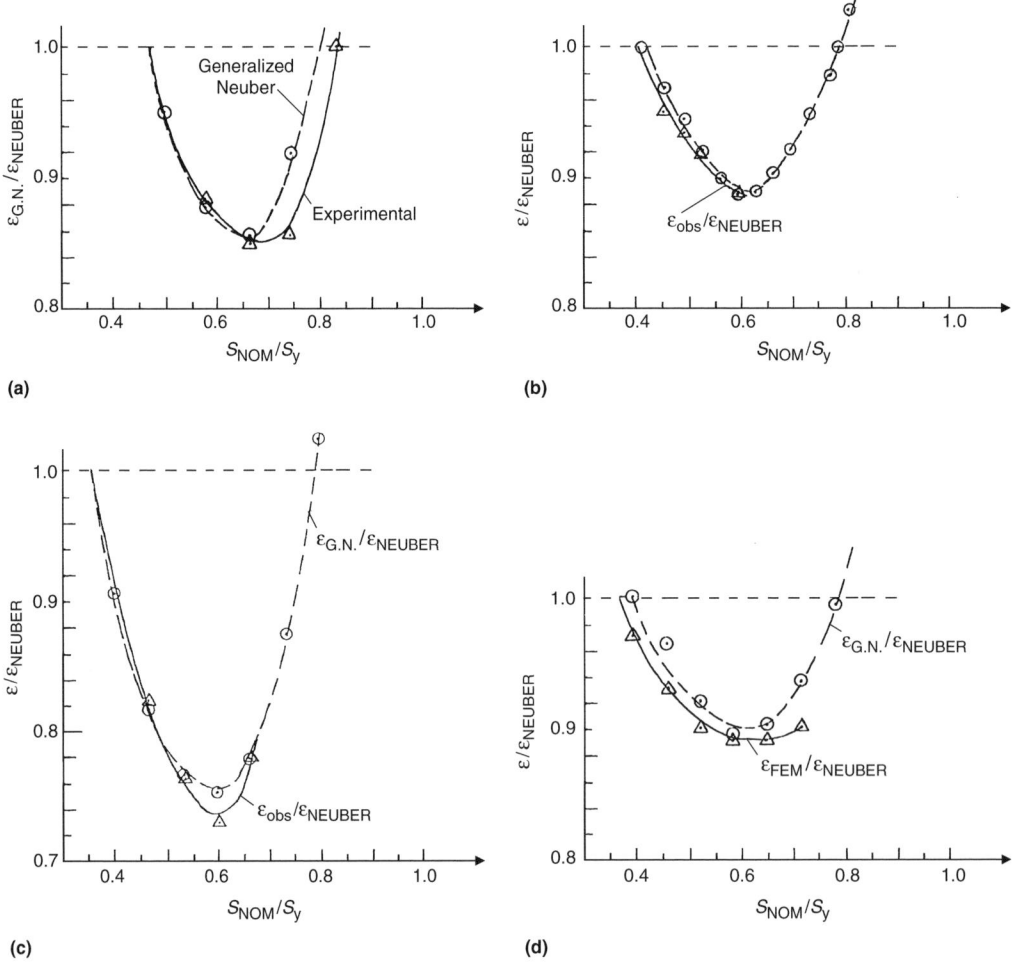

**Fig. 8.27** Comparison of experimental results with modified Neuber equations. (a) 316 stainless steel ($K_t = 2.1$, $S_y = 421$ MPa, or 61 ksi). (b) Ti-6Al-4V ($K_t = 2.37$, $S_y = 1000$ MPa, or 145 ksi). (c) 316 stainless steel ($K_t = 2.8$, $S_y = 524$ MPa, or 76 ksi). (d) PWA 1057 ($K_t = 2.32$, $S_y = 1069$ MPa, or 155 ksi)

solution of a practical problem is no more difficult by using the generalized Neuber equation than by using the ordinary Neuber equations. It is necessary to use the proper hyperbola only when calculating the intersection with the cyclic stress-strain curve. Usually, within the limited experience available, the strain as determined by the Neuber equation is greater than that measured (or determined by the generalized equation). However, the error is usually less than 15%. Usually, use of the Neuber equation is conservative because the error is on the side of designing for a higher strain. It is possible, however, that at high stresses, when the parabola of Fig. 8.23 turns up, the calculations by the Neuber hyperbola may actually become unconservative. A summary of the agreement between experimental results for a few materials and the equations for modified Neuber is shown in Fig. 8.27(a) to (d). These figures not only show reasonable agreement, they also show that the error is constrained to approximately 15% or less. It is also apparent that at high strains the strain can become greater than the strain predicted by the Neuber approach. A summary of the characteristics of the analysis is shown in Table 8.5. Furthermore, as discussed in Chapter 9, "Crack Mechanics," the strain concentration at the root of the notch governs only the initiation of the crack. The crack propagation phase is relatively unaffected, and this latter phase may be a large fraction of the total life (Ref 8.22).

## REFERENCES

8.1 W.C. Young and F.G. Budynas, *Roark's Formulas for Stress and Strain,* 7th ed., McGraw-Hill, New York, 2002
8.2 W.E. Pilkey, *Peterson's Stress Concentration Factors,* 2nd ed., John Wiley & Sons, 1997
8.3 G.N. Savin, *Stress Concentration around Holes,* Pergamon Press, 1961
8.4 J.A. Collins, *Failure of Materials in Mechanical Design,* John Wiley & Sons, 1993, p 419–425, 242–248. See also Ref 8.6, p 242–248.
8.5 H. Neuber, Theory of Notch Stresses—Principles for Exact Stress Calculation, Julius Spring, Berlin, 1937 (translated and published by J.W. Edwards, Ann Arbor, MI, 1946) and H. Neuber, *Theory of Notch Stresses: Principles for Exact Calculation of Strength with Reference to Structural Form and Strength,* 2nd ed., Springer-Verlag, Berlin, 1958 (translated and issued as AEC-TR-4547 by the U.S. Office of Technical Information, 1961)
8.6 R.C. Juvinall, *Engineering Considerations of Stress, Strain, and Strength,* McGraw-Hill, New York, 1967
8.7 P. Kuhn, "The Prediction of Notch and Crack Strength under Static or Fatigue Loading," SAE-ASME Paper 843C, presented April 1964
8.8 P. Kuhn and H.F. Hardrath, "An Engineering Method for Estimating Notch-Size Effect in Fatigue Tests of Steel," NACA TN 2805, National Advisory Committee for Aeronautics, Washington, DC, Oct 1952
8.9 R.E. Peterson, Notch-Sensitivity, Chapter 13, *Metal Fatigue,* G. Sines and J.L. Waisman, Ed., McGraw-Hill, New York, 1959, p 293–306
8.10 R.B. Heywood, *Designing against Fatigue of Metals,* Reinhold, New York, 1962
8.11 A.A. Blatherwick and B.K. Olsen, "Stress Redistribution in Notched Specimens under Cyclic Stress," A.S.D. Technical Report 61-451, 1961, Aero Systems Div., Wright-Patterson Air Force Base, Dayton, OH
8.12 J.H. Crews, Jr. and H.F. Hardrath, A Study of Cyclic Plastic Stresses at a Notch Root, *Exp. Mech.,* Vol 6 (No. 6), June 1966, p 313–320
8.13 R.M. Wetzel, Smooth Specimen Simulation of Fatigue Behavior of Notches, *J. Mater.,* American Society for Testing and Materials, Vol 3 (No. 3), Sept 1968, p 646–657
8.14 T.H. Topper, R.M. Wetzel, and J. Morrow, Neuber Rule Applied to Fatigue of Notched Specimens, *J. Mater.,* American Society for Testing and Materials, Vol 4 (No. 1), March 1969, p 200–209
8.15 J.H. Crews, Jr., "Effects of Loading Sequence for Notched Specimens under High-Low Two-Step Fatigue Loading," NASA TN-6558, Nov 1971
8.16 *SAE Fatigue Design Handbook,* 3rd ed., Publication AE-22, Society of Automotive Engineers (SAE), Warrendale, PA, 1997
8.17 *Fatigue under Complex Loading: Analyses and Experiments,* R.M. Wetzel, Ed., Vol 6, *Advances in Engineering,* Society of Automotive Engineers (SAE), Warrendale, PA, 1977

8.18 *Multiaxial Fatigue: Analyses and Experiments.* G.E. Leese and D. Leese, Ed., Publication AE-14, Society of Automotive Engineers (SAE), Warrendale, PA, 1989

8.19 *Recent Developments in Fatigue Technology,* R.A. Chernenkoff and J.J. Bonnen, Ed., Publication PT-67, Society of Automotive Engineers (SAE), Warrendale, PA, 1997

8.20 J. Walcher, D. Gray, and S.S. Manson, "Aspects of Cumulative Fatigue Damage Analysis of Cold End Rotating Structures," American Institute of Aeronautics and Astronautics preprint 79-1190, 1979

8.21 T.A. Cruse and T.G. Meyer, "Structural Life Prediction and Analysis Technology," Report FR-10896, Pratt & Whitney Aircraft Group, Nov 3, 1978

8.22 D.F. Socie, N.E. Dowling, and P. Kurath, "Fatigue Life Estimation of Notched Members," *Fracture Mechanics: Fifteenth Symposium,* ASTM STP 833, R.J. Sanford, Ed., American Society for Testing and Materials, 1984, p 284–299

# CHAPTER 9

# Crack Mechanics

## Introduction

A particularly exciting development that has been occurring over the past half century is the constant advancements in the treatment of specimens and structures containing cracks. While the importance of cracks in affecting structural strength has been recognized for hundreds of years, it was not until the 1950s that quantitative treatment was made possible through the development of what is commonly designated as *fracture mechanics*. Originally this description may have been appropriate because the early objective was to explain the then-mysterious phenomenon of "brittle fracture" of structures made from materials thought to be ductile. However, the derivative uses of the technology developed expanded to other applications such as "fatigue crack growth," even if "fracture" was not specifically involved. We shall, therefore, refer to this technology as *crack mechanics,* and include in the discussion both brittle fracture and fatigue crack propagation in ductile materials.

This chapter describes the mechanics of cracking. It is written for both novice students and practicing engineers who would like to gain greater insight into the cracking process without becoming deeply involved in the mathematical description of stress fields surrounding crack tips. Even for those with a higher level of experience using fracture mechanics, the information contained in the chapter should be of value. A list of current reference works and other pertinent publications is included at the end of the chapter for those interested in acquiring a greater depth in the knowledge of the subject.

## Brittle Fracture

The history of brittle fracture goes back many years. It has been long recognized that some materials such as glass and ceramics do not display ductility at ambient temperature and pressure and that a structure made from such materials will either sustain an imposed load or fracture with little or no evidence of plastic deformation. However, it was observed later that some structures failed in such a "brittle" fashion even if the material from which they were fabricated was basically not brittle, displaying perhaps considerable plasticity in a conventional tensile test. Why such behavior developed was not initially understood, and because the phenomenon occurred in many applications of great technological significance, intense interest developed in understanding the phenomenon both for philosophical reasons and for determining how to avoid the danger of its occurrence. We present here a brief history of the developments that led to the current highly sophisticated technology of this field.

**Some Prominent Examples of Brittle Fractures.** Before discussing the technology of fracture analysis, it is appropriate to relate some of the special cases involving brittle fracture. Some of these occurred well before the modern analytical methods were developed, others were the catalyst that sparked the intensive study that led to the modern understanding, and others yet were concurrent in the development of the modern analytical procedures.

An important example is that of the accident that occurred January 15, 1919 when a steel vessel containing 2.3 million gallons of molasses suddenly fractured at a processing plant on Commerce Street in Boston. Barsom and Rolfe (Ref 9.1) in an excellent text on fracture mechanics have provided a graphic description of the incident. They point out that the molasses flowed over many city streets, killing a dozen humans and numerous horses (and no doubt many other smaller creatures). Much litigation ensued in which liability was probed; was it faulty design, inferior material, or simply human error? No conclusion could be reached, and one of the court-appointed auditors admitted that if he could not understand what had happened, he was no more ignorant about the true cause of the accident than were over half of all the experts on both sides of the litigation. It seemed to be a total enigma.

It is ironic that it was in this very same year that A.A. Griffith (Ref 9.2) was pursuing his pioneering research on the mechanics of fracture of cracked bodies and drew the conclusion that

> The breaking load of a thin plate of glass having a sufficiently long straight crack normal to the applied stress is inversely proportional to the square root of the length of the crack.

Early the following year his classic paper was published on the brittle fracture of glass that could have been brought to bear on this accident. However, the litigants, Griffith, and even the whole engineering continuity did not see the connection for decades to come.

During World War II another mysterious set of disasters occurred. Numerous merchant ships simply broke in half for no apparent reason. Some broke while in harbor before they were ever commissioned or before they saw service. Of course, after Pearl Harbor, the United States entered the war hastily. The action was in Europe and in Britain, and it was necessary to build many sea vessels to carry material and personnel to these destinations. Previously, the experience in building such ships was on a much lower scale, and the mode of fabrication was by riveting plates together. To expedite construction, however, a welding procedure was adopted. Very rigid structures developed for which the steel used was not ideally suited, and furthermore, there were many welds with at least minor flaws. The result was a rash of failures. Figure 9.1 shows one such failure that occurred in harbor in 1941 (Ref 9.1). The cause was not understood until much later, but fortunately, because the U.S. Navy was involved, the Naval Research Laboratory became associated with the search for understanding. It was George Irwin, an administrator and researcher at the Naval Laboratory, who provided an insight that laid the framework for a new technology, designated *fracture mechanics,* which is discussed later in this chapter. Basically, however, it related to brittle behavior under some conditions for material that under other conditions might behave in a ductile fashion. In this case, it was the constraint due to large size in a structure that contained mechanical and microstructural defects.

**Fig. 9.1** Brittle fracture of a welded World War II ship

Another interesting case within our experience was the sudden fracture of a tank storing liquid natural gas at low temperature. The operator of the unit was the East Ohio Gas Company, Cleveland, Ohio. Figure 9.2 shows a Cleveland *Plain Dealer* photo of the disaster that occurred. The liquid natural gas that leaked out from a crack that developed in the tank was somehow ignited, and the holocaust that developed spread over many square blocks, devastating the entire area, killing at least 130 people, and burning hundreds of others. Here, too, it was a question of embrittlement, this time by the low temperature ($-160$ °C, or $-250$ °F liquefied natural gas) of a material that was otherwise ductile but apparently quite unsuitable for the purpose involved.

Finally, we can mention one more case within our direct experience. This was the 6600 mm (260 in.) diameter motor casing that was being developed by the U.S. Air Force and NASA to contain a solid fuel, the combustion of which would provide the propulsion for the first stage of a rocket engine. The issue under contention at the time was whether the casing should be made of 200 grade maraging material (tensile strength 1380 MPa, or 200 ksi), or whether a maraging 250 grade (tensile strength 1720 MPa, or 250 ksi) should be used to make the casing lighter. It was NASA's contention that the softer material would be preferable, even though it would require a slightly larger thickness to contain a given pressure. The stronger material could be used with lower thickness, but it would not be as tolerant of flaws because of its lower ductility. The vehicle under test at NASA was constructed of the 250 grade material and was pressurized with water rather than gas to limit the flying debris that could be expected if fracture occurred.

Indeed, a serious stress concentration did occur in the test vessel, and it caused an unanticipated catastrophic failure to occur at only 62% of the intended operating pressure. The stress concentration was a defect that was hidden by weld metal and was not detected by x-ray inspection prior to test. After fracture, the cracking origin became quite evident. After many days of searching and studying the fracture fragments, the investigation team was able to reassemble the test tank as shown in Fig. 9.3. Note the scale of the test part from the size of the 6 ft man standing adjacent to the fractured casing. The fracture surface features clearly pointed back to the origin of failure. This case, too, was one in which the material used was not suitable for the intended purpose because it lacked a quality of toughness that is best defined through fracture mechanics.

**Fig. 9.2** Devastation caused by brittle fracture of liquid natural gas tank in Cleveland, Ohio, Oct 1944

Over the years many user organizations came to be interested in fracture mechanics evaluation of materials. In the aeronautical field, designers of landing gear learned that brittle fracture can occur unexpectedly. Choosing stronger materials often accelerates failure! Designers of nuclear reactors had realized that nuclear radiation can cause embrittlement of metals, and that fracture could be disastrous. Fracture mechanics attracted much interest and research support. In this chapter we start with the role of fracture mechanics in explaining brittle fracture. Then we discuss another facet of the subject: crack mechanics. Finally, we discuss an application that combines these two facts in the study of materials that initiate and grow a crack through the fatigue process, finally fracturing when a fracture toughness condition is reached.

**Griffith's Study on Glass.** Griffith became interested in the brittle behavior of glass strictly for academic reasons; he had no particular application in mind. He simply wanted to understand why glass shows no ductility while most metals generally deform appreciably before they fracture. His method of reasoning and his conclusions became the basis of the science now known as *fracture mechanics*. Although his approach is still used as a basis for many studies, the modern procedures using his concept only differ in their practical implementation.

On the basis of his analysis, Griffith came to the conclusion that brittle fracture was associated with preexisting cracks. However, whether the cracks propagated suddenly and catastrophically depended both on the stress in the region at the tip of the crack and how this stress would change if the crack were to grow by a small increment. Such growth would change three factors: the stress at the tip of the crack, the mechanical energy stored in the specimen (structure) before and after the crack had grown, and the surface energy caused by the creation of new surface as the crack grew. If the surface energy required to extend the crack was greater than the stored mechanical energy given up by the structure in passing from the original crack length to the new crack length under its loading condition, then there was no driving energy to cause the crack to grow further; the crack would simply cease to grow. If, however, there was just enough stored energy given up by the structure in passing between the original and current crack

**Fig. 9.3** Catastrophic brittle fracture of a 260 in. diam rocket motor case during hydrotest. Failure occurred unexpectedly at about 50% of design pressure. Note size of case compared to the 6 ft tall men.

length to provide the new surface energy created by the extended crack, then extension of the crack was its own driving force; the crack would grow in a stable manner. Should more stored energy be given up than needed to form the new surface, then the growth of the crack would feed on itself and would continue in a self-propagating manner. That is, catastrophic failure would occur.

Thus, using Inglis's theory of stress distribution around a crack (Ref 9.3), Griffith was able to calculate the stored elastic energies involved in the body for both the original crack size and the extended crack. Applying his theory, for example, to an infinite plate of unit thickness containing a centrally located, through-the-thickness crack of length 2a and subjected to a uniform tensile stress, σ, applied to the surface edge, the resulting equation is written in the form:

$$\sigma\sqrt{a} = \sqrt{2\gamma_e E/\pi} \qquad \text{(Eq 9.1)}$$

The left side is the most common form of what has come to be known as the *stress-intensity factor K*. When the stress reaches a critical fracture strength $\sigma_c$, $K \rightarrow K_c$ (the *critical* stress-intensity factor). In Eq 9.1, $E$ is the elastic modulus and $\gamma_e$ is the energy required to create the new cracked surface. Thus, if the loading produces more stored elastic strain energy in the cracked plate than is required to create the new surfaces, the crack will continue to grow spontaneously.

Because Griffith's theory was directed to an explanation of the brittle behavior of glass, and because it involved an elastic surface property not familiar to engineers, it was not seized upon immediately by the engineering community for application to the brittle behavior of metals. However, the great importance of Griffith's theory lay in the form in which the final stability condition could be expressed. Equation 9.1 shows, for example, that the important loading parameter involving stress and crack length is $\sigma\sqrt{a}$ for the infinite plate, and the important material property parameter is $\sqrt{2\gamma_e E/\pi}$. Thus, even if $\gamma_e$ were a relatively unfamiliar property, it did not have to be determined from first principles. It could essentially be derived through Eq 9.1 by determining under what conditions of σ and $a$ the material *did behave* in a brittle fashion, and then this property would be usable for determination of brittle behavior under other loading conditions of stress and crack length. The concept could even be used for geometries other than an infinite cracked plate.

Studies have shown that the surface energy $\gamma_e$ required to extend the crack depends strongly on the environment. For example, in a *Scientific American* article (Ref 9.4), Michalske and Bunker discuss in laymen's terms their molecular modeling efforts. They show that a chemical interaction between water and the atomic bonds of the glass molecules at a crack tip causes a significant reduction in fracture resistance.

In 1948, Irwin (Ref 9.5), concerned with the brittle behavior of ship steels, because he was employed by the Naval Research Laboratory that was alarmed over the Liberty merchant ship failures such as shown in Fig. 9.1, proposed that Griffith's theory could also be applied to metals if the effective surface energies could be determined through appropriate testing. The fact that the energy associated with local plastic deformation in the vicinity of the crack tip could also be handled with appropriate corrections to Eq 9.1 made the new theory attractive. Orowan also proposed a similar concept in Ref 9.6; however, because of Irwin's more intense subsequent efforts in this field, his name has been linked more closely with the extension of Griffith's theory to brittle behavior of metallic materials. It is common to describe the applications to be discussed herein as the Griffith-Irwin theory.

## The Griffith-Irwin Approach for Treating Crack Mechanics

The innovation introduced by Irwin still relies on the same basis as that used by Griffith—namely, the stress field in the vicinity of the crack—but the formulation makes the treatment somewhat less eclectic and easier to implement. It also makes it possible to generalize more easily and to extend the treatment to cases that would be more difficult by the strict use of only the Griffith concept.

To understand the nuances, again consider Eq 9.1. In the form presented it states that the condition for rapid crack extension depends essentially on four parameters: the nominal stress σ in the vicinity of the crack (the stress that would be present in the region of the crack if the crack were not present); the crack length $a$; and two physical properties, $E$ and $\gamma_e$. We note, however, that these four parameters are really grouped as two: $\sigma\sqrt{a}$ and $\gamma_e E$. Thus, what is important is not $\sigma$ and $a$ separately, but their combined value $\sigma\sqrt{a}$. Similarly, the product $\gamma_e E$ is just another physical property, which does not really have to be thought of in relation to a surface energy, but that might be measured directly by some other way. Now, if we take a large plate of material, introduce a sharp crack of length $a_1$, gradually increase stress $\sigma_1$ until a rapid crack growth develops (indicating brittle behavior), then the value of $\sigma_1\sqrt{a_1}$ provides us with the value of $\gamma_e E$, according to Eq 9.1. If we designate the entire term $\sqrt{2\gamma_e E/\pi}$ as $P_c$ we can conclude that for any combination of $\sigma_n$ and $a_n$ such that $\sigma_n\sqrt{a_n} = \sigma_1\sqrt{a_1} = P_c$, brittle fracture can occur. Thus, we essentially replace the knowledge of a surface energy parameter (that is in the realm of the physicist) by a simple fracture test (that is in the realm of the engineering experimentalist).

Furthermore, Eq 9.1 suggests that we might better be able to treat more complex geometries such as cracks in finite structures. As it turns out, it also becomes easier to envision the effects of structure thickness and to extend treatment for metals where some plastic deformation may occur in a small region ahead of the crack.

**The Stress Field ahead of the Crack.** The study of the modern treatment of crack mechanics starts with an understanding of the stress distribution immediately at the tip of a crack in a body of isotropic material. For simplicity, we consider first that the material remains elastic, although it is possible to include plasticity and viscoplasticity resulting from the very high stresses immediately at the tip of the crack. Sneddon (Ref 9.7) was the first to present a solution for the stress fields for two particular problems, which later were generalized by Irwin (Ref 9.8, 9.9) and by Williams (Ref 9.10). In formulating these equations it was recognized that three cases exist as shown in Fig. 9.4:

- *Mode I:* wherein the surfaces move directly apart and therefore are designated the *opening mode*
- *Mode II:* wherein the surfaces slide over one another perpendicular to the leading edge of the crack and therefore are designated the *edge-sliding mode*
- *Mode III:* wherein the surfaces slide with respect to each other parallel to the leading edge of the crack and therefore also are called the *tearing mode*

Irwin (Ref 9.8, 9.9) derived the equations based on the method of Westergaard (Ref 9.11) as shown in Table 9.1.

Although the three pure modes of deformation and the resulting stresses are shown in Fig. 9.4 and Table 9.1, more general cases of crack-tip deformation can be treated by combination

**Table 9.1  Equations for the three basic modes of fracture**

**Mode I**

$$\sigma_x = \frac{K_I}{(2\pi r)^{1/2}} \cos\frac{\theta}{2}\left[1 - \sin\frac{\theta}{2}\sin\frac{3\theta}{2}\right]$$

$$\sigma_y = \frac{K_I}{(2\pi r)^{1/2}} \cos\frac{\theta}{2}\left[1 + \sin\frac{\theta}{2}\sin\frac{3\theta}{2}\right]$$

$$\tau_{xy} = \frac{K_I}{(2\pi r)^{1/2}} \sin\frac{\theta}{2}\cos\frac{\theta}{2}\cos\frac{3\theta}{2}$$

$$\sigma_z = \nu(\sigma_x + \sigma_y), \quad \tau_{xz} = \tau_{yz} = 0$$

$$u = \frac{K_I}{G}[r/(2\pi)]^{1/2}\cos\frac{\theta}{2}\left[1 - 2\nu + \sin^2\frac{\theta}{2}\right]$$

$$v = \frac{K_I}{G}[r/(2\pi)]^{1/2}\sin\frac{\theta}{2}\left[2 - 2\nu - \cos^2\frac{\theta}{2}\right]$$

$$w = 0$$

**Mode II**

$$\sigma_x = -\frac{K_{II}}{(2\pi r)^{1/2}} \sin\frac{\theta}{2}\left[2 + \cos\frac{\theta}{2}\cos\frac{3\theta}{2}\right]$$

$$\sigma_y = \frac{K_{II}}{(2\pi r)^{1/2}} \sin\frac{\theta}{2}\cos\frac{\theta}{2}\cos\frac{3\theta}{2}$$

$$\tau_{xy} = \frac{K_{II}}{(2\pi r)^{1/2}} \cos\frac{\theta}{2}\left[1 - \sin\frac{\theta}{2}\sin\frac{3\theta}{2}\right]$$

$$\sigma_z = \nu(\sigma_x + \sigma_y), \quad \tau_{xz} = \tau_{yz} = 0$$

$$u = \frac{K_{II}}{G}[r/(2\pi)]^{1/2}\sin\frac{\theta}{2}\left[2 - 2\nu + \cos^2\frac{\theta}{2}\right]$$

$$v = \frac{K_{II}}{G}[r/(2\pi)]^{1/2}\cos\frac{\theta}{2}\left[-1 + 2\nu + \sin^2\frac{\theta}{2}\right]$$

$$w = 0$$

**Mode III**

$$\tau_{xz} = -\frac{K_{III}}{(2\pi r)^{1/2}}\sin\frac{\theta}{2}$$

$$\tau_{yz} = \frac{K_{III}}{(2\pi r)^{1/2}}\cos\frac{\theta}{2}$$

$$\sigma_x = \sigma_y = \sigma_z = \tau_{xy} = 0$$

$$w = \frac{K_{III}}{G}[(2r)/\pi]^{1/2}\sin\frac{\theta}{2}$$

$$u = v = 0$$

Source: Ref 9.12

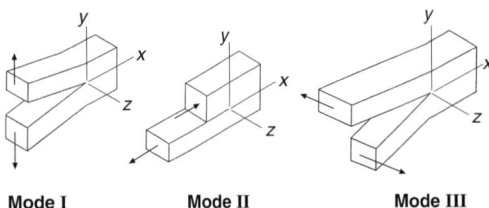

**Fig. 9.4** The three basic modes of crack surface displacement. Source: Ref 9.12

of these pure modes. To understand the implications of these equations and their use in the treatment of practical problems we shall consider only mode I. The conclusions we shall derive can also be extended to other modes (Ref 9.11, 9.12).

**Stress-Intensity Factor, Mode I.** The first important generality that can be derived from the equations in Table 9.1 is that there exists a single parameter $K_I$ that essentially establishes the whole stress field in the vicinity of the crack. Once $K_I$ is specified, all the stresses in the vicinity of the crack, $\sigma_x$, $\sigma_y$, and $\tau_{xy}$ are immediately established for all distances from the crack tip ($r$) and for all angles $\theta$ relative to the crack front. This is a remarkable observation because it establishes that once this quantity is known, all the stresses in the region of the crack, i.e., the entire stress field, are fully established. There are several second-order corrections to the equations, but they can generally be neglected for practical purposes. This quantity $K_I$ is known as the stress-intensity factor for mode I loading and it usually contains two factors: the nominal stress $\sigma$ that would be present in the region of the crack if the crack were absent, and a geometric factor that involves the crack length and its location within the structure. Extensive studies have been conducted, both analytical and experimental, to determine expressions for the stress-intensity factor for different geometries as outlined later.

**The Stress Singularity at the Crack Tip.** It is apparent from the equations that at the crack tip, where $r = 0$, the stresses become infinite. This fact would appear to be troublesome because it might imply that the material would immediately fracture, because the theoretical strength of the material would be exceeded long before the stress could get to infinity. Fortunately, extenuation occurs for two reasons. First, the derivation is based on the assumption that the radius of the crack at its tip is zero, i.e., the crack is infinitely sharp. Actually, this radius is finite; certainly it cannot be less than the lattice parameter for the material. Secondly, it is assumed that the material remains perfectly elastic, whereas some plastic flow always occurs, at least for ductile metals. We shall discuss this matter further, but for the present it is sufficient to indicate that the complication can be overcome by considering that a plastic region develops that can be accounted for in the analysis and the singularity does not render the analysis unworkable.

**Equal Biaxial Stresses at the Crack Tip.** In the x-y plane, i.e., where $\theta = 0$, the stress $\sigma_x$ becomes equal to the stress $\sigma_y$. This fact is easily deduced by setting $\theta = 0$ in the mode I equations (Table 9.1), resulting in $\sigma_x = \sigma_y = K_I/\sqrt{2\pi r}$. For all values of $r \neq 0$, finite stresses $\sigma_x$ and $\sigma_y$ develop, well into the depth of the material and they are always equal. For practical purposes, therefore, a condition of at least biaxial tension develops in the vicinity of the crack tip, and for thick structures a third stress $\sigma_z$ may also develop. As discussed in Chapter 5, "Multiaxial Fatigue," multiaxial tension has the effect of reducing the ductility of the material. Thus, a crack has the doubly dangerous effect of not only elevating the stress in its vicinity but also reducing the capability of the material to tolerate the elevated stress by reducing its ductility.

**Distinction between Plane Stress and Plane Strain.** An important distinction develops between plane stress and plane strain in the vicinity of the crack tip. For plane stress $\sigma_z = 0$ the material develops only a biaxial tensile stress $\sigma_x = \sigma_y$. For the case of plane strain, $\sigma_z = \nu(\sigma_x + \sigma_y) \cong (2/3)\sigma_x$, where $\nu$ is the elastic Poisson's ratio. Thus, the material is close to a condition of triaxial tension of equal magnitude. If total equitriaxiality occurred, the material would display zero ductility and would become totally embrittled even if it were quite ductile in the ordinary tensile test where $\sigma_y = \sigma_z = 0$. Fortunately, totally equal triaxial tension does not develop because $\sigma_z$ is only approximately $2/3$ of $\sigma_x$, or $\sigma_y$. However, highly embrittled it does become, and it is for this reason that brittle behavior can occur even in a normally fairly ductile material when plane-strain conditions develop at the crack tip.

**Some Illustrations of Stress-Intensity Factors.** Figure 9.5 shows some illustrative cases commonly encountered in practice. Where the body approaches an infinite size compared with the crack length, it is often possible to determine the expression for stress-intensity factor in closed form. Those for finite-size bodies are usually expressed as fractions of the stress-intensity factor for infinite-size bodies. In many cases the analytical forms are obtained as approximations. For example, the form of the equation is assumed by analogy to known solutions, but the coefficients are left as unknown constants that are then determined by equating to known values at specific points that are determined from finite-element solution or other approximation approaches.

**Fracture Toughness as a Critical Stress-Intensity Factor.** Fracture toughness is an important stress-intensity factor that has special engineering significance. If a specimen or structure is sufficiently thick for complete constraint at the tip of a crack, so that a plane-strain condition exists, then that stress intensity at which fracture occurs is designated the plane-strain fracture toughness, $K_c$. This parameter may be regarded as a physical property of the material because it represents the behavior of the material at a specific condition of principal tensile stress state in the ratio 1:1:2ν. In contrast, if near-complete constraint in the thickness direction does not exist, failure may still occur but the stress state is not unique because the stress in the thickness direction may have any value between 0 and $2ν\sigma_x$. Thus, the stress intensity at fracture indicates a property only for the circumstantial stress-state present. On the other hand, the plane-strain fracture toughness measured as the

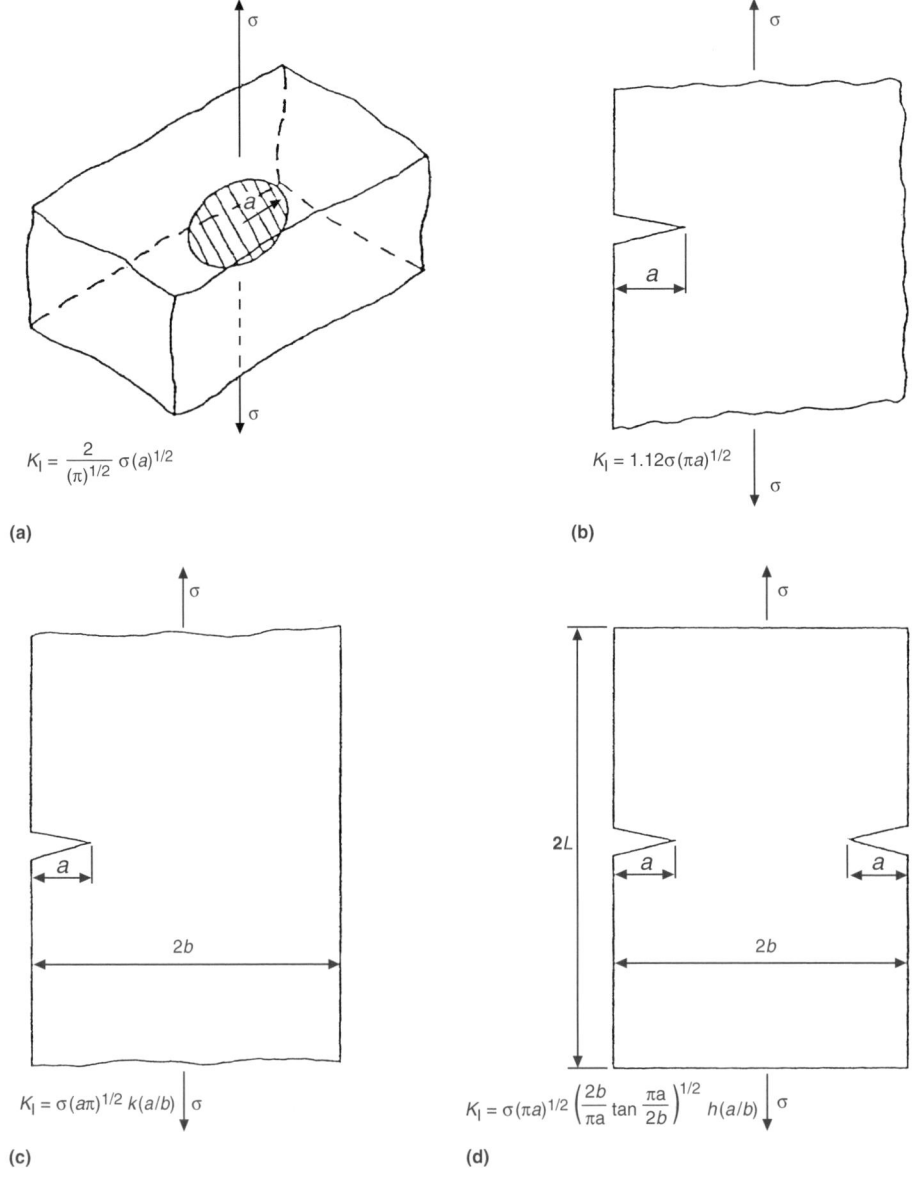

**Fig. 9.5** Stress-intensity factors for commonly encountered service conditions. Source: Ref 12. (a) Penny-shaped crack in uniform tension in infinite body. (b) Edge crack in a semi-infinite sheet subjected to tension. (c) Single edge-cracked strip in tension. (d) Symmetric double-edge crack loaded in tension. Source: Ref 9.12

failure condition for a unique stress state will always result in fracture when that stress state develops, i.e., any other plane-strain condition. Thus, the plane-strain fracture toughness is a true materials property, and, once it is determined for a particular material, it is useful for establishing the fracture condition whenever plane-strain conditions exist at the root of a natural crack.

It may be argued that failure at a condition of plane stress should also represent a materials property because the stress state here is also unique in the ratio 1:1:0. However, if the material has any ductility—and that is what we are concerned with here, brittle fracture of ductile materials—then it is very unlikely that brittle fracture will occur under plane-stress conditions. Examination of the equations in Table 9.1 shows that the shear type of failure is likely to dominate under this condition. When $\sigma_z$ is zero, the maximum shear stress is $\sigma_1/2$, and that is exactly what the maximum shear stresses are in the tensile test. Therefore, if the material displays ductility in the tension test, it will also tear at the crack tip when such cracks are present. Thus, ductile tearing is the common failure mode in this structure and plane-stress fracture toughness has little meaning for materials other than those that are normally recognized as brittle anyway and for which the issue of brittle failure is not a mystery.

**Inadequate Plane-Strain Fracture Toughness as an Explanation for Numerous Brittle Failures.** The development of the concept of plane-strain fracture toughness—in part resulting from inquiries aimed to understand mysterious cases of brittle fracture of structures constructed from ductile materials—did indeed lead to an explanation of these failures, as well as numerous others not significant enough to merit intensive study. It appears clear that the structures involved were very massive and did, indeed, lead to plane-strain conditions at the crack tip. If the plane-strain fracture toughness of the material involved was inadequate for whatever reason, rapid unstable fracture developed. The understanding developed was that although the materials were ductile in a uniaxial tensile test, they were lacking in ductility under the triaxial stress involved in the plane-strain condition at the tip of a natural crack; in other words, the materials did not have sufficient plane-strain fracture toughness for the conditions imposed. Accordingly, in recent years the focus has changed from the normal considerations of strength and ductility to a serious consideration of plane-strain fracture toughness as well. As discussed in another section of this chapter, fracture toughness depends on temperature, loading rate, composition cleanliness, heat treatment, environment, and other factors as well. Designers are careful today to include specification of fracture toughness especially when massive structures are involved. The measurement of this quantity has spawned an extensive discipline with many areas of expertise.

**Measurement of Plane-Strain Fracture Toughness.** The American Society for Testing and Materials (ASTM) has dedicated considerable study to establishing specifications for valid plane-strain fracture toughness testing. For a complete discussion, see Ref 9.13 and 9.14. Here, we shall outline the use of only two types of specimens (tensile and bending) commonly used.

Figure 9.6 shows the types of specimens and loading configurations that can be used when ample material is available. The specimen is sufficiently long so that four-point loading can be imposed to produce pure bending at the center of the span where the natural crack is located. This crack is produced by starting with a carefully machined notch and then subjecting the specimens to low amplitude loadings for many fatigue cycles until a crack develops at the root of the notch. Because the loading is light the growth of the crack can be controlled in a stable fashion until its depth reaches carefully specified limits. Because of the four-point bending configuration, the critical region (at the crack tip) can be kept remote from the loading points and associated secondary stresses. The load at which brittle fracture occurs is observed and the plane-strain fracture toughness is calculated.

A second type of specimen commonly used is shown in Fig. 9.7. This specimen is known as a compact tension specimen because the amount of material involved is kept as low as possible. The reasons for minimizing material usage are many. Sometimes it is because the available material is limited, such as when it is desired to conduct a test on material cut from an existing part, or when many different parameters (e.g., heat treatments) are to be studied and a large number of specimens are needed. Again, the preparation of the specimen consists of a machined starter notch (Fig. 9.7) followed by light fatigue loading to produce a natural crack within specified dimensions to ensure that a valid analysis can be conducted based on the requirements of plane-strain conditions. (It is also possible to create a tiny sharp crack in brittle materials by a simple compressive overload. See, for example, Ref 9.15.)

In conducting these tests the investigator must ensure that the important variables are incorporated, otherwise the results might be meaningless. For example, it must be ensured that the specimen be taken from the correct part of a structure. This is because composition and properties may vary considerably through the volume of a large structure due to segregation and service condition investigated. For example, the properties of the material may be considerably different in the three principal directions at any one location in roll-formed materials. The short-transverse direction as shown in Fig. 9.8, usually has the lowest fracture toughness so it is important to consider this factor seriously.

Table 9.2 lists representative fracture toughness values for materials of engineering interest determined experimentally by various investigators. Note the lack of correlation of toughness with $\sigma_y$. This is particularly true with the maraging steel and the titanium alloy.

## Application to Fatigue Crack Growth

Although the technology of crack mechanics was first developed with the aim of understanding sudden and catastrophic crack extension as discussed in connection with plane-strain fracture toughness, it soon became clear that the concepts involved could also be useful in studying cycle-by-cycle stable crack growth. After all, it is reasonable to expect that such stable crack growth should also depend on the stress field in the vicinity of the crack, which is the underlying basis on which unstable crack propagation is founded.

The study of stable fatigue crack growth did not, however, begin with the Griffith-Irwin crack mechanics development; it had a long prior history before the modern theory of crack mechanics entered the scene. First, Head (Ref 9.17), Frost and Dugdale (Ref 9.18), and Liu and Iinno (Ref 9.19) each developed crack growth laws based on the general form:

$$\frac{da}{dN} = C\sigma^m a^n \quad \text{(Eq 9.2a)}$$

where $\sigma$ is stress, $a$ is crack length, and $C$, $m$, and $n$ are constants fitted to a small amount of available data. However, there was disagreement among the early investigators as to the appropriate constants because their specimens were not identical; nor were the loading parameters. McEvily and Illg (Ref 9.20) interpreted crack growth data in terms of an effective crack-tip stress concentration factor and proposed a semi-empirical relation in the form:

$$\log\left(\frac{da}{dN}\right) = C_1 K_N \sigma_{net} - C_2$$
$$- \left[\frac{C_3}{(K_N \sigma_{net} - C_4)}\right] \quad \text{(Eq 9.2b)}$$

where $K_N$ is an effective stress concentration factor and $C_1$, $C_2$, $C_3$, and $C_4$ are empirical constants fitted to available data.

It was not until 1963, when Paris and Erdogan (Ref 9.21) published their ASME paper "A Critical Analysis of Crack Propagation Laws,"

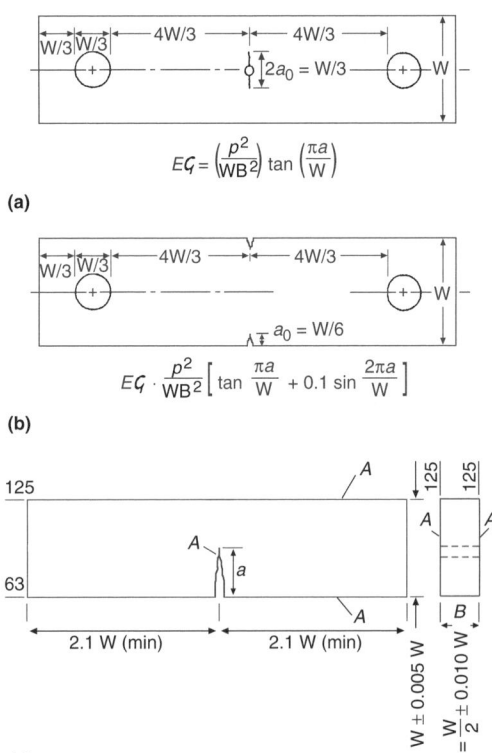

**Fig. 9.6** Practical fracture toughness specimens. (a) Symmetrical center-cracked plate. Source: Ref 9.13. (b) Symmetrical edge-cracked plate. Source: Ref 9.13. (c) Bend specimen. Source: Ref 9.14.

that it became clear what the trouble was. These investigators introduced the concept that crack propagation rate was associated with the stress-intensity factor at the tip of the propagating crack, as might be expected because the crack propagation rate should depend on the stress field in the vicinity of the crack tip. The stress intensity factors were not the same for all the specimen types and loading conditions for all the tests conducted by the early investigators. Thus,

not until the proper crack growth laws were based on local stress-intensity factor could consistency be obtained.

To prove their point, Paris and Erdogan analyzed the results of tests on a flat plate with a central crack loaded in two ways. In one, the loading was applied at the remote ends parallel to the crack; in the other, the loading was applied in the same direction, but on the crack surfaces. It happens that as the crack grows the stress-

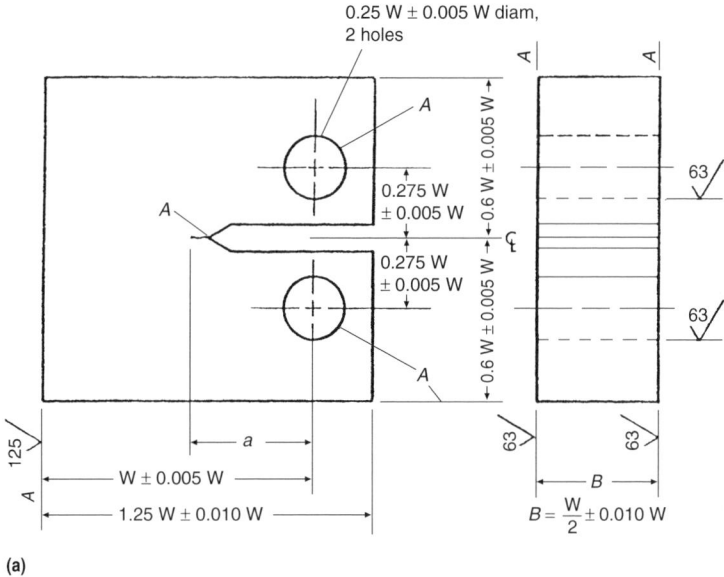

(a)

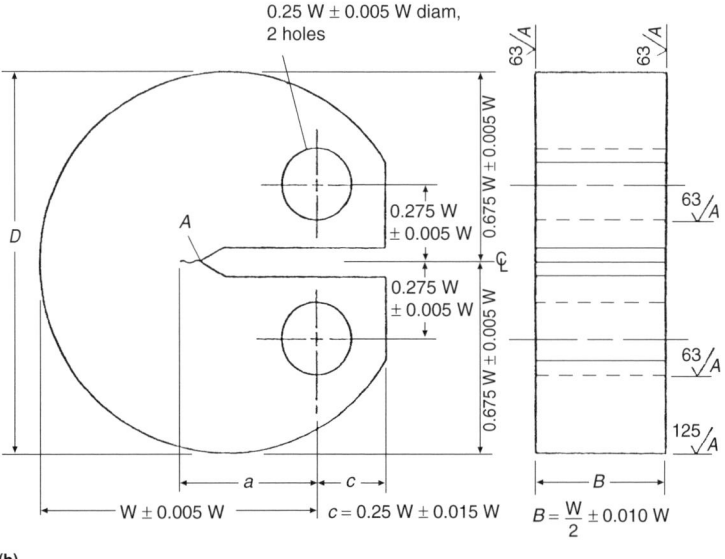

(b)

**Fig. 9.7** Compact tension specimen for fracture testing Source: Ref 9.14. (a) Rectangular. (b) Round disk

intensity factor increases in the first loading mode, but in the second loading mode the stress-intensity factor decreases. What this critical experiment demonstrated was that the crack growth rate followed the variation of stress-intensity in both cases. Thus, they effectively demonstrated that it is the stress-intensity factor at the tip of the crack that governs the crack growth rate. Their proposal for the relation was:

$$\frac{da}{dN} = A(\Delta K)^n \qquad \text{(Eq 9.3)}$$

where $a$ is crack length, $\Delta K$ is the range of stress intensity factor, and $A$ and $n$ are constants.

From their limited tests they were led to propose that $n$ is a universal constant equal to 4.0. While this was a seminal finding, and is still often used today, there are many modifications required based on later findings. These modifications are discussed next.

## Modifications to the Paris-Erdogan Law

### Consideration of the Broad Spectrum of Stress-Intensity Factor.
The tests conducted by Paris-Erdogan were over a limited range of stress-intensity factor, resulting in their power law relations, which produces a straight line on log-log coordinates of $(da/dN)$ versus $\Delta K$. If, however, the stress-intensity factor is greatly increased or decreased, distinct curvature in the relation becomes clear, as seen in Fig. 9.9. At very low $\Delta K$ the crack growth rate decreases substantially, reaching a near-zero value at a stress intensity designated as the threshold. Below this threshold the crack growth rate is insignificant. The numerical value of $\Delta K$ at the threshold depends significantly on specific material and environment. In fact, there is considerably greater difference in threshold intensity among materials and environment than there is among materials in the linear range.

As stress intensity increases to high values the curve also deviates from the straight line, swooping upward. It is thought that this occurs as the stress intensity begins to approach the

**Table 9.2 Fracture toughness values for structural engineering alloys**

| Alloy description | $\sigma_Y$, yield strength, ksi | $K_{Ic}$, fracture toughness, ksi$\sqrt{\text{in.}}$ |
|---|---|---|
| 4340 steel | 217–238 | 45–57 |
| D6AC steel | 217 | 93 |
| HP9-4-20 steel | 186–190 | 120–140 |
| 18Ni maraging steel | 277 | 45–58 |
| 2014-T651 aluminum | 63–69 | 21–25 |
| 2219-T851 aluminum | 50–52 | 33–37 |
| Ti-6Al-4V | 127 | 112 |

Source: Ref 9.16

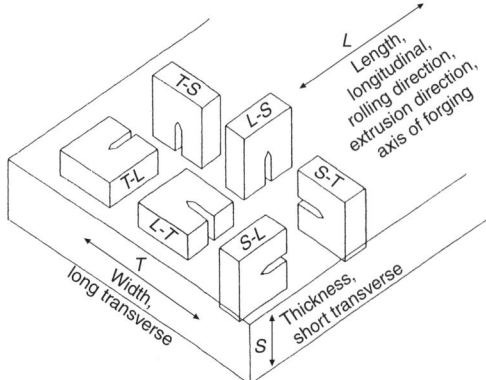

**Fig. 9.8** Orientation of crack in fracture specimen to major forming directions in plate stock. Source: Ref 9.14

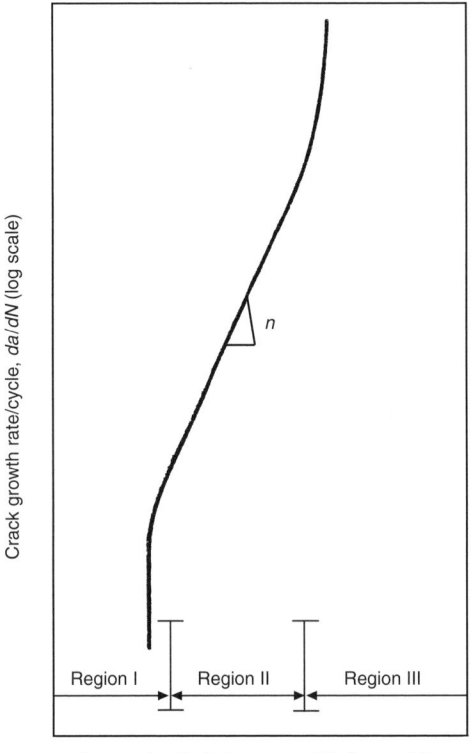

**Fig. 9.9** Fatigue crack growth beyond the regime of the Paris-Erdogan law

fracture toughness. However, this behavior occurs even for thin sheets; therefore, it is probably associated with the development of large plastic zones.

**Material Dependence of Exponent n.** Paris and Erdogan (Ref 9.21) recommended a universal value of $n = 4$. Figure 9.10 (Ref 9.22) justifies this choice for several engineering structural metals and alloys. However, subsequent tests by other investigators on many other materials under differing conditions have shown that the value of $n$ can be quite material dependent, with ranges of 2 to 5 being common (Ref 9.23).

**Mean Stress-Intensity Effects.** The first tests conducted by Paris and Erdogan subjected the plate to loading from zero to maximum. Thus, the nominal mean stress intensity was always one-half the maximum value. Later tests by other investigators demonstrated that the crack growth relationship depended significantly on the mean stress intensity as well as the range of stress-intensity factor imposed. Several formulations have been proposed. One of the more common is that of Forman (Ref 9.24):

$$\frac{da}{dN} = \frac{C\Delta K^m}{(1 - R)(K_{Ic} - \Delta K)} \quad \text{(Eq 9.4)}$$

Here, $\Delta K$ is the range of stress-intensity factor, and $R$ is the ratio of stress-intensity factors (minimum divided by maximum) so that for the usual zero-to-max loading, $R = 0$ and fully reversed loading, $R = -1$. $K_{Ic}$ is the fracture toughness, and $C$ and $m$ are empirical materials parameters.

For $R = 0$:

$$\frac{da}{dN} = \frac{C\Delta K^m}{(K_{Ic} - \Delta K)} \quad \text{(Eq 9.5)}$$

When $\Delta K$ approaches $K_{Ic}$, the denominator approaches zero, and the crack growth rate becomes very large. In this respect, the relation expresses the reasonable expectation that crack growth rates approach infinity at a stress-intensity range that approaches the fracture toughness. However, the equation doesn't degenerate properly to the Paris-Erdogan equation for the test condition used by these investigators. An alternate empirical equation has been proposed in the form:

$$\frac{da}{dN} = \frac{C\Delta K^m}{\sqrt{(1 - R)}} \quad \text{(Eq 9.6)}$$

where as $R$ approaches zero, the equation becomes the same as the Paris-Erdogan Eq 9.3. Of course, the constants $C$ and $m$ cannot be the same for the two formulations, and their choice must be tailored to the available data and the representation chosen. Equation 9.6, of course, uses the same values for $C$ and $m$ determined from conventional zero-to-max loading tests and has been used for steels for moderate positive values of $R$. However, it does not explicitly contain the fracture toughness as the stress range for which crack growth increases greatly.

Figure 9.11 shows the results obtained in a University of Illinois program on three materials tested at various mean stress-intensity ranges (Ref 9.25). Use was made of the Forman equations with the constants shown in Table 9.3.

All the data for different $R$ values tend to fall in a single narrow band (Fig. 9.11b) whereas without the Forman correction, the scatter is significant (Fig. 9.11a).

Figure 9.12 shows schematically the crack growth behavior as $R$ is varied. As $R$ is increased from zero, the threshold for crack growth occurs at lower stress-intensity factors, as does the onset of the linear range. The linear region has very nearly the same slope, although the crack growth rate is higher for a given stress intensity the

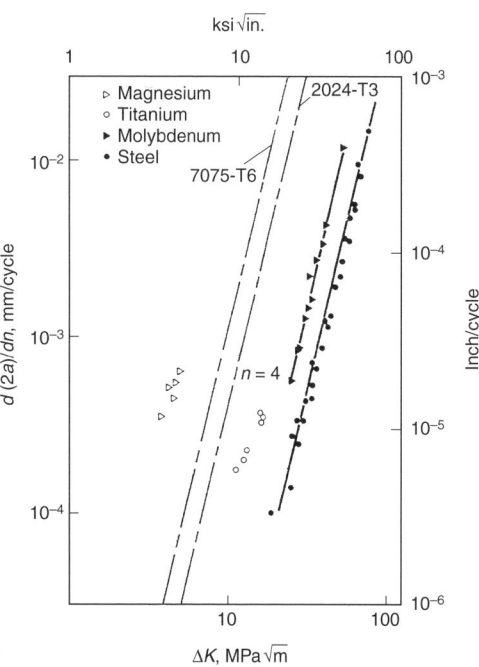

**Fig. 9.10** Fatigue crack growth behavior for various engineering alloys. Source: Ref 9.22

214 / Fatigue and Durability of Structural Materials

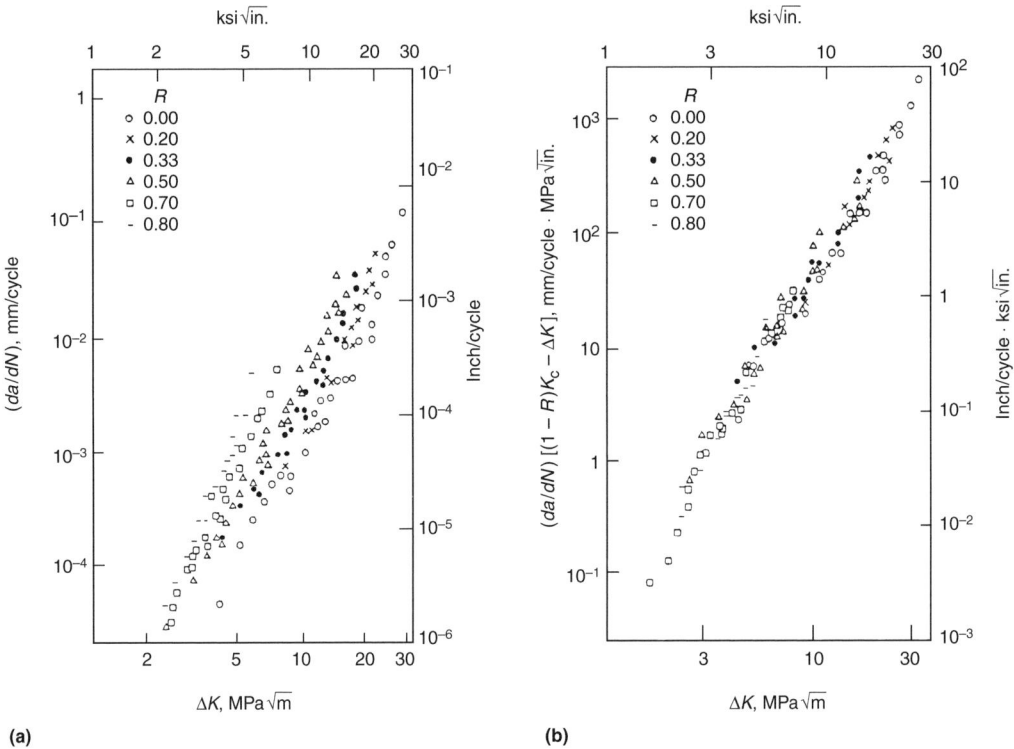

**Fig. 9.11** Effectiveness of Forman's formulation to collapse data obtained at various stress intensity factors. Source: Ref 9.25. (a) Individual crack growth data at various values of mean stress ratio, R. (b) Collapse of all data into narrow band by correlation according to Eq 9.4

higher the value of $R$. And the onset of increased crack growth rate occurs at lower stress intensities the higher is the stress-intensity range $R$. At all values of $R$ the crack growth rate eventually becomes nearly vertical at a sufficiently high stress-intensity range.

**An Alternate Graphical Representation of Mean Stress-Intensity Effects.** Nelson (Ref 9.27) has proposed an alternate type of diagram for representing mean stress-intensity effects, as shown in Fig. 9.13. This diagram is analogous to that for mean stress effects for fatigue crack initiation (Chapter 4, "Mean Stress"). The vertical axis is the mean stress-intensity amplitude

**Table 9.3 Paris-Erdogan law constants for three structural alloys**

| Constant | 7076-T6 aluminum alloy | Man-Ten steel | USS-T1 steel |
|---|---|---|---|
| C | $5 \times 10^{-7}$ | $1.6 \times 10^{-8}$ | $2.5 \times 10^{-8}$ |
| m | 3 | 3.11 | 3.08 |
| $K_c$, ksi$\sqrt{\text{in.}}$ | 68 | 121 | 153 |

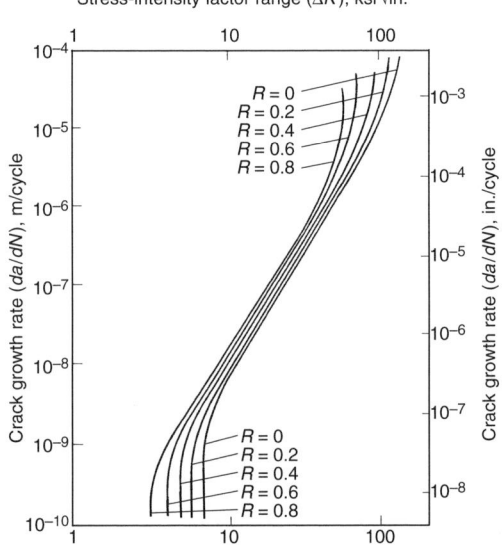

**Fig. 9.12** Schematic mean stress influence on fatigue crack growth rates. Source: Ref 9.26

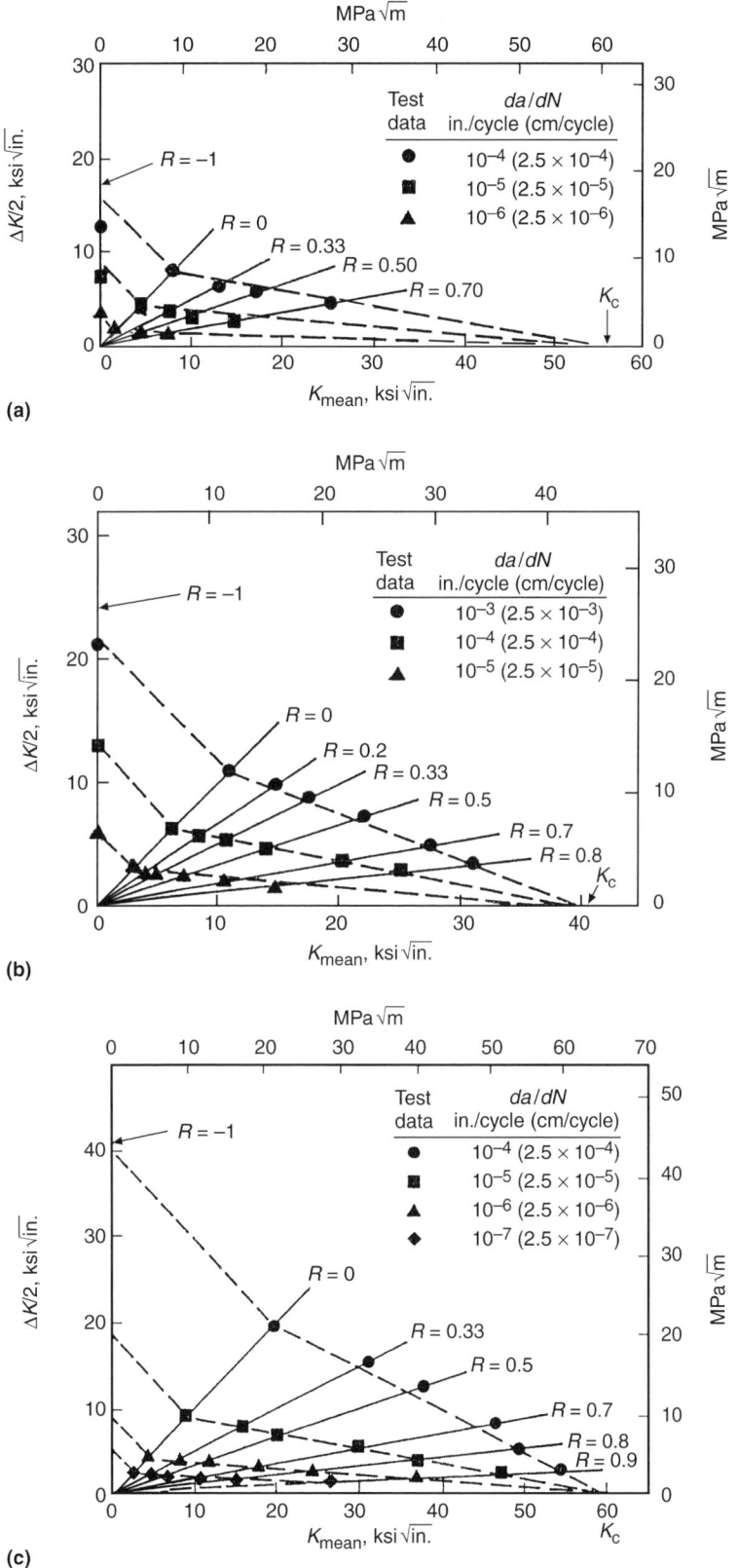

**Fig. 9.13** *R*-ratio effects for three lightweight structural alloys displayed on a mean stress intensity diagram proposed by Nelson (Ref 9.27). (a) Aluminum alloy 2024-T3. (b) Aluminum alloy 7075-T6. Source: Ref 9.28. (c) Titanium alloy Ti-6Al-4V. Source: Ref 9.29

(analogous to alternating stress amplitude) and the horizontal axis plots the mean stress intensity (corresponding to mean stress). Lines of constant crack growth rate are then shown in the diagrams (analogous to lines of constant life). The figure shows results for the aluminum alloys 2024-T3 and 7075-T6, and for the titanium alloy Ti-6Al-4V. Along the 45° line, where $R = 0$ points can be spotted for each chosen crack growth rate as determined from the Paris-Erdogan equation in standard tests (for which $R = 0$). Nelson then draws straight lines from these points to the common point at $K_{mean} = K_c$, the fracture toughness (analogous to drawing equal life curves to the point $\sigma_{mean} = \sigma_f$). For the region $-1 < R_1 < 0$, the lines have a slope of $-1$, so that the intercept on the vertical axis is twice the value as for $R = 0$. In this way it is represented that the inclusion of compression loading has little or no effect on crack growth rate. To construct such a diagram, therefore, the only data needed are from the Paris-Erdogan relation for standard $R = 0$ tests, and the fracture toughness. Reasonably good correlations between predictions and experimental data taken from independent sources is demonstrated in Fig. 9.13 for three alloys.

*The Threshold Stress Intensity.* Below a minimum value of stress intensity, crack growth does not occur at all. This minimum is called the *threshold stress-intensity factor*. As seen in Fig. 9.12, this threshold is very sensitive to mean stress-intensity effects (as well as several other effects). This might be expected because, for a given stress-intensity range, the maximum stress intensity is higher, the higher is the value of $R$. Thus, the sensitivity is really a reflection of the maximum stress intensity. The threshold is also very sensitive to environmental effects. This sensitivity is due to a wedging effect of the corrosion product in the nascent crack, as discussed in the next section.

*Environmental Effects.* Hostile environments will, of course, increase the crack growth rate for both chemical and mechanical reasons. If the environment attacks the surface chemically, the corrosion product will usually not be as strong as the base material, thus accelerating the fracture propagation. This is especially true for the crack tip where the local stress is very high. The corrosion product also forms a mechanical wedge upon release of load so that the stress at the crack tip can remain tensile (or perhaps even increase) even during the unloading. Figure 9.14 shows a comparison of the crack growth rates for a steel in air and in a 3% NaCl solution. In air all loading frequencies give essentially the same crack growth rate curve. In the sodium chloride solution, however, very large differences occur according to loading frequency. The lower the frequency, the longer is the time available for chemical attack to occur, and the higher is the crack growth rate. Similarly, if the amount of chemical reaction is increased by raising the temperature, the crack growth rate increases, as seen in Fig. 9.15, taken from Ref 9.25 using data from Ref 9.30. It is extremely important, therefore, in considering crack growth rate to account for chemical effects, which include the identification of the environment, temperature, and loading rate effects. The pattern of loading may also be important if it allows for longer time at stress to influence the crack growth.

*Typical Crack Growth Equations for Common Steels.* Although Paris and Erdogan suggested a universal value of 4 for the value of $n$ in the crack growth equation based on their limited early tests, subsequent more-extensive studies have shown that this exponent is quite material dependent. Barsom and coworkers have, for ex-

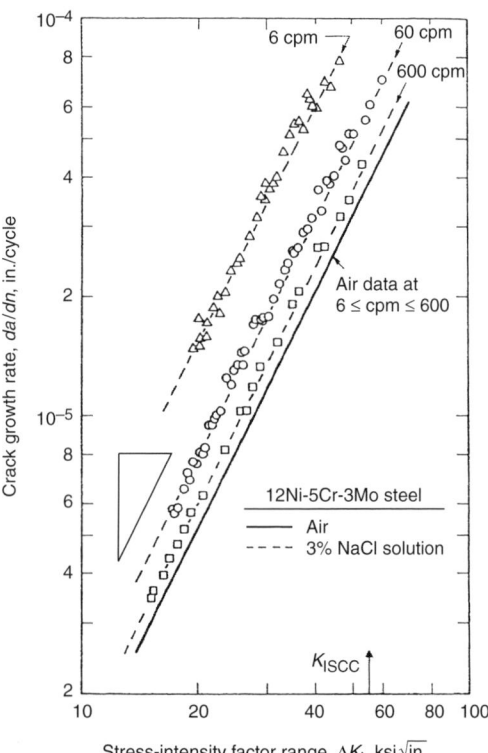

**Fig. 9.14** Detrimental influence of aggressive environment on crack growth rate. Source: Ref 9.1

ample, studied various classes of steels. Even within a given class, there was considerable scatter. However, scatter bands could best be drawn to separate the classes rather than including all steels together. Figure 9.16 shows Barsom's results (Ref 9.31) separating austenitic, martensitic, and ferritic-pearlitic steels as individual classes. Note that the austenitic steels fall in a very narrow band, while the other two steels exhibit a data spread of about 2 in crack growth rate for a given stress intensity. Equations 9.6 to 9.8 give the highest crack growth rates (i.e., the upper line of the scatter bands) for each of these classes.

For martensitic steels:

$$\frac{da}{dN} = 1.35 \times 10^{-10}(\Delta K)^{2.25}$$

(in units of m/cycle and MPa$\sqrt{m}$)   (Eq 9.6)

$$\frac{da}{dN} = 6.6 \times 10^{-9}(\Delta K)^{2.25}$$

(in units of in./cycle and ksi$\sqrt{in.}$)

For ferritic-pearlitic steels:

$$\frac{da}{dN} = 6.9 \times 10^{-12}(\Delta K)^{3.0}$$

(in units of m/cycle and MPa$\sqrt{m}$)   (Eq 9.7)

$$\frac{da}{dN} = 3.6 \times 10^{-10}(\Delta K)^{3.0}$$

(in units of in./cycle and ksi$\sqrt{in.}$)

For austenitic steels:

$$\frac{da}{dN} = 5.6 \times 10^{-12}(\Delta K)^{3.25}$$

(in units of m/cycle and MPa$\sqrt{m}$)   (Eq 9.8)

$$\frac{da}{dN} = 3.0 \times 10^{-10}(\Delta K)^{3.25}$$

(in units of in./cycle and ksi$\sqrt{in.}$)

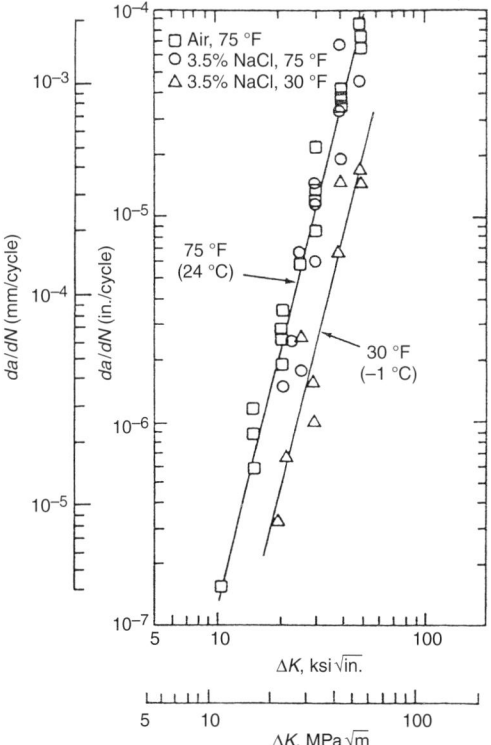

**Fig. 9.15** Comparison of fatigue crack growth rate in 3.5% aqueous NaCl and air at 30 and 75 °F. Data from Ref 9.30

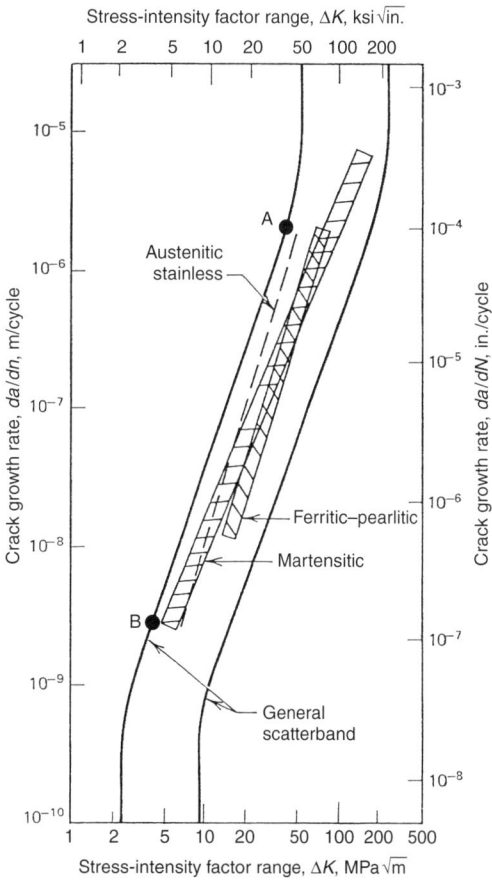

**Fig. 9.16** Superposition of Barsom's scatter bands (Ref 9.31) on the general fatigue crack growth scatter band for steels. Figure from Ref 9.26

Note that exponent values of 2.25, 3, and 3.25 are best for use with martensitic, ferritic-pearlitic, and austenitic steels, respectively. Of course, they all fall within a broader scatter band, and for all steels the upper line of this band could be used for conservative estimates. The equation for the line *AB* in Fig. 9.16 is:

$$\frac{da}{dN} = 5.5 \times 10^{-11}(\Delta K)^{2.8}$$

(in units of m/cycle and MPa$\sqrt{m}$) (Eq 9.9)

$$\frac{da}{dN} = 3.0 \times 10^{-9}(\Delta K)^{2.8}$$

(in units of in./cycle and ksi$\sqrt{in.}$)

However, the overestimate of crack growth rate can be more than 10 times for individual cases. Even within a given class the differences among specific materials can still be a factor of 4 times or more, as seen in Fig. 9.17 for various martensitic steels.

## Sequential Loading Effects

For continuous constant amplitude loading the crack growth rate that develops is as discussed previously. If amplitude changes, however, additional factors must be considered to determine how the crack growth rate is affected. Let us consider first the effect of periodic overloads.

Figure 9.18(a) shows an example loading history that includes an occasional tensile overload (points A and B) of a precracked specimen. Resultant crack growth response is displayed schematically in Fig. 9.18(b), where curve *OA* shows how the crack grows prior to the first overload, *OAA'* would be the crack growth if the loading continued at the same amplitude. At time $t_1$, a higher tensile amplitude load is applied for one or a few cycles. (This could be analogous to an occasional gust load encountered in an aircraft structure.) If higher amplitude loading is applied for only one cycle before returning to the original amplitude the subsequent crack growth rate is retarded, rather than increased as might be expected. In fact, total arrest may occur (that is,

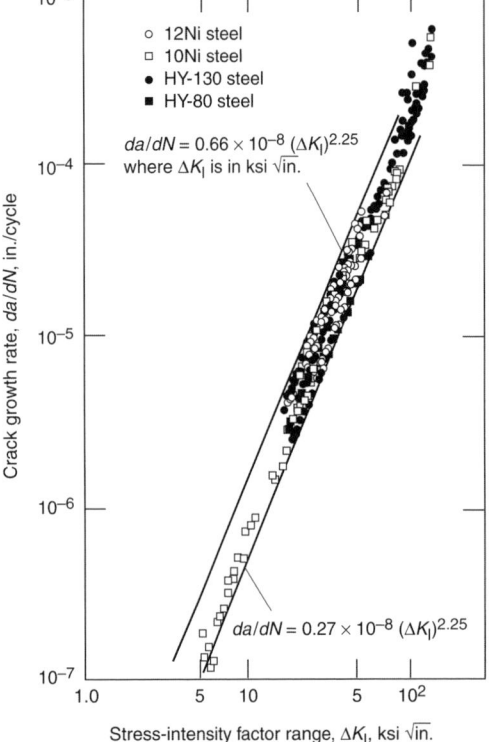

**Fig. 9.17** Summary of fatigue crack growth rate data for martensitic steels. Data from Ref 9.31; figure from Ref 9.26

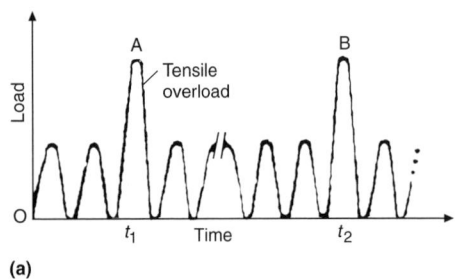

(a)

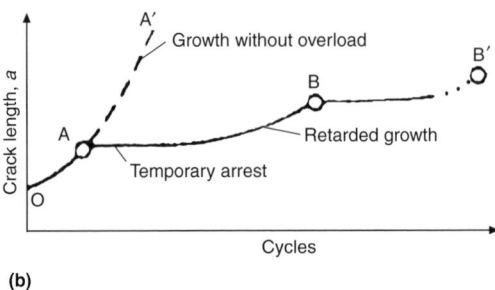

(b)

**Fig. 9.18** Crack growth retardation effects of periodic overloads (Ref 9.27). (a) Loading. (b) Crack growth

the crack doesn't grow at all) for a period of lower amplitude loadings. Eventually however, the crack starts growing again, and if the lower loading were indefinitely resumed, it would reach the growth rate associated with this loading with proper accounting for the extended crack size. If a second overload is applied at $t_2$, retardation ($BB'$) is again observed, and so on, for later overloads until the crack size reaches a critical dimension, and the specimen or structure fractures. Various explanations have been proposed for this phenomenon. McMillan and Pelloux (Ref 9.32) and Schijve (Ref 9.33) have suggested three possible factors: crack tip blunting due to the overload, strain hardening of the metal ahead of the crack tip, and residual compressive stresses in the metal ahead of the crack tip. Let us now consider some models that have been used for quantitative treatment of the problem of crack growth retardation due to overload. These models provide physical insight into the phenomenon.

**The Wheeler Model.** Wheeler (Ref 9.34) developed a model whereby the retardation is due to the large plastic zone ahead of the crack that is induced by the overload. Figure 9.19 illustrates the parameters as they are used in the model. The region $O$ is the plastic zone ahead of the growing crack for the steady loading at the lower level. The application of the higher load induces a larger plastic zone $O'$. When the loading returns to the lower level, the cracks, to progress, must first pass through $O'$. Wheeler proposed that the crack growth during this stage is retarded according to the equation:

$$\left(\frac{da}{dN}\right)_{i,ret} = (C_p)_i [C(\Delta K)_i^n] \qquad \text{(Eq 9.10)}$$

where $(C_p)_i$ is a retardation parameter for cycle i. This parameter is assumed to be in the form:

$$(C_p)_i = \left[\frac{(ry)_i}{(a_p - a)^m}\right] \qquad \text{(Eq 9.11)}$$

where: $(ry)_i =$ "current" plane-strain yield zone size $= (1/4)\sqrt{2\pi}(K_{max}/S_y)_i^2$; $(a_p - a) =$ distance from crack tip at cycle $i$ to the boundary of the yield zone caused by the last tensile overload in a loading sequence; and $m =$ a "shaping" factor chosen to force the data to fit the model

The application of this model to a complex loading history can become quite involved, but the method has been successfully applied. However, the method has two major drawbacks. First, it requires the analyst to determine by experiment the shaping factor $m$. While this is possible in a dedicated study of a particular problem, it is impractical for a designer to divert his or her attention when studying a variety of materials under a variety of loading conditions. Secondly, the model predicts essentially that the retardation is a maximum immediately after the overload and progresses monotonically to zero until the crack passes through the plastic zone produced by the overload. Actually, a much more complex pattern of retardation has been observed, which is best explained by the "crack closure" concept described in a subsequent section in connection with the Elber model (Ref 9.35).

**The Willenborg Model.** Willenborg et al. (Ref 9.36) also hypothesize that the crack retardation is due to the larger plastic zone produced by the overload. However, the effect on the

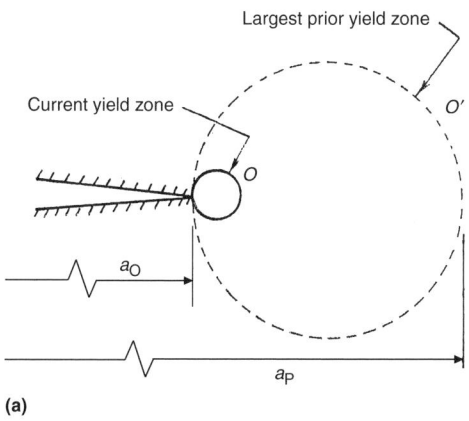
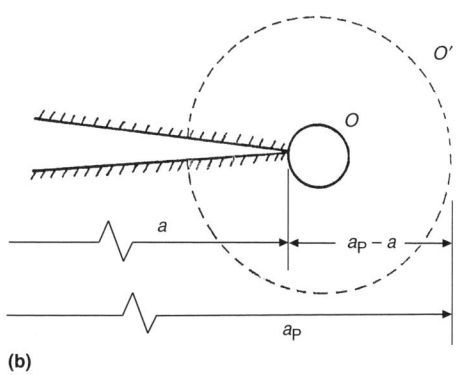

**Fig. 9.19** Relation of crack-tip yield zones in the Wheeler crack growth retardation model (Ref 9.34). (a) Initial condition at overload. (b) Progress of crack through plastic zone of overload

crack growth is presumed to be due to a residual stress field so that the growth is governed by the equations of mean stress-intensity effects discussed earlier. These effects are complex but can be dealt with through appropriate numerical computer programming. Thus, the method overcomes the disadvantage of the Wheeler model in avoiding the need for an experimentally determined shaping factor, but it requires the use of a complex computer program for its analyses. Furthermore, it also predicts a retardation pattern that is inconsistent with those typically observed experimentally.

**The Elber Crack Closure Model.** The introduction of the crack-closure concept by Elber in 1968 (Ref 9.35) provided an insight that has been very useful in interpreting crack growth understanding, especially the effects of nonsteady loading. His observations can be explained in connection with Fig. 9.20. This figure shows the pattern of opening and closing of a crack during tensile loading from zero to some maximum stress intensity. The normal concept would hold that the crack is closed during zero load, opens completely when the load is a maximum, and closes completely again when the load is removed. However, because of the plastic flow that occurs at maximum loading, the tip of the crack closes before the load is completely removed, as shown in Fig. 9.20(b). During the remainder of the unloading the crack closes further, but there is little effect at the tip of crack where the action takes place in relation to crack growth, as shown in Fig. 9.20(c). Similarly, during reloading, some of the range of stress-intensity factor has relatively little effect on the crack tip; the faces of the crack simply unload. For the remainder of the loading, the action at the crack tip is affected. In effect, the range that governs the crack growth is $AB = DA$ rather than $AC$ in Fig. 9.20. Elber (Ref 9.34) expressed the concept in the equation:

$$\frac{da}{dN} = C(\Delta K_{eff})^m \qquad (Eq\ 9.12)$$

where $(\Delta K_{eff})$ is the effective stress-intensity range (between the point where the crack opens and when it reaches its maximum value). Elber tested 2024-T3 aluminum to determine how $\Delta K_{eff}$ varied with stress ratio. His results are illustrated schematically in Fig. 9.21. For $R = 0$ the effective stress-intensity range is about 50% of the total range, whereas at $R = 0.5$, the effective range is 70% of the total range.

**Application of the Crack Closure Concept to Complex Loading.** The crack closure concept has been able to explain some behaviors better than other methodologies, and for this reason has gained a considerable following. One such example is illustrated in the upper portion of Fig. 9.22 for a low-high-low loading sequence. Here we have a low $\Delta K$ applied until a stable crack growth rate is achieved; a higher $\Delta K$ is then applied until another stable crack growth rate is achieved; and finally, the stress intensity is reduced to the original lower level. The dashed curve in the figure shows the expected pattern of when crack closure occurs within each cycle. The distance from the dashed curve to the

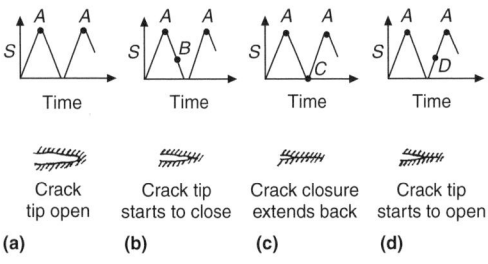

**Fig. 9.20** Illustration of crack closure. Source: Ref 9.25

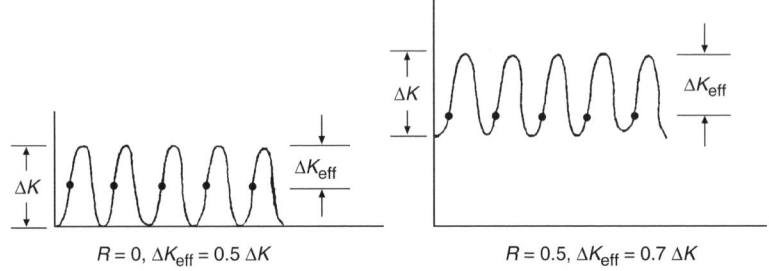

**Fig. 9.21** Schematic illustration of the results of Elber's closure model. Source: Ref 9.25

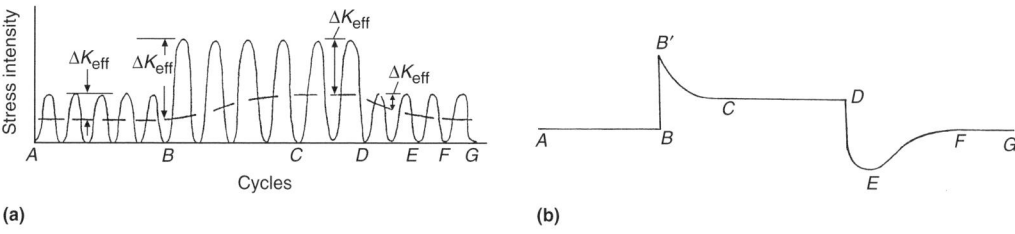

**Fig. 9.22** Illustration of effects of crack closure. (a) Applied sequence of low-high-low loading. Source: Ref 9.25. (b) Typical crack growth rate response

peak of the loading represents the region extent of crack opening. The changes of range for crack opening do not occur instantaneously, but rather change gradually as shown schematically in the figure. This is a result of the gradual buildup and reduction of residual stress in the plastic zone. Note, therefore, that in the region BC the effective $\Delta K$ is greater than the effective $\Delta K$ associated with the higher loading, and in the region DE the effective $\Delta K$ is lower than the effective $\Delta K$ associated with the smaller $\Delta K$ of the loading.

The observed crack growth rate pattern is shown schematically in the right half of the figure. In the region $AB$ the crack growth rate is that for the lower of the loadings. In the early cycles of application of the higher load, when the effective $\Delta K$ is actually higher than the steady-state $\Delta K$ in the larger loading, the crack growth rate jumps to a higher value shown by $B'C$, but eventually settles down to the steady-state crack growth rate of the larger loading, $CD$. The most interesting feature of the behavior is seen in the region $DE$. Here the crack growth rate drops dramatically because the effective $\Delta K$ is so small. However, during the region $EF$ it builds back up to the value associated with the lower of the two loading levels, and finally, during $FG$ it reaches the same crack growth rate as $AB$ associated with the initial lower loading level.

The behavior predicted by the crack closure model is in remarkably close agreement with experimental observations, contrary to the predictions of other models, such as those of Wheeler and Willenborg. Of course, the success of the crack closure model to indicate qualitative behavior does not mean that it can always be used quantitatively to predict behavior under various loadings because quantitative predictions of crack closure are not simple. But because crack closure can be observed experimentally, as can crack growth rate, this methodology has served as a very useful tool for studying general behavior.

In Ref 9.37, von Euw, Hertzberg, and Roberts have studied the crack retardation pattern during repeated loading immediately following an overload. The crack growth retardation pattern is shown in Fig. 9.23. It is consistent with the expectations of the crack closure model. If the overload is between 2.0 and 3.0 times the repeated lower loading level total crack arrest can occur (if the overload does not cause immediate fracture). Table 9.4 summarizes the overload (OL) results of investigations on the crack arrest (CA) behavior of several materials. The range of values shown for the variety of materials is small.

A large number of other investigations have been conducted to study the effects of complex loading patterns, including both tension and

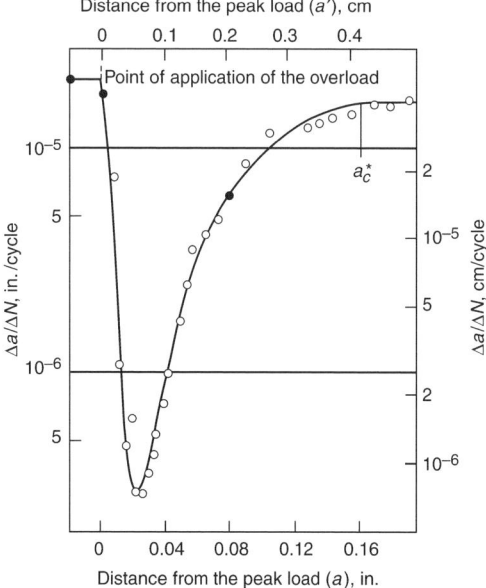

**Fig. 9.23** Crack growth rate retardation pattern after an overload. Source: Ref 9.37

compression. Reference 9.27 provides a good summary of the results and contains useful listings of publications on this subject.

## Crack Initiation, Crack Propagation, and Fracture in Circular Specimens

Most crack propagation and fracture studies are conducted on rectangular specimens that lend themselves to tracking of crack growth and to easy calculation of fracture toughness, i.e., the stress-intensity factor at fracture. On the other hand, fatigue crack initiation and total fatigue life have traditionally been studied with circular cross-section specimens. Circular specimens, however, are inherently ill-suited for tracking crack growth for two obvious reasons. Circular cross sections are not amenable to easy tracking of crack depths and lengths, and they would require prohibitively large diameter sections to accommodate the study of large cracks. Yet many engineering parts have circular sections, and direct study of this geometry is necessary for practical applications.

Two studies by Manson and coworkers (Ref 9.38, 9.39) have been conducted on circular specimens, and it is appropriate to review them at this juncture. One was conducted on ductile materials, and the aim was primarily to determine when the crack starts at the root of a surface stress concentration, and how many cycles are consumed in the crack propagation phase. The other study also involved the initiation and crack propagation associated with a surface stress concentration, but the materials involved had low fracture toughness so that the crack was only able to penetrate partially through the cross section before reaching a fracture condition based on the concept of plane-strain fracture toughness. These efforts provided a unique opportunity to study the relative roles of crack initiation and crack growth (qualitatively) over a range of ductile to brittle behaving materials.

**Table 9.4  Tensile overload ratios causing crack arrest**

| Metal | $(K_{OL}/K_{CA})$ for arrest |
|---|---|
| Ti-6Al-4V titanium | 2.7 |
| 1020 cold rolled steel | 2.5 |
| Austenitic manganese steel | 2.3 |
| 2024-T3 aluminum | 2.0–2.5 |
| 7075-T6 aluminum | 2.3–2.5 |
| 4340 steel | 2.4 (estimated) |

Source: Ref 9.27

**Study of Ductile Materials.** In Ref 9.38 Manson and Hirschberg studied 7075-T6 aluminum and annealed 4340 steel. The specimen used is shown schematically in Fig. 9.24. It is a standard 0.25 in. diameter hourglass specimen into which two diametrically opposed notches were machined. The two notches were used in the experiments so that the exact locations of crack initiation could be known and carefully observed visually. Specimens were machined with either of two notch root radii, 0.025 or 0.008 in., while a 0.010 in. root depth was held constant for all. The importance of the selected depth value will become obvious later in the discussion of this program. The intent was to create two stress concentrations, one of them approximately 2.0 and the other approximately 3.0, respectively, based on elastic calculations. To determine the elastic stress concentration, a computation was made assuming that a groove had been machined uniformly around the circumference of the specimen, yet retaining the cross-sectional shape shown in the figure. The effective fatigue stress concentration was also checked by conducting long-life fatigue tests on notched and smooth specimens at different external loads. The stress concentration factors are also shown in Table 9.5.

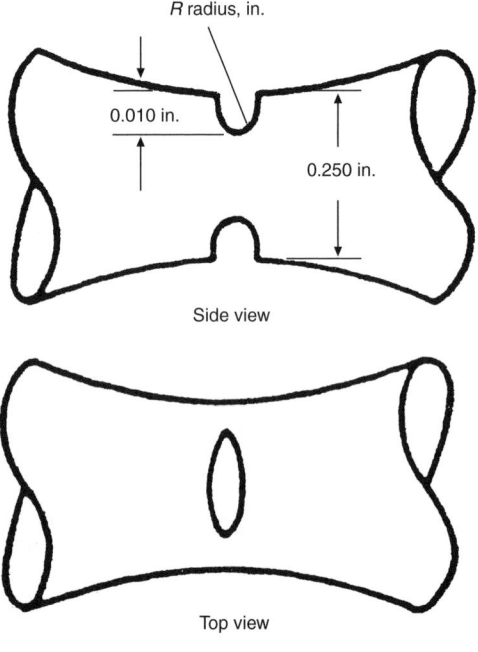

**Fig. 9.24** Test section of notched fatigue specimen used by Manson and Hirschberg. Source: Ref 9.38

*Calculation of Surface Strain.* Both the Neuber (Ref 9.40) and the Stowell, Hardrath, and Ohman (see, for example, Ref 9.41) approaches were used to determine the surface strain at the root of the notch. Because the results obtained were nearly identical, only the Neuber results are shown.

*Crack Initiation.* The early stages of fatigue constitute a period of complex microstructural events involving atomic rearrangement through dislocation motion. It is an intriguing subject that has attracted many highly qualified investigators. Volumes of theories have been proposed to explain fatigue behavior at this level of observation. However, for day-to-day design activities, engineers must generally resort to readily available equations based on macroscopic variables such as stress and strength, strain and ductility, readily identifiable cracks, and cyclic lives associated with major events of initiation and propagation. To help fulfill this practical need, a simplified formula has been proposed by Manson in Ref 9.42:

$$\frac{N_o}{N_f} = 1 - 2.5(N_f)^{1/3} \quad \text{(Eq 9.13)}$$

where $N_o$ is the number of cycles to initiate a crack at a given surface strain and $N_f$ is the fatigue life of a specimen subjected continuously to that strain until failure. This relation was obtained by fitting the results for many materials obtained in various ways, but mostly by counting striations back to the origin of a crack. Figure 9.25 shows the relations between the crack initiation lives predicted by this formula and the data used to generate the equation. In another treatment (Ref 9.39) an alternate equation was proposed:

$$\frac{N_o}{N_f} = 1 - 4(N_f)^{-0.4} \quad \text{(Eq 9.14)}$$

which also fit the data well. However, in connection with the latter equation it was proposed that it be used only for $N_f > 730$ cycles. In the analysis of the notch program under discussion here, Eq 9.14 was used, but the two equations would have given the same results because no lives less than 730 cycles were involved.

It is important to clarify what is meant by the term "crack initiation" because the earliest defects could be considered as cracks, and the term can be very ambiguous. For this purpose we introduced the concept of an "engineering size crack" by which we meant a crack that could be detected with the naked eye (assuming that the viewer had normal vision); a reasonable size for such a crack would be one whose surface length is about 0.010 in. (which is the dimension viewed). The depth of the crack would only be about 0.003 in. because we have observed that initial cracks are about ⅓ as deep as their surface length. Thus, for this study, if we calculated the surface strain, and knew the strain-life characteristics for smooth specimens, we could calculate the number of cycles required to develop a crack of 0.003 in. deep according to Eq 9.14.

*Crack Propagation.* In treating crack growth in a part with a notch, an important factor that must be considered is the effective crack depth and the nominally applied strain perceived by the material immediately in front of the developing crack. As an example, consider our case of a 0.010 in. deep notch with a crack depth from the root of 0.003 in. Imagine that material were added back to the faces of the notch such that a side view was made to look the same as a 0.013 in. deep crack in an unnotched part, both of which were subject to the same nominal remote strain range. This is shown schematically in Fig. 9.26. Because the original faces of the notch car-

**Table 9.5  Concentration factors for notched specimen**

| Material | Radius, R, in. | $K_t$ elastic | $K_f$ fatigue |
|---|---|---|---|
| 7075-T6 aluminum | 0.203 (0.008) | 3.0 | 2.9 |
|  | 0.635 (0.025) | 2.0 | 1.7 |
| 4340 steel (annealed) | 0.203 (0.008) | 3.0 | 2.0 |
|  | 0.635 (0.025) | 2.0 | 2.0 |

Source: Ref 9.38

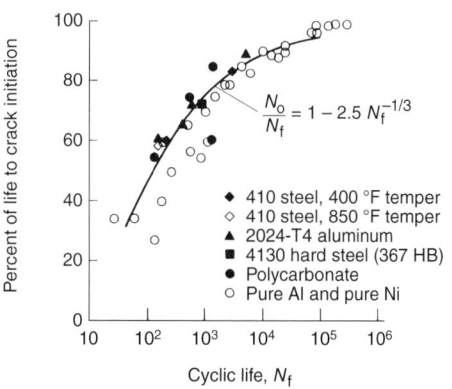

**Fig. 9.25** Proposal for representing crack initiation lives. Source: Ref 9.42.

ried no stress across them, the added material would not be expected to have appreciable effect on the stress fields ahead of the 0.003 in. deep crack emanating from the original notch root. Hence, the material ahead of the crack could be expected to respond nearly the same to either crack; a 0.003 in. crack from a 0.010 in. deep notch or a 0.013 in. crack from a smooth unnotched surface. This reasoning breaks down if we consider no crack at the root of the notch, because the stress and strain field at the uncracked root of the notch is known to be substantially different from the stress field at the tip of a 0.010 in. deep crack from an initially smooth surface. However, if it is assumed that the actual depth of the crack portion is at least 0.003 in. and has been caused by the concentrated strain at the notch root, then the crack growth beyond a total depth of 0.013 in. should be reasonably independent of whether the 0.010 in. deep notch was present. Thus, we can formulate a simple, yet quite useful procedure for partitioning the crack initiation and crack propagation stages of the total fatigue life for notched geometries.

To provide the basic information required we used the analysis developed by Manson in Ref 9.39. Figure 9.27 shows the relevant figure. What is shown here is the crack length in a ¼ in. diameter specimen versus the cycle ratio after the crack has once started. The figure does not concern itself with the cycles required to initiate the very first tiny crack due to metallurgical processes that occur during early plastic flow, nor is this information necessary for the present analysis.

This figure derives from simple analyses made in Ref 9.39, which combined two features: the assumption of a crack growth law consistent with then existing technology, and crack growth measurements made by striation counting on a number of specimens made of ductile materials.

Starting with a crack growth equation of the form:

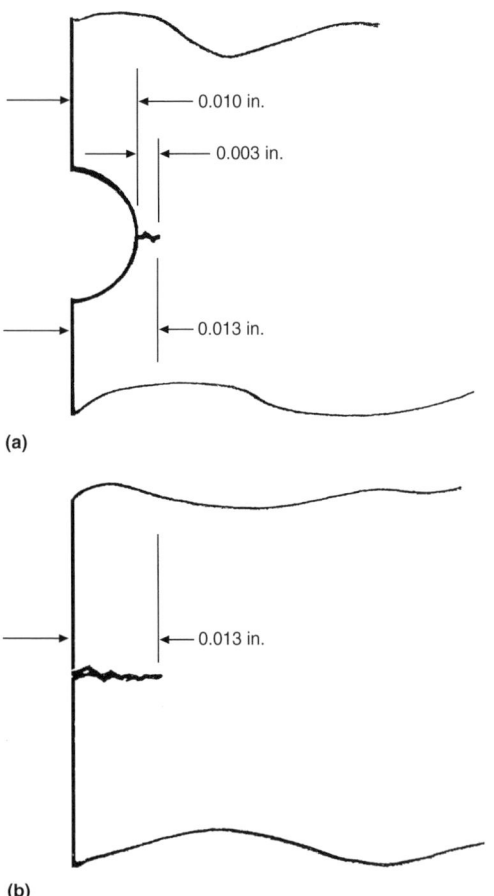

**Fig. 9.26** Crack initiation from (a) 0.010 in. notch depth to total depth of 0.013 in. compared with (b) crack depth of 0.013 in. without notch

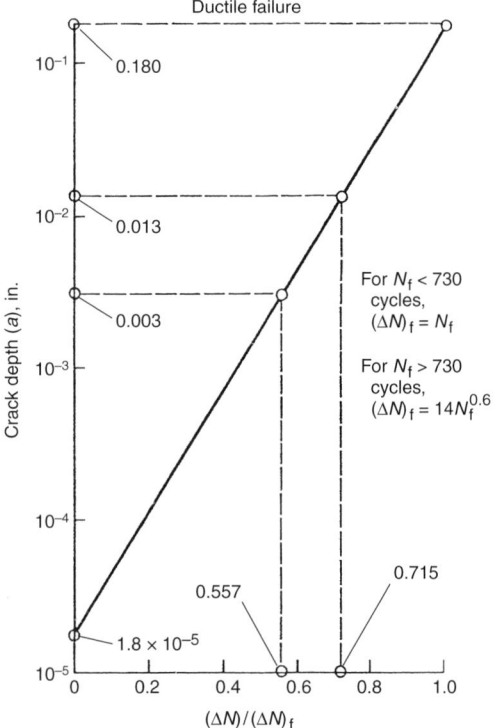

**Fig. 9.27** Manson's hypothesized crack growth relation. Source: Ref 9.39

$$\frac{da}{dN} = f(\Delta\varepsilon_p)(a) \quad \text{(Eq 9.15)}$$

where $a$ is crack depth, $N$ is the number of cycles required to propagate the crack, and $f(\Delta\varepsilon_p)$ is a function of plastic strain range. The equation is integrated and manipulated to the form:

$$\frac{\ln\left(\frac{a}{a_o}\right)}{\ln\left(\frac{a_f}{a_o}\right)} = \left(\frac{(N - N_o)}{(N_f - N_o)}\right) = \frac{\Delta N}{(\Delta N)} \quad \text{(Eq 9.16)}$$

where $a$ is the crack depth at any number of cycles $N$, $a_o$ is the equivalent crack size at $N_o$, $a_f$ is the crack size at fracture, and $N_f$ is the total life at failure. For purposes of the simplified analysis we can take $a_o = 1.8 \times 10^{-5}$ in. and $a_f = 0.18$ in. The hypothetical crack size at the start is, of course, academic, and derives from extrapolation of striation measurements, and the final crack length of 0.180 in. refers to a specimen of 0.25 in. diameter. Ideally, therefore, $a_f$ should be 0.25; however, the striation spacing is very coarse near the point of fracture, so the 0.18 in. length derived from extrapolation of the striation measurements is satisfactory, considering the logarithmic life scale involved.

Thus, using Eq 9.16 and selected values of $a$, we can determine the fraction of the fatigue life required to grow the crack to any chosen size. For the problem at hand the two important crack penetrations are 0.003 and 0.013 in. To simplify the discussion we show first in Fig. 9.28 a schematic of the life curves needed to make the calculation. (The curves are shown spread farther apart than actual for particular materials in order to be able to explain the procedure more clearly.) Curve $AA'$ is the strain-life relation of the test material, $BB'$ is the life to a crack length of 0.013 in., and $CC'$ the life to a crack length of 0.003 in. These are obtained by multiplying the life $N_f$ by 0.715 and 0.557, respectively, at each strain level, as dictated by Fig. 9.28.

Consider now a test in which the nominal strain $\Delta\varepsilon$ is applied to the test section. A Neuber calculation determines the actual strain concentration at the root of the notch $k_N$, so the actual strain at the root of the notch is $k_N \varepsilon_{app}$. The material then cycles until a detectable 0.003 in. deep crack develops (as measured by a 0.010 in. surface crack length, detected in our experiments by two optical microscopes focused at the roots of the two diametrical notches). The number of cycles in this phase is $PQ$ in Fig. 9.28. For purposes of later display we identify this number as crack initiation. The distance $RS$, between the curves for 0.013 in. crack and the failure curve, is designated as the crack-propagation phase.

*Test Results.* Figure 9.29 shows the cyclic stress-strain curves for the two materials tested in this program. When the aforementioned procedure was applied to these two materials for the two notch geometries represented, the results are as shown in Fig. 9.30. Quite good agreement is seen, considering all the approximations involved in the calculations and experiments. Figure 9.31 shows the comparison between prediction and experiment for the crack propagation phase.

Note that there is very little difference in the crack propagation lives whether the stress concentration factor is 2 or 3, while the initiation lives are significantly different. This is to be expected because the elastic stress concentration significantly affects the surface strain at the root of the notch, which governs crack initiation. The crack propagation period starts, however, once a natural crack has been developed beyond the notch. It makes relatively little difference how the natural crack was started because the geometry is about the same for either value of $K_t$; only the number of cycles required to establish the condition is different for each value of $K_t$.

This program brought out an important factor regarding the severity of a notch in affecting fatigue life: *while stress concentration will reduce the cycles to "start a crack," it doesn't significantly affect the propagation period.* Therefore, the total fatigue life is not as might be expected on the basis only of the higher strain developed at the root of the notch. In fact, in Ref 9.38 we made two calculations that bring out this point. The results are shown in Fig. 9.32. The material

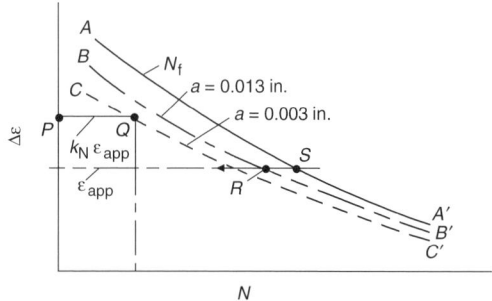

**Fig. 9.28** Schematic of the calculation procedure

analyzed was 7075-T6 aluminum that had been studied in the experimental program at stress concentrations of 2 and 3. Calculations are also shown for $K_t = 5.0$. On the basis of the method described previously, there is practically no difference in life between $K_t = 5.0$ and $K_t = 3.0$. But, on the basis of an approach proposed by R.E. Peterson (Ref 9.43), a prominent authority in the field of fatigue at the time, significant reduction in life can be expected for $K_t = 3.0$ and even greater reduction at $K_t = 5.0$. Peterson's calculations essentially reduce the entire life commensurably with the stress concentration because they do not distinguish between effects of the notch on crack initiation and crack propagation. Because the prediction for $K_t = 3.0$ already falls off the mark compared with the experimental results, it can be expected that the error would be even greater at $K_t = 5.0$.

The reasoning indicated in Fig. 9.32 provides some comfort to the engineer regarding the presence of high surface stress concentrations in engineering structures. In low- and intermediate-cycle fatigue, they are nowhere nearly as damaging as they might appear, because they only affect the crack initiation, not the propagation. However, in high-cycle fatigue, wherein the dominant portion of life is due to crack initiation, the effect of high $K_t$ can remain quite severe.

The aforementioned observation may also provide insight into the behavior of parts with shallow cracks that have been found to grow at vastly different rates than would normally be predicted on fracture mechanics-based crack growth equations for larger cracks. The region of concern is for stress-intensity values below the threshold regime of large crack growth.

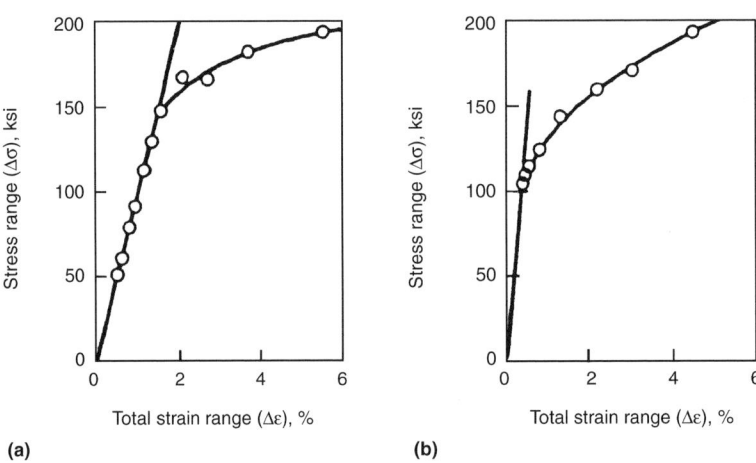

**Fig. 9.29** Cyclic stress-strain curves. (a) 7075-T6 aluminum alloy. (b) Annealed 4340 steel. Source: Ref 9.38

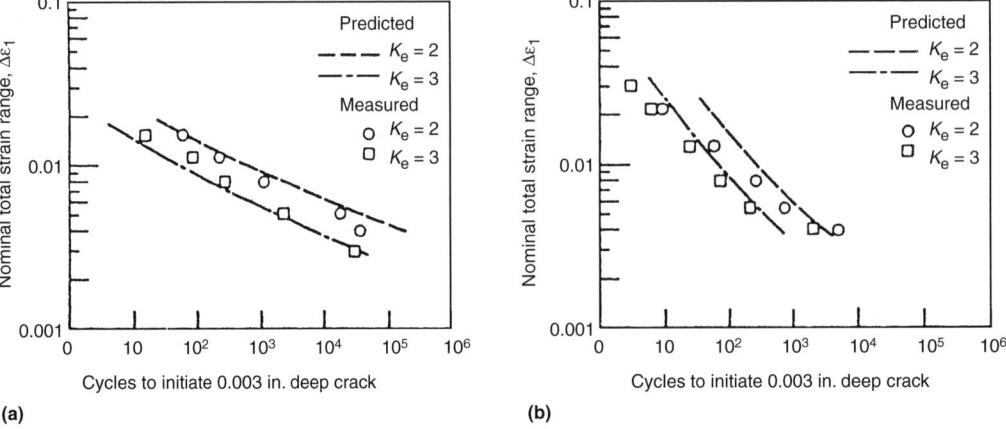

**Fig. 9.30** Crack initiation for notched circular specimens. (a) 7075-T6 aluminum alloy. (b) Annealed 4340 steel. Source: Ref 9.38

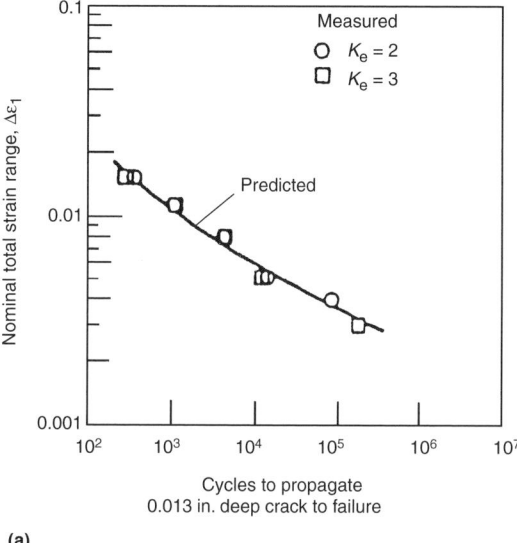
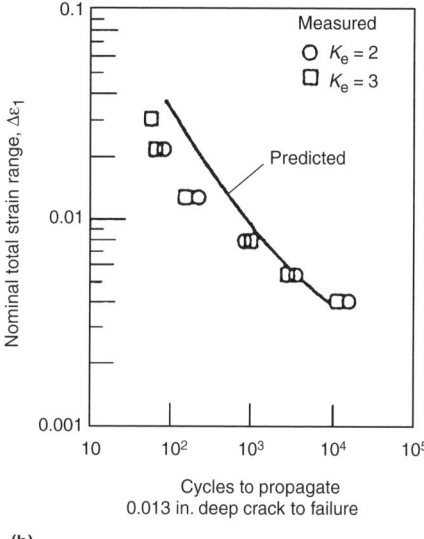

**Fig. 9.31** Crack propagation for notched circular specimens. (a) 7075-T6 aluminum alloy. (b) Annealed 4340 steel. Source: Ref 9.38

A significant literature has developed since the early 1970s that reflects the intensive probing of the cyclic growth resistance of what are known simply as "small cracks." Aside from the obvious problems of accurately defining the dimensions of small cracks and calculating their highly localized stress fields, it must be recognized that the assumptions of homogeneity and isotropy that are at the heart of most stress-intensity formulations are severely violated as the scale of crack sizes approaches the microscopic proportions of metallurgical features. These features with heterogeneous and anisotropically oriented characteristics are comparable in size to the small cracks of interest. More detailed discussion of small crack growth behavior is given in Chapter 10, "Mechanism of Fatigue."

**Study of Quasi-Brittle Materials.** The program on crack growth also involved testing on quasi-brittle materials in order to determine the extent to which fracture toughness affects fatigue life. Before we discuss the results of this program, let us briefly review previous findings regarding the role of fracture toughness on the fatigue of standard fatigue specimens (¼ in. diam). In expressing the Manson-Coffin-Basquin relations of fatigue, no discussion is included for fracture toughness. The universal slopes equation includes tensile strength and ductility but gives no consideration to fracture toughness.

In Ref 9.42 Manson provided an example evaluation of 410 stainless steel tested in two conditions of heat treatment, one to produce a ductile material, the other to produce considerable brittleness. In both cases the material was quenched from 1000 °C (1850 °F), but in one case it was tempered to 200 °C (400 °F) and in the other to 450 °C (850 °F). Both conditions resulted in approximately the same tensile strength and reduction of area, but the brittleness of the two conditions differed considerably. As shown in Fig. 9.33 the two conditions of temper

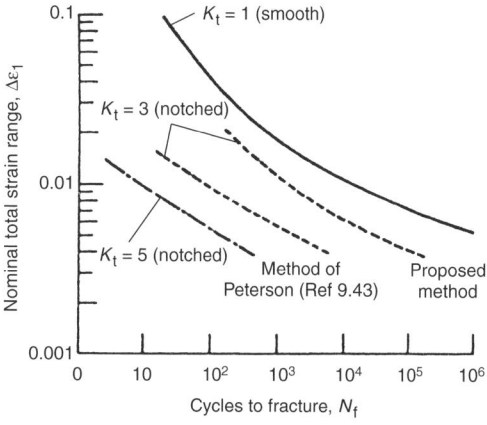

**Fig. 9.32** Comparison of methods for predicting cyclic life of notched specimens for 7075-T6 aluminum alloy. Source: Ref 9.38

produced about the same hardness, tensile strength, and ductility, but the Izod* impact energy absorption was greatly different for the two tempering conditions. No fracture toughness tests were made, but it was clear that fracture toughness would be greatly different for the two materials. The fatigue tests, also shown in Fig. 9.33, were almost identical, however. It was on the basis of these tests that the omission of fracture toughness was justified in the development of the universal slopes equations. This conclusion might have been surprising to some observers in the fatigue field. After all, is it not the final cycle of a fatigue test that is the one that terminates in fracture due to the development of a crack of critical size for the stress range to which the material is subjected. Why, then, should fracture toughness not enter into the representation of the fatigue behavior? The primary reason is that in the testing of small, unnotched specimens most of the fatigue life is consumed in initiating a crack and in the early stages of propagation. The later development of the crack from a moderate size to critical size takes only a small fraction of the total fatigue life because the rate of crack propagation per cycle at this stage is relatively high. Thus, the fraction of the total life involved in this stage is quite small and variations in this fraction may be indistinguishable within the scatter of the fatigue life data. In addition, the size of the ¼ in. diameter fatigue specimen does not permit much variation in critical crack sizes. The toughness of most engineering materials is sufficiently high that failure of such a specimen at a net section stress less than the yield strength cannot occur. Because the conditions for a linear elastic fracture mechanics toughness test are rarely realized in fatigue testing of small, smooth specimens, it should occasion no surprise that the fatigue life of a small specimen can be predicted without reference to the fracture toughness of the material.

In order to determine wherein the importance of fracture toughness lies, we must, therefore, resort to materials of low toughness so that brittle fracture will be manifest even in small specimens. We must also start with a notched specimen, to promote the process of crack initiation early in the cyclic loading process. As a vehicle

---

*Izod impact energy absorption is a measure of the notched-bar impact resistance of a material. A sharply notched cantilever bar is rigidly clamped at one end and an elevated pendulum is released, striking the exposed end, causing fracture to occur at the notched midsection. The kinetic energy of the pendulum is reduced by the fracture energy of the specimen. Izod testing is largely in disuse, having been replaced with its sister impact energy testing technique, Charpy (Ref 9.44). Notched-bar impact testing is still in use today, largely because of long-standing acceptance criterion written into manufacturing specifications. The testing procedure is inexpensive and gives quick results, but is giving way to evaluations of fracture toughness based on the much sounder discipline of fracture mechanics.

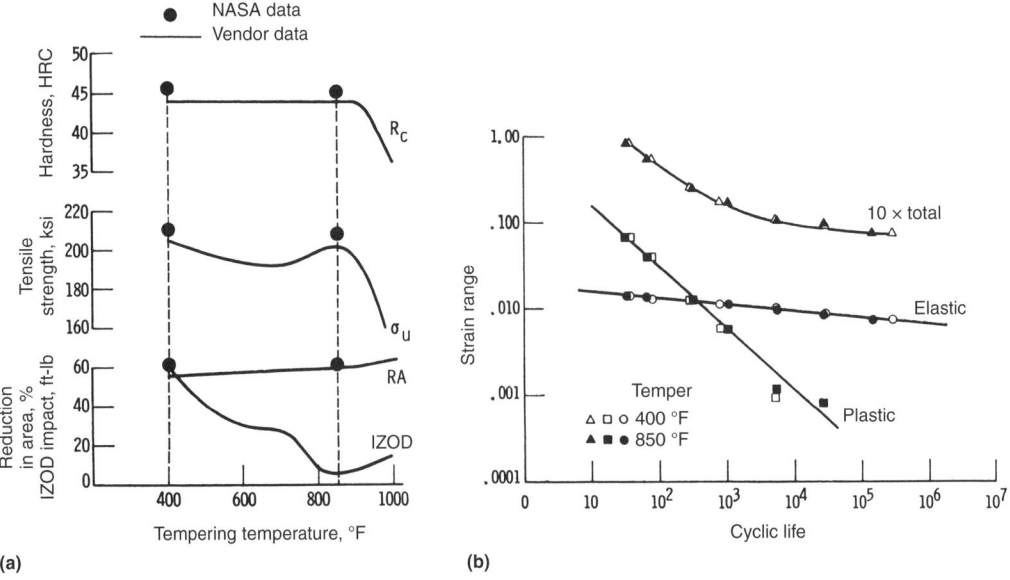

**Fig. 9.33** Effect of tempering temperature on 410 stainless steel on tensile properties and on fatigue properties when the tempering temperature was 400 and 850 °F. (a) Tensile properties. (b) Fatigue properties. Source: Ref 9.42

for discussing this interface between fatigue and fracture, we therefore choose to treat the behavior of notched specimens of quasi-brittle materials that have relatively low fracture toughness. We can then use the fatigue crack growth model of Fig. 9.27 to track the crack growth to the point of fracture.

Before presenting the results of the program that was conducted on quasi-brittle materials it should be pointed out that completely brittle materials, such as certain ceramics, would not have been appropriate for this purpose. Such materials tend not to involve enough crack growth to be able to reveal interesting results. The quasi-brittle materials studied do have some, but limited, ductility and can serve to study how a crack can grow before a fracture condition is reached and how properly to define the fracture condition. The materials studied were specially prepared to be suitable for this purpose. Unfortunately, there are few studies of this type available, and therefore the data are limited. The conclusions reached must therefore be regarded as tentative.

The material tested was AISI 52100 steel, very useful for applications requiring high hardness, such as rolling-contact bearings. Specimens of the material were machined to the notch geometry of Fig. 9.24, except that only one notch was used, whereas in the study of ductile materials two notches were used. Here we did not view the crack initiation and we wanted to avoid the complication of the possibility of growing two simultaneous cracks. The specimens were hardened and tempered at various temperatures to achieve several fracture toughness values $K_{Ic}$. The opening-mode fracture toughness value for each temper was measured using the techniques of Ref 9.13 on specimens of ½ in. square section in which fatigue cracks had been introduced. These fracture toughness values varied from 17 to 35 ksi$\sqrt{\text{in.}}$ as will be discussed in connection with Fig. 9.34. The round notched specimens of similarly heat treated materials were then fatigue tested at various strain levels. Observations were made of the stress amplitude σ, the number of cycles to failure, and the depth of penetration of the crack at failure. Figure 9.35 shows a typical fracture surface observed. Here, the 315 °C (600 °F) temper failed at 10,030 cycles, and the crack size was measured to the point where a clear transition occurred between the smooth growth of the crack and the sudden brittle fracture. Crack depth included the 0.010 in. dimension of the machined notch. The results are plotted in Fig. 9.34. Linear elastic fracture mechanics theory would predict that a logarithmic plot of $a_f$ versus $K_{Ic}/\sigma$ should produce a straight line of slope 2. The data lent themselves readily to such a correlation, the equation of the straight line of the figure being:

$$a'_f = \left(\frac{K_{Ic}}{\sigma}\right)^2 \text{ or } K_{Ic} = \sigma\sqrt{a'_f} \qquad \text{(Eq 9.17)}$$

where $a'_f$ is the crack depth at fracture for the quasi-brittle material (in contrast to the designation $a_f$ for the failure crack depth for com-

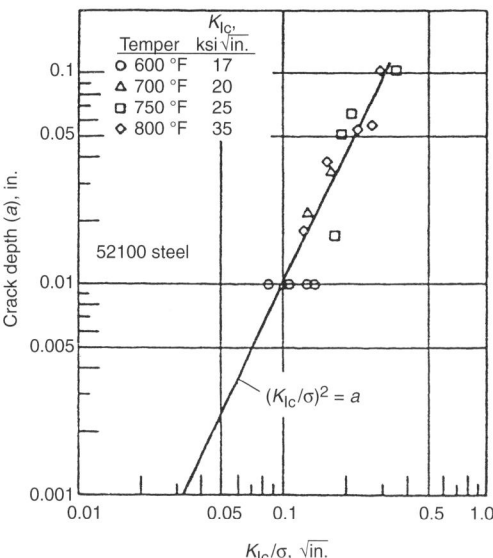

**Fig. 9.34** Fracture relation. Source: Ref 9.39

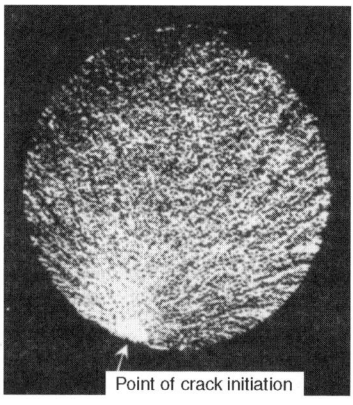

**Fig. 9.35** Fracture surface of 0.25 in. diam specimen of AISI 52100 steel (600 °F temper), $N_f$ = 10,300 cycles. Source: Ref 9.39

pletely ductile behavior, which is taken as 0.18 in. for all materials). While Eq 9.17 is manifestly derived for only one material, and for a limited range of $K_{Ic}$, its remarkable simplicity makes it attractive for general use. Further data are needed to justify its use for applications beyond the illustrations cited here. Equation 9.17, or its more general equivalent, when determined, then provides an expression for the crack depth at failure for a material of known value of $K_{Ic}$ cycled at a known stress amplitude $\sigma$. If the expression yields a value of $a'_f$ greater than 0.18 in., it may be assumed that completely ductile behavior (according to the definition previously cited) will characterize the material at the fatigue test conditions involved. If the value of $a'_f$ is less than 0.18 in., a quasi-brittle failure will occur when the crack depth becomes equal to $a'_f$.

**Application to Unnotched Specimens.** When a fatigue specimen behaves in a completely ductile manner throughout a test, the fatigue life $N_f$ can be predicted from its relation to the strain range and the tensile properties of the material as given by the method of universal slopes. It is of interest to predict the life $N'_f$ of a specimen that fails prematurely by brittle fracture and to determine to what extent the ratio $N_f/N'_f$ depends on $K_{Ic}$. This is accomplished by use of Eq 9.17 and the following equation (derived as Eq 5 in Ref 9.39):

$$\frac{\Delta N}{(\Delta N)_f} = \log_{10}[15.3(a)^{1/4}] \quad \text{(Eq 9.18)}$$

where $(\Delta N)_f$ is taken as $N_f$ for $N_f < 730$, and $14N_f^{0.6}$ for $N_f > 730$.

Because of the double condition placed on $(\Delta N)_f$ in Eq 9.18, it is necessary to examine separately the two cases $N_f < 730$, and $N_f > 730$.

*Case I: $N_f < 730$ cycles.* For this case, $N_o \approx 0$, and $\Delta N = N'_f$ and $(\Delta N)_f = N_f$. By Eq 9.17 and 9.18:

$$\Delta N = (\Delta N)_f \log_{10}[15.3(a)^{1/4}]$$
$$= (\Delta N)_f \log_{10}\left[15.3\sqrt{\frac{K_{Ic}}{\sigma}}\right] \quad \text{(Eq 9.19)}$$

Therefore:

$$N'_f = N_f \log_{10}\left[15.3\sqrt{\frac{K_{Ic}}{\sigma}}\right] \quad \text{(Eq 9.20)}$$

*Case II: $N_f > 730$.* Here, $(\Delta N)_f = 14N_f^{0.6}$, $N_o = N_f - (\Delta N)_f = N_f - 14N_f^{0.6}$. By Eq 9.17 and 9.18:

$$\Delta N = (\Delta N)_f \log_{10}\left(15.3\sqrt{\frac{K_{Ic}}{\sigma}}\right)$$
$$= 14N_f^{0.6} \log_{10}\left(15.3\sqrt{\frac{K_{Ic}}{\sigma}}\right) \quad \text{(Eq 9.21)}$$

But $N'_f = N_o + \Delta N$; therefore:

$$N'_f = N_f - 14N_f^{0.6} +$$
$$14N_f^{0.6} \log_{10}\left(15.3\sqrt{\frac{K_{Ic}}{\sigma}}\right) \quad \text{(Eq 9.22)}$$

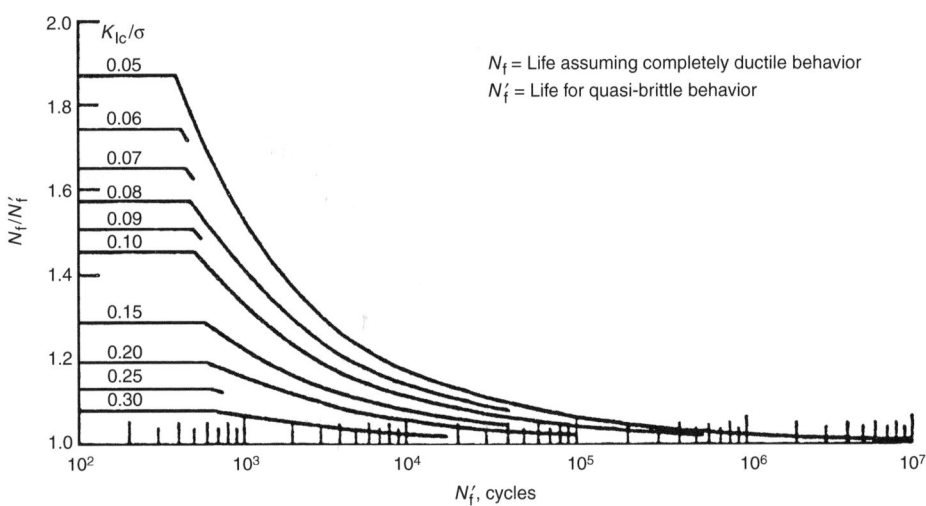

**Fig. 9.36** Effect of notch toughness on life of smooth specimens. Source: Ref 9.39

The relationships among $N_f$, $N'_f$, $K_{Ic}/\sigma$ expressed by Eq 9.19 and 9.21 are shown graphically in Fig. 9.37. It can be seen from the figure that at the very low cyclic lives and very low values of $K_{Ic}/\sigma$, $N_f$ may be appreciably larger than $N'_f$. As the life is increased, the effect becomes smaller and smaller. This result is to be expected on the basis of the relative duration of the crack initiation and crack propagation periods. For the low-cyclic lives the entire life consists of crack propagation. Interrupting the crack propagation by premature brittle failure before the crack has progressed across the entire section, therefore, results in a reduced life. Even so, the loss of life is only of the order of 50% or less for practical ranges of $K_{Ic}/\sigma$, and this amount of life reduction might not be clearly distinguishable from the data scatter common in fatigue measurements.

At the higher-cyclic lives, the effect becomes very small indeed and is surely to be hidden in data scatter. The reason for the small effect at the higher-cyclic lives is in the relative unimportance of the crack propagation phase. Most of the life is consumed in the crack initiation phase, which is unaffected by the fracture toughness, according to the approach adopted here. The only effect of the fracture toughness is to cause premature fracture at some critical crack depth during propagation of the crack. Because the crack propagation phase is a small part of the total life, removing even most of it does not have a large effect on total life.

Thus, it can be seen why fracture toughness does not enter as a major variable in the estimation of fatigue life of small, unnotched specimens. This observation has already been made in Ref 9.42. The preceding discussion and Fig. 9.37 serve mainly to make the explanation more quantitative.

**Application to Notched Specimens.** While fracture toughness is relatively unimportant in affecting the life of small unnotched specimens, the effect becomes more important when a notched specimen is used for material evaluation. The strain concentration at the notch root causes the crack to develop at a relatively early stage in the life of the specimen compared with an unnotched specimen, making the crack propagation period a larger percentage of the total life. Because low fracture toughness serves to cut short the crack propagation period, the effect

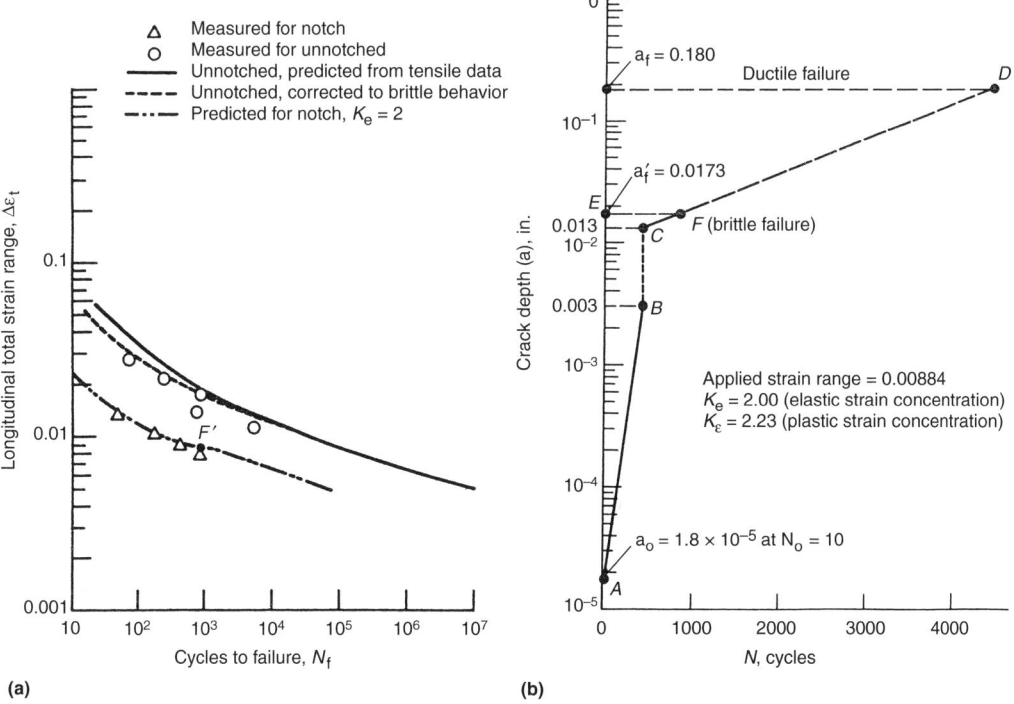

**Fig. 9.37** Fatigue behavior of AISI 52100 steel in 315 °C (600 °F) tempered condition ($K_{Ic} = 17$ ksi$\sqrt{\text{in.}}$). (a) Smooth and notched specimen behavior. (b) Hypothesized crack growth stages in notched specimen. Source: Ref 9.39

of low fracture toughness would be expected to be of greater importance in such notched specimens than in unnotched specimens. The method of computation again depends predominantly on the use of Eq 9.17 and 9.18 and their graphical equivalents, Fig. 9.27 and 9.34. We shall illustrate the approach by referring again to that AISI 52100 steel tempered at 315 °C (600 °F) (62 HRC), for which the value of $K_{Ic}$ was 17 ksi$\sqrt{\text{in}}$. An illustrative computation is presented subsequently. Here we shall indicate only the basis of the method and the results of the computations.

In Fig. 9.37(a) the continuous line represents the strain cycling fatigue behavior for unnotched specimens determined by the method of universal slopes. It is assumed that this curve would be obtained for the material knowing the measured tensile properties of the illustrative AISI 52100 steel if it behaved in a completely ductile fashion; that is, if the crack could progress in fatigue across the entire failure section. The dotted curve is then obtained from the continuous curve by the procedure indicated in the previous section. It is seen that the data points on smooth specimens for this material, as shown by the circles, agree well with the dotted curve. Alternatively, if the experimental data of the smooth material are well established over the entire life range (that is, the dotted curve is well established by experiment), the curve for completely ductile behavior (continuous curve) can be obtained from the experimental curve by inverting the process described in the previous section, again using Fig. 9.36 to determine $N_f$ from the measured $N'_f$.

Having established the life characteristics of the smooth material on the basis of completely ductile behavior, its first use is to determine the number of cycles required to start a crack at the root of the notch. To accomplish this, the strain range at the root of the machined notch is determined using the method described in Ref 9.37. Once the local strain at the root of the machined notch is known, the number of cycles at this strain level required to initiate the crack and cause it to grow to a size detectable with the unaided eye (assumed to be a depth of 0.003 in. from the root of the notch) can be determined from Fig. 9.27 and 9.38. At this point we start to take into account the 0.010 in. depth of the machined notch and regard the total crack as extending to a depth of 0.013 in., instead of 0.003 in. from the root of the notch. Concurrently, we revert to a consideration of crack growth as associated with the nominal strain range in the test section instead of the local strain range at the root of the notch (Ref 9.38) and analytically determine how the crack would continue to grow as a function of cycles of loads application if the material were completely ductile. From the value of $K_{Ic}/\sigma$, however, we can determine what the critical crack length is for brittle fracture. The number of cycles required to reach this critical crack length is the life of the notched specimen. This number of cycles can conveniently be determined with the aid of the auxiliary plot shown in Fig. 9.37(b). The use of this figure is discussed subsequently. The calculated results are shown by the dot-dash curve of Fig. 9.37(a) accompanied by the experimental data points obtained on notched specimens. Very good agreement is seen to exist between calculations and experiment.

**Illustrative Calculations for Notched Quasi-Brittle Specimen.** An illustrative example will be shown for an applied strain range of 0.00884 on a notched specimen of the material considered in Fig. 9.27. For this applied strain and a nominal elastic strain concentration factor of 2 (see Table 9.5), a Neuber analysis (see Appendix B of Ref 9.38) indicates an actual strain concentration factor of 2.23. Thus, the localized strain at the root of the notch is 0.01968, for which $N_f = 758$ cycles according to Fig. 9.37(a). At this value of $N_f$, the value of $(\Delta N)_f$ is $14(758)^{0.6}$, or 748 cycles. Thus, $N_o = N_f - 14(N_f)^{0.6} = 10$. This point is plotted at A in Fig. 9.37(b). Because the period from $N_o$ to a crack depth of 0.003 in. is $7.8(N_f)^{0.6} = 415$ cycles by Fig. 9.28, a crack depth of 0.003 in. beyond the root of the machined notch occurs at (10 + 415) = 425 cycles. This point is plotted at B in Fig.

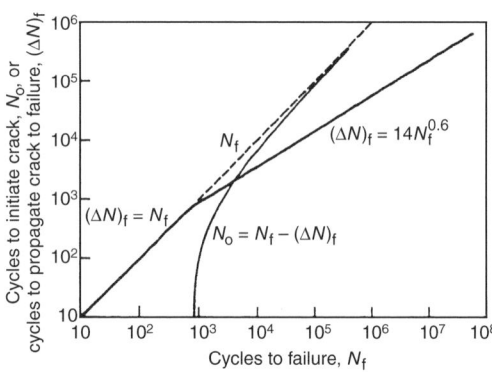

**Fig. 9.38** Relation of crack initiation to crack propagation and failure. Source: Ref 9.38

9.37(b). At this depth of crack the procedure is to change abruptly vertically in the figure from a crack depth of 0.003 to 0.013 in. (in order to take into account the 0.010 in. machined depth of the notch), and to change the applied strain level from 0.01968 back to the nominal strain range of 0.00884. The starting point of the abrupt change is shown at C in Fig 9.37(b). The next step is to determine how the crack would progress across the entire section of the specimen if it were not interrupted by brittle fracture. According to Fig. 9.37(b), the period of crack growth between 0.013 in. and 0.18 in. consumes $4(N_f)^{0.6}$ cycles, where $N_f$ now corresponds to the applied strain level of 0.00884. Thus, point D is plotted at $425 + 4(10^5)^{0.6} = 4425$ cycles at a crack length of 0.18 in. However, for a strain range of 0.00884 corresponding to a stress amplitude of 129.3 ksi, the critical crack length for brittle failure is $(K_{Ic}/\sigma)^2 = (17/129.3)^2 = 0.0173$ in., because $K_{Ic} = 17$ ksi$\sqrt{\text{in.}}$ for this temper of material. Thus, a horizontal line EF may be drawn in Fig. 9.37(b) at 0.0173 in. The intersection of this line with the crack growth line CD is at F where $N = 860$ cycles. Thus, the life dictated by quasi-brittle fracture is 860 cycles, which is plotted at point $F'$ in Fig. 9.37(a) at an applied nominal strain range of 0.00884. In this manner other points on the dot-dashed curve of Fig. 9.37(a) are determined for corresponding values of selected nominal strains.

## Discussion

Good agreement has been found between the predictions of the simplified method outlined and the experimental data obtained for both the ductile and quasi-brittle materials evaluated. As more data are accumulated, the following factors associated with the method may be reexamined.

**Consideration of the use of a normalized crack growth relation (Eq 9.18 and Fig. 9.28) for all materials and for all life levels.** Even if the assumption of such a relation is valid, this equation may be determined more accurately when more extensive data are available. It may also be observed that inherent in the use of a normalized crack growth relation is the assumption of the insensitivity of crack growth to notch toughness or other variables that do not directly manifest themselves in a determination of $N_f$ for a given strain. Thus, two materials having the same values of ductility, tensile strength, and elastic modulus would be expected to have the same value of $N_f$ at a given imposed strain range on a standard unnotched specimen (Fig. 9.39) in reversed strain cycling, regardless of the fact that the two may differ appreciably in fracture toughness. They would thus have the same crack growth per cycle at any stage of crack growth length. This feature is inconsistent with other analyses of the problem of crack growth in fatigue. Krafft (Ref 9.45) has, for example, suggested that the crack growth per cycle is inversely proportional to $K_{Ic}^2$. Should this or some other assumption be a better representation of materials behavior, it could be incorporated into the normalization of the crack growth relation.

**Reexamination of the Form of the Expression for $(\Delta N)_f$ as More Data Become Available.** It may be that the value of P in the expression $PN_f^{0.6}$, or the exponent 0.6, are material constants. It may also be that an expression for $N_o$ different from that used in the analysis, especially in the low ranges of life, might be required. Perhaps of even greater interest is the expression of $N_o$, which might be stated directly in terms of plastic strain rather than indirectly by first computing $N_f$ and $(\Delta N)_f$. The relation between $\Delta\varepsilon_p$ and $N_o$ would then be valid for geometric bodies other than the specimen of Fig. 9.39, and would therefore have general utility. Likewise, instead of using a crack propagation period relation such as $14N_f^{0.6}$, use could be made of more general expressions for crack growth in terms of plastic strain range per cycle, thereby again permitting the extension of the method to the treatment of more general geometries. The expression for $dl/dN$ in terms of $\Delta\varepsilon_p$ is discussed in detail in Appendix B of Ref 9.39 and serves as a suitable starting point, but it should be refined if more data become available.

**The assumption of a normalized fracture relation** (Fig. 9.34) may require revision both in form and in the magnitude of the parameters involved in the expression for other materials.

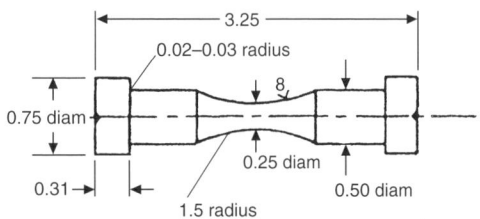

**Fig. 9.39** Fatigue specimen configuration (dimensions in inches.) Source: Ref 9.41

**Investigation of the Problem of Interpretation of Computations of $a_f$ Less Than 0.013 in.** Although a definite value of life does result from Fig. 9.37(b) for any value of $l_f$, it must be recognized that values that turn out to be less than 0.013 in. imply that brittle fracture occurs before the crack has penetrated a depth of 0.003 in. beyond the machined notch of 0.010 in. Because the method of analysis makes use of relations developed on limited data obtained at crack depths greater than these values, it is evident that special attention would have to be given to cases for which $K_{Ic}/\sigma < 0.36\sqrt{\text{in.}}$ and for which $l_f < 0.013$ in.

Despite the foregoing limitations, some of which might readily be overcome by changes in the quantitative relationships, the method is of interest because of its simplicity and because it quickly yields definite answers in cases where attempts at more exact analyses might greatly complicate the process. It may therefore be applied with some confidence because it yields good results for the case illustrated.

## REFERENCES

9.1 J.M. Barsom and S.T. Rolfe, *Fracture and Fatigue Control in Structures: Applications of Fracture Mechanics,* 3rd ed., American Society for Testing and Materials, 1999

9.2 A.A. Griffith, The Phenomena of Rupture and Flow in Solids, *Philos. Trans. R. Soc. (London),* Vol A221, 1920, p 163–198

9.3 C.E. Inglis, Stresses in a Plate due to the Presence of Cracks and Sharp Corners, *Proc. Inst. of Naval Architects,* Vol 55, 1913, p 219–241

9.4 T.A. Michalske and B.C. Bunker, The Fracturing of Glass, *Sci. Am.,* Dec 1987, p 122–129

9.5 G.R. Irwin, *Fracturing of Metals,* American Society for Metals, 1948, p 147–166

9.6 E. Orowan, *Fatigue and Fracture of Metals,* MIT Press, Cambridge, 1950

9.7 I.N. Sneddon, The Distribution of Stress in the Neighborhood of a Crack in an Elastic Solid, *Proc., R. Soc. (London),* Vol A187, 1946, p 229–260

9.8 G.R. Irwin, Analysis of Stresses and Strains Near the End of a Crack Traversing a Plate, *J. Appl. Mech. (Trans. ASME),* Vol 24 (No. 3), Sept 1957, p 361–364

9.9 G.R. Irwin, Fracture, *Handbuch der Physik,* Vol VI, Springer, Berlin, 1958, p 551–590

9.10 M.L. Williams, On the Stress Distribution at the Base of a Stationary Crack, *J. Appl. Mech. (Trans. ASME),* Vol 24 (No. 1), March 1957, p 109–114

9.11 H.M. Westergaard, Bearing Pressures and Cracks, *J. Appl. Mech. (Trans. ASME),* Vol 61, 1939, p A49–A53

9.12 C.P. Paris and G.C. Sih, "Stress Analysis of Cracks," *Fracture Toughness Testing and Its Applications,* ASTM STP 381, American Society for Testing and Materials, 1965, p 30–81

9.13 J.E. Srawley and W.F. Brown, Jr., "Fracture Toughness Testing Methods," *Fracture Toughness Testing and its Applications,* ASTM STP 381, American Society for Testing and Materials, 1965, p 133–196

9.14 ASTM Committee E08, "Standard Test Method for Plane-Strain Fracture Toughness of Metallic Materials," *Annual Book of ASTM Standards,* Vol 03.01, *Metals—Mechanical Testing; Elevated and Low-Temperature Tests; Metallography,* E 399, 2002

9.15 S. Suresh, *Fatigue of Materials,* 2nd ed., Cambridge Solid Science Series, Cambridge University Press, 1998

9.16 M.P. Blinn and R.A. Williams, Design for Fracture Toughness, *Materials Selection and Design,* Vol 20, *ASM Handbook,* ASM International, 1997, p 533–544

9.17 A.K. Head, The Growth of Fatigue Cracks, *Philos. Mag. series 7,* Vol 44 (No. 356), Sept 1953, p 925–938

9.18 N.E. Frost and D.S. Dugdale, The Propagation of Fatigue Cracks in Sheet Specimens, *J. Mech. Phys. Solids,* Vol 6 (No. 2), 1958, p 92–110

9.19 H.W. Liu and N. Iinno, A Mechanical Model for Fatigue Crack Propagation, *Fracture,* Chapman and Hall, 1969, p 812–824

9.20 A.J. McEvily, Jr. and W. Illg, "The Rate of Fatigue-Crack Propagation in Two Aluminum Alloys," NACA TN D-52, National Advisory Committee for Aeronautics, Oct 1959

9.21 P. Paris and F. Erdogan, A Critical Analysis of Crack Propagation Laws, *ASME J. Basic Engineering,* Vol 85, 1963, p 528–534

9.22 P.C. Paris, Fatigue—An Interdisciplinary Approach, *Proc., 10th Sagamore Conference,* Syracuse University Press, Syracuse, NY, 1964, p 107

9.23 R.W. Hertzberg, *Deformation and Fracture Mechanics of Engineering Materials,* John Wiley & Sons, 1976, p 465–539

9.24 R.G. Forman, V.E. Kearney, and R.M. Engle, Numerical Analysis of Crack Propagation in Cyclic-Loaded Structures, *ASME J. Basic Engineering,* Vol 89, 1967, p 459–464

9.25 D.F. Socie, M.R. Mitchell, and E.M. Caulfield, Fundamentals of Modern Fatigue Analysis, Fracture Control Program, FCP Report No. 26 (revised), College of Engineering, University of Illinois, Jan 1978

9.26 R.I. Stephens, A. Fatemi, R.R. Stephens, and H.O. Fuchs, *Metal Fatigue in Engineering,* 2nd ed., John Wiley & Sons, 2001

9.27 D.V. Nelson, Review of Fatigue-Crack-Growth Prediction Methods, *Exp. Mech.,* Vol 34, Feb, 1977, p 41–49

9.28 C.M. Hudson, "Effect of Stress Ratio on Fatigue Crack Growth in 7075-T6 and 2024-T3 Aluminum Alloy Specimens," NASA TN-D-5390, Aug 1969

9.29 J.H. Fitzgerald and R.P. Wei, A Test Procedure for Determining the Influence of Stress Ratio on Fatigue Crack Growth, *J. Test. Eval.,* Vol 2 (No. 2), March 1974, p 67–70

9.30 R.J. Bucci, "Environmentally Enhanced Fatigue and Stress Corrosion Cracking of a Titanium Alloy Plus a Simple Model for the Assessment of Environmental Influence of Fatigue Behavior," Ph.D. dissertation, Lehigh University, 1970

9.31 J.M. Barsom, Fatigue-Crack Propagation in Steels of Various Yield Strengths, *J. Eng. Ind. (Trans. ASME), series B,* No. 4, Nov 1971, p 1190

9.32 J.C. McMillan and R.M.N. Pelloux, "Fatigue Crack Propagation under Program and Random Loads," *Fatigue Crack Propagation,* ASTM STP 415, American Society for Testing and Materials, 1967, p 505–532

9.33 J. Schijve, "Significance of Fatigue Cracks in Micro-Range and Macro-Range," *Fatigue Crack Propagation,* ASTM STP 415, American Society for Testing and Materials, 1967, p 415–457

9.34 O.E. Wheeler, Spectrum Loading and Crack Growth, *ASME J. Basic Engineering,* Vol 94, 1972, p 181–186

9.35 W. Elber, "The Significance of Fatigue Crack Closure," *Damage Tolerance in Aircraft Structures,* ASTM STP 486, American Society for Testing and Materials, 1971, p 230–242. See also Fatigue Crack Closure under Cyclic Tension, *Eng. Fract. Mech.,* Vol 2, 1970, p 37–45

9.36 J. Willenborg, R.M. Engle, Jr., and R.A. Wood, "A Crack Growth Retardation Model Using an Effective Stress Concept," Air Force Flight Dynamics Laboratory Report AFFDL-TM-71-1-FBR, Jan 1971

9.37 E.F.J. von Euw, R.W. Hertzberg, and R. Roberts, "Delay Effects in Fatigue-Crack Propagation," *Stress Analysis and Growth of Cracks,* ASTM STP 513, American Society for Testing and Materials, 1972, p 230–259

9.38 S.S. Manson and M.H. Hirschberg, "Low Cycle Fatigue of Notched Specimens by Consideration of Crack Initiation and Propagation," NASA TN D-3146, National Aeronautics and Space Administration, Washington, D.C., June 1967

9.39 S.S. Manson, Interfaces between Fatigue, Creep, and Fracture, *Int. J. Fract. Mech.,* Vol 2 (No. 1), 1966, p 327–363

9.40 H. Neuber, Theory of Stress Concentrations for Shear-Strained Prismatical Bodies with Arbitrary Nonlinear Stress-Strain Law, *J. Appl. Mech.,* Vol 83 (No. 4), Dec 1961, p 544–550

9.41 H.F. Hardrath and L. Ohman, "A Study of Elastic and Plastic Stress Concentration Factors due to Notches and Fillets in Flat Plates," NACA TR-1117, National Advisory Committee for Aeronautics, Washington, D.C., 1953

9.42 S.S. Manson, Fatigue—A Complex Subject, *Exp. Mech.,* Vol 5 (No. 7), 1965, p 193–226

9.43 R.E. Peterson, Fatigue of Metals, Part III: Engineering Design Aspects, *Materials Research and Standards,* Vol 3, Feb 1963, p 122–139

9.44 Impact Toughness Testing, *Mechanical Testing and Evaluation,* Vol 8, *ASM Handbook,* ASM International, Oct 2000, p 596–611

9.45 J.M. Krafft, On Prediction of Fatigue Crack Propagation Rate from Fracture Toughness and Plastic Flow Properties, *ASM Trans. Quart.,* Vol 58 (No. 4), 1965, p 691–695

## SELECTED REFERENCES

- M.H. Aliabadi and D.P. Rooke, *Numerical Fracture Mechanics,* Computational Mechanics, Inc., Billerica, MA, 1991
- T.L. Anderson, *Fracture Mechanics Fundamentals and Applications,* 2nd ed., CRC Press, New York, 1995
- A.G. Atkins and Y.W. Mai, *Elastic and Plastic Fracture,* John Wiley & Sons, 1985
- *Ductile Fracture Handbook,* Novetech Corp., Gaithersburg, MD, Electric Power Research Institute, Palo Alto, CA Vol 1, 1989; Vol 2 and 3, 1991
- H.L. Ewalds and R.J.H. Wanhill, *Fracture Mechanics,* Edward Arnold, London, 1984
- E.E. Gdoutos, *Fracture Mechanics—An Introduction,* Kluwer Academic Publishers, Boston, MA, 1993
- K. Hellan, *Introduction to Fracture Mechanics,* McGraw-Hill, New York, 1984
- M.F. Kanninen and C.H. Popelar, *Advanced Fracture Mechanics,* Oxford University Press, New York, 1985
- V. Kumar, W.W. Wilkening, W.R. Andrews, H.G. deLorenzi, and D.F. Mowbray, "Advances in Elastic-Plastic Fracture Analysis," EPRI NP-1931, research project 1237-1, General Electric Company, R&D Center, Schenectady, NY, June 1987
- Y. Murakami, *Stress Intensity Factors Handbook,* Pergamon Press, Vol 1 and 2, 1987; Vol 3, 1992
- D.R.J. Owen and A.J. Fawkes, *Engineering Fracture Mechanics: Numerical Methods and Applications,* Pineridge Press Ltd., 1983
- H.S. Reemsnyder, "Fracture Mechanics of Mechanically Fastened Joints—A Bibliography," *Structural Integrity of Fasteners,* ASTM STP 1391, P.M. Toor, Ed., American Society for Testing and Materials, 2000, p 204–214
- D.P. Rook and D.J. Cartwright, *Compendium of Stress Intensity Factors,* HM Stationery Office, London, 1976
- G.C. Sih, *Handbook of Stress Intensity Factors,* Institute of Fracture and Solid Mechanics, Lehigh University, Bethlehem, PA, 1973
- R.N.L. Smith, *Basic Fracture Mechanics,* Butterworth-Heinemann Ltd., 1991
- H. Tada, P.C. Paris, and G.R. Irwin, *The Stress Analysis of Cracks Handbook,* 3rd ed., co-published by The American Society of Mechanical Engineers and ASM International, 2000
- D. Taylor, *A Compendium of Fatigue Thresholds and Growth Rates,* Engineering Materials Advisory Services, Ltd., Warley, England, 1985

# CHAPTER 10

# Mechanism of Fatigue

## Introduction

There has always been an aura of mystery regarding why metals, and materials in general, fail in fatigue. The impression seems to have developed that a part may function satisfactorily for many, many loadings, but when metal fatigue occurs, the failure is sudden and catastrophic. As early as 1839, Poncelet (Ref 10.1) in France described this phenomenon as "fatigue," presumably analogous to human fatigue that results when a motion is repeated successively. Fatigue comes on gradually in human endurance, and what frequently is overlooked is that fatigue of a metallic part also develops gradually. Failure is not really sudden but is the end result of progressive deterioration that eventually produces the failure "event." What the process consists of has been the subject of many studies over some 150 years, and we now understand reasonably well the nature of the fatigue mechanism. As with the fatigue process itself, our understanding did not develop suddenly or smoothly, and not without much controversy along the way.

## Some Early Theories

Because of the somewhat mysterious character of this newly observed type of failure, and because of its growing importance as new technological applications developed, many fertile minds directed their attention toward possible explanations. Unfortunately, the scientific tools available in those times were very limited, and much of the discussion was speculative, albeit quite insightful. A fairly complete discussion of the theories that developed in the 100 years following the first recognition of the problem was described by Gohn in 1963 (Ref 10.2). For the sake of brevity and because this discussion is available to the interested reader, we shall not include all the concepts described by him. However, it is appropriate to outline some of them that did indeed reflect some basis in truth, and to explain how more recent developments elucidate the underlying behavior that led to those early conclusions.

Consider, for example, the "crystallization" theory that was one of the first. The contention was that cyclic loading caused the initially amorphous metal to crystallize, and the resulting crystalline metal was somehow weaker than the amorphic (or "fibrous") material from which it evolved. The physical basis for this theory must have been that close examination of some fatigue failures did display some light-reflecting regions reminiscent of the facets of crystalline gems. After much controversy it was finally recognized that metals in general are crystalline and that the crystallinity did not derive from the cyclic loading.

The fatigue process involves slip on crystallographic planes of the metal, and the final fatigue fracture surface exposes some of these planes (especially in the stage I failure regime, as discussed later). Furthermore, we shall also observe that cyclic plastic deformation creates dislocations (see Appendix) within each grain, i.e., within each individual crystal. These dislocations become intertwined. In an attempt to seek a minimum energy level during repeated plastic deformation, the dislocations often tend to arrange themselves into what are known as low-angle boundaries between "subgrains" that begin to populate each grain as cyclic plasticity continues. The amount of angular misorientation between adjoining subgrains is only a few degrees. Fewer dislocations are needed to accommodate the misorientation between adjacent subgrains compared with the much larger misorientation of normal grain boundaries. In a sense, fatigue loading can produce changes in the detail of the crystallographic orientation within grains. However, such changes do not qualify as the "recrystallization" from amorphous material that was erroneously ascribed earlier to fatigue failures. Cyclic plastic flow may also cause other subtle crystalline changes, for example, in transformations from one crystalline structure to another, say from austenite to martensite. Therefore, although the concept that the fatigue process is due to crystallization is grossly inaccurate, there is a "grain of truth" in the original concept.

Another theory that developed early was that of "strain hardening." Gough and Hanson (Ref 10.3) noted that many materials harden under monotonic strain, until eventually a crack occurs at some local region where the hardness reaches the critical value and is unable to harden further. The concept was further developed by Orowan (Ref 10.4) to consider that in any material or structure a local region or enclave might be more susceptible to such hardening because of its location relative to the grain structure, or because of inherent flaws or inclusions. Many variations of the strain-hardening theme were also developed by other investigators, and it may, in fact, be an important component of the fatigue of cyclically strain hardening materials. However, it would be inadequate when considering cyclically strain-softening materials.

In a similar way many of the early theories described by Gohn can be evaluated as being either inapplicable to general materials behavior, or being unable to discriminate the true phenomenon causing the fatigue from attendant phenomena that accompany the process, but are not its cause. In this chapter we emphasize three underlying phenomena: plastic deformation, surface geometry degradation, and environmental chemical interaction as being the true demons that eventually are lethal to all materials in fatigue. Sometimes one factor is more important than the others, but together they can doom all cyclically loaded structures to eventual failure.

## Some Tools for Observation That Have Aided in Understanding

By far the most valuable tool that has been helpful to our understanding of the mechanism is the human mind: its ability to ask the proper question, its ability to fashion experiments that will shed light on these questions, and its ability to interpret the results—to separate the relevant results from the sideshows, and to reach insightful conclusions. In addition, however, there also may be physical tools—also fashioned by human minds. In the following we list briefly some of the tools that have enabled research to reach the conclusions we shall enumerate. The list is not all inclusive, of course, because each tool is developed from a progression of previous tools, and because some are too commonplace to require discussion. However, even this brief listing should demonstrate that the array of technology that has been brought to bear on this subject is substantial.

**Heat Generation Detection.** Because, as we discuss later, plastic deformation is generic to the fatigue process, it would be expected that hysteresis heat energy is generated during fatiguing. Several types of heat detection have been used in connection with fatigue studies. Putnam (Ref 10.5) has, for example, placed fatigue specimens in an autoclave to ensure that no external heat enters, or that heat generated during fatigue leaves the system. He then carefully measured the temperature of the surrounding fluid. When the applied fatigue stress was lower than the "endurance limit" (meaning that failure would not be expected until a very large number of cycles), no detectable heat generation was measured. As soon as some heat generation was detected, however, the explanation was that plasticity was occurring and failure would soon occur. Figure 10.1 shows the remarkably close correlation between the endurance limit pre-

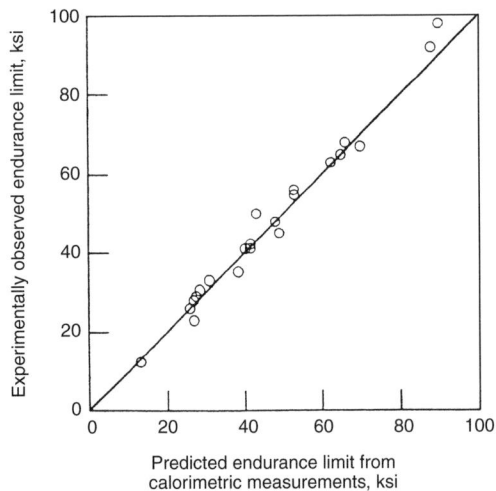

**Fig. 10.1** Endurance limit predictions for various steels based on calorimetric rise of temperature measurements. Data from Ref 10.5

dicted by this method and the experimentally determined value, validating the one-to-one relation between fatigue and plasticity.

Of course, many novel methods have been devised for measuring minute rises in temperature, and these methods could be applied in future fatigue investigations, for example, to localize the exact spots where fatigue initiates. The importance of all such methods is that plasticity is generic to fatigue, and we emphasize this point in our discussion.

**Bubble Raft Analogy Equipment.** A very interesting piece of equipment to illustrate how atoms (and especially flaws within the atomic structure) move under cycling loading was devised in 1955 by Sinclair and Corten (Ref 10.6). Basically, it was inspired by the work of W.L. Bragg (Ref 10.7) in which an array of soap bubbles representing the atoms within a crystal are acted upon by external stimuli to simulate what happens during such metallurgical processes such as recrystallization, dislocation movement, stacking fault formation, and so forth. Sinclair and Corten used it to study the effect of reversed shear in simulating the fatigue process. Their apparatus is shown in Fig. 10.2. Basically, the bubbles are generated by blowing an air jet into a special solution of propylene glycol, gelatin, and triethanolamine in water. The test chamber is a volume between two polymethylmethacrylate plates, the end faces of which could be displaced parallel to each other to represent shear strain. After creating a volume of bubbles, individual bubbles were burst by a hot wire to form the desired content of vacancies, dislocations, grain boundaries, and small and large impurity atoms. The faces were then cyclically displaced to represent shear strain, and the microstructural movements and formations were photographed. Very interesting conclusions were drawn regarding the mechanism of fatigue, as we shall discuss.

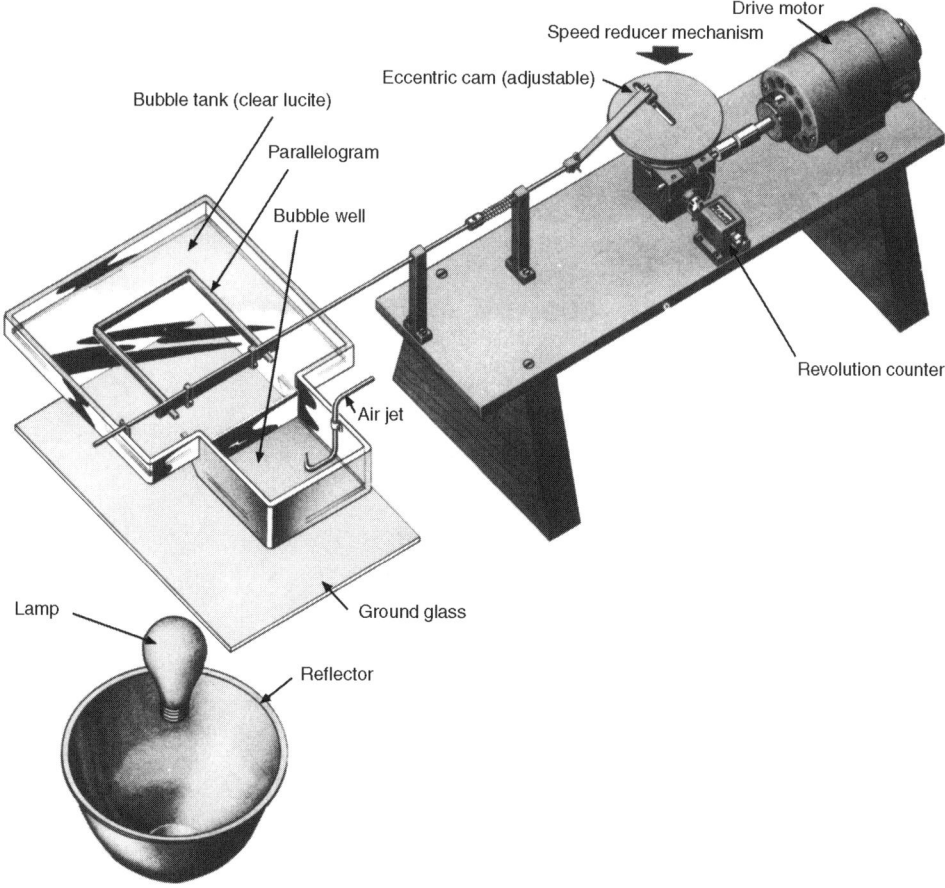

**Fig. 10.2** Bubble raft apparatus for study of "crystal" subjected to reverse shear. Source: Ref 10.6

**Photoelasticity as a Tool.** Because fatigue is so intimately related to the question of generation and propagation of cracks, and because the photoelastic technique really comes into its own when studying stress distribution in the vicinity of a crack, it is very logical that this research tool should have contributed to our understanding of fatigue. Figure 10.3 contains some results that have been presented by Gerberich (Ref 10.8) showing the strain distribution at the tip of a notch. The Moiré plastic-coating method is especially valuable here because it enables us to study the strain distribution in metallic materials. Because crack growth involves a plastic zone at the tip of the crack, knowing the shape of the plastic region has been helpful in formulating the assumptions that are most appropriate for analysis.

The phase-interference method has also been applied to study stress distribution in the vicinity of a notch tip. Figure 10.4 is taken from Oppel and Hill (Ref 10.9) in which they present their results using this technique. Briefly, light is directed through an optical system onto an initially flat surface of the specimen. In the undeformed condition of the specimen, the plate glass is par-

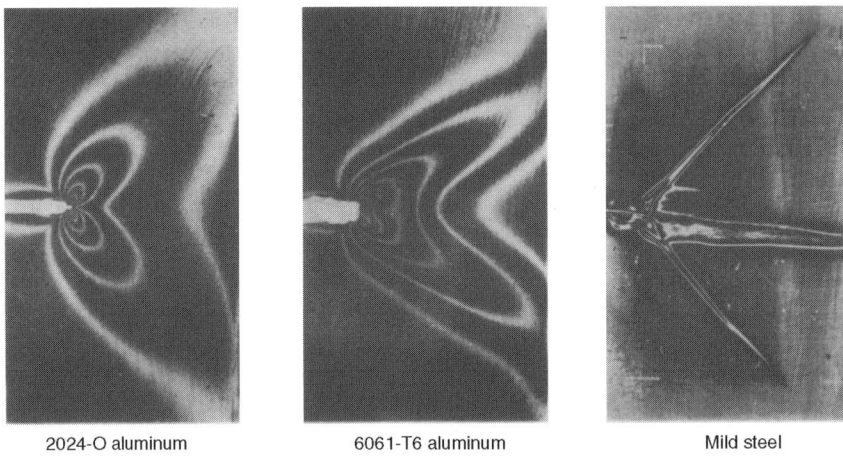

**Fig. 10.3** Photoelastic-coating method to determine strain distribution at notch tip. Source: Ref 10.8

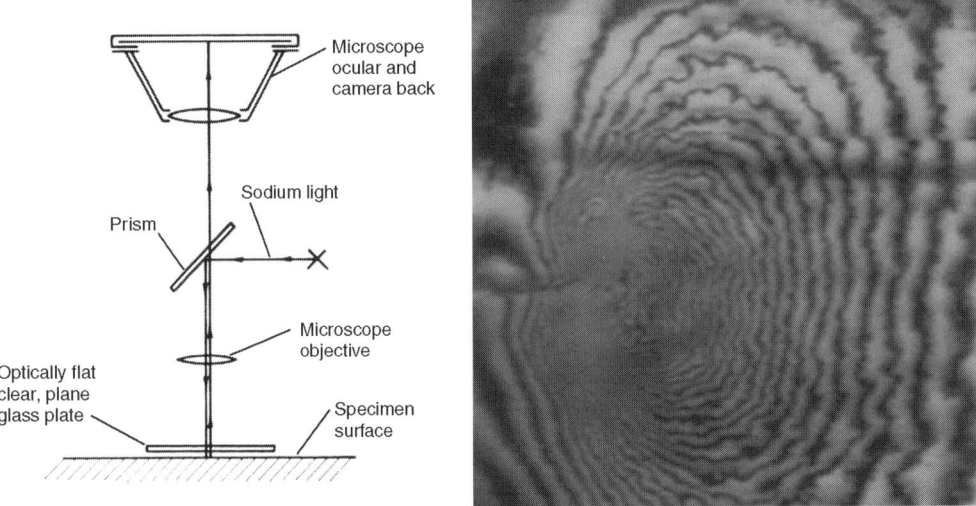

**Fig. 10.4** Phase-interference method to determine strain distribution at notch tip. Source: Ref 10.9

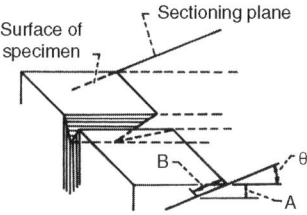
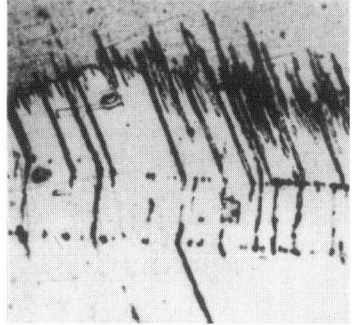

Taper magnification = cosecant θ
Notch depth:
on normal section = A
on taper section = B

Taper section of brass specimen in torsion

**Fig. 10.5** Observation of fatigue damage at specimen surface by taper sectioning. Source: Ref 10.10

allel to the specimen surface. Deformation of the specimen by loading causes a variation in the distance between the plate glass and the specimen at different points and results in the pattern shown on the right of the figure.

**Optical Magnification and Taper Sectioning.** The optical microscope is, of course, an extremely valuable tool for examining the surface of a stressed sample. Unfortunately, magnification beyond about 200 reduces the clarity of the field but it has been combined with taper sectioning to extend the range of magnification by another factor of 20 or so. As developed by Wood (Ref 10.10), the concept takes advantage of geometry. The surface is cut by a plane that makes a small angle with it, a shallow surface crack will be elongated in accordance with the cosecant relation of the angle between the cutting plane and the surface. Figure 10.5 shows the concept, and a typical result obtained by Wood and his coworkers (Ref 10.10). In this way, surface cracks that are hardly distinguishable in an ordinary microscope become very clear when "taper sectioning" is used. Of course, the magnification occurs only in the depth of crack, not in the width; thus care must be exercised in interpreting the results of such studies. However, they have been especially useful in observing surface deterioration.

**Laser Reflectivity.** Studying surface deterioration can also be accomplished by the use of lasers. Here, the surface of a fatigue specimen is exposed to an intense laser beam impacted at an angle. The intensity of the reflected beam depends on the surface smoothness. So long as the surface remains smooth, a good reflected beam is obtained. As the surface becomes jagged due to the fatigue process, however, the reflected beam intensity is attenuated. Figure 10.6 shows

in schematic form the type of results that can be obtained. Soon after the intensity of the reflected beam drops precipitously, the specimen develops a crack and fails in fatigue. It is expected that the reflected intensity would decease more gradually if the fatigue loading level were increased. Greater plasticity induces greater surface roughening at an earlier percentage of fatigue life, leading to a gradual drop in reflected intensity of the laser beam. Once a major crack forms and failure is imminent, the intensity would then fall rapidly.

**Electron Microscopy.** The development of the electron microscope has been a great boon to the study of fatigue mechanism. Basically, the concept of the electron microscope is analogue to the optical microscope, as seen in Fig. 10.7 (Ref 10.11). Instead of using a light source, the operating medium is a beam of electrons generated, for example, by a filament heated to a high temperature. The electron flow is then con-

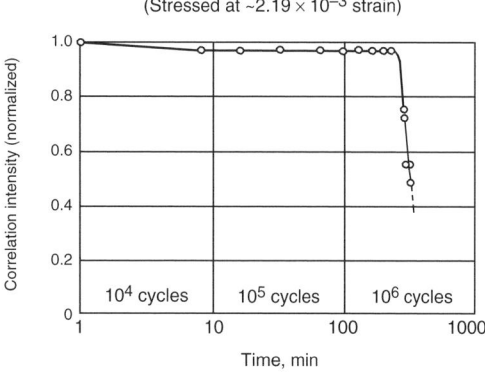

**Fig. 10.6** Schematic of decrease in reflected intensity (normalized) of a laser beam impinged at an angle to the surface of a metal being fatigued

trolled by magnets in an analogous way as lenses control the flow of light in an optical system.

The electron beam can then be used in one of two ways: reflection or transmission. In reflection, the electron beam is caused to impinge on the atoms of the test specimen, causing them to emit secondary electrons that are ejected from the surface and which then are observed by an appropriate image receptor. The condition of the atomic arrangement, which governs the reflected electrons, can then be observed. In the transmission electron microscope (TEM), the main beam passes through a thinned foil specimen, and the atomic arrangement in the specimen influences the electron path and the intensity as seen on the sensitive image receptor plate. Both methods are used, but the transmission method must use a very thin specimen so that the electron beam can penetrate it.

Both methods can be used with a stationary or scanning beam. The latter scans point-by-point locations of a desired field so that much more intense beams can be used because they remain only momentarily at any one point. The scanning reflection electron microscope system is referred to as SEM, while the scanning transmission system is termed STEM. Conventional SEM imaging can typically achieve magnifications up to 40,000 times. Recently introduced field emission SEM imaging can attain 100,000 times. TEM and SEM can attain 150,000 to 200,000 times.

The surface of any solid specimen can be studied by SEM* if a replica can also be made of the surface topography using a very thin material that can interact with the electron beam. Figure 10.8 shows one technique for creating

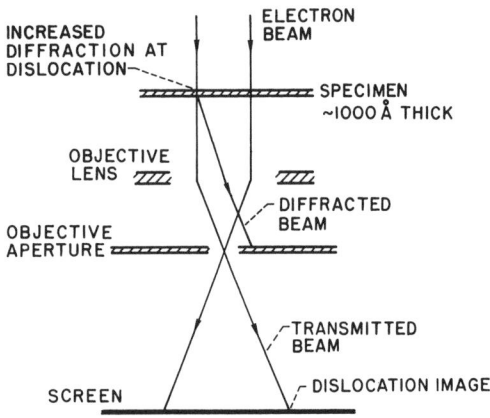

**Fig. 10.7** Formation of dislocation image in wafer-thin specimen by transmission electron microscope. Source: Ref 10.11

---

*Provided the surface is electrically conductive. Therefore, metallic specimens can be directly imaged with no preparation beyond cleaning. Insulators such as plastics and ceramics are usually first sputter-coated with a thin conductive layer of carbon or metal such as platinum.

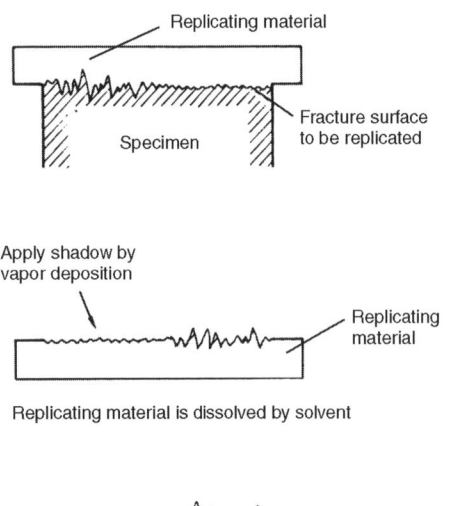

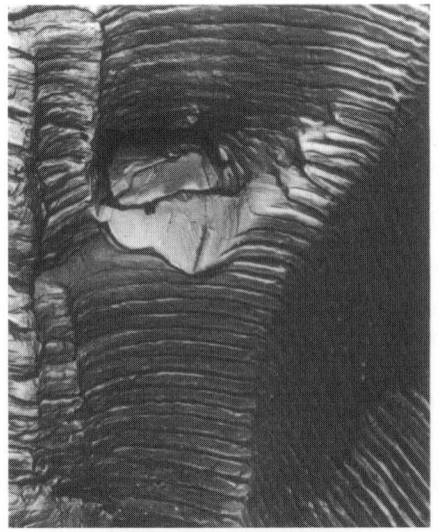

Striations on fracture surface characterizing fatigue crack growth

**Fig. 10.8** Scanning electron microscopy studies of fatigue-fracture surface by replication. Material: 7075-T6 aluminum alloy; fatigue life, 56,000 cycles. Source: Ref 10.31

such a replica. The surface of the specimen is covered with a plastic replicating material, typically cellulose acetate, as shown in the upper left of the figure. The plastic material is then carefully stripped off the specimen, leaving an exact replica of the surface topography. This surface is then coated or shadowed by the use of carbon or other material (e.g., gold, germanium) that forms an extremely thin (<0.1 μm) conductive film having the shape of the surface. Shadowing at an angle produces highlights that make it possible to distinguish regions of surface depression from those of elevation. The plastic replica is then dissolved, typically in acetone, and the remaining thin film is studied in a transmission electron microscope. The right side of Fig. 10.8 shows surface striations observed on an aluminum 7075-T6 alloy after it failed in fatigue at 56,000 cycles. The spacing of the striations here is of the order of 0.5 μm.

Figure 10.9 shows another approach involving transmission electron microscopy. Here, the ultra-thin specimen itself, rather than a replica of the surface, is placed in the microscope. To be penetrated by the electron beam, however, the specimen must be extremely thin, often less than 0.1 μm. The layers of atoms act as a diffraction grating, and any disarray in the lattice is indicated by a disturbed pattern generated by the electron beam on the screen. A typical pattern is shown in Fig. 10.9 taken from the work of Grosskreutz (Ref 10.12). The dark regions are dislocations—regions of atomic disarray—and it is seen from the photographs how the grains break up into a number of smaller regions called dislocation cells or *subgrains*, separated from each other by these dislocation networks. Because this method involves the use of very thin films of the specimen itself, it is sometimes questioned whether the dislocation structure observed in such films are typical of the behavior of bulk material. However, use of this technique is widespread and it is extremely useful.

A combination of the technique involving the use of the electron microscope and the Moiré method is illustrated in Fig. 10.10 (Ref 10.13). Here, an extremely thin layer of palladium was deposited on a correspondingly thin layer of gold. The local array of atoms within a single adjoining grain of each crystal constituted the grids that combined to form a moiré pattern. The interference patterns produced by the grids were observed with an electron microscope. In this figure, one of the grids is assumed to be perfect and the other to contain a dislocation. The moiré pattern that results is distinguishable by means of the electron microscope, whereas the actual atomic array of the materials themselves is too fine to be resolved in this manner. Thus, by this method, it is possible to observe

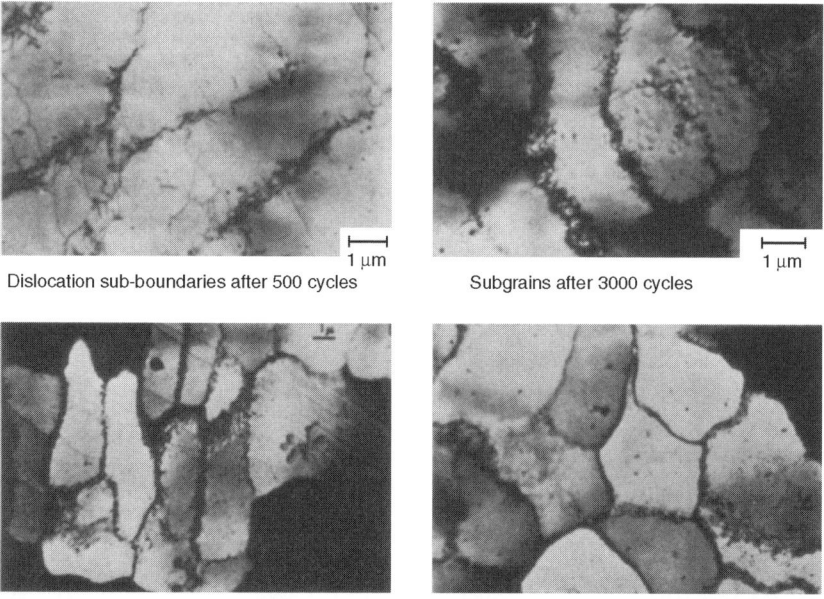

**Fig. 10.9** Use of transmission electron microscopy to observe formation of substructure in aluminum; total strain range = 0.004, life ≈ 500,000 cycles. Source: Ref 10.12

**244 / Fatigue and Durability of Structural Materials**

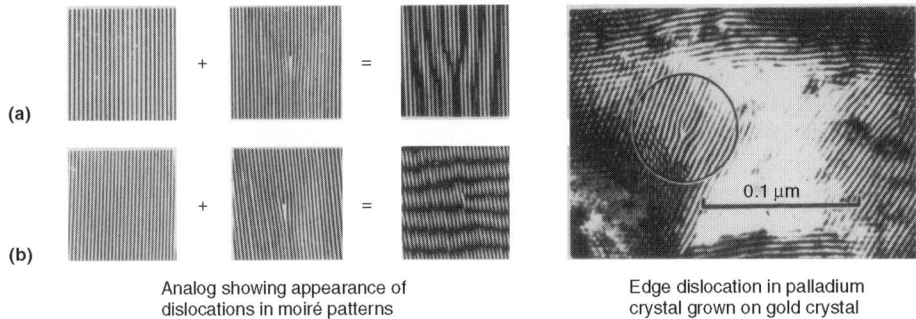

**Fig. 10.10** Moiré method used to observe dislocations in crystal lattice by electron microscopy. (a) Parallel case. (b) Rotation case. Source: Ref 10.13

disarrays in the atomic structure such as the circled edge dislocation shown on the right side of the figure.

**Decoration and etching** techniques have been used to provide photographic evidence of the existence of dislocations and their geometric properties. Figure 10.11 shows a classic photograph obtained by Dash (Ref 10.14) of a Frank-Read source using the copper decoration technique. The development of etch pits is now a well-established technique.

A very interesting innovation was developed by Hahn and Rosenfield (Ref 10.15). By using silicon-rich steel and a special etchant, they were able to bring out in bold contrast the regions of plastic flow in the vicinity of a notch. Figure 10.12 shows some of their results. A sheet with a sharp notch was subjected to loading, and the region of plastic flow was first observed by this technique at the surface. By machining away half of the specimen and then etching the midsection, the distribution of plastic flow in the midsection of the plate was made apparent. Note the difference in appearance of these regions. Such studies have been extremely useful in understanding the role of plastic flow and the effect of constraint of such plastic flow on the generation and propagation of cracks.

**X-rays** are an alternate means of detecting disarray in a crystal lattice, because the lattice will diffract x-rays in much the same manner as an electron beam. Thus, directing an x-ray beam onto very thin films of crystalline materials in which the lattice arrangements are not perfect can result in photographs similar to that shown in Fig. 10.13 (Ref 10.16).

For many years Professor Shuji Taira of Kyoto University in Japan developed methods for studying fatigue as well as creep by the x-

**Fig. 10.11** Observation of dislocations by decoration. Source: Ref 10.14

**Fig. 10.12** Etching technique to reveal plastic zone. Source: Ref 10.15

ray technique. Unfortunately, with his untimely death in 1979, much of the work was aborted. Using an x-ray diffraction probing technique, Taira and Honda (Ref 10.17) examined the effect of cyclic loading on the changes in the diffraction pattern from the fatigued surface layers of quenched and tempered low-carbon steel rotating bending specimens. Figure 10.14, taken from Taira and Honda's paper, shows the value of $b/\beta$ (ratio of the Burger's vector to the half value of the breadth of the diffraction peak as a function of the fractional fatigue life consumed). Clearly, as fatigue cycling progresses, a measure of the accumulation of damage is provided by the increase in line broadening. Physically, line broadening is a measure of the crystalline disorder, i.e., a departure from the originally ordered crystalline array of this body-centered cubic alloy. Fatigue cycling is responsible for the introduction of atomic level defects (vacancies, dislocations, subgrain boundaries, etc.) that cause a scattering of the diffracted x-ray beam, which in turn cause a more diffuse back reflection pattern from the rear surface layer of the tested specimens. An interesting observation is the fact that the correlation between line broadening and fatigue cycle ratio was independent of the stress amplitudes (and hence fatigue life level) over the range employed. If this observation is a general one, it implies an underlying fundamental basis for the accumulation of fatigue damage and hence a physical basis for understanding and predicting fatigue life.

**Ultrasonics.** Another approach that has been found useful in the detection of fatigue damage is an ultrasonic technique that is capable of detecting fatigue cracks prior to their observation with the unaided eye. The approach was used by Klima et al. (Ref 10.18) and is shown in Fig. 10.15. By this technique, it is possible to detect cracks of length less than 0.076 mm (0.003 in.) in notched sheet specimens. As a matter of interest, the cracks were detected well within the first 10% of the life of the sharply notched specimens.

## Role of Plasticity

Probably the most important factor in generating fatigue is the development of inelastic deformation. At normal temperatures the main type of deformation is plasticity—the relative sliding of atoms along the crystallographic slip planes of the crystals of the material. At high temperatures inelastic deformation of the time-dependent type—anelastic and creep—may also develop, but in the following discussion we emphasize plasticity, although some of the factors discussed may also be relevant to anelasticity and creep.

While dislocations are normally involved in facilitating plasticity, and it is commonly believed that the movement of a dislocation during plasticity may result in its annihilation, in actuality the main effect of the plasticity is to multiply dislocations. We show, for example, in the Appendix, "Selected Relevant Background Information," how one mechanism—the Frank-Read source—can generate many dislocations from just a few if subjected to plastic deformation, and in Fig. 10.11 it is seen that such Frank-Read sources do indeed exist. While the Frank-Read source is easy to understand and it serves as a simple teaching example, it is not the dom-

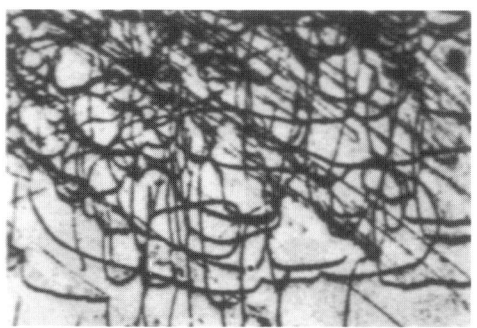

**Fig. 10.13** Observation of dislocations in silicon by x-ray diffraction. Source: Ref 10.16

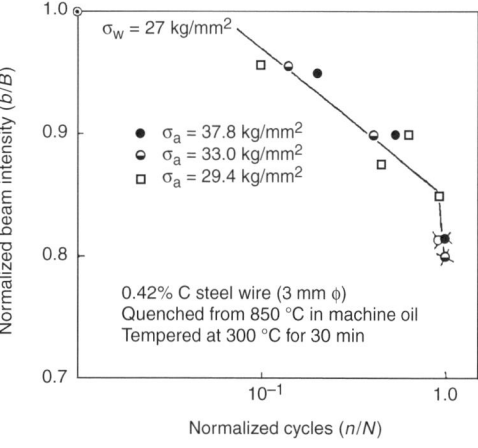

**Fig. 10.14** Diffraction line broadening during fatigue of quenched-and-tempered low-carbon steel. Source: Ref 10.17

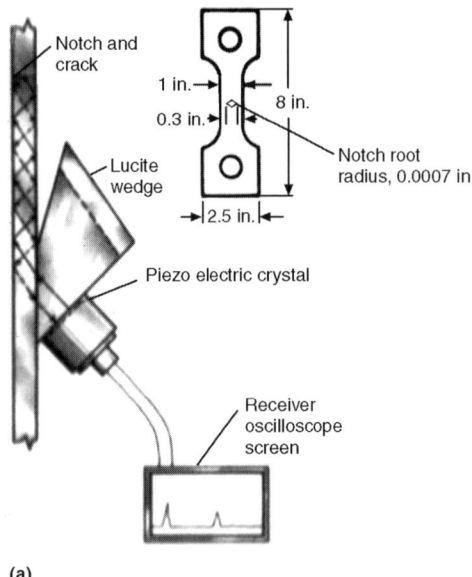
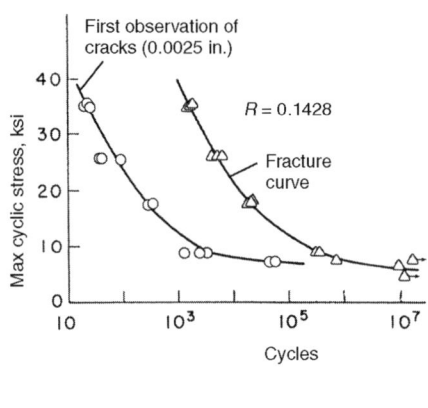

(a)          (b)

**Fig. 10.15** Ultrasonic technique for detecting early fatigue cracks. (a) Test configuration and specimen. (b) S-N curves for notched 2014-T6 0.60 in. sheet. Source: Ref 10.18

inant mechanism for dislocation multiplication. With progressive plastic deformation, a complex arrangement of dislocations can develop, such as illustrated in Fig. 10.13. Thus, we can predicate that a material subjected to plastic deformation will contain many dislocations. Repeated reversals of plastic deformation not only serve to generate new dislocations, but also serve to agglomerate the imperfections associated with the dislocations into groupings that are more damaging structurally than when they exist individually. There are several ways in which this effect can occur, and we discuss them in the following.

**Agglomeration of Dislocations.** The proliferation of dislocations and their to-and-fro motion enables their agglomeration to cause local structural imperfections, that start small but eventually destroy the cohesive integrity of larger regions of the material.

Figure 10.16 shows a very simple case of how a localized region can develop a microstructural flaw. Dislocations A and B, independently moving back and forth may eventually arrive at the configuration shown. The field of atomic cohesion is weakened, and the line of atoms along a-a and b-b may spring apart, establishing a vacancy, the first level of atomic imperfection. Or, as shown in Fig. 10.17, two, three, or even more dislocations can coalesce along the slip plane, and react to establish even larger atomic imper-

fections. Further plastic flow can either cause an imperfection to grow locally or cause independently developed imperfections to agglomerate into larger imperfections.

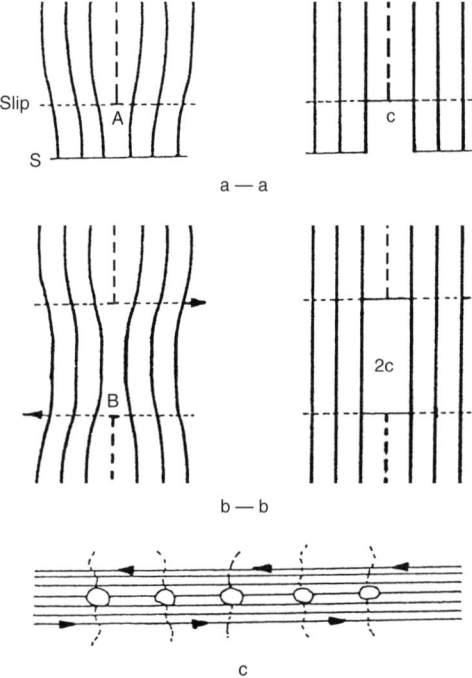

**Fig. 10.16** Sketch of dislocations coalescing to form a crack nucleus. Source: Ref 10.10

The bubble raft analogy described in a previous section helps to explain how the discrete imperfections agglomerate. Figure 10.18 shows results of an experiment conducted wherein individual bubbles were burst throughout the bubble area, and then the flat edges displaced back and forth to simulate reversed shearing plasticity. These imperfections migrated in a somewhat random manner until they agglomerated into a single larger imperfection. By analogy, the individual vacancies move around, seeking other vacancies with which to merge and produce larger imperfections (pores). Figure 10.19 shows specifically how a bubble arrangement (a) involving a pair of dislocations would move (b) with cycles toward a large blown-out region, and eventually falling into the sump (c), enlarging it. We can thus hypothesize that plastic deformation mobilizes the dislocations, and in their motion brings them together to cause larger flaws. However, while bubble raft experiments can help illustrate such concepts, they do not account for the actual three dimensional stress fields surrounding edge and screw dislocations. The mechanics governing how various dislocations and their stress fields interact to produce new dislocations and other crystalline defects have been extensively considered (Ref 10.20).

Because the notion of the dislocation is within the slip planes, it is the slip planes that are most damaged by plastic deformation. Different slip planes develop different amounts of damage according to the happenstance distribution of dislocations in its vicinity and the resolved shear stress on each slip plane and associated slip direction. A highly damaged slip plane can go on to complete fracture. Figure 10.20 shows what Forsyth (Ref 10.19) called a "persistent slip band." By this, he meant that the slip band had so many deep grooves that attempts to level the surface topography in its vicinity by etching or electropolishing were unsuccessful. The damage "persisted" even through appreciable depth was removed. The eventual complete fracture of a slip plane across the whole grain dimension is illustrated in Fig. 10.21. Here it is seen that *AB* is a slip plane (it is parallel to the other slip bands in this vicinity) and that its larger width implies local cracking. The crack extends across the whole grain (at its ends the slip bands, or slip planes, run in other directions). Other cracks, such as *CD*, form independently along the direction of the slip planes of their own crystals. The growth of the jagged crack *ABCD* is not by extending *AB*, but rather by the merging of the independent cracks *AB* and *CD*. At this point the ligament *BC* breaks because it can no longer sustain the displacements of its terminals *B* and *C* after the independent cracks have formed.

**Breakup of Grains into a Substructure.** Another important factor served by plastic deformation is to break up the individual grains into smaller entities called *subgrains*. While the grain size for most materials is commonly measured in terms of fractions of a millimeter, the sizes of the subgrains are commonly measured in microns. As the many generated dislocations are moved about by plastic flow, they tend to agglomerate, forming boundaries between the subgrains. Adjacent subgrains have crystalline orientation slip planes that are only at a small angle to each other, but the effect is to permit more dispersed slip rather than have only a few discrete slip planes participate in the plastic deformation as would be required if the grains did not break up. This applied strain is more easily absorbed without inducing excessive stress among the atoms in the grain. The breakup of the grain into subgrains starts very early in the life and in itself does not represent severe

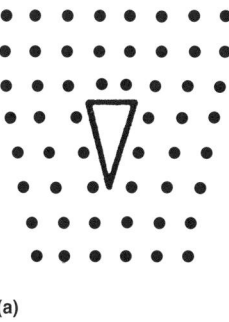

(a)

(b)

**Fig. 10.17**   Vacancy formation. Coalescence of (a) two dislocations and (b) three dislocations

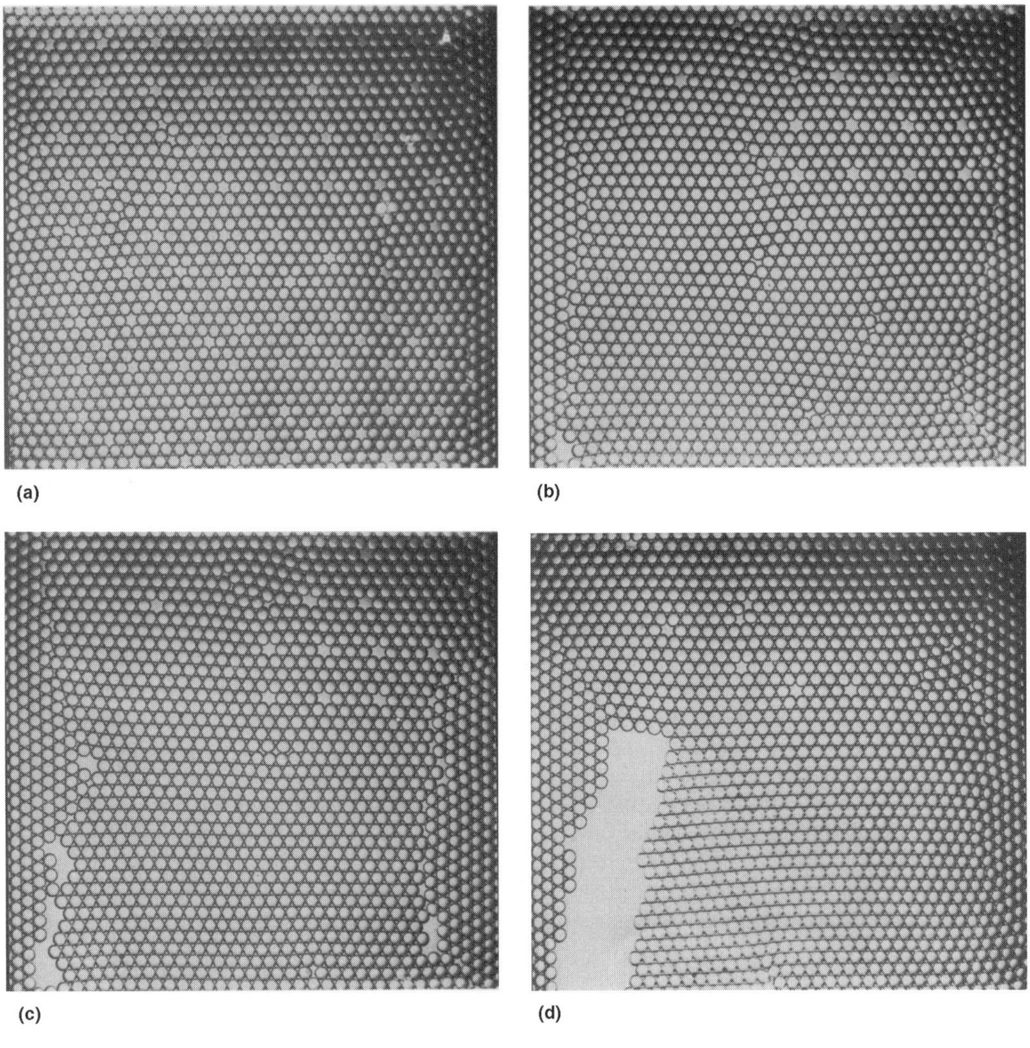

**248 / Fatigue and Durability of Structural Materials**

**Fig. 10.18** Bubble raft with initial 7% vacancies. (a) 0 cycles. (b) 2 cycles. (c) 3 cycles. (d) 9 cycles. Source: Ref 10.6

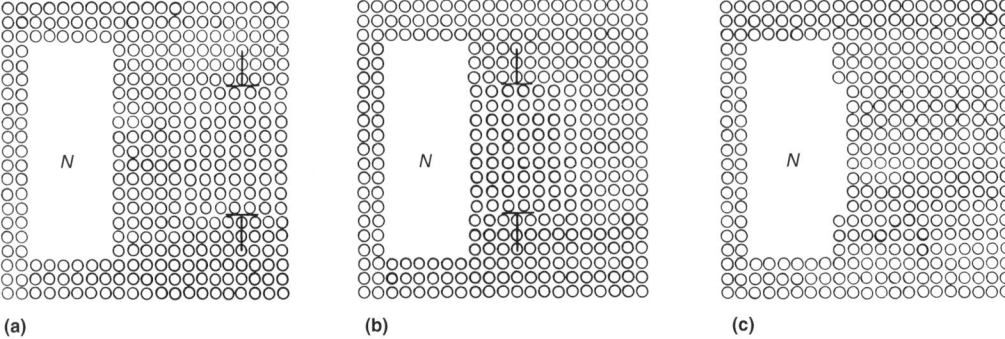

**Fig. 10.19** Edge dislocations of opposite sign (shown by ⊥ shaped symbol) moving along glide planes to condense and increase size of nucleus (N). Source: Ref 10.6

Chapter 10: Mechanism of Fatigue / 249

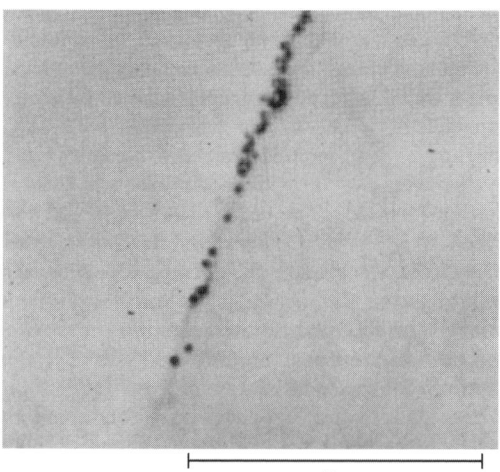

Fig. 10.20  Persistent slip band. Source: Ref 10.19

damage; in fact, as already noted if may be beneficial to the material in distributing, more evenly, the imposed strain. Observation of the subcell formation is easily made using transmission electron microscopy, as discussed earlier and shown in Fig. 10.9. The imposed strain range in this case was 0.004, giving a life of 500,000 cycles; yet such boundaries were clearly visible at 0.1% of the life. These boundaries change size and orientation and may reach a stable geometry after 1%, or a few percent at most, of the life.

That the initial formation of subgrains is not in itself microstructurally damaging is attested to by the fact that if the strain range is changed in early life, the dislocations simply rearrange themselves to be optimal for the new strain range. Each strain range has its own optimum cell size. Figure 10.22 shows results obtained in torsion by Pratt (Ref 10.22) on oxygen-free, high-conductivity (99.997% pure) copper. Some tests were conducted near room temperature, others near liquid nitrogen temperature, but the results were basically the same: the cell size was inversely related to strain. Stress is used in the figure, but essentially the same relation would be evident if strain were used. Nor does temperature affect the results; if suitable, small corrections are made to account for temperature changes on modulus of rigidity and Burgers vector (atomic spacing). Cell size depends only on current strain, and not on past history provided that the past history was not sustained for sufficient cycles to start cracking, and that the current strain is sustained long enough to achieve saturation. Here, the prior stress of 9.9 ksi for 1500 cycles was then changed to 8 ksi (Fig. 10.23). According to Pratt, the same cell structure was obtained by applying only 8 ksi without the prior loading.

What is the purpose of cell-size formation, and what effect does it serve? It facilitates the absorption of the imposed internal strain with a minimum of atomic disruption. It allows the material to trend toward a more natural or equilibrium state. There seems to be a natural hardness state for each material, for each strain level. If prior thermal and mechanical history has left the material harder than this natural state associated with that strain level, the rearrangement of the

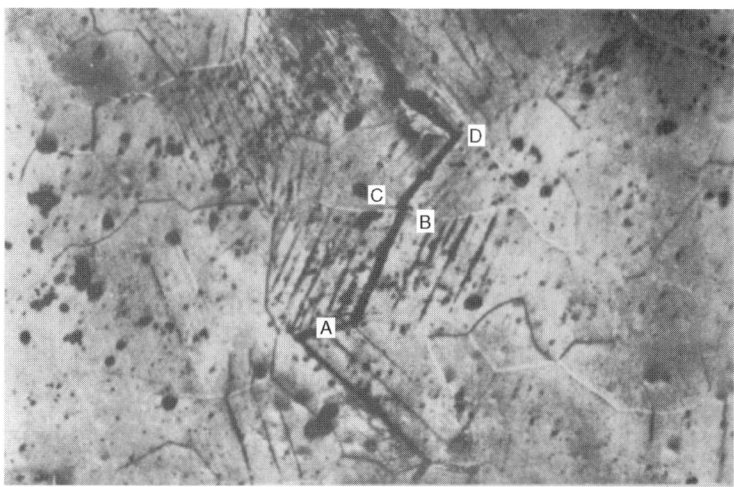

Fig. 10.21  Crack propagation along slip lines in aluminum. Source: Ref 10.21

dislocations through the mechanism of cell formation will soften it, and plastic flow can occur at lower stress levels. Conversely, if for the imposed strain level the dislocation substructure makes flow too easy, the dislocations arrange themselves in another substructural form to harden it toward its natural state. We discuss this tendency for copper in Chapter 2, "Stress and Strain Cycling." Thus it would appear that there are tendencies toward changes in cell formations associated with nonequilibrium between the current value of imposed stress (at a given temperature) and the saturation stress for that strain range. Once the saturation stress has been achieved, the cell structure stops changing and it too becomes saturated. That is why, in the work of Pratt discussed previously, each imposed strain range produces both a saturation cell size and corresponding saturation stress.

There is a price to pay for the benefit of deformation facilitation, however. The dislocations that form the subcell boundary walls are, of course, concentrated imperfections. And the large number of dislocations in close proximity renders them susceptible to flaw formation, as discussed in the section "Agglomeration of Dislocations." Eventually, then, the cell boundaries could become the origin of local cracks that could merge and start the disintegration of the microstructure. Thus it can be seen that dislocation formation and movement can serve a temporary beneficial function, but in the end it is the nemesis of structural integrity.

**Macroeffects.** In addition to the macroeffects of plastic deformation discussed previously, plasticity also contributes to several macroscopic effects that induce fatigue. It is appropriate to discuss these factors separately in the following sections.

## Surface Deteriorations

While reversed plastic flow can eventually cause fatigue in any interior volume of a material, it is the surface where the fatigue most usually originates. True, even when the fatigue starts at a surface, plasticity is an important factor, but other factors are present at the surface not present in the interior volume, and these factors aggravate the effects of plasticity.

The surface, by its very nature, is weaker than interior metal in the same state. It is possible, of course, to strengthen a surface mechanically, chemically, or metallurgically relative to interior volume, but it is then in a different state. First, the atoms at the surface do not have neighbors, and their constraint, in the region on the other side of the surface in the same way as interior atoms have a three-dimensional network of atoms surrounding them. It is likely, therefore, that the stress required to cause plastic flow for these atoms is at least slightly lower that at interior atoms. Secondly, a machined surface has invariably been plastically deformed. A higher number of resulting dislocations, twins, and other crystallographic defects are usually present near the surface, which can encourage enhanced fatigue damage there. Furthermore, the surface is inevitably not perfectly smooth. Whether it be

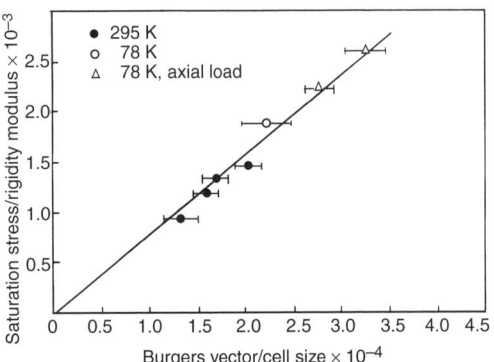

**Fig. 10.22** Dislocation cell size corresponding to saturation stress achieved at various cyclic strain ranges for 99.997% pure oxygen-free, high-conductivity copper. Source: Ref 10.22

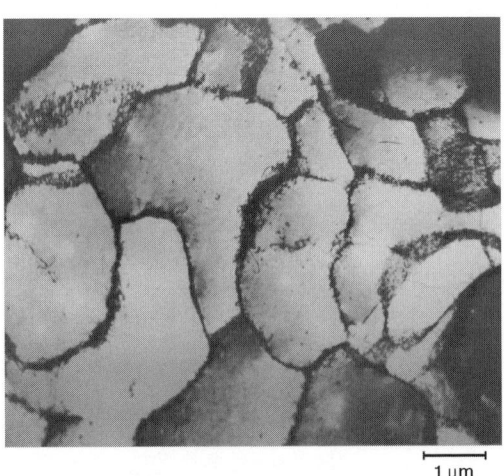

**Fig. 10.23** Dislocation cell structure (≈15,000×) for 99.997% pure oxygen-free, high-conductivity copper after strain cycling at a saturation stress of 9.9 ksi for 1500 cycles and then 8 ksi at 295 K. Source: Ref 10.22

by the initial surface preparation ranging from extremely smooth electropolishing to rough hogging, perfect smoothness is never achieved. Therefore, irregularities introduce stress and strain concentrations that amplify surface strain. Finally, the surface is usually in contact with an environment that more often than not is chemically hostile to the metal involved. In this section let us consider the geometric factor first.

An interesting experiment reflecting the role of surface roughness is shown in the already discussed Fig. 10.6. Here, a specimen was first polished to a very fine finish and then subjected to alternating strain. A high-intensity laser beam was directed onto the surface and the intensity of the beam reflected from the surface was carefully measured. Initially, the reflected beam was very bright, but at some point its intensity became highly attenuated, as seen in the figure. Irregularities on the surface scattered the incoming light energy, leaving less to be uniformly reflected. Shortly after the reflected beam intensity started to decrease, the specimen fractured in fatigue. Either the rumpling of the surface was concurrent with the development of fatigue, or it was the very factor that led to the failure. It is likely that the latter is the case, but in any event, it is clear that the surface roughening was a precursor to the fatigue.

Another interesting experiment associating surface with fatigue was described by Raymond and Coffin (Ref 10.23). In their tests they cycled a specimen to greater than half of its life, and then machined enough material from the surface to eliminate any cracks they may have formed. The remaining specimen was then loaded at the same strain range for the same number of cycles as the initial specimen. If the sum of the two cyclic loadings had been imposed without machining, the specimen would have failed, but with the machining, no failure occurred. Many more sequences of machining and reloading were applied, bringing the total number of strain reversals to many times the sustainability of a single specimen without remachining. In fact, in this particular test, enough material was removed so that the testing was terminated for lack of material rather than because of the onset of fatigue. While questions could be directed at the generality of this finding, because such generality would imply an over-important role to surface compared with plasticity alone, the test clearly points to a greater importance of surface in inducing fatigue.

Wood et al. (Ref 10.10) have studied surface damage by taper sectioning, as already discussed in this chapter (Fig. 10.6). Much of the recognition of the importance of surface degradation is due to his work with this tool. Figure 10.24(a) (Ref 10.24) shows a schematic of how sliding slip planes can establish a surface configuration that results in *extrusions* (projection of material above the surface) and *intrusions* (missing material below the surface). Experimental evidence, attributed to W.A. Wood, is shown in Fig. 10.24(b).

The net effect of all the surface considerations is to cause many surface cracks to develop. Lipsitt (Ref 10.25) has made a careful study of the number of cracks that develop at the surface and the depth to which they penetrate, as shown in Fig. 10.25. He determined depth by etching

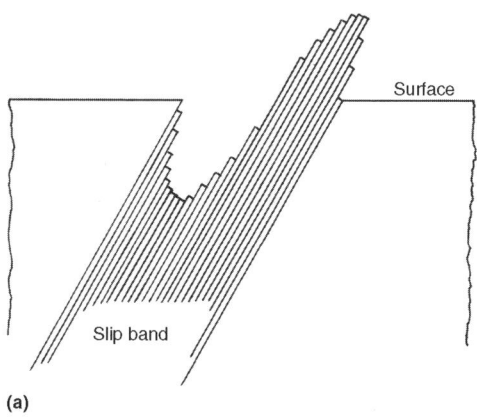
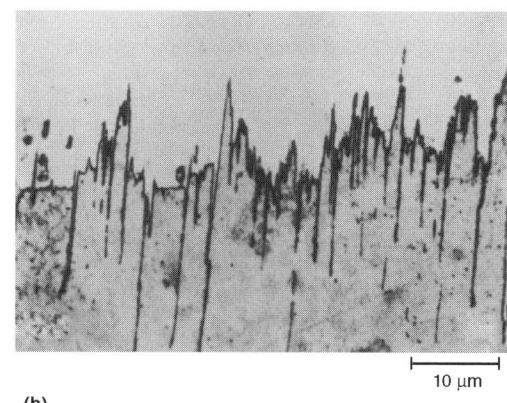

**Fig. 10.24** Slip-band geometry develops at slip band. (a) Schematic. (b) Experimental evidence attributed to taper-section work on copper of W.A. Wood; see Ref 10.24

away surface gradually and noting the remaining number of cracks on the surface after each etch. The high crack density is truly remarkable.

## Environmental Effects

The presence of a hostile environment aggravates both the effects of plasticity and surface degradation by adding chemical phenomena to the process. Examining Fig. 10.24, it is clear that as the surface notch develops, planes of atoms that were formerly interior now became exposed to whatever environment may exist on the other side of the surface. The nascent surface is initially very clean, and therefore very susceptible to chemical reaction. Even without such enhancement a hostile environment can attack the surface; with enhancement the attack can be worse.

Many environments are chemically hostile in fatigue. Even air, which is almost always present, contains oxygen, which oxidizes most materials. The resulting oxide is often foreign to the inherent metal. It is usually brittle and does not adhere well once formed. It is likely to spall off, removing surface metal, further roughening the surface and exposing fresh substrate material to the environment in a vicious spiraling cycle of deterioration. If moisture is present in the air, the attack can be much more severe.

A good example of the importance of moisture can be found in the work of Holshouser and Bennett (Ref 10.26). They attached a piece of transparent pressure-sensitive tape to an aluminum specimen that they were testing in bending under normal laboratory atmospheric conditions. To their surprise they noted that bubbles developed under the tape during stressing. To determine the nature of the bubbles they first made a chemical analysis and found them to be hydrogen—a further surprise. After considering a number of other possible explanations for the development of hydrogen bubbles, they concluded that the most probable one related to the dissociation of water vapor trapped in the air under the tape. The bubbles formed during surface deterioration when nascent clean interior at-

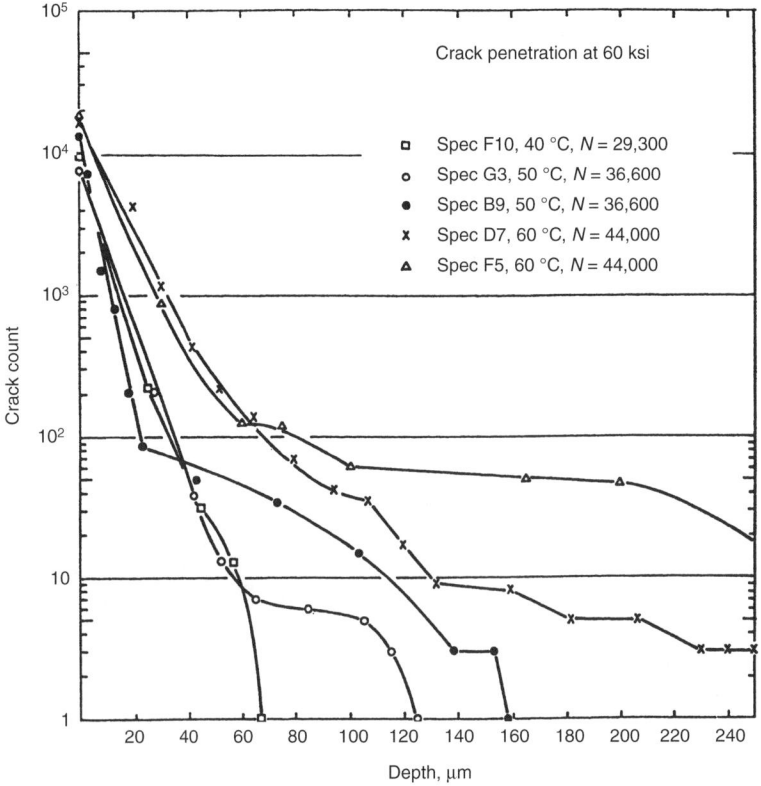

**Fig. 10.25** Surface cracking results of Lipsitt (Ref 10.25)

oms became exposed to the water. Because of the high affinity toward oxidation of the freshly formed surface, the latter actually dissociated the water into its oxygen and hydrogen components, seized the oxygen to form a surface oxide, and left the unused hydrogen as bubbles. They also found, interestingly, that the growth of surface cracks in the untaped region of the specimen was more rapid then in the taped region. Very likely the taped region had only a limited amount of water vapor from which the oxide could be formed, while the untaped region could use more of the surrounding air moisture to continue the oxidation as needed. In other tests they found that if the entire test section was covered with tape, the fatigue life increased, even though the structural strengthening due to tape was negligible. Obviously, by limiting the surface moisture to that existing after the tape was applied, the surface oxidation possible was less than if the moisture in the entire volume of air surrounding the specimen were available. Without protection, fresh subsurface material becomes oxidized and the brittle oxide spalls from the surface, thus exposing new surface for the process to repeat.

Figure 10.26 (Ref 10.27) shows a schematic of a typical extruded tongue (only 1000 Å thick) of nascent material that is produced during fatigue of an aluminum-copper alloy. Because aluminum has a high affinity for oxygen, oxidation occurs rapidly on both sides of the thin sliver of material. Being brittle, the oxide fractures readily and the tongue easily spalls from the surface of the sample.

How environmental pressure can affect fatigue life is clearly indicated in Fig. 10.27 (Ref 10.28). Here, aluminum specimens were fatigue tested in air at various pressures from extremely low values encountered in space applications (where $10^{-6}$ torr is typical), through normal atmospheric pressure (approximately 1 torr) up to $10^3$ torr. Two testing frequencies were used, 25 and 50 Hz, and various strain amplitudes were applied. The figure shows an abrupt change in fatigue life at about $10^{-2}$ to $10^{-3}$ torr. Below this transition the amount of oxygen present in the environment is inadequate to cause much oxidation, and going to pressures even 1000 times lower than the vicinity of the transition value did not improve fatigue life much, because in all cases there was inadequate oxygen to do much damage. Above the vicinity of the transition value the fatigue life was not reduced by much even if pressure was increased a thousandfold because the oxidation is paced by the amount of surface aluminum available to oxidize. In the transition range, however, the amount of oxygen was just right to maximize the oxidation damage. At the lowest strain range tested, the effect of the oxidation was enough to reduce life by a factor of 10. Note also that testing frequency also affected the results. At the higher frequency the time required to generate a given number of cycles is lower, and the oxidation damage is lower, increasing the cyclic life. These tests are typical of a vast literature that attests to the importance of environment in affecting the mechanism of fatigue. We give further evidence in discussing methods of avoiding fatigue in Chapter 11, "Avoidance, Control, and Repair of Fatigue Damage," where we cite various surface protection methods.

## Role of Foreign Particles

Another factor that significantly influences the mechanism of damage development is the

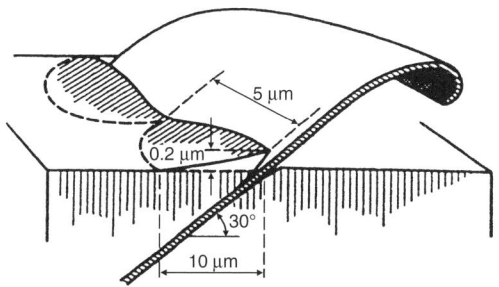

**Fig. 10.26** Tongue of extruded nascent material being formed prior to oxidation. Source: Ref 10.27

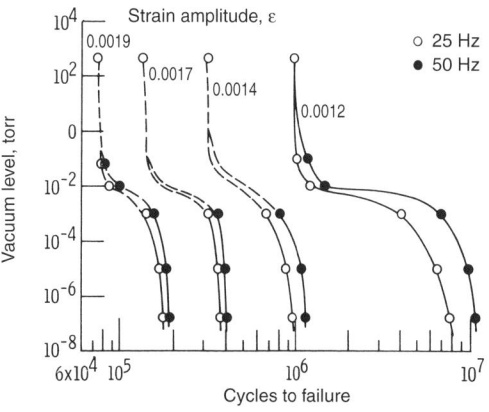

**Fig. 10.27** Pressure dependence of the fatigue life of 1100-H14 aluminum with strain amplitude and cyclic frequency. Source: Ref 10.28

presence of foreign particles in the microstructure. Such foreign particles may be inadvertent, for example, nonmetallic inclusions. Or, they may be purposely dispersed throughout the material (at least in some region where they are desirable) for strengthening the material both in monotonic and in cyclic loading. We shall not dwell on the latter dispersion strengthening materials because this subject has so many facets, and metallurgists have spent much time studying the principles of particle composition and spacing for various metallic systems. Suffice it to say that many successes have been achieved in imparting strength by this method. The particles are usually very small, and their spacing very close, so that for fatigue purposes the material can be considered as a continuum in the same sense that ordinary engineering materials are considered as homogeneous for purposes of analysis. It is the foreign particles that require special consideration from a fatigue point of view. We shall refer to them as nonmetallic inclusions.

Nonmetallic inclusions may be of various size and shape and are usually not uniformly distributed throughout the material of interest. They disturb the smooth transmission of stress and strain in the material because their elastic and plastic properties differ from those of the base material; they may be of jagged geometry, producing local stress concentration; they may be weak and brittle so that they can fracture at relatively low stresses; and the interface between the particle and base metal may be weak so that it is already debonded or can easily debond under strain.

If a crack starts at an inclusion, the local crack size can correspond to that which would require many reversals to achieve by the stage I process; this may change naturally as described later in the section "Distinction between Crack Imitation and Crack Propogation Phases." Thus, the failure of an inclusion either by fracturing or debonding can essentially bypass many of the crack-initiation cycles normally required to start a crack, especially in the high-cycle fatigue range. It is not uncommon to find that cracks in this range almost always start near a surface inclusion. Figure 10.28 shows such a crack that started after only 5% of the fatigue life in a 2024-T4 aluminum alloy. The implication is that the total fatigue life has been reduced by perhaps as much as a factor of 10 to 20 of the life without a surface inclusion. At high strain ranges associated with low fatigue life, where high plastic strain is already required throughout the test section volume, inclusions are not as damaging as in the high-cycle range. We highlight this distinction in Chapter 11, "Avoidance, Control, and Repair of Fatigue Damage."

## Summary: Factors Controlling Crack Initiation

Although the aforementioned four factors are in some ways interrelated—for example, surface irregularities develop because of plastic flow, internal discontinuities enhance local plastic flow—they must be considered as separate factors because they produce damage in independent ways as well and because in considering remedial or avoidance measures, it is appropriate to examine ways to suppress each without regard to the interrelating factors. Thus, it is more apparent that improving cleanliness would lead to improved fatigue in the high-cyclic life range even if considerations of plasticity might lead to the same conclusion in a more roundabout way.

## Crack Growth Factors

The development of microstructural and surface flaws does not necessarily mean that the fatigue life of a specimen or test piece is over. Sometimes the flaws develop early in life and the remaining life is very long, sometimes the

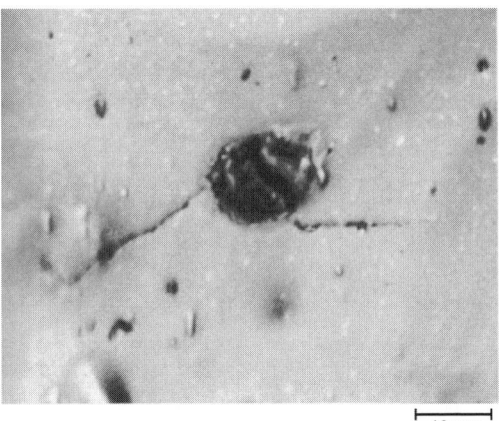

**Fig. 10.28** Fatigue crack initiation at a surface inclusion in 2024-T4 aluminum after 5% of total fatigue life. Source: Ref 10.24

critical flaws do not develop until most of the life has already been used up, and sometimes flaws develop but do not enlarge, as in, for example, so-called nonpropagating cracks. It is important, then, to consider the growth of these flaws from their first formation to the point of final fracture, or at least to the point when good engineering judgment dictates the wisdom of retiring the part from service.

**Realistic Model for Crack Growth in Ductile Materials.** In developing the crack growth laws of fracture mechanics (see Chapter 9, "Crack Mechanics") little consideration was given to the micromechanism by which the cracks actually grow. The treatment is empirical, and tests are set up wherein previously initiated fairly large-size cracks do grow, but the equation used to track the growth are based on equations of elasticity: The growth of the cracks must develop by either brittle fracture at the head of the crack or by some mechanism involving plastic flow. In general, the plastic flow mechanism in a ductile metal involves a blunting of the crack tip in tension and, on unloading, the resultant local compressive stress resharpens the tip, causing it to advance. The telltale sign of this mechanism is a ridge/valley feature, or striation, that is perpendicular to the direction of crack growth. For some materials, such as low-strength steels or high-strength superalloy single crystals, and for some cycling conditions, especially those involving high compressive forces, the striations become obliterated during crack closure, leaving the mechanism hidden. Various proposals have been made to explain the mechanism. The model proposed by Laird (Ref 10.29) is simple, and it fits the observations well. Let us thus adopt it for description of the growth process.

Initially, let us reexamine Fig. 10.3 to show the strain distribution at the head of the crack as determined by attaching a photoelastic coating on metallic materials. Also, examine Fig. 10.12, which reveals the plastic zone as determined by etching a material for which the etched microstructure is sensitive to plastic flow. It is evident from these figures that when load is applied to a part with a very sharp notch or a natural crack, the shear stresses and plasticity are highest in the directions that make an angle of anywhere from about $\pm 45°$ to about $\pm 70°$ with the direction of the crack line. Theoretical plasticity analysis (Ref 10.30) predicts an angle of $\pm 69°$. Laird drew on this observation and proposed a model that is depicted schematically in Fig. 10.29.

In Fig. 10.29(a) we show schematically the crack prior to load application. For reference we note that as shown, there are five striations included in the figure. Now start to apply vertical load. The crack opens and high shear stresses develop in the $\pm 69°$ directions, causing plastic flow at the lobes of the fifth striation at the head of the crack, as shown in Fig. 10.29(b). The plasticity continues, and the material ahead of the last striation rounds out as much as is necessary to increase the radius as is needed to reduce the stress concentration to stop plastic flow (Fig. 10.29c). In other words, by extending the crack by the needed length (depending on the load or strain applied), equilibrium is reached between the applied load and the stress concentration at the crack front, and crack growth stops. Next, a compressive load is applied. The material cannot, of course, simply reverse the process of plastic flow in an identical manner to the forward flow. Instead, the bulge at the nose is collapsed to the shape shown in Fig. 10.29(d) because of reverse shear stresses in the $\pm 69°$ directions. When the load is reversed, the configuration becomes as shown in Fig. 10.29(e). The tip of the crack now looks just like the tip in Fig. 10.29(a), except that a sixth striation has been added. In the next loading the configuration becomes as shown in Fig. 10.29(f), which is just like Fig. 10.29(b), except that the opening displacement is higher because the crack is longer.

Laird's model accomplishes several desirable goals:

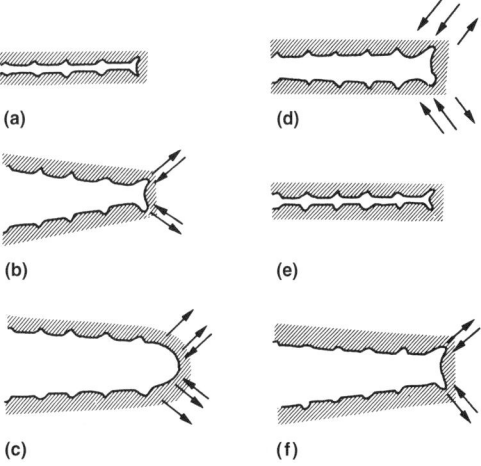

**Fig. 10.29** Laird's crack growth model. See text for explanation of various views. Source: Ref 10.29

- It explains how striations form
- It attributes the striations to plasticity
- It explains why the crack stops growing at the maximum load, answering the dilemma of why the crack does not just keep getting larger at the maximum load since stress intensity is obviously higher.

Most importantly, it explains the geometric feature of the striations that might otherwise be puzzling. Note that a trough in the upper surface lies opposite a trough in the lower surface, and a peak lies opposite a peak. It is not that the metal missing in a trough on one surface is matched by a mating peak on the other surface, as might be expected if local fracture were the mechanism of crack extension. Of course, for somewhat brittle materials some extension could be a combination of both mechanisms.

In general, striations often have been shown to develop on a one-for-one basis with load cycles as illustrated in Fig. 10.30. Here, a program of ten moderate amplitude loadings was followed by a single large load, and the program repeated. The result is 10 striations of moderate spacing followed by one of high spacing. Counting of striations can be a very useful tool in failure analysis because it can provide insight into loading orders that actually occurred when ambiguity might otherwise exist.

In some studies it has been found that tests in vacuum did not appear to produce striations; however, careful examination showed that striations were indeed present but not as pronounced. This finding reflects an emphasis on the role of oxidation in the fatigue process, even during crack growth. If plastic flow occurs at the head of a crack and exposes freshly formed surface material from previously protected interior material, local oxides may form. Such oxides interact with the plastic flow process itself and with the optical reflections from the cracked surface, so it is not surprising that striations formed in vacuum should have a different appearance from those formed in air. Yet, it is not likely that the basic mechanism is different in the two environments.

**Distinction between Crack Initiation and Crack Propagation Phases.** We are now in a position to make a clear distinction between two phases of the fatigue process, commonly called stage I and stage II, as defined by Forsyth (Ref 10.31). These are shown clearly by Laird in Fig. 10.31. In stage I the mechanism of defect formation is associated with dislocation and other microstructural defect agglomeration as discussed in the section in this chapter entitled "Agglomeration of Dislocations." In any case, the

**Fig. 10.30** Fatigue fracture surface of 7075-T6 aluminum showing the striations produced by a program consisting of a severe overload followed by ten constant amplitude load cycles. Source: Ref 10.24

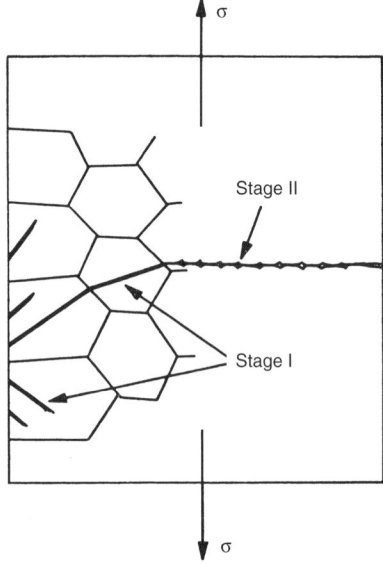

**Fig. 10.31** Illustration of fatigue stages I and II. Source: Ref 10.29

damage accumulates primarily in the slip planes in the near-surface grains. Eventually, a network of cracks in three or four grains forms a continuous crack. When viewed on the fractured surface they reflect a crystalline appearance and, as already described, may be the basis for one of the early theories of fatigue predicting the formation of a crystalline structure. This early cracking is commonly called stage I. In this region the crack is mainly featureless, that is, no striations are detected. Eventually, however, the crack becomes long enough to develop a stress intensity at its tip sufficient to start the cracking by the striation-forming mechanism described previously. This phase is now called stage II. Alternatively, these two stages are often described as the crack initiation phase and the crack propagation phase.

**Quantitative Separation of Crack Initiation and Propagation.** Because the crack initiation phase is mainly featureless, but the crack propagation phase shows distinct striations, it is easier to measure the striation phase and to determine the initiation phase by subtraction. Such a study was made by Manson in 1964 (Ref 10.32). Several materials were used, including a polycarbonate polymer that gave very good quantitative results. The clear striations for one of these tests are shown in Fig. 10.32, which gave a life of 115 cycles. Other specimens were studied as well. Admittedly, this material is a polymer, and there is no certainty that it follows the same crack growth laws as metals; nevertheless, it was quite useful in gaining an insight into crack growth, and, as will be shown in Chapter 12 the behavior is not inconsistent with the results from several metals also included in the study. Figure 10.33 shows how precisely the striations define the crack growth, especially when a semilogarithmic scale is used. The crack growth work requires much analysis before it can be used in a general way because stress intensity changes so drastically once the crack penetrates deeply into the specimen, and because it is not clear how the applied load is distributed once the cross section is partly separated. However, for determining initiation, i.e., when a crack depth of about 0.051 or 0.076 mm (0.002 or 0.003 in.) develops, they are very useful. For the 115 cycle life about 60% of the life is used up at this crack depth, and for the 1300 cycle life, it is about 90%. These points are plotted as the solid circles in Fig. 10.34 together with other points for other materials. The open circles are taken from Laird and Smith (Ref 10.33) in which they also counted striations on pure aluminum and pure nickel; data for other materials were taken from the studies of Manson and NASA colleagues. For the aluminum alloy and the two steels, initiation was approximated

**Fig. 10.32** Crack-growth striations in fracture surface of 6 mm (¼ in.) diam polycarbonate-resin specimen; 115 cycles to fracture. Source: Ref 10.32

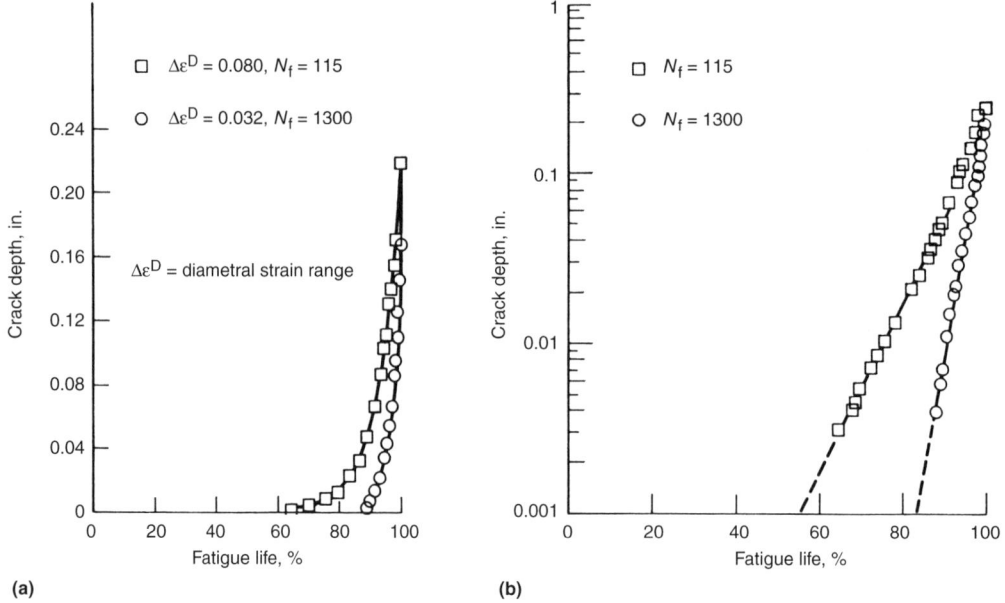

**Fig. 10.33** Striation crack growth in polycarbonate. (a) Rectangular coordinates. (b) Semilogarithmic coordinates. Source: Ref 10.32

from surface observations. When the surface crack length was about 0.254 mm (0.010 in.), it was deduced that the crack depth was of the order of 0.076 mm (0.003 in.) because experience had taught that depth was about ⅓ of surface length.

The remarkable observation that can be made from Fig. 10.34 is that in a smooth round specimen (in this case about 6 mm, or ¼ in. diam), most of the life is consumed in initiating a crack. This was observed for several different materials, and could be the case for many materials. When the life is $10^6$ cycles, over 99% of the cycles are required to initiate the crack. It may appear that the failure is sudden, which is why the impression has developed that fatigue is a sudden, unanticipated form of failure. But even 1% of life, in this case, is 10,000 cycles; so once the crack starts, it can take many cycles for it to propagate to failure. For lower lives a larger percentage of the life is used for crack initiation, but even at the 1000 cycle life, about 80% of the cycles is required to start the crack, while only 20% of the cycles propagate it.

**The Manson-Coffin-Basquin Equations as a Measure of Crack Initiation.** At this point it is appropriate to comment on the use of the strain-based analyses involving the Manson-Coffin-Basquin equation to calculate fatigue life. It has become common to call such analyses *crack initiation*, whereas another school of analysts who use the crack growth equations of fracture mechanics term their methodology *fatigue crack growth*. In the early days of the development of modern fatigue analysis practice, these two schools were almost mutually exclusive; the Manson-Coffin-Basquin advocates wondered why fracture mechanics was needed, and the fracture mechanics advocates sought to apply crack growth equations for cracks of the submicron size, and in fact felt that the Manson-Coffin-Basquin equation could be derived by in-

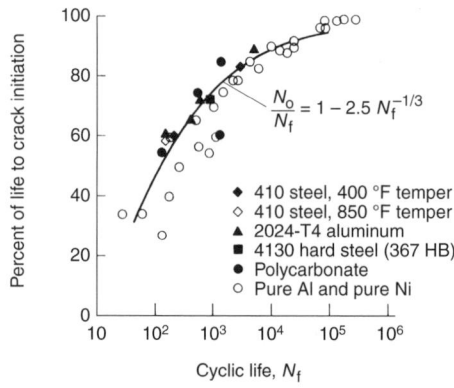

**Fig. 10.34** Relation between percent of life to crack initiation and fatigue life data from Ref 10.32. Source: Ref 10.32

tegrating the fracture mechanics crack growth equations starting with near-zero-size crack lengths. Actually, both schools were wrong in their perceived basis, for each method provides a niche that the other cannot fill, and the current enlightened view is that each method is useful in the treatment of a complete problem.

First, consider the Manson-Coffin-Basquin equation. The constants for it are determined from smooth specimens, usually of small size, cycling them at a fixed strain range until failure occurs. Here it is important to define what "failure" means. Some early investigators monitored the surface of the specimen until a visible crack occurred normally about 0.25 mm, or 0.01 in., which could be detected with the unaided eye. Obviously this is a subjective decision, and can seriously influence and confuse the results. Another school, which included our NASA laboratory, took the view that it was too much trouble to monitor the whole surface continuously. Rather, the specimen was loaded until fracture of the cross section occurred, defining the event as failure. In this way cycle counting was easy, and if the specimen broke in the absence of the investigator, an interrupting device would stop the loading, and life would be easily measured anyway. Nearly all the data in our laboratory were generated in this way, and probably also most of the published literature from other laboratories as well. Thus, data defining the Manson-Coffin-Basquin properties of materials refer to actual fracture of the specimens.

Figure 10.34 shows, however, that although most of the life as defined by these numbers is used to initiate the crack, it is not the entire life. Some of the cycles propagate the crack, and as the crack grows the stress intensity changes in a complex manner, so it is cumbersome to attempt fracture mechanics crack growth analysis once the crack has started in this type test. The Manson-Coffin-Basquin data are a good measure of crack initiation for a known surface strain, but they must be corrected according to Fig. 10.34 if they are really to refer to crack initiation.

**The Usefulness of the Paris Crack-Growth Equation and its Limitations.** Consider how crack growth data are normally generated. Usually the test specimen is a flat plate. Initially, a long crack is generated by first machining a deep chevron notch and then sharpening the tip by applying numerous fatigue cycles of very low amplitude, thereby developing a natural crack. The test then proceeds by applying an alternating load, and observing the growth of the essentially long crack. Figure 10.35 shows schematically the typical results from such a test. In region BC the curve is approximately straight, leading to the familiar Paris power law ($da/dN = C(\Delta K)^m$) wherein the crack growth clearly occurs by the striation mechanism. A threshold stress intensity, $\Delta K_{th}$ is indicated at point A. Below the threshold, cyclic crack growth is presumably absent, or at least very low. At point D, growth rate has increased dramatically and fracture is imminent.

**Small Crack Behavior.** Now consider the application of this type of data to a crack developing from a smooth specimen in a standard crack initiation-type fatigue test. So long as no crack develops, the Paris crack growth law is not applicable at all. Even if small cracks somehow develop in the early stages the curve may still not be applicable. Tiny cracks may produce very small stress-intensity factors (below the threshold) and crack growth would not be predicted. But the curve relates to the case where the low $\Delta K$ is generated by a very low stress associated with a very long crack artificially induced in the crack growth specimen, while for the small crack initiation fatigue specimen, the stress intensity derives from a very small crack size and whatever moderate stress is present. It has been recognized since the mid 1970s that "small cracks" do not follow the same crack growth law

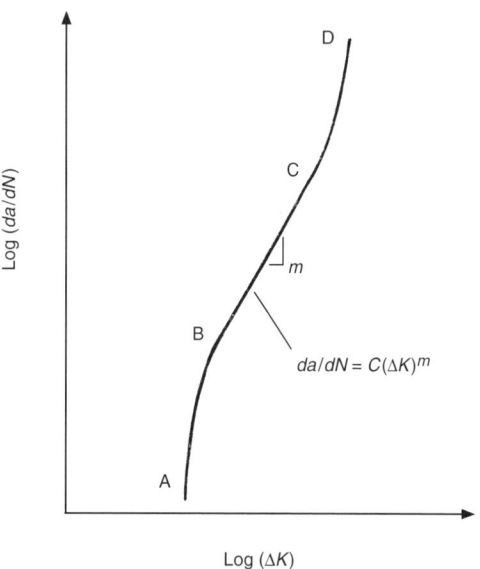

**Fig. 10.35** Schematic representation of fatigue crack growth rate versus range of stress-intensity factor, $\Delta K$

as long cracks. In the intervening years many researchers have compared, in the vicinity of the threshold regime, fatigue crack growth curves for short cracks with those generated with long cracks. It is readily apparent from the example shown in Fig. 10.36 (Ref 10.34) that not only do cracks grow at stress-intensity ranges below the threshold, but the growth rate can be significantly greater than rates for long cracks at the same $\Delta K$. Similarly, to treat small cracks associated with sizes in the crack initiation stage by the Paris power law in the region where the crack growth curve is linear may likewise be misleading because the curves are generated from long crack data while the types of stress fields generated by the small cracks differ substantially from those of the long cracks.

It would seem appropriate to conclude, therefore, that the crack initiation period should not be treated as a crack growth phenomenom based on crack growth curves for large cracks. It is wiser to treat initiation from data generated for Manson-Coffin-Basquin application, with proper correction as shown in Fig. 10.34, and then to apply the conventional crack growth laws for the region after crack initiation has already occurred and the small individual cracks of the crack initiation phase have already consolidated into a single dominant crack.

**Limitations of the Laird Crack Growth Model and Some Attempts to Quantify It.**

While the Laird crack growth model incorporates many of the features that agree with experimental observation, it oversimplifies the process. According to the model as described in the section in this chapter entitled "Realistic Model for Crack Growth in Ductile Materials," the entire crack extension process is due to plastic flow. No fracturing is involved in the crack growth, and no consideration is given to materials with limited ductility that are unable to sustain the large plastic strains implied in the model. Many analyses have attempted to improve the modeling and to quantify it. Pelloux (Ref 10.35, 10.36) has provided insight into potential differences in crack propagation mechanism between air and vacuum as shown by the model of Fig. 10.37. The crack tip deforms plastically while the remainder of the sample behaves elastically. In theory, plastic deformation should be reversible, but there are several reasons why slip can be only partly reversible during unloading. Pelloux's proposal is that the newly formed oxide layer provides the necessary obstacle. Thus, striations are produced in air where oxide formation restricts reversed plasticity, but striations are absent in a vacuum where such restriction is not present. Actually, striations are formed in vacuum, but to a lesser extent than in air as discussed earlier.

Another model, due to Tomkins (Ref 10.37) and Tomkins and Biggs (Ref 10.38), is shown

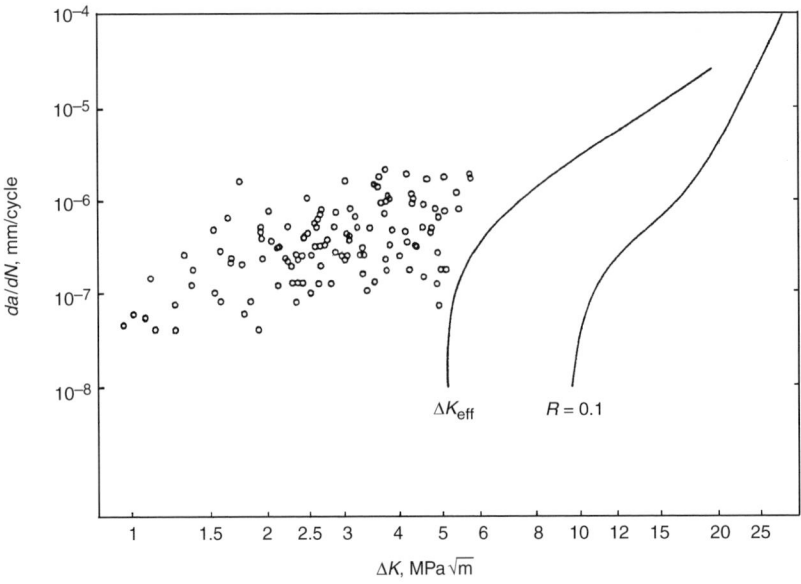

**Fig. 10.36** Small crack growth behavior doesn't follow long-crack growth behavior; Nimonic 901 at room temperature (curves are from long-crack data, points are short crack data). Source: Ref 10.34

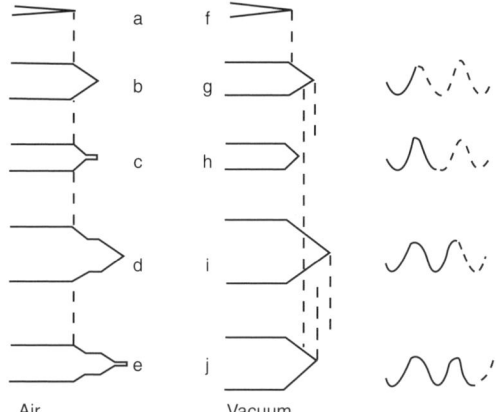

**Fig. 10.37** Alternate shear mechanism of crack tip advance for air and vacuum proposed by Pelloux (Ref 10.35, 10.36)

in Fig. 10.38. In this model tensile loadings initiate plastic flow in two narrow shear bands at 45° to the crack tip. Continued loading to the tensile strain limit produces a new crack surface that is formed by shear decohesion along the inner edges of the flow bands where the shear gradient is the largest. If strain hardening occurs in the flow bands before the strain limit is reached, rupture can be expected at the tip of the blunt crack. Rupture at the crack tip exposes new flow bands, allowing the process to repeat. Unloading reverses shear flow, and the crack closes without any significant recohesion.

Other methods of crack growth that consider the quantitative plastic flow and relate it to fatigue fracture have been proposed. A summary of many methods is discussed by Baïlon and Antolovich (Ref 10.39) but they cannot all be discussed here for reasons of brevity. One, in particular, may be mentioned here because it attempts to tie crack growth with the Manson-Coffin life equation. Antolovich, Saxena, and Chanani (Ref 10.40) have developed a theory based on the idea that a low-cycle fatigue process occurs in some region one unit ahead of the crack tip, and that the crack advances $N$ units in $N$ cycles, which represents the number of cycles to crack initiation at an average plastic strain range $\Delta\varepsilon_p$ in the process zone ahead of the crack. In this way the crack growth becomes interrelated with the crack initiation life as defined by the Manson-Coffin plastic strain-versus-life equation. On this basis they concluded that the Paris equation exponent should be equal to $2/\beta$, where $\beta$ is their designation for the Manson-Coffin exponent. Figure 10.39 shows remarkably good agreement between these two parameters, which may give some credence to their concept. However, it may be noted in the figure that $2/\beta$ ranges in value from 1.5 to 5, implying Manson-Coffin exponents from 0.4 to 1.33, which is quite broad, based on our experience. Also, this method might have difficulty explaining the striation spacing.

**Concluding Remarks Regarding Treatment of Fatigue as a Two-Stage Process.** While it is a noble goal to attempt to relate the parameters of the crack initiation stage to that of crack propagation because valuable insight into the mechanism of fatigue results, it fortunately does not seriously affect the ability to treat engineering problems even if such a goal is not accomplished. The constants for the Manson-Coffin equation and for the Paris crack growth law can both be simply obtained experimentally, and with the proper interpretation can be used to treat the two stages separately. The method is especially useful when treating cracks starting from stress concentrations. It is important to recognize, however, that inclusion of both stages is

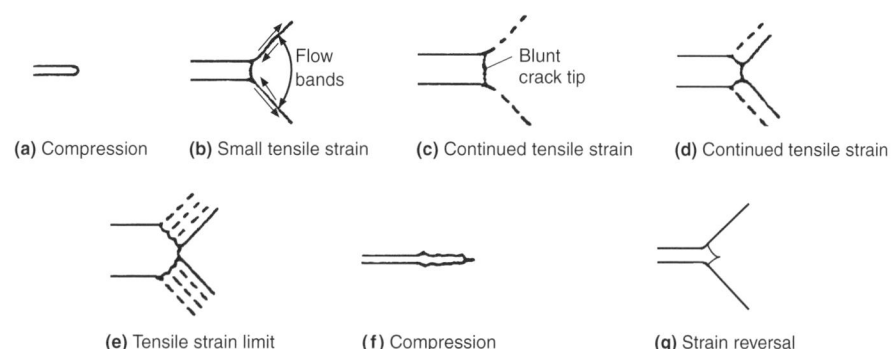

**Fig. 10.38** Fatigue crack tip advance by plastic flow proposed by Tomkins (Ref 10.37) and Tomkins and Biggs (Ref 10.38)

necessary and that neither needs be related to the other.

## Implications of Rolling Contact Applications Regarding the Fatigue Mechanism

Rolling contact parts—ball and roller bearings, for example—provide an excellent application for examining the fatigue mechanism. A tremendous amount of progress has been made in recent years toward improving fatigue life of these components, and we briefly review some of this progress in Chapter 11, "Avoidance, Control, and Repair of Fatigue Damage." Here, it is appropriate to examine the principles that come into play in relation to the fatigue mechanism.

If calculations are made of stresses that develop in a modern ball bearing, the numbers are surprisingly high. It appears that a million pounds per square inch or more are being supported by materials that, based on common engineering properties, should be able to support in the vicinity of 1380 to 1725 MPa (200 to 250 ksi) for the achieved lifetimes. Part of the discrepancy is artificial because the usual calculations do not take into account plastic flow and realistic contact areas that actually develop in service. However, much of the apparent increase in service strength is real and is a reflection of some of the factors discussed previously regarding mechanism, and others that we can now discuss.

**Factors Associated with Calculation Procedure.** Stresses as calculated by the Hertzian equations are based on the elastic stress-strain curve and take into account only elastic distortions at the point of contact. Thus, for example, if a sphere is loaded onto a flat plate, the Hertzian calculations would take into account the elastic deformation at the point of contact of both the ball and the plate. Because the contact area is small, the calculated stress turns out high. But some plastic deformation always occurs because of the high stresses. Consequently, the contact area increases, and the stresses become lower both because of the higher contact area and because the cyclic stress strain curve is lower than the extrapolated elastic line when plasticity occurs. Thus, the Hertzian calculations are high. In addition, the rolling (together with some sliding) of the ball on the plate will wear a track, so the contact area is even higher. Thus, when taking into account plasticity and increased track width, the real stresses would be reduced relative to the Hertzian calculations, perhaps by 20% in an extreme application. This accounts for only a small part of the seeming discrepancy, however.

**Subsurface Location of the Critical Stress.** An additional factor is that the most highly stressed volume of material is below the surface. Figure 10.40 schematically shows the stress distribution in the subsurface of a plate loaded by

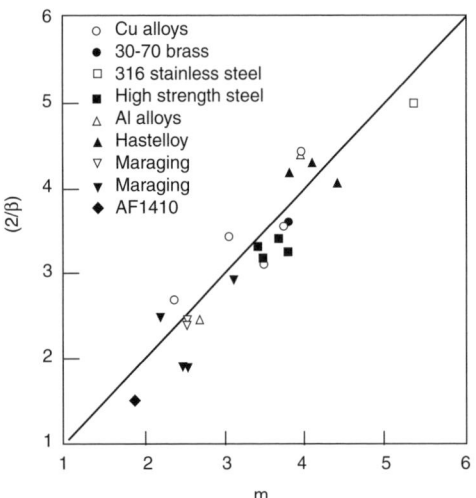

**Fig. 10.39** Predicted 2/β versus observed m Paris equation exponent. β is the Manson-Coffin exponent. Source: Ref 10.39

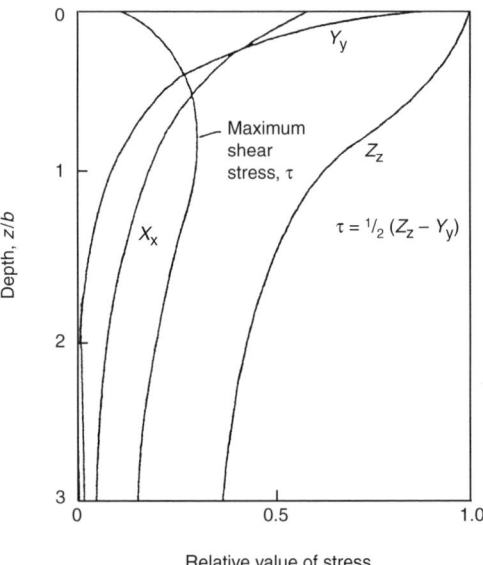

**Fig. 10.40** Relative value of three principal stresses and of shear stress with depth from surface. Source: Ref 10.41

a ball. The highest stresses occur below the surface, and they are triaxially compressive. Three benefits derive from this fact: surface irregularities are not involved, environmental attack is absent, and the compressive triaxiality improves fatigue life. This latter subject is discussed in Chapter 5, "Multiaxial Fatigue."

**Volumetric Effects.** As investigated by Kuguel (Ref 10.42), the amount of material subjected to the maximum stress significantly affects the load-carrying capacity. For this reason flexural bending can support more stress than rotating bending, which in turn can support a higher stress than axial loading, for the same fatigue life as discussed in Chapter 7. In rolling contact fatigue, the volume subjected to the highest stresses are small compared with the volume of specimens used in conventional axial fatigue testing.

**Fine Grain Size and Material Cleanliness.** Bearing materials have a very fine grain size and are exceptionally free of undesirable inclusions. As discussed in Chapter 11, "Avoidance, Control, and Repair of Fatigue Damage," both of these factors significantly improve fatigue life in the very-high cyclic life range. For bearings, life requirements are usually $10^8$ cycles or greater, and therefore these factors are especially controlling in this application. As noted in the section "Role of Foreign Particles" in this chapter, high-cycle fatigue failure is very often associated with the stress concentrations of inclusions, and fatigue fracture often initiates at inclusions. The initiating inclusions may be subsurface or surface connected. Bearing steels are melted and remelted multiple times in vacuum in order to clean the microstructure, increasing fatigue life immensely. The melt may also be center tapped, removing molten metal from the center of the heat, in order to avoid foreign material from the free surface or the walls of the melt vessel. Similarly, the refined grain size, achieved by working and heat treatment, provides many stage I crack stoppers, so fracture in many more grains is necessary before the crack length becomes large enough to develop a stress intensity high enough to propagate a stage II crack. In fact, the nature of the failure in rolling contact differs appreciably from that in axial or bending fatigue. Because of the subsurface maximum stress condition, the crack starts subsurface. But, because of the complex stress state, the growth of the crack is such as to form a pit as the separated metal spalls off. Thus, no large crack growth develops until the pits themselves form a network that produces a long crack.

**A Quantitative Study to Verify the Importance of Some of the Dominant Variables.** Moyar (Ref 10.43) and Moyar and Morrow (Ref 10.44) have made a thorough study of the importance of some of the aforementioned factors. They reconciled the apparent high stresses calculated in rolling contact of an AISI 52100 bearing steel that would be expected to support a much lower stress conventionally tested in rotating bending or torsion. We cite only a portion of the results of their extensive study:

- Assume that the apparent stress in the rolling contact application is 6900 MPa (1000 ksi) for a life of $10^7$ cycles to failure.
- Correcting for plastic deformation, the stress really supported for the $10^7$ life is about 5500 MPa (800 ksi).
- Correcting for distortion and track alteration, the real stress is of the order of 1400 MPa (200 ksi).
- Correcting for the small volume in the rolling contact application compared with that in a torsion specimen, the torsion specimen would be expected to support 690 MPa (100 ksi).
- Correcting for the multiaxial compression effect, the corresponding stress supportable in a torsion test is 520 MPa (75 ksi).
- In an actual torsion test of a moderate-size specimen, the measured stress for $10^7$ cycles was 520 MPa (75 ksi).

Thus, what appeared to be a stress of 6900 MPa (1000 ksi) in the rolling contact application was partly fictitious because the stress was calculated elastically and did not take into account plastic flow or wear at the contacting surfaces that would reduce the stress. However, it was also a real difference because of the small volume of material subjected to the high stress and because of the multiaxial compression present in the critical region. Yet, the researchers didn't even take into account the fact that in the critical region, the most highly stressed material is subsurface and protected from the environment. Nor did they correct for frequency effects.

## Putting It All Together: The Development of a Fatigue Failure

It is appropriate to conclude this chapter by discussing the probable sequences of events that lead to a fatigue failure. As noted in the introduction of this chapter, fatigue is not a failure event; it is a sequence of processes.

A clue to the fatigue process is the Manson-Coffin equation that relates fatigue to the amount of plastic deformation; it is the *cyclic plastic deformation* that is lethal to the material and causes it to fracture in the last loading even though earlier loadings of the same magnitude were readily withstood. As noted in this chapter, plastic deformation results in several different types of changes in the microstructure. It may, in some cases, but not always, affect the grain structure. For example, initially annealed metals become populated with small subgrains separated by low-angle boundaries, whereas initially cold-worked metals with a high density of intragranular dislocation tangles within grains undergo annihilation of dislocations as well as rearrangement and the metal loses strength. On rare occasions, the crystalline structure may change. For example, AISI 316 stainless steel may convert from austenite to martensite due to cyclic plasticity. Such effects are more or less global in the bulk of the microstructure. Even in the absence of such global changes, however, local changes always attend plastic deformation. The dislocation movement during plastic flow always changes the microstructure locally. What may have been a random distribution of isolated vacancies in the microstructure can be rearranged by the plastic flow into agglomerated larger imperfections. Each loading cycle enlarges the imperfection by bringing smaller imperfections together, as suggested by the work of Sinclair and Corten (Ref 10.6) in their bubble raft analogy. Eventually, the microstructure may become riddled with large imperfections. This behavior is much more pronounced at elevated temperatures when thermally activated processes can take place.

Identifying the fatigue process with plastic deformation pinpoints the slip planes so that the imperfections accumulate in selective areas of the microstructure and are not uniformly distributed throughout the entire volume. This enables fewer dislocations to cause more intense damage. Eventually, a small region, or several similar small regions, becomes weakened enough to start a small crack. When the crack is large enough to cause its own stress concentration, as discussed in Chapter 9, "Crack Mechanics," it starts growing according to the laws described there and eventually growing large enough to weaken the bulk cross section so that it can no longer support the peak applied loads. Then and only then does fatigue fracture occur. While the fracture may be sudden, it is the climactic termination of a series of processes.

## REFERENCES

10.1 S.P. Timoshenko, *History of Strength of Materials,* McGraw-Hill, New York, 1953
10.2 G.R. Gohn, "The Mechanism of Fatigue," *Materials Research & Standards,* American Society for Testing and Materials, Vol 3 (No. 2), 1963, p 106–116
10.3 H.J. Gough and D. Hanson, The Behavior of Metals Subjected to Repeated Stresses, *Proc. R. Soc. (London) A,* Vol 104, 1923, p 538–565
10.4 E. Orowan, Theory of the Fatigue of Metals, *Proc. R. Soc. (London) A,* Vol 171, 1939, p 79–105
10.5 W.J. Putnam and J.W. Harsch, Rise of Temperature Method of Determining Endurance Limit, *An Investigation of the Fatigue of Metals,* H.F. Moore and J.B. Kommers, Ed., University of Illinois Engineering Experiment Station Bulletin 124, Oct 1921, p 119–127
10.6 G.M. Sinclair and H.T. Corten, "Bubble Raft Model of Crystal Fatigue," T. & A. M. Report 86, Department of Theoretical and Applied Mechanics, University of Illinois, Urbana, IL, June 1955
10.7 W.L. Bragg and J.F. Nye, A Dynamical Model of a Crystal Structure, *Proc. R. Soc. (London), A,* Vol 190, 1947, p 474
10.8 W.W. Gerberich, "Plastic Strains and Energy Density in Cracked Plates, Part I: Experimental Techniques and Results," GALCITSM 63-23, Graduate Aeronautical Labs, California Institute of Technology, Pasadena, CA, June 1963
10.9 G.U. Oppel and P.W. Hill, Strain Measurements at the Root of Cracks and Notches, *Exp. Mech.,* Vol 4 (No. 7), 1964, p 206–211
10.10 W.A. Wood, S. McK. Cousland, and K.R. Sargent, Systematic Microstructural Changes Peculiar to Fatigue Deformation, *Acta Metall.,* Vol 11, 1963, p 643–652
10.11 A. Howie and M.J. Whelan, Diffraction Contrast of Electron Microscope Images of Crystal Lattice Defects, *Proc., R. Soc. (London), A,* Vol 263, 1961, p 217–237

10.12 J.C. Grosskreutz, "Research on the Mechanisms of Fatigue," Tech. Documentary Report WADD-TR-60-313, Part II, WPAFB Contract AF 33(616)-7858, Midwest Research Institute, Kansas City, MO, Dec 1963

10.13 G.A. Bassett, J.W. Menter, and D.W. Pashley, Moiré Patterns on Electron Micrographs, and Their Application to the Study of Dislocations in Metals, *Proc., R. Soc. (London) A,* Vol 246, 1958, p 345–368

10.14 W.C. Dash, The Observation of Dislocations in Silicon, *Dislocations and Mechanical Properties of Crystals,* John Wiley & Sons, Inc., New York, 1957, p 57–67

10.15 G.T. Hahn and A.R. Rosenfield, "Local Yielding and Extension of a Crack under Plane Stress," Report published by the Battelle Memorial Institute, Columbus, Ohio, 1964

10.16 A.R. Lang, Studies of Individual Dislocations in Crystals by X-Ray Diffraction Microradiography, *J. Appl. Phys.,* Vol 30, 1959, p 1748–1755

10.17 S. Taira and K. Honda, X-Ray Investigation of Fatigue Damage in Metallic Materials, *Trans. Jpn. Inst. Met.,* Vol 1 (No. 1), July 1960, p 43–48

10.18 S.J. Klima, D.J. Lesco, and J.C. Freche, "Ultrasonic Technique for Detection of Fatigue Cracks," NASA TN D-3007, National Aeronautics and Space Administration, Washington, DC, 1965

10.19 P.J.E. Forsyth, Slip Band Damage and Extrusion, *Proc. R. Soc. (London) A,* Vol 242, 1957, p 198. See also Fig. 7.1 in E.R. Parker, Theories of Fatigue, *Mechanical Behavior of Materials at Elevated Temperature,* J.E. Dorn, Ed., McGraw-Hill, New York, 1961, p 130.

10.20 J.P. Hirth and J. Lothe, *Theory of Dislocations,* 2nd ed., John Wiley & Sons, Inc., New York, 1982.

10.21 J. Schijve, "Analysis of the Fatigue Phenomenon in Aluminum Alloys," NLR-TR-M2122, National Aeronautical and Astronautical Research Institute, Amsterdam, 1964

10.22 J.E. Pratt, Dislocation Substructure in Strain-Cycled Copper as Influenced by Temperature, T. & A.M. Report 663, Department of Theoretical and Applied Mechanics, University of Illinois, Urbana, IL, March 1966. See also J.E. Pratt, Dislocation Substructure in Strain-Cycled Copper, *J. Materials,* Vol 1 (No. 1), ASTM, March, 1966, p 77–88.

10.23 M.H. Raymond and L.F. Coffin, Jr., Geometric and Hysteresis Effects in Strain-Cycled Aluminum, *Acta Metall.,* Vol 11, July 1963, p 801–807

10.24 J.C. Grosskreutz, "Fatigue Mechanisms in the Sub-Creep Range," *Metal Fatigue Damage: Mechanism, Detection, Avoidance, and Repair,* ASTM STP 495, American Society for Testing and Materials, 1971, p 5–60

10.25 H. Lipsitt, Crack Propagation in Cumulative Damage Fatigue Tests, *Proc., 11th Annual Air Force Science and Engineering Symposium* (Aerospace Medical Division, Brooks Air Force Base, Texas), Oct 20–22, 1964

10.26 W.L. Holshouser and J.A. Bennett, Gas Evolution from Metal Surfaces during Fatigue Stressing, *Proc. American Society for Testing and Materials,* Vol 62, 1962, p 683–693

10.27 P.J.E. Forsyth, *The Physical Basis of Metal Fatigue,* Elsevier, New York, 1969, p 50

10.28 M.J. Hordon, Fatigue Behavior of Aluminum in Vacuum, *Acta Metall.,* Vol 14 (No. 10), Oct 1966, p 1173–1178

10.29 C. Laird, "The Influence of Metallurgical Structure on the Mechanisms of Fatigue Crack Propagation," *Fatigue Crack Propagation,* ASTM STP 415, American Society for Testing and Materials, 1967, p 131–168

10.30 I.S. Tuba, A Method of Elastic-Plastic Plane Stress and Strain Analysis, *J. Strain Anal.,* Vol 1, 1966, p 115–122

10.31 P.J.E. Forsyth, A Two-Stage Process of Fatigue-Crack Growth, *Proc., Symposium on Crack Propagation* (Cranfield, UK), 1961, p 76–94

10.32 S.S. Manson, Fatigue: A Complex Subject—Some Simple Approximations, *Exp. Mech.,* Vol 5 (No. 7), 1965, p 193–226

10.33 C. Laird and G.C. Smith, Crack Propagation in High Stress Fatigue, *Philos. Mag.,* Vol 7, 1962, p 847–857

10.34 P. Newman and C.J. Beevers, Growth of Short Fatigue Cracks in High Strength

Ni-Base Superalloys, *Small Fatigue Cracks,* R.O. Ritchie and J. Lankford, Ed., The Metallurgical Society, Warrendale, PA, 1986, p 97–116

10.35 R.M.N. Pelloux, Mechanisms of Formation of Ductile Fatigue Striations, *Transactions Quarterly,* American Society for Metals, Vol 62, 1969, p 281–285

10.36 R.M.N. Pelloux, Crack Extension by Alternating Shear, *Eng. Fract. Mech.,* Vol 1, 1970, p 697–704

10.37 B. Tomkins, Fatigue Crack Propagation: An Analysis, *Philos. Mag.,* 8th series, Vol 18, Nov 1, 1968, p 1041–1066

10.38 B. Tomkins and W.D. Biggs, Low Endurance Fatigue in Metals and Polymers, Parts I, II, III, *J. Mater. Sci.,* Vol 4, 1969, p 532–553

10.39 J.-P. Baïlon and S.D. Antolovich, "Effect of Microstructure on Fatigue Crack Propagation: A Review of Existing Models and Suggestions for Further Research," *Fatigue Mechanisms: Advances in Quantitative Measurement of Physical Damage,* ASTM STP 811, J. Lankford, D.L. Davidson, W.L. Morris, and R.P. Wei, Ed., American Society for Testing and Materials, 1983, p 313–349

10.40 S.D. Antolovich, A. Saxena, and G.R. Chanani, *Eng. Fract. Mech.,* Vol 7, 1975, p 649–652

10.41 E.E. Bisson and W.J. Anderson, "Advanced Bearing Technology," NASA SP-38, National Aeronautics and Space Administration, Washington, DC, 1964, p 150

10.42 R. Kuguel, "The Highly Stressed Volume of Material as a Fundamental Parameter in the Fatigue Strength of Metal Members," T. & A.M. Report 169, Department of Theoretical and Applied Mechanics, University of Illinois, Urbana-Champaign, IL, June 1960

10.43 G.J. Moyar, "A Mechanics Analysis of Rolling Element Failures," Ph.D. thesis, Department of Theoretical and Applied Mechanics, University of Illinois, Urbana, IL, 1960. See also T. & A.M. Report 182, Department of Theoretical and Applied Mechanics, University of Illinois, Urbana, IL, Dec 1960.

10.44 G.J. Moyar and JoDean Morrow, "Surface Failure of Bearings and Other Rolling Elements," Engineering Experiment Station Bulletin 468, University of Illinois Bulletin, Vol 62 (No. 35), University of Illinois, College of Engineering, Urbana, IL, Nov 1964, p 14–24

# CHAPTER 11

# Avoidance, Control, and Repair of Fatigue Damage*

## Introduction

The most effective way to avoid fatigue is by attention to design details. All the preceding chapters are relevant to this precaution. Understandings of the mechanism and of the quantitative approach to design are, of course, expected to be incorporated in safe and successful designs. There are aspects, perhaps of a more qualitative nature, however, of which the designer must be aware in order to avoid pitfalls. For example, if a change of cross section is to be accommodated through a fillet radius that has already been chosen, it may matter less how accurately the stress distribution is calculated than to recognize that the stress can be lowered significantly by using a larger radius. Thus, it is important to include certain generalities in the designer's mind so that an efficient approach may be the starting point rather than an afterthought.

In addition to design there are many factors in materials specification, fabrication process (including the nature of the microstructure), and service parameters specification that are under the control of the designer and user that can greatly aid in improving service life. It is good for the designer to bear these factors in mind. Also, there is the question of inspection and possible repair that can greatly extend fatigue life.

If design features are incorporated to make it easy to inspect critical regions at specified intervals, or after some unexpected service conditions have been encountered, the need for remedial measures can be decided more effectively. Finally, there is the question of which remedial measures are available. Part replacement is usually possible, but only if the designer has anticipated this eventuality and has allowed for easy execution.

In discussing design and remedial factors it is desirable, wherever possible, to describe the fundamental basis underlying the approach. It is also important to draw distinctions that may render the approach meaningful for one set of conditions, while much less meaningful under other conditions. Method of loading, for example, load cycling as compared with strain cycling, may constitute an important difference. Whether the part is notched or smooth, whether it performs at high or low temperature, whether it is expected to last 100 cycles or 100 million cycles can make important differences in the efficacy of a given approach to the solution of a fatigue problem. Thus, we shall make use of the factors discussed in Chapter 10, "Mechanism of Fatigue." Because the subject of fatigue has been studied by engineers for well over 100 years, it is natural that many remedial approaches have been attempted, some with encouraging results. It is also to be expected that new approaches should be the reward of new understanding. In this chapter we do not try to cover all approaches that have been successfully applied in the past; a fine literature exists on this subject, and several references are cited later. Nor do we attempt to speculate on as-yet-untried approaches suggested by the basic mechanism. Rather, we cite selective approaches that have proved successful

---

*This chapter includes much of the content of a manuscript written by the senior author over three decades ago (Ref 11.1). To bring our discussion up-to-date, we have deleted some outmoded fatigue mitigation practices and new ones have been added. Occasional reference is made herein to high-temperature behavior when of direct pertinence to the materials being covered. A planned second volume will be devoted exclusively to high-temperature material and structural durability.

in application, or which merit further study on the basis of practical considerations and experience. It is convenient to categorize the principal approaches according to the following headings.

**Choice of Materials or Material Combinations of Inherently High Fatigue Resistance.** Although such a sweeping generalization almost defeats itself by begging the question, there nevertheless are some concrete guidelines that are very useful. The option of this category lies largely with the discretion of the designer, whose choice is therefore of fundamental importance. If the designer makes a bad choice of material, the user may have relatively little opportunity for rectifying the error.

**Avoidance of Strain Concentration.** Because fatigue usually is associated with high localized strains, any feature that reduces such localization, whether by initial design or by service operation, can be extremely helpful. The option here is, again, largely the designer's, but service operation and maintenance practices are also of extreme importance. It probably can be said with reasonable assurance that more fatigue failures occur because of neglecting the importance of strain concentrators than because of any other single factor.

**Surface Protection.** While fatigue may start in the interior of a body, it more commonly starts, in the absence of specialized surface treatments, at the surface. As discussed in Chapter 10, "Mechanism of Fatigue," the absence of corresponding atoms on the outer side of the surface alters the cohesive forces, removes constraints, and thereby renders the surface "weaker" than the interior. The surface often contains geometric stress-strain concentrations incorporated by the designer for functional reasons, or introduced inadvertently by rough handling.

Also, the surface may be attacked chemically by a pernicious environment, or mechanically by wear (gear teeth, bearings, journals, etc.). Thus, protection of the surface from the deleterious effects of each of these influences by coatings or surface treatments such as carburizing and nitriding (for steels) may also result in improved fatigue performance.

**Favorable Residual Stress.** Several methods are used commonly for improving fatigue life of components by the introduction of compressive residual stresses in the fatigue-prone surface layer. In all methods, only the surface material is stretched by the imposed localized plastic deformation. Upon removal of the deforming forces, the large elastic constraint of the bulk material works to "push" the stretched surface back toward its original configuration, thus leaving the surface material in a state of high compressive residual stress. These stresses serve as beneficial mean stresses that prolong fatigue life.

Because of its relatively low cost and ease of use, the most commonly used method is *shot peening*. Additional methods that offer distinct superior advantages over shot peening are available, but they are invariably less versatile and more expensive. These methods include *autofrettage, hole-expansion,* and two relatively new technologies: *laser shock peening (LSP)* and *low-plasticity burnishing (LPB)*. These more sophisticated methods can produce residual stresses to a much greater depth (0.635 mm, or 0.025 in.) than shot peening (0.127 mm, or 0.005 in.) while introducing less mechanical cold working to the material. Each method is referenced and discussed in the subsequent section in this chapter "Shot Peening and Similar Surface Treatments."

The efficacy of any of the methods depends on the ability of the material to retain the residual stresses during service. High local monotonic or cyclic strains regardless of sign tend to reduce, or even annihilate, the residual stresses. Likewise, high temperature may anneal away the stresses associated with the residual strains. It is here, especially, where a distinction must be made between high-cycle and low-cycle fatigue. Thus, in order to exploit the opportunities involved in this method it is necessary not only to outline the variants available, but the mechanism involved, and under what conditions the method may be expected to be applicable.

It is also worthwhile to mention that surface treatments such as carburizing and nitriding of steels, normally thought to assist only wear resistance, can also, in many cases, induce beneficial compressive residual stresses.

**Property Conditioning and Restoration.** Many processes designed to improve fatigue resistance can be considered under the first heading regarding the basic fatigue properties of the material. Thermomechanical processing, in which the material is subjected to a progressive history of cold and hot working together with a suitable schedule of heat treatment, is one such process. However, there are others, such as coaxing, which must be regarded as a separate class of treatment, not only because it may involve a fundamentally different mechanism, but because it is unique to certain materials. Thus, it is desirable to identify a separate category relating to fatigue property conditioning, which is intended to restore fatigue resistance once damage has occurred.

**Safe-Life and Fail-Safe Design.** When the designer is guarding against fatigue failures, two commonly accepted design philosophies are available: *safe-life* and *fail-safe*.

*Safe-life-designed* structures are those whose failure would eventually result in catastrophic loss of structure provided no intervention was taken.

*Fail-safe-designed* structures, however, retain functional load and deflection capabilities despite the fact that a single member is damaged or fails. The philosophy typically incorporates multishared load paths that provide structural redundancy. It also suggests use of high-toughness and tear-resistant materials that exhibit slow crack propagation rates that enable rigorously scheduled crack length inspections so that maintenance and repairs can be achieved. In a sense, fail-safe design can be regarded as very common because the concept of inspection for cracks and attendant part replacement are certainly not new. When parts are designed purposely for finite life, certain additional quantitative approaches become significant and require close attention to detail. In-depth discussion is found in a later section of this chapter. The fail-safe concept has been adopted in a number of applications.

**Distinction between High-Cycle and Low-Cycle Fatigue.** In dealing with remedial measures it is important to distinguish between high- and low-cycle fatigue. In many respects the distinction is arbitrary; some authors prefer to designate "low cycle" as less than $10^5$ cycles. From the point of view taken herein the distinction does not relate so much to the number of cycles to failure as to the mechanism involved. We shall regard "low cycle" as implying that a high plastic strain exists over a *macroscopic* region (many grain diameters), while "high cycle" implies that the strain is nominally elastic and involves very tiny amounts of cyclic plasticity only within *microscopic* regions. In high-cycle fatigue, therefore, a material discontinuity, such as a scratch or an inclusion, may cause plastic strain but only in a highly localized region. The presence of this discontinuity is inherently necessary to cause the plastic flow that initiates or propagates the fatigue crack. In the low-cycle region, on the other hand, the plastic flow is imposed globally, whether or not there are geometric strain concentrations. This distinction takes on importance in considering mitigation approaches relating to the significance of inclusions or other strain concentrators. It is also important in considering the significance of residual compressive stresses. Such stresses can be beneficial as long as they can be retained, but under high plastic strain, retention is difficult. Thus, remedial measures such as those associated with material cleanliness and the introduction of favorable residual stresses that are extremely useful in low-strain range applications, may not have corresponding utility in the high-strain range. Although it is strain range that really governs the utility of the approach, it is often more meaningful to the analyst to refer to the life range in discussing it. This reference to high-strain range may be taken as synonymous with low-cycle fatigue and low-strain range as equivalent to high-cycle fatigue. In many industrial applications there is often a combination of several events involving both the high-strain, low-cycle fatigue and low-strain, high-cycle fatigue regimes. This makes it all the more difficult to apply mitigation approaches that are beneficial across the entire spectrum of loading.

**Additional Mechanisms to Be Considered at High Temperature.** The mechanism that contributes to fatigue at moderate temperatures may potentially be active in fatigue at high temperatures. However, high temperature introduces additional mechanisms not normally encountered at lower temperatures. Although mentioned in passing in this chapter, we reserve detailed discussion of these mechanisms for a planned second volume. The prominent mechanisms are:

- The process of time-dependent, thermally activated straining called *creep,* a term that implies the damage is related to the type of damage incurred during a creep test.
- Material instabilities, which can cause, for example, new metallurgical phases to precipitate. Such phases not only influence the properties of the material, but also may affect the stresses that develop, which in turn may influence life. Among other instabilities that may occur are grain growth, and solutioning of precipitates that affect fatigue resistance.
- Surface interactions of a mechanical, chemical, or metallurgical nature such as corrosion, erosion, oxidation, sulfidation, decarburization, etc. In most cases such interactions are deleterious to fatigue life. Surface compounds produced by interaction with the environment may be weak and brit-

tle, or the interaction may cause surface discontinuities that lead to strain localization resulting in early fatigue failure.
- The presence of high temperature leads to thermal strains and stresses. Invariably, high-temperature parts will undergo nonuniform temperature changes both as they are heated to their operating temperatures and when subsequently cooled. The resultant thermal gradients cause different parts of the body to expand differentially, thereby introducing thermal strains and stresses that algebraically add to existing mechanical strains and stresses.

## Choice of Materials of Inherently Favorable Fatigue Resistance

We shall now review various factors to be considered in the choice of materials, material combinations, and processing variables that can be very helpful in improving fatigue life.

**Optimum Balance among Strength, Ductility, and Fracture Toughness.** A valuable guide in the choice of a material can be obtained by examining the method of universal slopes equation discussed in Chapter 3, "Fatigue Life Relations." From the equation it can be readily seen that if a material is subjected to a given strain range $\Delta\varepsilon$, its life $N_f$ depends on its ductility $D$ (related to reduction in area, RA, in a tensile test, by $D = \ln[1/(1 - RA)]$), its ultimate tensile strength $\sigma_u$ and elastic modulus $E$. A careful study of this relation reveals that when the imposed strain is high (of the order of 1 to 5%, or more, for most materials) the governing property is ductility; the higher the ductility, the longer will be the life. Below this strain range most materials benefit more from improved tensile strength than from increased ductility. The question of trade-off between strength and ductility has been discussed in Ref 11.2 where consideration was given to the determination of optimum heat treatment for fatigue life. It was assumed that a number of alternative heat treatments were available, each producing a different tensile strength and ductility. Generally, the higher the tensile strength, the lower is the ductility. A simple graphical procedure was suggested for determining the optimum heat treatment for maximizing fatigue life.

The role of fracture toughness is not as clear, however. In Ref 11.3 it was shown that two materials having the same ductility and tensile strength gave the same fatigue life regardless of the fact that their toughness values were far different. However, in none of the tests involved in this program were the geometrical and materials factors such that fatigue failure occurred in a brittle fashion. The same materials, tested either with sharp notches, or using specimens of large size, would undoubtedly have shown large differences in fatigue life; the tougher material very likely would have outlasted the less-tough material by a sizable factor. Thus, fracture toughness enters as an important material variable whenever an appreciable part of the life is in the crack growth stage. Certainly the length to which the crack can grow before catastrophic fracture occurs is dependent on the fracture toughness; even the rate at which the crack grows possibly is influenced by toughness in some materials, although quantitative relations are not completely established. The control-mode imposed on fatigue cracking has a great deal to do with the length of the crack at final failure, especially in high-cycle fatigue. A low-toughness material under force control will have a relatively small fatigue crack length at the point of residual strength overload. On the other hand, under displacement control, this material will have a longer fatigue crack length at failure. In the former case, the crack-driving force increases with crack length, while in the latter case, the driving force decreases with crack length.

Figure 11.1, taken from Ref 11.4 shows some indication of the importance of fracture toughness on fatigue. It is observed that the very high-cycle fatigue strength, for steels of moderate hardness, is directly proportional to the hardness. Beyond a critical hardness, however, the fatigue strength no longer increases as the hardness is further increased; in fact, the long-life ($10^7$ cycles-to-failure) fatigue strength becomes lower beyond this critical hardness. The "breakaway" hardness depends on the carbon content, the higher the carbon content the harder the material can be made before a loss in high-cycle fatigue strength sets in. Very likely, this behavior is related to loss in fracture toughness associated with increased sensitivity to inclusions as the material is hardened beyond a critical value.

Related information on the effect of fracture toughness is shown in Fig. 11.2 for low-cycle fatigue. Here, fracture toughness measurements were made (Ref 11.5) for a steel wherein hardness was varied by heat treatment. Beyond a hardness of approximately 300 Brinell, fracture toughness was found to decrease. Fatigue data (Ref 11.6) for this material at a large strain range of 2% under completely reversed loading are shown in Fig. 11.3. Above a Brinell hardness of 300 the life falls substantially. The reason for the drop in life is related undoubtedly to the drop in fracture toughness. For a large strain range, which produces large cyclic plasticity and a very low life, the number of cycles required to initiate a small crack is a relatively small proportion of the total life. The crack lengthens, of course, with increased cycling. When the crack length reaches a critical value, the laws of fracture mechanics take over and the specimen fractures. Thus, as the fracture toughness is lowered the life will decrease because the crack can grow only as far as the fracture toughness condition permits. Low fracture toughness tolerates a significantly lower critical crack length, thus taking fewer cycles to reach the critical crack length as shown in the figure. The above-mentioned results for low-cycle fatigue can differ considerably from high-cycle fatigue because at large strain ranges, the crack driving force does not diminish with crack length to the same extent as it does in nominally elastic cycling that is associated with high-cycle fatigue. We also discuss this factor in Chapter 9, "Crack Mechanics," in connection with a mild notch in a standard fatigue specimen.

From the foregoing semiquantitative discussion it can be seen that strength, ductility, and fracture toughness are inherent and often controllable materials properties that interact to govern fatigue life. Whether high ductility or tensile strength is the desired property, or whether fracture toughness is the governing parameter, depends on the loading conditions, and an analysis is very desirable to determine which property to feature. Further in-depth discussion on this subject can be found in Ref 11.5, 11.7, and 11.8.

**Material Cleanliness Factor.** Material soundness and cleanliness have long been recognized as major factors in governing fatigue. However, it is important to recognize that the role of cleanliness is very much greater in low-strain (high-life) fatigue than it is in high-strain (low-life) fatigue. As pointed out earlier, the high-cycle fatigue mechanism is more dependent on the presence of a localized strain concentration for the introduction of plastic flow than is the low-cycle fatigue mechanism wherein gross plastic flow occurs as a result of externally imposed high stress. Thus, localized imperfections in the form of pores or inclusions are highly instrumental in producing strain concentrations. The geometric nature of the discontinuity, decohesion at the interface between matrix and particle, or particle fracture itself, may be the cause of the localized strain. When the nominal strain is low, the strain concentrations associated with the particles become the governing factor in initiating plastic flow, and therefore cleanliness strongly influences fatigue strength. If the nominal strain is high, however, plastic

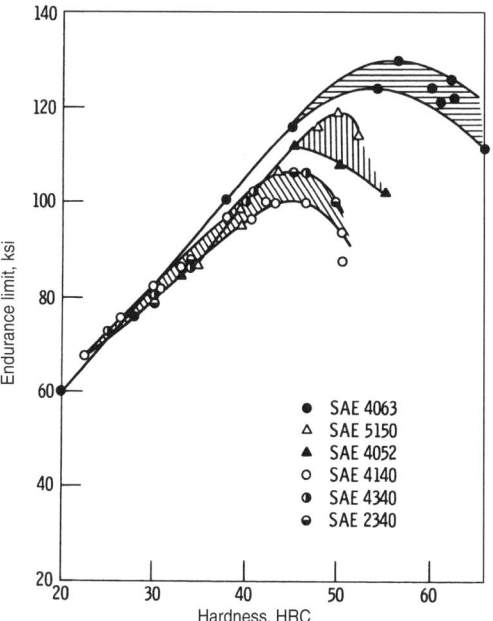

**Fig. 11.1** Optimum hardness for maximum fatigue strength of steels. Source: Ref 11.4

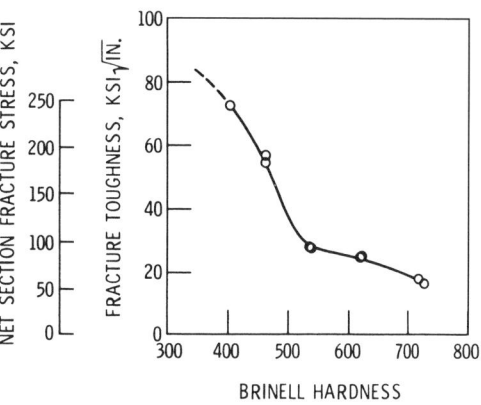

**Fig. 11.2** Decrease in fracture toughness at high hardness, 1.27 cm (1/2 in.) SAE 1045 steel specimens, V-notched to 0.953 cm (3/8 in.), precracked in fatigue, then heat treated. Source: Ref 11.5

flow occurs, particles or no particles, and life is influenced very little by cleanliness. Thus, we must expect the cleanliness factor to be most important in the high-cycle fatigue regime, and of less importance in the low-cycle life regime.

Evidence of the importance of material cleanliness is shown in Fig. 11.4 (Ref 11.9). Here the high-cycle fatigue strength is plotted against the inclusion count, and it is seen that as the inclusion count is decreased the long-life fatigue strength is increased substantially. The figure also shows, incidentally, one of the ways of determining a measure of cleanliness without expensive equipment. The density of the material, of course, will reflect the degree of inclusion count because inclusions (or voids, especially) are generally less dense than the metal they displace. Thus, an accurate determination of density can be used as a measure of fatigue resistance, as suggested in the figure.

A more conventional approach to the determination of cleanliness is the use of various nondestructive testing techniques, of which ultrasonic detection is a typical example. Figure 11.5 shows the results (Ref 11.10) of an investigation of a high-strength steel using an ultrasonic detection technique to determine a cleanliness index parameter. While the data are scattered, there is a definite trend to show higher fatigue life as the cleanliness is increased. Additional evidence from the same investigation is shown in Fig. 11.6. Here, the $L_{50}$ life (mean life at which 50% of the specimens of a given inclusion index have failed, sometimes referred to as $B_{50}$ in the bearing industry) is shown for loading in both the longitudinal and transverse directions relative to the working direction of the material. Results are shown for both unidirectional and completely reversed loading. Again, it is obvious that the lower the inclusion count, as determined by ultrasonic equipment, the greater is the life. The longitudinal direction produces a greater life at a given inclusion count and loading condition, as would be expected, because the inclusions are probably elongated with the major axis in the rolling direction; the strain concen-

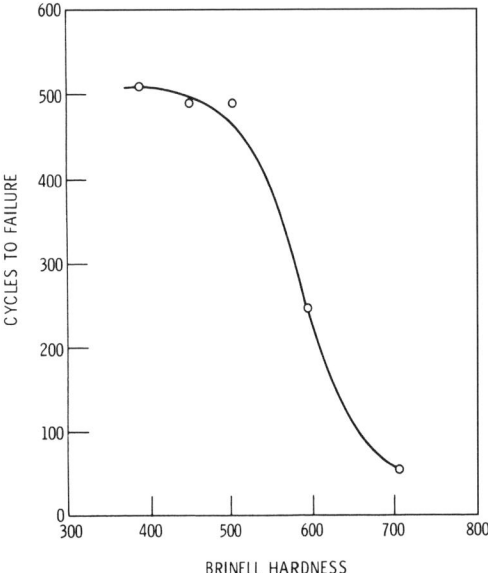

**Fig. 11.3** Fatigue life of SAE 1045 steel as a function of hardness for 2% strain range. Source: Ref 11.6

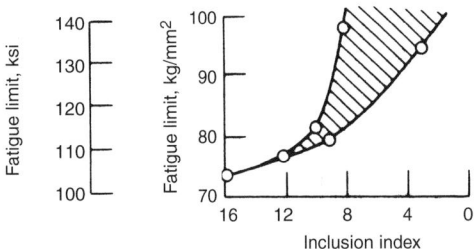

(a) Total average index of nonmetallic inclusions

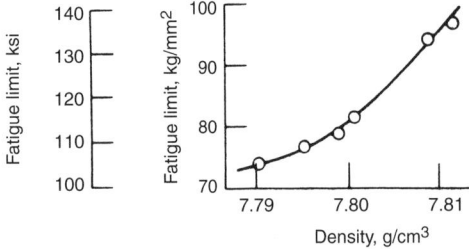

(b) Density of steel

**Fig. 11.4** Fatigue strength of ball bearing steel ShKh15 as a function of inclusion content. Source: Ref 11.9

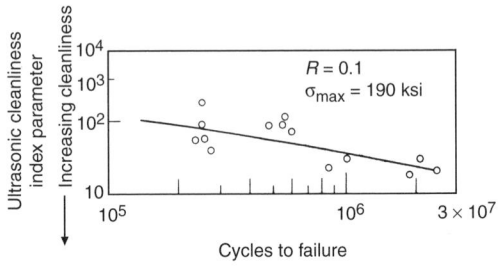

**Fig. 11.5** The effect of cleanliness on longitudinal fatigue specimens of SAE 4340 steel. Source: Ref 11.10

tration in the longitudinal direction is thus lower. Because fatigue life is dependent on inclusion count, it is clear that one way of improving performance is to separate lots of a component according to their inclusion count, using the higher-quality material for only the more demanding applications. Another approach is to control the cleanliness of the entire lot by increasing the cleanliness of the constituent materials, or by successive vacuum remelting to separate out impurities. Figure 11.7 shows some evidence of the success of this approach. Here, bearing material was successively vacuum remelted (Ref 11.11) to improve cleanliness, and it is seen that mean life kept improving even up to the fifth remelt. Note, however, that the life values are measured in the millions of cycles, which implies low nominal stresses relative to the yield point of the material. For the low-cycle fatigue range the effect of cleanliness becomes much less important because of the relative unimportance of the inclusions in governing plastic flow. The upward sweep of the curves at the lower lives in Fig. 11.6 demonstrates this behavior. As the life is decreased, inclusion count (and inclusion shape) remains important, but not as important in governing high-cycle fatigue life. Bilinear log (plastic strain range) versus log (life) curves that exhibit a steeper slope in the longer-life regime is believed to be due to material cleanliness.

**Grain Size Effects.** It has long been known that grain size influences mechanical properties. Thus, yield stress $\sigma_y$ for steels increases with decreasing grain size $d$ according to the well-known Hall-Petch relation (Ref 11.12) given by Eq 11.1:

$$\sigma_y = A + B(d)^{-0.5} \quad \text{(Eq 11.1)}$$

Fatigue life also is influenced considerably by grain size. There are several possible reasons for this effect. Fine grain size influences the amount of plastic constraint adjacent grains impose on each other as their yield strengths are exceeded individually. Also of importance, however, is the role of the grain boundary in impeding the progress of microcracks once they start in the slip planes of individual grains. Growth of a crack within a single grain is much easier because it involves plastic flow along a single set of slip planes. For the crack to penetrate an adjacent grain boundary, however, usually requires plastic flow along another set of slip planes, which may not be as favorably oriented with respect to the applied load as the crack that already has arrived at the boundary. Thus, grain boundaries can be regarded, in a sense, as crack arrestors. Fine grain size means more crack arrestors and smaller lengths of the cracks, if any, have initiated.

Evidence of the importance of grain size in governing fatigue life (Ref 11.13) is shown in Fig. 11.8. For smooth unnotched material, the fatigue strength at $10^8$ cycles can vary by a fourfold factor as the equivalent grain size is varied from about 0.0001 to 0.010 in. As a strain concentration factor is introduced, the significance

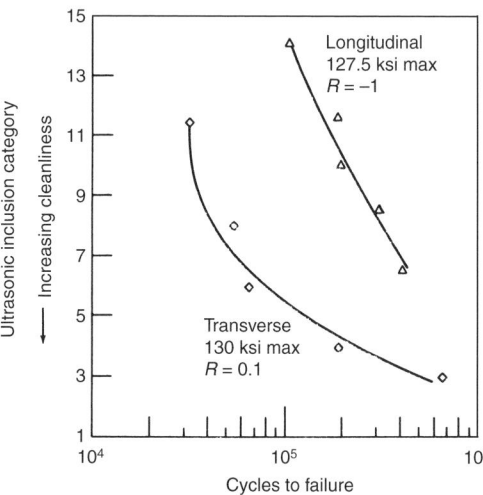

**Fig. 11.6** Axial fatigue properties of SAE 4340 steel (unidirectional and reversed loading) as a function of material cleanliness for $L_{50}$ life. Source: Ref 11.10

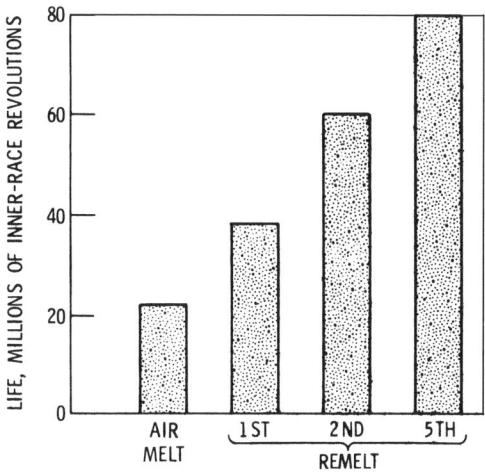

**Fig. 11.7** Increase in life of bearing inner races from successive consumable remelts of the same heat of steel (6309 size bearing). Source: Ref 11.11

of the grain size is reduced in magnitude. But some improvement in long-life fatigue strength always is obtained by reduction in grain size for all three curves shown in Fig. 11.8.

A more complete investigation on the effect of grain size was made by Vitovec in Ref 11.14. He chose a magnesium alloy because its hexagonal structure results in relatively few slip systems, thus emphasizing the effect of constraint of plastic flow resulting from the different orientation in adjacent grains. His results are shown in Fig. 11.9. Here are plotted the fatigue strengths at $10^7$ cycles for a series of specimens of different sizes wherein the theoretical stress concentration was the same. Three grain sizes in each specimen size were studied. Analysis of the results is intimately related to a number of complicating factors, including grain size effect on tensile strength, stress gradient effect for different stress concentrations, volumetric effects for specimens having proportional geometry but different absolute size, plastic constraint of adjacent grains, and so forth. However, one characteristic is clear from the figure: for any magnitude of stress concentration and physical size of specimen, the greater fatigue strength is displayed by the material of smaller grain size. The effect becomes greater as the size of the part increases. Thus it is generally true that, at room temperature at least, fine grain size is beneficial. The magnitude of the superiority of one grain size over another may depend, however, on the material and its basic crystalline structure, the size of the specimen, and the stress concentration factor.

In the preceding discussion the main emphasis is on the high-cycle life range of $10^7$ and $10^8$ cycles. Even in the low-cycle fatigue range, however, there can be a grain size effect. Table 11.1 shows some results from Boettner, Laird, and McEvily (Ref 11.15). In these tests aluminum alloy specimens were prepared to achieve average grain diameters ranging from about 0.127 to 2.667 mm (0.005 to 0.105 in.). When subjected to a stress level of 10.5 ksi, the lives varied from 1530 to 860 cycles; the larger the grain size, the lesser was the life. The most interesting aspect of these results, however, lies in the breakdown of the total lives into crack initiation and crack propagation periods. The crack propagation period is identified by the number of striations observed in a post-mortem examination of the fractured surface; the crack initiation period is determined by subtraction of the crack propagation cycles from the total life. It can be seen from the last two columns of Table 11.1 that the major difference in the lives of the various specimens was in the number of cycles required to *initiate* the crack; the finer the grain size, the higher the number of cycles for this stage of life. Again, therefore, even in the low-cyclic life range of about 1000 cycles, fine grain size is advantageous.

While the fine grain size is desirable at ambient and at moderately elevated temperatures, coarser grains become more advantageous in the *creep* range. It is well known that grain boundaries, which can be regarded as agglomerations of dislocations, contribute substantially to creep at temperatures above approximately half the absolute melting temperature. As discussed by Wells, Sullivan, and Gell (Ref 11.16), alternating creep deformation contributes substantially to high-temperature fatigue disintegration of materials. Thus, it might be expected that at temperatures well into the creep range the material of coarser grain would display the superior fatigue performance. In fact, if grain boundaries can be eliminated entirely, or at least kept in the longitudinal direction relative to the loading direction, fatigue performance might be improved.

It was as a result of this type of reasoning that directionally solidified alloys were investigated for improved performance both in static creep and in high-temperature fatigue. Thus, VerSnyder and Guard (Ref 11.17) showed that a normally brittle nickel-chromium-aluminum (Ni-Cr-Al) alloy had both higher ductility and longer creep-rupture life when a unidirectional columnar grain structure with a preferred ori-

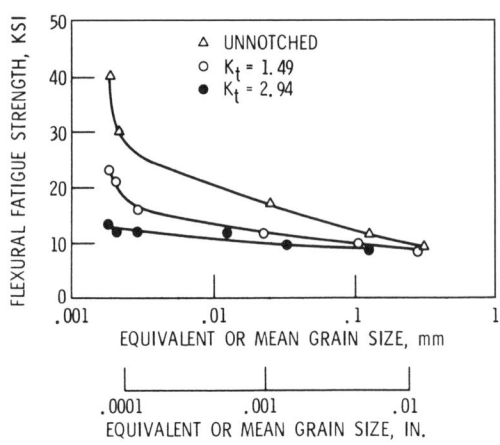

**Fig. 11.8** Influence of mean grain size on the fatigue strength of alpha brass at $10^8$ cycles. Source: Ref 11.13

Chapter 11: Avoidance, Control, and Repair of Fatigue Damage / 275

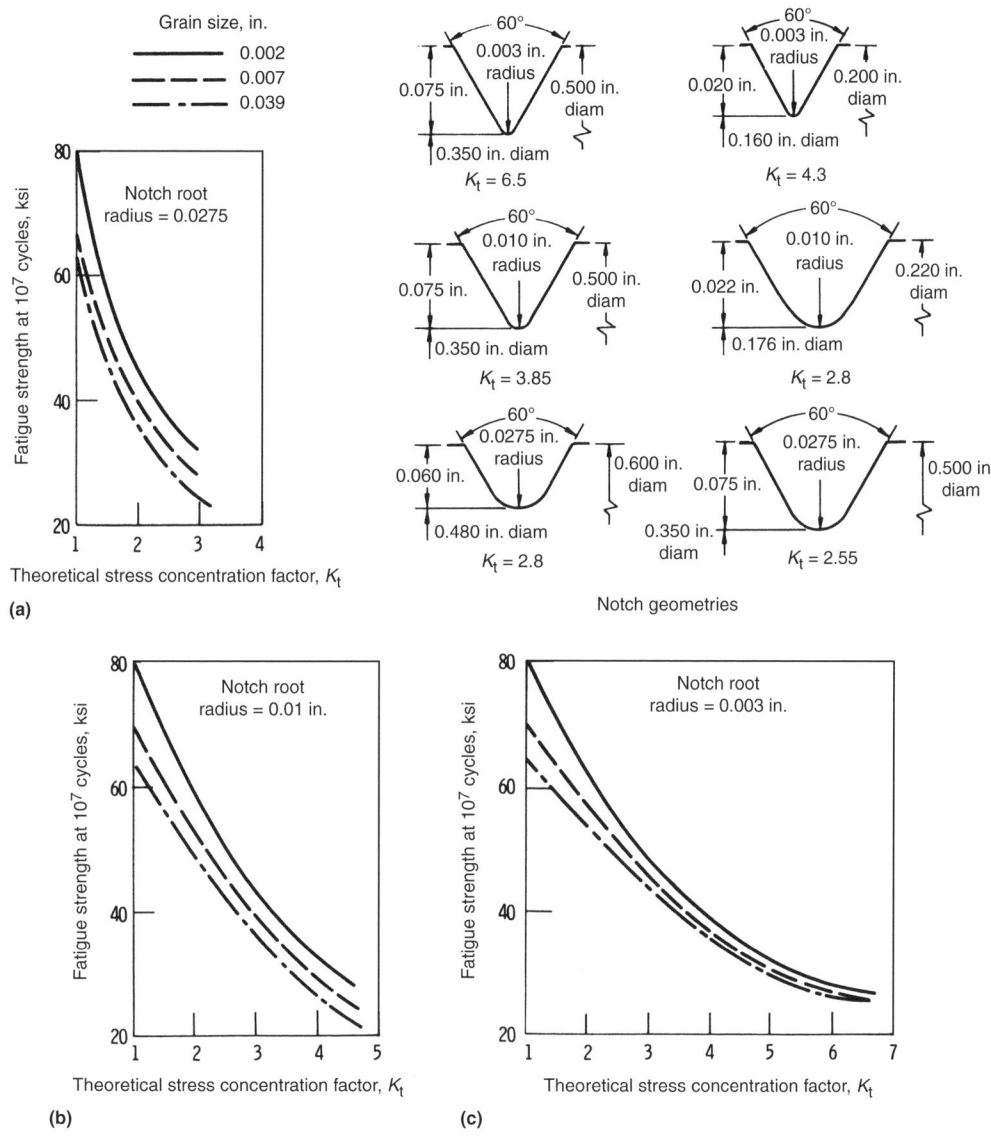

**Fig. 11.9** Fatigue strength as a function of theoretical stress concentration factor for an aluminum-magnesium alloy in several grain sizes. Source: Ref 11.14

**Table 11.1  Effect of grain size on low-cycle fatigue of aluminum alloy**

| | $\Delta\sigma = 10.5$ ksi, $R = -1.0$ | | |
|---|---|---|---|
| Specimen grain size diam, mm (in.) | Number of cycles to failure | Number of striations | Number of cycles for initiation |
| 0.127 (0.005) | 1530 | 250(a) | 1280 |
| 0.254 (0.010) | 1180 | 375 | 805 |
| 0.005 (0.020) | 1210 | 350 | 860 |
| 1.4 (0.055) | 1010 | 410 | 600 |
| 2.667 (0.105) | 860 | 405 | 455 |

(a) Accidentally low value caused by simultaneous crack propagation from two sites on the specimen surface. Source: Ref 11.15

276 / Fatigue and Durability of Structural Materials

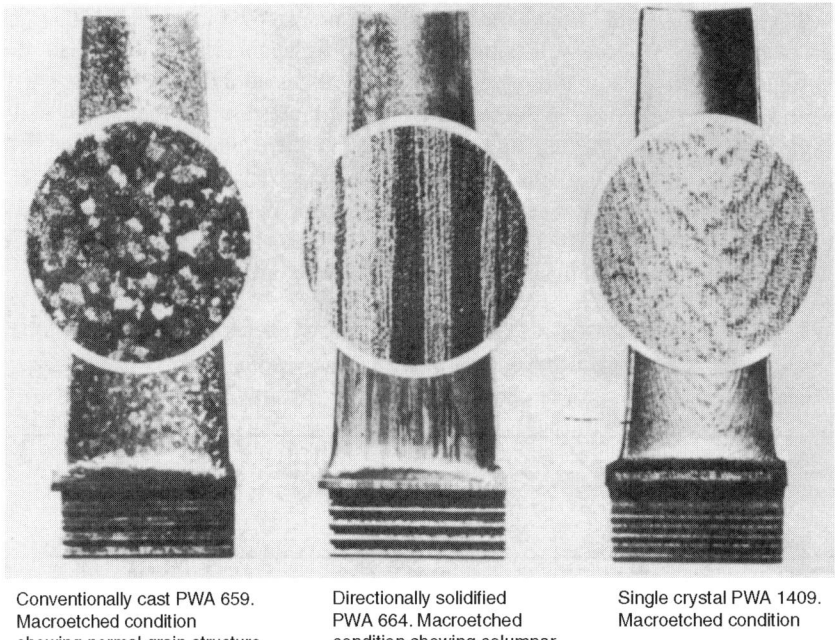

Conventionally cast PWA 659. Macroetched condition showing normal grain structure

Directionally solidified PWA 664. Macroetched condition showing columnar grain structure

Single crystal PWA 1409. Macroetched condition

**Fig. 11.10**   Pratt & Whitney developments of controlled grain structure in shroudless JT4 turbine test blades. Source: Ref 11.18

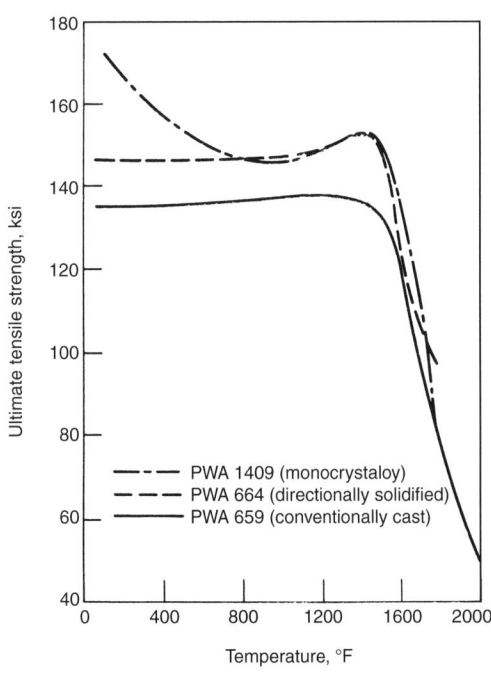

**Fig. 11.11**   Effect of test temperature on ultimate tensile strength of a nickel-base alloy in conventionally cast, directionally solidified, and single-crystal (monocrystaloy) form. Source: Ref 11.18

entation was obtained by directional solidification. The advanced materials development staff of Pratt and Whitney Aircraft (PWA) pioneered the area of controlled solidification. Much of their early work was done with the nickel-base alloy MAR-M 200. Figure 11.10 shows a turbine blade and enlarged sections of the microstructure for each of three forms of this alloy (Ref 11.18). The random polycrystalline form (PWA 659) is a relatively fine-grained material, the diameter of the grains being approximately equal along all three axes. In the directionally solidified form (PWA 664) the blade is "grown" from its base by control of an axial temperature gradient in a furnace, which causes progressive solidification onto already solidified material. Elongated grains are thus grown, and grain boundaries are all in the longitudinal direction, nominally parallel to the major stress axis of the blade. In the extreme case, solidification can be controlled to grow single crystals with no grain boundaries (PWA 1409). Refinements in chemistry and processing parameters of the single crystal alloy have resulted in subsequent designations of PWA 1480 and PWA 1484.

Figure 11.11 shows the effect of controlled solidification on the tensile strength of MAR-M 200. Comparing the ultimate tensile strength of the alloy in the random polycrystalline form (PWA 659), and in the directionally solidified form (PWA 664), the latter is seen to have higher strength up to at least 1700 °F. In the single crystal form (PWA 1409), higher strength is also achieved at room and low temperatures (Ref 11.18). Even greater superiority of the directionally solidified alloys is shown in their deformation characteristics. Figure 11.12 shows the tensile elongation of the three forms of the alloy (Ref 11.18), and it is seen how the progressive removal of grain boundaries increases the ability to elongate—a desirable characteristic for fatigue performance, especially when stress concentrations are present. For resistance to thermal stress, a low elastic modulus is desirable because a given strain (imposed by a thermal constraint) then manifests itself in lower induced thermal stresses. Figure 11.13 shows that both forms of directionally solidified material have a lower elastic modulus (in the [001] crystallographic orientation) than the conventionally cast alloy (Ref 11.18). In service the main advantage of the directionally solidified material would be expected to accrue because of its higher creep rupture and thermal fatigue properties. Figure 11.14 shows a comparison (Ref 11.18) of the rupture ductility and creep-rupture life of PWA 659, 664, and 1409, at three temperatures: 760, 870, and 980 °C (1400, 1600, and 1800 °F) and three stress levels. Significant improvements in rupture life and rupture ductility were obtained with the directionally solidified material at all three test conditions. The difference in creep-rupture life among the three materials decreases with increasing temperature.

It is interesting to note that significant increases in room-temperature strength and ductility have been obtained by means of controlled solidification as compared with the random polycrystalline form. Hence, the elimination of grain boundaries and the proper selection of crystallographic orientation of the remaining single crystal or columnar grains have brought about an improvement in properties. The prior

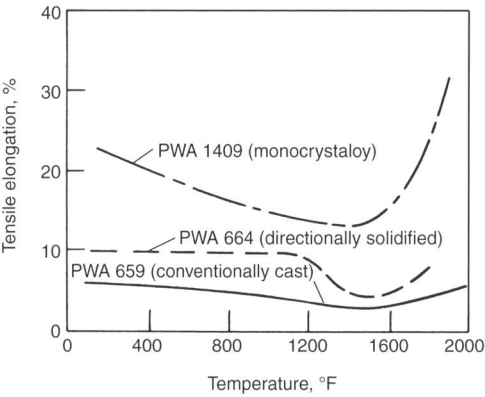

**Fig. 11.12** Effect of test temperature on elongation of a nickel-base alloy in conventionally cast, directionally solidified, and single-crystal (monocrystaloy) form. Source: Ref 11.18

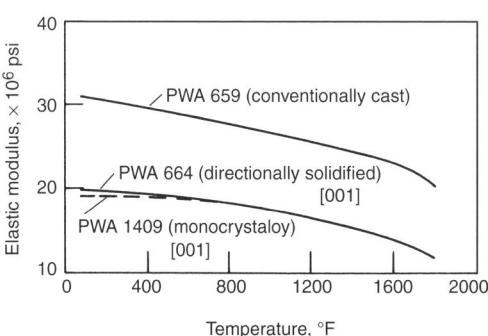

**Fig. 11.13** Effect of test temperature on elastic modulus of a nickel-base alloy in conventionally cast, directionally solidified, and single-crystal (monocrystaloy) form. Source: Ref 11.18

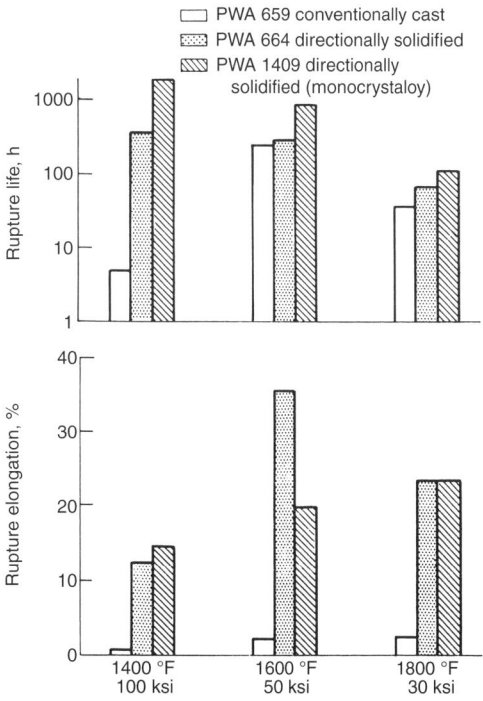

**Fig. 11.14** Creep-rupture properties of conventionally cast (PWA 659), directionally solidified (PWA 664), and single crystal (directionally solidified monocrystaloy) (PWA 1409). Source: Ref 11.18

discussion that attributed a strengthening character to grain boundaries at low temperatures would appear to be in conflict with these results. However, there is no real conflict when it is recognized that up to intermediate temperatures grain boundaries can strengthen grains that are weak as a result of an unfavorable crystallographic orientation with respect to the direction of applied tress. Of course, it must be further recognized that these weak grains would not be present were it not for the random grain orientations associated with conventionally cast polycrystalline structures.

Thermal fatigue results obtained at NASA (Ref 11.19) with conventionally cast and directionally solidified MAR-M 200 are shown in Fig. 11.15. Here, specimens simulating turbine blades were subjected alternately to a high-velocity, hot-gas stream and to air cooling, as designated in the figure. The plot shows the number of uncracked specimens as a function of applied thermal cycles. Of the seven random polycrystalline specimens tested, six were cracked after 20 cycles; all were cracked after 40 cycles. The seven specimens of directionally solidified material ([001] orientation) showed no cracking even after 100 cycles, when the test was discontinued.

The early work by Pratt & Whitney with directional solidification was later followed by NASA studies with some materials of their own development (Ref 20, 21). In general it has been found that for nickel-base alloys the greatest improvements due to directional solidification occur in the range from room temperature to approximately 980 °C (1800 °F) and that improvements of about a factor of 2 in creep rupture lives can be expected for many materials as shown (Ref 11.21), for example, in Fig. 11.16. However, some materials with relatively high ductility do not respond to directional solidification in that their properties are not improved by this process. Also the process is relatively expensive (although improved techniques have reduced cost greatly).

**Mechanical and Thermomechanical Processing.** While this chapter deals with the mechanical behavior of materials at temperatures below the high-temperature regime, many of the materials are processed at temperatures well into the high-temperature regime. Only cursory discussion is given herein of germane aspects of the high-temperature processing. For greater details, the reader is referred to the Appendix, "Selected Background Information."

One of the most effective ways to influence the mechanical properties of a material is by means of suitable heat treatment. Another is by working it, or deforming it under external force. More effective yet is the combination of mechanical working and heat treatment in proper sequence, commonly termed *thermomechanical processing*. Fatigue is among the properties most sensitively influenced by such processing.

An important way in which mechanical working improves the fatigue properties of a metal is by introducing a preferential grain flow direction. Not only are the grains elongated in the direction of the plastic flow, but material imperfections such as segregated impurities and voids also are elongated in the flow direction. Their effective stress concentrations thus are reduced in the longitudinal direction but are increased correspondingly in the transverse direction. When mechanically worked metal is used in fa-

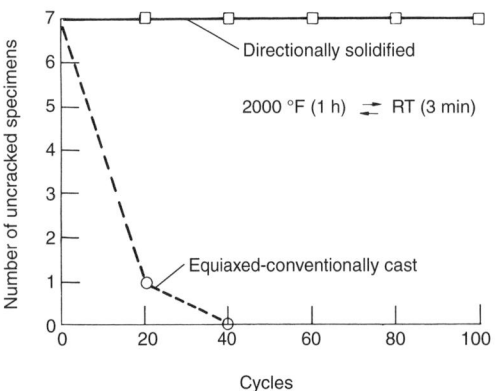

**Fig. 11.15** Thermal fatigue performance of conventionally cast and directionally solidified [001] nickel-base alloy MAR-M 200. Source: Ref 11.19

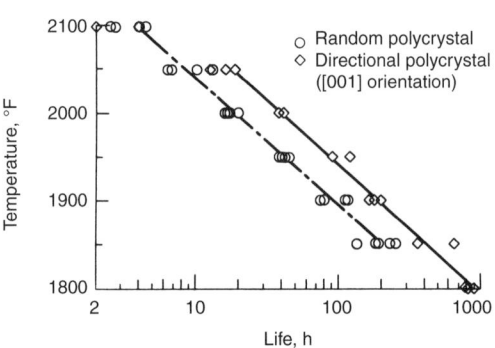

**Fig. 11.16** Stress rupture properties of random polycrystalline (conventionally cast) and directional polycrystalline (directionally solidified) WAZ-20 at 15 ksi. Source: Ref 11.21

tigue applications it becomes important that the "flow lines" be oriented in the proper direction relative to the maximum principal stress direction. Figure 11.17 shows the flow line directions for two process methods for forming bolt heads. In Fig. 11.17(a) the bolt head is made out of conventional bar stock (Ref 11.22). Flow lines are longitudinal, but the stress concentration (an important parameter that should always be minimized, independent of the other variables) associated with the bolt head causes principal stresses that are transverse to these flow lines, resulting in low fatigue life. If the head is produced by upset forging, as shown in Fig. 11.17(b), the flow lines become favorable to the direction of imposed stresses, and a higher fatigue life is achievable. In bearing races, similarly, proper grain flow direction is important. Figure 11.18(a) shows a construction in which the flow lines are not oriented properly relative to the applied stresses, while in Fig. 11.18(b) the relation between grain flow and stress direction is favorable because the principal tangential stress is parallel to the flow lines. Improved fatigue resistance (Ref 11.11) is also displayed by material with this grain flow structure.

While in most cases the flow line direction is inherent in the material from which the part is constructed, sometimes it is associated with small details of the construction itself. Figure 11.19 shows (Ref 11.22) the flow line orientation of threads ground from standard stock as well as those developed by rolling the threads with a machine tool. The rolling process causes plastic flow and develops flow lines tangential to the surface of the root radius. Because the principal stress at the root is also tangential, fa-

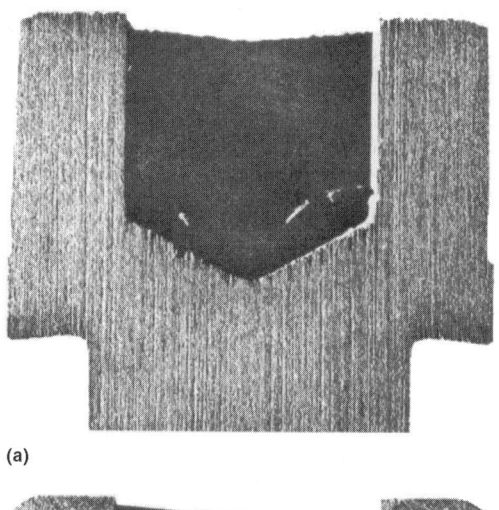

(a)

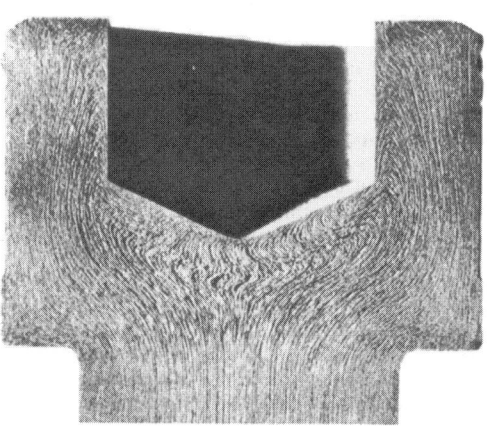

(b)

**Fig. 11.17** Flow lines resulting from two methods of forming bolt heads. Forged bolt had three times the fatigue strength of the machined bolt (Ref 11.22). (a) Machined. (b) Forged. Courtesy of *Machine Design*

(a)

(b)

**Fig. 11.18** Control of grain flow in bearing races (Ref 11.11). (a) Fiber flow nearly perpendicular to race. (b) Fiber flow parallel to race. Courtesy of *Machine Design*

vorable matching is accomplished in the rolled thread, improving fatigue life. Rolling has been found to be one of the most effective methods of improving fatigue strength of threaded fastenings. In part, this effect is also due to favorable residual stresses, which are discussed later.

An important way in which thermomechanical treatment can influence the fatigue properties is to provide a favorable dispersoid. In some metal systems mechanical working provides nucleation sites for the precipitation of very fine particles upon subsequent thermal aging. These particles can have a strengthening and ductilizing effect, both of which are desirable for good fatigue performance. The metallurgy of each system requires individual study. The application of thermomechanical treatments to a high-temperature cobalt-base alloy, L-605, is described in Ref 11.23. Figure 11.20 shows how various degrees of cold reduction of wrought material, followed by aging, substantially increased the tensile strength and ductility of the alloy as compared with material in the aged condition without prior working. In this case maximum strength was obtained with material that was cold reduced 30% and subsequently aged 200 h at 870 °C (1600 °F). Elongation was increased simultaneously from 6.5 to 9%.

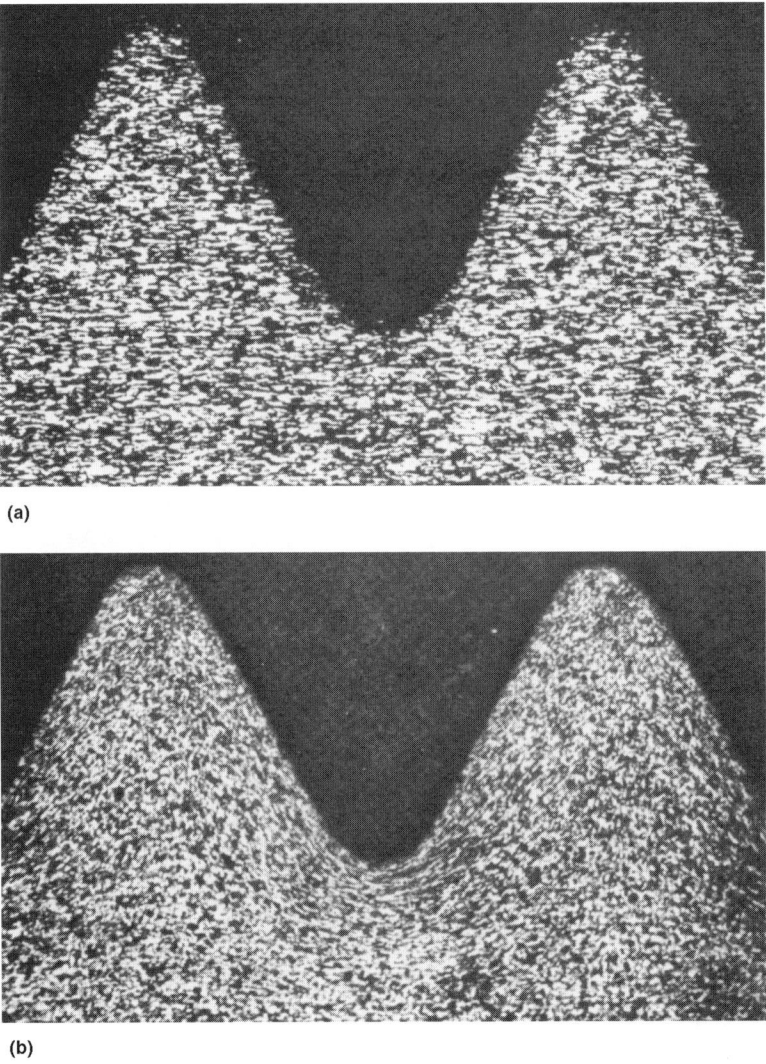

**Fig. 11.19** Chromatographs showing grain flow for two methods of forming threads (Ref 11.22). (a) Ground. (b) Rolled. Courtesy of *Machine Design*

Also among the systems amenable to thermomechanical working are the ausformed steels. Here, the steel is worked while it is at intermediate temperatures within the unstable austenitic range. Upon cooling to the martensitic range a large number of very fine carbide particles precipitate on the sites nucleated by the working in the austenitic range. The mechanism involved in the strengthening is actually much more complicated than simply the development of a very fine stable precipitate (Ref 11.24); however, much of the benefit is derived from this factor together with improved homogeneity of the microstructure and cleanliness. The principal benefits that accrue are high strength together with high ductility and fracture toughness; all these factors contribute to improved fatigue resistance. Figure 11.21(a) shows the consequence (Ref 11.25) of ausforming the tool steel H-11. Here is plotted the ratio of the high-cycle fatigue strength to the ultimate tensile strength. The shaded band shows results for a series of high-strength steels, indicating that most steels fall within a fairly well-defined scatterband. The filled circles show the ausformed H-11 results. Not only does it have an extremely high ultimate tensile strength, but also the ratio of its high-cycle fatigue strength to tensile strength falls well above the scatterband for conventionally heat treated steels. Further results for ausformed H-11 steels are shown in Fig. 11.21(b) where a comparison is made of the undeformed (conventional) and deformed (ausformed) conditions. Here, enough data were available (Ref 11.26) to make a statistical analysis, and it is clear that at any survival level the increase in strength is of the order of 25% higher due to ausforming. The improvement carries into the low-cycle fatigue range, as would be expected on the basis of the method of universal slopes, Chapter 3, "Fatigue Life Relations," because tensile strength, ductility, and fracture toughness are all increased by the ausforming. The dramatic improvement of fatigue life due to ausforming in bearing applications has been verified by Bamberger (Ref 11.27).

The potential for thermomechanical working as a means of improving fatigue resistance has been barely tapped. Fortunately, interest in this approach is expanding. Ostermann and Reimann (Ref 11.28) have reviewed fundamental processes involved in fatigue of aluminum alloys and possibilities of increasing fatigue resistance by thermomechanical processing.

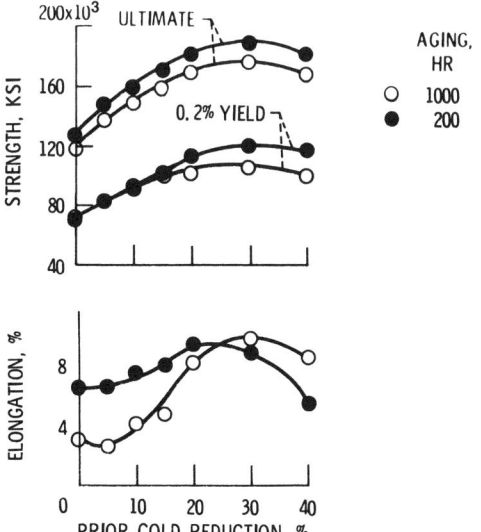

**Fig. 11.20** Effect of prior cold reduction on room temperature tensile properties of cobalt-base alloy L-605 after aging for either 200 or 1000 h at 871 °C (1600 °F). Source: Ref 11.23

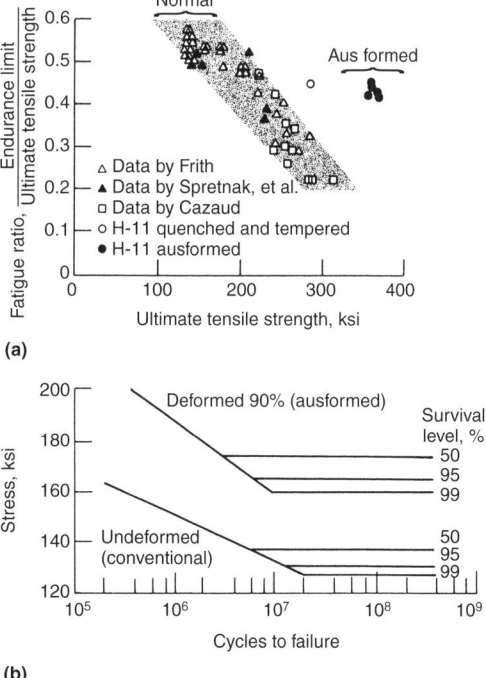

**Fig. 11.21** Examples of benefits derived from ausforming H-11 tool steel. (a) Relationship between high-cycle fatigue and the ultimate tensile strength. Sources listed are given in Ref 11.25. (b) Fatigue behavior of conventional and ausformed H-11 steel. Source: Ref 11.26

**Materials Compatibility and Matching.** The choice of an optimum material or material condition for a given application often depends not only on the loading to which it is to be subjected, but also on materials with which it is to interact during the loading. Environmental effects, for example, corrosion, obviously depend on the chemical compatibility of the material with the fluid environment. Similarly, fretting involves frictional forces between rubbing surfaces, and compatibility here involves the interaction of the two solid materials as well as the separating medium. This subject also is touched upon later in the discussion in connection with surface protection.

There are some situations, however, when a beneficial effect results from an interaction of the mechanical properties of mating materials rather than from a chemical effect. It is well to bear such opportunities in mind when selecting materials. An interesting example is the differential hardness effect associated with the life of ball bearing races. It has been found in Ref 11.29 and 11.30 that the life of the race can be increased substantially when the balls are harder than the race by about 2 points on the Rockwell C scale (when both are in the range of about 60 HRC). Either harder or softer balls than that associated with this optimum relation causes lower race life and higher life scatter. Figure 11.22 (Ref 11.11) shows the $B_{10}$ life (life when 10% of those specimens being tested have failed) versus the hardness differential (ball hardness minus race hardness). It can be seen that if the race is harder than the ball, the $B_{10}$ is very low. When the balls are about two points harder than the race, the life is optimized. If the balls are made harder than the race by more than 2 points life falls again, but not as rapidly as when the races are harder than the balls by the same differential. Factors of at least 10 in life are involved in small differentials in hardness. Figure 11.23 shows (Ref 11.11) that scatter is also minimized when the hardness relation between balls and races is optimized.

The reason for the large effect on fatigue life and life scatter, arising out of small differences in relative hardness, is not completely understood. It is thought to be related to residual stresses introduced in the races, or possibly to the amount of retained austenite. However, the reliability and economic implications are clear. Sorting of balls according to hardness, and matching hardness of ball lots with race lots, can produce most satisfying rewards.

## Strain Concentration Effects

Metal fatigue is due largely to the concentration of strain in a localized region. If the strain

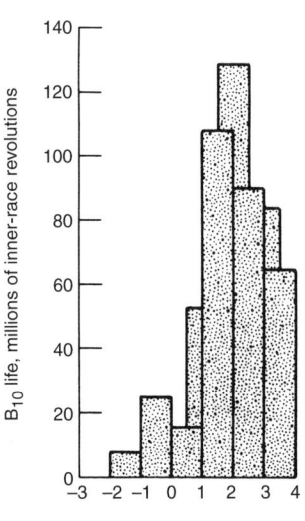

**Fig. 11.22** Bearing life at various values of differential hardness between balls and races, (radially loaded, 207 size ball bearings). Source: Ref 11.11

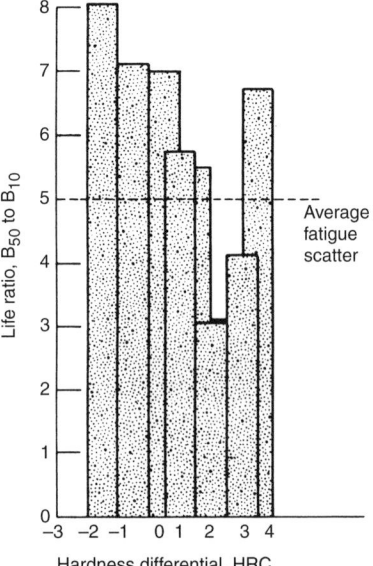

**Fig. 11.23** Bearing fatigue scatter (indicated by ratio of $B_{50}$ life to $B_{10}$ life) for various values of differential hardness between balls and races (radially loaded, 207 size ball bearings). Source: Ref 11.11

could be distributed over a larger volume of metal, the fatigue life could be greatly extended. But, because the strain is concentrated, large plastic strains may be introduced, even though the average strain is low. The strain concentration may occur on a microscopic scale; being localized, for example, in a few slip planes of a discrete region, or it may be localized on a macroscopic scale, as for example, within a limited region of a machined notch. Avoidance of such strain concentration is not always possible, but wherever it is possible, premium benefits result in the extension of fatigue life. The subject is extremely broad and cannot be covered completely in a limited discussion of this type. Therefore, we shall treat it briefly, breaking it down according to whether a particular aspect is within the province of the designer, the processor, or the user.

**Design Aspects.** Mechanical equipment inherently involves the use of holes, fillets, threads, and a myriad of other geometric discontinuities that cause strain to be concentrated in a limited volume of metal. Often the designer, while facing up to the need for these geometric discontinuities, can specify geometric details that mitigate their adverse effects. Keeping radii as large as practical, avoiding the cumulative superposition of piggyback strain concentrators, and adding metal or removing metal so as to cause smoother flow of stress from one region to another are but a few ways open to the designer to minimize the problem. Many good treatises have been prepared to help him (Ref 11.31–11.35), and almost every good book on fatigue contains numerous suggestions along this line. Figure 11.24 shows an example taken from Ref 11.34, which shows several design suggestions. Figure 11.25 is an example of the consequences, measured in fatigue life, of apparently minor design details (Ref 11.22). The large radii at the root of the teeth did not in any way interfere with the efficient functioning of the threaded part, yet a fourfold increase in life resulted from relatively minor design modifications. Part of the life improvement was due to the residual stresses introduced by rolling the threads, which are discussed later in connection with residual stresses, but part of it was also due to removing the sharp radii. Examples of this effect can be cited almost without limit.

Sometimes the stress concentrator is not a real design necessity, just an oversight, and can be avoided completely. Such a case is shown in Fig. 11.26, where a diagonally drilled hole was used in a turbine disk as part of a blade-retention mechanism (Ref 11.32). Fatigue failures resulted from the stress concentration. An alternate retention scheme, involving a wire that was merely bent over the edges of the disk rim, served equally well without introducing any stress concentrations.

In addition to avoiding stress concentration due to geometric discontinuities, the designer should always be alert to the question of force distribution. Providing many force-carrying elements is always helpful. For example, in a threaded fastening the number of threads among which the force is distributed should be as large as possible. Figure 11.27 shows the results of some fatigue tests (Ref 11.36) on threaded fastenings. Several strength levels were studied, and for each strength level three nut heights were tested. In all cases, the taller the nut, providing more threads, the longer was the fatigue life.

Distribution of the force among the force-carrying elements is also very important. Again, threaded fastenings provide examples wherein poor force distribution results in low fatigue life. The thread closest to the point of application of the force tends to absorb the major part of the force; other threads only absorb force transmitted to them by the deformation of intervening threads. Arranging the deflection characteristics of the assembly so as to distribute the force can be very important, as shown by the two examples (Ref 11.22) in Fig. 11.28. Here, alternatives are offered; in the first, tightening the nut causes it to flex, thereby reducing the pressure on the lower threads and transmitting force to the upper ones; in the second, progressively smaller undercuts in successive threads also allow gradual transfer of force among threads.

Specifying the correct material and material combinations is also within the province of the designer. This factor has been already discussed in a more general context earlier in this chapter, but specific illustrations in connection with threaded fastenings (Ref 11.36) are shown in Fig. 11.29 and 11.30. Figure 11.29 shows the effect on fatigue life of a nut as a function of its strength level. There appears to be an optimum strength that is about the same for all nut heights. Below the optimum, it is the inherent low strength that limits the fatigue resistance; above the optimum, material cannot take advantage of the greater strength because of the low fracture toughness of the material in the regions of high stress concentration. Figure 11.30 shows a second example of the importance of material matching in relation to service application. Here the fatigue tests were conducted using washers of progressively increasing hardness. The greatest lives were obtained for the hardest washer. The harder the washer, the more likely the initial tightness will be uniform and the tightness be maintained throughout the loading.

These illustrations are but a few of the factors within the control of the designer in affecting fatigue life. Obviously, there is also a role in relation to the fabrication and processing, as well as in recommending service practice. These are discussed in the following sections.

**Fabrication and Processing.** Many of the factors that govern the fatigue life are associated with the manner in which the part is translated from an engineering drawing to a working mechanism. In discussing this aspect of the fatigue problem, we shall not deal with material integrity because this already has been discussed; we shall also omit consideration of residual stresses that are, in fact, intimately associated with fabrication and processing because this subject is treated elsewhere in this book.

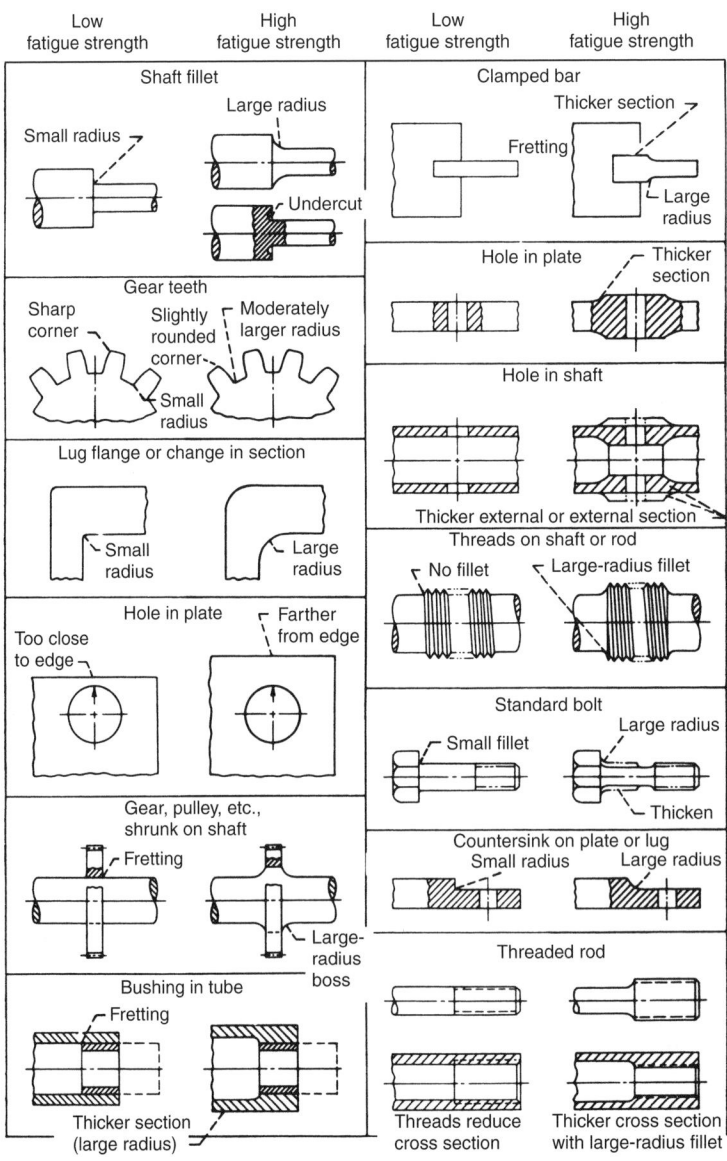

**Fig. 11.24** Some design techniques for improving fatigue resistance (Ref 11.34). Courtesy of *Machine Design*

Chapter 11: Avoidance, Control, and Repair of Fatigue Damage / 285

It has long been known that the fatigue resistance depends on the process by which the part is shaped from raw material. Figure 11.31 shows (Ref 11.37) the variation of fatigue strength at $10^7$ as a function of tensile strength for aluminum alloys and steels. It is seen that as-forged parts are very poor relative to machined or ground parts.

Figure 11.32 shows (Ref 11.38) more detail of the effect of processing for a steel of medium strength. Here it is seen that the detrimental effect of forming procedures such as forging and hot rolling depends on the life level sought. In the low-cycle fatigue range all methods of shaping give approximately the same strength levels, but in the high-cycle range there can be a factor of as much as $3\times$ difference in fatigue strength depending on the manner of forming. These differences are related to surface discontinuity effects as well as residual stresses, both of which are diminished in importance as the externally applied stress (or amount of plastic flow) is increased.

The metal deformation process affects not only the surface finish and residual stress state, but also its surface metallurgy. Among the factors that are identified with surface metallurgy is grain flow. Figure 11.33 shows (Ref 11.39) some results for a gear produced by high-velocity forging. The grain flow is seen to be favorably oriented relative to the stress concentration associated with the gear teeth; a sevenfold increase in fatigue life occurred

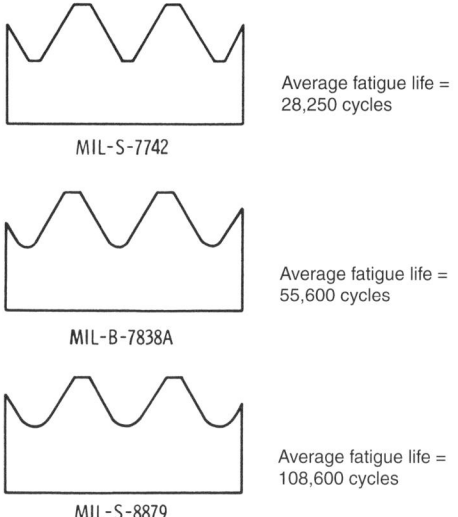

Fig. 11.25  Effect of thread shape on fatigue life of heat treated and rolled threads. Source: Ref 11.22

Fig. 11.26  Cracks started at locking-pin hole in turbine disk. Source: Ref 11.32

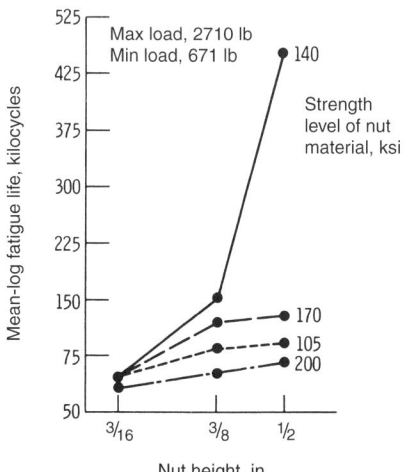

Fig. 11.27  Effect of nut height on mean-log life of aircraft fastening. Source: Ref 11.36

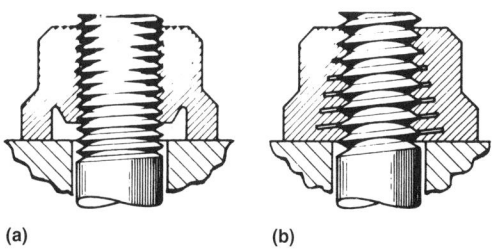

Fig. 11.28  Two methods for distributing force among threads. (a) Large undercut at base to permit flexing. (b) Distributed small undercuts at several threads. Source: Ref 11.22

The machining process itself governs both the surface roughness and the amount of residual stress. Figure 11.34 shows some results (Ref 11.40) of different machining methods for a titanium alloy. At the two extremes of the spectrum are electric discharge machining and ultrasonic machining, and in the $10^7$ life range there can be as much as a factor of $5\times$ difference in strength level by the two methods. Electric discharge machining produces local heating and affects the material both mechanically and metallurgically. High-surface stress raisers, together with deteriorated metallurgical properties, wreak havoc with the long-life fatigue properties. Different materials respond differently to this machining process, and titanium alloys are very seriously affected. Ultrasonic machining, on the other hand, is very gentle. The particles removed are very small, and a fine surface finish results. Fatigue properties are correspondingly high.

Note, however, that the differences in fatigue life due to machining method diminish as life is reduced, as discussed previously.

As already indicated, surface finish is not the only factor that is associated with the machining process. A process such as grinding, which produces residual compressive stresses, can result in a higher fatigue life than electrochemical machining or electropolishing, which result in a smoother surface but one without residual stress. This point is illustrated in Fig. 11.35 for the nickel-base alloy Inconel 718. This figure is taken from Ref 11.41, which provides a rather extensive compendium of the manner and degree to which fatigue properties of various materials are altered by surface effects.

The machinist also must be cautious not to introduce stress concentrations not intended by the designer, nor to defeat the designer's efforts toward minimizing stress concentrations. Some-

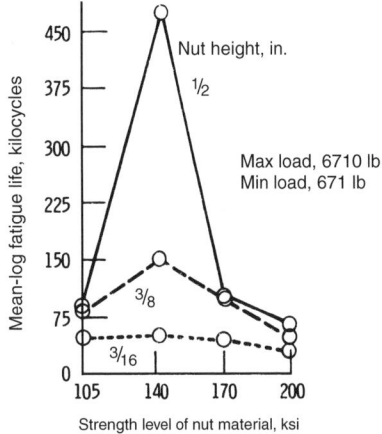

**Fig. 11.29** Effect of strength level of nut material on mean-log fatigue life of threaded fastening. Source: Ref 11.36

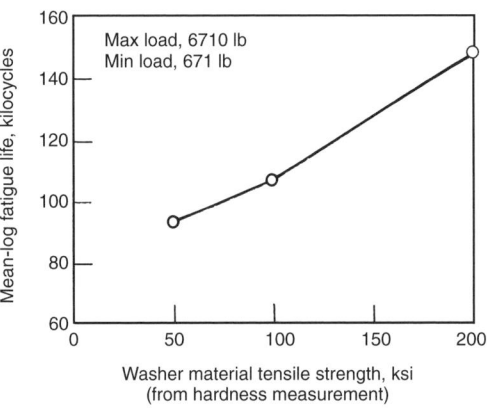

**Fig. 11.30** Effect of washer hardness on fatigue life of threaded fastening. Source: Ref 11.36

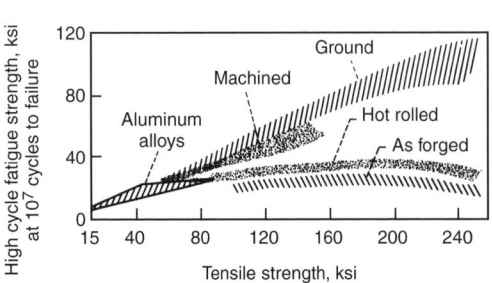

**Fig. 11.31** Effect of tensile strength level and processing treatments on reversed bending high-cycle fatigue strength at $10^7$ cycles-to-failure of aluminum alloys and steels. Source: Ref 11.37

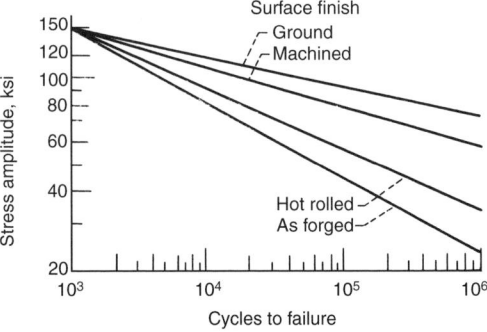

**Fig. 11.32** Surface finish effect on fatigue of medium strength steel. Source: Ref 11.38

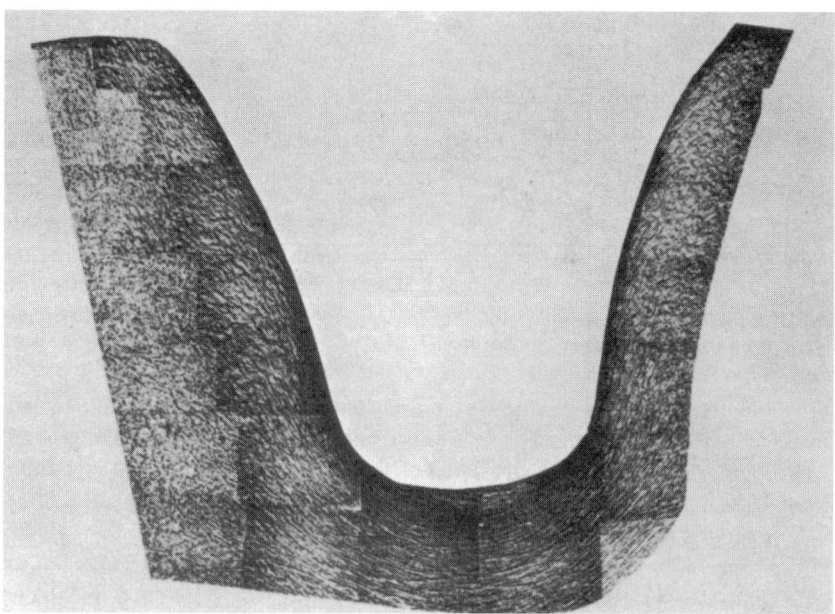

**Fig. 11.33** Improved grain flow in gear teeth resulting from high-velocity forging (sevenfold increase in fatigue life) (Ref 11.39). Courtesy of *Machine Design*

times, seemingly minor details may be incorporated by the designer to improve fatigue life. These details might be bypassed by the machinist who may feel that he or she is producing an "equivalent" rendering. An example (Ref 11.22) is shown in Fig. 11.36. The compound-fillet would double the fatigue life compared with the single-radius fillet, which might be substituted by an inventive machinist.

Assembly by welding is perhaps one of the more treacherous of the factors contributing to weakness in fatigue loading. A weld can usually be expected to be a metallurgically inferior region of metal. Impurities, inhomogeneities, and a large variety of imperfections can be expected to be concentrated in weld regions. Certainly it cannot be expected that a region in which there has been localized melting and solidification, and in which dissimilar metals have participated, will be as sound metallurgically as the remainder of the material, which has been consolidated under much more desirably controlled conditions. Tensile residual stresses created by the shrinkage of the weld metal as it cools can also contribute to decreased fatigue resistance. Because fatigue, especially in the high-cycle range, is sensitive to localized imperfections, welds very commonly are regions of fatigue failure origin. The manner of coping with this problem involves many factors that require close attention to detail. It should be emphasized that not all welds are bad. Quite frequently, it is the local geometry of a weld that introduces stress concentration due to the presence of a weld bead. For example, in steels, the removal of the raised weld bead of a butt weld, to make it flush, results in a fatigue

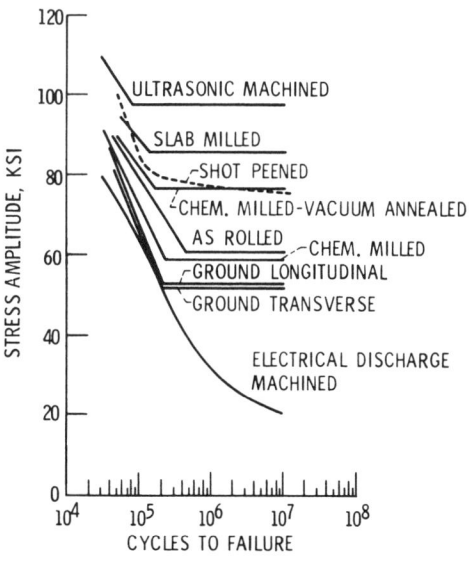

**Fig. 11.34** The effect of various machining processes on the bending fatigue strength of Ti-5Al-2.5Sn alloy. Source: Ref 11.40

performance nearly equal to that of the parent metal. Unfortunately, the same cannot be said about welding of heat treated aluminum alloys. Gurney's classic reference book (Ref 11.42) is devoted entirely to the subject of conventional welding. More recent volumes on practical aspects of welding are by Radaj (Ref 11.43) and Maddox (Ref 11.44).

The basic approaches for improving the fatigue performance of weldments can be classified into several distinct groups, as described in the following paragraphs.

*Favorable Geometry.* Choice of geometry so as to avoid placing the weld in a critical region of stress, as for example Ref 11.42 in Fig. 11.37; filleting or stress-relief drilling after welding (Fig. 11.38).

*Material Improvement by Heat Treatment.* Either by initial choice of welding technique, or by the subsequent heat treatment of the welded region, its soundness may be improved. Welding technique is a very specialized subject for the welding engineer and depends highly on the individual metallurgy of specific materials. Heat treatment after welding can relieve unfavorable residual stresses and introduce favorable metallurgical factors thus overcoming some of the problems associated with the welding process.

*Mechanical Working.* After welding, mechanical working can do three things: (a) modify the mechanical properties of the surrounding material, (b) improve the soundness of the joint (by closing voids), and (c) introduce favorable residual stresses. Residual stresses are covered elsewhere; a typical example of the effects of (a) and (b) is illustrated in Fig. 11.39. The process demonstrated here is termed *hydrodynamic compression treatment* (Ref 11.45) and has been used to improve the notoriously low fatigue resistance of spot welds. In this process the spotweld region is subjected to a force normal to the sheet surface, while a ring prevents the material from flowing in the radial direction in the plane of the sheet. The spotweld region is thus sub-

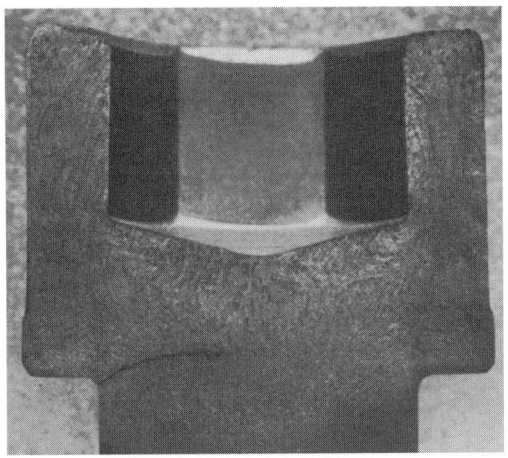

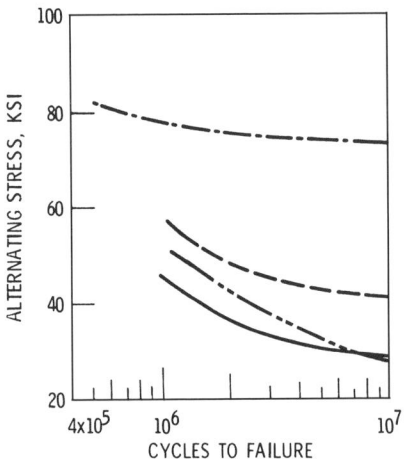

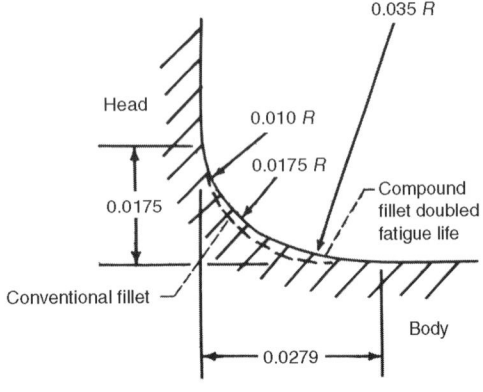

**Fig. 11.35** Effect of surface conditions on fatigue properties of nickel-base alloy Inconel 718, cantilever bending, zero mean stress, 24 °C (75 °F). Source: Ref 11.41

**Fig. 11.36** Effect of fillet geometry on fatigue life of bolt (Ref 11.22). Courtesy of *Machine Design*

Chapter 11: Avoidance, Control, and Repair of Fatigue Damage / 289

jected effectively to a high hydrostatic pressure, which improves the bonding and mechanically works the material. Improvement in strength at the $10^7$ life level can be as high as a factor of 3 or more. The figure shows the results for a range of test temperatures, and sizable improvements are obtained at all temperature levels.

*Residual Stresses.* Numerous methods have been used to introduce favorable residual stresses in weld regions, thereby improving fatigue resistance immensely. Because these methods are similar to those associated with fatigue resistance improvement in general, they are discussed elsewhere in this book. Numerous advanced permanent metal joining techniques have evolved that permit superior welds under certain conditions or for certain metals. Among these are cold welding, diffusion bonding, forge welding, friction welding, friction stir welding, fu-

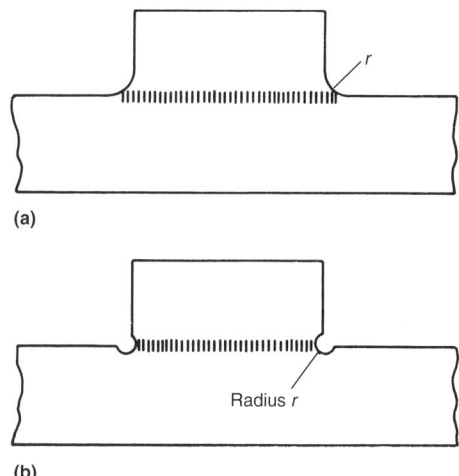

**Fig. 11.38** Methods of reducing the stress concentration at the ends of discontinuous longitudinal butt welds. (a) End of weld radiused. (b) Drilled hole at weld end. Source: Ref 11.42

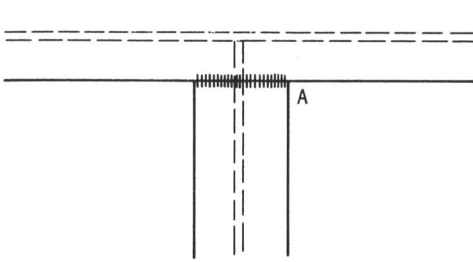

(a) Badly designed square frame corner with severe stress concentration at the corner A.

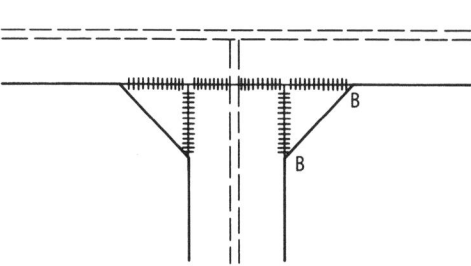

(b) The addition of triangular gussets produces little improvement. There are still severe stress concentrations at the corners B.

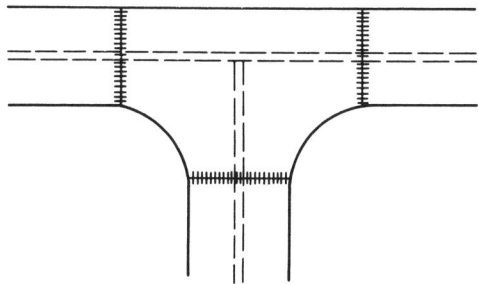

(c) Improved design with reduced stress concentration and with welds separated from it.

**Fig. 11.37** Improvement of the design of the corner of a welded frame. Source: Ref 11.42

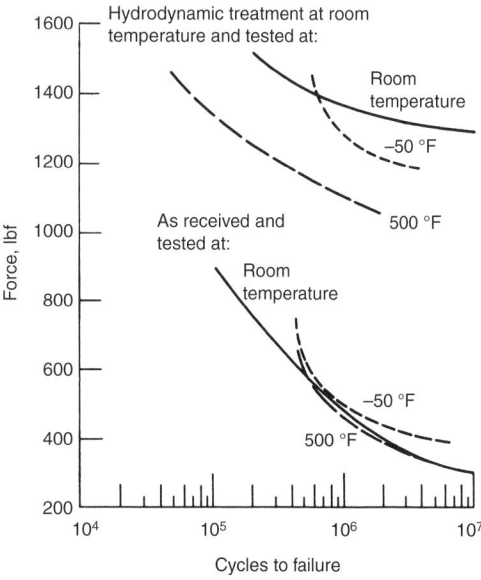

**Fig. 11.39** Illustration of effects of hydrodynamic compression treatment and test temperature on the fatigue life of single spotwelds in 0.064 in. type 301 stainless steel. Source: Ref 11.45

sion welding, resistance welding, electric-arc welding, thermal welding, and high-energy-beam welding. For in-depth information, the interested reader should consult The Welding Institute (TWI), The Welding Research Council (WRC), and ASM International.

**Assembly and Use.** Both the mechanic and the operator significantly influence the fatigue life of a machine part. Dents and scratches are obvious stress concentrations and can seriously reduce fatigue resistance. Other influences are more subtle. Figures 11.40 to 11.42 show some interesting results (Ref 11.36) relating to threaded steel fasteners. In Fig. 11.40 it is seen that tightening torque can have more than a five-fold effect on life.

If the nut is too loose the mean stress due to assembly is low, but the dynamic stresses upon application of alternating loads are high due to sloppy fit. If the nut is too tight, the mean stress due to assembly is high, again reducing life. There is obviously an optimum nut tightness to give peak fatigue resistance. Two of the most common points of weakness in tight bolt-nut attachments are: at about one turn in from the loaded, bearing-face of the nut (the first 25% of the engaged threads carry the bulk of the force); and at the end thread of the bolt where it transitions into the shank when a nut is run up against the shank. Although fatigue failure of nuts is rare, in certain cases, wherein laps or discontinuities are rolled into the nut blank, the weakened material affects the strength of the subsequently formed female threads. The reduced strength combined with the hoop stress due to tightening can result in fatigue failure of the nut.

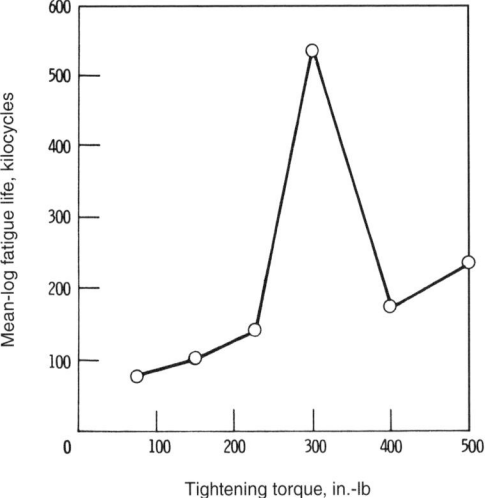

**Fig. 11.40** Effect of tightening torque on fatigue life of threaded steel fastening. Minimum load, 671 lbf; maximum load, 6710 lbf. Source: Ref 11.36

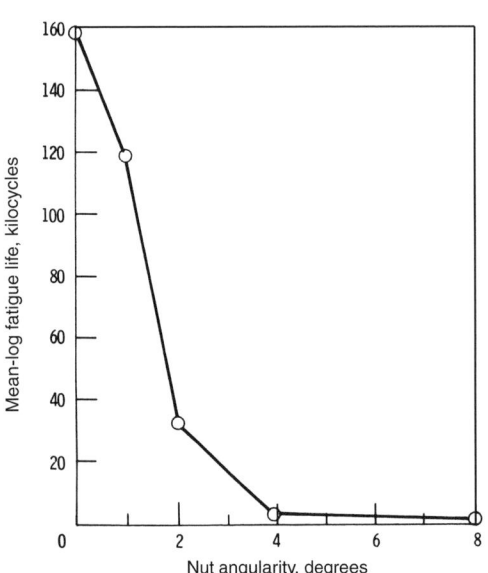

**Fig. 11.41** Effect of nut surface angularity on life of threaded steel fastening. Minimum load, 671 lbf; maximum load, 6710 lbf. Source: Ref 11.36

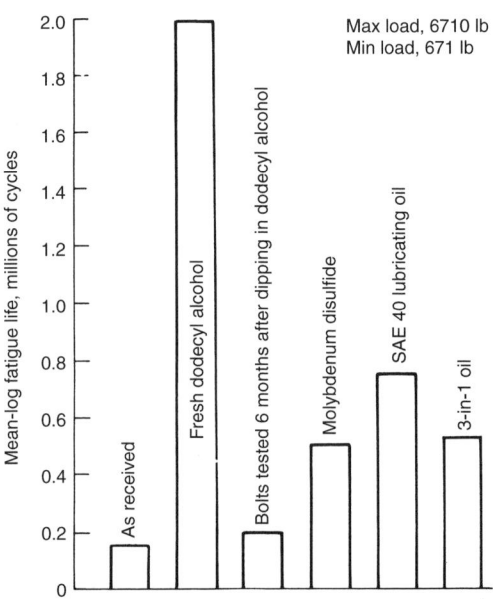

**Fig. 11.42** Effect of lubricant on fatigue life of threaded steel fastening. Minimum load, 671 lbf; maximum load, 6710 lbf. Source: Ref 11.36

Increased fatigue durability of bolted connections can usually be achieved by reducing the preload in each bolt by the following techniques:

- Threads of bolt and nut must be smooth and continuous, clean, and adequately lubricated
- Where permitted, decrease the bolt diameter and increase the number of bolts
- Ensure the clamped surfaces are flat and clamped with a force greater than the tensile service force
- Avoid soft inserts between the clamped surfaces that could collapse with time and reduce the original clamping force, i.e., use as thin and as hard a gasket as possible.

Figure 11.41 shows the results of angularity between nut and mating surface produced by the intentional use of tapered washers in the tests. When an angularity exists not only is the load localized and therefore high, but the eccentricity of the load introduces unnecessary bending moments. A few degrees of angularity can reduce the life a hundredfold. The importance of providing a lubricant, so that the mating surfaces can set properly, avoid undue friction, and protect the surface is illustrated in Fig. 11.42. The type of lubricant is also important, and for this application fresh dodecyl alcohol gave the best results. The question of lubricants will come up again in connection with surface protection and fretting.

Proper maintenance is also within the province of the user. Removing stress concentrations that arise out of service conditions can greatly extend fatigue life. A good example is cited by Manson (Ref 11.32, p 390) relating to combustion liners. Reaming, sanding, and vapor blasting of the punched edges of these liners improved their fatigue life. Undoubtedly, periodic treatment of this type would have been even more beneficial. It also is pointed out in this reference that in one investigation on turbine blades periodic polishing and etching improved life by removing service-induced irregularities. This matter also is discussed later in connection with material reconditioning.

Sometimes the service conditions can be altered to avoid unnecessary strain concentration. Again, an illustration in connection with turbine blades can be found in Ref 11.32. The rapid temperature escalation associated with rapid starting of a turbine engine produces high strain concentration due to thermal stress in regions of high temperature. Starting the engine more gradually, in this case by programmed sequential burner ignition, greatly reduced strain concentration and improved thermal fatigue life.

The operator has many opportunities to improve fatigue life by avoiding or mitigating strain concentrations through careful attention to detail. It is important that he be made aware of his responsibility and opportunity.

**Surface Protection.** Although fatigue is associated largely with localized plastic flow, and therefore may be initiated whenever such plastic flow is induced, the surface is the most vulnerable region. The reasons are many and some are discussed in Chapter 10, "Mechanism of Fatigue." In this section we emphasize two of them: the surface is susceptible to mechanical damage by rubbing or other physical interaction, and fatigue is accelerated by chemical interaction with environmental gases or liquids in contact with the surface. Both the mechanical and chemical interactions therefore should be minimized if fatigue resistance is to be enhanced. In the following sections we consider several methods of protecting the surface.

**Environment Neutralization or Exclusion.** Air, particularly the moisture in the air, is a very hostile environment for many materials in relation to fatigue. Aluminum, copper, steel, and magnesium are all known to be sensitive to atmospheric pressure (principally the partial pressure of oxygen) and to humidity. In fact, much of the variability in fatigue results on the same alloy tested in different laboratories, or tested in the same laboratory during different seasons, is known to be essentially a "humidity effect." Figure 3.11 in Chapter 3, "Fatigue Life Relations," shows some interesting results on a steel to demonstrate the effect of environment. Plastic strain is plotted against cyclic life for data obtained in air and in vacuum. A straight-line relation is expected. It is seen that the line drawn through the vacuum data is straight up to 4 or 5 million cycles. In air, however, not only is the life for a given plastic strain somewhat lower than in vacuum, but deviation from linearity is seen to start at about 100,000 cycles life.

Another significant set of data has already been discussed in connection with Fig. 10.27 in, Chapter 10, "Mechanism of Fatigue." Here the material is aluminum, and the plot is that of environmental pressure against fatigue life for several selected applied strain levels. The effect of frequency also is indicated by showing results for 25 and 50 Hz. As discussed in Chapter 10, "Mechanism of Fatigue," it is seen that pressure has an effect on fatigue life, and that the lower the pressure the higher the life. The increase is not, however, linear; there is a rather sharp discontinuity at about $10^{-2}$ to $10^{-3}$ torr. Also, the higher frequency results in the higher life, the largest effect of frequency occurring at the discontinuity pressure. Based on these observations, the authors of this study proposed that the mechanism involved is related to the amount of oxygen present, and to the time available for the oxygen to penetrate cracks formed by fatigue. As the oxygen penetrates it interacts chemically with the nascent surface exposed as the crack grows. Above about $10^{-2}$ torr there is adequate oxygen at all pressures, and little decrease in life results by further increases in pressure. Below $10^{-4}$ torr all pressures involve inadequate oxygen, and little increase in life results from further decreases in pressure. In the region from $10^{-2}$ to $10^{-3}$ torr, however, pressure is critical because the amount of oxygen involved is just enough to penetrate the crack and react; also frequency is critical because it governs the time required to reach a given number of cycles, and therefore influences the amount of time available to the oxygen for penetration.

Operation in a vacuum, or at least in an inert environment, has been shown repeatedly to result in improved fatigue resistance. One such investigation for a nickel- and a cobalt-base alloy is described in Ref 11.46. An example of the generally improved fatigue life observed in vacuum is shown in Fig. 11.43 for the cobalt-base alloy, S-816. Here, fatigue life was increased by a factor of two to four times that in air depending on stress level, thus showing the potential benefit of protection from the atmosphere.

Oxygen and moisture are only two of the aggressive environments encountered in fatigue applications. In gas turbines these components are present in the combustion gas, but there are other even more aggressive constituents in the products of combustion, especially if the sulfur content of the fuel is high. Of course, little can be done to neutralize the atmosphere in cases requiring specific working fluids. The lesson here is the value of avoiding particularly damaging constituents. If that is not possible, protection of the surface from the environment by coatings is indicated. Discussion of coatings is presented later.

**Minimization of Mechanical Damage.** In some cases damage, at least in part, is caused by mechanical rather than chemical effects. Erosion by solid particles in a gas stream, and fretting due to frictional forces acting during small relative displacements of mating parts, provide two examples. The mechanical damage here is but a prelude to the development and growth of a fatigue crack, but it is nevertheless important because the suppression of this damage will delay the development of the crack. One approach in each case is to remove or reduce the origin of the mechanical damage. In the case of erosion, removal of the erosive particles is the objective. Screens and inertia devices may be useful. The subject of fretting is very broad, and many detailed studies have been made, among them Ref 11.47 to 11.49. From these reports and from the papers cited within them, it is clear that the mechanism of fretting corrosion involves a number of factors:

- Mechanical contact of two materials involving frictional interaction
- Oscillating relative motion of small magnitude
- An atmosphere that can react chemically with nascent surfaces formed by the abrasion between the two materials or by the release of pitted particles from the fretting surface

One way of minimizing fretting corrosion, therefore, is to reduce friction, thereby reducing

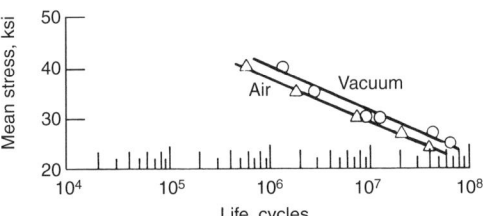

(a) Ratio of cyclic to mean stress, 0.125

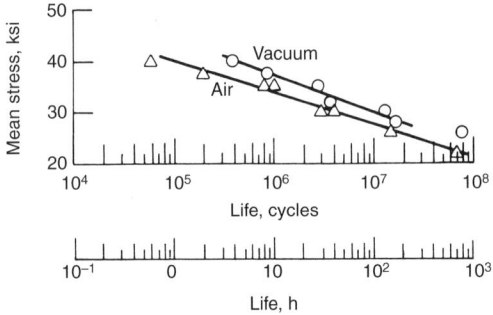

(b) Ratio of cyclic to mean stress, 0.667

**Fig. 11.43** Effect of environment on axial tensile fatigue properties of S-816 at 816 °C (1500 °F) in both vacuum and air. Source: Ref 11.46

the basis for mechanical abrasive damage. Figure 11.44 shows some results of a study (Ref 11.50) on fretting corrosion. One of the most effective ways found to improve the fatigue performance was to control the frictional characteristics of the mating surfaces. Magnesium pads on steel surfaces gave, for example, good results, but high-hardness steel pads gave poor fatigue results. Other ways of minimizing fretting fatigue relate to surface protection by lubricants and by the introduction of compressive residual stresses.

**Fluid Coatings.** A fluid coating, whether a thin film of lubricant or a bath of oil, can significantly enhance fatigue resistance. One reason, associated with improved load distribution and part functioning, is illustrated in Fig. 11.42. A second reason is the reduction of friction and associated mechanical damage such as fretting. A thin film of lubricant can improve fretting fatigue life, but here caution must be exercised to ensure that the film is replaced if it is squeezed out by the rubbing action of the two surfaces. If not replaced, the benefit will accrue for only a short period of time. A third reason for the beneficial effect of a fluid coating is the protection it might afford against the potential chemical attack of the environment.

Figure 11.45 shows the results (Ref 11.51) of fatigue tests on high-hardness steel piano wire. The fatigue life is shown on the vertical axis, and at different positions along the horizontal axis are shown the fatigue results for each of several conditions in which the material was tested. At the lowermost right, indicated by the solid squares, is the basic fatigue behavior of the material in an air environment after having been wiped clean with acetone. At a maximum stress of 1210 MPa (175 ksi) ($R \approx 0.1$), the average life was somewhat less than 20,000 cycles. Applying a thin coating of oil (as long as the oil is free of corrosive elements) to the cleaned material raised the average life to about 100,000 cycles, as shown by the open symbols. The effect of platings is shown by the solid symbols on the lower left of the figure. In most cases, the platings studied had a detrimental effect on life. What is of interest here is the effect of a coating of oil on each of the materials with the various platings. In all cases, as shown by the open circles, life was extended on the average by a factor of 5 or more relative to the clean surface.

The beneficial effect of coatings extends to materials with stress concentrations (Ref 11.52) as seen in Fig. 11.46. A life increase of a factor of about 2 is seen for both the smooth and notched specimens for quenched and double-tempered 4340 steel.

The nature of the surface coating and its possible chemical or physical interaction with the

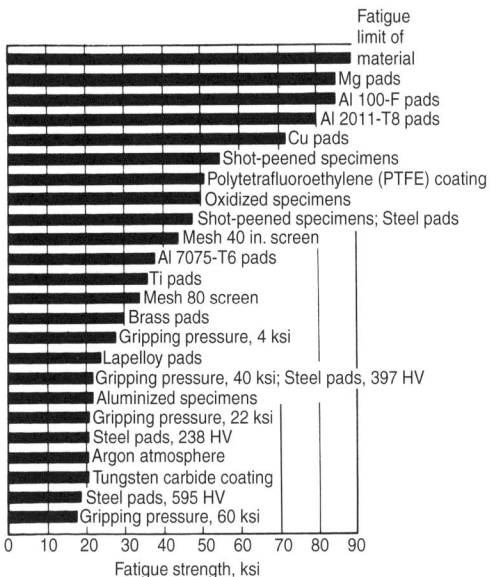

**Fig. 11.44** Fretting fatigue strength of RC 130B under various conditions (Ti-4Al-4Mn, an early titanium alloy). Source: Ref 11.50

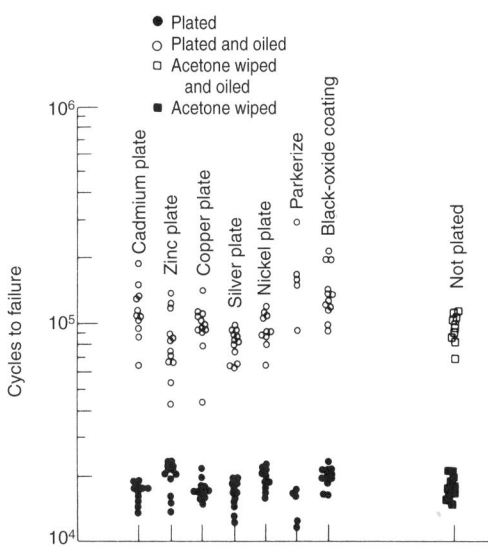

**Fig. 11.45** Effect of various platings and of oil protection on fatigue resistance of 0.051 cm (0.020 in.) diam, high-hardness, steel piano wire (ASTM A228); max stress = 175 ksi ($R \approx 0.1$). Source: Ref 11.51

substrate must also be considered. In some cases the effect may be detrimental. Figure 11.47 shows how various alcohols affect the high-cycle fatigue strength of a high-strength steel (Ref 11.53). The alcohols have the general chemical formula $C_nH_{2n+1}OH$, but for a given value of $n$ several atomic arrangements are possible, resulting in different chemical compounds of the same molecular weight. In particular, if the OH radical is bonded to a carbon atom, which in turn is bonded to only one other carbon atom, the alcohol is called a primary alcohol; if the corresponding carbon atom is bonded to two other carbon atoms, it is called a secondary alcohol. In Fig. 11.47 a plot is made of the life of a notched specimen of a high-strength steel under a nominal stress of 12.5 ± 4.37 ksi. In an air environment the life of such a specimen under this loading is beyond $10^7$ cycles. But in the presence of the alcohols, the life can fall below $10^6$ cycles. The degree of embrittlement depends on the value of $n$ in the molecular formula (being greater the lower the value of $n$), and on the carbon-coordination number (being greater for the secondary alcohols than for the primary alcohols). Rostoker (Ref 11.53) makes the interesting point that distilled water, $H_2O$, falls on both the curves, for primary and secondary alcohols at a value of $n$ of zero, and that water therefore causes an embrittling action characteristic of the limiting case of the primary and secondary alcohols. Of course, water is known to produce embrittling effects in high-strength steels. Figure 11.48 shows the effect of humidity on fatigue of magnesium (Ref 11.54). In this case, low and high humidity refers to 3% and 85% relative humidity, respectively. Extensive research on aluminum also has demonstrated an adverse susceptibility to moisture.

**Mechanical coatings** can influence the fatigue of the substrate material in several ways:

- Providing a desirable inherent property of the coated material, such as hardness, to the

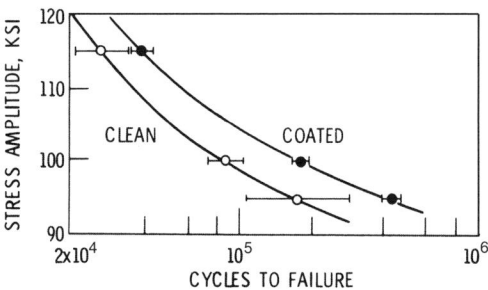

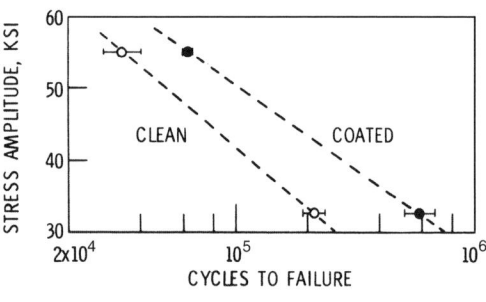

**Fig. 11.46** R.R. Moore rotating-beam S-N curves for (a) smooth and (b) notched (0.013 cm, or 0.005 in., root radius) specimens of quenched and double-tempered SAE 4340 steel (hardness, 302 HB) with and without polar organic liquid coating. Source: Ref 11.52

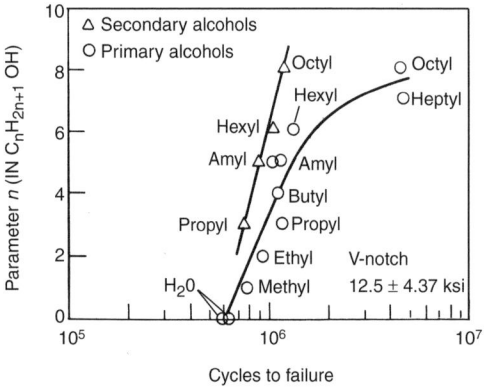

**Fig. 11.47** Fatigue of high-strength steel in primary and secondary alcohols. Source: Ref 11.53

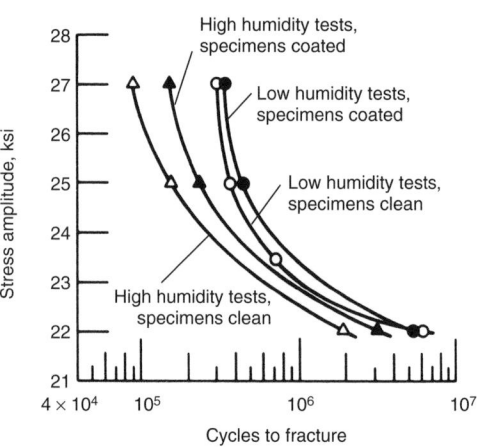

**Fig. 11.48** Effect of environment on fatigue of magnesium alloy AX61A. Source: Ref 11.54

surface, which might be subject to mechanical abrasion. Even apart from the aspect of abrasion, the hard surface may be desirable because of its inherent fatigue resistance, especially in the low-strain, long-life range.
- Metallurgical interaction with the substrate to form a new compound or alloy on the surface that has desirable fatigue properties beyond that of the coating itself.
- Introduction of favorable residual stresses.

In some cases the beneficial effect of the coating clearly arises from one or another of these functions; in others, two or more functions may be combined. Figure 11.49 shows a case in which the fatigue improvement is due primarily to isolation of hostile environment (Ref 11.55). Here, the plastic coating on a magnesium substrate produced the same fatigue properties as did a vacuum environment.

Figure 11.50 shows results (Ref 11.56) for the titanium alloy Ti-6Al-4V wherein the net effect is probably due to several factors. A nitride coating increased the fatigue strength at $10^6$ cycles over 60% and a diffused chrome coating together with pickling produced an improvement of over 200%. Nitriding is known to be associated with the introduction of residual compressive stresses. Chromium produces a hard, fatigue-resistant surface, and also may alloy with the substrate and probably introduces some residual compressive stresses.

A very early development in the coatings field was metalliding (Ref 11.57). In this process the part to be coated is immersed in a fluid bath at high temperature, and the coating material is deposited at a rate equal to its rate of diffusion into the substrate material. Because of the high temperature and diffusion, the surface becomes an alloy involving the substrate and the deposited material. Unusual coatings that are not possible by conventional coating procedures were made possible by this process.

It should be pointed out, however, that coatings are not always beneficial for fatigue applications. Possible adverse effects of cracking of a brittle coating at high-strain range have already been mentioned. Adverse metallurgical reaction with the substrate or the introduction of tensile residual stresses either as a result of deposition, or upon temperature cycling, may also result in detrimental fatigue properties. Finally, side effects of the plating process may be undesirable. For example, nickel plating usually results in lower fatigue life, the reason being embrittlement due to hydrogen that diffuses into the material during the plating process.

Thus, mechanical coatings can be good or bad, and care must be exercised to take into account the nature of the substrate and the coating material before selecting a coating for fatigue application.

## Residual Stresses

Perhaps the most effective way of prolonging fatigue life is by the introduction of compressive residual stress. Several methods, as discussed later, may be used to accomplish this purpose, the main ones being localized tensile plastic flow and metallurgical transformations that involve a

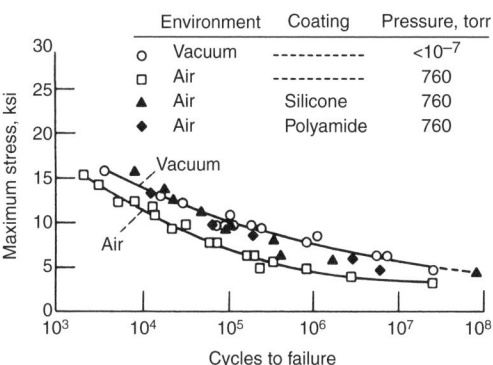

**Fig. 11.49** Comparison of fatigue behavior of coated and uncoated magnesium specimens tested at room temperature at 30 Hz. Source: Ref 11.55

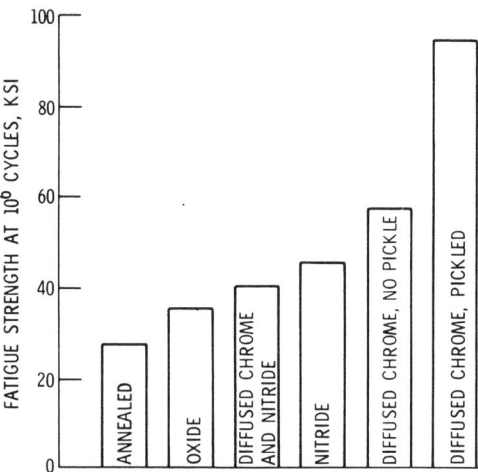

**Fig. 11.50** Effect of various coatings on fatigue strength of Ti-6Al-4V alloy. Source: Ref 11.56

volume change. In either case the beneficial effect arises because the critical region relating to fatigue is placed in compression. This compression is balanced by smaller tensile stresses distributed over a larger, less critical, subsurface region. Because the residual stresses are caused by elastic interaction of adjacent regions, subsequent plastic flow in the critical region can annihilate the residual stresses. Therefore, low-cycle fatigue, which involves gross section plastic flow, is essentially not improved by initial compressive residual stresses because these can be quickly relaxed during cycling*. Relaxation of residual stresses can also occur even in high-cycle fatigue by minute increments of localized plasticity, thus eventually negating their intended benefit. Compressive residual stresses are more beneficial when external loading conditions impose significant tensile stresses. A combination of significant compressive residual stresses and active compressive loading, however, could cause undesirable compressive yielding and resultant tensile residual stress that could actually reduce fatigue life. For example, this would be the case when a notched geometry, which had been treated to induce compressive residual stresses, was subjected to a compressive overload. If the sum of the compressive overload stress and the residual compressive stress exceeded the compressive yield strength, the end result could be a detrimental tensile residual stress for subsequent cyclic loading. Caution must therefore be exercised when specifying surface treatment to induce compressive residual stresses.

The importance of stress relaxation is indicated in Fig. 11.51 from the results of a classic investigation in 1958 by Morrow and Sinclair (Ref 11.58). They imposed an initial uniform tensile mean strain of 0.0033 on smooth axial specimens of a soft (29 HRC) steel. This resulted in an initial stress of 100 ksi based on a nominally elastic response with $E = 30{,}000$ ksi. Subsequently, each specimen had various levels of alternating strains superimposed and the relaxation of the initial 100 ksi stress was recorded as a function of applied cycles. These tests simulated salient features of fatigue testing of specimens containing a large tensile residual stress.

The advantage of their testing procedure was in the ability to measure the axial load response of the specimen, which provided a direct cycle-by-cycle measure of the stress relaxation taking place during cyclic straining. The general behavior of such stress relaxation for cyclic hardening and softening and cyclically stable materials is demonstrated and discussed in Chapter 2, "Stress and Strain Cycling." Had actual residual stresses been imposed by some surface deformation or thermal treatment, the tensile residual stresses of concern would have been confined to a very thin surface layer of material, and they would have been balanced internally by much smaller compressive residual stresses over the remainder of the specimen cross section in a complex pattern of stress distribution. It would not have been possible to detect the cycle-to-cycle stress relaxation without resorting to ill-suited and laborious techniques such as x-ray diffraction. As shown in Fig. 11.51, the mean stress (shown as a ratio to the initial mean stress) decayed rapidly when high alternating strains (represented as stresses, $\sigma_a = \varepsilon_a E$) were imposed. For low strains, however, relatively little relaxation of residual stresses occurred. The possibility of such stress relaxation must always be considered, especially under the high strains involved in low-cycle fatigue.

In evaluating the significance of residual stress as a means of increasing fatigue life it is important to determine whether the residual stress is equivalent to an externally applied stress of the same magnitude. A number of investigators have answered this question in the affirmative. Rowland (Ref 11.59), analyzing lit-

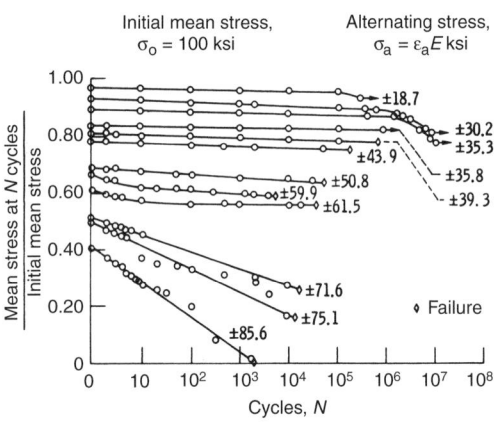

**Fig. 11.51** Relaxation of initially imposed mean stress during strain cycling of annealed SAE 4340 steel (hardness, 29 HRC). Source: Ref 11.58

---

*During mission loading, it may be possible to sustain a significant residual (mean) stress for a cycle or more, leading to fatigue damage being accumulated at a much different rate than if no mean stress were present. See Chapter 4, "Mean Stress," and Chapter 6, "Cumulative Fatigue Damage," for discussion relative to this set of circumstances.

erature data, has for example, presented the results shown in Fig 11.52. Here, high-cycle fatigue strength is plotted against mean stress, similar to a modified Goodman diagram. The mean stress for each of the data points consists of a residual stress produced by shot peening (the peak value for which is shown in parenthesis adjacent to each point) together with that due to an external load. Because the line representing the data points is very similar to a modified Goodman diagram, Rowland concluded that residual stresses are the equivalent of externally applied mean stresses. For some limited situations this conclusion may be valid, but there are significant differences between a locked-in residual stress and stress due to externally applied loads to justify extreme caution against this general assumption. Among the most important differences are:

- Cyclic relaxation: as discussed previously
- Limited area of action of residual stress: In contrast to the large area of action of an externally applied uniform mean stress, in consequence of which the residual stress loses its significance when the crack passes its region of presence.
- Variability of location of peak residual stress relative to the surface, complex stress gradient effects, and concurrent presence of balancing levels of tensile and compressive stresses.

Sheratt (Ref 11.60) presented information that mean compressive stresses could, under certain circumstances, be more influential in resisting crack propagation than in preventing crack initiation, and that some of the anomalous results associated with the behavior of shot-peened specimens can be explained in terms of this factor. It is nevertheless an established fact that compressive residual stresses can be very beneficial in improving high-cycle fatigue life, and we shall now discuss various methods of introducing such residual stresses to advantage.

## Shot Peening and Similar Surface Treatments

Several surface cold working processing treatments are available for inducing compressive residual stresses. Among these, shot peening (SP) is by far the most common and least costly. Other established processes include rolling (R), autofrettage (AF), and hole expansion (HE). New processes such as laser shock peening (LSP) (Ref 11.61) and low-plasticity burnishing (LPB) (Ref 11.62) are currently being developed due to their promising technical advantages.

**Shot peening (SP)** is a multirepeated surface impact process. Hard spherical steel shot are air blasted against the critical surface region of a component (Ref 11.63, 11.64). The indentation at each point of impact is the result of local plastic deformation in a shallow surface layer. Because the plastically stretched surface layer wants to expand and the adjacent elastically responding material immediately around and below the impact restrains the expansion, a compressive residual stress field is generated in the near surface layers. High compressive residual stresses are confined to a shallow surface layer (up to 0.127 mm, or 0.005 in. depth) while low-tensile residual stress spread deeper through the cross section to maintain static equilibrium. Unfortunately, the degree of cold work due to SP can be exceptionally high (up to 30–40%) and can reduce the available ductility of the material within the affected surface layer. It should be cautioned that SP of thin sections may lead to undesirable warping. The reader is also cautioned not to confuse shot peening with shot blasting; the latter is a surface stripping/cleaning process and utilizes sharp edges of fractured shot to speed the process. Shot blasting may even reduce fatigue resistance of the treated surface.

**Rolling (R)** is another very practical way to induce surface residual stresses. The concept is the same: compressive stress normal to the surface causes tensile plastic flow in the surface plane, but change of dimension in these directions is constrained by the bulk subsurface, thereby inducing compressive residual stress. When such residual stresses are introduced in regions of stress concentrations, such as at the root of threaded fastenings, they counteract the

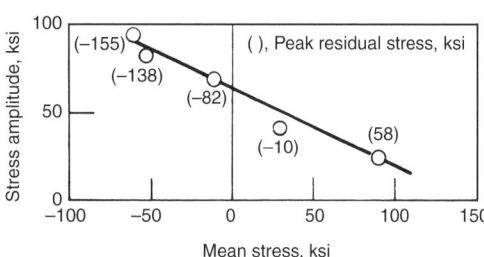

**Fig. 11.52** Combination of mean and alternating stresses at $10^6$ to $10^7$ cycles to failure for steel subjected to residual stress and superimposed external loading. Source: Ref 11.59

corresponding concentration of tensile stresses resulting from external loading, and thus contribute substantially to fatigue resistance in the high-cycle life range. Figure 11.53 (Ref 11.65) shows the effect of cold-rolling threads of a high-strength steel. When the rolling was performed after the heat treatment, a significant increase in fatigue resistance was achieved; heat treating after cold rolling removed the residual stresses, destroying their beneficial effect. The difference between heat treating before and after rolling is greater the higher the hardness of the steel. For soft steel, plastic flow is induced readily by external loading, thus washing out the benefit of the residual stress even if produced by rolling after heat treating. Thread rolling, however, also has a beneficial geometric effect, producing a larger root radius that is retained whether the heat treatment is affected before or after rolling. It also introduces an improved grain flow, such as illustrated in Fig. 11.33.

**Autofrettage (AF)** is very similar to rolling, except that the rolling is done internally on thick-walled tubes such as gun barrels. The rolling tool is typically rolled in a spiral manner down the length of bore as it tries to expand the inside diameter. The localized surface deformation results in large compressive residual stresses in the circumferential and longitudinal directions. The increased cold work also strengthens the surface to make it more wear resistant.

**Hole expansion (HE)** is basically a variant of autofrettage except that the length of the hole is much, much shorter. This technique is used frequently with thin plates that are to be connected by bolting or riveting. The technique is widely used in the aircraft airframe industry to produce favorable compressive residual stresses that prolong fleet life of aircraft.

**Low-plasticity burnishing (LPB)** is the newest of the techniques for mechanically introducing favorable compressive residual stresses. Low-plasticity burnishing is a moderate-cost process, developed for aircraft engine and other high-performance components by Paul Prevey of Lambda Research, Cincinnati, OH (Ref 11.62). The process has commonality with rolling, autofrettage, and hole expansion but goes far beyond in its applicability to almost any geometric shape. Low-plasticity burnishing uses a hydraulically floated ball (similar to a ballpoint pen) to press and roll freely along the surface of the component so that near surface layers are deformed plastically. A series of overlapping passes are applied until the intended surface area has been treated. Great depths of residual compressive stress can be generated. Because the rolling ball minimizes plasticity (no shearing due to sliding), less cold work is generated at the surface. These relatively deep (0.025 to 0.035 in.) compressive residual stresses prolong fatigue crack initiation and thus help to prolong the fatigue life of treated components. As discussed later, the degree of cold work (c.w.) left in the surface layer by LPB is considerably lower than for SP and LSP. Research by Telesman, Kantzos, and Gabb at NASA-GRC and Prevey at Lambda (Ref 11.66) has clearly shown that the LPB process also significantly retards crack propagation through the compressive residual stress layer, thus further enhancing service life. Dramatic photographic evidence is shown in Fig. 11.54 (Ref 11.67) of the influence of LPB on fatigue crack growth behavior of a turbine disk alloy. An axially loaded fatigue specimen was subjected to electric discharge machining (EDM) to produce a small crack starter of approximately 0.006 to 0.007 in. depth. The specimen was then cyclically loaded at 1200 °F until the crack had grown to the depth shown (about 0.020 in.). Testing was suspended, the specimen removed from the machine and sent to Lambda where the surface containing the crack was subjected to LPB. The resultant residual stress pattern in this specimen was measured in duplicate by Prevey, utilizing x-ray diffraction, and is shown in the upper graph. Upon its return to NASA, fatigue testing of the specimen was resumed. The resultant severe constriction from the highest compressive residual stresses in the region from about 0.005 to 0.015 in. depth is clearly evident. Crack growth was retarded within this region but not beyond. Note the shallow surface cracking and the mushroom-shaped growth as the crack front progressed beyond the constriction zone.

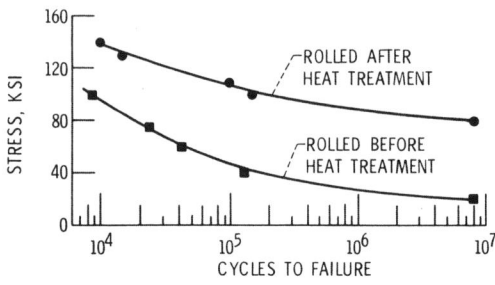

**Fig. 11.53** Effect of rolling threads before and after heat treatment on 220 ksi steel bolts. Source: Ref 11.65

**Laser shock peening (LSP)** is perhaps the highest-tech and highest-cost process used for generating compressive residual stresses. It uses a high-energy pulsed-laser beam (diameter of about 0.25 in.) to vaporize a thin opaque coating on the surface to be treated. The explosive nature of the vaporization generates a powerful shock wave that impacts and propagates into the component so that a relatively deep surface layer of the component is plastically deformed, resulting in a deep penetration of high-compressive residual stress (0.025–0.035 in.) (Ref 11.61). Although LSP costs are currently high, efforts are being made to reduce these by the current chief proponents, Clauer and coworkers at LSP Technologies, Dublin, OH in conjunction with the U.S. Air Force at Wright-Patterson Air Force Base.

Example distributions of residual stress and cold working into the depth of a component are shown in Fig. 11.55 for SP, LPB, and LSP. Both distributions were measured by an automatic x-ray diffraction technique developed by Paul Prevey (Ref 11.62). The technique measures the residual stresses from the determination of the shift in the x-ray diffraction peak position, caused by the change of material lattice spacing due to the presence of residual elastic strain (stress). Simultaneously, the technique can measure the broadening of x-ray diffraction peaks to quantify the degree of cold work in the material. Based on these techniques, residual stress and cold-work distributions as functions of depth can be measured after surface layers are removed by electropolishing or chemical machining and by an associated correction technique. As can be seen from the figure, both LPB and LSP are able to generate deeper and higher-magnitude compressive residual stress fields with much less cold work when compared with SP. Most importantly, for high-temperature aircraft engine components, surfaces treated by LPB and LSP will have higher thermal relaxation resistance (due to lower cold work and lower propensity for annealing or recrystallization) than those treated by SP. Significant improvements in high-cycle fatigue life over conventional shot peening have been reported for these advanced treatment methods (Ref 11.61, 11.62).

### Commonality of Surface Treatments for Compressive Residual Stresses

Although each of the aforementioned processes has a unique means for imparting the sur-

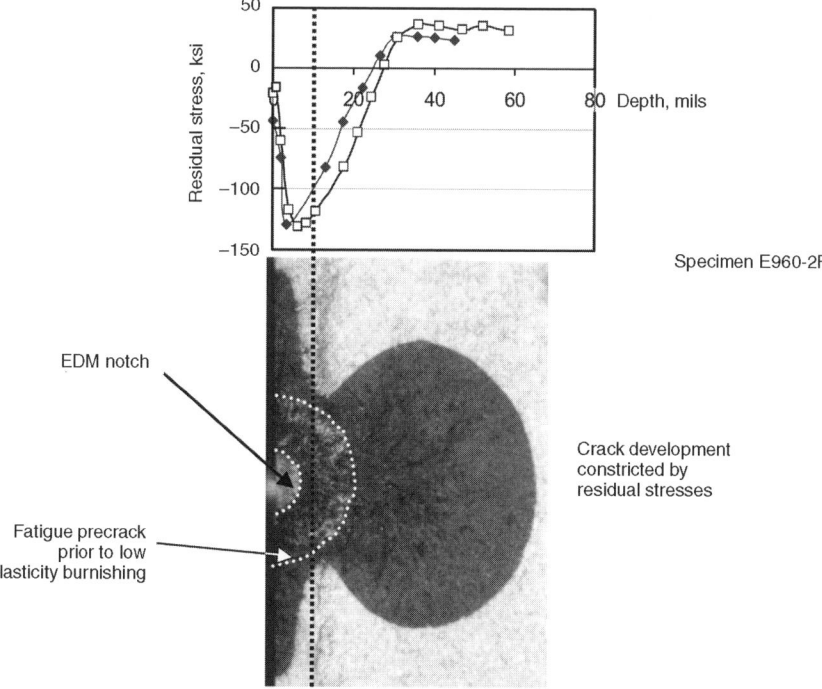

**Fig. 11.54** Dramatic effect of low plasticity burnishing (LPB) on fatigue crack growth in a nickel-base turbine disk alloy. Source: Ref 11.67

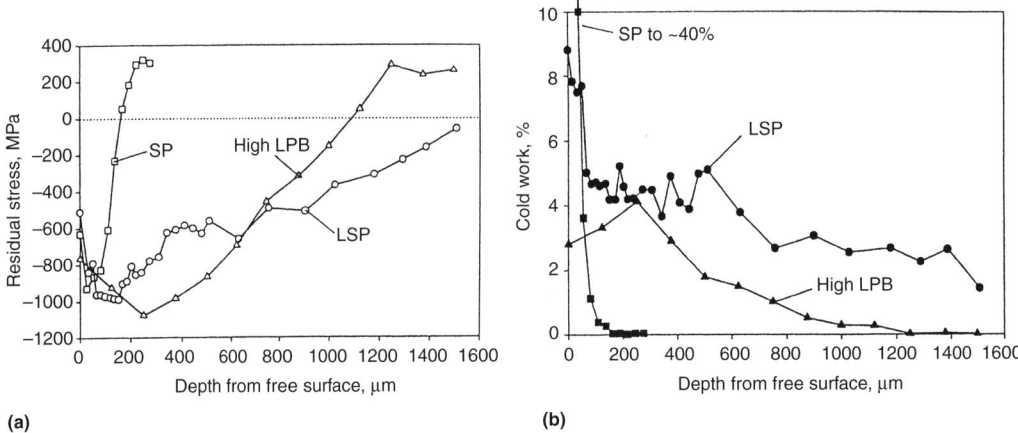

**Fig. 11.55** Residual stress and cold-work distributions in Inconel 718 after various surface treatments. (a) In-plane residual stress distribution. (b) Cold work-distribution. SP, shot peening; LPB, low-plasticity burnishing; LSP, laser shock peening. Source: Ref 11.68

face cold working that gives rise to compressive residual stresses, all of the processes share the commonality of plastically stretching the surface material that is constrained by the much larger elastically behaving volume of material around and beneath the treated area. As the plasticity imparting forces are removed, the elastic bulk overwhelms the rather small volume of plastically deformed material and pushes it back toward its initial position, leaving it with a biaxial compressive residual stress. Figure 11.56 shows schematically (Ref 11.60) a one-dimensional view of the general stress distribution that develops from any of the aforementioned surface treatment methods. The highest compressive residual stress is actually at a small depth below the surface. The balancing tensile stresses, which, for thick sections, are considerably lower in magnitude than the peak compressive stresses, lie farther yet from the surface. Each application that would seem to benefit from surface-induced compressive residual stresses should also be examined for the potential deleterious effects due to the balancing tensile residual stresses. One example that has been brought to the attention of the authors is of fir-tree attachments of highly loaded turbine blades for which the failure initiation site was shifted from surface to subsurface where the residual stress was tensile. Shot peening was used for inducing the compressive residual stress in the fir-tree region. Although the subsurface initiation life was a little greater than observed for the nonpeened condition, a potentially serious downside to subsurface crack initiation is that cracks are not observable using direct surface observations until they have grown to the extent that they break through to the surface.

Quantitative analyses of beneficial or detrimental results of surface working treatments become very difficult because of the following reasons:

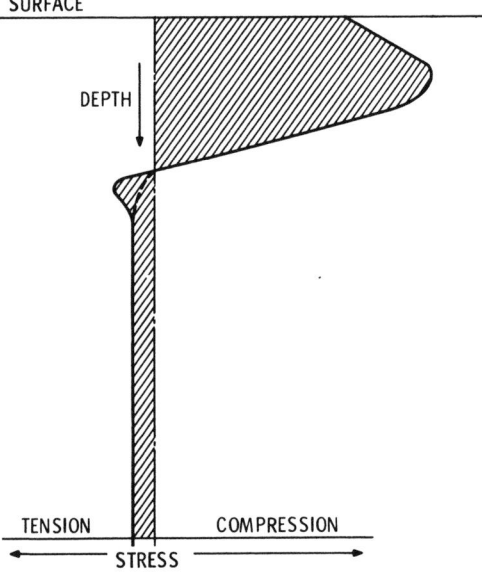

**Fig. 11.56** General form of residual stress distribution induced by shot peening or similar surface cold working treatments. Source: Ref 11.60

- Complexities of stress distribution and attendant alteration of mechanical properties of surface material due to working
- Uncertainties regarding the quantitative role of the residual stresses in affecting crack initiation and crack propagation
- Geometrical surface effects, that is, irregularity of surface, and perhaps even cracking caused by the peening process itself

However, it is irrefutable that beneficial results can be obtained. The following are applications that have been found amenable to this process of fatigue strengthening.

**For General Part Strengthening and Mitigation of Stress Concentrations.** Figure 11.57 shows some results (Ref 11.69) on the crack propagation rates in the very early stages (crystallographic stage I, often considered to be a portion of the crack initiation life) of fatigue of Udimet 700, a nickel-base superalloy, resulting from glass bead blasting (a process similar to shot peening, except that glass instead of steel beads are used; in either case, fractured beads are not recycled). The specimens contained no stress concentrations and were subjected to bending fatigue loadings. It is seen that the glass-beaded specimen showed a stage I crack propagation rate 10,000 times lower than the electropolished specimen. The same investigation showed that crack initiation also is retarded by the residual stresses associated with the glass beading process.

For notched parts, in which the major fraction of the fatigue life normally is the crack propagation phase, nearly all investigators agree that shot peening can be beneficial, although some materials can benefit more than others from this process. The best practice is to try it on a simulated part under loading that reasonably simulates service conditions, in order to determine empirically how much benefit can be derived.

Figure 11.58 shows some results obtained by Fuchs (Ref 11.70) showing the very substantial increase in life that is obtainable by shot peening of internal combustion engine connecting rods. This part had a number of stress concentrations, and it is probable that the extreme effectiveness in this case is related to this factor. It is interesting to note that without peening, the life of the connecting rod is so low as to imply considerable plastic flow at the point of stress concentration. With peening, the life is raised in the high-cycle fatigue regime, suggesting that residual stress can retard crack initiation as well as crack propagation. Apparently, also, plastic flow was retarded by the presence of the cold work imposed to achieve the residual stresses.

The fact that gross plastic flow can cause relaxation and nullification of the residual stresses is seen in Fig. 11.59. The point made here is that shot peening of the generator shaft can substantially improve fatigue life at a torsional twist of ±10°. If a twist of ±13° is applied, the plastic strain relaxes the residual stresses, and little or no benefit occurs. The figure also emphasizes the importance of simulating true service conditions when laboratory tests are conducted to evaluate the merit of a process. Accelerated testing, by increasing the torsional twist, certainly would have masked the benefit of residual stresses in this case. Proposed accelerated testing should be scrutinized carefully to ensure that

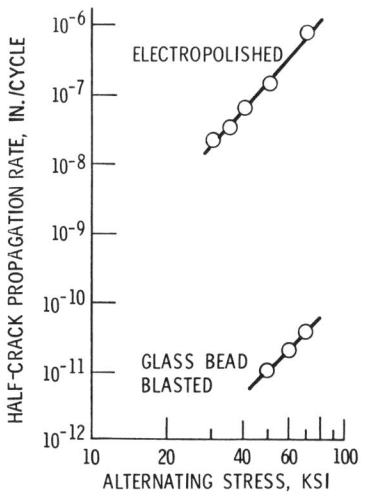

**Fig. 11.57** Stage I half-crack fatigue propagation rates at various alternating stresses in electropolished and glass bead-blasted material tested in bending. Source: Ref 11.69

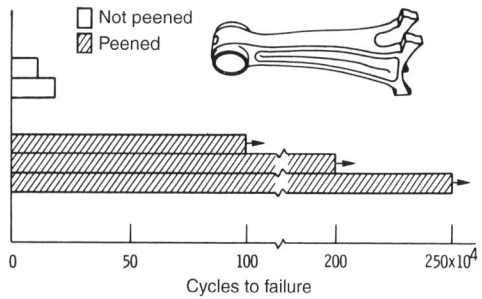

**Fig. 11.58** Fatigue durability of polished versus rough-finish and shot peened forked connecting rods. Source: Ref 11.70

new phenomena aren't being introduced that could overshadow the effects being accelerated.

Shot peening can be used even for materials subjected to high-temperature service provided the temperatures are not so high as to relieve the residual stress. As seen in Fig. 11.60, results for a nickel-base alloy show benefits from peening, even at 538 °C (1000 °F). The data are for 2.5 mm (0.100 in.) thick sheet Inconel 718 tested in fully reversed bending at 538 °C (1000 °F) and room temperature. Because no parent metal data were reported at 538 °C (1000 °F), a 538 °C (1000 °F) curve was estimated for the alloy in the same condition (same grain size: ASTM 6.5–7; heat treatment, etc.) based on data from other programs. The data show significant improvement in the fatigue strength for shot-peened material over both room temperature and elevated temperature parent metal strengths.

Thus, shot peening may have numerous high-temperature applications if applied to the proper materials. However, it is obvious that residual stresses can anneal out if the temperature is sufficiently high. Each material and loading condition should be evaluated before use in a service application. The question of annealing of residual stresses due to high-temperature exposure certainly merits further study. Recent research by Prevey (Ref 11.62) addresses the issue of the degree of cold working involved in generating compressive residual stresses by surface working. Residual stresses imposed by small amounts of cold work are expected to be retained at higher temperatures than equal values of residual stresses imposed with higher amounts of cold work.

**For Counteracting Detrimental Surface Treatment.** Sometimes it is necessary to apply surface treatments that, unfortunately, degrade resistance. Machining and electroplating are among these treatments. Introducing residual stresses after the surface treatment has been performed may serve to restore the fatigue resistance. Figure 11.61 shows (Ref 11.70) how shot peening can compensate for the detrimental effects of grinding, which can leave the surface with tensile residual stresses and/or minute cracks. Figure 11.62 suggests that shot peening also can compensate for the detrimental effects of nickel plating on steel, which normally leaves the surface in residual tension (Ref 11.70).

Another demonstration of the value of shot peening in restoring damaged parts is seen in Fig. 11.63. Here, turbine blades of the nickel-base alloy Inconel 713C were fatigue tested. The basic fatigue life curve is *A*, and that for shot peened blades without prior service is *B*, show-

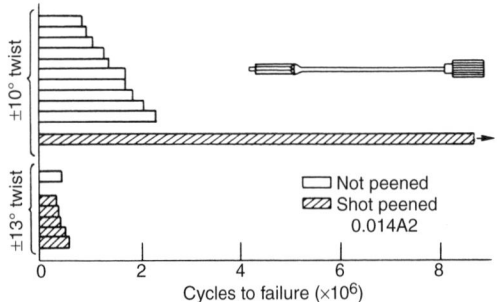

**Fig. 11.59** Loss of effectiveness of shot peening at high strain levels for 6150 steel (34–38 HRC). Source. Ref 11.70

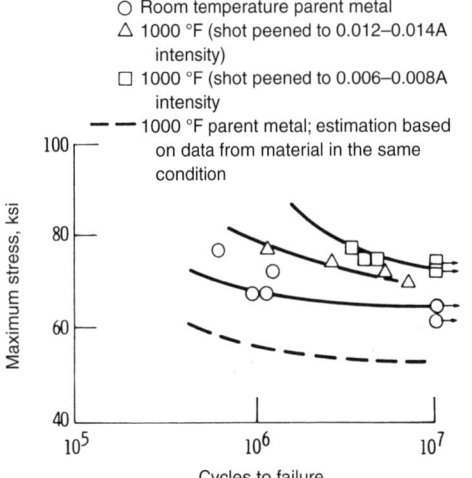

**Fig. 11.60** Effect of shot peening on Inconel 718 as shown by fatigue curves obtained in fully reversed bending tests. Source: Ref 11.71

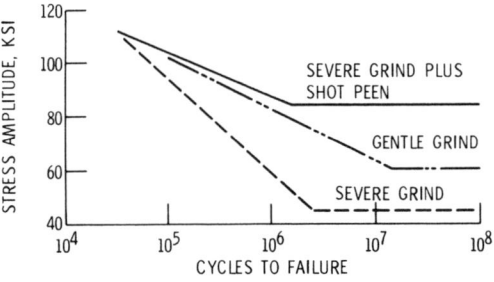

**Fig. 11.61** The effect of grinding and peening on reverse-bending fatigue strengths of flat steel bars (hardness 45 HRC, or 421 HB). Source: Ref 11.70

ing a substantial benefit of shot peening. The S-N curve for blades showing "grinding indications" in the shanks is shown as C, but when these blades were shot peened prior to testing, the curve D was obtained. Thus, the shot-peened blades with shank indications were a little stronger even than the standard blades without shot peening.

**To Compensate for Fatigue Damage due to Service Loading.** Minute cracks representing fatigue damage due to service loading also can be counteracted by the compressive residual stresses of shot peening. Figure 11.64 shows results obtained on 4340 steel specimens tested in rotating bending. Curve A is the basic S-N curve for the material, and B the S-N curve for the shot-peened condition, showing a substantial improvement obtainable by shot peening of standard smooth geometries. Also shown in the figure is curve C, obtained on specimens that had been prestressed for approximately half their fatigue lives at 100 ksi, and curve D, which refers to material that has been fatigue damaged (as for curve C), but peened after damage, and then retested. It is seen that the peening has more than restored the material to its basic fatigue life. Thus, peening can overcome fatigue damage. But Viglione (Ref 11.73) also found that if the damage had progressed too far (say about 80% of the fatigue life at high stress had been used up), shot peening was no longer beneficial. In fact, the process might be detrimental, although this effect as observed might have been really a reflection of data scatter. He also found that if the material had originally been shot peened, reshot peening after fatigue damaging was not effective in restoring life. Viglione's results are consistent with the fact that shot peening is capable of introducing compressive residual stresses to a maximum of only 0.005 in. If a fatigue crack had progressed to a greater depth (for example, 0.020 in.) the tip of the crack would then be in a tensile residual stress field, albeit small, with an attendant higher driving force for propagation and therefore subsequent shorter fatigue life. Even if his specimens had been double peened (peened twice), the depth of penetration would not have been appreciably more than 0.005 in., and the surface would have suffered considerably more cold working, with attendant loss of ductility.

Other techniques for generating compressive residual stresses to depths of the order of 0.040 in. (LSP and LPB), however, would be expected

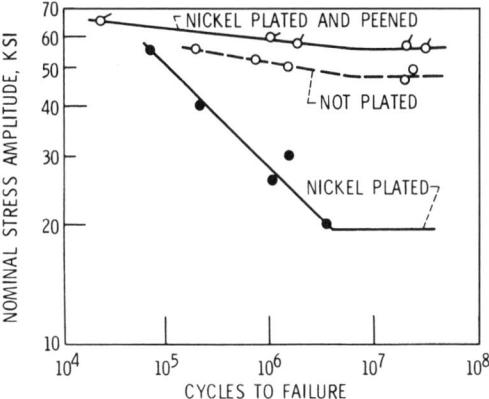

**Fig. 11.62** Comparison of fatigue strengths of unplated, nickel-plated, and nickel-plated and peened steel rotating-beam specimens. Source: Ref 11.70

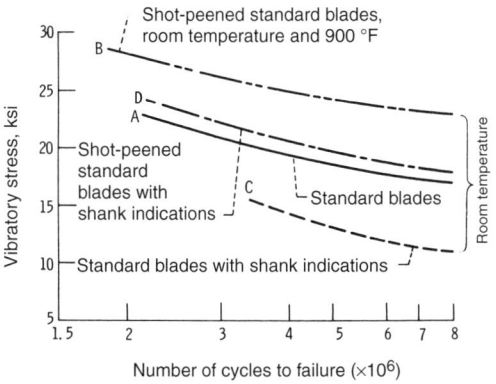

**Fig. 11.63** Suppression of fatigue damage of Inconel 713C turbine blades by shot peening. Source: Ref 11.72

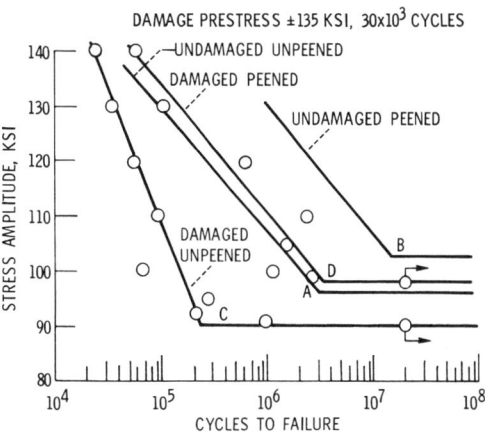

**Fig. 11.64** Shot peening as a means of overcoming prior fatigue damage with SAE 4340 steel tested in rotating bending. Source: Ref 11.73

to restore appreciably the fatigue damage of cracks up to and beyond 0.040 in. In fact, in thin sections such as found on leading and trailing edges of gas turbine engine fan and compressor blades (wherein the thickness does not exceed two depths, i.e., 0.080 in.), the entire leading and trailing edges could be totally in compression for distances of up to 0.5 in. from the actual edge. This is provided that both surfaces are simultaneously burnished. Even under such circumstances, a severely fatigue-damaged blade could have its fatigue resistance raised above the original untreated resistance. It is for this reason that the U.S. Air Force has had an intense interest in pursuing advanced surface treatment techniques.

## Cold Rolling

As discussed in an earlier section, shot peening is but one of several ways to induce residual stresses by working the surface. Another very practical way is to roll the surface. The concept is the same: compressive stress normal to the surface causes tensile plastic flow in the surface plane, but change of dimension in these directions is constrained by the bulk subsurface, thereby inducing compressive residual stress. When such residual stresses are introduced in regions of stress concentrations, such as at the root of threaded fastenings, they counteract the corresponding concentration of tensile stresses resulting from external loading, and thus contribute substantially to fatigue resistance in the high-cycle life range. For example, Fig. 11.53 clearly demonstrates the benefit of cold-rolling threads of a high-strength steel. When the rolling was performed after the heat treatment, a significant increase in fatigue resistance was achieved; heat treating after cold rolling removed the residual stresses, destroying their beneficial effect. The difference between heat treating before and after rolling is greater the higher the hardness of the steel. For soft steel, plastic flow is induced readily by external loading, thus washing out the benefit of the residual stress even if produced by rolling after heat treating. Thread rolling, however, also has a beneficial geometric effect, producing a larger root radius, which is retained whether the heat treatment is applied before or after rolling. It also introduces an improved grain flow, such as already discussed in connection with Fig. 11.33.

## Interference Fits

Interference fits have been found to be beneficial to fatigue life, the principal factor being the introduction of residual stresses. But the problem is somewhat more involved than merely the introduction of residual compressive stresses by plastic deformation. To understand the nature of the benefit, we consider several specific applications of this process.

Figure 11.65 shows the procedure involved in the use of a Taperlok (Wilmad, Buena, NJ) fastening, a patented device intended to improve fatigue resistance (Ref 11.74). The process involves plastic flow due to forcing a pin of a given diameter into a hole of smaller diameter drilled between the two parts to be fastened. If the pin were removed, there would remain residual compressive stresses in the mating parts, just as if the plastic flow had been caused by shot peening or rolling. However, the pin remains within the structure in order to perform its fastening function; therefore, the pressure between pin and structure is not removed. As it is first fabricated, the residual stresses in the structure are tensile.

Even the tensile prestress, in this case, is beneficial. The mechanism involved is analogous to a bolt that bonds two members together, especially when the rigidity of the bolt is much lower than the rigidity of the remaining members of the assembly. As the assembly loads the bolt in the axial direction, the load on the bolt does not increase until the load stress exceeds the prestress. Prestressing thus has the effect of reducing the stress range, while having relatively little effect on the maximum stress developed; therefore, it results in increased fatigue life. Actually, the problem is very complex, and a careful analysis really is required to take into account changes in localized stresses as affected by deformations, which involve rigidity calculations.

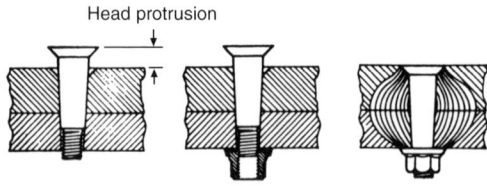

**Fig. 11.65** Geometry of taper pin fastening. Installation sequence: 1. Following hole preparation with tapered drill, tapered shank bolt is inserted in hole and firmly seated by hand pressure. 2. Full contact along entire shank of bolt and hole prevents rotation of bolt while tightening washer-nut. During tightening, nut spins freely to the locking point, but washer remains stationary and provides a bearing surface against structure. 3. Torquing of washer-nut by conventional wrenching method produces a close tolerance interference fit, seats the bolt head, and creates an evenly balanced prestress condition within the bearing area of the structural joint. Source: Ref 11.74

But qualitatively it is easy to see that an interference fit may improve fatigue performance as a result of a combination of the following factors:

- Filling the hole, which itself reduces local stresses and strains, compared with an open hole (which, of course, would in most cases not be present at all if it were not included)
- Cold working of the material in the vicinity of the hole, which may increase the strength of the critical region
- Reducing the stress range for a given external loading
- Introduction of residual compressive stresses

Figure 11.66 shows the effect of taper pin interference in the fatigue life of small lugs. It is seen that a substantial increase in fatigue life can result from an interference fit, but that beyond a certain point further increases in interference do not produce corresponding increases in life. The fatigue behavior is a reflection of the complex interplay of the aforementioned factors involved. Figure 11.67 shows fatigue results for aluminum strips with open holes, conventional bolts, and tapered bolts. It is seen that simply filling the holes improves the fatigue life considerably compared with a strip with open holes. Tapered bolts improve the life further. Figure 11.67, however, does not reflect the behavior of structures involving load transmittal from one member of the assembly to another through the bolts. But it does indicate that under certain conditions tapered bolts can be beneficial.

Another application of interference fits for which improved fatigue resistance is claimed is the Sine-Lok (Valley-Todeco, Inc., Sylmar, CA) fastener. Figure 11.68(a) shows the geometry of such fasteners, the shank profile having the shape of a modified sine wave with a pitch of 1.6 mm ($^1/_{16}$ in.) and a peak-to-valley height of 0.08 mm (0.003 in.) The fastener is used to bolt together assemblies in which the holes are made smaller than the outside diameter of the bolt, and therefore a forced fit is required to affect the assembly. The producers of the fastener claim, however, that the forces involved are considerably lower than corresponding forces for straight-sided interference-fit bolts because contact is made in limited areas, and the metal in the vicinity of the contact area has a place into which to distort. They also claim that dimensional tolerances in the structural holes are less critical, and that there is less of a tendency to gall or seize during assembly. Improved fatigue resistance also is claimed.

Figure 11.68(b) shows some results in fatigue tests of aluminum alloy plate containing a hole. It is seen that if the hole is filled with a straight shank without pressed fit, life is increased by a factor of about 3. Filling the hole with an interference-fit Sine-Lok fastener improves the fatigue life by a factor of 10. Most of the tests available to date do not involve load transfer from one structural member to another through the Sine-Lok fastener, which is, of course, the main function of such devices.

### Mechanical Overload

The application of a mechanical overload, defined as a load higher than commonly encountered in service, often can be beneficial in prolonging fatigue life. If this overload causes tensile plastic flow at the root of notches or cracks that already have been started by the fa-

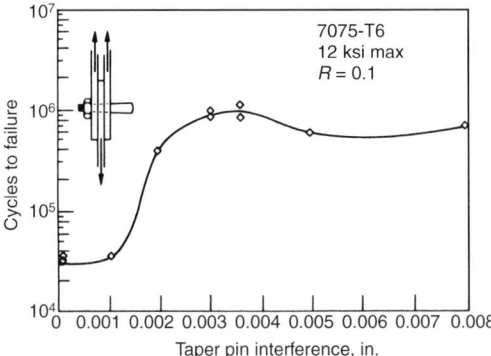

**Fig. 11.66** Effect of taper pin interference in fatigue life of small lugs. Source: Ref 11.75

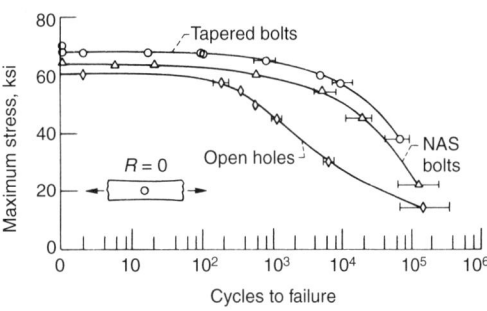

**Fig. 11.67** Fatigue life of 2.5 cm (1 in.) wide strips of aluminum 7075-T6 with 0.476 cm ($^3/_{16}$ in.) diam 100° countersunk holes compared with lives when filled with unloaded NAS-333 bolts and tapered bolts. Source: Ref 11.73

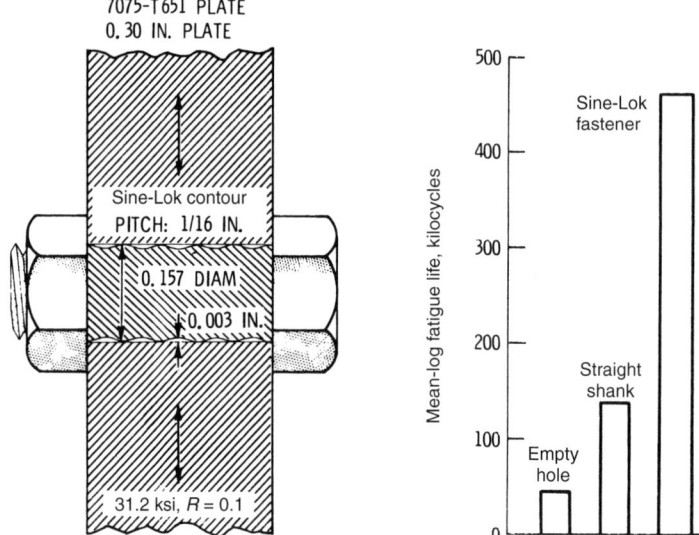

**Fig. 11.68** Fatigue performance of aluminum plate containing hole or with hole filled with either straight shank or Sine-Lok fastener. (a) Sine-Lok assembly. (b) Fatigue life. Source: Ref 11.75

tigue process, the residual stresses upon release of load will be compressive. Conversely, if the overload is in compression, the residual stresses will be tensile. Thus, a tensile overload will generally produce beneficial effects in fatigue, while compressive overloads often produce adverse effects.

Figure 11.69 shows the effects of axial overloads on notched bars of aluminum 7075-T6, evaluated in rotating bending after the overload. The effect is dramatic: tensile overloads increase the fatigue strength, in the life range above $10^6$ cycles, by a factor of 2 or more; compressive overloads decrease the fatigue strength by about the same factors.

Single overloads, while improving fatigue performance, are not effective as periodic repetitions of the overload. The residual stresses associated with the overload tend to become relaxed, as already discussed in relation to Fig. 11.51. Thus, overload repetition may restore the residual stresses and further improve fatigue life. Figure 11.70 shows that this is the case in a typical service application. Here, the effect on life is plotted against the ratio of overload to the 0.1% proof stress. The shaded region is the scatterband for parts subjected to a single overload, and it is again evident that tensile overloads increase life while compressive overloads reduce life. Of special interest, however, is the scatterband enclosed by the dotted curves. This region refers to results wherein periodic repetitions of overload are applied. It is seen that even larger increases in life can be achieved in this manner if the overload is repeated. Of even more importance is that the overloads can be kept lower while still achieving large life improvement. Because the practice of overloading is a dangerous

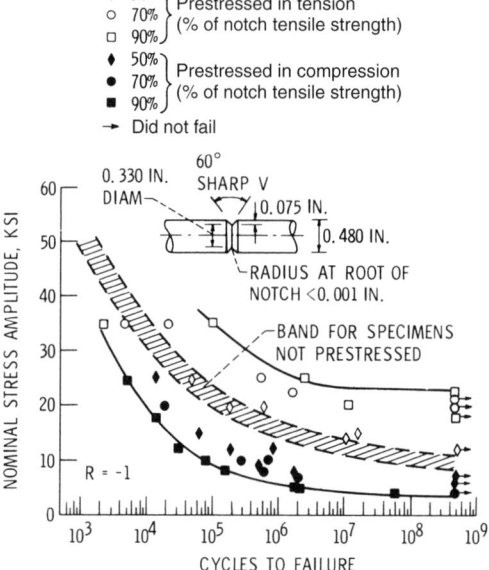

**Fig. 11.69** Rotating-beam fatigue curve for aluminum 7075-T6 rod, 1.9 cm (3/4 in.) diam rolled and drawn with various amounts of prestress. Source: Ref 11.76

one anyway (considering the possibility of fracture during overload), it is advantageous to be able to keep the overloads as low as possible.

A dramatic illustration (Ref 11.78) of the potential benefit of overload is shown in Fig. 11.71. The figure shows the failure cross section of an R.R. Moore rotating beam specimen. A crack was initiated at the root of the machined notch A, and this crack was propagated across region B by application of a relatively high stress of 65 ksi. The high-load sequence was terminated with the application of a tensile stress on the side of the specimen that contained the crack. The load was then reduced to a nominal value of 20 ksi, which was maintained until the specimen failed. It can be seen from the figure that final fracture did not occur by continued propagation of the crack already formed in region B. Rather, a new crack started from the opposite smooth side of the specimen, and this crack propagated across region D until the cross section became reduced sufficiently by cracks B and D to cause final static crack growth across region C. Because the fatal crack did not propagate from the root of region B it can be inferred that the compressive residual stress due to the last application of the tensile load was sufficient to prohibit further crack propagation—striking example of the power of residual stress in arresting crack growth.

Figure 11.72 shows a final example of how overloading can be used to strengthen parts in fatigue loading. Illustrated here is a fillet weld deposited on a tension specimen and the scatterband fatigue curve for a series of such specimens at loading condition $R = 0$. Overloading to cause plastic flow at the stress concentration associated with the weld improves the fatigue resistance at subsequently applied lower stress levels. The figure also shows that in this case,

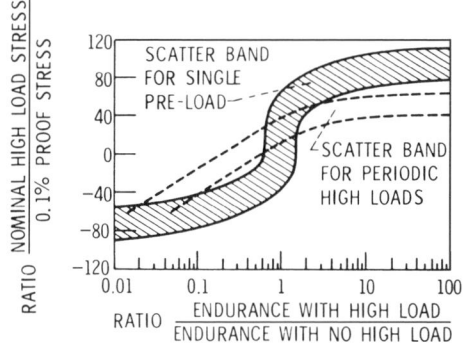

**Fig. 11.70** The effect of one preload and of periodic high loads on endurance of aluminum alloy parts subjected to fluctuating tension. Source: Ref 11.77

**Fig. 11.71** Illustration of crack growth arrest by residual stress due to overload. A, slotted notch; B, crack developed at 65 ksi; C, tensile portion of fracture; D, crack developed at 20 ksi. Source: Ref 11.78

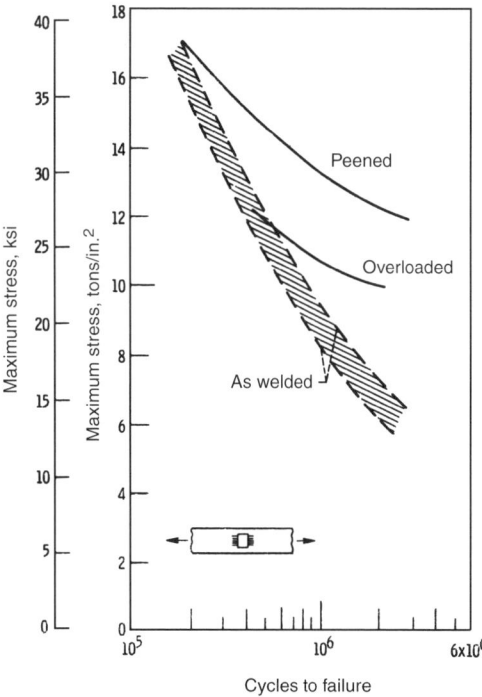

**Fig. 11.72** Several improvement methods for mild steel specimens with transverse nonload-carrying fillet welds. Source: Ref 11.42

shot peening was more effective than the particular overload studied.

It should be emphasized that while overloading can be used effectively to increase fatigue life, great care must be exercised in choosing the overloading conditions to be applied. If an excessive load is applied, cracks might be started that would otherwise not be initiated. For some materials and loading conditions the presence of these cracks could serve to weaken the structure compared with its strength in the absence of the applied overload. Each case must be studied individually to determine the wisdom of applying the overload, taking into account material, structure, and experience with similar situations. In an extensive study of the merit of the practice of overloading for fatigue protection, it was determined, however (Ref 11.79), that it was worth the attempt in a large number of applications. Caution should be taken in considering potential benefit of overloads that they do not introduce a detrimental low-cycle fatigue-high-cycle fatigue interaction as discussed at length in Chapter 6, "Cumulative Fatigue Damage."

### *Prestress Coatings*

Coatings, as a means of surface protection, have already been discussed. Another purpose such coatings can serve is to introduce favorable residual stresses. Various processes can be used to introduce the stresses, among them differential thermal expansion of the coating and substrate material and metallurgical phase change.

The use of nitride coatings in the bearings industry constitutes an example of the application of phase changes. In the marstressing technique (Ref 11.80) the material is heated to the austenitic range in a nitrogen environment. As the material cools below the transformation range, the core material transforms to martensite prior to transformation of the surface material because the presence of nitrogen retards this transformation. The transformation is accompanied by an increase in volume, but because the surface material is at a high temperature while the core transforms, the strains imposed on the surface by the expansion of the core can be readily accommodated by surface creep and plastic flow. Eventually, as the temperature is further lowered during the quench, the point of martensitic transformation of the surface is reached. This transformation also is accompanied by expansion, but now, because of the lower temperature and because of geometric effects, this expansion cannot cause easy creep and plastic flow on the interior material. The stronger, more massive interior material thus constrains the expansion of the surface material, imposing on it a residual compressive stress that is beneficial to fatigue life.

Figure 11.73 shows the residual stresses in ball-bearing inner rings introduced by the prenitriding process. Compared with shot peening, the residual stresses are lower, but sufficiently high, according to the investigators, to produce a beneficial effect. A sample of the results is shown in Fig. 11.74. It is seen that life is ap-

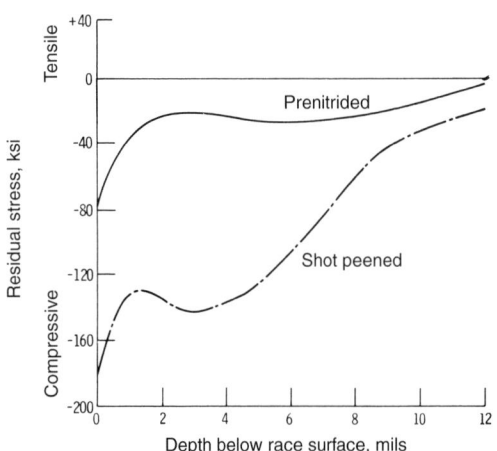

**Fig. 11.73** Residual stress in prenitrided and shot-peened ball bearing inner rings. Source: Ref 11.80

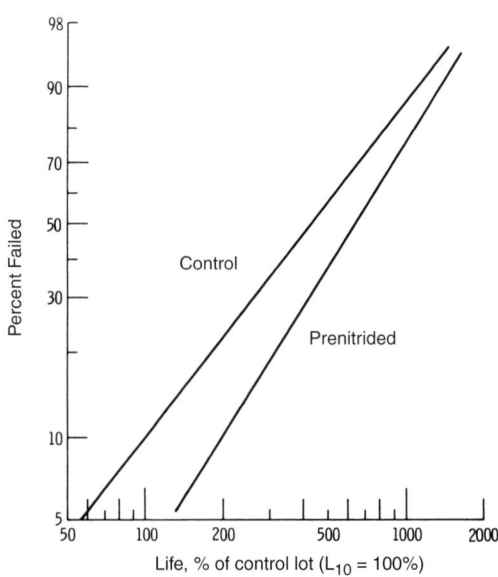

**Fig. 11.74** Relative endurance of prenitrided and control-lot bearing inner rings. Source: Ref 11.81

proximately doubled for the early failures, which are the important ones governing overall bearing life.

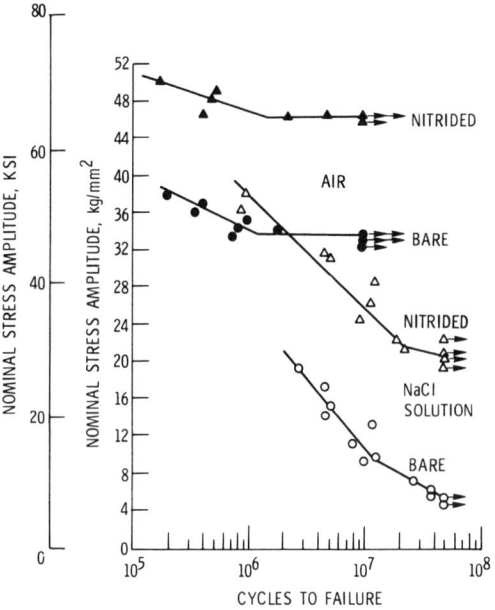

**Fig. 11.75** Fatigue curves for smooth rotating beam specimens of steel 45 tested in air and in 3% NaCl solution before and after nitriding. Source: Ref 11.81

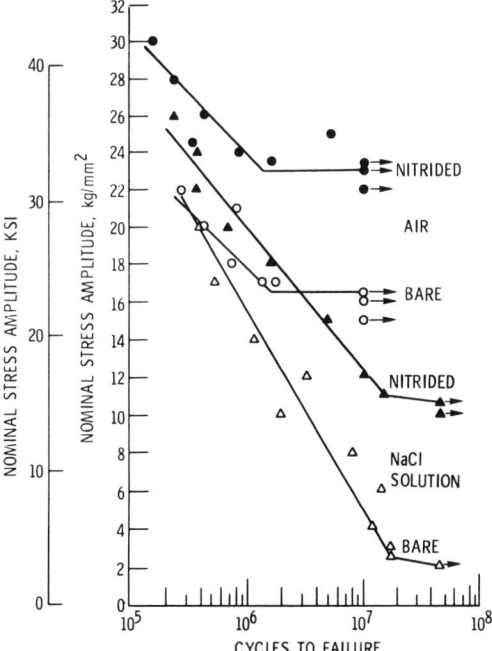

**Fig. 11.76** Fatigue curves for notched rotating beam specimens of steel 45 tested in air and in 3% NaCl solution before and after nitriding. Source: Ref 11.81

The benefits of nitriding in introducing residual compressive stresses are to be added to the protective effect of the coating against hostile environments. Figure 11.75 shows the effect of nitriding on steel when tested with smooth specimens. In air the fatigue strength is increased by about one-third in the life range of $10^7$ cycles. In a 3% sodium chloride (NaCl) solution the fatigue strength is substantially reduced in both the nitrided and unnitrided conditions. But the fatigue strength is over 2½ times as high for the nitrided as the unnitrided material in the life range of $10^7$ cycles.

Figure 11.76 shows the result of rotating beam fatigue tests of notched steel specimens. The beneficial effects of the nitriding are even greater for this case than for smooth material. Nitriding has been found to be beneficial to fatigue life for other steel applications, among them maraging steels (Ref 11.82).

### Thermal Treatment

Subjecting a structure or a specimen to a controlled temperature distribution that causes plastic flow, thereby producing a favorable state of residual stress, also can be beneficial to fatigue life. Figure 11.77 shows one way in which the approach has been used successfully to strengthen welded structures in fatigue. Region

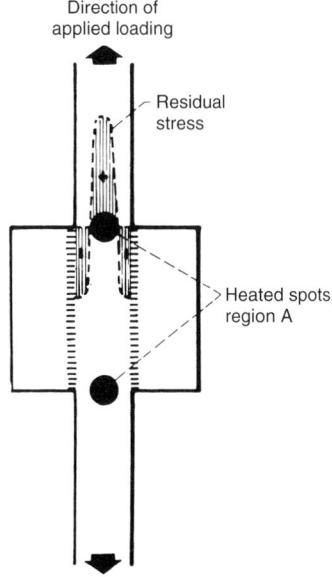

**Fig. 11.77** Method of improving fatigue strength of welded regions by inducing residual stresses as a result of spot heating. Source: Ref 11.42

A is heated locally to a high temperature. The temperature of the adjacent area is kept lower. The thermal stresses associated with such a temperature distribution are, of course, tensile in the regions of low temperature, and compressive in the vicinity of region A where the temperature is high. Plastic flow caused by the thermal stress then causes a residual stress distribution of mirror image relative to the stress distribution when the temperature subsides to room value; that is, the residual stress is compressive on either side of the heated spot. The welds are thus in the region of compressive residual stress, which is beneficial to fatigue resistance. Figure 11.78 shows that fatigue resistance is thereby improved considerably.

### Property Conditioning and Restoration

Because fatigue is a process of gradual, cycle-by-cycle deterioration, the question arises as to whether such deterioration can be minimized by proper conditioning of the material so that it may better be able to withstand the loading conditions. Alternatively we may ask whether there are practices that will aid in restoring the initial fatigue resistance after the material already has been subjected to partial deterioration by prior loading. Some of the opportunities along this line have already been outlined in connection with residual stresses: periodic overloading, shot peening after prior use, and so forth. Others will now be discussed briefly. Unfortunately, this field has not been explored extensively, and much remains that can yet be studied to exploit these directions.

### Coaxing and Strain Dispersal

That the long-life fatigue resistance of some materials can be increased substantially by subjecting them to a programmed sequence of loading has long been known under the heading of *coaxing*. Most of the studies, however, have concentrated on the mechanism involved and the magnitude of the effect on a limited number of materials rather than on the potential applicability to structural fatigue improvement.

Figure 11.79 shows the results of a study of coaxing on steel and illustrates the phenomenon. The dotted curve is the ordinary S-N curve of the material and shows a high-cycle fatigue strength of approximately 43 ksi in the $10^7$ to $10^8$ cyclic life range. The line *ABCD* shows a special sequence of loading wherein a series of stress levels are applied to the material, each for several million cycles. The first stress is at 40 ksi, below the horizontal portion of the initial fatigue curve, and is repeated for $5 \times 10^6$ cycles. The stress is then raised to 41 ksi, and $2 \times 10^6$ cycles of this stress level are applied, until point *C* is reached. The stress is then again increased to 42 ksi, and another $2 \times 10^6$ cycles applied. In this manner, progressively increasing the stress for limited numbers of cycles, the material is gradually "coaxed" to be able to with-

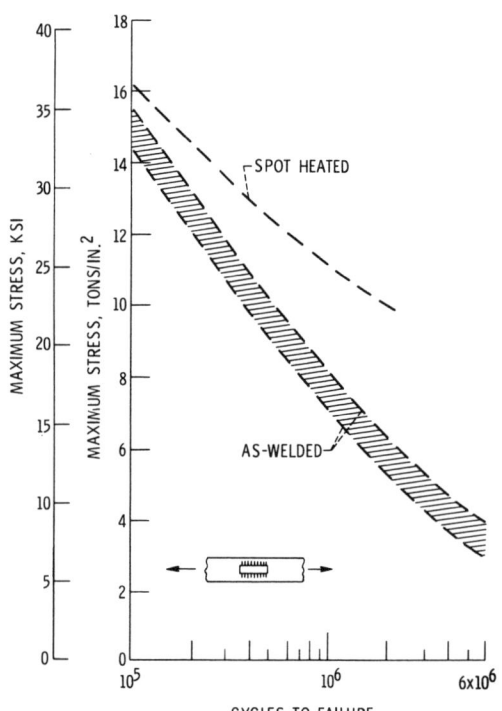

**Fig. 11.78** Effect of spot heating on fatigue of simulated weld. Source: Ref 11.42

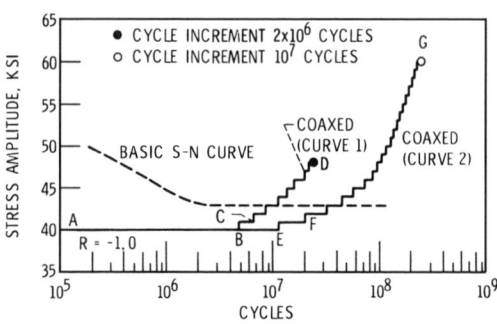

**Fig. 11.79** Improvement of fatigue life of 1045 steel by coaxing. Source: Ref 11.83

stand stress levels above the initially determined 43 ksi level for many more cycles than would have been possible (see basic S-N curve) had the stress levels below this level not been applied first. The curve ABEFG shows even more dramatic strength increases due to coaxing than ABCD. Here, the number of cycles applied in each block of progressively increasing stresses in $10^7$ cycles, starting below the 43 ksi initial level. It can be seen, therefore, that the material was able to withstand $10^7$ cycles of each unit stress level from 44 to 60 ksi, whereas it would not have been able to withstand $10^7$ cycles of even one of these stress levels had this stress been applied solely rather than as part of the coaxing program.

*Coaxing* is the progressive improvement in the fatigue strength of a material induced by applying a progressively increasing stress, starting below the originally determined "fatigue limit" and increasing the loading to a value well above this limit. How much increase in fatigue capacity can be achieved depends on the coaxing program and on the material. One of the most complete studies of this phenomenon is described by Belyaev (Ref 11.84). He studied hard and soft steels, as well as some nonferrous metals, and found coaxing opportunities in them all, although some materials were more susceptible to this type of improvement than others. He found, moreover, that the optimum loading sequence was rather crucially dependent on material. The softer materials required more cycles to build up the fatigue resistance than the harder steels. In fact, if the cycles required to coax a soft steel were applied to a hard steel, damage, rather than benefit, occurred.

Belyaev also considered the practical matter of coaxing in actual machine components. He concluded that it is well worth the try and that in some cases the practice was almost inherent in the manner by which components were loaded in service. Bearings, for example, are initially run at low loads (below what might be considered a "fatigue limit"). As parts wear and clearances increase, the loading becomes progressively more severe, as implied in a coaxing loading program. Figure 11.80 shows some of the results on a bearing coaxing program (Ref 11.84). The curve AB is the S-N curve for the bearing. Coaxing results are indicated by CD and CDE (two tests, one failing at point D and one at E). Obviously, a benefit accrues from the controlled "run-in" program. It was pointed out, in fact, that the failures at points D and E were due to cage fatigue rather than race fatigue represented by curve AB. If the cages had not failed, the life of the bearing races would have been much longer than D and E.

Belyaev found, however, that components with notches, cracks, or other severe stress raisers cannot be coaxed to higher fatigue life by a progressively increasing loading sequence. This is attributed to the high strains associated with the discontinuities. His reasoning was that coaxing achieves its beneficial result by progressively strengthening points of weakness in the material. A weak slip plane, for example, slips first, but is strain hardened during a large number of cycles of low-stress application, which is not in itself damaging. As the weak regions are strengthened, the strains associated with the later higher loads are distributed more uniformly over the entire surface, thereby avoiding the overconcentration of strain in weak regions as is common in conventional single-load fatigue testing.

The mechanism whereby coaxing achieves its beneficial effects has not been completely established. It may, as Belyaev contended in the previously mentioned studies, have to do with progressive strengthening of weak regions. Or, it may be due to favorable subgrain structure

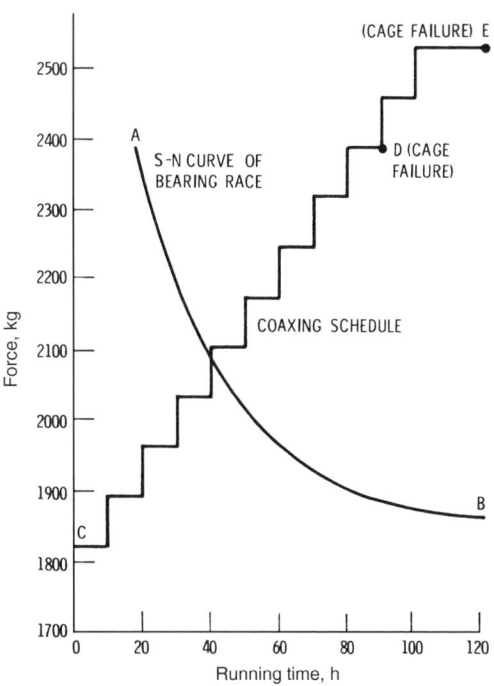

**Fig. 11.80** Improvement in fatigue life of bearing resulting from coaxing (run-in) program. Source: Ref 11.84

formed at the progressively increasing stresses, as described by Grosskreutz (Ref 11.85), which again, permits more uniform absorption of the strain among many grains rather than being so highly localized. Whatever the mechanism, however, the concept has yet to be exploited to its fullest extent.

Other investigators have examined related processes, somewhat similar to coaxing, as a means of improving fatigue life. Waisman and his coworkers (Ref 11.86), for example, have found that a material is better able to sustain a high fatigue stress level if periods of low stress level are interspersed during the loading cycle. Again, it may be simply a matter of "conditioning" the microstructure to distribute the required amount of slip more uniformly over a large volume of material rather than causing severe localization or strain inherent in fatigue failure. Nine and Wood (Ref 11.87) also have emphasized the role of dispersed slip in improving fatigue life. These ideas merit further study.

## Surface Removal

Two factors associated with the mechanism of fatigue lead us to consider surface removal as an important means of structural rejuvenation. First is the fact that in most cases fatigue damage starts at the surface, whether because strains are usually highest at the surface, or because of surface exposure to damaging external influences, or because of the inherent weakness of a surface. Second, a large fraction of the total fatigue life is consumed in initiating cracks at the surface (at least in the absence of high-strain concentration factors such as geometric notches or preexisting flaws, and when the cyclic life is appreciable, say over $10^4$ cycles, as is common in most applications). Thus, periodic surface removal may result in the elimination of most of the fatigue-damaged material. How often to remove surface, and how much to remove, is a matter of matching fatigue fundamentals to the economics and convenience of a particular application.

In one of the earliest studies (Ref 11.88) designed to determine whether life can be improved by surface removal, it was found that an increase of a factor of 6 or more in life could be obtained by removal of about 0.002 in. after every million cycles of loading (50% of the nominal life at the loading conditions involved). Other investigators have found similar improvements due to surface removal. Thompson and Wadsworth (Ref 11.89) found that removal of approximately 0.001 in. after each 25% of life prolonged life substantially. After 2.25 times the normal life, the specimen was unbroken and looked as good as new.

A more engineering-oriented approach to the problem of surface removal is described by Christensen (Ref 11.90). Figure 11.81 shows some of the results. The test specimen contained a central hole constituting a source of strain concentration where the crack nucleated. The $S$-$N$ curve for the test specimen without surface removal is shown as the cross-hatched band. Two specimens were tested by removing 0.015 in. of surface at intervals of 50,000 cycles, the results of which are shown by the line $PQRSTUVW$. (The stress variation as shown by this line is partly due to material removal and partly to changes in external loading conditions.) It can be seen from the figure that life was increased appreciably by the surface removal. Based on the original fatigue curve, the summation of the cycle ratios (applied cycles at each stress level divided by fatigue life at the stress level) is about 3.5, suggesting that surface removal was beneficial.

Figure 11.82 illustrates, on the other hand, one of the cautions (Ref 11.90) that must be exercised in seeking to extend life by this method, namely, to ensure that cracks present in the surface are completely removed—a daunting task. The $S$-$N$ curve for the filleted specimens is shown by $AB$. Five specimens were first loaded at 25 ksi. for 100,000 cycles, half the life that would have caused failure at that stress level. Two of these specimens were then machined to

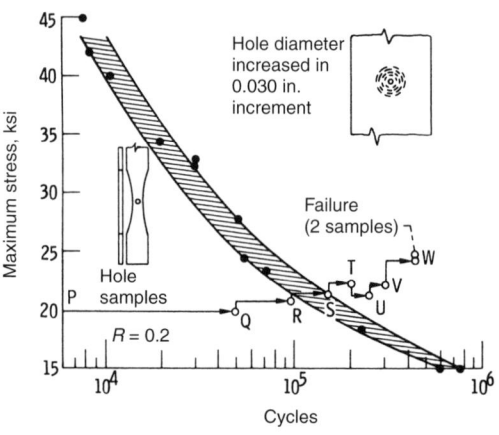

**Fig. 11.81** Effect of surface removal at intervals of 50,000 cycles on fatigue of specimen with circular hole. Source: Ref 11.90

remove 0.0015 in. as well as remachining the fillet to a radius of 1 in. When subsequently tested at 42.4 ksi., the life was 18,000 cycles, whereas the life at this stress in the original S-N curve was about 25,000 cycles. Considering the reduction in stress concentration due to the enlarged fillet radius, the surface removal in this case actually reduced the life. The other three specimens were machined to a 0.003 in. depth, ensuring that all fatigue-damaged material was removed. Subsequent tests at a higher stress level of 42.5 ksi. resulted in a life about equal to that of virgin material machined to the larger fillet radius. Apparently, the 0.003 in. surface removal did essentially remove the crack-damaged material while the 0.0015 in. cut did not.

Most of the foregoing investigations related to loading applications that resulted in long lives, of the order of several millions of cycles. Beneficial results also have been obtained in the low-cyclic life range. Bentele and Lowthian (Ref 11.91) investigated the thermal-shock resistance of nozzle blades of gas turbines and found that if the test blades were polished and etched after every 250 cycles in a mild chemical solution, the crack resistance was vastly improved. They attributed this improvement largely to the removal of surface irregularities by the mild chemical change that occurred without an appreciable attack on the grain boundaries. Raymond and Coffin (Ref 11.92) also have studied the effect of surface removal on the low-cycle fatigue life of aluminum. They found that by periodic removal of several thousandths of an inch they were able to prolong life by a large factor (possibly indefinitely, they believed, or at least until most of the material was removed by machining). In each machining operation the surface-rumpled material was removed, this material being in general the origin of crack nucleation.

A very important example of life extension in the low-cycle fatigue range resulting from surface removal in conjunction with other remedial measures is the case (Ref 11.93) of the General Electric CJ805 engine used in commercial jet engines in service during the 1960s. The turbine blades were made from cast Udimet 500, a common high-temperature nickel-base alloy of the era. When production was first started, the time between overhauls (TBO) was only about 1200 h. Thermal fatigue cracking was the major cause for overhaul. As experience accumulated, several major changes were instituted. The starting sequence of the combustors was changed to reduce the thermal stresses, as discussed earlier in this report in connection with strain concentration effects. Also, rather than using the blades in the as-cast condition, 0.004 in. was machined from the as-cast surface to remove casting imperfections. These changes alone increased the TBO from 1200 to 4000 h. After 4000 h of service, thermal fatigue cracks again formed. However, during overhaul after 4000 h, 0.010 in. was machined from the surfaces of the leading and trailing edges. Two such surface removals were permitted during two successive overhauls. Thus, by the three modifications indicated, two

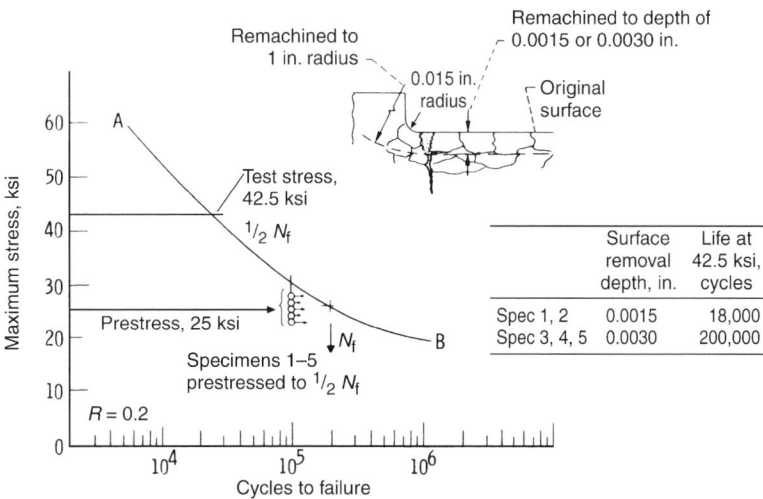

**Fig. 11.82** Tests to illustrate the importance of complete removal of fatigue crack if surface removal is to be effective in improving fatigue. Source: Ref 11.90

of which are basically surface removal, service life was extended from 1200 to 12,000 h of life.

## Property Restoration by Heat Treatment

The hope of removal of fatigue damage by periodic exposure to high temperature has been expressed many times and studied by a number of investigators. From these studies the most common conclusion reached has been that the hope is a rather futile one. In almost all cases periodic heat exposures, even if performed frequently, did not significantly increase life, and sometimes the effect was even detrimental (for a summary of early work in this field see Ref 11.89 and 11.94). Generally it has been agreed that once cracks start, even minute ones not detectable under the ordinary microscope, such cracks will not be repaired by high-temperature exposures. If the damage is in the form of microscopic flaws rather than discrete cracks, then high-temperature exposure might make these flaws more mobile, allowing them to diffuse and agglomerate to produce an even more damaging larger flaw than were the individual microscopic imperfections. In such cases, the effect of high-temperature annealing is to reduce fatigue life.

In a few instances, however, beneficial effects of annealing have been found. Wright and Greenough (Ref 11.95) found beneficial effects on gold when it was annealed at intervals of either one-half or one-quarter of its life (when this life was in the range of $10^7$ cycles). Ivanova and coworkers (Ref 11.96) studied reheat treatment of a steel after various prior cycles at a stress where life was about $2.5 \times 10^5$ cycles-to-failure. Unfortunately, they evaluated the effect of the heat treatment by testing the material at a lower stress level corresponding to the virgin material's "fatigue limit." Repair of the damage was interpreted to mean that life will remain indefinitely long when tested at this lower stress level. On this basis they concluded that if the heat treatment were performed prior to the introduction of submicroscopic flaws, the damage could be removed by heat treatment, but if submicroscopic flaws were induced by prior fatigue loading, reheat treatment did not repair the damage. The evidence of this conclusion given in Ref 11.96 is sparse and moot, but the possibility that such a conclusion, depending on material, could be correct can be accepted on the basis of logic alone. Just where in the fatigue process the submicroscopic flaws develop and how to "heal" them remains a problem of the individual metallurgy of each material.

Sandor (Ref 11.94) has made systematic study of the effect of heat treatment on the healing of steel. He found a major improvement in fatigue life in SAE 4140 steel by frequent 1 min anneals at 316 °C (600 °F). In this case, the specimen was force cycled with a zero mean stress. Because the material was cyclically strain softening, the repetitive loading caused a progressively increasing plastic strain from cycle to cycle. Upon each brief heating, the previously generated dislocations provided numerous additional sites for the small carbon atoms in the steel to diffuse to. The carbon then pinned the dislocations, making their further glide more difficult; that is, the alloy was strengthened by what is know as *strain-age hardening*. Upon cooling back to room temperature and continuing the fatigue cycling, the dislocations could again become unpinned, causing cyclic strain softening with attendant increases in the plastic strain range. Repetitions of this schedule of heating, cooling, and fatiguing resulted in greater cyclic life. Sandor concluded that the Manson-Coffin relation accurately predicted the prolongation of fatigue life. The results are shown in Fig. 11.83. The line *AB* shows the basic *S-N* behavior of the steel for completely reversed loading. At a stress range of 150 ksi the material ordinarily withstands about 100 to 200 cycles. Specimens heated every 10 cycles resulted in median lives of about 400 cycles, and reheating every cycle prolonged the life to thousands of cycles, with some specimens still running after 8000 cycles. It is important, of course, to note that the benefit in this case arose because the material was cyclically strain softening while the specimen was force cycled. Had the material been tested under strain-cycling, this improvement would proba-

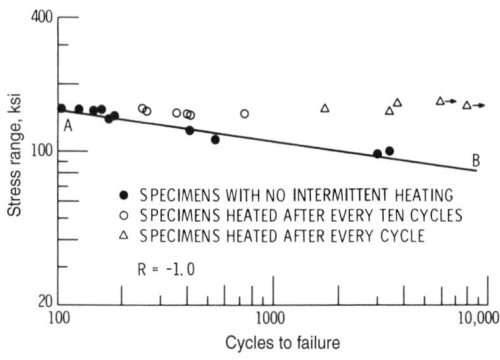

**Fig. 11.83** Effect of intermittent heating on fatigue life of cyclic strain softening steel subjected to load cycling conditions. Source: Ref 11.94

bly not have been obtained. It also should be observed that the frequency of reheat treatment required to greatly increase life renders the process impractical for many applications (unless, of course, the heating is somehow inherent in the service history).

The question of "healing" by reheat treatment is an interesting one and deserves more study than has been devoted to it. A better understanding of the fundamentals of the fatigue process, together with possible practical procedures might be the reward of such studies. However, it is to be expected that the individual metallurgy of each material will enter into the results, as well as the loading conditions encountered.

## Safe-Life and Fail-Safe Design and Surveillance

In many cases the appearance of a visible surface crack implies that the major part of the fatigue life has already been consumed. It is frequently cited, for example, that in a smooth specimen loaded so as to have a life of about a million cycles the appearance of a visible crack means that well over 90% of the life has been used up. This is very true for smooth specimens in the long-life range. But even in this instance, 2 or 3% of the life may still represent a large number of actual cycles. If the crack could be detected shortly after it occurs, it would still be possible to remove it before catastrophic failure occurs. Or, if inspections are scheduled to ensure that the number of intervening cycles does not exceed the number of cycles between detection and fracture, then it can be reasonably assumed that if a crack does not appear during a given inspection, then no catastrophic failure will occur before the next inspection.

The smooth, polished specimen, loaded for a life in the range of a million or more cycles, is the uncommon example of practical service simulation. More common is the part that has a stress concentration in the form of a natural imperfection, an unintended surface flaw, or a machined geometric discontinuity. In these cases the appearance of a crack occurs at a relatively low percentage of the total life, and, in fact, the major part of the life may be crack growth. This is especially true in the low-cycle fatigue range, and even more so at high temperature where cracks may be generated by a creep mechanism. Under these conditions, the fail-safe concept of mandatory part removal after some preestablished point of crack growth may be especially relevant.

The fail-safe and safe-life philosophies have been undergoing considerable study, and the practicality of their implementation has been augmented by the development of fracture mechanics laws for crack growth. The philosophies can be expressed in several simple rules (Ref 11.97):

- The structure must have an adequate life, either as a crack-free period, or one during which the growth rate of cracks is sufficiently low as to escape detection by inspection procedures.
- The structure should be able to carry a predetermined load under a given amount of damage, that is, to have adequate residual strength.
- Visual inspection of all critical areas must be possible, either in service or after tests.
- Repair of damaged elements must be accomplished either by replacement or by use of doublers. (The large amount of labor and time plus the high probability of accidental damage in replacing a major member in a complex structure often leads to use of doublers or conventional patching.)

While this philosophy is simple in principle, applying it requires invoking elegant sophistication. For example, the analysis may require the use of a crack propagation law to establish how rapidly the crack will grow once it is formed. Fortunately, fracture mechanics devotees have been studying the problem intensively. To date, the greatest progress has been made in the regime wherein the nominal stress in the vicinity of the crack is elastic (Ref 11.98). However, rapid progress is being made in treating problems within the plastic range, for example, Ref 11.99. Figure 11.84 shows the commonly used correlation between cyclic crack growth rate $da/dN$ and range of stress intensity $\Delta K$, in this

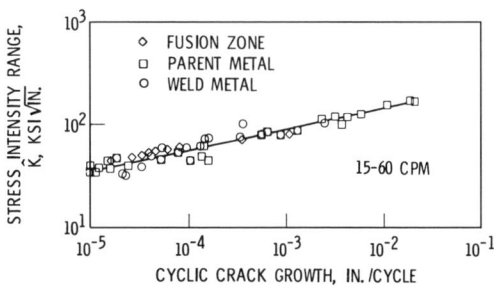

**Fig. 11.84** Crack growth rates in Inconel 718 weld components at 538 °C (1000 °F). Source: Ref 11.100

case, for Inconel 718 weld constituents (Ref 11.100). A straight line relation results on log-log coordinates. This power law is known as the *Paris-Erdogan relation* (Ref 11.101) and the curve is normally shown with the axes reversed relative to the figure shown. More sophisticated analyses also take into account the effect of mean value of stress intensity, as well as linearity deviations of such plots at the extremities of the curve. At the low end of the curve, $\Delta K$ approaches $\Delta K_{threshold}$ where $da/dN$ approaches zero more rapidly than implied by the linear Paris-Erdogan relation. At the high end of the curve, $\Delta K$ approaches $K_{IC}$ (the critical stress-intensity factor for fracture) where $da/dN$ sweeps vertically upward. Most of the past studies of crack growth have been conducted at a sufficiently low temperature to make unnecessary the need for inclusion of creep effects in the analysis. For the hot parts of gas turbine engines, however, this factor certainly must be considered. Time effects become very important, as seen in Fig. 11.85, also taken from Ref 11.100. Here it is seen that crack growth rate is a very strong function of hold time at high temperature. For example, the crack growth rate for a 2 h hold time is about 100 times higher than when no hold time is allowed.

The question of when crack extension ceases to be gradual "growth" and merges into "rapid fracture" is of intimate concern to the fail-safe philosophy. Fracture mechanics uses the concept of "fracture toughness" to provide the conversion factor, and much has been done in recent years to study this property and to provide methods of valid measurement. A special property, "plane-strain fracture toughness," associated with the fracture mechanism when the crack is in a region of plane strain has been designated, and standards have been set (Ref 11.102) for measurement of this quantity. Many applications in gas turbines, for example, turbine disks, are thick enough to conform to the requirements of the condition of plane strain. Other components, not so thick, do not satisfy the requirements of plane strain, yet require fracture criteria for determining allowable crack growth.

Figure 11.86, also taken from Ref 11.100, shows the effective fracture toughness at 538 °C (1000 °F) for 0.056 in. Inconel 718 sheet as a function of crack length $2a$ and sheet width $W$. It is seen that considerable variation in apparent fracture toughness can occur depending on the dimensions of the sheet, demonstrating that the fracture toughness as so determined is not a true materials property (although *plane strain fracture toughness* may be). However, such data are directly applicable for computing fracture for specific geometries, and if available for specific materials, temperatures, geometries, and loading conditions, can be very useful in calculating residual strengths-load carrying capacities after a given crack geometry has been developed by flaw growth.

The emergence of fracture mechanics as an analytical tool has given impetus to the fail-safe concept. By designing so as to be able to detect cracks, by developing procedures for crack detection as discussed in the literature, and by setting limits of permissible crack growth based on sound engineering judgment determined from fracture mechanics analysis, it may well be possible to extend fatigue life beyond its present limits. The appearance of a crack would then lose its role as a source of mystery and anxiety, or as a cause for automatic removal of the part. We would learn to deal with cracks

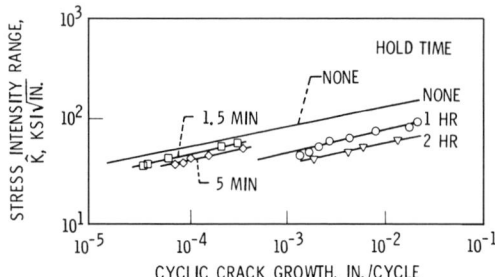

**Fig. 11.85** Hold-time effects on crack growth at 538 °C (1000 °F) for Inconel 718. Source: Ref 11.100

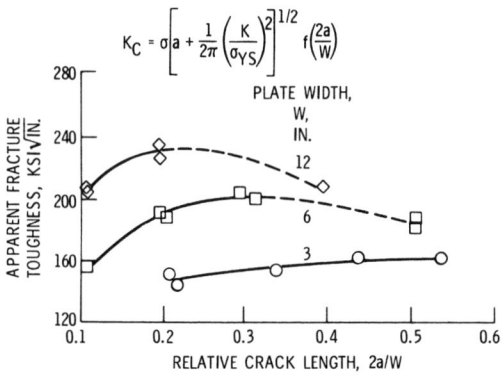

**Fig. 11.86** Apparent fracture toughness of 0.056 in. Inconel 718 sheet at 538 °C (1000 °F). Source: Ref 11.100

realistically, resulting in economic advantage, and perhaps even in improved overall reliability.

*Proof Loading*

Proof loading is a procedure commonly used to ensure subsequent safe operation of slings, hooks, cranes, and other lifting devices whose safety is clearly force controlled. Pressure vessels are also important structures that qualify for proof loading. Overpressurization proof testing of a pressure vessel is a well-controlled procedure often used to demonstrate "proof" of subsequent safe operation and to identify unsafe vessels. If the vessel fails due to excessive deformation or rupture, it is discarded as unsafe. All similar vessels would be suspect and likely would be removed from service for reworking or be scrapped. If the design is new, better quality control in manufacturing or a redesign typically would be called for. While the procedure serves to detect undesirable, potentially catastrophic defects, it also can create, in ductile-metal vessels, beneficial compressive residual stresses at critical locations. Compressive residual stresses generally impart greater durability due to a reduction of the peak and mean stresses during service. This is true whether service loading is static or cyclic (fatigue). However, if the proof load is too great, excessive plasticity could occur and impart undesirable damage in localized areas. Excessive proof pressurization damage can take the form of surface roughening and localized cracking. This is especially true in the presence of aggressive environments. Small-scale local cracking can reduce service fatigue resistance. Such cracking is virtually impossible to detect using standard nondestructive crack detection techniques. The potentially damaging aspects of proof loading have not been recognized by the engineering community to the same extent that beneficial aspects have been appreciated. The benefits, and potential pitfalls, of proof loading are illustrated for two cases of pressure vessels described next.

**Example 1: Heavy Vessel.** We presented earlier an example in Fig. 9.3 in Chapter 9, "Crack Mechanics," of a large rocket motor case that failed unexpectedly and catastrophically during the early stages of proof pressurization. The burst pressure was *only 50% of its intended* operating pressure. Fortunately, under the well-controlled conditions of proof testing, no human casualties were encountered. Had the proof test not been conducted and the motor case been put directly into service, the results would have been cataclysmic. A spectacular failure would have occurred on the launch pad resulting in casualties and serious injuries, as well as extensive damage to the launch facility. As pointed out in Chapter 9, "Crack Mechanics," the fracture initiation site was a concealed welding flaw that was undetected during nondestructive inspection (NDI). The applied loading level, coupled with the flaw size, was large enough to cause the stress-intensity factor at the flaw to reach its critical plane strain fracture toughness value for this high-strength, moderate-toughness, alloy steel. The most important general characteristic of this pressure vessel was its relatively thick-walled construction that led to plane strain conditions and a requirement for multipass welding. The left-hand column of Table 11.2 summarizes the features of pressure vessels of this class.

**Example 2: Light Vessel.** An early design for electric power storage for a U.S. space station considered arrays of $NiH_2$ batteries in the form of pressure vessels that would charge and discharge with an orbital solar exposure/eclipse period of 90 min. A 20 year design life corresponded to about $10^5$ fatigue cycles. Peak vessel pressure of 6.9 MPa (1000 psi) corresponded to a full charge while 4.5 MPa (650 psi) represented maximum discharge in each 90 min cycle. Nominal dimensions of the vessel were 9 cm (3.5 in.) diam, 24 cm (9.4 in.) overall length, and a wall thickness of only 0.06 cm (0.02 in.). The middle 10.5 cm (4.1 in.) long cylindrical section was circumferentially welded to nominally hemispherical end caps with a disk (0.06 cm, or 0.02 in., thick by 9-plus cm, or 3.5-plus in., diam) separating and sealing off the cylindrical portion from each end cap. The only detail of importance herein is the fillet area of the weld as shown sche-

**Table 11.2 Characteristics of extreme classes of pressure vessels**

| Heavy vessels | Light vessels |
|---|---|
| Thick walled | Thin walled |
| Plane strain | Plane stress |
| Multiple-pass welds | Single pass welds |
| Moderate quality control | High quality control |
| Difficult for NDI | Amenable to NDI |
| Moderate fracture toughness | High fracture toughness |
| Service cycles $< 10^3$ | Service cycles $> 10^3$ |

NDI, nondestructive inspection

matically in Fig. 11.87 (Ref 11.103). The material of construction was Inconel 718, in a solution-treated and aged condition. No filler weld metal was used because the slightly oversized circumference of the disks provided the excess material to form the desired weld fillet upon cooling. Prototype vessels exhibited excess metal in the weld fillet. This gave rise to a sharp stress concentration at the juncture (toe of the weld) of the relatively hard weld metal and the surrounding soft heat-affected-zone metal.

While proof loading has traditionally served as a defect inspection tool for large ground-based pressure vessels with the characteristics shown in the left-hand column of Table 11.2, this proofing procedure does not necessarily lend itself to the example lightweight, space-based vessel. In the latter, a cracklike defect such as an imperfect circumferential weld must penetrate 90% of the wall thickness before a commonly used 50% over-pressurization would expose the flaw. This is so because of the plane stress condition of the thin-walled vessel and the alloy's high-fracture toughness. By contrast, however, the critical depth of a fatigue life-limiting defect is less than about 0.0025 cm (0.001 in.) (<4% of the wall thickness). A defect this tiny cannot be detected by over-pressurization proof testing of such a thin-walled vessel. In fact, gross yielding, bulging, and rupture of the vessel would occur without exhibiting any detectable influence of the tiny flaw. Thus, it is not possible to use an over-pressurization proof test as a nondestructive inspection technique for fatigue life limiting flaws in thin-walled vessels. Obviously, a more appropriate NDI technique that is capable of detecting small manufacturing flaws is needed for this purpose.

Despite the aforementioned rationale to forego proof testing of the $NiH_2$ battery vessels, 50% overpressure proof loading was applied to sacrificial prototype vessels. In fact, as many as five repetitions of pressurization were applied in the belief that if one proof test is good, then four more must be even better (Ref 11.104). The result was not beneficial, rather it was quite damaging. Destructive examination of the critical toe of the weld revealed 0.0025 cm (0.001 in.) deep cracking due to the first proof-load application. Each subsequent proof loading incrementally extended the crack to a total depth of 0.010 cm (0.004 in.) (16% of the way through the vessel wall thickness!). This extent of proof loading severely damaged the vessel rather than proving that it was fit to be placed into service to withstand a minimum design lifetime of $10^5$ cycles. The conclusion was that proof loading afforded no benefit to the safety of the vessels of this example. In fact, damage resulted from the proof loading, and that damage accumulated with repeated proof loading. The lesson learned is that each specific situation must be examined carefully to ascertain the beneficial and detrimental effects of proof loading.

## Concluding Remarks

In this chapter the attempt is made to outline a number of approaches that can be pursued to improve fatigue life of materials. The principal aspects that must be considered involve:

- Proper choice by the designer of materials or materials combinations that give the best possible fatigue resistance
- Avoidance of strain concentrations in design, fabrication, and service
- Provision for surface protection against inadequate design, rough handling, and pernicious environments
- Introduction of beneficial residual compressive stresses by various means, taking into account such factors as design, fabrication, and usage requirements that have a bearing on the ability of the material to retain these stresses during service
- Property conditioning and restoration such as reheat treatment, surface removal, and any other procedure that can restore fatigue resistance once damage has already incurred

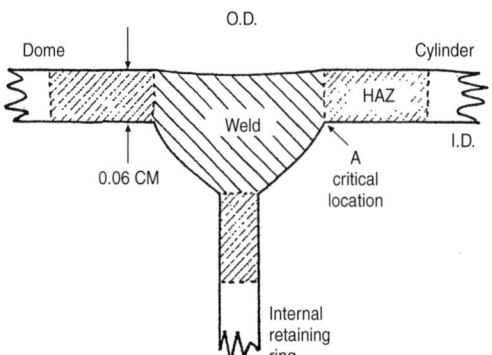

**Fig. 11.87** Cross section of a thin-walled pressure vessel at a circumferential weld. HAZ, heat-affected zone. Source: Ref 11.103

- Safe-life design approach together with the concept of continuous surveillance.

Although there are many approaches, most of them based on some facet of understanding of the fundamental mechanism of fatigue, it must be emphasized that more work is needed in many of these areas in order that the concepts set forth can be exploited to their fullest extent. For example, a few outstanding areas where further research is needed and can be particularly rewarding are the following: To achieve materials with improved fatigue resistance, further development of controlled solidification techniques promises to hold a great potential for enhancing intermediate and high-temperature fatigue resistance; this should be especially useful for applications such as turbo jet engine components. The potential of thermomechanical processing as a means of improving fatigue resistance is being tapped, and further research on this approach in various metal systems have yielded dramatic improvements. In the area of surface protection much remains to be done to properly "marry" coatings to the metal substrate so as to achieve a suitable materials combination for good fatigue resistance. In this connection, metalliding is a particularly beneficial technique. Property conditioning affords many opportunities for improved fatigue life, and coaxing and related processes merit further study. Further exploitation of material restoration by techniques such as surface removal of thin layers that have proved beneficial in turbojet engine practice is indicated clearly, as is the study of healing-by-reheat treatment for various metal systems. A more sophisticated application of fracture mechanics as an analytical tool to establish limits of permissible crack growth promises to augment the practicality of the fail-safe design philosophy.

Finally, although a number of approaches have been suggested for improving fatigue life, it also is clear that any approach used must be applied cautiously, with full consideration of all effects that might be expected. All materials do not respond in the same way to a given treatment; what is good for cyclic strain-hardening materials may not be as effective for, or may even be bad for, cyclic strain-softening materials; thermal treatments may be good for one material, bad for another. The range of fatigue life involved in the application is also important; what is good for the low-cycle fatigue life range may not be good for the high-cyclic-life range, and vice versa. In all cases it is important to take into consideration the type of loading, the part geometry, basic material behavior, environmental effects, reliability requirements, and a host of other factors that govern fatigue life. Many of these are identified in this chapter and in Chapter 10, "Mechanism of Fatigue." In some cases the effects have been studied extensively and are reasonably well understood; in others, much more work is needed.

## REFERENCES

11.1 S.S. Manson, "Avoidance, Control, and Repair of Fatigue Damage," *Metal Fatigue Damage: Mechanism, Detection, Avoidance, and Repair*, ASTM STP 495, American Society for Testing and Materials, 1971, p 254–346

11.2 S.S. Manson, Discussion of Experimental Support for Generalized Equation Predicting Low Cycle Fatigue, *J. Basic Eng.*, Vol 84 (No. 4), Dec 1962, p 537–541

11.3 S.S. Manson, Fatigue: A Complex Subject—Some Simple Approximations, *Exp. Mech.*, 1965, p 193–226

11.4 M.F. Garwood, H.H. Zurburg, and M.A. Erickson, Correlation of Laboratory Tests and Service Performance, *Interpretation of Tests and Correlation with Service*, American Society for Metals, 1951, p 1–77

11.5 J. Morrow, G.R. Halford, and J.F. Millan, Optimum Hardness for Maximum Fatigue Strength of Steel, *Proc. First International Conference on Fracture*, Sendai, Japan, T. Yokobori, T. Kawasaki, and J.L. Swedlow, Ed., Japanese Society for Strength and Fracture of Metals, Vol III, 1965, p 1611–1635

11.6 C.C. Nelson, Effect of Strain Cycling on a Hard Steel, *Sixth Student Symposium on Engineering Mechanics*, T&AM Report 289, Department of Theoretical and Applied Mechanics, University of Illinois, Urbana, IL, June 1966

11.7 S.S. Manson, Interfaces between Fatigue, Creep, and Fracture, *Int. J. Fract. Mech.*, Vol 2 (No. 1) March 1966, p 327–363

11.8 S.S. Manson and M.H. Hirschberg, The Role of Ductility, Tensile Strength, and Fracture Toughness in Fatigue, *J. Frank-*

*lin Inst.,* Vol 290 (No. 6), Dec 1970, p 539–548

11.9 G.V. Karpenko, A.B. Kuslitskii, and Y.I. Babei, Effect of Density on Fatigue Strength of Electroslag and Vacuum Remelted Ball-Bearing Steel, Translation 5911, Henry Brutcher, 1965

11.10 R.A. Cellitti and C.J. Carter, "Ultrasonic Measurement and Influence of Nonmetallic Inclusions on Fatigue and Engineering Behavior of Medium and High Strength Steels," Paper 690049, Society of Automotive Engineers, Jan 1969

11.11 E.V. Zaretsky, The Changing Technology of Rolling-Element Bearings, *Mach. Des.,* Vol 38, Oct 13, 1966, p 206–216, 218–291, and 222–223

11.12 N.J. Petch, The Fracture of Metals, *Progress in Metal Physics,* Vol 5, B. Chalmers and R.B. King, Ed., Pergamon Press, 1954, p 1–52

11.13 R.W. Karry and T.J. Dolan, Influence of Grain Size on Fatigue Notch-Sensitivity, *Proc. American Society for Testing Materials,* Vol 53, 1953, p 789–801

11.14 F.H. Vitovec, "Grain Size Effects on Fatigue and Their Relationship to Notch Geometry, Stress Gradient and Size of Specimens," WADC-TR-59-539, AD-206665, University of Minnesota, Minneapolis, MN, Dec 1958

11.15 R.C. Boettner, C. Laird, and A.J. McEvily, Jr., Crack Nucleation and Growth in High Strain-Low Cycle Fatigue, *Trans. American Institute of Metallurgical Engineers,* Vol 233 (No. 2), Feb 1965, p 379–387

11.16 C.H. Wells, C.P. Sullivan, and M. Gell, "Mechanisms of Fatigue in the Creep Range," *Metal Fatigue Damage: Mechanism, Detection, Avoidance, and Repair,* ASTM STP 495, American Society for Testing and Materials, 1971, p 61–122

11.17 F.L. VerSnyder and R.W. Guard, Directional Grain Structures for High Temperature Strength, *Trans. American Society for Metals,* Vol 52, 1960, p 485–493

11.18 B.J. Piearcey and F.L. VerSnyder, "Monocrystaloys, a New Concept in Gas Turbine Materials, The Properties and Characteristics of PWA 1409," Report 66-007, Pratt & Whitney Aircraft, Feb 2, 1966

11.19 J.R. Johnston and R.L. Ashbrook, "Oxidation and Thermal Fatigue Cracking of Nickel- and Cobalt-Base Alloys in a High Velocity Gas Stream," NASA TN D-5376, 1969

11.20 J.C. Freche, W.J. Waters, and R.L. Ashbrook, The Application of Directional Solidification to a NASA Nickel-Base Alloy (TAZ-8B), *Trans. Society of Automotive Engineers,* Vol 77, 1988, p 1744–1753

11.21 W.J. Waters and J.C. Freche, "A Nickel Base Alloy, WAZ-20, with Improved Strength in the 2000° to 2200 °F Range," NASA TN D-5352, 1969

11.22 R.A. Walker and G. Meyer, Design Recommendations for Minimizing Fatigue in Bolts, *Mach. Des.,* Vol 38 (No. 21), Sept. 15, 1966, p 182–186

11.23 G.D. Sandrock and L. Leonard, "Cold Reduction as a Means of Reducing Embrittlement of a Cobalt-Base Alloy (L-605)," NASA TN D-3528, 1966

11.24 S. Dunleavy and J.W. Spretnak, "Soviet Technology on Thermal Mechanical Treatment of Metals," DMIC memo 244, AD-69874, Battelle Memorial Institute, Nov. 15 1969

11.25 F. Borik, W.M. Justusson, and V.F. Zackay, Fatigue Properties of an Ausformed Steel, *Trans. American Society for Metals,* Vol 56 (No. 3), Sept 1963, p 327–338

11.26 W.M. Justusson and V.F. Zackay, Engineering Properties of Ausformed Steels, *Met. Prog.,* Vol 82 (No. 6), Dec 1962, p 111–114

11.27 E.N. Bamberger, "The Effect of Ausforming on the Rolling Contact Fatigue Life of a Typical Bearing Steel," Paper 65, LUB-9, American Society of Mechanical Engineers, Oct 1965

11.28 F. Ostermann and W.H. Reimann, "Thermo-Mechanical Processing and Fatigue of Aluminum Alloys," Report AFML-TR-69-262, AD-863211, Air Force Systems Command, Aug 1969

11.29 E.V. Zaretsky, R.J. Parker, and W.J. Anderson, "Effect of Component Differential Hardness on Rolling-Contact Fatigue and Load Capacity," NASA TN D-2640, 1965

11.30 E.V. Zaretsky, R.J. Parker, W.J. Anderson, and D.W. Reichard, " Bearing Life and Failure Distribution as Affected by

Actual Component Differential Hardness," NASA TN D-3101, 1965
11.31 C.R. Smith, "Tips on Fatigue," Report GD/C-63-091, General Dynamics/Convair, 1963
11.32 S.S. Manson, *Thermal Stress and Low-Cycle Fatigue,* McGraw-Hill, New York, 1966
11.33 R.B. Heywood, *Designing against Fatigue of Metals,* Reinhold, New York, 1962
11.34 F.B. Stulen, H.N. Cummings, and W.C. Schulte, Preventing Fatigue Failures, Part 1: Basic Factors, *Mach. Des.,* Vol 33 (No. 8), April 27, 1961, p 116–120; Part 2: Geometric Stress Concentration, Vol 33 (No. 10), May 11, 1961, p 191–195; Part 3: Effect of Heat Treatment and Surface Hardening on Fatigue Life, Vol 33 (No. 11), May 25, 1961, p 146–170; Part 5: Calculations for Determining Fatigue Life, Vol 33 (No. 13), June 22, 1961, p 159–165; Part 6: Effect of Biaxial Stresses on Fatigue Life, Vol 33 (No. 14), July 6, 1961, p 154–158
11.35 A.F. Madayag, Ed., *Metal Fatigue: Theory and Design,* Wiley, New York, 1969
11.36 J. Viglione, "The Effect of Nut Design on the Fatigue Life of Internal Wrenching Bolts," Report NAEC-AML-1910, AD-44647, Naval Air Engineering Center, June 17, 1964
11.37 J.L. Waisman, Factors Affecting Fatigue Strength, *Metal Fatigue,* G. Sines and J.L. Waisman, Ed., McGraw-Hill, New York, 1959, p 7–35
11.38 G.C. Noll and M.A. Erickson, "Allowable Stresses for Steel Members of Finite Life," presented at the Annual Meeting of the Society for Experimental Stress Analysis (Detroit, MI), 1959
11.39 F.J. Lavoie, High-Velocity Forging: New Force in Gear Technology, *Mach. Des.,* Vol 40 (No. 28), Dec 5, 1968, p 146–151
11.40 C.L. Harmsworth, "Design Criteria and Test Techniques," Air Force Materials Symposium, Report AFML-TR-65-29, AD-463572, Air Force Systems Command, May 1965, p 831–853; R.J. Rooney, "The Effect of Various Machining Processes on the Reversed-Bending Fatigue Strength of A-110AT Titanium Alloy Sheet," Report WADC-TR-57-310, AD-142118, Wright Air Development Center, Nov 1957
11.41 W.P. Koster, M. Field, L.J. Fritz, L.R. Gatto, and J.F. Kahles, "Surface Integrity of Machined Structural Components," Report AFML-TR-70-11, Metcut Research Association, Inc., March 1970
11.42 T.R. Gurney, *Fatigue of Welded Structures,* 2nd ed., Cambridge University Press, 1979
11.43 D. Radaj, *Design and Analysis of Fatigue Resistant Welded Structures,* Abington Publishing, Abington Hall, Cambridge, England, 1990
11.44 S.J. Maddox, Fatigue Strength of Welded Structures, 2nd ed., Abington Publishing, Abington Hall, Cambridge, England, 1991
11.45 J.A. Choquet, Improvement of the Fatigue Life of Spotwelds, *Weld. Res. Counc. Bull.,* No. 112, Feb 1966, p 1–16
11.46 A.J. Nachtigall, S.J. Klima, J.C. Freche, and C.A. Hoffman, "The Effect of Vacuum on the Fatigue and Stress-Rupture Properties of S-816 and Inconel 550 at 1500 °F," NASA TN D-2898, 1965
11.47 D.H. Rimbey, " An Experimental Study of Fretting Corrosion and Its Influence on Fatigue Strength," Ph.D. thesis, Department of Theoretical and Applied Mechanics, University of Illinois, Urbana, IL, 1967
11.48 W.D. Milestone, " An Investigation of the Basic Mechanisms of Mechanical Fretting and Fretting-Fatigue at Metal-to-Metal Joints, with Emphasis on the Effects of Friction and Friction-Induced Stresses," Ph.D. thesis, Ohio State University, Columbus, OH, 1966
11.49 D.A. Draigor and G.I. Valchuk, *Size Effect in Wear and Fatigue of Steel,* Daniel Davey & Co., New York, 1963
11.50 H.W. Liu, H.T. Corten, and G.M. Sinclair, Fretting Fatigue Strength of Titanium Alloy RC 130B, *Proc. American Society for Testing and Materials,* Vol 57, 1957, p 623–642
11.51 P.C. Clarke and M.M.B. Kay, Effect of Surface Films on Fatigue Life of Steels, *Mater. Research & Standards,* Vol 5 (No. 12), Dec 1965, p 600–606
11.52 W.L. Holshouser and H.P. Utech, Effect of Oleophobic Films on Metal Fatigue, *Proc. American Society for Testing and Materials,* Vol 61, 1961, p 749–756

11.53 W. Rostoker, "Embrittlement of Metals by Organic Liquids," Report IITRI-B183B2-4, AD-426964, IIT Research Institute, Chicago, IL, Jan 2, 1964

11.54 T.R. Shives and J.A. Bennett, The Effect of Environment on the Fatigue Properties of Selected Engineering Alloys, *J. Mater.*, Vol 3 (No. 3), Sept 1968, p 695–715

11.55 H.T. Sumsion, Vacuum Effects on Fatigue Properties of Magnesium and Two Magnesium Alloys, *J. Spacecr. Rockets*, Vol 5 (No. 6), June 1968, p 700–704

11.56 R.D. Weltzin and G. Koves, "Surface Treatment of Ti-6Al-4V for Impact-Fatigue and Wear Resistance," *Applications Related Phenomena in Titanium Alloys*, ASTM STP 432, American Society for Testing and Materials, 1968, p 283–298

11.57 N.C. Cook, Metalliding, *Sci. Am.*, Vol 221 (No. 2), Aug 1969, p 38–46

11.58 JoDean Morrow and G.M. Sinclair, "Cycle-Dependent Stress Relaxation," *Basic Mechanisms of Fatigue*, ASTM STP 237, American Society for Testing and Materials, 1958, p 83–103

11.59 E.S. Rowland, Effect of Residual Stress on Fatigue, *Fatigue: An Interdisciplinary Approach*, J.J. Burke, N.L. Reed, and V. Weiss, Ed., Syracuse University Press, Syracuse, NY, 1964, p 229–244

11.60 F. Sheratt, "The Influence of Shot-Peening and Similar Surface Treatments on the Fatigue Properties of Metals," Part 1: Report S&T-Memo-1/66, Ministry of Aviation, Feb 1966; "Appendix on Selected Case Histories," Part 2: Report S&T-Memo-2/66, Ministry of Aviation, Feb 1966

11.61 A.H. Clauer, Laser Shock Peening for Fatigue Resistance, *Surface Performance of Titanium*, J.K. Gregory, H.J. Rack, and D. Eylon, Ed., TMS, Warrendale, PA, 1996, p 217–230

11.62 P. Prevey, J. Telesman, T. Gabb, and P.T. Kantzos, FOD Resistance and Fatigue Crack Arrest in Low Plasticity Burnished IN718, *Proc. of 5th National Turbine Engine High Cycle Fatigue (HCF) Conference* (Chandler, AZ), March 2000

11.63 J.O. Almen and P.H. Black, *Residual Stresses and Fatigue in Metals*, McGraw-Hill Book Co., New York, 1963

11.64 M. Kobayashi, T. Matsui, and Y. Murakami, Mechanism of Creation of Compressive Residual Stress by Shot Peening, *Int. J. Fatigue*, Vol 20 (No. 5), 1998, p 351–357

11.65 T.C. Baumgartner and R.L. Sproat, "Basic Design and Manufacturing of Aircraft Fasteners for Use Up to 1600 °F," presented at the Fastener Symposium, 1957–1958

11.66 J. Telesman, P.T. Kantzos, T. Gabb, and P. Prevey, "Effect of Residual Stress Surface Enhancement Treatments on Fatigue Crack Growth in Turbine Engine Alloys," presented at The Third Joint FAA/DoD/NASA Conference on Aging Aircraft, (Albuquerque, NM), Sept 1999

11.67 I. Telesman, P.T. Kantzos, and T.P. Gabb, NASA Glenn Research Center, personal communication, Aug 2003

11.68 W.Z. Zhuang and G.R. Halford, Investigation of Residual Stress Relaxation under Cyclic Loading, *Int. J. Fatigue*, Vol 23, supplement, *Proc. International Conference on Fatigue Damage of Structural Materials III*, A.K. Vasudevan, T. Nicholas, J.T. Cammett, and P.C. Paris, Ed., 2001, p S31–S37

11.69 L.H. Burck, C.P. Sullivan, and C.H. Wells, Fatigue of a Glass Bead Blasted Nickel-Base Superalloy, *Metall. Trans.*, Vol 1 (No. 6), June 1970, p 1595–1600

11.70 H.O. Fuchs, Techniques of Surface Stressing to Avoid Fatigue, *Metal Fatigue*, G. Sines and J.L. Waisman, Ed., McGraw-Hill, New York, 1959, p 197–231

11.71 R.M. Niemi and R.E. Johnson, Private communication, General Electric Co., Materials and Process Technical Laboratories, Aircraft Engine Group, Cincinnati, OH, 1972

11.72 J.W. Faber, " Experience with Inconel 713C Alloy," Report A-2667, Westinghouse Aviation Gas Turbine Division, Feb 19, 1960

11.73 J. Viglione, "The Effects of Shot Peening on the Fatigue Properties of Damaged Bare and Damaged Chrome Plated High Strength 4340 Steel," Report NAEC-AML-2187, AD-470417, Naval Air Engineering Center, July 30, 1965

11.74 C.R. Smith, The Effect of Tapered Bolts

11.75 C.R. Smith, Interference Fasteners for Fatigue-Life Improvement, *Exp. Mech.,* Vol 5 (No. 8), Aug 1965, p 19A–23A

11.76 J.A. Scott, The Influence of Processing and Metallurgical Factors on Fatigue, *Metal Fatigue: Theory and Design,* A.F. Madayag, Ed., Wiley, New York, 1969, p 66–106

11.77 R.B. Heywood, Effect of High Loads on Fatigue, *Colloquium on Fatigue of Metals,* Springer-Verlag, Berlin, 1956, p 92

11.78 S.S. Manson and C.R. Ensign, "Discussion of Cumulative Damage Analysis in Structural Fatigue," *Effects of Environment and Complex Load History on Fatigue Life,* ASTM STP 462, American Society for Testing and Materials, 1970, p 69–72

11.79 R.W. Nichols, The Use of Overstressing Techniques to Reduce the Risk of Subsequent Brittle Fracture, Part 1, *Br. Weld. J.,* Vol 15 (No. 1), Jan 1968, p 21–42; Part 2, Vol 15 (No. 2), Feb 1968, p 75–84

11.80 A.J. Gentile, and A.D. Martin, "The Effects of Prior Metallurgically Induced Compressive Residual Stress on the Metallurgical and Endurance Properties of Overload Tested Ball Bearings," Paper 65-WA/CF-7, American Society of Mechanical Engineers, Nov 1965

11.81 V.B. Dalisov, V.S. Zamikhovskii, and V.I. Pokhusrkii, Effect of Nitriding on the Fatigue and Corrosion-Fatigue Strength of Steel 45, *Fiz. Khim. Mekh. Mater.,* Vol 2 (No. 2), 1966, p 173–176 (in Russian)

11.82 A. Graae, How to Nitride Maraging Steels, *Met. Prog.,* Vol 92 (No. 1), July 1967, p 74–76

11.83 T.J. Dolan, Basic Concepts of Fatigue Damage in Metals, *Metal Fatigue,* G. Sines and J.L. Waisman, Ed., McGraw-Hill, NY, 1959, p 39–67

11.84 V.I. Belyaev, *Problems of Metal Fatigue,* Daniel Davey & Co., New York, 1963

11.85 J.C. Grosskreutz, "Fatigue Mechanism in the Sub-Creep-Range," *Metal Fatigue Damage: Mechanism, Detection, Avoidance, and Repair,* ASTM STP 495, American Society for Testing and Materials, 1971, p 5–60

11.86 J.S. Waisman, A. Shrier, and V. Greenhut, "Fatigue of Metal Crystals, Part 1: Extension of Fatigue Life of Crystals through Control of Substructures," Report AFML-TR-66-123, Part 1, AD-809533, Rutgers University, New Brunswick, NJ, Aug 1966

11.87 H.D. Nine and W.A. Wood, "Improvement of Fatigue Life by Dispersal of Cyclic Strain," Technical report 36, Institute for Study of Fatigue and Reliability, Columbia University, New York, July 1966

11.88 E. Siebel and G. Stahli, Experiments on the Proof of Damage Stress and Work Hardening in Fatigue Tests (on Steels and Duraluminum), *Arch. Eisenhüttenwes.,* Vol 15 1942, p 519–527

11.89 N. Thompson and M.J. Wadsworth, Metal Fatigue, *Adv. Phys.,* Vol 7 (No. 25), Jan 1958, p 72–170

11.90 R.H. Christensen, Fatigue Cracking, Fatigue Damage, and Their Detection, *Metal Fatigue,* G. Sines and J.L. Waisman, Ed., McGraw-Hill, New York, 1959, p 376–412

11.91 M. Bentele and C.S. Lowthian, Thermal Shock Tests on Gas Turbine Materials, *Aircr. Eng.,* Vol 24 (No. 276), Feb 1952, p 32–38

11.92 M.H. Raymond and L.F. Coffin, Geometrical Effects in Strain Cycled Aluminum, *J. Basic Eng.,* Vol 85 (No. 4), Dec 1963, p 548–554

11.93 H.G. Popp, Private communication, General Electric Co., Cincinnati, OH, 1970

11.94 B.I. Sandor, "Metal Fatigue with Elevated Temperature Rest Periods," *Achievement of High Fatigue Resistance in Metals and Alloys,* ASTM STP 467, American Society for Testing and Materials, 1970, p 254–275

11.95 M.A. Wright and A.P. Greenough, The Effect of High-Temperature Intermediate Annealing on the Fatigue Life of Copper, *J. Inst. Met.,* Vol 93, 1964–65, p 309–313

11.96 V.S. Ivanova, P.A. Antikain, and N.S. Sabitova, Healing of Defects in Steel Accumulated during Cyclic Overloading, *Metalloved. Term. Obrab. Met.,* No. 1, Jan 1965, p 7–9

11.97 C.C. Osgood, Damage-Tolerant Design, *Mach. Des.,* Vol 41 (No. 25), Oct. 30, 1969, p 91–95

11.98 *Fatigue Crack Propagation,* ASTM STP 415, American Society for Testing and Materials, 1967

11.99 A.J. McEvily, Fatigue Crack Growth and the Strain Intensity Factor, *Proc., Air Force Conference on Fatigue and Fracture of Aircraft Structures and Materials,* AFFDL 70-144, H.A. Wood, R.M. Bader, W.J. Trapp, R.F. Hoener, and R.C. Donat, Ed., Air Force Flight Dynamics Laboratory, Sept 1970, p 451–459

11.100 H.G. Popp and A. Coles, Subcritical Crack Growth Criteria for Inconel 718 at Elevated Temperatures, *Proc., Air Force Conference on Fatigue and Fracture of Aircraft Structures and Materials,* AFFDL 70-144, H.A. Wood, R.M. Bader, W.J. Trapp, R.F. Hoener, and R.C. Donat, Ed., Air Force Flight Dynamics Laboratory, Sept 1970, p 71–86

11.101 P.C. Paris and F. Erdogan, A Critical Analysis of Crack Propagation Laws, *Trans. American Society of Mechanical Engineers, J. Basic Eng.,* Dec 1963, p 528–534

11.102 "Standard Test Method for Plane-Strain Fracture Toughness of Metallic Materials," ASTM E399-90 (re-approved 1997), American Society for Testing and Materials

11.103 G.R. Halford, Over Pressure Proof Factors for Fatigue Loaded Pressure Vessels, *Material Durability/Life Prediction Modeling: Materials for the 21st Century,* PVP, Vol 290, S.Y. Zamrik and G.R. Halford, Ed., ASME, 1994, p 49–56

11.104 *NASA Standards for Launch Packages,* NASA, 1993

# CHAPTER 12

# Special Materials: Polymers, Bone, Ceramics, and Composites

## Introduction

In this chapter we consider several classes of materials that are of special interest: polymers, bone, ceramics, and composites. In addition to bone (a particular type of ceramic), each has its counterpart in nature. Wood is probably the most familiar natural polymer. Composites, which combine several desirable mechanical features of their individual material components, are found in virtually every living plant and animal. Rocks and other natural formations are basically ceramics and, of course, bone is endowed by nature with its special features so useful in supporting living structures and serving as tools in life functions. Nature has had eons to optimize the properties of these materials for their intended functions, and it is a goal of engineering to emulate nature in developing materials for structural applications and carrying the development a step further toward perfection.

Considerable engineering studies have been made on all of these types of materials. It is not our purpose to review all the available knowledge. Here we shall focus only on the fatigue aspects of the behavior, and in particular we shall explore the applicability of the fatigue technology developed in this book. Specifically, we shall question whether the types of relations developed for metals apply to the crack initiation and propagation of these nonmetallic materials as well, or where modifications are required.

## Polymers

### Introduction

Before discussing the applicability of fatigue technology of metals to polymers, it is important to recognize the differences in basic microstructure between the two systems and how such differences might impact fatigue behavior.

### General Comparison of Polymers with Metals

**Microstructure.** First, we must recognize that the microstructure of polymers is entirely different from that of metals. While metals are crystallographic, polymers are long chains or strands of atoms (typically carbon and hydrogen) of several sizes. These strands may be mechanically or chemically linked where they cross each other. In some regions, for certain polymers, some of the atomic strands line up in a manner somewhat analogous to the atomic array in metals and are said to be "crystalline." The bonding, however, is not the same as in the atomic planes of a metallic crystal. Thus, while inelastic deformation occurs in metals through the movement of dislocations along crystallographic planes, polymeric inelastic deformation occurs as sliding of adjacent strands past each other or changing the crisscross arrangement of these strands where they tangle. Generally, therefore, we must be aware that the strain mechanisms are considerably different in the two systems.

**Proximity of Test Temperature and Melting Point.** We also must be aware that the proximity of common test temperatures to their melting points is different for the two systems. For most structural metals room temperature is a small fraction of the melting point; it requires temperatures above 540 °C (1000 °F) to reach one-half the melting point, where creep considerations become important. For polymers, even room temperature is a high fraction of the "melting" point (temperatures must, of course, be in degrees absolute). In fact, in polymers the critical temperature is not the actual melting temperature, but rather the *glass transition temperature,* which is where the bonds due to cross linking or other tangling become substantially reduced—for example, a polycarbonate for which the melting point is 260 °C (500 °F) and the glass transition temperature is 150 °C (300 °F). Room temperature represents about 55% of the melting point, while it represents 70% of the glass transition temperature. The latter would represent a test temperature of nearly 980 °C (1800 °F) to produce the same homologous temperature for steel (depending on carbon content).

The fact that even at room temperature a polymer is so close to its glass transition temperature results in inelastic (anelastic and creep) strains that normally are neglected in treating metal fatigue at room and moderately elevated temperatures. Thus, measurement of the inelastic strain that is most closely related to fatigue must be given close attention.

**Temperature increases associated with testing** can have severe effects normally neglected in metal studies. Because large hysteresis loops can develop in polymers, the energy dissipated per cycle as heat may be considerably larger than in metals. And, also because of polymer's low thermal conductivity, temperatures tend to become higher in localized regions, especially at cracks developed by fatigue. Even small increases in local temperature can so elevate the proximity to glass transition that the results can be made meaningless, or at least very difficult to interpret. Thus, often tests on polymers are conducted at much lower temperatures, not necessarily because they are used at these temperatures, but in order to avoid the complications of temperature distortions. Care must also be exercised in choosing testing frequencies. To avoid time-dependent strains, the strain rate should be high, but high frequencies also generate higher heat in a given time. Therefore, a balance must be found to compromise the two effects. In interpreting literature data this factor must be given special consideration.

**Crazing and Other Phenomena.** Some phenomena may occur in polymers that are very uncommon or absent in metals. These phenomena may influence fatigue results substantially. Some polymers are susceptible to craze development that forms normal to applied tensile stress. While these crazes do not produce much reduction in cross-sectional area, they provide a preferential path for cracks to propagate. They may also substantially reduce the ability of the material to deform inelastically. Thus, we shall see in a later discussion that crazing can render an essentially ductile polymer semiductile (much lower ductility). If the crazing can be suppressed (for example, by inducing surface compressive residual stresses), ductility could be enhanced substantially. Therefore, in considering fatigue of polymers, the existence or absence of such extraneous factors as crazing must be meticulously considered.

**Relationship between Tensile and Compresive Yield Strength.** Another important difference exists between polymers and metals; while it is rare for metals to show large differences between tensile and compressive yield strengths (cast iron is an uncommon example), polymers usually do show such large differences. It is usual for compressive yield stress to exceed tensile values by 25% or more. The result is that in cyclic straining between equal tensile and compressive strain limits, compressive mean stresses develop, while cycling between equal tensile and compressive stresses or forces causes tensile strain ratcheting.

**Monotonic Loading.** A typical tensile stress-strain curve, for example, for the thermoplastics polyethylene or nylon, is shown in Fig. 12.1 (Ref 12.1). In the region *OA* linear elasticity develops if the loading rate is relatively high, but if the loading rate is slow some anelasticity or creep may be evident.

At point *B*, an upper yield stress may develop, similar to many steels. Probably the tangled fibers develop internal stresses, similar to those developed at blocked dislocations in metals. But when the stress exceeds a critical value, an avalanche of fiber sliding develops similar to the

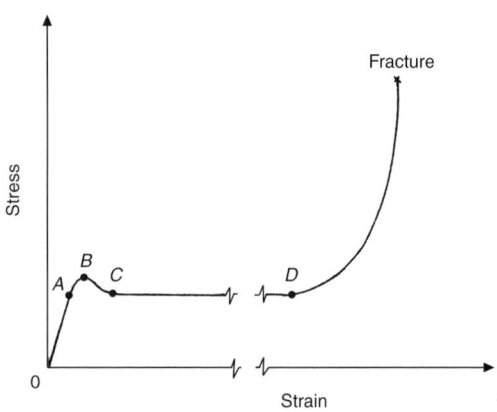

**Fig. 12.1** Tensile stress-strain curve for a ductile polymer. Source: Ref 12.1

avalanche of dislocation motion in metals. A very large amount of plastic flow can occur at the lower yield stress, CD, as the fibers straighten out. When fully aligned the stress to induce further deformation increases substantially from D to fracture. It should be pointed out that the mechanical properties of polymers depend strongly on their molecular weight.

Figure 12.2 shows the early part of the monotonic tensile and compressive stress-strain curve for a polycarbonate at room temperature (Ref 12.2). The upper and lower yield stress is clearly evident, even though it does not quite resemble the familiar yield curve of mild steels. Even the compressive stress-strain curve shows a reversal of curvature as the long chains start to move over one another with increasing compressive strain. Eventually, of course, at higher strains than shown in the figure, both the tensile and compressive curves resume their positive curvature.

Polymers can be divided into three basically different types of behavior in monotonic loading according to polymer type, test temperature, and strain rate. Ductile behavior is characterized by the curve of polyethylene or nylon as shown in Fig. 12.1. Ductilities up to 100% can characteristically be achieved. Some materials such as polymethyl methacrylate (PMMA) can act in a semiductile manner. The deformation in tension is limited by the development of crazes and very fine surface cracks that limit the ductility. In compression the crazing is inhibited and much higher deformation is possible. Figure 12.3 (Ref 12.2) shows this behavior. Up to about 5% tensile strains the behavior is like a ductile metal in tension, but above this value the stress concentration effect of the crazes produces brittle fracture. Compression continues on to much higher strains because crazes are inhibited. In brittle polymers, typified by highly cross-linked, thermosetting Bakelite, even small amounts of plasticity are inhibited. Such materials fracture in tension with little or no ductility. This characteristic can also occur for otherwise more ductile polymers when tested at low temperatures, well below the glass-transition temperature, especially at high loading rates. Each polymer class has its own special characteristics.

Thus, in summary, for uniaxial loading, high-quality polymers behave similarly to metals, except that anelasticity and creep can occur if the temperature is near or above the glass transition value. The amount of plasticity possible can well exceed that developed in most metals (unless the metals are either very unusual or are tested in a

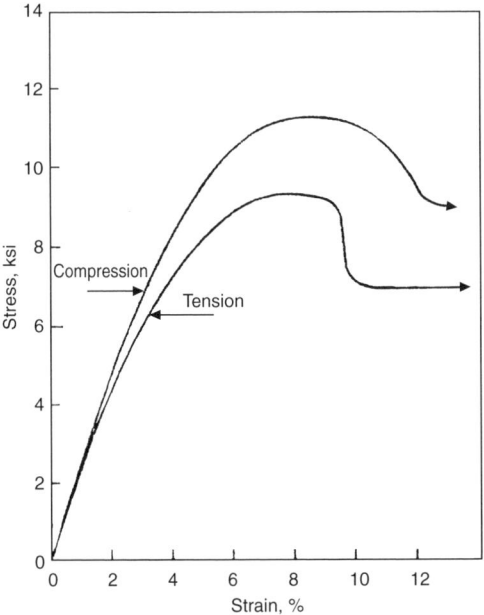

**Fig. 12.2** Tensile and compressive stress-strain curves for polycarbonate at room temperature. Source: Ref 12.2

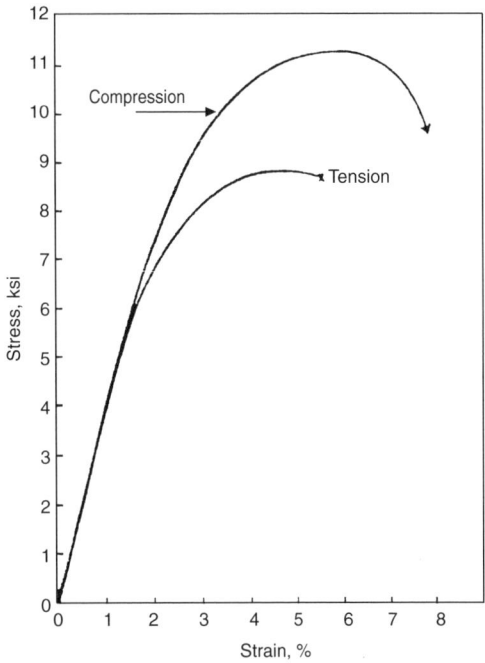

**Fig. 12.3** Semiductile behavior of polymethyl methacrylate at room temperature (strain rate = 2 × $10^{-4} sec^{-1}$). Source: Ref 12.2

very special range). On the other hand, the ductility of polymers can be very limited, or even absent, depending on composition (molecular weight), layout of the polymer chains, cross-linking, temperature, and strain rate. Given the sometimes ductile nature of polymers, it is understandable that there is an interest in the strain cycling, i.e., low-cycle fatigue, resistance of these materials.

**Cyclic Loading.** Polymers differ from common metals under cyclic loading in several ways. Because the homologous temperature under test conditions for polymers is usually higher than for metals (at room temperature, at least) some behavior is displayed by polymers at room temperature that are shown by metals only at high temperatures—namely, anelastic deformation and creep. Figure 12.4 illustrates the phenomenon and some of the propeller shapes of the hysteresis loops generated by this effect. The hysteresis loop shown is typical for polycarbonate at room temperature for a strain amplitude of ±4% (Ref 12.2). It has a propellerlike shape rather than that commonly displayed by metals. Time-dependent inelastic strain occurs during tension $OA$, causing the curve to deviate farther from the elastic line $OO'$ than would be expected on the basis of plasticity alone. Some of this strain is recovered on the return $AB$, causing it to develop an inflection. The length $OB$ represents a measure of the plasticity, and it is considerably lower than the total inelastic strain imposed. Analysis of such a loop becomes complicated if the time-dependent inelastic strain is to be separated from the total inelastic strain. In order to minimize the time-dependency effect, it is common to conduct cyclic tests at a high strain rate, but this approach introduces other complications. Heat is generated by the rapid strain rate; but because conductivity of polymers is low, the temperature increases, usually more in the central volume of the test part than at the surface where convection cooling occurs. Thus, in addition to generating a high temperature, there is also a temperature gradient, which complicates interpretation. It is not uncommon for the hysteresis loop (and the fatigue process, discussed later) to be dictated largely by the temperature and temperature gradient effects than by the flow and fatigue characteristics of the material at the nominal test temperature. To obtain meaningful tests the conditions must be carefully chosen to achieve a balance between the nature of the strains generated and the thermal factors involved. It is quite possible that some of the published properties are spurious because of failure to control test conditions carefully.

A second factor to consider is the large amount of inelastic deformation that polymers can support. The ±4% shown in Fig. 12.4 is quite large for metals, but it is small for polymers. Strains of ±10% are commonly used in polymer tests, as we shall show, and such high strains are beyond the common experience in metals.

A third factor of importance is the cyclic softening common in most polymers. While metals may either harder or soften under cycling, polymers usually only soften, and the amount of softening is usually greater than commonly observed in metals. Although the ratio of ultimate to yield strength would be greater than 1.4, and therefore it might be expected that cycling would cause hardening (Chapter 2, "Stress and Strain Cycling"), the common characteristic is for softening to occur.

The softening that occurs follows various patterns. In Fig. 12.5(a) the pattern for polycarbonate is shown. There is an incubation period $AB$ during which the stress range falls only slightly; a transition period $BC$ where a rapid reduction of stress occurs; a stable region $CD$ where the stress range changes little, but at the end of

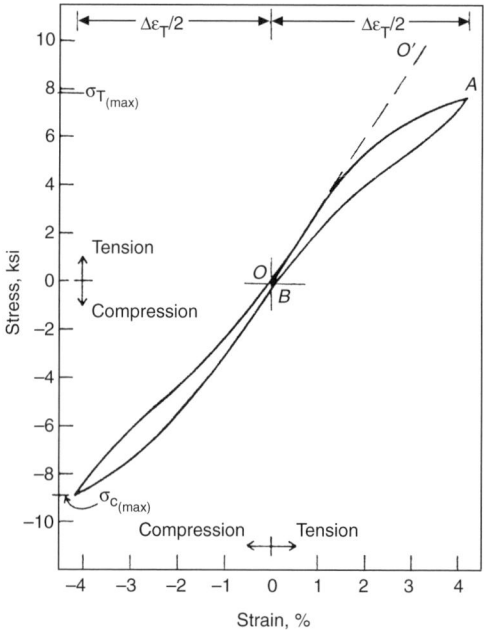

**Fig. 12.4** Initial hysteresis loop for polycarbonate at room temperature ($F = 0.1$ Hz triangular waveform, giving a strain rate of $1.6 \times 10^{-2} \text{sec}^{-1}$ for a strain range of 0.080). Source: Ref 12.2

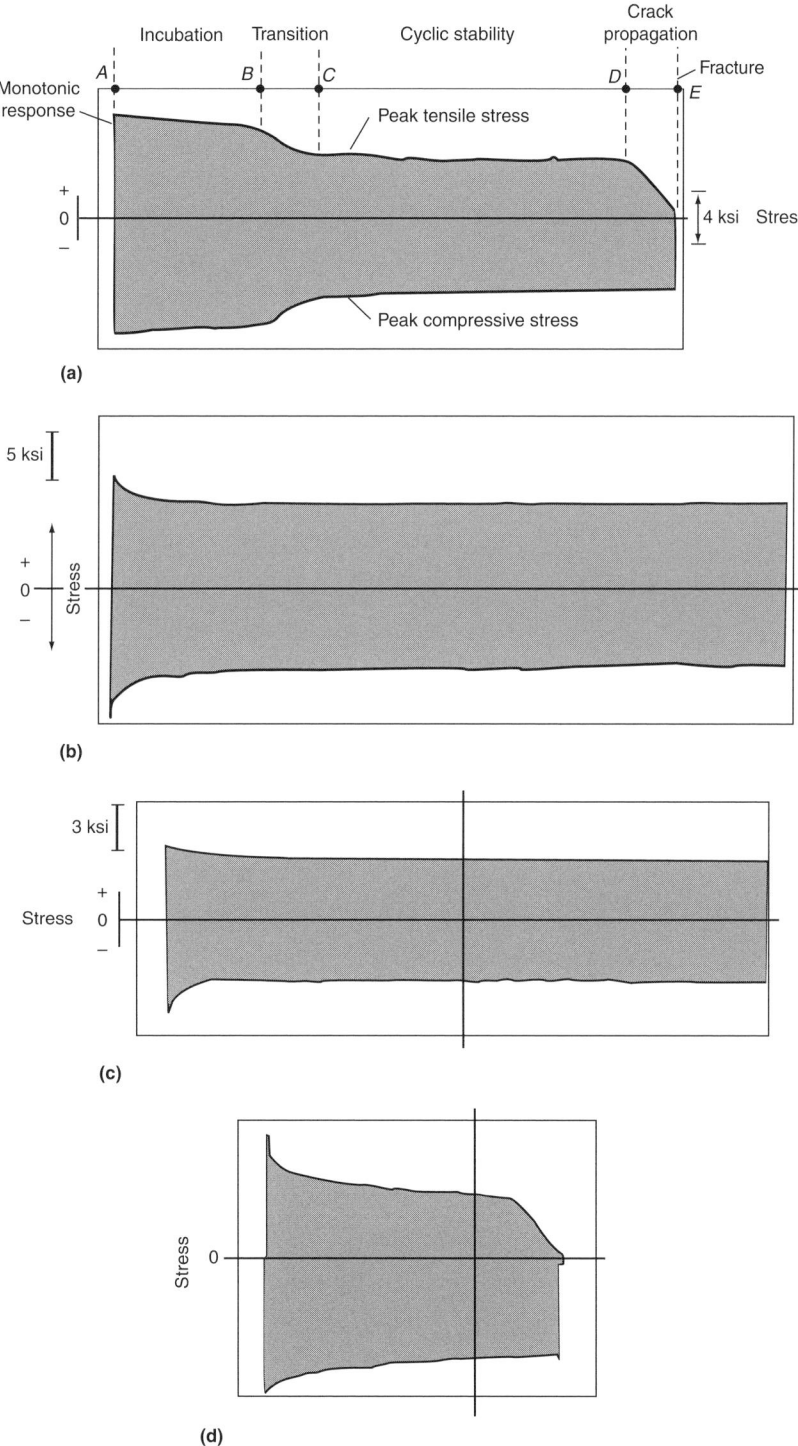

**Fig. 12.5** Softening patterns for various polymers tested under strain control at room temperature. (a) Polycarbonate (Ref 12.2). (b) Nylon (Ref 12.2). (c) Polypropylene (Ref 12.3). (d) ABS plastic (Ref 12.3)

which a crack starts. During the crack propagation period *DE* the apparent stress range (nominal stress as determined by load required to maintain the nominal strain range) falls rapidly.

Another type of behavior is represented by nylon as shown in Fig. 12.5(b). Here the softening starts almost immediately, but after relatively few cycles a stable pattern develops, which changes very little until a crack starts. A similar behavior is shown by polypropylene in Fig. 12.5(c) (note, however, that the compression softens more in the early loading than the tension and that the stress developed in tension and compression is about equal). Finally, in Fig. 12.5(d) is shown the behavior of acrylonitrile-butadiene-styrene (ABS) plastic, which softens appreciably in the early cycles and continues to soften throughout the cycling up to crack development. Notice, also, that the compressive stress developed is substantially higher (by more than 50%) than the tensile stress.

Thus, it is evident that many different types of cycling behavior can be displayed by different polymer systems. All, however, are of the softening type.

Cyclic stress-strain curves can be constructed in several ways similar to those used for metals. Stabilized stress values (at ½ life) for selected strain ranges would be a direct way of obtaining the cyclic stress-strain curve, but the incremental stress tests, for which test results are shown in Fig. 12.6, have been used by Beardmore and Rabinowitz (Ref 12.3). On the basis of such tests, plotting stress versus strain, hysteresis loops can be obtained such as shown in Fig. 12.7 for polycarbonate.

Using such techniques the cyclic stress-strain curves were generated. For polycarbonate, the

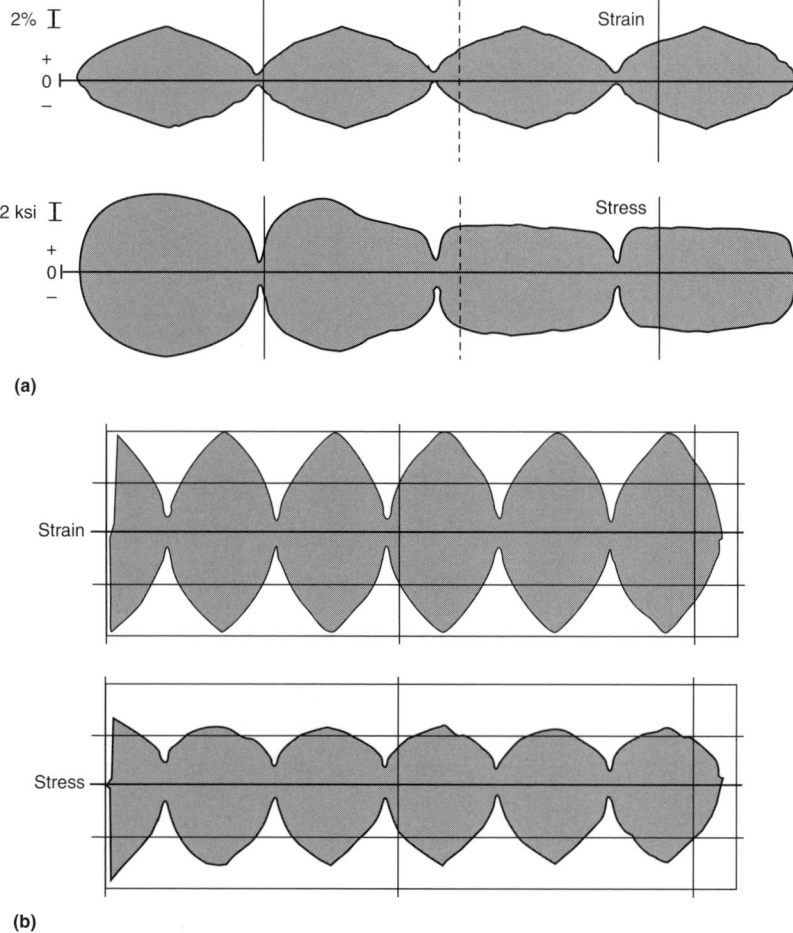

**Fig. 12.6** Incremental step test results for determining cyclic stress-strain curves of polymers at room temperature. Source: Ref 12.3 (a) Polycarbonate. (b) Polypropylene

results are shown in Fig. 12.8. Note that, as with metals, the irregular behavior of the monotonic tension test does not carry over into cyclic behavior. Note also that the cyclic compression stress response is of the order of 25% greater than cyclic tension. Other cyclic stress-strain curves generated by Beardmore and Rabinowitz are shown in Fig. 12.9.

## Fatigue: Conventional Tests

Both conventional *S-N* curves, in which stress is varied from zero to a maximum, and strain control tests, in which strain is varied between equal positive and negative values, have been conducted on several classes of polymers by Beardmore and Rabinowitz (Ref 12.3). They distinguished behavior according to whether crazing developed in the low-cycle, high-strain (or stress) regime. Beardmore and Rabinowitz

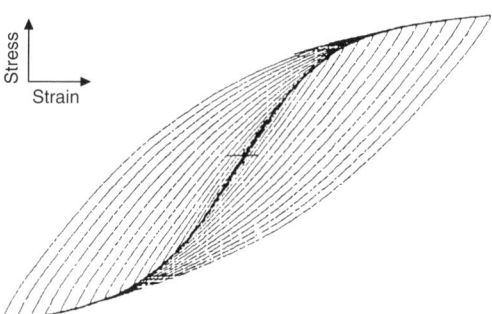

**Fig. 12.7** Hysteresis loops generated in step tests of Fig. 12.6(a) for polycarbonate at room temperature. Source: Ref 12.3

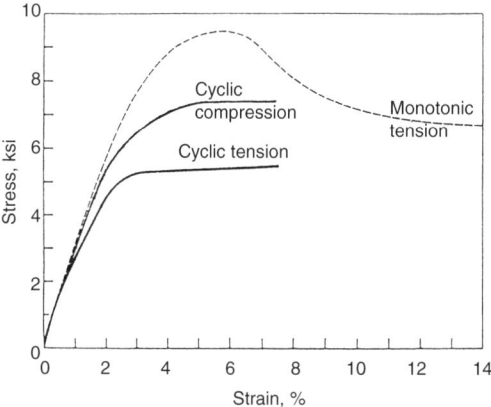

**Fig. 12.8** Monotonic tension and cyclic tension and compression stress-strain curves for polycarbonate at room temperature. Source: Ref 12.4

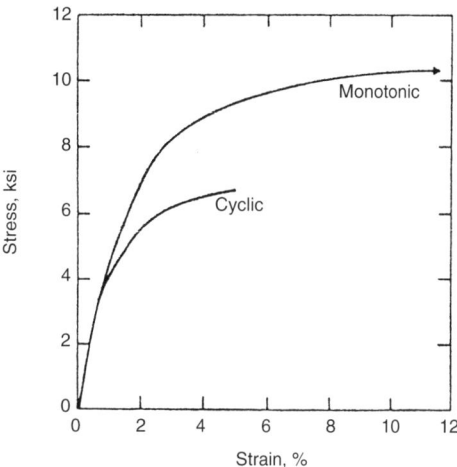

(a)

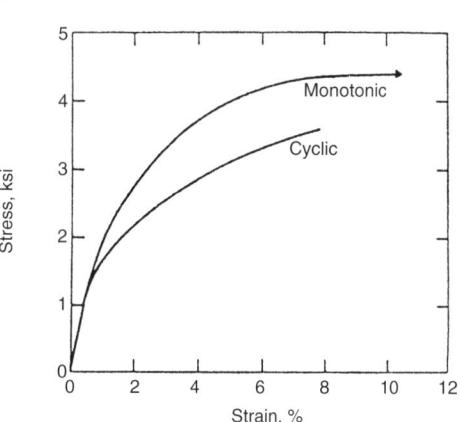

(b)

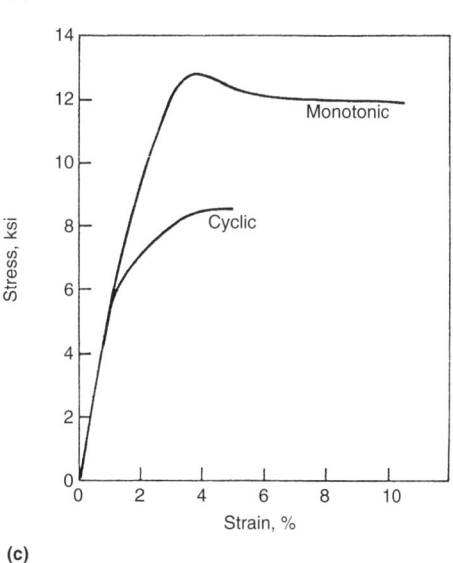

(c)

**Fig. 12.9** Monotonic and cyclic stress-strain curves for several polymers at room temperature (298 K). Source: Ref 12.3 (a) Polyoxymethylene. (b) Polypropylene. (c) Nylon

made no attempt to separate elastic and plastic components of strain; rather, for strain-cycling fatigue, they considered only total strain range. Figure 12.10 shows schematically how they interpreted their results. Normally they expect a behavior $AA'BC$ where $AA'B$ is linear, and $BC$ is linear and horizontal, representing a strain level below which life is essentially infinite. Actually, they found that in real material behavior the life could be divided into three regions. For some materials that tend to produce crazing in region I, the line $AA'$ is replaced by $A''A'$, but if no crazing occurs, the behavior remains as $AA'$. Region II is where the life curve is straight. Whether or not the material tends to craze at high strains, it does not craze in region II. Region III is the fatigue limit region. Figure 12.11 shows some of the results they obtained to illustrate specific behavior of various polymers. In Fig. 12.11(a) and (b) we see polycarbonate at room temperature where no crazing occurs, and at liquid nitrogen temperature (78 K) where crazing occurs. The figures imply two different behaviors. The room-temperature polycarbonate is shown on a strain range basis while the liquid nitrogen behavior is shown on a stress basis. Polystyrene and PMMA, both of which craze for high strains at room temperature, clearly display the three-region characteristic, as seen in Fig. 12.11(c) and (d).

## Strain-Cycling Study and Opportunities for Accelerated Evaluation

While a good start has been made to apply the principles discussed in this book for strain-cycling fatigue evaluation of polymers, these studies have been limited. Total strain range has been used as a parameter, but separation of the strain range into its elastic and inelastic components, so useful a concept for metals, has not been attempted for polymers. Partly this is due to difficulty in separating inelastic strain into its time-independent and time-dependent components. Usually the attempt is to avoid generation of time-dependent inelastic strain by using high strain rates, but heat generation becomes a problem. However, application of the data generated for one set of special conditions to more general conditions can become problematical. Here, then, is a problem demanding further study. There are a large number of polymers available, each with variations of properties. Furthermore, there is difficulty in setting test parameters in order to obtain reliable data within reasonable test times. It becomes readily apparent that accelerated testing technologies discussed herein would be quite useful for the enormous task of developing fatigue data bases for current as well as for future polymer systems. This goal is especially valuable if consideration is given to the rapidly growing use of polymers in high-technology applications.

To this end, a student of our fatigue course at Case Western Reserve University, Fred Lisy, took on a special project (Ref 12.4) of applying various testing techniques to several polymers of interest. Three materials, polycarbonate, nylon 6/6, and polyethylene, were examined in a limited program. These materials had been studied previously by Beardmore and Rabinowitz (Ref 12.3), making cross-checking possible. While several different approaches were studied, we shall limit our discussion here to only three: the Manson-Hirschberg Method of Universal Slopes (MUS) (Ref 12.5), which requires no experimental fatigue data, and two methods requiring a single fatigue test, as described subsequently.

Table 12.1 shows the physical and mechanical properties of the three materials as compiled by Lisy, some of which were estimates deduced from various sources. Table 12.2 lists various mechanical properties determined by Lisy for polypropylene as a function of strain rate. This program was conducted to determine a suitable testing frequency to minimize the testing time. It was decided to use 500% per min even though at this rate the surface temperature rise was 47 °C (117 °F) above room temperature, which was reduced to only a 4 °C (39 °F) rise by blowing air across the surface. Still, the final failure

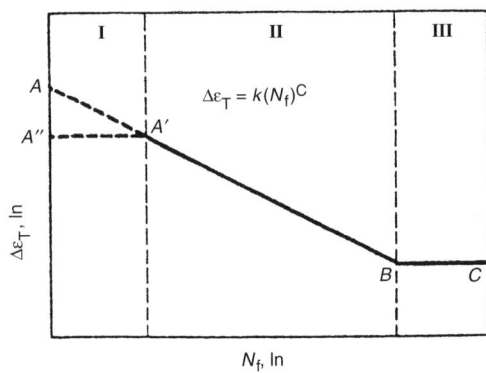

**Fig. 12.10** Schematic representation of the general fatigue response of polymers in terms of strain range versus number of cycles to failure. Source: Ref 12.3

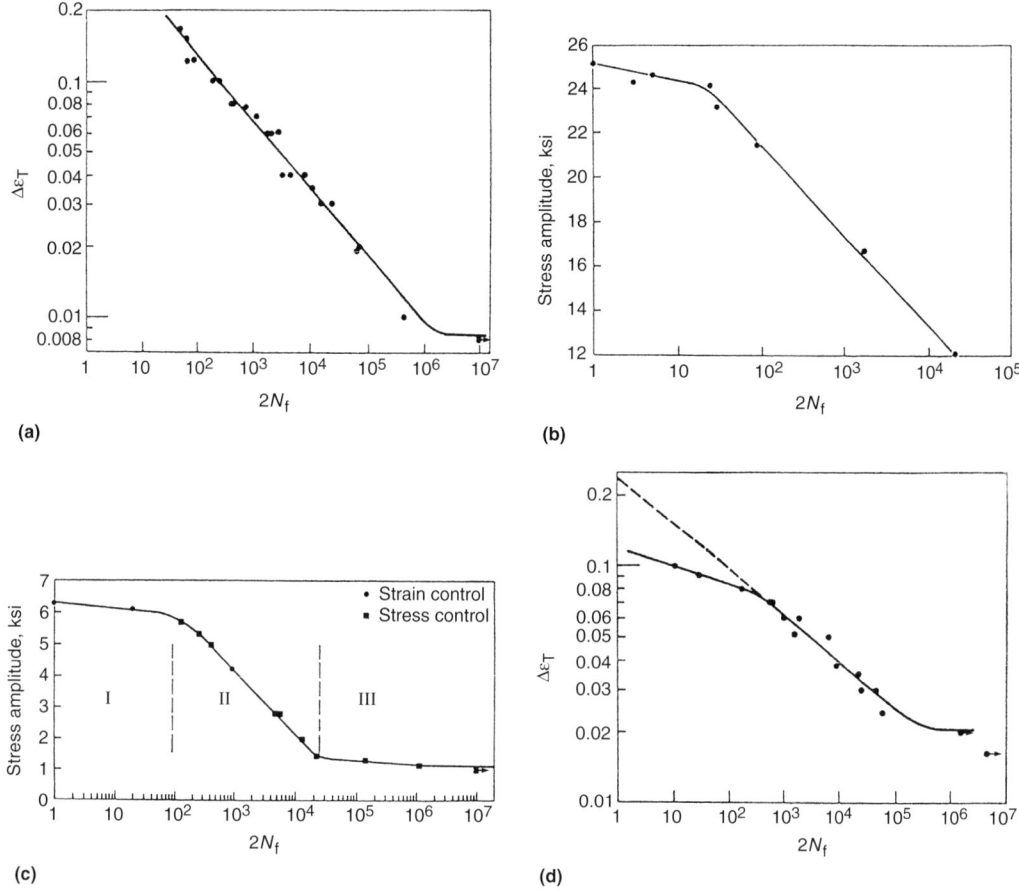

**Fig. 12.11** Completely reversed strain cycling fatigue of various polymers. Source: Ref 12.3. (a) Polycarbonate at room temperature. (b) Polycarbonate at 78 K. (c) Polystyrene at room temperature. (d) Polymethyl methacrylate at room temperature.

**Table 12.1 Physical and mechanical properties of polymers used in accelerated fatigue evaluation study**

| | Polymer | | |
|---|---|---|---|
| Property | Polypropylene | Nylon 6/6 | Polycarbonate |
| Polymer structure | Semicrystalline | Semicrystalline | Amorphous |
| Glass transition temperature, $T_g$, °C (°F) | 0 (32) | 60 (140) | 150 (302) |
| Melting temperature, $T_m$, °C (°F) | 165 (329) | 280 (536) | Amorphous |
| Modulus, $E$, MPa (ksi) | 2344 (340) | 3723 (540) | 388 (395) |
| Yield strength, $S_y$, MPa (ksi) | 42.7 (6.2) | 90 (13) | 66.2 (9.6) |
| Ultimate tensile strength, $S_u$, MPa (ksi) | 44.1 (6.4) | 90 (13) | 70 (10) |
| Tensile reduction of area (RA) | 0.79 | 0.70 | 0.70 |
| True fracture strain, $\varepsilon_f = \ln[1/(1 - RA)]$ | 1.56 | 1.20 | 1.20 |
| True fracture strength, $\sigma_f \approx S_u(1 + \varepsilon_f)$, MPa (ksi) | 112.3 (16.4) | 197.2 (28.6) | 151.7 (22.0) |

Source: Ref 12.4

**Table 12.2 A comparison of the elastic modulus, yield strength, and strain to fracture of polypropylene samples tested at six strain rates**

| Strain rate, %/min | Modulus, MPa (ksi) | Yield strength, MPa (ksi) | Strain to fracture, % |
|---|---|---|---|
| 500 | 2344 (340) | 42.7 (6.2) | 21 |
| 400 | 2103 (305) | 40.7 (5.9) | 21 |
| 300 | 1999 (290) | 39.3 (5.7) | 21 |
| 200 | 1875 (272) | 39.3 (5.7) | 23 |
| 100 | 1793 (260) | 37.2 (5.4) | 49 |
| 25 | 1696 (246) | 1.8 (5.1) | 480 |

Source: Ref 12.4

showed some localized overheating, so there may be some question as to the validity of the test results.

Hysteresis loops for the three materials at specific strain ranges are shown in Fig. 12.12. In Lisy's tests on polypropylene, four strain ranges were used: 3, 4, 6, and 8%. The two lowest range tests were discontinued because negligible plastic deformation developed, even after 48 hours of cycling. The 6 and 8% strain ranges produced analyzable results, the hysteresis loop for the 8% strain range for polypropylene being shown in Fig. 12.12(a). Data were taken from the literature of Beardmore and Rabinowitz (Ref 12.3) for nylon 6/6 and polycarbonate.

We can now discuss the application of three methods of analyses to accelerate fatigue property estimation.

**Method I: Application of the Manson-Hirschberg Method of Universal Slopes.** Here no fatigue data are required; the only properties needed are those obtained in a tensile test. Elastic modulus $E$, ultimate tensile strength $\sigma_u$, and reduction in area (which defines ductility, $D$) are the only properties required (Ref 12.5) (see also Chapter 3, "Fatigue Life Relations"). Of concern is the strain rate dependency of the monotonic tensile stress-strain behavior compared with the strain rate behavior of the cyclic stress strain, and its effect on cyclic life. The authors are unaware of any studies wherein the strain rates of testing were held constant at discrete values over a range of this variable to help address this concern. Another question arises as to what value to use for elastic modulus because of the anelastic and time-dependent strains that develop in the early part of the tensile loading. In this analysis we chose $E$ as the slope in the very initial part of the stress-strain curve, as displayed in Table 12.1. The results are plotted in Fig. 12.13. For the polycarbonate the agreement between the predictions and experiments are especially good; they are reasonable for nylon 6/6 but not very good for the polypropylene. For the polypropylene the data are very limited, however, and it should be recognized that the surface of the propylene did show evidence of overheating. Thus, the surface where the fatigue starts probably has more ductility than attributed to the material in the calculation. Furthermore, the MUS equation was derived from test results for metals under completely reversed loading (nominally zero mean stress). For polymers a mean compressive stress develops under completely reversed straining. Strictly, some account should be taken of the mean stress. A full accounting, however, is unlikely. This is an area for future research. But for the present it can be accepted that the MUS equation can be useful in estimating fatigue properties of polymers.

**Method II: Assuming That the Fracture Stress and Strain in the Tensile Tests Provides the Intercepts on the Vertical Axis for the Elastic and Plastic Lines and Incorporating One Fatigue Data Point.** This method is attractive because the tensile test provides one point each on the elastic line and the plastic line.

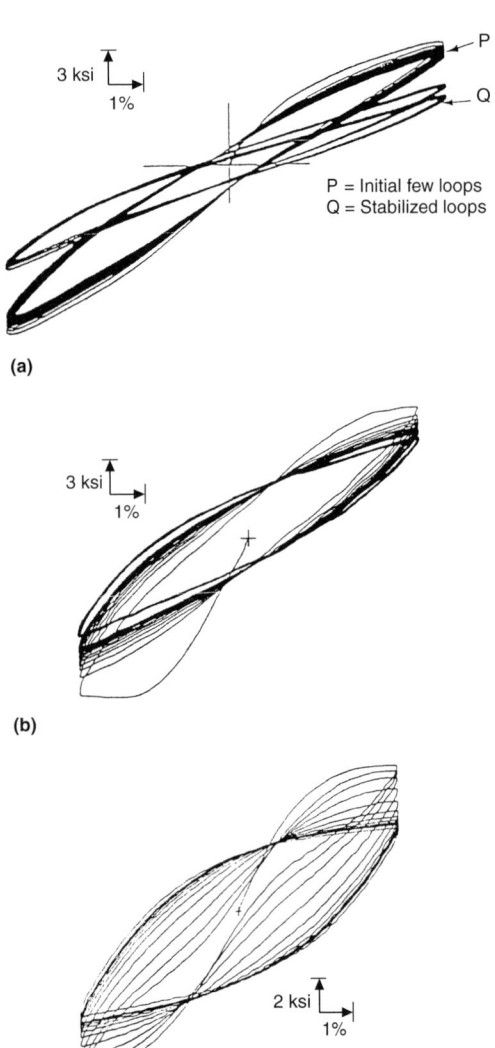

**Fig. 12.12** Hysteresis loops for three polymers cycled at various strain ranges. (a) Polypropylene at 298 K, $\Delta\varepsilon_t = 8\%$ (Ref 12.4). (b) Nylon 6/6 at 298 K, $\Delta\varepsilon_t = 12\%$ (Ref 12.3). (c) Polycarbonate at 298 K, $\Delta\varepsilon_t = 10\%$ (Ref 12.3)

The relations are that $\varepsilon'_f \approx \varepsilon_f$ and $\sigma'_f \approx \sigma_f$. The other point on each of the two lines is established at the single test point. Thus, the elastic and plastic lines are both determined, and total strain range versus life is established as the sum of the elastic and plastic components. Figure 12.14 shows the results for this approach. For polypropylene the 8% strain range test conducted by Lisy (Ref 12.4) was used, point A. The 6% point B is shown in the figure but was not used to establish the lines. Of course, it falls close to its predicted value but does not inspire confidence in the method because it is so close to A. However, the point brought out here is that the method does force the predictions to realistic values at least in one region, and therefore probably over a reasonably wide range.

For the nylon 6/6 and polycarbonate materials, considerable fatigue data were available from the tests of Beardmore and Rabinowitz (Ref 12.4). A single test result was hypothesized to be known, as shown by the Xs in the figure, and the constructions were made to pass through

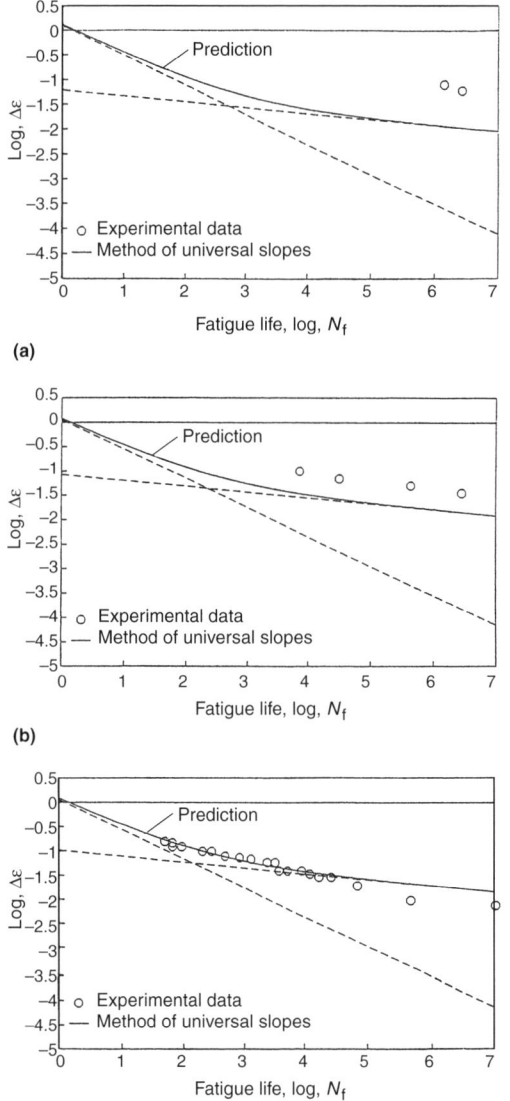

**Fig. 12.13** Application of the Manson-Hirschberg Method of Universal Slopes (MUS) equation to estimate the fatigue resistance of three polymers. (a) Polypropylene data (Ref 12.4). (b) Nylon 6/6 data (Ref 12.3). (c) Polycarbonate data (Ref 12.3)

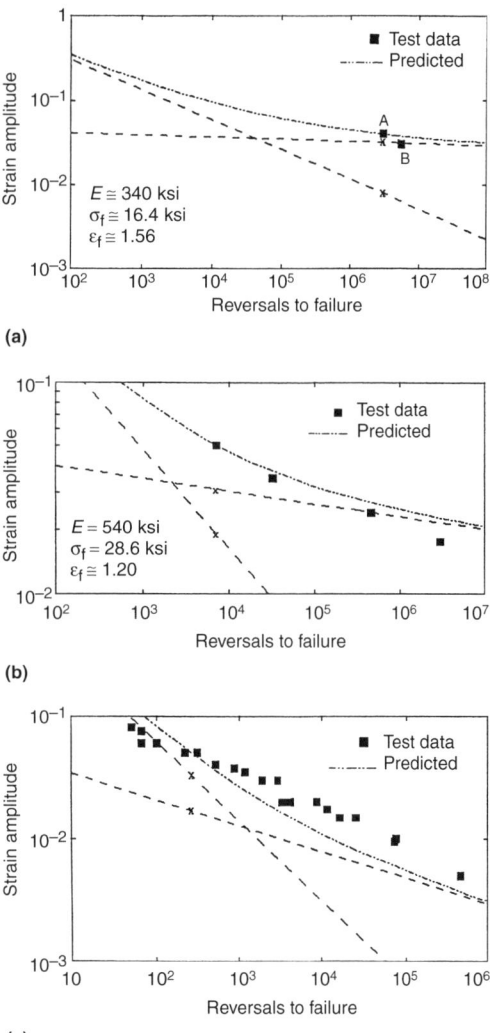

**Fig. 12.14** Application of method whereby $\varepsilon'_f \approx \varepsilon_f$ and $\sigma'_f \approx \sigma_f$ are determined from tensile test and a single fatigue experiment is used to establish an additional point on the elastic and plastic lines. (a) Polypropylene data (Ref 12.4). (b) Nylon 6/6 (Ref 12.3). (c) Polycarbonate data (Ref 12.3)

this point. Fatigue at all other strains could then be computed and compared with the experimental points. The agreement is remarkably good for both materials over quite a range of the life values.

**Method III Use of the Cyclic Stress-Strain Relation Inherent in the Experimental Hysteresis Loops to Refine the Life Relations.** Because the approximations $\varepsilon'_f \approx \varepsilon_f$ and $\sigma'_f \approx \sigma_f$ are not always valid, the slopes of the elastic and plastic lines are not always accurate when determined by method II. The slope of the plastic line is not quite so sensitive to the point of intersection of the plastic line with the vertical axis because the experimental point will generally have a much lower value than the ductility of the material. Thus, the slope of the plastic line as established by method II can be accepted as reasonably accurate. Because there is only a small difference in stress between the intersection of the elastic line on the vertical axis and its value in the experiment, however, error in this intercept can change the slope appreciably. Thus, in method III, the value of slope of the plastic line is accepted as determined by method II, but the value of the slope of the elastic line $b$ is determined from the hysteresis loop of the experimental point. By plotting plastic strain against stress, the cyclic hardening exponent $n'$ can be determined, and because $b/c = n'$, the value of $b$ can be determined from $c$ and $n'$. The analysis by method III is shown in Fig. 12.15 and 12.16. In Fig. 12.15 $\varepsilon_p$ is plotted versus $\sigma$, and the slope determines $n'$ for each of the three materials. In Fig. 12.16 the plastic line is drawn as in method II, and its slope $c$ determined. The value of slope of the elastic line is determined, however, by multiplying $c$ by $n'$, and the line of slope $b$ is drawn through the experimentally determined point of the single experiment. Once the elastic and plastic lines are known, the total strain curve is drawn as their sum. The points experimentally determined by Lisy and by Beardmore and Rabinowitz are then superimposed and show quite good agreement.

## Discussion of Strain Cycling

The agreement between prediction and experiment shown in Fig. 12.13, 12.14, and 12.16 is very encouraging and suggests that further research on these and other accelerated testing procedures is justifiable. The MUS equation yields only coarse predictions but provides ballpark figures that can be useful in some cases. Also, they can perhaps be improved if some empirical correlations can be introduced to allow for the existence of mean compressive stress. The other two methods make some allowance for the mean stress effect because they involve experimental values in a test, which includes mean stress, so the mean stress effect does not distort the predictions as much. In any case, the results are encouraging and suggest the value of their study. All three methods also suggest that

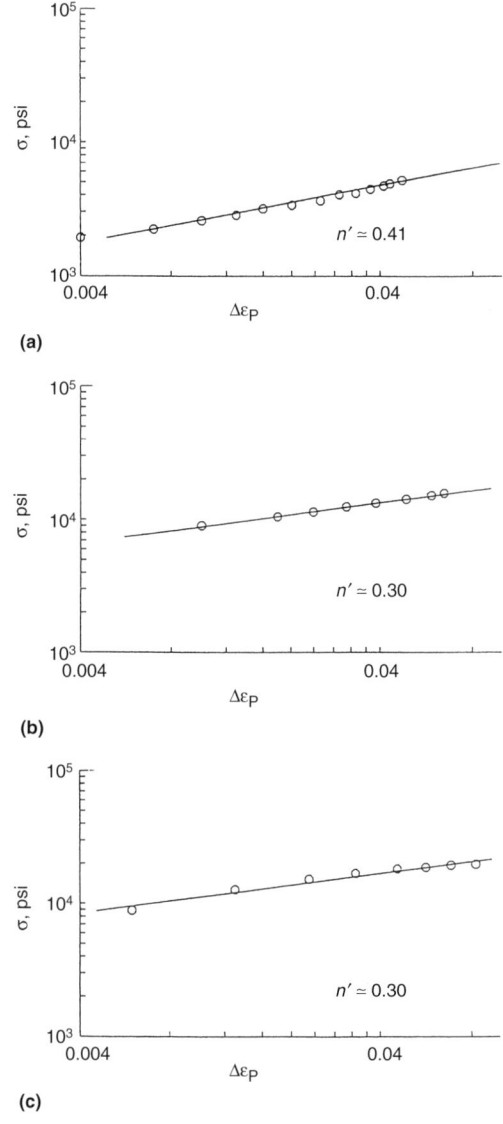

**Fig. 12.15** Determination of cyclic strain-hardening exponents for three test materials for which the slope of the elastic line is calculated. (a) Polypropylene data (Ref 12.4). (b) Nylon 6/6 (Ref 12.3). (c) Polycarbonate (Ref 12.3)

the effect of testing frequency can be estimated because strain rate affects ultimate tensile strength and ductility. Thus, if the effects are known from uniaxial testing alone, they can be built into analysis, especially if the simple experiment is also conducted at the appropriate frequency of interest. Many more materials should be evaluated at room temperature and at other temperatures of interest. The approach is relatively quick and easy and could yield useful results using only a fraction of the testing that would normally be used to evaluate fatigue curves.

The issue of crazing complicates the picture and should be considered. For materials that craze, such behavior is bound to appear in the tensile test and to reduce the tensile ductility. Thus, the ductility of a material that tends to craze will be low, and computations of the fatigue life at higher strain ranges, which are based on tensile ductility, may tend to give conservative results. The question is quite complicated and needs to be studied in detail. Beardmore and Rabinowitz have provided interesting studies of the role of crazing in the three regions they characterize for fatigue (Fig. 12.10, 12.11), but how the factors they discuss would influence the accelerated fatigue testing described previously requires extensive study.

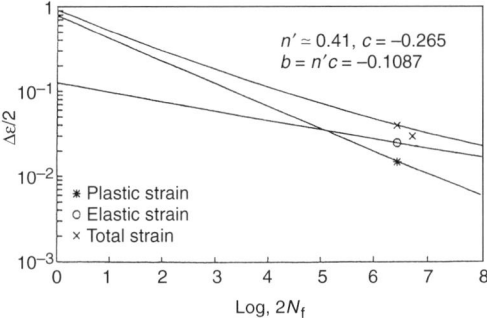

(a)

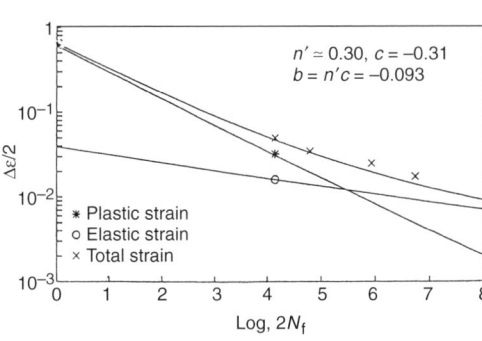

(b)

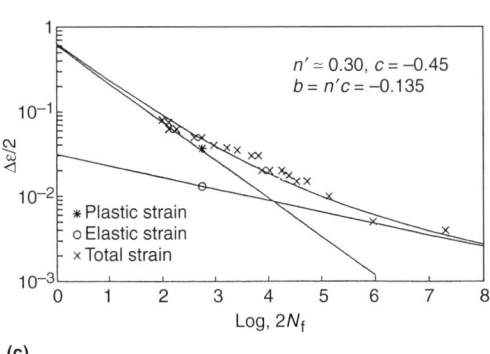

(c)

**Fig. 12.16** Determination of elastic and plastic lines by method III wherein slope $c$ of the plastic line relies on the assumption that $\varepsilon'_f \approx \varepsilon_f$, and slope $b$ of the elastic line are determined from $c$ and the cyclic strain-hardening exponent $n'$. (a) Polypropylene data (Ref 12.4). (b) Nylon 6/6 (Ref 12.3). (c) Polycarbonate (Ref 12.3).

## Fatigue Crack Growth

The concepts developed for fatigue crack growth in metals seem to be valid for polymers. As early as 1965, McEvily and coworkers (Ref 12.6) presented the concept of crack growth illustrated in Fig. 12.17. Their model describes how striations are developed defining the amount of growth from cycle to cycle. They indicate communication with Laird, who was then developing his model for crack growth in metals. This model is discussed in Chapter 10, "Mechanism of Fatigue." In 1964, Manson (Ref 12.5) showed that crack growth striations developed in polycarbonate specimens in much the same manner as in metals (Fig. 12.18). As pointed out by Manson and McEvily, the use of polymers for studying crack growth is easier because of the large deformations sustained by polymers compared with metals. The nature of the crack growth curves in general were also illustrated in Ref 12.5, as reproduced in Fig. 12.19, by studying a polycarbonate resin. And the relation between percent of life to crack initiation and fatigue life, as discussed by Manson and reproduced in Fig. 12.20, clearly shows that the behavior of the polycarbonate falls well within the realm of behavior of metals.

Later, various investigators studied the crack growth rate of numerous polymers according to its relation to stress-intensity factor, used for studying metals. The eight polymers (Ref 12.7) shown in Fig. 12.21 display the linearity commonly exhibited by metals.

**338 / Fatigue and Durability of Structural Materials**

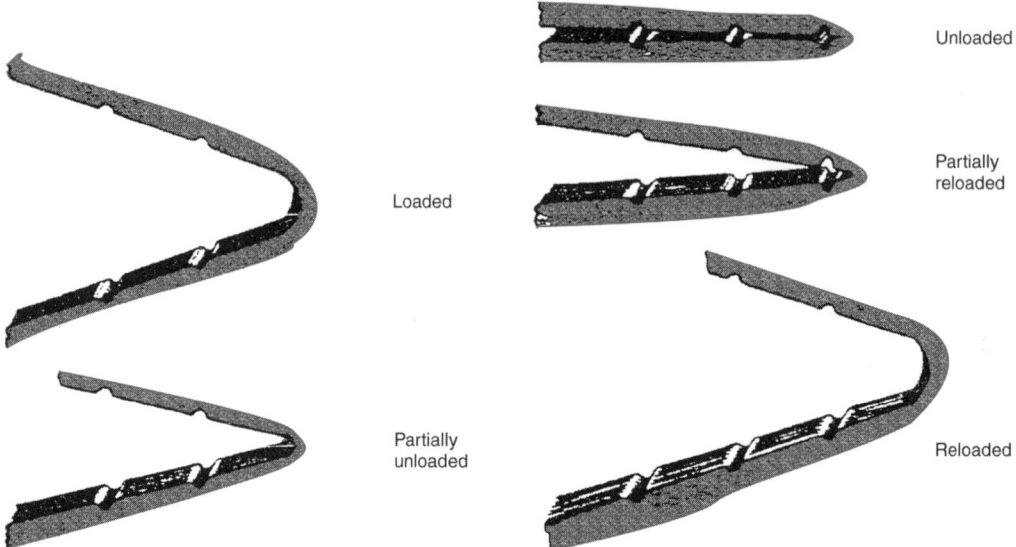

**Fig. 12.17** Schematic illustration of deformation at crack tip during a loading cycle. Source: Ref 12.6

**Fig. 12.18** Crack growth striations on fracture surface of ¼ in. diam polycarbonate-resin specimen; 115 cycles to fracture. Source: Ref 12.5

In conclusion it appears that polymer behavior in crack growth can be studied according to the technology developed for metals. But, of course, polymers will be more sensitive to temperature and loading rate than are metals, and each polymer system must be exclusively studied and the appropriate quantitative relations must be developed.

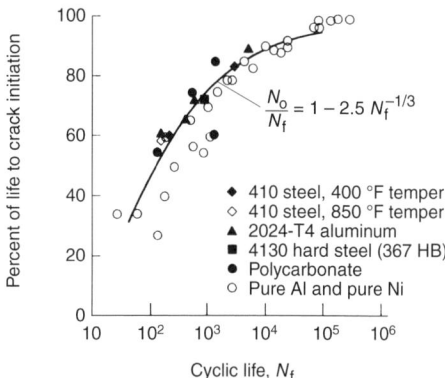

**Fig. 12.20** Relation between percent of life to crack initiation and fatigue life. Source: Ref 12.5

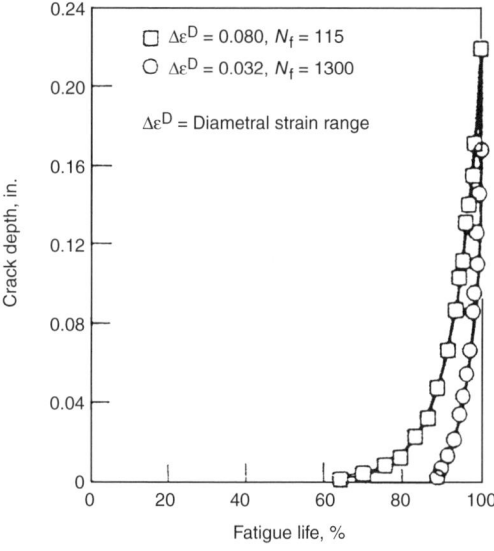

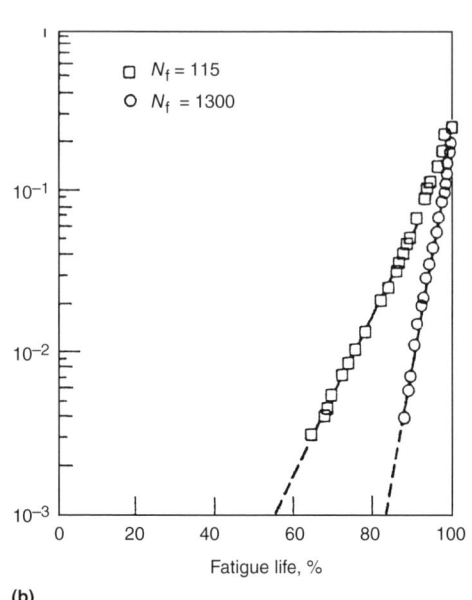

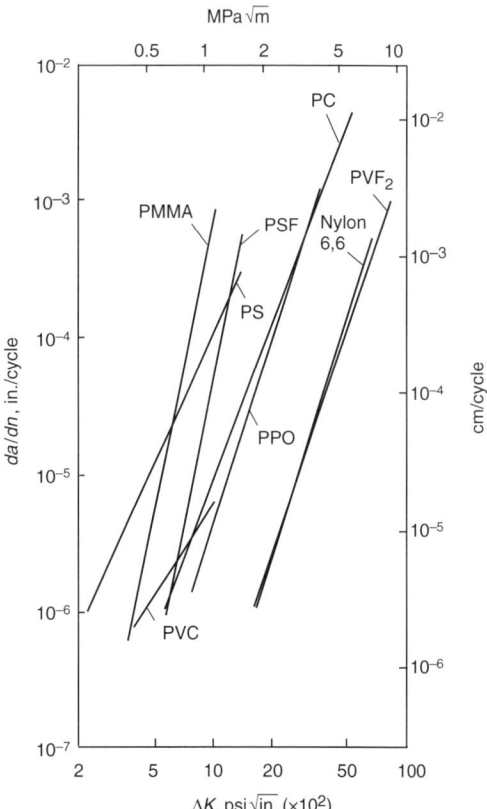

**Fig. 12.19** Crack growth curves for polycarbonate-resin. Source: Ref 12.5. (a) Rectangular coordinates. (b) Semilogarithmic coordinates

**Fig. 12.21** Fatigue crack growth rates of eight polymers (Ref 12.7); figure from Ref 12.8

# Bone

## Introduction

Bone is a natural material that has been in development over millions of years. Natural selection has therefore resulted in a high degree of optimization for its intended structural purpose. Of course, the main engineering interest is in human bone, and the study of animal bone is largely as a convenient means of learning indirectly what to expect for human bone.

Interest in the properties of bone has existed for a long time, but special attention has been given in recent times because of human sports. These have provided the need for applications and have also provided the funding for research. In addition, recent improvements in life extension have provided the incentive for understanding how bone properties change with aging.

It is very difficult to study the fatigue properties of living human bone. Obviously, specimens cannot easily be extracted for studies under exactly known loading conditions and responses determined without destroying the life in the specimen itself. Even qualitative studies in living animals and humans are difficult because of humane considerations and because loading conditions and responses are very inexact. While a few in vivo conditions have been studied, the largest numbers of tests have been conducted on devitalized in vitro bone extracted after the death of the subject. Most of the tests have been conducted on animal bone, but some on human subjects as well. By testing at various times after death (in most cases on different specimens), some idea has been gained about in vivo properties by extrapolation. But the properties of living bone, which can heal imposed damage, are very elusive. Still, because of the strong motivation, a considerable number of programs have been conducted in recent years on the behavior of devitalized bone in the hope of discovering fundamental principles that can be useful in dealing with their problems.

## A Typical Study of Fatigue Behavior of Bone

The literature on mechanical properties of bone, including fatigue, is vast. Most of the fatigue studies do not conform to the discipline of the methods covered in this book and will not be covered here. A doctoral dissertation by a colleague, Cheryl Pattin (Ref 12.9), is very instructive. We use her results, and some appropriate research to which she refers, to illustrate our interpretation of recent bone fatigue studies and how some of the concepts described in this book might advantageously be adapted in the future.

Specimens of the shape shown in Fig. 12.22 were machined from cadavers, aged 21 to 42 years, prematurely killed by heart attack or motor accident. The bones were kept in dry ice at $-20\ °C$ ($-4\ °F$) from time of extraction to time of testing, except during machining when they were also kept wet.

Testing was conducted in either of three ways. Force was varied from zero to a positive maximum (Fig. 12.23), zero to a negative maximum (Fig. 12.24), or completely reversed (Fig. 12.25). Strain was measured with an extensometer allowing the display of hysteresis loops. Stress was calculated based on gross cross-sectional area.

Typical results of such tests are described subsequently and, to the extent that it is possible, we show the data for a single type of bone used in the experiments. However, because of the limited supply of any particular kind of bone, various types were tested, taken from animals of different age at death, different locations in the body, and so forth. Not all types of tests that we shall show were from the same corpse. However, the behaviors described are typical of all the bone specimens tested.

**Tension-Only Fatigue Testing.** Figure 12.26 shows the hysteresis loops obtained from a sample in force cycling between zero and maximum

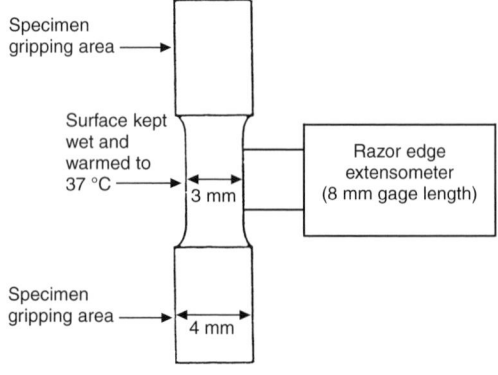

**Fig. 12.22** Cylindrical specimen shape with 1 mm (0.04 in.) root radius. Source: Ref 12.9

stress. Several characteristics of these hysteresis loops are clear.

*The Strain Ratchets from Loop to Loop.* A quantitative display of the ratcheting is shown in Fig. 12.27, where the strain is displayed as a function of time. The ratcheting is very rapid in the early cycling, reaches a steady state for most of the life, and accelerates at the very end. The fatigue life for this case is 7688 cycles, but after only 0.2% of life (approximately 15 cycles), the minimum strain has already ratcheted to about 20% of the ratcheting that occurs during the entire lifetime. The range of strain, however, does not change appreciably during the life.

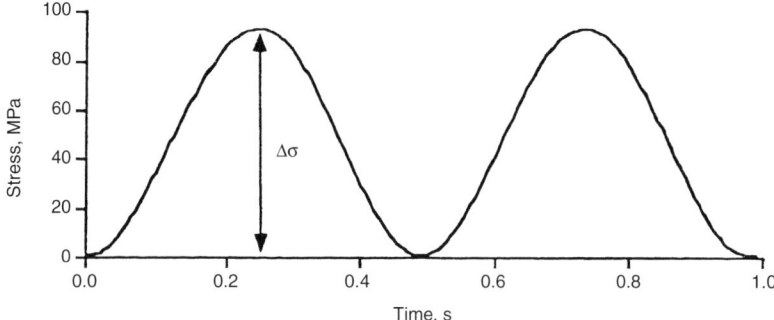

**Fig. 12.23** Tensile stress versus time waveform for two cycles. Source: Ref 12.9

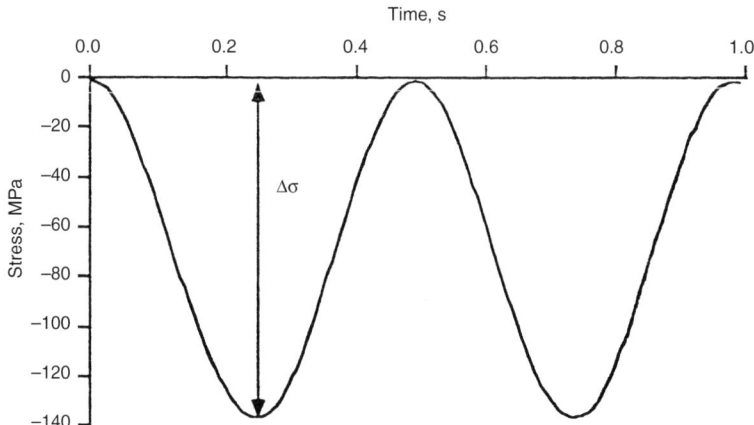

**Fig. 12.24** Compressive stress versus time waveform for two cycles. Source: Ref 12.9

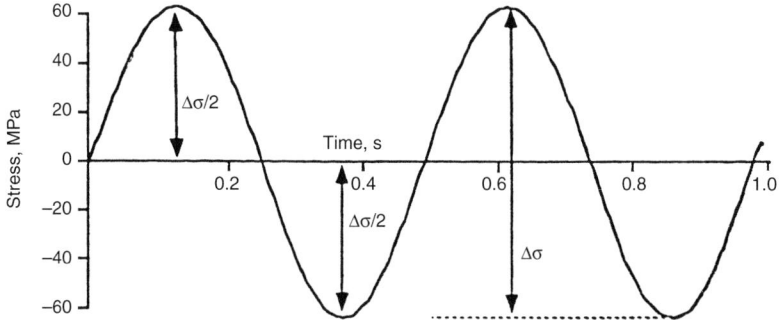

**Fig. 12.25** Completely reversed stress versus time waveform for two cycles. Source: Ref 12.9

342 / Fatigue and Durability of Structural Materials

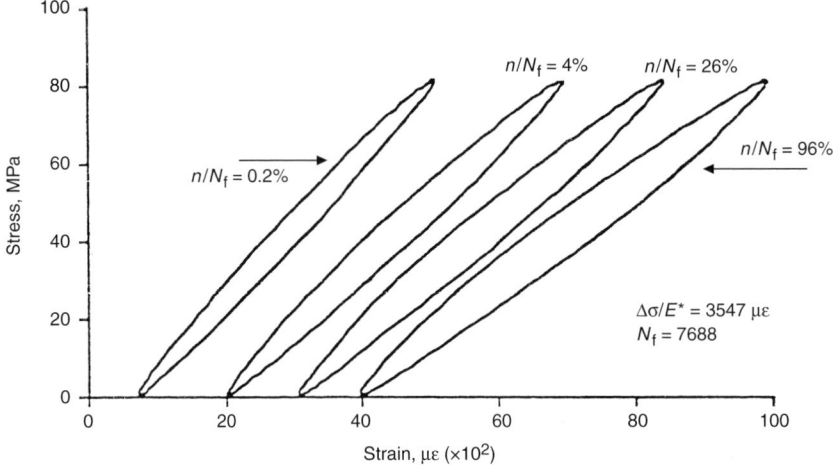

Fig. 12.26  Stress-strain hysteresis loops for tensile fatigue specimen TH. Source: Ref 12.9

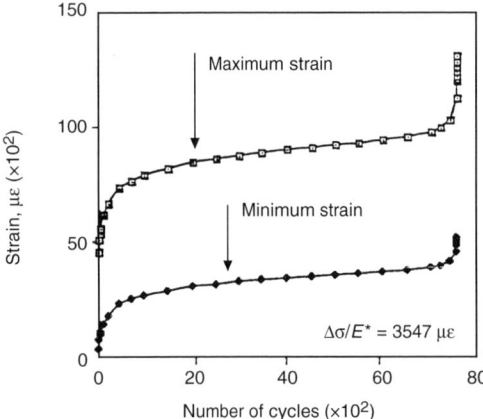

Fig. 12.27  Maximum and minimum strain curves showing tensile ratcheting for specimen TH. Source: Ref 12.9

*The elastic modulus* E* *(tangent to the loading curve at the beginning of the loading of each loop) tends to remain constant for all loops.* However, the constant slope remains at its value for only a small strain. Nonlinearity occurs very soon, due to inelasticity (plasticity, creep, etc.), as we shall discuss. Figure 12.28 shows the loading part of the hysteresis loops at various cycles representing the displayed percentages of the life of 7688 cycles.

*The tip-to-tip slope of the hysteresis loops, i.e., the secant modulus, progressively decreases,* rapidly at first, reaching a steady-state rate of decline for most of the life and decreasing very rapidly near the end of life. This characteristic can be seen quantitatively in Fig. 12.29.

*The width of the hysteresis loops increases progressively as the specimen is cycled.* It is

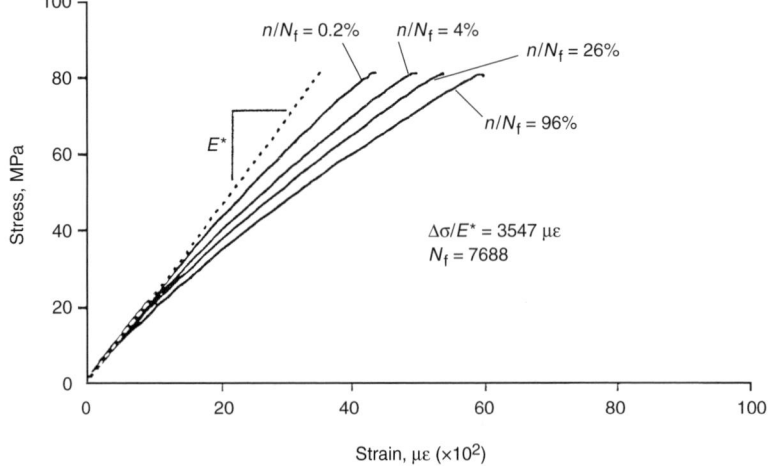

Fig. 12.28  Tensile loading curves at various life fractions for specimen TH. Source: Ref 12.9

common in this type of work to study loop width through the calculation of hysteresis energy, i.e., area of the hysteresis loop. While hysteresis energy is sometimes used in fatigue analyses of metals also, we have not used it extensively in this book. For this reason we include a brief discussion of this approach in the Appendix, "Selected Relevant Background Information." The hysteretic strain energy dissipated per cycle is shown in Fig. 12.30 to increases slightly from cycle to cycle over the vast majority of the life. The deviations occurring at the very beginning and the very end of life do not contribute appreciably to the total integrated energy. For purposes of life prediction, the energy dissipated per cycle could be assumed constant at its value at half-life (3844 cycles). This assumption is consistent with the observation in Fig. 12.31 that the integrated energy dissipated accumulates in a nearly linear manner with applied cycles. The line labeled TH refers to the accumulation of strain energy for the bone type shown in the previous figures; the other two lines refer to other specimens, for reference. Remarkably, if the total accumulated strain energy dissipated data are plotted against life on log-log coordinates, the straight line slope is 0.62. We discuss this finding later when the results of other loading conditions are presented.

**Compression-Only Fatigue Testing.** Tests in compression yielded the same qualitative behavior, although there were, of course, quantitative differences. As we would expect, the initial modulus of elasticity in compression was greater than in tension ($\approx 25\%$). Figure 12.32 shows the compressive hysteresis loops corresponding to the tensile loops of Fig. 12.26. Again, there is ratcheting from cycle-to-cycle, the width of the

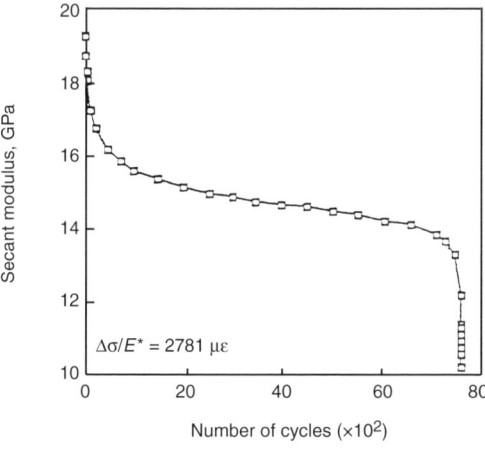

**Fig. 12.29** Secant modulus degradation with applied cycles for specimen TH. Source: Ref 12.9

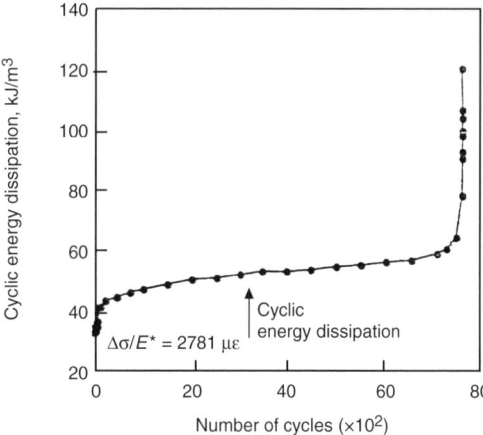

**Fig. 12.30** Cyclic energy dissipation per cycle with applied cycles for specimen TH. Source: Ref 12.9

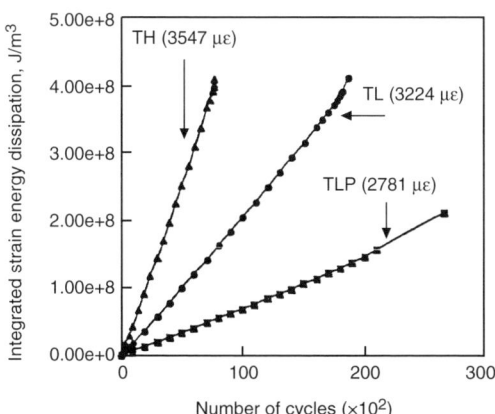

**Fig. 12.31** Integrated strain energy dissipation with accumulated cycles of tensile fatigue. Source: Ref 12.9

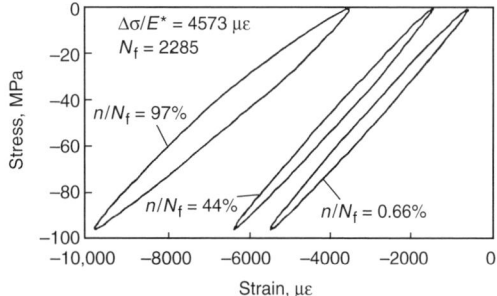

**Fig. 12.32** Stress-strain hysteresis loops for zero-to-compression fatigue loading. Source: Ref 12.9

loop broadens in the later cycling, and the secant modulus decreases progressively. The compressive ratcheting shown in Fig. 12.33 corresponds to the tensile ratcheting of Fig. 12.27. The initial ratcheting is less for compression than for tension, but the ratcheting progresses reasonably uniformly, and only near the end of the life does it accelerate.

Figure 12.34 shows the stress-strain response curves representing only the compressive loading portion of the cycle. These correspond to the curves for tension shown in Fig. 12.28. The quantitative loss in compressive secant modulus with applied cycles is shown in Fig. 12.35, corresponding to Fig. 12.29 for tension. The rapid loss of secant modulus seen in tension in the early cycles now disappears in compression, but for most of the life the secant modulus in compression continues to deteriorate, as in tension, and at the very end of life, the deterioration is rapid. Figure 12.36 shows the integrated energy dissipation versus cyclic life for compression of all the specimens so tested. Figures 12.37 and 12.38 display the total lifetime hysteresis energy dissipated as a function of the number of cycles to failure for compressive and tensile fatigue, respectively. The data fall on approximately linear log-log coordinates with slopes of 0.55 (compressive fatigue) and 0.62 (tensile fatigue).

**Completely Reversed Fatigue Testing.** Finally, there are the tests in completely reversed loading. Figure 12.39 shows the hysteresis loops for one specimen (others were tested and showed less severe results; however, we concentrate on this specimen for brevity). The hysteresis loops showed increases in both the tension and compression extremes, as quantified by Fig. 12.40. Secant modulus of both the tensile and compressive halves decreased as shown in Fig.

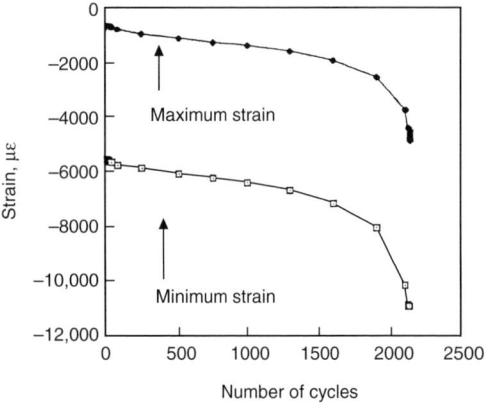

**Fig. 12.33** Maximum and minimum strain curves showing compressive ratcheting for specimen CH. Source: Ref 12.9

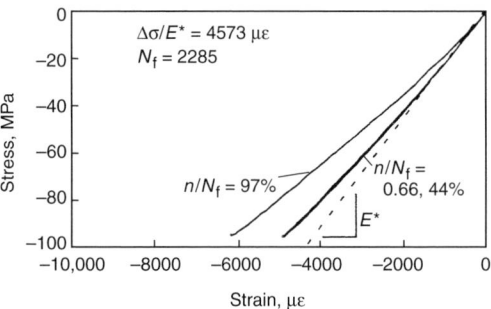

**Fig. 12.34** Compressive loading curves at various life fractions for specimen CH. Source: Ref 12.9

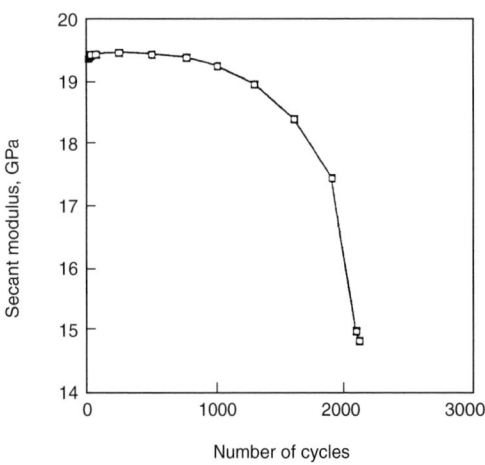

**Fig. 12.35** Secant modulus degradation with applied cycles during compressive loading only for specimen CH. Source: Ref 12.9

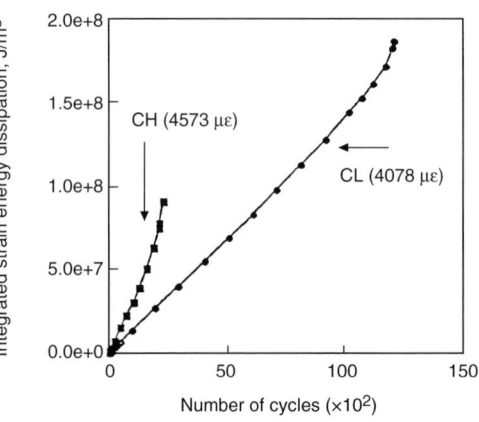

**Fig. 12.36** Integrated strain energy dissipation with accumulated cycles of compressive fatigue. Source: Ref 12.9

12.41. Also shown is the decrease in the average secant modulus (loop tip to loop tip). The deterioration is monotonic during the entire lifetime, although the loss of the tensile modulus is greater than that of the compressive modulus. Figure 12.42 shows the relative linearity of hys-

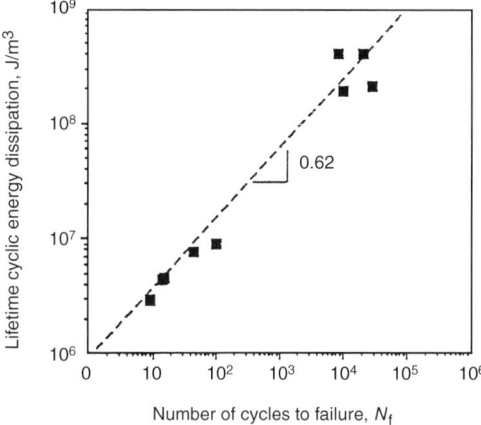

**Fig. 12.37** Total lifetime hysteresis energy dissipation as function of cycles to failure in compressive fatigue. Source: Ref 12.9

**Fig. 12.38** Total lifetime hysteresis energy dissipation as function of cycles to failure in tensile fatigue. Source: Ref 12.9

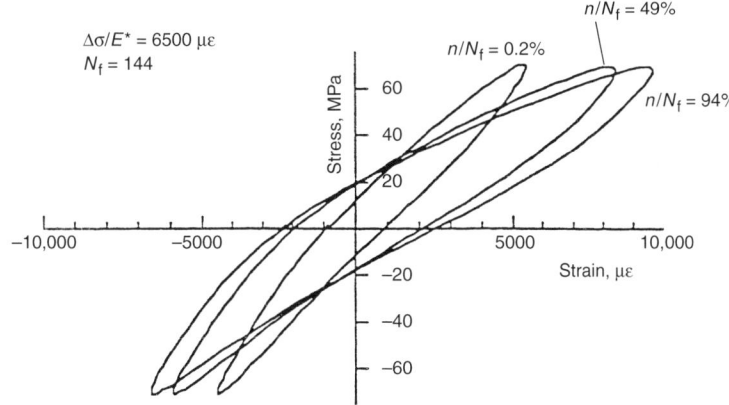

**Fig. 12.39** Stress-strain hysteresis loops for fully reversed fatigue specimen TCH. Source: Ref 12.9

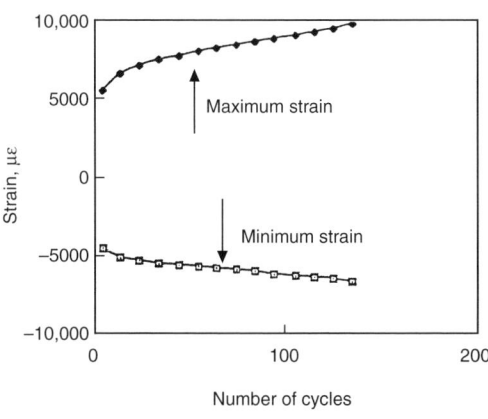

**Fig. 12.40** Maximum and minimum strain curves showing maximum and minimum strain increases during fully reversed fatigue for specimen TCH. Source: Ref 12.9

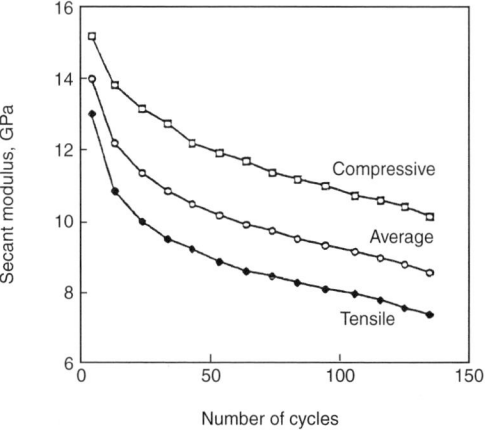

**Fig. 12.41** Secant moduli degradation during fully reversed fatigue of specimen TCH. Source: Ref 12.9

teresis energy accumulation with cycles, while Fig. 12.43 shows how cyclic energy dissipation varies with cyclic life for the case of completely reversed load cycling. Only four data points are available for this type of testing, but the slope of the best line through these points produces a slope of approximately 0.55.

A summary of all the data for cyclic energy dissipation is shown in Fig. 12.44. Included are cyclic tensile loading, compressive loading, and completely reversed loading. While there is some scatter, it is remarkable that all points fall along a single straight line of slope 0.60. That all types of loading fall on a single line is surprising because each type of loading produced such a different response. However, as Pattin (Ref 12.9) points out, those conditions that produce more energy dissipation per cycle for a given strain range (e.g., tensile loading) also produce a shorter fatigue life, so the total energy dissipation over the lifetime is no different from compressive loading in which the energy per cycle is lower, but the number of life cycles is greater. These results, therefore, point to the validity of a hysteresis energy criterion for fatigue failure for this type of material.

## Expected Behavior of Metallic Materials under the Types of Loading Used in Tests of Bone

As pointed out in the introduction to this chapter, one of our goals for discussing the special materials addressed here is to determine whether the technology presented in this book for metals is directly transferable to other materials such as polymers, ceramics, composites, and bone. Bone, in particular, is a polymer, as well as a ceramic, so it is an especially interesting material to examine from this perspective. In the tests conducted by Pattin, three major types of loading were imposed, so we shall start by considering the expected behavior of various types of metals under the same types of loadings.

### Modeling of Stable Metal

Consider first an ideally stable metal, which neither hardens nor softens under cycling loading. The same three types of loading patterns used by Pattin are considered. For each type, two

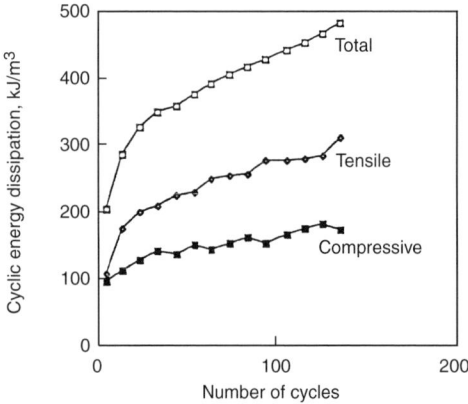

**Fig. 12.42** Integrated strain energy dissipation with accumulated cycles of fully reversed fatigue loading of specimen TCH. Source: Ref 12.9

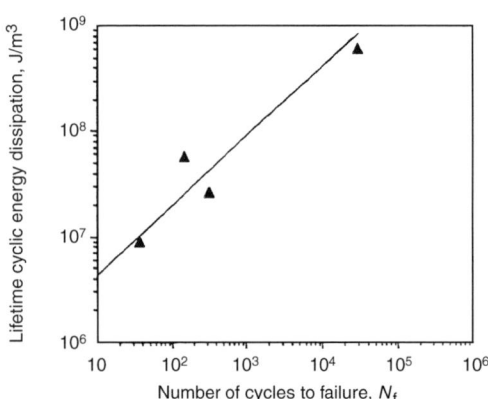

**Fig. 12.43** Total lifetime hysteresis energy dissipation as function of cycles to failure in fully reversed fatigue loading. Source: Ref 12.9

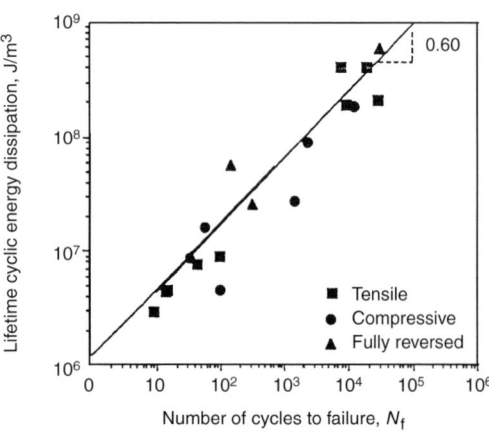

**Fig. 12.44** Total lifetime hysteresis energy dissipation as function of cycles to failure for tensile, compressive, and fully reversed fatigue loading. Source: Ref 12.9

different loading levels are analyzed and their corresponding fatigue lives predicted. For quantitative purposes, we have chosen the constants for the stress-strain and life relations as shown in Fig. 12.45. As is typical, these relations are based on data from fully reversed loading tests.

**Zero-to-Tension Fatigue Loading.** We examine two cases of loading. For the first, we force-control cycle between zero and 57 ksi tensile stress (points A to B in Fig. 12.45a) which is nominally within the elastic regime. The total strain range is computed to be 0.00195. Although not readily observable at the macroscopic scale, a tiny amount of cyclic plasticity ($\approx 0.00005$) is calculated to be present and contributes to damage. Continued cycling replicates the very narrow hysteresis loop AB until fatigue failure is imminent. Because a tensile mean stress (equal to half the controlled stress range) is present in the zero-to-tension fatigue test, it is necessary to account for its effect on fatigue life. To do this we assume that the modified Morrow-type equation (Ref 5.5) correctly models the effect of mean stress on fatigue life (Chapter 4, "Mean Stress"). The calculated life for this case thus is 172,000 cycles to failure.

Now, we consider the second case with a greater maximum tensile stress of 75.3 ksi (point C as shown in Fig. 12.45a). A hysteresis loop ABCA develops that includes a detrimental tensile mean stress that is also equal to half the applied stress range. The total and plastic strain ranges for this case are calculated to be 0.00271 and 0.00020, respectively. Because a totally stable material under load control will neither ratchet nor cyclically relax the mean stress, continued cycling will replicate this loop until failure occurs at a calculated life of 8980 cycles to failure. The secant modulus, as represented by an imaginary line connecting points A and C, is slightly less ($\approx 7\%$) than the elastic modulus of 30,000 ksi.

**Zero-to-Compression Fatigue Loading.** For the fatigue loading condition of zero-to-compression at the same two stress and strain ranges just noted, the hysteresis loops are the same as for zero-to-tension fatigue loading, except that the cyclic stress-strain loops lie below rather than above the horizontal strain axis. The corresponding computed fatigue lives (accounting for compressive mean stresses) for cases (a) and (b) are $N_f = 4,280,000$ and 647,000 cycles, respectively. These lives are significantly longer (by factors of 25 and 72, respectively) than that in tension because of the beneficial effect of compressive mean stress compared with tensile. The secant modulus is the same as for zero-to-tension cycling.

**Completely Reversed Fatigue Loading.** For completely reversed fatigue loading at the same stress range, the hysteresis loops for cases (a) and (b) again will be identical to those for tension-only and compression-only fatigue loading. The compression portions of these hysteresis loops are symmetrical to those in tension, and

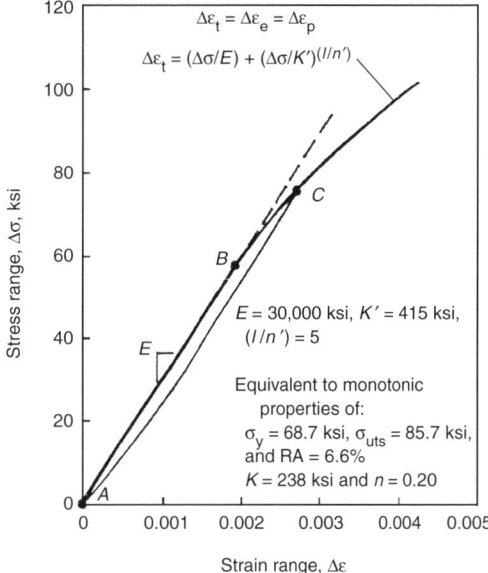

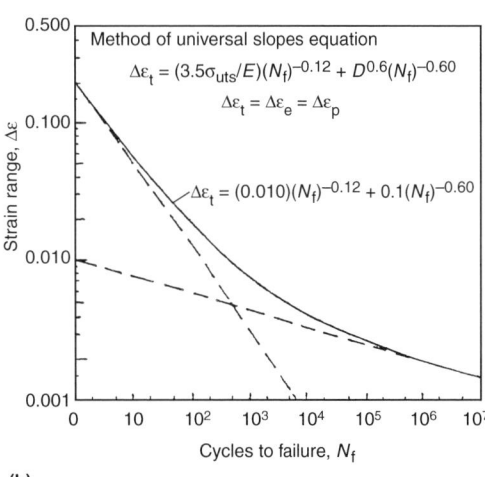

**Fig. 12.45** Typical stress-strain and life relations for a cyclically stable metal. (a) Cyclic stress-strain relation based on ranges. RA, reduction in area. (b) Completely reversed strain-life relation based on the method of universal slopes. Source: Ref 12.5

the loops repeat until failure intercedes at computed lives of 1 million and 100,000 for the two cases respectively. The secant modulus is always lesser the greater the plastic strain range. The secant modulus remains fixed in value until fatigue failure occurs.

Now we can consider the hysteresis energy factor for these cases. Of course, because for any given type of loading, the hysteresis loop remains constant, the integrated plastic strain energy accumulates linearly with cycles, analogous to that found for bone. However, the plots of the total energy expended versus life are quite different from that of bone, as shown in Fig. 12.46. The equations used in calculating the hysteresis energy are presented in the Appendix, "Selected Background Information." Each type of loading imposed on the metal produces a unique curve with different slopes. This behavior is due entirely to the mean stress effects on the calculated fatigue lives.

We can now summarize the differences between bone behavior and that expected from an ideal, *stable* (nonsoftening and nonhardening) metal. The tensile and compression *rheological* behavior is the same for metal but differs appreciably in bone. Ratcheting occurs in bone both in tension, and compression, as well as completely reversed loading (better described as softening), while no such ratcheting occurs, by the aforementioned definition of *stable*, in the metal under any of the loading types. Finally, for metals, there is a substantial mean stress that significantly affects the total hysteresis energy absorption during the fatigue lifetime. It can be concluded that the overall fatigue behavior of bone cannot be modeled using the behavior of a stable conventional metal. However, because bone cyclically softens, we now examine the behavior of a cyclically strain-softening metal.

### Modeling of Cyclic-Softening Metal

We ask whether the behavior of a cyclically softening metal would be of assistance in modeling the cyclic behavior of bone, which we earlier showed to exhibit cyclic softening and ratcheting behavior. In Ref 12.5 we discussed the cyclic strain-softening behavior of hardened 4340 steel. Very early in the strain cycling the softening was almost by a factor of 2 in strength, so this metal might give us some insight as to how to model bone. Unfortunately, force-controlled cyclic tests were not performed for 4340 steel in its hardened condition. Consequently, it is necessary to make educated guesses as to its behavior had such tests been conducted. As discussed in general terms in Chapter 2, "Stress and Strain Cycling," a cyclically softening alloy under force control with a tensile mean load will ratchet significantly in the tensile direction. Furthermore, if the loads are high enough, a premature tensile necking instability could occur and significantly shorten the cyclic life below that for fatigue-induced cracking and failure. Even under completely reversed force control, a cyclically softening alloy will also ratchet in the tensile direction, but much more slowly than when subjected to a tensile mean load. The reason for this behavior is subtle. First, the cyclic strain range increases significantly during softening. Then, as the strain range gets larger, the true stresses (force/instantaneous cross-sectional area) developed at the force extremes differ significantly (greater in tension and less in compression). Thus, there is a tensile mean stress that in turn promotes tensile strain ratcheting, although not at the same rate as for zero-to-maximum tension loading. At the other extreme of force-controlled cycling with a compressive mean load, the cyclically softening 4340 steel will strain ratchet toward compression, causing the cross-sectional area to increase progressively. The rate of ratcheting and hence area change, however, diminishes as the area gets larger and the true stresses diminish. At some point, the specimen will either shake down to a repeatable cycle with a compressive mean ratchet strain or will fail by low-cycle fatigue before stability is reached.

In comparing the behavior of the cyclically softening metal, it is noted as having similarity

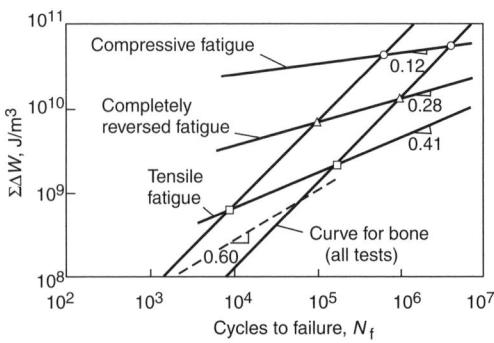

**Fig. 12.46** Total lifetime hysteresis energy dissipation as function of cycles to failure for tensile, compressive, and fully reversed fatigue loading for a typical metal compared with the general behavior of bone

in stress-strain response characteristics as that of bone for each of the loading modes. Because experimental data on cyclic lives were not determined for the softening 4340 steel, it is not possible to draw hard conclusions with respect to the trend in fatigue resistance when compared with bone. However, it would be expected that the integrated hysteresis energy expended by the metal would not fall in line with that for bone. While a single curve represents all of the bone fatigue data, regardless of loading in tension only, compression only, or in both with completely reversed loading, it is highly unlikely that a similar result would be found for the 4340 steel. It is reasonable to expect that a plot of integrated expended energy versus cyclic life for the softening alloy would show an even more exaggerated separation of the results of the three different types of loading than was shown for the stable alloy. Unfortunately, no fatigue data on alloys are available to quantify the differences between these material systems.

## Brief Concluding Remarks Regarding Fatigue of Bone

It is concluded that while the fatigue behavior of metals provides some guidance as to critical fatigue tests appropriate for bone, it gives but limited insight into the expected behavior of bone. The closest similarity in behavior of the two vastly different materials is in their cyclic stress-strain response when a cyclic strain-softening alloy is used for comparison. Obviously, considerably more fatigue research on bone and cyclically softening alloys is required to develop trends and a better understanding of their behavior under cyclic loading.

## Ceramics

### Introduction

The term *ceramics*, to the layman, conjures images of bricks and mortar, dinnerware, decorative figurines, bathroom fixtures, glass, and even gemstones. However, to a new breed of engineers and designers, the word *ceramics* suggests a whole new solution to achieving critical needs not met by conventional metallic materials. New, man-made structural-grade ceramics offer thermal, physical, and mechanical properties unachievable by conventional metallic materials. We will concentrate on these engineered ceramics because of their unique role in structural design of high-performance, high-temperature systems. Examples of applications include heat exchangers, diesel piston caps, poppet valves, turbine rotors, airfoils, combustor liners, protective coatings, and even ultrahigh-speed ball bearings.

Most of this section originates not from the authors' research activities, but rather from their colleagues at NASA-Glenn, Case Western Reserve University, and elsewhere. The reasons for preparing this section are limited to assessing any similarities in mechanical behavior between this unique class of materials and metallic materials. Topics such as stress-strain response, fatigue crack initiation, fatigue crack propagation, and fracture characteristics are of dominant concern. There is no intent to review the sizeable literature that has been written over the past three decades on the mechanical behavior of ceramics.

**Advantages.** The foremost desirable characteristics of engineered ceramics are their ability to withstand very high temperatures while retaining a large fraction of their lower-temperature strength. A few of the more advanced ceramics retain useful load-carrying capability to 1500 °C (2700 °F) and higher, well above the melting point of many high-temperature alloys. Combined with their high-temperature strength retention, they also offer much higher resistance to aggressive environments. Most of these ceramics are thermally stable and resistant to oxidation and chemical attack. They are resistant to long-term creep and relaxation and have high hardness and wear resistance. They are also excellent electrical and thermal insulators, have low coefficients of thermal expansion, and high moduli of elasticity. Coupled with their relatively low density, ceramics offer a high strength-to-density ratio—a very desirable characteristic for aerospace applications where flight weights are of paramount concern. Table A.7 of the Appendix to this book contains a listing of physical, thermal, mechanical, and electrical properties of representative monolithic engineered ceramics.

**Disadvantages.** The notorious disadvantages of engineered ceramics for structural application is their brittleness and extreme sensitivity to growth, at ambient temperatures, of small flaws undetectable by the unaided eye. These factors are responsible for a great deal of the scatter in measured mechanical properties. To deal with this scatter in design, statistical representation usually is required. Weibull statistics (Ref 12.10) are used frequently (See the Appendix, "Selected Background Information," to this book). In fact, the Weibull modulus $m$ for tensile strength is listed in Table A.7 of the Appendix. The lower the value of $m$, the greater is the scatter. The strongest ceramic listed also has the lowest value of $m$ ($=7$) and hence the greatest scatter.

Brittleness is synonymous with poor fracture resistance (See Chapter 9, "Crack Mechanics"). Fractures initiate from flaws or geometric discontinuities introduced during manufacturing. Even in the absence of readily recognizable discontinuities, fractures originate from tiny flaws that are detectable only with the aid of high-resolution instruments. Unlike metals, the fracture resistance of engineered ceramics is low, despite inherently strong atomic bonds. This is so because the material is unable to relieve high concentrations of stress through plastic deformation. Thus, even relatively low forces can result in high, but localized, stresses that initiate fracture. Once localized fracturing commences, the stresses at the tip of the crack continue to increase with no relief as there is no energy-absorbing mechanism to relieve the high stress such as there is for metals and polymers. Hence, once a crack in a ceramic has initiated and starts to grow, only one energy dissipation mechanism is available to help slow the growth. That dissipation mechanism is the energy required to create the new surfaces of the crack. Unfortunately, this energy is no match for the crack-driving energy associated with the release of elastic strain energy as the crack grows. Thus, fracturing becomes a runaway process that occurs extremely rapidly. To guard against such undesired fractures, intensive research and development efforts have been expended over the past half century to increase the "toughness" of ceramics. Efforts have centered on decreasing the already-microscopic flaw sizes through densification processes, improved bonding characteristics, and increasing the effectiveness of fracture energy absorption/dissipation mechanisms. Fracture and fatigue characteristics are discussed further in a subsequent section.

A less obvious, but nonetheless important, disadvantage of engineered ceramics is the difficulty of being able to shape these materials once they are in their fully hardened state. Products must be formed to near-net shape while still in an unfired (unsintered), *green state*. Once sintered, machining options are limited to grinding, ultrasonic erosion, and abrasive water-jet techniques. Material removal must be done carefully and at low rates to avoid mechanical or thermally induced stresses and hence lower the probability of damaging the defect-sensitive surface.

**Composition and Manufacturing of Engineered Ceramics.** Most engineered ceramics have compositions involving a single metallic element (e.g., aluminum and titanium) or nonmetallic element (e.g., cesium and silicon) that chemically bond to small-diameter, low-atomic weight elements such as oxygen and carbon. The latter form oxides, carbides, borides, nitrides, and silicides. Mechanical blending of fine powders of the elemental ingredients is used to create most of the engineered ceramics. These powders are held together with a polymer binder and are compacted prior to any major machining. The binder burns off and dissipates during the hot sintering cycle, ideally leaving no voids in the process. Parts are sintered under very high temperatures and hydrostatic pressures, i.e., hot isostatic pressing. This process must be accomplished under high quality control to avoid introduction of tiny voids or flaws that will limit static fracture and fatigue resistance.

The nature of the atomic bonding in ceramics covers a broad range from purely ionic to purely covalent. The inherent high strength, but brittle nature, of ceramics derives from these atomic bonding characteristics. Ceramics do not possess the same forms of bonding as found in the close-packed crystalline structures of ductile metallic materials. It is much more difficult for an atomic bond in a ceramic, once broken, to be remade, thus the motion of a dislocation through one atomic distance in a ceramic material results in a local, unhealed fracture. This is in sharp contrast to metals wherein broken atomic bonds can be remade instantaneously as the dislocation moves through an integral number of atomic distances.

**Fiber Reinforcement.** Compensations for the low energy absorption characteristics of monolithic ceramics can be made by incorporating reinforcing ceramic fibers into a ceramic matrix. The fibers may be of similar or dissimilar composition to the existing ceramic matrix. Such artificially constructed materials are known as fiber-reinforced ceramic matrix composites, or simply CMCs. The fibers are often made stronger than the matrix, using appropriate processing techniques. The fibers serve two critical functions, as illustrated schematically in Fig. 12.47. They act as barriers to propagating cracks because of the discontinuity between fiber and matrix and because the stronger fibers tend to hold the cracks closed. The latter is called *crack bridging*. The fibers carry some of the force across the open crack, thus redistributing force away from the critical crack tip, reducing the stress-intensity factor. To increase the energy absorption during fracture, it is intended that the fiber/matrix bond be weak so that a certain amount of interfacial sliding occurs between the fiber and matrix. The sliding friction dissipates some of the energy supplied from the fracturing process, analogous to, but far less effective than, the role of plasticity in dissipating fracture energy in metals. Fiber-reinforced ceramic matrix composites are discussed further in the subsequent major section on composite materials.

## Strength and Durability of Ceramics

Withstanding all of the mechanical and thermal forces poses a significant design challenge for monolithic ceramics owing to their low toughness and high degree of inherent scatter. Because of the high degree of scatter in mechanical properties, deterministic approaches to fatigue and fracture are of limited value. Instead, considerable use is made of statistical analysis. A computer code, CARES (Ref 12.11–12.14), written by researchers at NASA-Glenn, is available for performing probabilistic design of ceramic components. The code takes into account the highly statistical nature of the behavior of ceramics.

### Fracture Strength

An example of ceramic mechanical material behavior analyzed for input into the CARES code is given by the Weibull statistical distribution for silicon carbide (SiC) bend bars displayed in Fig. 12.48. Note the fanning out of the 90% upper and lower confidence bounds at the ends of the distributed data. The measured fracture stress varied by a factor of three within this set of data. Sources of this variation are tiny, nearly undetectable voids and defects. Their size and distribution must be controlled with great care during the basic processing of the ceramic into its final form. In simplest terms, the problem is analogous to that of the weakest link in a long chain causing the failure of the entire chain. This analogy implies that the larger the volume of material subjected to the highest stresses, the greater is the probability of encountering strength-reducing flaws. Hence, the concept of the highly stressed volume of material becomes an important consideration in assessing durability of engineered ceramics.

### Bending versus Axial Strength

Ceramics are so brittle it is especially difficult to conduct a successful conventional tensile test owing to undesired stresses induced in the specimen by improper gripping and slight misalignment. Consequently, tensile strength measurements are usually determined from three-point or four-point bend tests of rectangular cross-section specimens. Conventional elastic strength-of-materials bending calculations ($\sigma = Mc/I$) are used to assess the ultimate tensile strength in the outer surface from the bending moment at the instant of fracture. This value is sometimes referred to by the terms *modulus-of-rupture, bending fracture strength,* or more simply, *fracture stress* (as used in Fig. 12.48).

### Thermal Shock

Another manifestation of the brittleness of ceramics is that they are unable to withstand what

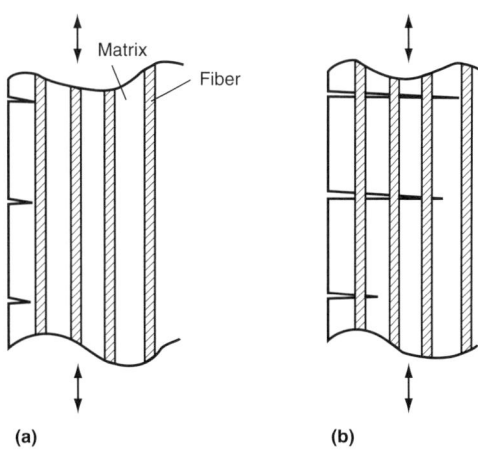

**Fig. 12.47** Fiber barrier to rapid fracture. (a) Internal interface deflects and slows crack propagation. (b) Bridging crack with fibers reduces stress intensity at crack tip.

has been conventionally referred to as *thermal shock*. When a severe thermal gradient is imposed—either by a rapid-heating (up-shock) or a rapid-cooling (down-shock)—on a brittle material, large transient thermal stresses are introduced. If the maximum tensile thermal stress exceeds the critical fracture strength, fracture initiates and rapidly propagates to complete fracture. Up-shocking induces tensile thermal stresses on the usually stronger interior of a sample, whereas down-shocking causes tensile thermal stress at the usually weaker surface. The latter is the most devastating to ceramics. Ductile materials do not experience this mode of failure, although they can fail by thermal fatigue if heating and cooling is repeated many times.

### Fracture Toughness

The concept of fracture toughness (See Chapter 9, "Crack Mechanics"), a measure of the ability of a material to withstand loadings in the presence of cracks, seems ideally suited for use in describing the fracture resistance of brittle ceramics. However, insertion of cracks of well-controlled size into test coupons is particularly difficult. Because crack growth resistance is very low, it is difficult to control and thereby stop a crack from growing beyond a desired limit. Novel techniques have been developed to achieve appropriate crack-arrest during sample preparation. In ductile metals, cracks can be introduced through high-cycle fatigue crack propagation, but this technique is not as viable for brittle ceramics. Instead, a crack is created using a machined notch that is lightly loaded until a crack starts to propagate. However, the propagation is arrested at a desired crack length soon after beginning by an imposed compressive stress field (Ref 12.15). Upon removal of the compressive stresses, the crack remains stable and the specimen is ready for fracture toughness testing. Deviations in measured fracture toughness are comparable to the deviations in the tensile fracture strength; i.e., the Weibull moduli are nearly the same.

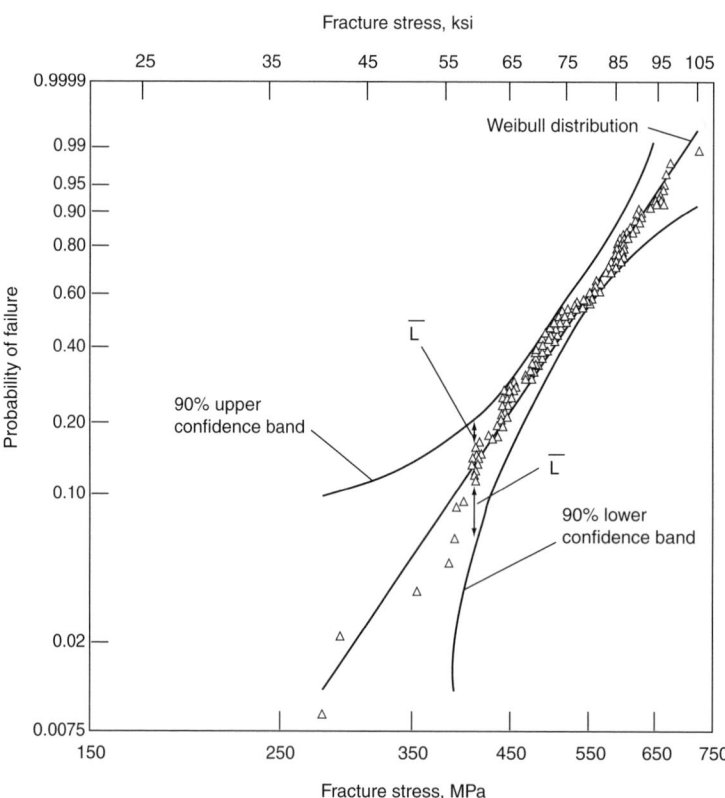

**Fig. 12.48** Weibull distribution and 90% confidence bands determined for fracture stress of hot isostatically pressed silicon carbide. Source: Ref 12.14

## Fatigue Crack Initiation and Propagation

Based on these large variations in fracture strength, it would be quite reasonable to expect equal or even greater scatter in fatigue resistance. However, compared with metallic materials, few fatigue crack initiation curves for ceramics have been generated. As a case in point, the 1200 page ASM International 1991 *Handbook on Ceramics and Glasses* (Ref 12.14) does not list a single citation for fatigue of ceramics in its lengthy 42 page index. *Static* fatigue of glasses is mentioned, but the term in that field implies time-dependent, environmentally induced fracture while under constant tensile loading, not cycle-dependent loading induced cracking and fracture.

**Crack initiation.** The fatigue crack initiation behavior of ceramics is vastly different from that of metallic materials. At temperatures below the creep range, ceramics lack the ability to create their own crack-initiating defects due solely to cyclic loading. They do not undergo cyclic inelastic deformation that produces the reversed crystallographic slip (plasticity) induced intrusions and extrusions that are possible in metals (See Chapter 10, "Mechanism of Fatigue"). Hence, classical microscopic fatigue crack initiation is not of particular concern as a failure mode in ceramics. Cyclic loading, however, can contribute to strength degradation and fracture. If other degradation mechanisms such as oxidation or creep at very high temperatures are operating, cyclic loading can interact with these prime mechanisms. This gives the impression of cycle-dependent behavior, whereas the failures are truly time dependent. Figure 12.49 illustrates "cyclic crack initiation" data generated in bending (Ref 12.16) for two hot-pressed silicon nitride materials designated HS-110 and HS-130. At the relatively low temperature of 250 °C (480 °F), the effect of cyclic loading is negligible; i.e., fatigue doesn't appear to occur, resulting in a horizontal distribution of data points. For HS-110 at 1000 °C (1830 °F) and 1200 °C (2190 °F), cycle dependency of life appears to govern the life. However, small amounts of inelastic deformation in the form of time-dependent creep were identified from metallographic examination of the tested specimens. Small degrees of oxidation were also observed. Similar results have been reported for silicon carbide (SiC) in 1995 (Ref 12.17). However, these researchers also found that when high-temperature cyclic loading tests were performed in vacuum, the time-dependency of failures was not observed. This implies that thermally activated oxidation, rather than creep, plays a significant role in the failures of SiC. At the highest temperatures, ceramics begin to behave somewhat like metals do at high temperatures. That is, cyclic inelastic deformation, in the form of time-dependent creep, and oxidation interactions create microscopic discontinuities that grow and link to form macroscopic discontinuities. These, in turn, cyclically propagate. However, it must be recognized that the extent of oxidation and cyclic inelastic strain per cycle is orders of magnitude smaller than that experienced in ductile metals. Note that when cyclic loading tests are conducted at a single frequency (30 Hz, in Fig. 12.49), it is impossible to determine whether the results are truly cycle dependent or time dependent. Because the only damaging mechanisms involved are time dependent, it is reasonable to conclude that the cyclic loading test results are also time dependent, not cycle dependent. The consequence is that while, in principle, a time-dependent inelastic strain range versus cyclic life relation (like the Manson-Coffin law) might exist for a ceramic at high temperatures, this life relation falls far below the corresponding elastic strain range versus life re-

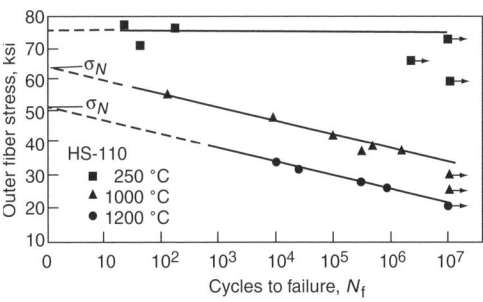

(a)

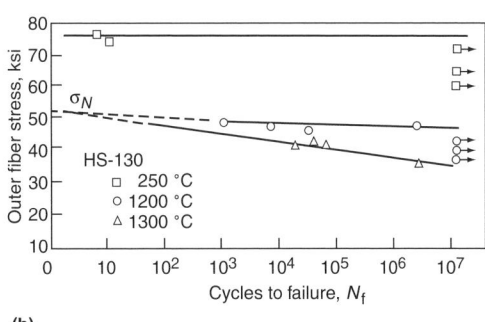

(b)

**Fig. 12.49** Reversed bending cyclic crack initiation behavior for (a) HS-110 and (b) HS-130 hot-pressed silicon nitride at 30 Hz for various temperatures. Source: Ref 12.16

lation and contributes very little to the total strain range versus cyclic life relation. This condition can be estimated using the MUS equation (See Chapter 3, "Fatigue Life Relations") wherein the creep ductility $D$ of the ceramic might be on the order of 0.1% or less. Presuming an ultimate tensile strength of 300 MPa and an elastic modulus of 300 GPa, the predicted fatigue equation would appear as below and in Fig. 12.50:

$$\Delta\varepsilon = (3.5\sigma_{UTS}/E)(N_f)^{-0.12} + (D^{0.6})(N_f)^{-0.60}$$
(Eq 12.1)

or

$$\Delta\varepsilon = (0.0035)(N_f)^{-0.12} + (0.0158)(N_f)^{-0.60}$$
(Eq 12.2)

It is important to caution that the MUS equation was developed based on the known fatigue and tensile test behavior of a large number of monolithic metals that typically exhibit far greater ductility than do ceramics. Even a very low ductility metal may have 10, 100, or even 1000 times greater ductility than a ceramic. This fact is important relative to the term representing ductility, i.e., $D^{0.6}$, in the MUS equation, because raising the value of ductility (for ceramics, much, much less than 1.0) to a power less than 1.0 (0.6) results in a numerical value many times greater than the original ductility value. In the current example with $D = 0.001$, the value of $D^{0.6}$ becomes 15.8 times larger than $D$ and the value may not properly represent the behavior of very brittle materials. The case may be moot, however, because even with the consequently exaggerated values associated with the plastic strain term in the MUS equation, the relative contribution of the plastic strain range term may be extremely small, compared with the elastic strain range term, to be of practical concern. That is, the plastic strain range term remains negligible even though it is exaggerated by the MUS equation. For this reason, only the elastic strain range versus life term needs to be considered. However, the slope of the elastic line for metals is in the region of $-0.12$ as indicated by the MUS. However, as seen from the essentially horizontal fatigue curves in Fig. 12.49, if there is a slope, it would be on the order of $-0.01$. Thus, the elastic strain range at $10^6$ cycles to failure would be only 13% less than the strain range at 1 cycle to failure. Note that the scatter in the tensile strength properties can be quite high ($\pm 40\%$), which causes the scatter in the strain range at one cycle to failure to be equally high. Thus, the scatter at one cycle is three-times greater than the mean effect that is to be expected. This combination virtually excludes the possibility of conducting a valid fatigue testing program. Scatter of results would overwhelm the ability to measure the fatigue curve.

It should be apparent that the MUS, or any of the other life prediction fatigue models developed for metals, is inapplicable for predicting the fatigue life of ceramics at low temperatures.

**Crack Propagation.** Pure fatigue crack growth in ceramics occurs at such rapid rates that it is quite difficult to conduct valid $da/dN$ tests. However, under certain elevated-temperature testing conditions cyclic crack growth can occur without relying on metallic-like cyclic plasticity at the crack tip. Three potentially responsible mechanisms can be involved. First is moisture adsorption by the ceramic material at the crack tip where the chemical potential is the highest. Second is high-temperature oxidation at the crack tip, and thirdly is crack opening wedging as the asperities on the newly formed crack faces interfere as they attempt to close as the tensile load is reduced to zero. Moisture adsorption and local oxidation lowers the local bonding strength, resulting in small amounts of crack extension on each cycle. Imperfect mating of the crack faces causes a wedging action at the crack tip, thus locally increasing the crack tip stresses and causing a small amount of crack extension due to brittle fracture.

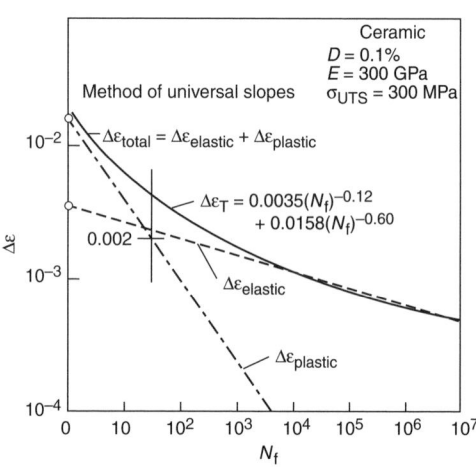

**Fig. 12.50** Method of universal slopes prediction of fatigue for a hypothetical ceramic

Until very recently, cycle-dependent failure of ceramics has been of limited engineering interest for the following reasons:

- Limited service experience
- Ceramics don't fatigue in the same classical mechanistic fashion as ductile metals.
- The effects of cyclic loading alone on degradation of a ceramic's strength are not pronounced; i.e., the slope of the fatigue crack initiation curve is either zero or very near zero and the $\log(da/dn)$ versus $\log(\Delta K)$ curve is very steep—on the order of 25 to 30.
- Ceramics exhibit far more scatter in tensile strength than metals do and hence large factors of safety (or very low levels of failure probability) would dictate the acceptance of only very low structural loading levels for ceramics. These loading levels would be sufficient to encompass the high-cycle fatigue endurance strength (even if such exists, it is not appreciably lower than the tensile strength).

### Other Aspects of Fatigue of Ceramics

In addition to the aforementioned factors, there are a number of features of brittle ceramics that would be expected to cause these materials to differ significantly for metallic behavior characteristics. Some of the most important of these characteristics are mentioned briefly here.

**Cyclic Stress-Strain and Cyclic Hardening and Softening.** Because of the limited ductility exhibited by high-strength ceramics at ambient temperatures, there are essentially no concerns for their cyclic stress-strain behavior. For engineering purposes, ceramic stress-strain response is linearly elastic. Thus, there is no concern for cyclic hardening or softening. At very high temperatures, ceramics may gain some ductility and begin to exhibit characteristics of metals at temperatures in the creep regime.

**Hysteresis Energy.** In addition, hysteresis energy considerations would have little to offer for the description of the fatigue resistance of ceramics because any inelasticity is too small to reliably quantify at all levels of cyclic loading.

**Multiaxial Stress-Strain States.** Multiaxial stress-strain influences on the fatigue and fracture behavior of brittle materials should be easier to predict than for ductile metals. This is so because brittle behavior tends to be governed simply by the maximum principal stress, not a function of the shear stresses as is the case for ductile metals. As the temperature of application for ceramics becomes quite high and some small amount of ductility is evident, more metalliclike behavior would be expected.

**Mean Stress Effects.** It is not expected that the classical mean stress effects observed for ductile metals will have applicability to brittle ceramics. Based on the knowledge that cycle-dependent failures can occur in ceramics, but principally only at high temperatures, and are due to a combination of environmental degradation, creep, and some indirect aspect of cycling, we would expect the maximum tensile stress in the cycle would dictate behavior. This would be in sharp contrast to ductile metal behavior for which a range of stress coupled with the mean stress (or maximum, or minimum—it matters not as there is an interrelationship between the various measures) is well known to be responsible for fatigue failures.

**Cumulative Damage.** To our knowledge, there is no experimental information available on cumulative fatigue damage for ceramics. Furthermore, little guidance is available from theoretical grounds.

## Brief Summation Regarding Fatigue of Ceramics

Few, if any, of the metallic-based fatigue models have direct applicability for describing or predicting the fatigue behavior of brittle ceramic materials. The observed cycle-dependencies of failure behavior of ceramics appear to be more fundamentally related to time-dependent phenomena such as creep (only at very high temperatures) and environmentally assisted interaction than to reversed plasticity that underlies the fatigue behavior of ductile metals (See Chapter 10, "Mechanism of Fatigue").

The mechanical behavior of composited ceramics materials are discussed briefly in the next section of this chapter, "Introduction to Composite Materials."

# Composite Materials

## Introduction

Here we concentrate on force-bearing structural composite materials engineered by humans. Natural-appearing composites, while interesting, and providing insight for developing engineered composite architectures, seldom possess the structural characteristics achievable through human intervention.

A composite material can be interpreted as being a structural subelement with constituents of two or more distinct monolithic materials. Typically, each constituent possesses one or more highly desirable properties but suffers from at least one inferior property. The purpose in assembling the constituents is to create a new entity with properties, as a whole, that is superior to those of the individual constituents. The predominant structural composite properties that have been achieved are higher tensile strength, greater stiffness, lower density, and greater resistance to catastrophic brittle fracture. Other desirable characteristics are also attainable. For example, composite electric power transmission wire with a strong steel core and high-thermal conductivity copper cladding offers a combination of strength and low electrical resistance not achievable from either monolithic metal alone.

Quite naturally, relatively high costs are encountered in producing composite materials. The original base cost of the constituents is only one of the many factors. In fact, the constituent costs may be higher still if their form must be precisely controlled and protected prior to fabrication into the resultant composite. Fabrication costs are considerably higher than for monolithic materials because the constituents must remain in proper proportion and, as necessary, in proper alignment to one another. Final machining costs may also increase because of the specialized practices that must be employed to deal with the nonuniform machining characteristics of the individual constituents. Repair of damage and scrap recovery costs are also much higher than for monolithic materials. Also, the mechanical testing and evaluation costs are considerably higher for composites than for monolithic materials. Consequently, composites are relegated to roles wherein overall production costs are not the dominant deciding factor. Several industries are currently taking advantage of the benefits of composite materials. These are the high-technology, high-cost consumer industries (sports equipment manufacturers being one of the dominant users) and aerospace industries. The consumer products are performance-driven (golf clubs, skis, boat hulls, and specialty automotive components*, etc.). Cost, as long as it is not prohibitive, is of secondary importance. Notable common exceptions to high-cost composites are plywood and steel bar-reinforced (rebar) concrete used in buildings, dams, bridges, and nearly every mile of concrete highway throughout the world. The authors' principal association and that of our colleagues, however, has been with composites for aerospace applications.

High strength and stiffness, combined with relatively low density (and hence even higher ratios of strength to density), have made composites highly attractive for construction of high-performance aircraft and spacecraft. Significant government funding from both the Department of Defense (DoD) and the National Aeronautics and Space Administration (NASA) has been in place for several decades to develop the necessary technology for routine utilization of composites of all sorts. Example aeronautical applications for airframes include wings, doors, and vertical stabilizers. Engine applications include such components as fan blades (used, for example, in the General Electric GE90 high-bypass gas turbine engine), fan blade containment hardware, and other relatively low-temperature applications involving polymeric composites. A massive effort was initiated in the late 1980s and the 1990s to develop high-temperature, polymer matrix composites (PMCs), ceramic fiber-reinforced metal matrix composites (MMCs), and ceramic matrix composites (CMCs) for compressor and turbine blades, vanes, combustors, exhaust nozzles, and so forth. The advances as well as the economic and inherent technological barriers that dogged the programs will be discussed in a chapter on high-temperature applications of MMCs in a planned future volume.

While the initial attraction to composite materials in aerospace application was due to the potential achievement of high monotonic strength-to-weight and stiffness-to-weight characteristics, such materials must also exhibit commensurately high resistance to fatigue and fracture. In this section we assess the applicability of the fatigue and fracture methods that

---

*For example, all Chevrolet Corvette (General Motors Corp., Detroit, MI) bodies (1953 to present) are formed of chopped, randomly oriented glass fibers in an epoxy matrix. Since the early 1990s, Corvettes have also sported a graphite fiber-reinforced composite single-leaf transverse-rear spring with stiffness characteristics tailored for desired performance. The result is significantly lower weight than conventional steel springs.

have been developed on the basis of metal technologies to the characterization and life prediction of engineered composites. First, we must review some of the basics of composites.

## Basic Types and Architectures of Composites

Structural composite materials are classified as either particulate- or fiber-reinforced. Particulate reinforcement may be viewed simply as the equivalent of fiber reinforcement wherein the fiber length is reduced to equal its diameter. Particulate composites are nominally isotropic in their properties, and they are produced in both sheet and solid three-dimensional form. Fibrous composites may be either two- or three-dimensional. The fiber layout may be either random or oriented. If the fibers are short, they are normally randomly oriented. Long fibers, however, are used in a variety of architectures. Most often, long fibers are aligned in parallel directions or are continuously wrapped around axisymmetric forms (e.g., pressure vessels and tubes), usually in a crisscross pattern. They are also used in more complex architectures such as two-dimensional woven sheets (like cloth), or stitched into three-dimensional configurations. Figure 12.51, from Ref 12.18, illustrates several of the fiber-reinforcing architectures.

**Particulate Composites.** Unless we are dealing with pure elemental materials, many of the monolithic engineering materials and alloys we use on a daily basis are, in the truest sense, composite materials. At their microscopic level, combinations of two or more elements may bond to form a new material and a synergistic improvement in overall properties. Some of the more sophisticated alloys (to be discussed in a planned second volume) owe their beneficial strength to second-phase particles that can precipitate within the alloy (via appropriate heat treatment) to form tiny, hard, and uniformly dispersed particles. The precipitates act as barriers to dislocation motion and hence improve yield and tensile strengths, although ductility may suffer. Small particle composites (particles on the order of just a few microns in size) are highly effective because of the strong interfacial bonds at this scale. Metallurgically generated composites of this type are referred to as *in situ* particulate composites. Unfortunately, much larger macroscopic particles do not possess as great an interfacial bonding strength, as do the small particles. This is principally because of the greater dimensions involved. The particles and the matrix have considerably different atomic characteristics. As the volume of a particle displaces an equal volume of matrix, the result is a greater degree of atomic mismatch at the interface. Thus, the atomic bonds across the interface are strained to a greater degree and the overall bond strength is reduced. Furthermore, because of the differences in stiffness and strength characteristics of the constituents, internal stress concentrations are created that further reduce the overall strength of the system. However, if these larger particles are stiffer and stronger than the matrix, there may remain a net benefit, strengthening and stiffening of the composite. A prime example of large particulate composite is common concrete. The matrix is cement and the reinforcing particles are small stones of varying diameter. In this case, the primary direction of the increase in strength is in compression.

The fabrication of engineered large-particulate composites creates some difficult challenges to ensure that the particles are uniformly distributed throughout the matrix because local particle volume fraction is important in imparting strength to the composite. Particulate composites, however, do not possess the same potential for high stiffness and strength, as do oriented fiber-reinforced composites. The principal attributes of particulate composites are moderate strength and isotropy of mechanical and thermal

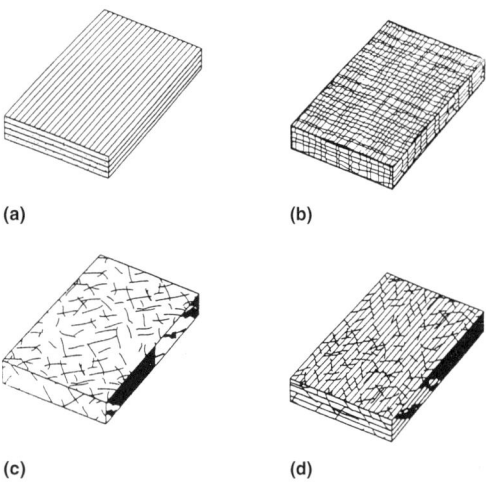

**Fig. 12.51** Various forms of fiber-reinforced composites. (a) Continuous fiber. (b) Woven fiber. (c) Chopped fiber. (d) Hybrid. Source: Ref 12.18

properties. They also have relatively low production costs.

**Fiber-Reinforced Composites.** Fibers made of glass, graphite, ceramics, and (occasionally) metals are the most common and effective reinforcements available for high stiffness and high strength. Sheets, or plies, of fiber-reinforced composites may be constructed using the fibers in different orientations:

- Randomly oriented short fibers
- Oriented short- and intermediate-length fibers
- Aligned continuous fibers (including continuous filament-wound fibers)
- Woven continuous fibers (2-D and 3-D)

The broadest classifications of matrix materials are:

- Polymeric
- Metallic
- Ceramic

Aside from concrete, the most widely used composite materials have a random, chopped-fiber architecture owing to the ease and consequent low cost of fabrication. (A fiber is usually considered to be effective as a fiber—rather than as an elongated particle—if its aspect ratio [length to diameter] is on the order of 10:1 or greater [see, for example, Ref 12.19]). A sketch of a three-dimensional random chopped fiber composite is shown in Fig. 12.51(c). The dominant application of random fiber composites is in sheets or layers of the material that are formed into three-dimensional shapes. The latter possesses nearly isotropic in-plane mechanical characteristics and is appropriate for structural loading that might occur in almost any direction. Thus, design methodologies for use of these composites are not appreciably different from those applied to monolithic materials. A significant deviation from metal-fabrication procedures involves additional reinforcement to the composite at attachment points. Another deviation is the necessity to deal with exposed fibers at cut edges and drilled holes. The machined surfaces must be protected from the environment to avoid attack of the exposed fiber/matrix interface. Polymeric matrix composites, for example, can absorb considerable moisture along the length of the interface. This is of little concern unless the exposure temperature exceeds the boiling point of water. Such exposure creates interior pockets of steam that expand rapidly and extensively, causing virtual explosions within a PMC. Interfacial bonds are destroyed and the PMC suffers local loss of strength. The overall strength advantages are thereby lost. Protective moisture barriers are required to prevent such catastrophic strength loss.

**Ply Layup Architectures.** Once plies have been manufactured, they may be stacked and bonded to form thicker material known as *laminates*. Alignment of the stacking can be controlled to achieve desired stiffness, strength, and other mechanical or thermal properties in preferred directions. For example, greatest unidirectional stiffness and strength are achieved by stacked plies with all fibers aligned parallel to the direction of anticipated loading. This architecture, however, has the highest degree of anisotropy in strength. It is quite weak in the directions perpendicular to the fiber direction. The degree of anisotropy in stiffness is shown in Fig. 12.52(a) from Ref 12.20. Similarly, the dramatic anisotropy in tensile strength is shown in Fig. 12.52(b) (Ref 12.21). Here, the angle between the fiber and loading direction was varied from 0 to 90°. Over this range, the tensile strength dropped to only 2.5% of its 0° strength when tested in the 90° direction. Even at an orientation of only 10° from zero, the strength was reduced to just 20%. These results clearly demonstrate the extreme care that must be taken to ensure accurate alignment of the fibers to achieve the maximum unidirectional strengthening benefit of the fibers.

A "quasi-isotropic" stacking involves plies stacked with certain fixed angles between the plies so that no single direction within the planes parallel to the plies has appreciably more strength than any other. The properties in the direction normal to the surface, however, would be essentially as low as the transverse direction in a unidirectional ply layup.

The stacking of plies to resist in-plane loading must involve symmetry of stacking to avoid undesired out-of-plane bending (warping). This is envisioned by considering a laminate composite with just two identical plies of differing orientation. As an in-plane load is applied, the stiffer-oriented ply would not stretch as far as the more compliant one, thus causing the laminate to curl concave toward the stiff side. Symmetric ply orientation about the midplane removes the problem. For this reason, most laminate composites use a symmetric layup. An example of a symmetric and a nonsymmetric layup is given in

Fig. 12.53, taken from Ref 12.22. From top to bottom, the symmetric layup shown is designated by the following sequence:

$$0°/0°/+45°/-45°/-45°/+45°/0°/0°$$

A corresponding nonsymmetric layup is:

$$-45°/+45°/+45°/-45°/0°/0°/0°/0°$$

A shorthand notation has been established to aid in more precise descriptions of the stacking sequence for specifying the most common architectures (see, for example, Ref 12.23). A symmetric layup is written as:

$$[0°/\pm 45]_S$$

The subscript "S" denotes symmetric layup, and only the first half need be written. If subscripts are inappropriate, a colon (:) precedes the full-size alphanumeric, i.e., for the preceding example, $[0:2/\pm 45]:S$. The degree signs are absent but understood, and the $\pm$ implies the $+$ orientation first and vice versa. For ply layups with a nonrepeating ply in a symmetric laminate with an odd number of plies, the odd ply is designated

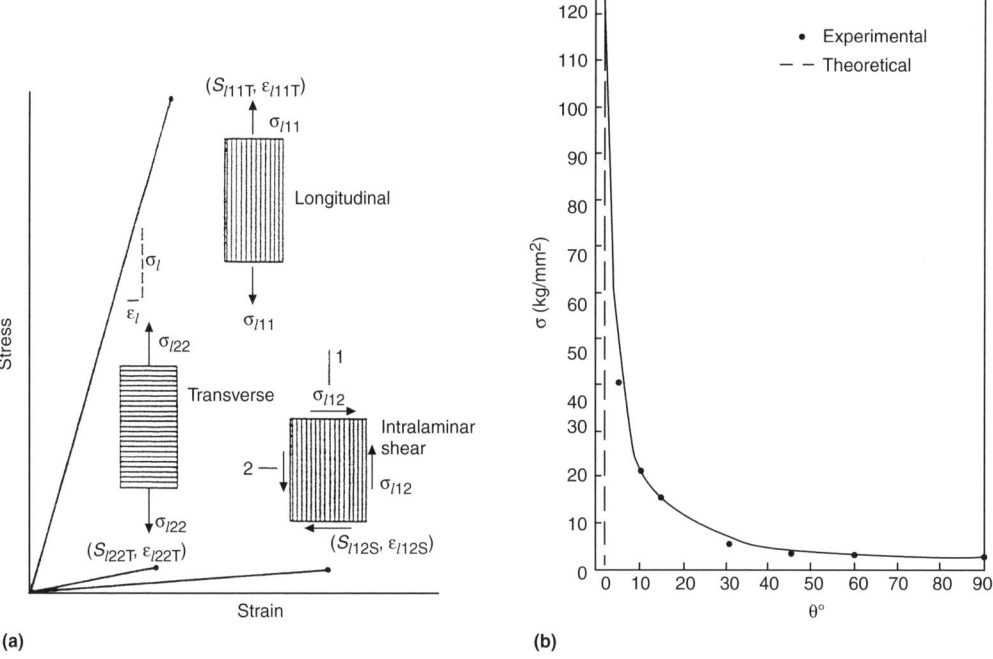

**Fig. 12.52** Degree of anisotropy in properties of unidirectional fiber-oriented composites. (a) Stiffness (Ref 12.20). (b) Tensile strength (Ref 12.21)

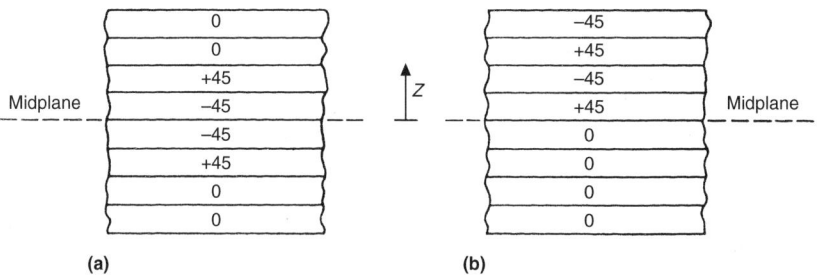

**Fig. 12.53** Two 8-ply laminates. (a) Symmetric. (b) Nonsymmetric

with a bar over the orientation of that ply or by use of a slash (/) after the ply. If the composite is a sandwich construction, the core location is given by the word *Core* as indicated in the designation:

[(0/45/90/Core/0/45/90)]

An example of a 50-ply stacking made up of exact repetitions of:

0°/0°/+45°/−45°/90°

is written as:

[(0₂/±45/90)₅]ₛ

Note that +90 and −90 are the same, so no distinction is made. A simple unidirectional 8 laminate composite is designated: [0₈]

Multiple laminate composites sometimes are referred to by the percentage of each orientation present. For the three respective examples, the descriptions are:

50% 0s, 50% ±45s

40% 0s, 40% ±45s, 20% 90s

100% 0

Before dealing with the mechanical and physical properties of ply layups, it is first necessary to understand the mechanics of a single ply, i.e., how the properties of the matrix and constituent fibers contribute to the properties of a ply.

**Mechanics of Composites.** The macroscopic structural and physical characteristics of composite plies (elastic stiffness, Poisson's ratio, strength, thermal expansion, thermal conductivity, etc.) can be computed directly using the principles of mechanics, heat flow, and knowledge of the principal properties of the constituents. The density of packing of the fibers (volume ratio fraction of fibers, $k_F$) is also very important and is usually expressed as a decimal, e.g., 0.25. If a stiff and strong fiber is high and is used in high volume, the stiffness and strength of the ply will be proportionately high in the fiber direction. The complementary volume fraction of the matrix, $k_M$, would correspond, in this case, to 0.75, i.e., $k_F + k_M \equiv 1.0$. Computed ply properties typically are referenced to the direction parallel to the direction of the fibers. Figure 12.54 (Ref 12.24) defines the *x-y* coordinate system (*z* is the mutually perpendicular direction) as well as the first and second principal directions (1 is parallel to the fiber in each ply). The general assumptions include linear elastic behavior of both constituents as well as perfect bonding between fiber and matrix.

The basic equations of mechanics for a composite ply have been presented, for example, by Chamis (Ref 12.20) and are reproduced here for completeness. Table 12.3 contains a summary of the more important equations. Equation nomenclature is given in Table 12.4. A sampling of constituent mechanical and thermal properties for several polymeric matrices and appropriate fibers are listed in Tables 12.5 and 12.6, respectively.

## Fatigue Modeling of Composites

In this book, we restrict our discussion of fatigue of composites primarily to room temperature behavior. Discussion of elevated-temperature behavior, especially under thermal cycling conditions, will be addressed in a planned future volume.

We learned earlier in this chapter that polymers exhibit fatigue behavior not unlike that of metals, so it might be expected that polymeric composites behave in a similar manner. However, while there are numerous similarities between metallic and polymeric matrix composites, the relation of monolithic constituent properties to their composited properties is not as straightforward. This is so because a composite has characteristics of a miniature structure whose component materials properties are intentionally varied from one location to the next. It is far from being uniform, homogeneous, and isotropic, as is the usual case for monolithic materials. In addition, the mechanical properties of the matrix and fiber are intentionally different. They are so different, in fact, that the two constituents will inevitably have widely different fatigue characteristics.

**Fatigue Resistance of Constituent Components.** For the MMCs of greatest current interest (and all CMCs), fibers are ceramic. In the previous section, reasoning was presented that ceramics do not exhibit classical characteristics of fatigue crack initiation and propagation. Furthermore, because fatigue involves highly localized, weakest-link processes, the "weakest" constituent invariably will be the weak link of the composite. For fatigue, the weakest link is the metallic matrix because its fatigue resistance is considerably lower than the maximum tensile strength of the nonfatiguing ceramic fiber. The matrix is not necessarily the weakest link in the case of monotonic composite stiffness and tensile strength, which are the characteristics most emphasized when designing a composite. With this as a starting point, a fatigue life prediction model was proposed (Ref 12.25) for metal matrix composites with ceramic reinforcing fibers. The model deals principally with the fatigue and cyclic stress-strain resistance of the matrix and how that resistance is altered by the presence of ceramic fibers. Although the model was created for predicting composite lives under thermal and mechanical cycling at high temperatures, it is equally applicable to isothermal fatigue resistance at room temperature.

The model separates the total fatigue life (macrocrack initiation, $N_f$) of the matrix into microcrack initiation, $N_i$, and microcrack propagation, $N_p$. The macrocrack initiation equation (Ref 12.25) retains the form of the Manson-Hirschberg Method of Universal Slopes equation (MUS) (Chapter 4, "Mean Stress"). In fact, the slopes of the elastic and plastic life relations remain the same, but the coefficients are taken to be general values $B$ and $C$:

$$\Delta\varepsilon_t = B(N_f)^{-0.12} + C(N_f)^{-0.60} \qquad \text{(Eq 12.3)}$$

If $B$ and $C$ are not known, they are approximated from the MUS equation ($B = 3.5\sigma_{UTS}/E$ and $C = D^{0.60}$).

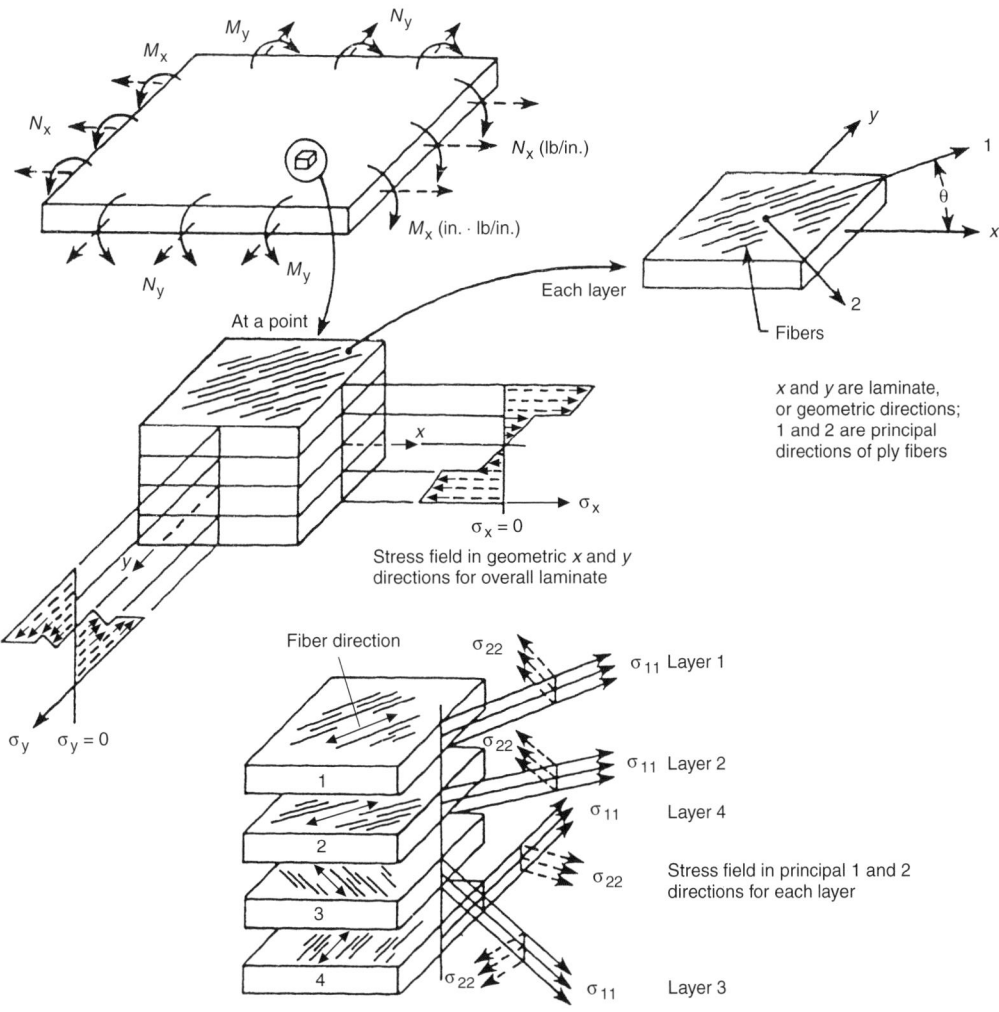

**Fig. 12.54** Defining plies, orientation, loads, and stresses and strains in laminates of continuous fiber-reinforced matrix material. Source: Ref 12.24

There is widespread acceptance that microcracks initiate very early in low-cycle fatigue, leaving the vast majority of the life to be spent in microcrack propagation. In high-cycle fatigue, however, microcracks initiate late in life and the microcrack propagation life is a small fraction of the total life. For the sake of a numerical example, it is not unreasonable to assume that, at the lowest possible life of $N_f = 1$:

$$N_i = 0.1 N_f \qquad \text{(Eq 12.4)}$$

$$N_p = 0.9 N_f \qquad \text{(Eq 12.5)}$$

whereas in a high cyclic fatigue at a life of $N_f = 10^7$:

$$N_i = 0.9 N_f \qquad \text{(Eq 12.6)}$$

$$N_p = 0.1 N_f \qquad \text{(Eq 12.7)}$$

By selecting the two points (on log-log coordinates) given by Eq 12.4 at $N_f = 1$ and Eq 12.6 at $N_f = 10^7$ and connecting a straight line between them, an approximate equation is obtained for the microcrack initiation fatigue life relation. Similarly, the two points given by Eq

### Table 12.3 Mechanical and thermal ply properties

Symbols are defined in Table 12.4.

| | |
|---|---|
| Partial volumes: | $k_f + k_m + k_V = 1$ |
| Ply density: | $P_l = k_{fpf} + k_m p_m$ |
| Matrix volume ratio: | $k_m = (1 - k_V)/[1 + (p_m/p_f)(1/\lambda_m - 1)]$ |
| Fiber volume ratio: | $k_f = (1 - k_V)/[1 + (p_f/p_m)(1/\lambda_f - 1)]$ |
| Weight ratios: | $\lambda_f + \lambda_m = 1$ |
| Ply thickness(a): | $T_l = \tfrac{1}{2} N_f d_f \sqrt{\pi/k_f}$ |
| Interply thickness: | $\delta_l = \tfrac{1}{2} [\sqrt{\pi/k_f} - 2] d_f$ |
| Interfiber spacing(a): | $\delta_s = \delta_l$ |
| Contiguous fibers(a): | $k_f = \pi/4 \sim 0.785$ |
| Longitudinal modulus: | $E_{\ell 11} = k_f E_{f11} + k_m E_m$ |
| Transverse modulus: | $E_{\ell 22} = \dfrac{E_m}{1 - \sqrt{k_f}(1 - E_m/E_{f22})} = E_{\ell 33}$ |
| Shear modulus: | $G_{\ell 12} = \dfrac{G_m}{1 - \sqrt{k_f}(1 - G_m/G_{f12})} = G_{\ell 13}$ |
| Shear modulus: | $G_{\ell 23} = \dfrac{G_m}{1 - \sqrt{k_f}(1 - G_m/G_{f23})}$ |
| Poisson's ratio: | $\nu_{\ell 12} = k_f \nu_{f12} + k_m \nu_m = \nu_{\ell 13}$ |
| Poisson's ratio: | $\nu_{\ell 23} = \dfrac{E_{\ell 22}}{2 G_{\ell 23}} - 1$ |
| Heat capacity: | $C_\ell = \dfrac{1}{p_\ell}(k_f p_f C_f + k_m p_m C_m)$ |
| Longitudinal conductivity: | $K_{\ell 11} = k_f K_{f11} + k_m K_m$ |
| Transverse conductivity: | $K_{\ell 22} = (1 - \sqrt{k_f}) K_m + \dfrac{K_m \sqrt{k_f}}{1 - \sqrt{k_f}(1 - K_m/K_{f22})} = K_{\ell 33}$ |
| For voids: | $K_m = (1 - \sqrt{k_v}) K_m + \dfrac{K_m \sqrt{k_v}}{1 - \sqrt{k_v}(1 - K_m/K_V)}$ |
| Longitudinal thermal expansion coefficient: | $a_{\ell 11} = \dfrac{k_f a_{f11} E_{f11} + k_m a_m E_m}{E_{\ell 11}}$ |
| Transverse thermal expansion coefficient: | $a_{\ell 22} = a_{f22} \sqrt{k_f} + (1 - \sqrt{k_f})(1 + k_f \nu_m E_{f11}/E_{\ell 11}) a_m = a_{\ell 33}$ |

(a) Square array of fibers. Source: Ref 12.20

12.5 at $N_f = 1$ and Eq 12.7 at $N_f = 10^7$, define the straight line describing the approximate microcrack fatigue propagation life relation.

For microcrack initiation, we can describe the fatigue life relation by:

$$\Delta\varepsilon_t = (0.80B)(N_i)^{-0.10} + (0.34C)(N_i)^{-0.50}$$

(Eq 12.8)

Correspondingly, the complementary microcrack propagation is described by:

$$\Delta\varepsilon_t = (1.30B)(N_p)^{-0.14} + (1.14C)(N_p)^{-0.70}$$

(Eq 12.9)

The separation of life was done in order to deal with individual influences on each stage caused by the presence of fibers and their flaws. While alternate equations representing the two phases of fatigue have been proposed for cumulative fatigue damage (See Chapter 6, "Cumulative Fatigue Damage"), they are not as convenient to work with as are Eq 12.8 and 12.9. The latter retain the familiar simple form of the total strain range versus life relation for macrofatigue crack initiation used throughout this book.

**Influences of Fibers on Matrix Properties and Behavior.** One of the most influential factors affecting the fatigue resistance of unidirectional fiber-reinforced composites is the angle between the fiber and the loading direction. The influence of angle on tensile strength was shown in Fig. 12.52(b). The experimental fatigue results of Ref 12.21 are shown in Fig. 12.55 for a unidirectional E-glass fiber-reinforced epoxy composite. Loading was tension-tension so that the ratio of mean stress to stress amplitude for the composite is $\bar{V}_\sigma = 1.2$ ($\bar{R}_\sigma = +0.1$). Note that the bar-over notation herein refers to composite variables and constants. Fatigue curves

**Table 12.4  Nomenclature for equations of Table 12.3**

| | Definition |
|---|---|
| $c$ | Heat capacity |
| $D$ | Diffusivity |
| $d$ | Diameter |
| $E$ | Modulus of elasticity |
| $G$ | Shear modulus |
| $K$ | Heat conductivity |
| $k$ | Volume ratio |
| $N_f$ | Number of filaments per roving end |
| $P$ | Property |
| $S$ | Strength |
| $s$ | Filament spacing |
| $T$ | Temperature |
| $t$ | Thickness |
| x,y,z | Structural Reference Axes |
| 1,2,3 | Ply material axes |
| $\alpha$ | Thermal expansion coefficient |
| $\delta$ | Interfiber, interply spacing |
| $E,\varepsilon$ | Fracture strain, strain |
| $\Theta$ | Ply orientation angle |
| $\lambda$ | Weight percent |
| $\rho$ | Density |
| $\delta$ | Stress |
| **Subscripts** | |
| f | Fiber property |
| C | Compression property |
| L | Ply property |
| m | Matrix property |
| S | Shear |
| T | Tension |
| v | Void |
| 0 | Reference property, temperature |
| ∞ | Saturation |
| 1,2,3 | Direction corresponding to 1,2,3 ply material axes |

Source: Ref 12.20

**Table 12.5  Matrix properties**

| Name | Symbol | Units | LM | IMLS | IMHS | HM | Polyimide | PMR |
|---|---|---|---|---|---|---|---|---|
| Density | $p_m$ | lb/in.³ | 0.046 | 0.046 | 0.044 | 0.045 | 0.044 | 0.044 |
| Tensile modulus | $E_m$ | 10⁶ psi | 0.32 | 0.50 | 0.50 | 0.75 | 0.50 | 0.47 |
| Shear modulus | $G_m$ | 10⁶ psi | – | – | – | – | – | – |
| Poisson's ratio | $v_m$ | – | 0.43 | 0.41 | 0.35 | 0.35 | 0.35 | 0.36 |
| Heat capacity | $C_m$ | Btu/lb/°F | 0.25 | 0.25 | 0.25 | 0.25 | 0.25 | 0.25 |
| Heat conductivity | $K_m$ | Btu/hr/ft²/°F/in. | 1.25 | 1.25 | 1.25 | 1.25 | 1.25 | 1.25 |
| Thermal Exp. Coef. | $\alpha_m$ | 10⁻⁶ in./in./°F | 57 | 57 | 36 | 40 | 20 | 28 |
| Diffusivity | $D_m$ | 10⁻¹⁰ in.²/sec | 0.6 | 0.6 | 0.6 | 0.6 | 0.6 | 0.6 |
| Moisture exp. coef. | $\beta_m$ | in./in./%M | 0.33 | 0.33 | 0.33 | 0.33 | 0.33 | 0.33 |
| Tensile strength | $S_{mt}$ | ksi | 8 | 7 | 15 | 20 | 15 | 3 |
| Compression strength | $S_{mc}$ | ksi | 15 | 21 | 35 | 50 | 30 | 16 |
| Shear strength | $S_{ms}$ | ksi | 8 | 7 | 13 | 15 | 13 | 8 |
| Tensile fracture strain | $G_{mt}$ | in./in. (%) | 8.1 | 1.4 | 2.0 | 2.0 | 2.0 | 2.0 |
| Compr. fracture strain | $G_{mc}$ | in./in. (%) | 15 | 4.2 | 5.0 | 5.0 | 4.0 | 3.5 |
| Shear fracture strain | $G_{ms}$ | in./in. (%) | 10 | 3.2 | 3.5 | 4.0 | 3.5 | 5.0 |
| Air heat conductivity | $K_V$ | Btu/hr/ft²/°F/in. | 0.225 | 0.225 | 0.225 | 0.225 | 0.225 | 0.225 |
| Glass trans. temp. (Dry) | $T_{GD}$ | °F | 350 | 420 | 420 | 420 | 700 | 700 |

LM, low modulus; IMLS, intermediate modulus low strength; IMHS, Intermediate modulus high strength; HM, High modulus. Thermal, hygral, compression, and shear properties are estimates only; $G_m = E_m/2(1 + v_m)$. Source: Ref 12.20

are shown for off-axis angles of 0, 5, 10, 30, and 60 degrees. Even small off-axis angles of 5 to 10 degrees cause a drastic reduction in the fatigue strength compared with 0 degree loading. With the off-axis angle of 60 degrees, the fatigue strength at $10^5$ cycles to failure has decreased to about 2.5 kg/mm$^2$ from a value of about 80 kg/mm$^2$ at 0 degree orientation.

The fact that fibers are present in a composite imparts changes in both the flow and failure response of the matrix. A fully developed MMC life prediction method must deal with these induced changes. A listing of the most significant mechanical influences of a fiber on the surrounding matrix is given in Table 12.7 for isothermal fatigue. Each factor is identified according to whether it influences the microcrack initiation phase or microcrack propagation phase of the macrocrack initiation life. A few of the influences can be treated analytically; others require future development. Detailed discussions of these factors, and others, are given in Ref 12.25. There are also mechanical influences of the matrix on the behavior of the fibers as discussed subsequently.

**Mechanical Effects of Matrix on Fibers.** It is necessary to examine the mechanical influence of the matrix on the response characteristics of the reinforcing fibers. Fibers of primary concern are elastic and brittle. They fail progressively throughout the fatigue life of the MMC as a result of the continual transferring of tensile stresses from the matrix material as it cyclically deforms and cracks. Cyclic relaxation of mean (residual) stresses and strain hardening or softening of the matrix results in various scenarios of behavior depending on the combination of operative conditions.

Selecting the specific force-controlled condition of zero-to-maximum tension is instructive to demonstrate a qualitative micromechanics analysis of how the fibers are influenced by the matrix behavior. Cyclic tensile mean stress relaxation, as well as cyclic strain softening of the matrix, causes a transfer of additional tensile stress (and hence, strain) to the fibers, loading them closer to their critical fracture strain. If the matrix cyclically strain hardens, it will tend to carry a larger portion of the total peak tensile load. This effect is counteracted, and possibly overshadowed, by the cyclic relaxation of any initial tensile residual stress in the matrix. All else being equal, cyclic strain softening of the matrix will tend to strain the fibers in tension to a greater extent than cyclic strain hardening. In addition to cyclic hardening, softening, and relaxation of mean stress (either by cyclic or time-dependent means), fibers are forced to carry a

**Table 12.6  Fiber properties**

| Name | Symbol | Units | Boron | Hms | AS | T300 | kEV | S-G | E-G |
|---|---|---|---|---|---|---|---|---|---|
| Number of fibers/end | $N_f$ | ... | 1 | 10,000 | 10,000 | 3000 | 580 | 204 | 204 |
| Fiber diam. | $d_f$ | in. | 0.0056 | 0.0003 | 0.0003 | 0.0003 | 0.00046 | 0.00036 | 0.00036 |
| Density | $P_f$ | lb/in.$^3$ | 0.095 | 0.070 | 0.063 | 0.064 | 0.053 | 0.09 | 0.090 |
| Longitudinal tensile modulus | $E_{f11}$ | 10$^6$ psi | 58 | 55.0 | 31.0 | 32.0 | 22 | 12.4 | 10.6 |
| Transverse tensile modulus | $E_{f22}$ | 10$^6$ psi | 58 | 0.90 | 2.0 | 2.0 | 0.6 | 12.4 | 10.6 |
| Longitudinal shear modulus | $G_{f12}$ | 10$^6$ psi | 24.2 | 1.1 | 2.0 | 1.3 | 0.42 | 5.17 | 4.37 |
| Transverse shear modulus | $G_{f23}$ | 10$^6$ psi | 24.2 | 0.7 | 0.8 | 0.7 | 0.22 | 5.17 | 4.37 |
| Longitudinal Poisson's ratio | $v_{f12}$ | ... | 0.20 | 0.20 | 0.20 | 0.20 | 0.35 | 0.20 | 0.22 |
| Transverse Poisson's ratio | $v_{f23}$ | ... | 0.20 | 0.25 | 0.25 | 0.25 | 0.35 | 0.20 | 0.22 |
| Heat capacity | $C_f$ | Btu/lb/°F | 0.31 | 0.20 | 0.20 | 0.22 | 0.25 | 0.17 | 0.17 |
| Longitudinal heat condition | $K_{f11}$ | Btu/hr/ft$^2$/°F/in. | 22 | 580 | 580 | 580 | 1.7 | 21 | 7.5 |
| Transverse heat condition | $K_{f22}$ | Btu/hr/ft$^2$/°F/in. | 22 | 58 | 58 | 58 | 1.7 | 21 | 7.5 |
| Longitudinal thermal expansion coefficient | $\alpha_{f11}$ | 10$^{-6}$ in./in./°F | 2.8 | −0.55 | −0.55 | −0.55 | −2.2 | 2.8 | 2.8 |
| Transverse thermal expansion coefficient | $\alpha_{f22}$ | 10$^{-6}$ in./in./°F | 2.8 | 5.6 | 5.6 | 5.6 | 30 | 2.8 | 2.8 |
| Longitudinal tensile strength | $S_{ft}$ | ksi | 600 | 250 | 300 | 350 | 400 | 600 | 40 |
| Longitudinal compression strength | $S_{fc}$ | ksi | 700 | 200 | 260 | 300 | 75 | ... | ... |
| Shear strength | $S_{fc}$ | ksi | 100 | ... | ... | ... | ... | ... | ... |

Note: Transverse, shear, and compression properties are estimates inferred from corresponding composite properties. Source: Ref 12.20

greater portion of the imposed tensile load because of matrix cracking perpendicular to the fibers. As the matrix fatigues due to microcrack initiation and microcrack propagation, more and more of the applied load is transferred to the fibers contained within the plane of the cracks. Furthermore, as fibers begin to crack, the load they had been carrying is transferred to the remaining unbroken fibers, thus increasing their peak tensile stresses and strains, pushing them closer to ultimate fracture. Failure of the composite into two pieces occurs when the remaining axial fibers are required to carry stresses and strains beyond their critical value. This important fact forms the basis for the "fiber-dominated" approach to composite life prediction proposed by Johnson (Ref 12.26, 12.27). An integral part of the MMC life prediction method is the inclusion of an upper strain limit that will cover those conditions wherein the life of the material is clearly governed by the tensile strain capacity of the fibers. This is represented in a fatigue curve as a horizontal line at the value of the critical tensile fracture strain of the fiber. To place this strain value on a strain range axis requires that a micromechanics strain analysis of the MMC be used to determine the initial mean strain in the fibers.

Because of the importance of mean stresses and strains on the matrix and fiber failure behavior, a quantitative analysis is required to determine how the initial mean or residual stresses cyclically relax during fatigue loading. Mean stress effects on fatigue life of composites must be evaluated as well. These analyses are described qualitatively next.

**Mean Stress Effects in MMCs.** The need to understand mean stress effects on the fatigue behavior of MMCs is at least as important as it is for any other structural material. It is perhaps more so for three important reasons; first, every test sample of an MMC will have residual stresses in both the matrix and fibers that are an inherent result of cooling down from a high consolidation temperature. The different coefficients of thermal expansion of the matrix and the fibers are the cause of these unavoidable initial residual stresses. Second, most MMC fatigue testing to date has been conducted under tension-tension loading, leading to concern as to

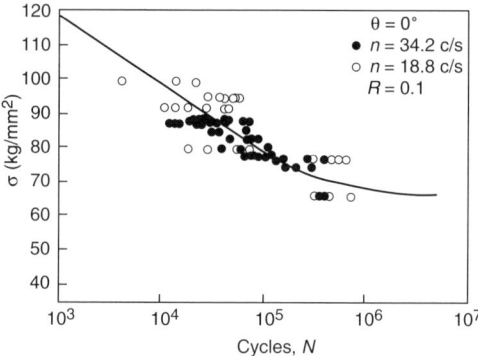

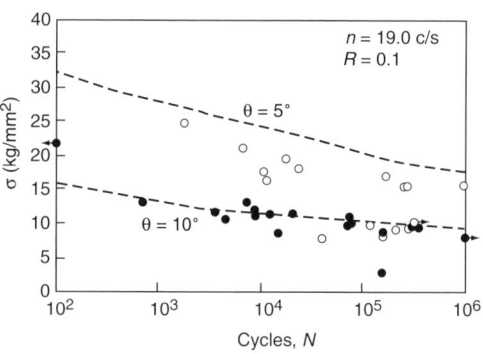

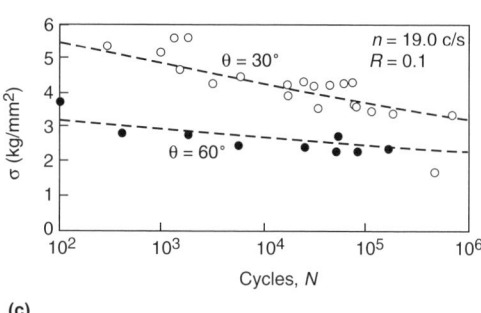

**Fig. 12.55** Influence of off-axis loading on fatigue strength of fiberglass-epoxy unidirectional composite. Source: Ref 12.21. (a) 0°. (b) 5 and 10°. (c) 30 and 60°

**Table 12.7 Factors associated with fibers mechanically influencing matrix fatigue response**

| Influencing factor | Is there an effect? | |
| --- | --- | --- |
| | Microinitiation | Micropropagation |
| Mean and residual stresses | Yes | Yes |
| Multiaxial stress state | Yes | Yes |
| Off-axis fibers | Yes | Yes |
| Internal stress concentrations | Yes | No |
| Multiple initiation sites | Yes | Yes |
| Nonuniform spacing | Yes | Yes |
| Interfacial layers | Yes | No |
| Fractured fibers | Yes | No |
| Fiber debonding | No | Yes |
| Fiber crack retardation | No | Yes |
| Fiber bridging | No | Yes |

Source: Ref 12.25

how to generalize the fatigue behavior to tension-compression and compression-compression loadings, as is routinely done for monolithic metals. Third, because of the inherent structure of an MMC, local inelasticity can occur in the matrix at peak loads, resulting in redistribution of internal stresses (between fiber and matrix) while not affecting the composite stress. Such internal redistribution alters the mean stresses in both the matrix and the fibers, and hence affects the fatigue resistance of the MMC.

Because anticipated service loadings for components made of MMCs will invariably involve a multitude of operating conditions, cyclic mean stresses undoubtedly range from large to small and from tensile to compressive. Here we address the need to model adequately the mean stress effects in MMCs over a range of practical operating conditions.

Mean stress effects on fatigue of MMCs are relatively unexplored. The sparse data are limited generally to tensile mean stresses under tension-tension loading. Compressive mean stress has largely been unexplored, principally due to the thin, buckle-prone specimen configurations most commonly employed. Thin specimens have been preferred because they can be produced at a fraction of the cost of thick specimens. By contrast, costs of monolithic engineering metal specimens are comparatively unaffected by section thickness. Consequently, mean stress effects for monolithic metals have been studied extensively over the past century under tensile and compressive loading. Numerous mean stress models for monolithic materials have resulted, as discussed in Chapter 4, "Mean Stress." Mean stress modeling for MMCs, however, is in its infancy owing to the lack of experimental guidance.

We now present analysis of some of the first systematically generated tension-compression mean stress results. Data were generated at NASA's Glenn Research Center using [0] continuous-fiber reinforced MMCs (Ref 12.28, 12.29). The intent is fourfold:

- To document mean stress effects over a broad range for a representative MMC, silicon carbide fiber (SCS-6) imbedded in a titanium matrix (Ti-15-3)
- To evaluate applicability of monolithic material mean stress models to this MMC
- To demonstrate analytically the direct applicability of certain monolithic mean stress models to MMCs loaded in the elastic regime
- To outline and discuss an approach for dealing with mean stress effects in the presence of inelastic deformation of the matrix in MMCs

**Brief Review of Mean Stress Effects in Monolithic Metals.** As documented in Chapter 4, "Mean Stress," and in Ref 12.30 and 12.31, the effect of mean stresses in fatigue on the lifetime of monolithic metals has been studied for well over a century. Extensive mean stress data have been generated, and numerous mean stress models have been proposed. Each model has enjoyed a modicum of success in representing selected sets of data. A common feature of these models is their reflection of the general observation that tensile mean stresses are detrimental, and compressive mean stresses are beneficial, compared with a base of zero mean stress. No single model* has been able to represent the disparate behavior exhibited by various monolithic metals. The models differ quantitatively in the extent to which detrimental or beneficial effects are ascribed.

Mean stress effects have been studied most extensively in the nominally elastic, high-cycle fatigue regime. Here, fatigue life, at constant stress amplitude, can vary over orders of magnitude depending only on the magnitude and sign of the mean stress. In these cases mean stresses are stable and not prone to cyclic relaxation. In the low- to intermediate-cycle fatigue range there is diminished concern for mean stress effects due to inelasticity (plasticity or creep) that causes cyclic relaxation of initial mean stresses under conditions of locally imposed strain control. However, under conditions of force control in the low- to intermediate-cycle fatigue regime, large cyclic life losses are possible. The intervention of a radically different damaging mechanism, *strain ratcheting*, is possible. Ratcheting is the gradual cyclic increase of strain in one direction that, if not held in check, can lead to excessive localization and further intensification of damage. If a tensile mean load is maintained, the deformation is unconstrained, and localized necking can ensue, with immediate failure by rupture. A compressive

---

*A unified mean stress model was proposed by the senior author in the early 1980s that mimics the desirable features of numerous models. The first published application of this model appeared in the doctoral thesis of Prof. Manson's graduate student, Kurt Heidmann (Ref. 12.32). Since that time further efforts have produced the promising comprehensive model described in Chapter 4 "Mean Stress."

mean load can lead to localized cycle-dependent buckling and the potential collapse of the structure. While they are cyclically induced, ratcheting and buckling failure mechanisms are *not* in the same category as classical fatigue failure mechanisms (crack initiation and propagation). Integrity of structural configuration is not maintained in the presence of excessive ratcheting. Thus, in the consideration of mean cyclic loading effects, four distinct behavioral patterns must be considered:

- Classical mean stress effects during nominally elastic loading
- Cyclic relaxation of initial residual stresses and actively imposed mean stresses
- Ratcheting in tension with potential for excessive deformation and localized necking
- Ratcheting in compression with potential for excessive deformation and buckling

During cyclic deformation-controlled structural or coupon sample loading, only the first two patterns of behavior are of importance. For cyclic force-controlled structural or specimen testing, all four responses are possible. For elastic material response, the very circumstances under which most mean stress models had been developed, only the first is important. We address the classical elastic mean stress conditions in MMCs before advancing to the more complex patterns.

**Mean Stress Modeling for MMCs.** An elastic loading condition for both the matrix and the fibers is our starting point for adopting a fatigue mean stress model for MMCs. Most current high-performance MMCs of interest use continuous ceramic fibers that are not prone to cycle-dependent cracking, particularly at temperatures well below their creep and oxidation regime (Ref 12.33). The ceramic fibers thus can be considered to respond elastically. They fracture in a brittle fashion when a maximum critical tensile strain (or stress) is reached. Hence, only the metallic matrix undergoes fatigue in the classical sense of plasticity-governed, cycle-dependent, crack initiation and propagation (See Chapter 10, "Mechanism of Fatigue"). It follows that if the metallic matrix controls the cycle-dependent fatigue response of the composite, it is reasonable to expect mean stresses to exert their fatigue effect on composite life via their effect on the matrix fatigue life. This expectation will be demonstrated analytically. Thus, we can adopt the position that if a mean stress model is known to apply to a monolithic metal, that model is expected to work equally well for an MMC with a matrix of that metal. Two conditions, however, must be met for this expectation to be demonstrated analytically:

- The failure mechanism is indeed classical fatigue crack initiation and propagation in the metallic matrix.
- The stress-strain conditions in the matrix are nominally elastic so there is no shift, during testing, in the relation between the value of mean stress in the matrix and that in the composite.

To assess reasonableness of these assumptions, we examine the mean stress results reported in Ref 12.28 and 12.29 for SCS6/Ti-15-3 at 427 °C (800 °F). The test temperature is somewhat above room temperature. However, it is low enough and the cyclic frequency is high enough to preclude time-dependent effects from operating in either constituent. The response of the matrix was nominally elastic for the test conditions studied. The Halford-Nachtigall Modified Morrow (HNMM) mean stress model (Ref 12.34) discussed in Chapter 4, "Mean Stress," is compared with the experimental results in Fig. 12.56. Note that for conditions of elastic loading, the HNMM model is identical to the original Morrow model (Ref 12.35). Furthermore, because of the low ductility of the composite (nominally one percent), the true fracture strength $\sigma_{TFS}$ used in the original Morrow model is only 1% greater than the ultimate tensile strength $\sigma_{UTS}$ used in the Modified Goodman model (Ref 12.36). (Pertinent tensile properties

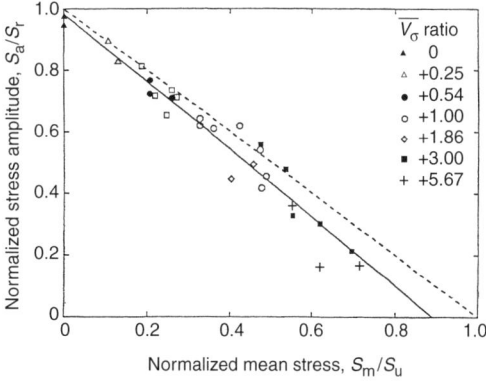

**Fig. 12.56** Halford-Nachtigall Modified Morrow, original Morrow, and Modified Goodman correlation of mean stress results for SCS6/Ti-15-3 MMC at 427 °C (800 °F). The dashed line represents all three models, while the solid line is the best curve fit to the data. Source: Ref 12.29

at 427 °C (800 °F) were measured at NASA-Glenn Research Center to be: $\sigma_{UTS} = 1450$ MPa and true ductility, $D \approx 0.01$. Since $\sigma_{TFS} \approx \sigma_{UTS}(1 + D)$, $\sigma_{TFS} \approx 1465$ MPa.) Consequently, there is only a 1% difference in the normalized mean stresses between the three models.

The three-stress normalized equations are:
H-N Modified Morrow:

$$(\sigma_a/\sigma_R) + (\sigma_m/\sigma_{TFS}) = 1 \quad \text{(Eq 12.10)}$$

Morrow (original):

$$(\sigma_a/\sigma_R) + (\sigma_m/\sigma_{TFS}) = 1 \quad \text{(Eq 12.11)}$$

Modified Goodman:

$$(\sigma_a/\sigma_R) + (\sigma_m/\sigma_{UTS}) = 1 \quad \text{(Eq 12.12)}$$

where $\sigma_a$ is stress amplitude with mean stress, $\sigma_R$ is stress amplitude with zero mean stress at same life as for $\sigma_a$, $\sigma_m$ is mean stress corresponding to $\sigma_a$, $\sigma_{TFS}$ is true fracture strength, and $\sigma_{UTS}$ is ultimate tensile strength.

It is observed in Fig. 12.56 that the predicted mean stress results from Eq 12.10 to 12.12 are reasonably accurate up to a mean stress ratio $V_\sigma = 1.0$ ($R_\sigma = 0$). As shown by the table in the figure, the data at higher mean stress ratios fall below the dashed predicted line. A plausible explanation for the discrepancy is that at the highest mean stress ratios, the maximum tensile stresses are the greatest, and hence there are greater propensities for yielding of the matrix as well as some degree of mean stress relaxation. If that is the case, the basic assumption of elastic matrix response is violated. Thus, the relation between mean stress in the composite and the mean stress in the matrix also is violated. Reference 12.29 showed that in the region of $V_\sigma > 1.0$ ($R_\sigma > 0$) the Normalized Soderberg model* (Ref 12.37) gave the best correlation with the data, but only if the 0.02% offset yield strength was used. This correlation is indicative of some matrix yielding at the highest mean stress levels. It is noted that the Normalized Soderberg model is identical in form to the three models discussed previously. As shown in Fig. 12.57, two other popular mean stress models, the Smith-Watson-Topper parameter (Ref 12.38) and the Walker parameter (Ref 12.39), did not adequately correlate the mean stress fatigue results for this composite under these test conditions.

**Basis for Applicability of Monolithic Mean Stress Models to MMCs.** In the following, conditions are examined under which a mean stress model that is applicable for a monolithic metal can be applied directly to a ceramic fiber-reinforced MMC using that metal as its matrix. The following assumptions are imposed:

- Fibers are of a typical high-strength, high-modulus ceramic.
- Fibers are aligned with the direction of uniaxial loading, i.e., [0] orientation.
- Fibers are uniformly distributed and are fully bonded to the in situ matrix.
- Fibers behave in a linear-elastic manner up to their fracture point.

The in situ matrix deformation behavior is taken as linearly elastic, although extension to inelastic behavior is possible. The underlying premise is that the fatigue life of the composite is governed by the in situ metallic matrix, the only fatigue-prone constituent. However, it must be understood that in situ matrix fatigue behavior may differ from stand-alone behavior because of the influence of the fibers (see Table 12.7). To distinguish between the characteristics of the composite, in situ matrix, and stand-alone matrix, the following notation is adopted for the variables:

$\bar{x}$ overbar for the composite
$|x|$ vertical lines for the in situ matrix
$x$ no markings for stand-alone matrix

*Linear Elastic Matrix Behavior.* It is shown subsequently that the Halford-Nachtigall Modified Morrow (HNMM) model (and by analogy for linear elastic behavior, the original Morrow and the Modified Goodman models) applies equally well to the composite as to the matrix. Thus, credence is added to the notion that the MMC fatigue life is indeed governed by the in situ matrix fatigue behavior and the mean stress as it is applied to the matrix. We have selected the HNMM model because of its simplicity in expressing fatigue lives with or without mean stress. As presented in Chapter 4, "Mean Stress," the original Morrow equation for the stand-alone matrix was modified (Ref 12.34) and presented in the following useful form:

$$(N_{fm})^{-b} = (N_f)^{-b} - V_\varepsilon \quad \text{(Eq 12.13a)}$$

---

*The Normalized Soderberg model has the same form as the Modified Goodman equation. However, Soderberg uses the yield strength $\sigma_Y$ to normalize the mean stress instead of the ultimate tensile strength, i.e., $(\sigma_a/\sigma_R) + (\sigma_m/\sigma_Y) = 1$.

where $N_{fm}$ is fatigue life in the presence of mean stress $\sigma_m$ and alternating stress $\sigma_a$; $N_f$ is fatigue life when no mean stress is present and the alternating stress is $\sigma_a$; $b$ is fatigue strength exponent (log-log slope of elastic strain range versus fatigue life), known as the Basquin exponent (a number generally in the range of $-0.05$ to $-0.20$); and $V_\sigma$ is the ratio of mean to alternating stress, $\sigma_m/\sigma_a$ ( = inverse of the traditional fatigue $A$ ratio, also related to the stress ratio $R_\sigma$ ( = $\sigma_{min}/\sigma_{max}$), by the relation, $V_\sigma = (1 + R_\sigma)/(1 - R_\sigma)$

It is assumed that the form of Eq 12.13(a) is directly applicable to the in situ matrix material and consequently to the composite:

$$(|N_{fm}|)^{-|b|} = (|N_f|)^{-|b|} - |V_\sigma| \quad \text{(Eq 12.13b)}$$

$$(\overline{N_{fm}})^{-\bar{b}} = (\overline{N_f})^{-\bar{b}} - \overline{V_\sigma} \quad \text{(Eq 12.13c)}$$

The fatigue curves in Fig. 12.58 (Ref 12.25) show $\bar{b} \cong b$, implying $|b| = \bar{b}$. No further distinction is made and the overbar and vertical

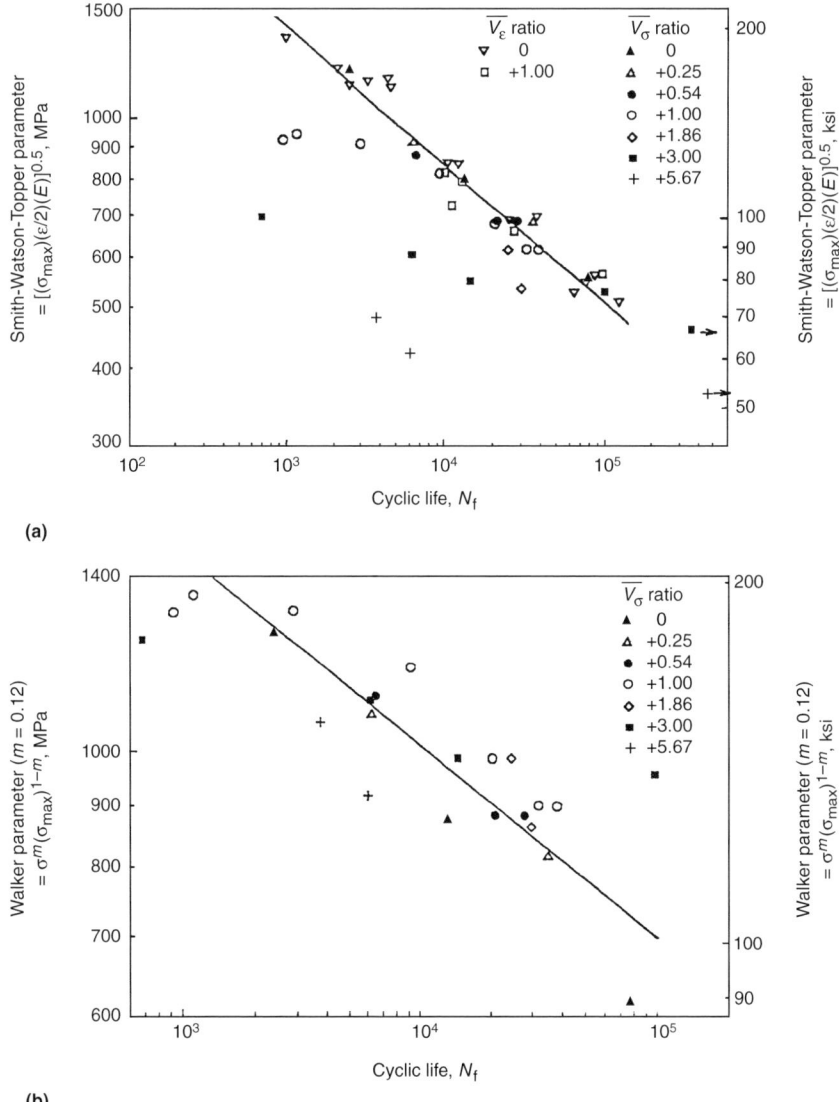

**Fig. 12.57** Correlation of mean stress results for SCS6/Ti-15-3 MMC at 427 °C (800 °F). (a) Smith-Watson-Topper parameter (Ref 12.38). (b) Walker parameter (Ref 12.39)

bars will be omitted from this material constant in the following.

It can be shown, based upon the *rule of mixtures*, equilibrium, strain compatibility, and linear elasticity, that the axial stress in the matrix $|\sigma|$ is related to the axial stress in the composite $\bar{\sigma}$ by the simple linear equation:

$$|\sigma| = \lambda \bar{\sigma} + |\sigma_r| \qquad \text{(Eq 12.14a)}$$

where the subscript $r$ denotes the residual, or mean, stress in the in situ matrix. Correspondingly, the mean and alternating stresses in the in situ matrix and composite are related by:

$$|\sigma_m| = \lambda \bar{\sigma}_m + |\sigma_r| \qquad \text{(Eq 12.14b)}$$

$$|\sigma_a| = \lambda \bar{\sigma}_a \qquad \text{(Eq 12.14c)}$$

where:

$$\lambda = \left[ \frac{1}{k_F \left( \frac{E_F}{E} - 1 \right) + 1} \right]$$

and $k_f$ is the volume fraction of fibers in composite, $E_f$ is the elastic modulus of fiber, $E$ is the elastic modulus of matrix, and $\sigma_r$ is the initial residual stress in the matrix due to cooldown from high composite processing temperature (usually tensile, because the coefficient of thermal expansion of the stiff fibers is invariably less than that of the matrix). A numerical example for $\lambda$ is 2/3, for $k_f = 1/4$ and $E_f/E = 3$.

For completely reversed loading of the composite, $\bar{V}_\sigma = 0$ and the fatigue life is $(\overline{N_f})$. Because the in situ matrix is assumed to govern the life of the composite and the in situ matrix life is $|N_{fr}|$ in the presence of the tensile residual stress $|\sigma_r|$, it follows that these two lives are equal:

$$(\overline{N_f}) \equiv (|N_{fr}|) \qquad \text{(Eq 12.15)}$$

Substituting Eq 12.15 into Eq 12.13c with:

$$(\overline{N_{fm}})^{-b} = (|N_{fr}|)^{-b} - \bar{V}_\sigma \qquad \text{(Eq 12.16)}$$

Recognizing that for the in situ matrix material with residual (mean) stress $|\sigma_r|$, the fatigue life $(|N_{fr}|)^{-b} = (|N_f|)^{-b} - |V_{\sigma_r}|$. Substituting into Eq 12.16:

$$(|N_{fr}|)^{-b} = (|N_f|)^{-b} - |V_{\sigma_r}| - \bar{V}_\sigma \qquad \text{(Eq 12.17)}$$

Noting that

$$|V_{\sigma_r}| = \frac{|\sigma_r|}{|\sigma_a|} \text{ and } \bar{V}_\sigma = \frac{\bar{\sigma}_m}{\bar{\sigma}_a}$$

and utilizing Eq 12.14(b) and (c):

$$-|V_{\sigma_r}| - \bar{V}_\sigma = -\frac{|\sigma_r|}{|\sigma_a|} - \frac{\bar{\sigma}_m}{\bar{\sigma}_a} = -\frac{|\sigma_r|}{|\sigma_a|} - \frac{\frac{|\sigma_m| - |\sigma_r|}{\lambda}}{\frac{|\sigma_a|}{\lambda}} = -\frac{|\sigma_m|}{|\sigma_a|} = |V_\sigma|$$

(Eq 12.18)

Thus, Eq 12.17 becomes:

$$(|N_{fr}|)^{-b} = (|N_f|)^{-b} - |V_\sigma| \qquad \text{(Eq 12.19)}$$

However, the right-hand side is known from Eq 12.15 to be $(\overline{N_f})^{-b}$. Thus, it is seen that the shift in life of the composite due to the mean stress in the composite is the same as the shift in life of the in situ matrix due to the corresponding shift in its mean stress. Another way of interpreting this result is to state that if the experimental results show that the composite obeys the monolithic metal mean stress equation, the implication is that the composite fatigue life is indeed governed by the fatigue resistance of the in situ matrix.

*Inelastic Matrix Behavior.* Similar analyses have been applied to cases involving cyclic re-

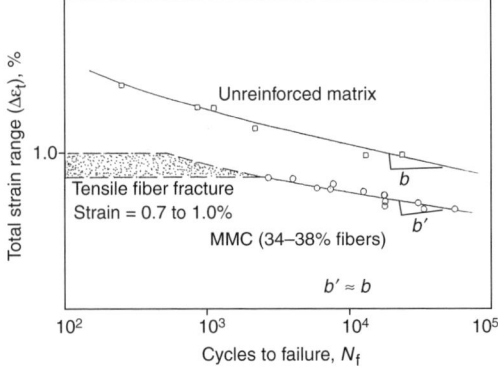

**Fig. 12.58** Fatigue curves for the matrix alloy Ti-15-3 and the SCS6/Ti-15-3 MMC tested at 427 °C (800 °F). Note similar slopes ($b' \cong b$) of the two curves at the higher lives. Source: Ref 12.25

laxation of the mean stress in both the composite and the in situ matrix. Under certain conditions, when the relaxation of the mean stress in the in situ matrix can be tracked analytically or even experimentally, the same conclusion can be drawn; i.e., the mean stress effect on the matrix life is indeed shown to control the corresponding mean stress effect on the composite life. We have also analyzed cases wherein ratcheting of the matrix material due to inelasticity causes the composite to ratchet under force-controlled testing. If sufficient ratcheting occurs, the elastic fibers are strained to higher and higher tensile strain values until brittle fracture of the fibers occurs. An example of such behavior is given in the following. Such behavior is particularly prevalent under cyclic thermal and mechanical loading at temperatures within the creep regime.

**Comparison of MMC Fatigue Results for Strain-Controlled and Force-Controlled Tests.** Attention is now directed toward experimentally observed differences between strain-controlled and load-controlled testing of the MMC SCS6/Ti-15-3. The principal concern is for the differences in behavior when the matrix undergoes inelastic deformation. The experimental results are plotted in Fig. 12.59. As expected, there is no difference between the stress- and strain-controlled $\overline{V} = 0$ results; therefore, a single line is fitted to the combined data. It is observed that for any cyclic strain range, the fatigue lives are significantly reduced in the presence of tensile mean stresses. Under force control, for $\overline{V}_\sigma = +1$, the mean stress on the composite is forced to be constant, and the life is reduced by nearly a factor of 20. Under strain control there is the opportunity for mean stress relaxation in the matrix and hence lower effect on life is observed. Additionally, the stress-controlled results at this mean ratio show a region of life at the highest strain ranges (on the order of 0.7 to 0.8% as seen in Fig. 12.59) that is very sensitive to applied strain range. Such behavior has been observed in titanium matrix composites when tested under force-controlled conditions (see, for example, Ref 12.26 and 12.27). It is a reflection of premature failures induced by cyclic tensile strain ratcheting of the matrix. The matrix ratcheting gradually transfers more and more of the force and strain to the fibers until they fracture prior to the development of fatigue cracks in the ductile matrix. Figure 12.60 shows an example of tensile ratcheting of the composite caused by ratcheting of the matrix while under force control with $\overline{V}_\sigma = +1$.

The fatigue results shown in Fig. 12.59 are replotted in Fig. 12.61 in terms of stress range

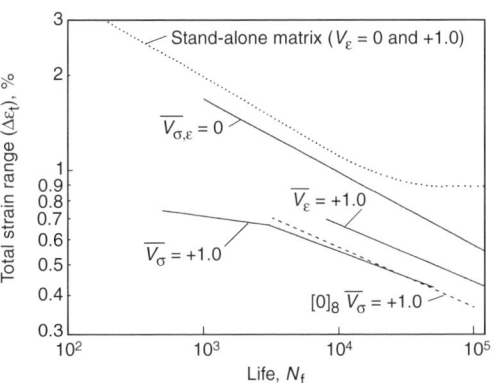

**Fig. 12.59** Summary of total strain-range fatigue curves for SCS6/Ti-15-3 MMC at 427 °C (800 °F). Source: Ref 12.28

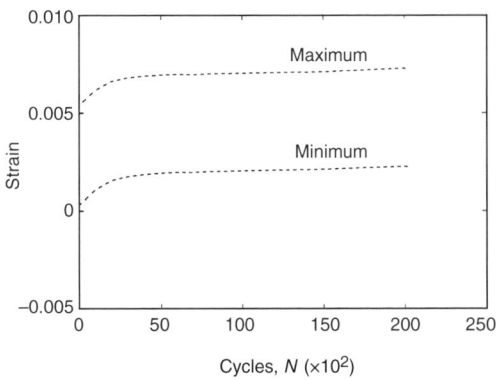

**Fig. 12.60** Maximum and minimum strain response for force-controlled test with $\overline{V}_\sigma = 1$, SCS6/Ti-15-3 MMC at 427 °C (800 °F), $\Delta\sigma = 940$ MPa (140 ksi). Source: Ref 12.28

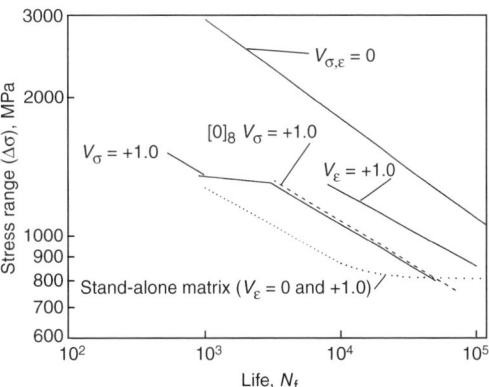

**Fig. 12.61** Summary of stress range fatigue curves for SCS6/Ti-15-3 MMC at 427 °C (800 °F). Source: Ref 12.28

versus life. Again, and as expected, there is no distinction between strain-controlled and force-controlled results for completely reversed loading. The difference in the appearance of these two figures is slight. Thus, the stress ranges and the strain ranges are essentially linearly related by the modulus of elasticity.

**Summary of Applicability of Monolithic Metal Fatigue Modeling to Composite Materials.** We have examined the applicability of metals-based fatigue models in assessing the cyclic durability of metal matrix composites reinforced with long ceramic fibers that are strong and stiff, yet brittle. In our discussion we have emphasized mean stress effects because of their importance in governing fatigue behavior of such materials. For purely elastic loading, the existing metal fatigue mean stress formulations appear to apply reasonably well, and the composite can be treated as though it were a monolithic metallic material. However, when plasticity is encountered in the matrix, it becomes necessary to treat the composite as a miniature structure with materials properties varying from point to point. Such treatment is basically the same as that employed when applying material fatigue models to geometrically complex structural components made of metals.

Because of the success of being able to apply metallic fatigue models to polymeric materials, as described in the earlier section on polymers, it is expected that polymer-based fibrous composites could be addressed with as much success as metal matrix composites.

All ceramic matrix composites, and all ceramic constituents in other composites, should not be expected to follow the metallic fatigue models because the nature of the inelastic deformation in ceramics differs so greatly, both in kind and magnitude, from that of ductile metals.

**Some Future Trends.** Current composite materials are generally made of constituents at the macroscopic scale. Future composites may involve property enhancement by constituents at the molecular level. As research on the use and manufacture of nanomaterials and functionally graded materials (FGM) is given greater emphasis in the future, microcomposites should play an increasingly important role in structural force-bearing applications. Two examples of functional grading of the architecture are shown schematically in Fig. 12.62 (Ref 12.24). Factors such as interfacial bonding, thermal expansion mismatch between constituents, internal stress concentrations, and so forth might be tailored to remove the difficulties that have hampered macrocomposites in gaining widespread use, both in performance and analysis. Generic computational thermal and mechanical modeling of FGMs has advanced considerably in the past decade with the view toward enhancing structural durability (Ref 12.40, 12.41).

Some FGMs currently are used in nonforce-bearing applications. Two widely different examples are:

- Coatings in thermal protection systems for high-temperature components in aerospace propulsion systems. The coatings decrease the maximum metal temperature of the component so that it can withstand larger mechanical and thermal loadings. The thermal conductivity, stiffness, and strength are controlled in the thickness direction to provide the properly tailored end result.
- A molecular scale FGM that has been in existence for decades is the electronic micro-

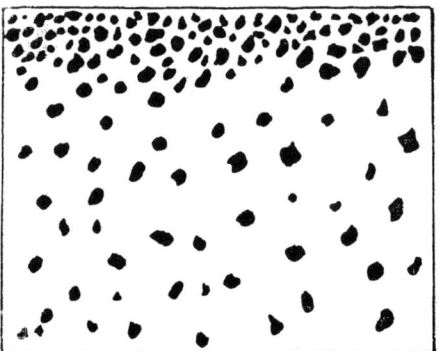
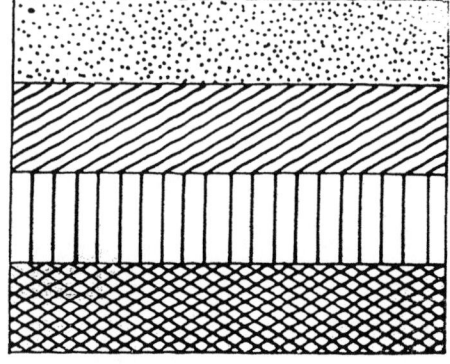

**Fig. 12.62** Two examples of functionally graded materials. Source: Ref 12.24

chip. Its purpose, however, is strictly for electronic applications.

The task of achieving structural, force-bearing, microcomposite materials using nanotechnology and functional grading concepts will be far more difficult than that required of the nonstructural (thermal and electronic) materials.

**REFERENCES**

12.1 M.F. Ashby and D.R.H. Jones, *Engineering Materials 2, An Introduction to Microstructures, Processing and Design, International Series on Materials Science and Technology,* Vol 39, 1986, p 229. (See Fig. 23.10)

12.2 S. Rabinowitz and P. Beardmore, Cyclic Deformation and Fracture of Polymers, *J. Mater. Sci.,* Vol 9, 1974, p 81–99

12.3 P. Beardmore and S. Rabinowitz, Fatigue Deformation of Polymers, *Treatise on Materials Science and Technology,* Academic Press, San Diego, CA, Vol 6, 1975, p 267–331

12.4 F.J. Lisy, "Fully Reversible Fatigue of Thermoplastics," Student Special Projects Paper, EMAE 689, Case Western Reserve University, Cleveland, OH, 1991

12.5 S.S. Manson, Fatigue: A Complex Subject—Some Simple Approximations, *Exp. Mech.,* Vol 5 (No. 7), 1965, p 193–226

12.6 McEvily et al., *10th Sagamore Conference,* 1964, p 101. (See Fig. 5)

12.7 R.W. Hertzberg, J.A. Manson, and M. Skibo, Frequency Sensitivity of Fatigue Processes in Polymeric Solids, *Polymer Engineering Series,* Vol 15 (No. 4), April 1975, p 252–260

12.8 R.W. Hertzberg, *Deformation and Fracture Mechanics of Engineering Materials,* John Wiley & Sons, New York, 1976, p 527. (See Fig. 13.45)

12.9 C.A.G. Pattin, "Cyclic Mechanical Property Degradation in Bone during Fatigue Loading" Ph.D. thesis, Department of Mechanical Engineering, Stanford University, 1991

12.10 W. Weibull, *Fatigue Testing and Analysis of Results,* Pergamon Press, London, 1961

12.11 J.P. Gyekenyesi, SCARE: A Postprocessor Program to MSC/NASTRAN for the Reliability Analysis of Structural Ceramic Components, *J. Eng. Gas Turbines Power,* Vol 108 (No. 3), July 1986, p 540–546

12.12 J.P. Gyekenyesi and N.N. Nemeth, Surface Flaw Reliability Analysis of Ceramic Components with the SCARE Finite Element Postprocessor Program, *J. Eng. Gas Turbines Power,* Vol 109 (No. 3), July 1987, p 274–281

12.13 N.N. Nemeth, J.M. Manderscheid, and J.P. Gyekenyesi, "Ceramics Analysis and Reliability Evaluation of Structures (CARES) User's and Programmer's Manual," NASA TP-2916, National Aeronautics and Space Administration, Washington, DC, 1990

12.14 N.N. Nemeth and J.P. Gyekenyesi, Probabilistic Design of Ceramic Components with the NASA CARES Computer Program, *Ceramics and Glasses,* Vol 4, *Engineered Materials Handbook,* ASM International, 1991, p 700–708

12.15 A.M. Calomino, "Damage Mechanisms and Controlled Crack Propagation in a Hot Pressed Silicon Nitride Ceramic," NASA TM-106595, National Aeronautics and Space Administration, Washington, DC, May 1994

12.16 R. Kossowsky, Cyclic Fatigue of Hot-Pressed Si3N4, *J. Am. Ceram. Soc.,* Vol 56, Oct 1973, p 531–535

12.17 T. Machida, H. Miyata, S. Usami, and H. Ohta, Fatigue Behavior of Structural Ceramics at Elevated Temperatures, *Cyclic Fatigue in Ceramics,* H. Kishimoto, T. Hoshide, and N. Okabe, Ed., Elsevier and the Society of Materials Science, Japan, 1995, p 33–58

12.18 R.L. Gibson, *Principles of Composite Material Mechanics,* McGraw-Hill Book Company, New York, 1994

12.19 C.R. Barrett, W.D. Nix, and A.S. Tetelman, *The Principles of Engineering Materials,* Prentice-Hall, Inc., Englewood Cliffs, NJ, 1973

12.20 C.C. Chamis, Simplified Composite Micromechanics Equations for Hygral, Thermal, and Mechanical Properties, *SAMPE Q.,* April 1984, p 14–23

12.21 Z. Hashin and A. Rotem, A Fatigue Failure Criterion for Fiber Reinforced Materials, *J. Compos. Mater.,* Vol 7, Oct 1973, p 448–464

12.22 *Composite Materials for Aircraft Structures,* B.C. Hoskin and A.A. Baker, Ed.,

AIAA Education Series, J.S. Prezemieniecki, Ed.-in-Chief, American Institute of Aeronautics and Astronautics, Inc., New York, 1986

12.23 *Annual Book of ASTM Standards,* Section 15, Vol 15.03, *Space Simulation; Aerospace and Aircraft; High Modulus Fibers and Composites,* 1997

12.24 M.M. Schwartz, *Composite Materials Handbook,* McGraw-Hill, New York, 1984, p 3.13

12.25 G.R. Halford, B.A. Lerch, J.F. Saltsman, and V.K. Arya, "Proposed Framework for Thermomechanical Fatigue (TMF) Life Modeling of Metal Matrix Composites (MMCs)," *Symposium on Thermo-Mechanical Fatigue Behavior of Materials,* ASTM STP 1186, H. Sehitoglu, Ed., 1993, p 176–194

12.26 W.S. Johnson, S.J. Lubowinski, and A.L. Highsmith, "Mechanical Characterization of Unnotched SCS$_6$/Ti-15-3 Metal Matrix Composites at Room Temperature," *Thermal and Mechanical Behavior of Metal Matrix and Ceramic Matrix Composites,* ASTM STP 1080, J.M. Kennedy, H.H. Moeller, and W.S. Johnson, Ed., American Society for Testing and Materials, 1990, p 193–218

12.27 W.S. Johnson, "Modeling Stiffness Loss in Boron/Aluminum Laminates below the Fatigue Limit," *Long Term Behavior of Composites, Proceedings of the Symposium,* ASTM STP 813, T.K. O'Brien, Ed., American Society for Testing and Materials, 1983, p 160–176

12.28 B.A. Lerch and G.R. Halford, Effects of Control Mode and R-Ratio on the Fatigue Behavior of a Metal Matrix Composite, *Mater. Sci. Eng.,* Vol A200, 1995, p 47–54

12.29 B.A. Lerch and G.R. Halford, Fatigue Mean Stress Modeling in a [0]$_{32}$ Titanium Matrix Composite, *1995 HITEMP Conference,* NASA CP 10178, NASA-Lewis Research Center, Oct 23–25, 1995, p 21-1 to 21-10

12.30 W.Z. Gerber, Bestimmung der Zulossigne Spannugen in Eisen Constructionen, *Bayer. Arch. Ing. Ver.,* Vol 6, 1874, p 101 (in German)

12.31 J.B. Conway and L.H. Sjodahl, *Analysis and Representation of Fatigue Data,* ASM International, 1991, p 149

12.32 K.R. Heidmann, "Technology for Predicting the Fatigue Life of Gray Cast Iron." Ph.D. thesis, Department of Mechanical and Aerospace Engineering, Case Western Reserve University, Cleveland, OH, May 22, 1985

12.33 Cyclic Fatigue in Ceramics, H. Kishimoto, T. Hoshide, and N. Okabe, Ed., *Current Japanese Materials Research,* Vol 14, The Society of Materials Science, Japan, Elsevier, 1995

12.34 G.R. Halford and A.J. Nachtigall, The Strainrange Partitioning Behavior of an Advanced Gas Turbine Disk Alloy, AF2-1DA, *J. Aircr.,* Vol 17 (No. 8), 1980, p 598–604

12.35 J. Morrow, in *Fatigue Design Handbook,* Vol 4, Society of Automotive Engineers, 1968

12.36 J. Goodman, *Mechanics Applied to Engineering,* Longman, Green & Company, London, 1899

12.37 C.R. Soderberg, Factor of Safety and Working Stress, *Trans.,* American Society of Mechanical Engineers, Vol 52 (Part 1), 1930, APM 52-2

12.38 K.N. Smith, P. Watson, and T.H. Topper, A Stress-Strain Function for the Fatigue of Metals, *J. Mater.,* Vol 5 (No. 4a), American Society for Testing and Materials, 1970, p 767–778

12.39 K. Walker, "The Effect of Stress Ratio during Crack Propagation and Fatigue for 2024-T3 and 7075-T6 Aluminum," *Effects of Environment and Complex Load History on Fatigue Life,* STP462, American Society for Testing and Materials, 1970, p 1–14

12.40 J. Aboudi, M.-J. Pindera, and S.M. Arnold, Higher-Order Theory for F Functionally Graded Materials, *Composites Part B: Engineering,* Vol 30, 1999, p 777–832

12.41 M.-J. Pindera, J. Aboudi, and S.M. Arnold, "Analysis of Plasma-Sprayed Thermal Barrier Coatings with Homogeneous and Heterogeneous Bond Coats under Spatial Uniform Cyclic Thermal Loading," NASA TM-2003-210803, Dec 2003

# APPENDIX

# Selected Relevant Background Information

**Preface**

This book is mainly about fatigue. But limiting the discussion to this subject introduces some gaps in understanding for some of its likely readers or students. The fatigue content is mainly an outgrowth of either short courses or full-length courses presented by the senior author to various groups over a period of three decades. Initially, the presentations were one- or two-week courses at the Pennsylvania State University. Attendees were usually sophisticated members of the engineering community whose interest was to learn the state of the art. Most had a fairly good background in materials technology, but the occasional use of a word or secondary subject brought forth questions that made digression necessary. Similarly, in later presentations senior and graduate mechanical engineering students who also were well versed in the fundamentals and the required terminology, but lacking in some details, again raised questions resulting in digressions. Finally, there has been a class of students whose major interest has been outside of mechanical engineering, but who were taking the course to gain breadth in knowledge. For them especially, such digressions were necessary in order to make the presentation understandable. It appeared that clearly there was a need to collate the kind of useful information to which they could refer when necessary, limiting therefore the frequency of digression and providing a written record for later reference. This Appendix is intended to fill that need.

The Appendix is not intended to break new ground. Most of the information presented herein can be found in classic books devoted to their subject in much more sophisticated and inclusive form. References A.1 to A.24 are but examples of such books, and the reader wishing more complete background on any one of the subjects discussed is referred to these books and to numerous others that can be found in libraries. Our intent is to cover at least as much of each of the subjects undertaken to make back reference possible without too much inconvenience to the reader.

**Introduction**

We begin with the periodic table of elements, Fig. A.1. It is of fundamental usefulness in understanding the "whys" of materials behavior. From basic courses in chemistry and physics, nearly all readers are familiar with the background of this table—how the elements align in relation to their atomic weight, how elements bond to themselves to form solids, why some are inert and others so reactive with other elements, and so on. It is clear from the table, for example, why polymers, containing only elements of low atomic numbers (hydrogen, nitrogen, fluorine, and carbon) are the low-density materials (yet, as discussed later in the text, can be combined with other elements to make high-strength alloys); why the superalloys, containing mainly nickel, chromium, cobalt, and so forth, are heavy; and why the ultrahigh-temperature alloys containing tantalum and tungsten become extremely heavy. Ceramics (based on silicon, aluminum, boron, nitrogen, or carbon) are relatively lightweight materials; yet they can be made strong at very high temperatures. Unfortunately, the periodic table alone does not dictate how to build "strength" into materials. Many other factors enter into this technology, which constitutes the art and science of the materials scientist. But the atomic table contains clues, and when properly used by the scientist and technologist, makes their task easier.

While the atomic and nuclear physics of the elements is well beyond the scope of this book, it should be pointed out that the basic structure of the electron shells surrounding the nucleus of protons and neutrons control many of an elements mechanical, thermal, electrical, and optical properties. In each element, there is one electron (negative charge) for each proton (positive charge) there are usually more neutrons (no charge, but the same mass as a proton) than there are protons. Isotopes of an element are caused by having an excess of one, two, or more neutrons. This leads to the atomic weight being slightly more than twice the atomic number. Obviously, as the atomic number increases, the number of electrons increases. As the number of electrons increase, they eventually fill four subshells designated $s$, $p$, $d$, and $f$, with capacities of 2, 6, 10, and 14 electrons, respectively. Once the last ($f$) subshell is full, a shell is completed and a new set of subshells and shells begins. This periodic progression continues to the highest atomic number known. Armed with this limited information, it is easier to appreciate the periodicity of the table of elements.

The electrons that occupy the outermost subshell and shell of an element are known as the *valence electrons,* and they are the electrons that actively participate in the bonding between atoms and molecules. Fully filled shells and subshells are much less interactive than unfilled ones. Thus, it becomes easier to understand how elements might interact with themselves to form a gas, a liquid, or a solid. Correspondingly, it is much easier to envision why an element interacts with other elements to form polymers, alloys, oxides, etc. A more fundamental background is found by consulting some of the earlier references (see, for example, Ref. A.20).

Having the periodic table available is more than a convenience. It is a powerful metallurgical tool that suggests, for example, elements effective for alloying purposes. Groupings of elements denote the nonmetals, the noble metals, and even the refractory metals. The metals with high affinity for oxygen and other smaller elements can also be identified from their placement in the table. Even the crystallographic structure of many of the elements is indicated by their placement in the table.

Largely, the electrons in the outermost shell of the atoms, i.e., the valance electrons, govern atomic bonding of the structural materials. The three principal types of atomic bonds are ionic,

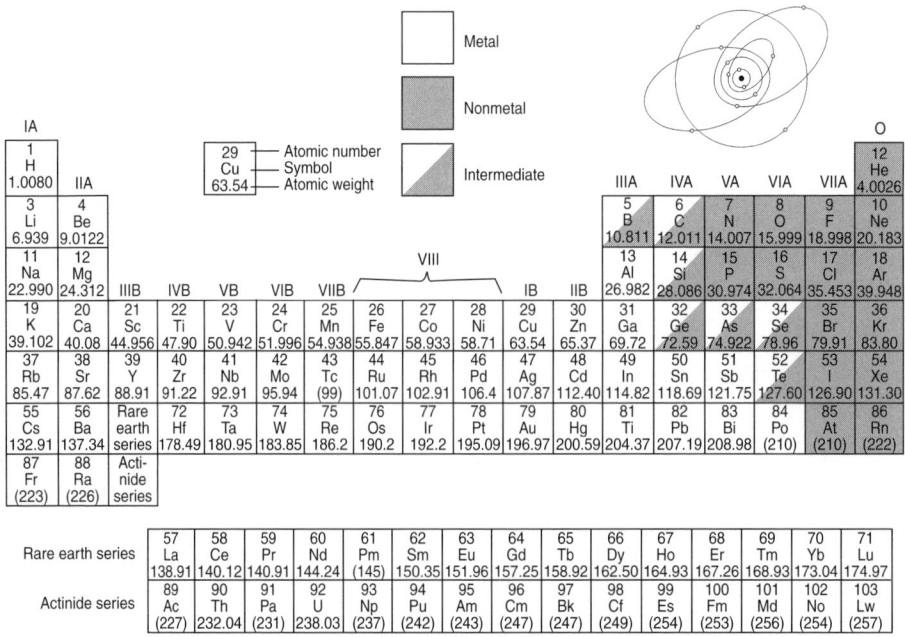

**Fig. A.1** Periodic table of elements. Source: Ref A.20

covalent, and metallic. Our greatest concern is for metallic bonding because the major focus of this book is on metallic material behavior.

Ionic bonding, as the name implies, involves the mutual attraction of ions, i.e., elements that are positive ions (missing an electron) with elements that are negative ions (having an extra electron). Ionic bonding is always found in molecules that are composed of metallic and nonmetallic elements. The elements most commonly bonded are at the horizontal extremes of the periodic table. Perhaps the most commonly known ionic-bonded molecule is NaCl, Sodium Chloride (table salt). Ionic bonding is isotropic, equal in all directions. Examples of ionic bonding are given later in the Appendix as well as in Chapter 12 (see section on "Ceramic Materials").

Covalent bonding is aptly named because of the co-sharing of a valence electron between the bonding elements. Multiple bonds frequently exist when the number of valence electrons is larger for one of the elements than for the others. Bond strengths may be quite weak (polymeric materials), or exceptionally strong (diamond) depending upon the elements. Covalent bonding is highly anisotropic and is the strongest in the direction between the two atoms that are sharing electrons. Most covalent-bonded compounds also involve a mixture of ionic bonding. The greater the horizontal and vertical separation (lower left to upper right) in the periodic table relative to Group IVA, the more ionic is the bond. Correspondingly, according to this layout, the closer the atoms are in the periodic table, the more covalent is the bonding. A few examples of covalent bonding are given later in the Appendix as well as in Chapter 12 (see section on "Polymers").

Metallic bonding is the primary bonding type for the engineering structural building materials for this book. It defines the bonding found in all metals and alloys. In fact, metals are the dominant elements in the periodic table of the elements. Metallic materials have one to three nominally free (valence) electrons per atom. They spread freely among all of the atoms and are shared equally by all the atoms in metal. They may be thought of as belonging to the metal as a whole, or forming a "sea of electrons" or an "electron cloud." The nonparticipating (nonvalence) electrons and their nuclei are called ion cores. They possess a net positive charge that is equal in magnitude to the total valence electron charge per atom. Figure A.2 is a detailed schematic diagram of metallic bonding. The free-valence electrons shield, or neutralize, the positive ion cores from the repulsive electrostatic forces. The result is an isotropic force field that acts in every possible direction. In effect, the free-floating electrons actually hold the ion cores together by preventing their mutual repulsion. Metallic materials are good conductors of both electricity and heat, because of the free-valence electrons.

The well-trained material scientist utilizes much of the inherent information contained in the periodic table of the elements to guide the development and processing of polymeric, ceramic, and metallic engineering materials; as well as provide designers with better materials of construction. However, a great deal more in-depth information is necessary than can be provided herein.

Mechanics of solids concepts of stress and strain in the elastic and plastic range are then discussed, and the very rudimentary tests that are conducted to provide mechanical material characterization upon which property performance may be judged. Common hardness tests and their inter-relationships are mentioned along with material strength tests. The latter include tensile, compressive, torsional, and bending tests. More sophisticated testing for fatigue, creep-fatigue, thermomechanical fatigue, and thermal fatigue are presented. Cyclic crack growth and fracture toughness testing are discussed in Chapter 9, "Crack Mechanics." Creep and creep-rupture testing are to be described in

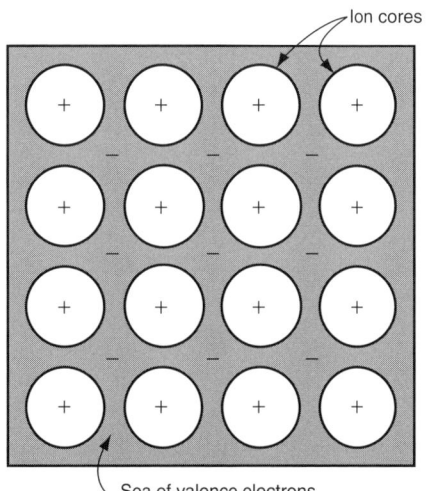

**Fig. A.2** Schematic illustration of metallic bonding. Courtesy, John Wiley & Sons, Inc.

Chapter 1 of the proposed second volume on high-temperature structural fatigue.

All mathematical equations of a general nature appearing in this book are developed in the chapter in which they are introduced with the exception of one: the inversion of an equation of the form $y = Ax^a + Bx^b$, which has no closed-form solution. The inversion technique is an original development by the senior author and provides very accurate approximate solutions to several practical problems. It is discussed here rather than where applied because of the several uses that are made of the results. Although initially developed for inverting the strain-life equation (Chapter 3, "Fatigue Life Relations"), its applications were later extended to the stress-strain equation (Chapter 2, "Stress and Strain Cycling"), mean stress effects (Chapter 4, "Mean Stress"), and cumulative fatigue damage (Chapter 6, "Cumulative Fatigue Damage"). We considered it best, therefore, to present it in the Appendix in a form suitable for generalized application.

## Microstructural Factors

Study of the mechanical behavior of materials starts with the recognition that they are composed of atoms arranged in some systematic manner, including defects in the ideal structure, that govern the properties of interest to us. The atoms are held together by attractive forces—being the net of attraction due to gravitational forces, and electronic bonding, and repulsion—due to the similar and opposite charges of electrons and protons within the atoms. Applied forces can distort the arrangement of the atoms. For moderate forces the atoms can be displaced while still retaining their basic relative positions. This type of displacement is referred to as *elastic* because removal of the forces allows the atoms to be restored to their original positions. If the forces become high enough they may temporarily overcome the attractive forces among some of the atoms, causing them to slip over each other, resulting in permanent distortion in a process referred to as *plastic deformation*. Finally, if the forces are high enough, the net attractive forces may be overcome, and the atoms may separate completely, causing *fracture*. The basic structure that governs the forces of attraction and repulsion among the atoms, as well as the atomic defects present, greatly influence the elastic stiffness and the forces required to cause plasticity and fracture.

In this brief review we consider some rudiments that elementary students can apply to understand the many concepts and terms used in this book.

**Crystallography.** Alloys of metals are composed of atoms bonded to each other so as to achieve the desired properties sought for engineering materials. The starting point governing the properties of a material is the crystallographic, or lattice, structure of the unit that makes up the microstructure—the scheme according to which the atoms are arranged relative to each other.

The simplest lattice is the cubic structure shown in Fig. A.3. Here the elemental structural cell consists of eight atoms at the corners of a cube. The simple cubic lattice is not a common atomic arrangement for structurally significant materials because it leads to very low ductility and hence low fracture resistance. However, many of the metals of engineering structural interest possess a structure that is based on slightly more complex cubic structures—the body-centered cubic (bcc) and the face-centered cubic (fcc). Closely related to the fcc structure is the hexagonal close-packed (hcp) metallic atomic arrangement. The three metallic structures are illustrated in the following figures.

Figure A.4 shows the bcc structure. Portions of eight atoms are used to define the cube, one-eighth at each corner. In addition, there is a complete atom at the center of the cube, accounting for the designation. In a pure metal all the atoms are, of course, the same. In a compound or an

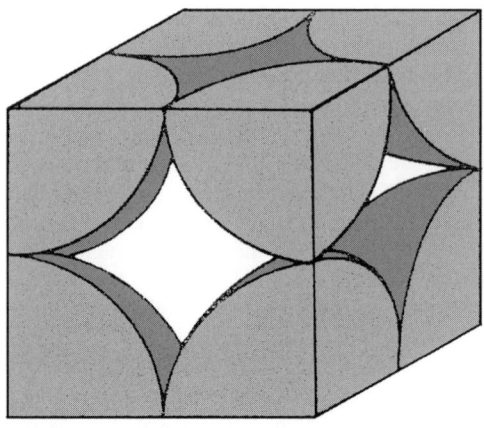

**Fig. A.3** Simple cubic structure. Source: Ref A.20

alloy, however, the atoms may be different but still are arranged systematically according to the scheme of the figure.

Figure A.5 shows the fcc structure. Again, portions of eight atoms define the corners of the cube, but one additional half-atom is located at the face of each of the planes enclosing the cube. As in the body-centered cube, the atoms at each of the corners, or in the center of the faces, may differ when more than one element is present.

Figure A.6 shows the atomic arrangement for the hcp structure. The basal planes are six sided, with a portion of an atom at each apex of the hexagon and a half-atom at its center. However, similarly arranged basal planes are alternated with other planes in which the atoms are rotated in position so as to be able to stack more closely to the basal planes. The three-dimensional arrangement of successive planes is such that the peaks of the spheres in one plane fit into the hollows between the spheres in the adjacent plane. For this reason, this crystallographic arrangement is called *close-packed*.

Four of the most common lattices found in metals and various other materials have been shown and briefly described in the foregoing discussion. There are ten more unit cell packing structures, making a total of 14. These 14 arrangements are known as the Bravais lattices and are shown in Fig. A.7. Table A.1 provides a brief description of the geometric characteristics of these crystal systems and lists example elements and compounds of each. Note that the 14 Bravais lattices can be consolidated into seven more generalized crystal systems. For example, the general cubic lattice (with all axes equal in length and all angles of 90°) includes the simple cubic, face-centered cubic, and body-centered cubic.

**Packing of Atoms and Stacking of Successive Planes.** The easiest way of envisioning how the atoms must arrange themselves, and how

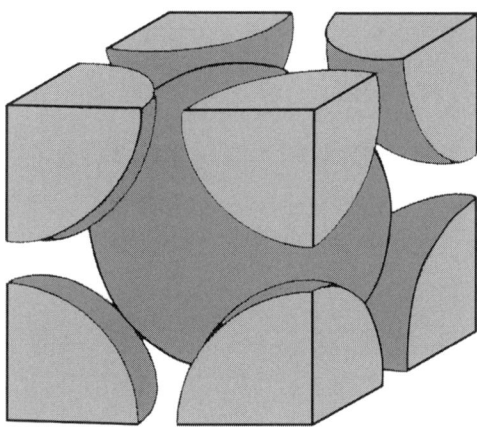

**Fig. A.4** Body-centered cubic structure. Source: Ref A.20

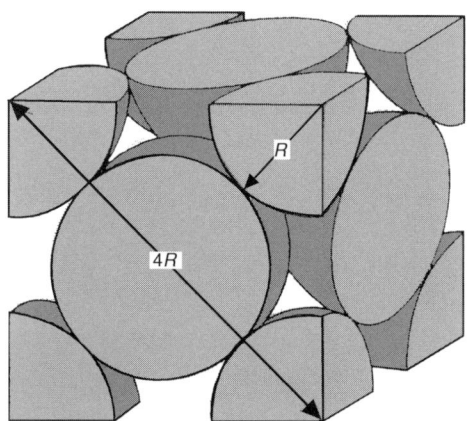

**Fig. A.5** Face-centered cubic structure; also cubic close-packed structure. Source: Ref A.20

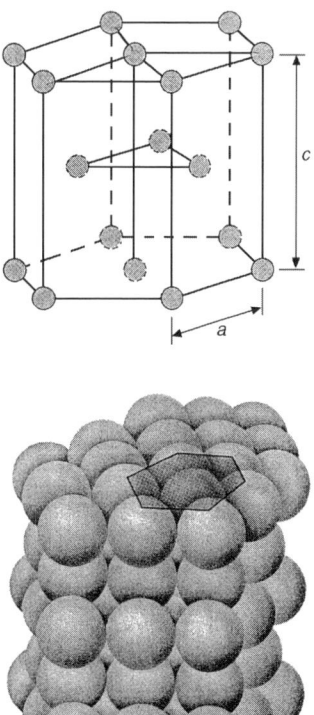

**Fig. A.6** Hexagonal close-packed structure. Source: Ref A.20

successive planes stack relative to each other, is to regard them as hard spheres, similar to billiard balls. In this way each atom is allocated a private space, which may not be invaded by a neighbor. However, these billiard balls may touch each other at their periphery. A simple cubic system has atoms stacked according to the unit cell of Fig. A.8(a). As unit cells are added to form a larger three-dimensional solid, each atom touches four in an orthogonal array. In any of the orthogonal planes passing through the centers of the atoms, the array is square. As explained later, this arrangement does not lend itself readily to sliding of one plane over the other, thus resulting in low ductility. The "density" of the simple cubic crystal is the least of any stack-

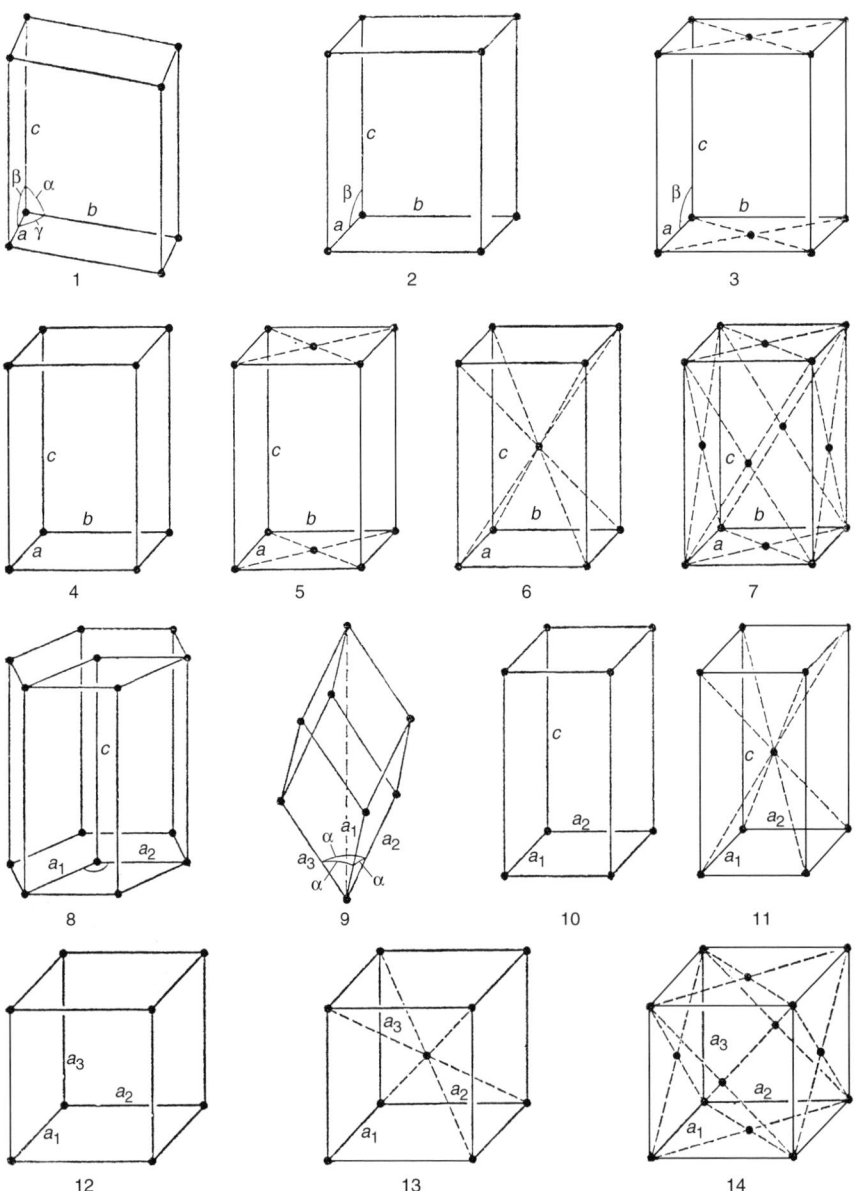

**Fig. A.7** The 14 Bravais lattices. (1) Triclinic, simple. (2) Monoclinic, simple. (3) Monoclinic, base-centered. (4) Orthorhombic, simple. (5) Orthorhombic, base-centered. (6) Orthorhombic, body-centered. (7) Orthorhombic, face-centered. (8) Hexagonal. (9) Rhombohedral. (10) Tetragonal, simple. (11) Tetragonal, body-centered. (12) Cubic, simple. (13) Cubic, body-centered. (14) Cubic, face-centered. Source: Ref A.2

## Table A.1  Characteristics of the seven crystal systems most frequently encountered

In this table, $\neq$ indicates "not necessarily equal to and generally different from."

| Bravais No. | System | Axes and interaxial angles | Examples |
|---|---|---|---|
|  | **Triclinic** | Three axes are not at right angles, of any lengths. | $K_2CrO_7$ |
| 1 | (simple) | $a \neq b \neq c$    $\alpha \neq \beta \neq \gamma \neq 90°$ |  |
|  | **Monoclinic** | Three axes, one pair not at right angles, of any length | $\beta$-S and $CaSO_4 \cdot 2H_2O$ (gypsum) |
| 2 | (simple) | $a \neq b \neq c$    $\alpha = \gamma = 90° \neq \beta$ |  |
| 3 | (base-centered) |  |  |
|  | **Orthorhombic** | Three axes at right angles; all unequal | $\alpha$-S, Ga, and $Fe_3C$ (cementite) |
| 4 | (simple) | $a \neq b \neq c$    $\alpha = \beta = \gamma = 90°$ |  |
| 5 | (base-centered) |  |  |
| 6 | (body-centered) |  |  |
| 7 | (face-centered) |  |  |
| 8 | **Hexagonal** | Three axes coplanar at 120°, equal fourth axis at right angles to these | Zn, Cd, and NiAs |
|  |  | $a_1 = a_2 = a_3 \neq c$    $\alpha = \beta = 90°, \gamma = 120°$ |  |
|  |  | (or $a_1 = b \neq c$) |  |
|  | **Rhombohedral** | Three axes equally inclined, not at right angles; all equal | As, Sb, Bi, and calcite |
| 9 | (triagonal) | $a = b = c$    $\alpha = \beta = \gamma \neq 90°$ |  |
|  | **Tetragonal** | Three axes at right angles; two equal | $\beta$-Sn (white) and $TiO_2$ |
| 10 | (simple) | $a = b \neq c$    $\alpha = \beta = \gamma = 90°$ |  |
| 11 | (body-centered) |  |  |
|  | **Cubic** | Three axes at right angles; all equal | Cu, Ag, Au, Fe, and NaCl |
| 12 | (simple) | $a = b \neq c$    $\alpha = \beta = \gamma = 90°$ |  |
| 13 | (body-centered) |  |  |
| 14 | (face-centered) |  |  |

Source: Ref A.2

ing arrangement, having only one complete atom per unit cell (⅛ atom at each of the eight corners). The unit cell has equal edge lengths of one atomic diameter, $D$. The bcc arrangement shown in Fig. A.8(b) also has ⅛ atom at each of the eight corners, but in addition, has another complete atom at the center of the unit cell for a "density" of two atoms per unit cell. This arrangement requires that the unit cell dimension be greater than one atomic diameter. The bcc unit cell dimension is $(2/\sqrt{3})D$, resulting in a volume of 1.54 $D^3$. There being 2 bcc atoms of 0.5236 $D^3$ volume per atom in the unit cell gives rise to an *atomic packing factor* (density) of 0.68. For the simple cubic, this factor is only 0.52.

In dramatic comparison we look next at the hcp arrangement of Fig. A.9. In the basal plane each atom is surrounded by a hexagonal arrangement of 6 atoms, and they are packed as closely as spheres can be packed. In the second plane the spheres are arranged such that their peaks fit into one of two sets of "triangular hollows" created by the spheres of the first layer. Similarly, the spheres of the third plane fit into the "hollows" of the second plane. In the figure, the spheres of the third plane are arranged exactly as those in the first in a repeating *ABAB* stacking pattern. The spheres are packed together as closely as possible, resulting in the terminology "close-packed." For this hard-sphere model it can be shown that the atomic packing

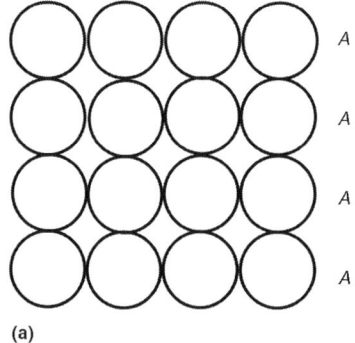
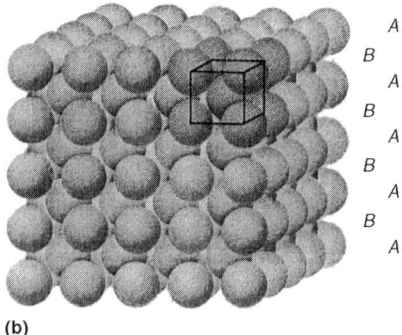

**Fig. A.8** Atomic stacking. (a) Simple cubic crystal. (b) Body-centered cubic. Source: Ref A.20

factor is 0.74 compared with only 0.52 for the simple cubic cell of Fig. A.3, and 0.68 for bcc packing. Because there are numerous planes of high atomic packing, plasticity that, in simplistic terms, is the slip of one plane over the other, is greatly facilitated. Pure metals with a close-packed structure are therefore relatively ductile.

The fcc crystalline structure, shown in Fig. A.10, is also close-packed with 74% of the volume taken up by the spheres. To appreciate the difference between hcp and fcc structures, one must concentrate on the third plane in the stacking sequence for hcp. Note that the third plane in the hcp stacking differs from both the first and second planes, resulting in a three-plane repeating sequence, *ABCABC*. The plastic flow properties of hcp and fcc structures differ therefore because successive stacking of atomic planes is different.

**Relation of Crystallographic Structure to Plastic Flow.** Because plastic flow is achieved by sliding of one crystallographic plane over another, it is clear that crystallography affects the ability to produce such sliding, affecting ductility. The planes most likely to slide are those parallel to the ones containing the most densely packed spheres. A greater distance between the most densely packed planes reduces the attractive forces between adjacent atoms, and gives the planes more freedom to break bonds and to slide. The direction of sliding is along the line containing the most densely packed atoms. Therefore, for each crystallographic structure there are definitive sets of crystallographic planes and directions along which slip is most likely to take place.

In the hcp metals there is only one plane of highest atomic density, the basal plane, and there are three directions within the plane of the same least atomic spacing. Thus, there are only three easy-slip systems, partially explaining the limited ductility of such metals. In the fcc structure there are eight closely spaced planes, but they are arranged in four sets parallel to each other, so there are really only four effective sets of closely spaced planes. Each has three closely spaced directions, so there are 12 possible easy-slip systems. Fundamentally, therefore, the fcc metals should have higher ductility than the hcp metals. However, there are other factors that govern ductility, which are discussed later. Therefore, this generality may be overridden. The bcc structure is not close packed, so there is no plane of dominant atomic density. Mathematically, of course, there is one set that does have the highest density of packing, but this plane does not differ much from other planes. The net effect is that there are 48 possible slip systems, but because the spacing of the slip planes is closer than those in close-packed structures, higher shear stresses are required to cause the slip. When slip occurs, its direction can move readily from one slip system to another because of the near-equality of shear stress required for slip in different systems. Thus, the slip lines are often observed as wavy, reflecting the changes of slip direction as the slip moves from one system to the next. Wavy slip has been studied extensively in relation to metal fatigue.

While slip systems have a considerable effect on material ductility, other factors also contribute to plastic flow, primarily defects in the

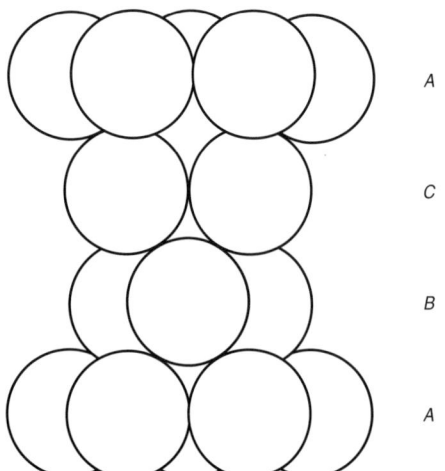

**Fig. A.9** Packing of hexagonal structure. Source: Ref A.16

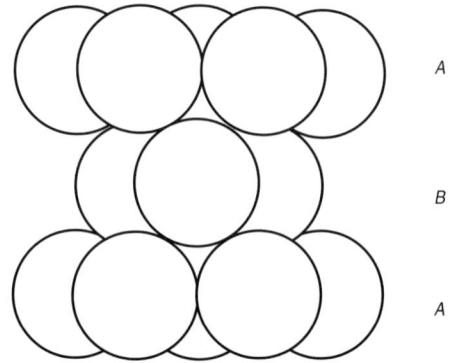

**Fig. A.10** Packing of face-centered cubic structure. Source: Ref A.16

atomic structure. We outline this subject in the following sections, "Defects in Crystalline Structure" and "Dislocations." Greater detail is further provided in Chapter 10, "Mechanism of Fatigue."

In summary, there are three crystalline structures common in metals: the body-centered cubic, the face-centered cubic, and the hexagonal close-packed. Other crystalline structures exist, for example, the simple cubic, but no metals crystallize in this configuration. Other metals, such as mercury, gallium, zinc, and cadmium crystallize in complex form but are not of special interest as structural materials.

Annealed fcc materials, such as aluminum, copper, gold, lead, nickel, austenitic stainless steels, platinum, and silver, have relatively low yield strengths and high ductility, but they strain harden rapidly. They are malleable and thus can be rolled into very thin sheets without fracturing (e.g., aluminum foil, copper foil, gold leaf, etc.). The bcc metals and alloys (steel being the most common) are usually associated with higher strengths and are frequently ductile but may exhibit large strain hardening in the annealed state. They often are associated with a response known as the *yield point phenomenon*. This trait is brought about by the presence of small atoms such as carbon, oxygen, and nitrogen—even boron—that find energetically stable positions in the interstices (triangular hollows) between the atoms. When dislocations, as discussed later, move through the lattice they can be "pinned" by these interstitial atoms and temporarily become blocked from further motion. Eventually, however, with added loading the dislocations "break away" from the pinning points and a limited amount of plasticity takes place abruptly, and the stress drops. This behavior is analogous to static and dynamic sliding friction. Also associated with the yield point phenomenon is a similar distinct break in the fatigue curve exhibited by some bcc metals in the life range of $10^5$ to $10^7$ cycles to failure.

With the exception of beryllium, which possesses a number of unusual features, the hcp metals are often low in strength, low in elastic modulus, and of medium to high ductility. Titanium is an exceptional hcp metal. It stands alone in having a relatively high elastic modulus, high strength, high melting temperature, but thermally softens at only 30% of its absolute melting temperature compared with 50% for most other metals. Titanium, despite a melting temperature of over 1650 °C (3000 °F), exhibits small amounts of creep even at room temperature.

**Twinning as an Alternate Form of Permanent Deformation.** While not as common in service, or as important in our discussion of plasticity, the term *twinning* is often encountered in the literature. It is appropriate for the reader to be at least aware of the meaning of this term in relation to our discussion of plastic deformation.

In twinning, a permanent deformation is introduced in the crystal by movement of a portion of the atoms to a new stable configuration in which the atoms on one side of a plane, known as the *twin plane*, form the mirror image of the atoms on the other side. The process is illustrated in Fig. A.11. Figure A.11(a) represents a section perpendicular to a polished surface in an

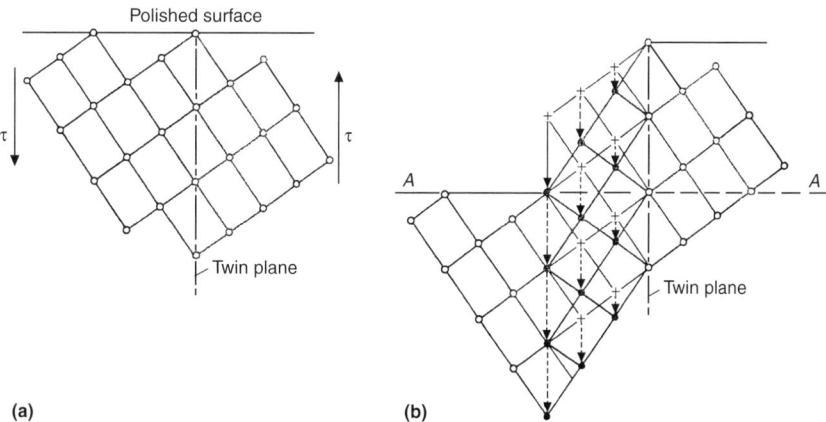

**Fig. A.11** Illustration of the process of permanent deformation by twinning. See text for discussion. Source: Ref A.22

undeformed cubic lattice. The potential twinning plane is shown perpendicular to the paper as well as to the polished surface. If shear stress is applied vertically as shown, the crystal will twin about the twinning plane, Fig. A.11(b). The region to the right of the twinning plane remains undeformed. The deformed region to the left of this plane has sheared such as to make the lattice a mirror image across the twin plane. Each atom in the twinned region in this sketch moves by a sheared amount that is proportional to its distance from the twin plane. Undeformed atomic positions are shown by open circles, plus signs indicate the original positions in the lattice of those atoms that change position, and solid circles are the final positions of these atoms within the twinned band. Note that a twin band will be visible on a subsequently polished surface due to the difference in crystallographic orientation between the deformed and undeformed regions. Twinning deformation does not involve dislocations and can be viewed as a nearly elastic "snap-through" event analogous to mechanical oil canning, although not reversible. The deformed state, which is quite stable, is at a higher energy level than the original, thus requiring expenditure of energy. Twinning contributes to the overall permanent deformation of the aggregate of crystals participating in the process. Twinning is most prevalent in bcc and hcp metals, although fcc metals have been observed to twin when loaded at very low temperature or under shock loading. Sometimes twinning occurs with an audible clicking (e.g., "tin cry," the noise emanations from pure tin as it is deformed). It also occurs more readily if the material is highly cold worked and then annealed. The release of stored energy upon annealing can produce local twinning called *annealing twins*.

**Defects in Crystalline Structure.** While overall crystallography tends toward one of the available structures, there may also be considerable deviation from repetitive placement of atoms on a very local basis. These deviations may restrict the ability of atoms to slide over one another, or alternatively they may, in fact, improve this ability. Defects are classified as point, line, or planar. Examples are discussed next.

*Point Defects.* Figure A.12 shows, in a two-dimensional array, three types of so-called point defects commonly encountered. Each point defect has a higher internal energy state than the perfect repetitive lattice. In Fig. A.12(a) is shown a vacancy—simply one of the atoms is missing. In Fig. A.12(b) an extra atom is shown in one of the interstices of the lattice; this type of defect is called an *interstitial*. Figure A.12(c) illustrates one of the atoms in the regular lattice being replaced by a foreign atom—an atom of a different element. Point defects play an important role in the mechanical and electrical behavior of metallic crystals.

Vacancies are always present in all metals. As temperature is increased, the equilibrium density of vacancies increases. This is a direct result of thermal energy or what we call heat. Random vibrational motion associated with temperature has each and every atom in a constant state of flux. This creates the possibility that at any given instant, there will be atoms missing from their equilibrium atomic site. These vacant sites and their thermally induced mobility are what are responsible for self diffusion. Vacancies contribute significantly to the mechanisms of creep, i.e., thermally activated, time-dependent, diffusion-controlled, deformation associated with an applied stress.

Interstitial atoms are smaller atoms than the parent metallic atoms. This allows them to fit within the naturally small spaces between the parent atoms of the metal. Obviously, the electron bonding characteristics between an interstitial atom and its surrounding parent atoms are different from the parent-to-parent bond. Depending on the atoms involved, the bonding may be better or worse than for parent atoms only. Thus, a material's strength can be increased or decreased by intentionally selecting atom pairings. Interstitial atoms can also play a significant role in how other atomic defects (dislocations) affect the behavior of a metallic material. Because interstitial atoms affect the electronic bonding between the atoms, they also affect electrical resistance. The more interstitial atoms present, the greater is the electrical resistivity.

Impurity (foreign) atoms are substitutional atoms for the parent metallic atoms; i.e., they simply replace the parent atoms at their normal sites. For substitutional atoms to bond to parent atoms, they must have a similar atomic size (weight)

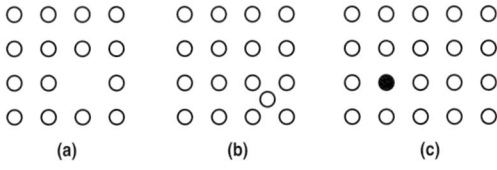

**Fig. A.12** Point defects. (a) Vacancy. (b) Interstitial. (c) Foreign atom. Source: Ref A.22

and a correct number of free electrons in their outer shell. So-called binary solid solutions are examples of substitutional atoms being added to parent atoms to form an alloy whose strength properties are generally better, but their thermal and electrical conductivity are poorer, than either pure metal constituent. The substitutional atoms differ slightly from the parent atoms, thus distorting the local bonding and impeding the motion of line defects (dislocations) and the flow of heat and electrical current.

*Line and Planar Defects.* In addition to the lattice point defects, there may be coarser surface defects such as imperfection in the stacking of the alternate planes in a close-packed structure or in the twinned region of a crystal. Also, if two crystals in a polycrystalline material have nearly the same crystallography, i.e., orientation, but not exactly, the low-angle boundary between the two may be regarded as a planar defect in the structure. Of course, in adjacent crystals of distinctively different orientation there must be an abrupt transition from one crystallographic arrangement to the other, which necessarily involves defects in what might otherwise be ideal atomic arrangement in each crystal. For the purpose of explaining these defects, as well as numerous other crystallographic defects, it has been necessary to recognize a special type of line defect known as a dislocation. Along this line, which may be straight or curved, the atoms are arranged slightly differently from those in the rest of the atoms in the crystal. This difference causes important changes in mechanical behavior of materials. Of interest is the influence of dislocation motion in plastic deformation, which is responsible for metal fatigue.

## Dislocations

Use of dislocation theory is so common in explaining deformation and metallurgical phenomena that at least a rudimentary knowledge is extremely useful in discussing fatigue theory as well. Most students today are exposed to dislocation theory, although not always in the special connection that is most useful in clarifying the fatigue process. Also, for some readers it may be useful to state in an abbreviated form those concepts that are most helpful in understanding the terminology that often creeps into the discussion in this book.

All engineering metals of interest to us (except, perhaps, whiskers of very small diameter)

contain microstructural imperfections. In general, most of the useful metals consist of crystals (grains) in which the atoms are arranged in a regular order. Commonly, there are a great many grains in the cross section of an engineering part, each grain having the same basic crystallographic arrangement, but the crystals are rotated relative to each other. Although the atoms within a crystal normally are arranged quite regularly, there are several types of imperfections that are of importance to us, the most important of which are dislocations.

**Edge Dislocations.** That atomic irregularities must exist within the grains was long known by consideration of the stress necessary to break the atomic bonds of a perfect crystal. As calculated by physicists, the stresses necessary to break these bonds, as would be required for fracture, or even for plastic flow, would have to be of the order of one-tenth the elastic modulus. Thus, for steels, the fracture stress would have to be of the order of $3 \times 10^6$ psi, whereas even the highest strength steels show only about one-tenth this strength. However, it was not until 1934 when three investigators—G.I. Taylor (Ref A.25), E. Orowan (Ref A.26), and M. Polanyi (Ref A.27)—simultaneously, but independently, identified these imperfections in the form of dislocations. Two major types of dislocations exist: one called *edge,* the other, *screw.*

Consider first the edge dislocation shown in Fig. A.13(a). Here the atoms, shown as small circles, are, in general, arranged in a very regular manner. However, the line *AP* (representing a plane of atoms perpendicular to the plane of the figure) represents an extra plane that breaks up the regularity of the array. The bonding between atom $A'$ and atom $B$ is thus of lower intensity than that between the more regularly placed atoms, say $P$ and $P'$.

Imagine, now, a shear stress $\tau_s$ applied to the upper and lower halves of a one atom layer thick crystal. Because of the weakened bond between $A'$ and $B$, it is possible for this stress to break the bonds between $A$ and its neighbors, causing atom $A$ to move over and bond with atom $B$. The result is the extra atomic plane appears to have moved one atomic spacing (a Burgers vector) to the left. Breaking and remaking only one weakened atomic bond accomplished this apparent movement. Continued application of the shear stress causes a series of similar weakened atomic bonds that are broken and remade. This mechanism continues until the extra plane of atoms reaches the edge of the crystal, leaving a slip step of one Burgers vector that is opposite in sign to the step shown originally on the right-hand side of the crystal. Figure A.13(b) shows schematically the successive movements that would take place under this circumstance. It will be noted in this figure that the "sliding" for an edge dislocation takes place normal to the plane associated with the dislocation.

An alternate way of describing an edge dislocation is to identify it as the separation of a region of the crystal, which has already slipped from that which has not yet slipped. In Fig. A.14(a), we start first with a perfect crystal and imagine that a slit ABCD is made along one of the atomic planes only through part of the crystal. The upper atoms in the slit region are displaced one Burgers vector (one atomic spacing) relative to the lower atoms. The atoms along AB, which is remote from the end of the slit (CD), are displaced one atomic dimension toward CD.

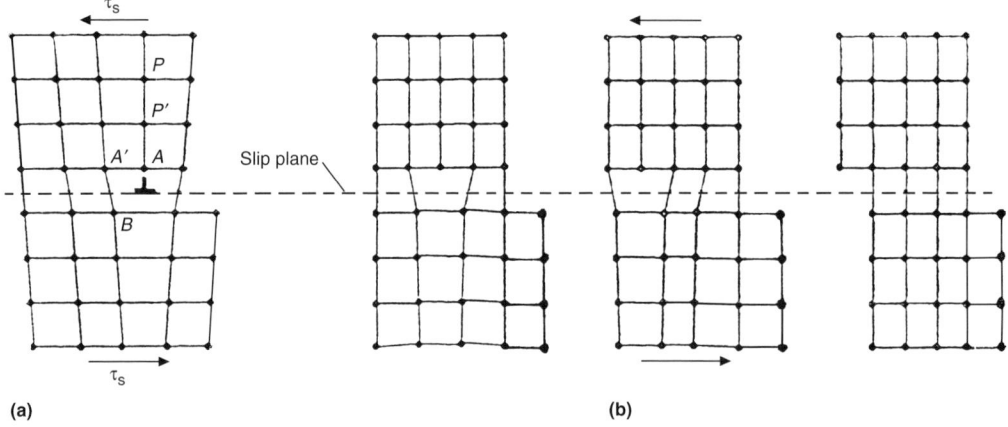

**Fig. A.13** Edge dislocation. (a) Atomic arrangement in a plane normal to edge dislocation. (b) Movement of an edge dislocation

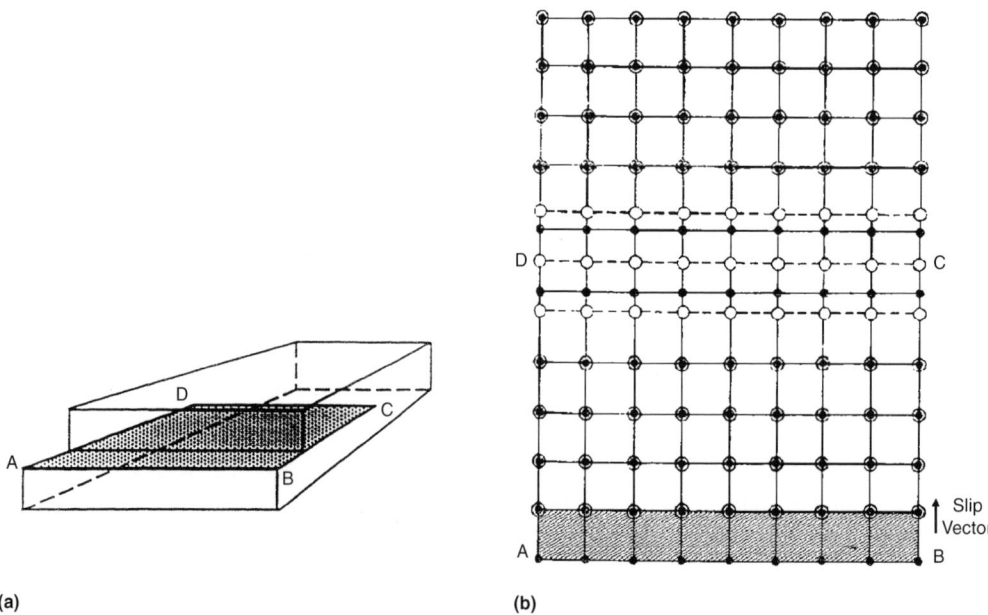

**Fig. A.14** Edge dislocation perceived as the region separating the region of the crystal in which slip has taken place from the region that has not yet slipped. (a) The slip that produces an edge dislocation. Unit slip has occurred over the area ABCD. The boundary CD of the slipped area is the dislocation; it is perpendicular to the slip vector. (b) Arrangement of atoms around the edge dislocation shown in (a). The plane of the figure is parallel to the slip area and CD the dislocation. The open circles represent atoms in the atomic plane just above the slip plane, and the solid circles represent atoms in the plane just below the slip plane. Source: Ref A.28

Although the displacement is a whole atomic spacing in the remote region, the displacement is less and less at the atoms that are closer to the point at which the slit stopped. If we now imagine that the atomic plane that was originally slit is atomically rebonded in its displaced condition, the atomic arrangement will be similar to that near a dislocation line, as shown in Fig. A.14(b). The dislocation line can therefore be regarded as the separation between the slipped and unslipped region of the crystal. In the case of an edge dislocation, the displacement vector is clearly perpendicular to the dislocation.

**Screw Dislocations.** If the slip displacement occurs parallel to the dislocation, it is called a *screw dislocation* (because the dislocation occurs locally as a spiral). Figure A.15(a) shows a crystal that is initially perfect, but which one can imagine to be slit from the surface *BC* to the depth *AD*. If now a slip vector is applied to cause the region above the plane *ABCD* to slide parallel to *AD*, the resulting atomic arrangement is a screw dislocation. Figure A.15(b) shows the atomic arrangement around a screw dislocation. The open circles represent atoms in the plane just above the slip plane, whereas the solid circles represent the atoms in the plane just below. These atoms are arranged in the form of a screw about the dislocation line, giving the dislocation its designation.

More complex types of dislocations can exist, such as shown in Fig. A.16(a). Here the line separating the slipped from the unslipped region is the curved line *AC*. Thus, *AC* is the dislocation. If the slip vector is parallel to *BC*, the direction of slip is parallel to the dislocation in the region *A* while it is perpendicular to the dislocation in region *C*. Figure A.16(b) shows the arrangement of the atoms above the slip planes (open circles) and those in the plane below the slip plane (solid circles). It is clear that near *A* the dislocation is of the screw type, near *C* it is of the edge type, and between *A* and *C* it is a more complex combination of the two. Numerous complicated types of dislocations can exist (for example, Ref A.29), but are beyond the scope of this Appendix.

**Generation of Multiple Dislocations.** Although a finite number of dislocations may already be present in a material in its initially formed state, such dislocations would soon be displaced to the edge of the crystal and would vanish. However, it turns out that plastic flow, in addition to annihilating dislocations, also generates new dislocations by several different processes. In fact, orders of magnitude more dislocations can be generated than are annihilated by plastic flow. For example, an annealed sample of metal may have only $10^7$ dislocation lines, on average, piercing each $cm^2$, but after heavy

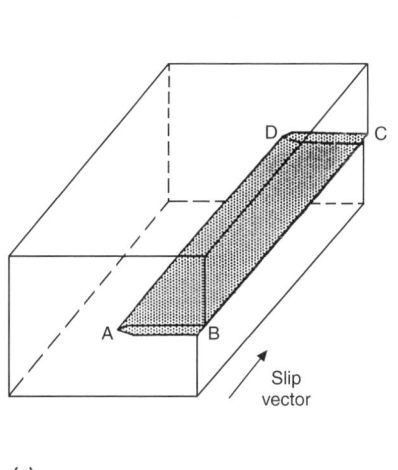
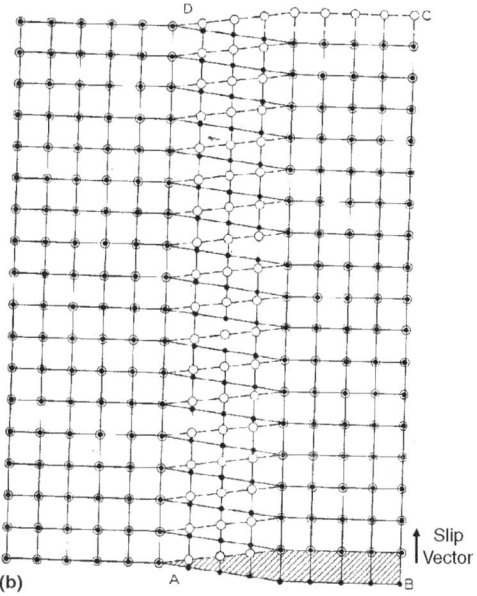

**Fig. A.15** Screw dislocation. (a) The slip that produces a screw dislocation. Unit slip has occurred over *ABCD*. The screw dislocation *AD* is parallel to the slip vector. (b) Arrangement of atoms around the screw dislocation shown in (a). Open circles represent the atomic plane just above the slip plane; solid circles represent the atomic plane just below. Source: Ref A.28

cold working can have as many as $10^{10}$ dislocations/cm$^2$. This is a multiplication ratio of $10^3$:1.

One of the first postulated dislocation generation mechanisms is the classical Frank-Read source (Ref A.30), illustrated in Fig. A.17 (Ref A.22). Although it is a seldom-observed mechanism of dislocation multiplication, it is one of the easiest to visualize. The ends of a short edge dislocation line in Fig. A.17(a) are locked from moving, perhaps by impurity atoms or by the dislocation continuing, although out of the current slip plane. As a shear stress is applied, the line is forced to bulge as in Fig. A.17(b). Continued shearing occurs and the dislocation line curls around Fig. A.17(c) and (d). When the bulging line meets itself, annihilation occurs along that length because the two mating incre-

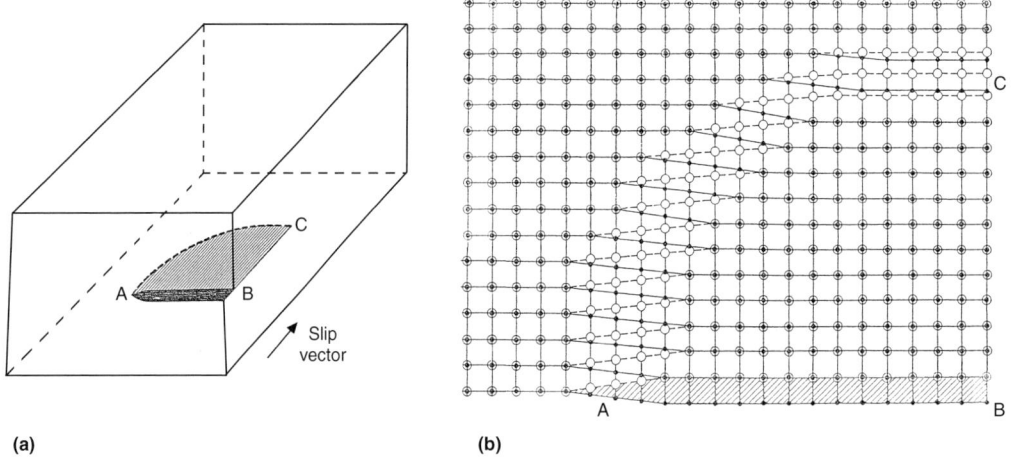

**Fig. A.16** Complex dislocation. (a) Unit slip in the area *ABC* produces a curved dislocation *AC* lying in a single slip plane. (b) Arrangement of atoms. Open and solid circles represent atoms just above and just below the slip plane. Source: Ref A.28

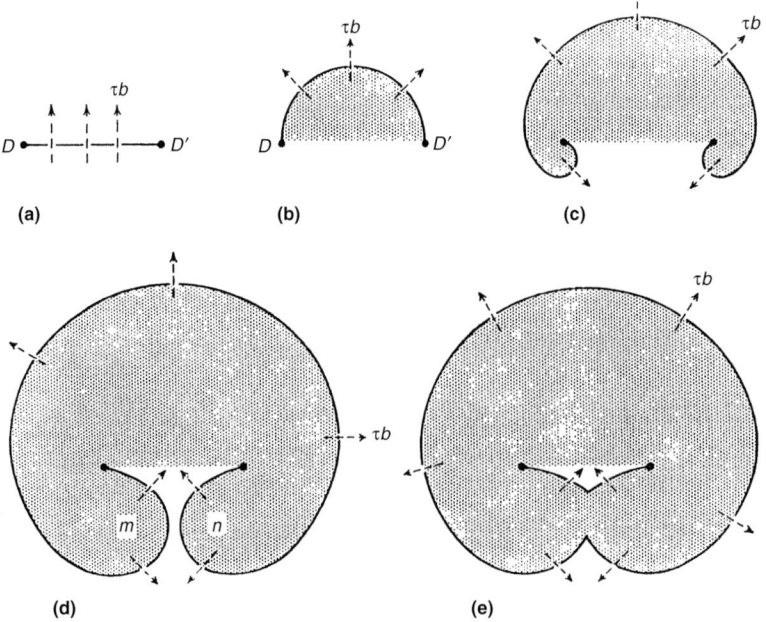

**Fig. A.17** Multiplication of dislocations by the Frank-Read source. Source: Ref A.22

ments are of opposite sign, creating the cancellation process. The outer ring of dislocation can continue to expand until it meets an obstacle, whereas the inner short-length portion reforms Fig. A.17(e) and the entire process can be repeated many times. While the Frank-Read source can generate many more dislocations, they could only be within the plane of the original pinned dislocation. Transmission electron microscopy, however, has revealed that nearly all of the newly created dislocations are complex tangles that cut across slip planes, thus relegating the Frank-Read source to more of an academic exercise than a realistic model. The model, however, did spawn the development of numerous and more viable theories of dislocation multiplication. These are beyond the scope of this Appendix.

Figure A.18 shows an example of the complex tangle of dislocations that can result from plastic flow on numerous parallel and intersecting slip planes.

## Heat Treatment as a Metallurgical Tool

One tool available to the metallurgist to influence the properties of an alloy is heat treatment. One of our most prevalent structural alloys is steel, an iron-base alloy. Steel is a remarkably complex material metallurgically. By subjecting it to programmed heat history, the microstructure can be altered significantly and properties desirable for special applications can be imparted.

Heat treatment is a complex subject, with the details best left to the professional specialist among the metallurgists. There are two relevant metallurgical societies in the United States that are valuable sources of subject detail: ASM International and the American Institute of Mining, Metallurgical, and Petroleum Engineers (AIME). However, even the elementary student can grasp the principles upon which procedures are based. We attempt here to provide the rudiments, which are useful in understanding some of the concepts and terminology, found in the literature, especially in reports relating to metal fatigue. We start with an introduction to equilibrium phase diagrams of metals. For simplicity we use the *binary* iron-carbon system not only because of its importance in steel technology, but because it provides an opportunity to introduce the terms so often encountered in the literature for other systems as well. The reader interested in the equilibrium phase diagrams of other common engineering alloys such as aluminum, titanium, and nickel should consult the voluminous literature (see, for example, Ref A.17 and A.21).

**The Equilibrium Phase Diagram.** Figure A.19 shows what phases (a uniform atomic mixing of the elements involved) will develop if a given weight percentage of carbon (horizontal axis) is added to molten iron and the temperature lowered and held at a particular value. It is important to emphasize that this is an *equilibrium* diagram, meaning there is no time factor involved; any given composition must be held at temperature long enough, whatever time it takes, to achieve the equilibrium phases shown. In such a case it makes no difference what the past temperature history has been. Otherwise, temperature history is very important and other phases (nonequilibrium) may develop, which could be desirable or undesirable. In fact, heat treatment often relies for its effectiveness on avoiding equilibrium conditions as discussed later. To understand the diagram, we need to define at least some of the terms shown:

**liquid (liquidus).** In this region at the top of the figure, the iron is molten and the carbon is dissolved in the iron.

**alpha-iron.** A solid solution of carbon in an iron phase ($\alpha$Fe or ferrite) in which the atoms are arranged in a bcc structure. It is a relatively soft phase having 0 to 0.022 wt% carbon.

**delta-iron.** A relatively unimportant bcc phase ($\delta Fe$) existing over a narrow high-temperature range

**austenite.** A solid solution of carbon in an iron phase ($\gamma$Fe, gamma iron) in which the atoms are arranged in an fcc structure

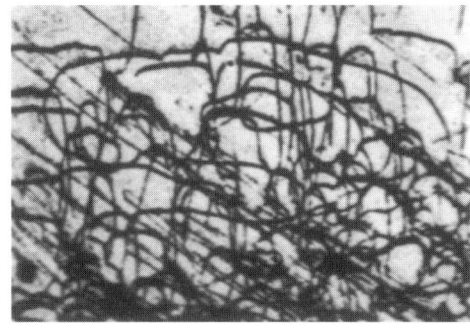

**Fig. A.18** A complex jungle of dislocations observed in silicon by x-ray diffraction. Source: Ref A.31

**cementite.** A compound of iron and carbon of composition Fe₃C, corresponding to ≈6.7 wt% carbon (well into the cast iron range). It is a relatively hard and brittle phase with a complicated lattice.

**pearlite.** A lamellar phase composed of alternate platelets of ferrite and cementite.

**martensite.** Of great importance to obtaining high-strength properties of steel. However, it does not appear in the iron-carbon equilibrium diagram because it is a metastable version of austenite and carbon that is obtained as a result of rapid cooling to near room temperature. The fcc lattice of the austenite experiences a virtually time-independent polymorphic transformation to a bcc tetragonal lattice involving no diffusion. The transformation is competitive with pearlite and bainite; these phases might form under other appropriate conditions. Presence of diffusion would result in formation of ferrite and cementite.

**bainite.** A mixture of ferrite supersatuarated with carbon and cementite, but in a nonequilibrium structure quite different from pearlite

**eutectic.** A mixture formed at the lowest possible freezing point of all combinations of constituents. The term can also refer to the temperature at which this occurs, i.e., point $C$ in Fig. A.19

**eutectoid.** A mixture transformed at the lowest possible transformation temperature for all combinations of constituents involved

**hypoeutectoid.** A mixture involving a lower concentration of the added element (carbon in this case) than that of the eutectic added to the base constituent (ferrite)

**hypereutectoid.** A mixture involving a higher concentration of the added element (carbon in this case) than that of the eutectic that is added to the base constituent (ferrite in this case)

Note that, depending on the carbon content, it requires a different temperature to cause the ma-

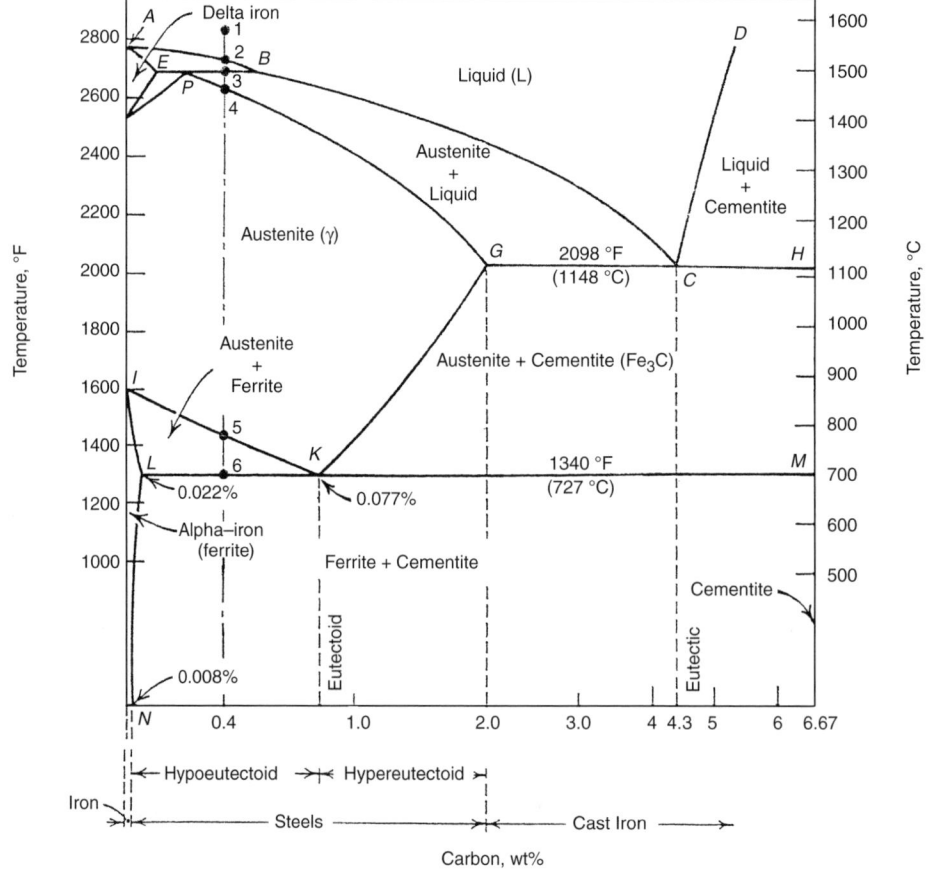

**Fig. A.19** Iron-carbon equilibrium phase diagram. Source: Ref A.17

terial to become totally liquid. The line *ABCD* essentially defines the liquidus temperature, which can vary from about 1130 to 1520 °C (2066 to 2770 °F). As the temperature drops from its liquidus, various transformations take place, according to carbon content. To illustrate, we can discuss the 0.4% carbon composition designated by the vertical line 1-2-3-4-5-6 in Fig. A.19.

At point 1 the material is totally liquid. As the temperature is lowered to about 1500 °C (2730 °F), point 2, the melt solidifies into delta-Fe + liquid. But at point 3 the phase becomes austenite plus liquid. Upon reaching point 4, the new phase becomes solid austenite. Proceeding down to point 5 (at about 790 °C, or 1450 °F), yet another solid transition occurs and austenite + ferrite is formed. Finally, beyond point 6, transformations conclude with an equilibrium phase of pearlite and ferrite. Further lowering of the equilibrium temperature doesn't change the microstructure.

In this way, the transformations expected for any composition of iron and carbon can be analyzed by following along the vertical line representing this Fe-C composition. But it is important to emphasize that the predestined structure, as temperature is reduced, is premised on the assumption that sufficient time is always allowed to achieve an equilibrium structure. The transformations are controlled by the dynamics of thermally activated diffusion and require time to take place. If the temperature is traversed rapidly, and the transformations are not given adequate time to take place, other compositions, such as martensite, can be achieved. These may not be equilibrium compositions, and they may thus not be stable. However, they may be stable enough for engineering purposes and their properties may be far superior to the equilibrium structures to justify themselves for a particular application. This is the underlying philosophy behind heat treatment: to impose a temperature history that will produce a desirable structure and associated mechanical properties that are best suited for a given application. The metastable structure may not be permanent and it may change with time, but this change, sometimes very minor, may be acceptable considering the benefits received.

**Quenching as the Starting Point in Heat Treatment.** The first step in heat treating an alloy is often the quenching (extremely rapid cooling) from a high temperature to a lower one in order to obtain a metastable structure that is allowed to develop because of the nonequilibrium conditions. To illustrate the principles without introducing complexities that are best handled by the metallurgist, we shall discuss the eutectoid composition steel containing 0.77% carbon, shown in Fig. A.19 as the vertical dashed line down from point *K*. As seen in the figure this composition is special because in cooling from the solid-solution austenitic state at 1340 °F, the transformation starts at the lowest possible temperature among the various compositions and there are no other potential equilibrium states that must be encountered below this temperature. Metallurgists refer to this specific composition as the *eutectoid* composition. To understand the nonequilibrium circumstances existing at this composition below 1340 °F, reference is made to Fig. A.20. This figure is the cooling transformation diagram for this eutectoid steel. This type diagram is commonly referred to as a *time-temperature-transformation* (T-T-T) *curve*. It is clear that above point *K* (1340 °F) the material will exist as austenite, and below 1340 °F it will transform to some other phase (in the case of long-time equilibrium conditions, the new phase would have been pearlite + cementite). However, as seen in this figure, if sufficient time is allowed at various isotherms below 1340 °F, a wide range of metastable, yet highly desirable, compositions can be achieved. Two C-shaped curves are shown: one giving the time it takes at temperature for the transformation to begin, the other when the transformation is completed. If the material is held very close to 1340 °F, the transformation is very sluggish; it takes of the order of $10^5$ seconds (28 h) to start and it isn't complete even after a million seconds (280 h). But if the temperature is allowed to fall below 1340 °F, the transformation starts earlier and ends earlier, as seen in the figure. The earliest it will start is at a temperature of around 1050 °F where, if the temperature is held constant, the transformation begins in about 1 second and ends after 5 seconds. At lower temperature there are further delays in the start and completion of the process because of slower reaction rates at lower temperature. So, if we wish to avoid the reaction entirely we must cool at a very rapid rate to miss the nose (the minimum time) of the initiation curve. Cooling according to the curves labeled $Q_1$, $Q_2$, and $Q_3$ will accomplish such a purpose. The critical cooling rate curve $Q_1$ is just tangent to the *C* curve for transformation start and is the slowest cooling rate that will prevent the pearlite transformation. Quenching this eutectoid steel in water or in oil can accomplish faster rates and are suitable for avoiding such a transformation.

Assume, for example, curve $Q_3$ is followed in the quench. No transformation will proceed to produce pearlite, but what does happen? Instead, an entirely new phase, martensite, forms starting when the temperature drops to about 475 °F. The line $M_s$ designates the start of the martensitic transformation. Martensite is a metastable iron phase supersaturated in carbon that is the product of a diffusionless transformation from austenite and can serve, with proper further heat treatment, to develop tough, high-strength steel. This transformation occurs almost instantaneously, but the degree to which it goes to completion depends on how much lower the temperature is reduced. For this particular material the completion is at about room temperature, shown as the $M_f$ (martensitic finish) line. Note that martensite is not an equilibrium structure, and as noted before, does not appear in the equilibrium phase diagram. It can decompose either spontaneously or as a result of further exposure to higher temperature. The responsibility of the

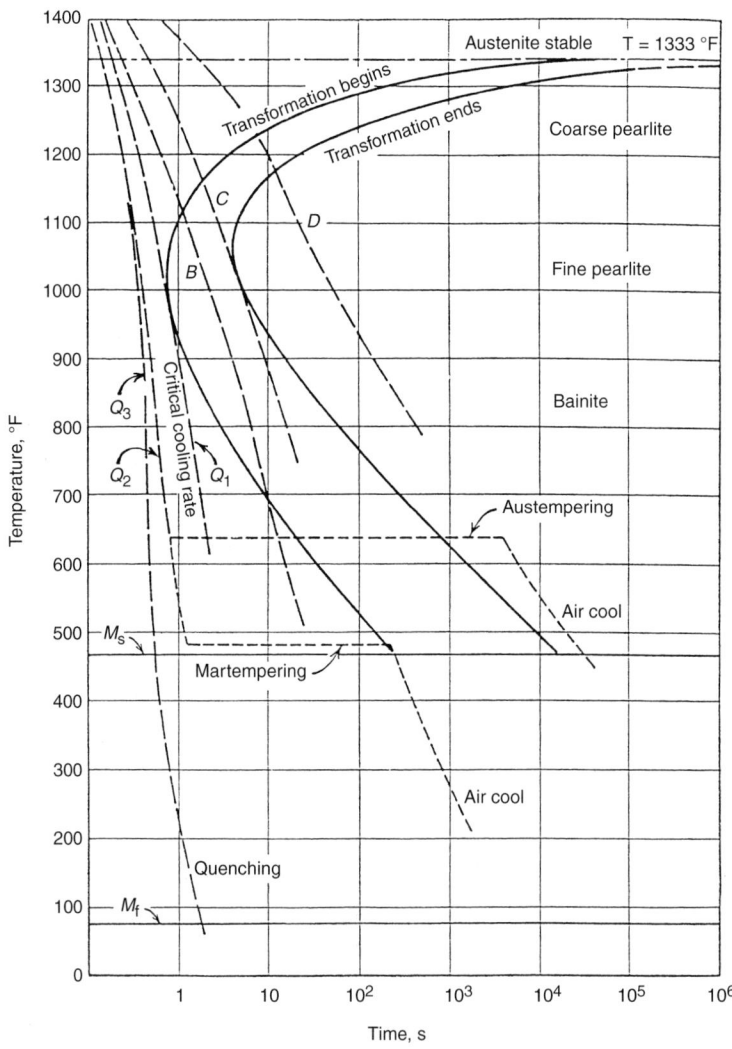

**Fig. A.20** Cooling transformation diagram for a eutectoid carbon steel (0.77% C). Curve B results in the microstructure composed of pearlite and martensite. Curve D corresponds to the rate of cooling during annealing. Curve C corresponds to normalizing. Cooling rates (curves $Q_1$, $Q_2$, $Q_3$) higher than the critical cooling rate result in martensitic structure on quenching to room temperature, or in bainite (austempering) or martempering according to the temperature of quenching. Source: Ref A.4

metallurgist is to provide further treatment to optimize the properties for his particular material.

**Common Heat Treatments.** Quenched steel, if it contains mostly martensite, is not suitable for many engineering applications. It is too hard and displays little toughness (resistance to crack extension). It also is riddled with internal residual stresses developed because of volume expansion as the martensite spontaneously grows from numerous initiation sites. Thus, the residual stresses must be removed and the hardness reduced to a tolerable level without undue reduction in strength. The metallurgist usually resorts to a tempering process, which means exposing the part to higher temperatures between approximately 400 °F and about 1350 °F for 2 h to only 10 min (in general, the higher the temperature, the lower the time), depending on the size of the part (the heavier the part, the higher the temperature and/or longer the time). Local creep then relieves the residual stresses, and some Fe$_3$C (cementite) starts to precipitate out, softening the martensite. A part treated in this way is referred to as *quenched and tempered*. But the metallurgist has numerous other choices as well. Among other terms that are often encountered in this connection are:

- *Normalizing*, in which the part is heated to a temperature above the upper critical range (IKG in Fig. A.19), wherein some transformation to austenite can take place. In this way the structure is made more uniform. It is then cooled in still air.
- *Annealing*, wherein the part is heated to a temperature above the critical range, and then cooled to below this range at a predetermined slow rate in order to allow some of the equilibrium products to precipitate and to reduce locked-in quenching stresses. In this way improved ductility and a host of other desirable properties are imparted to the material.
- *Spheroidizing*, wherein the steel is heated to a temperature of approximately 1300 °F (just below the *eutectoid* temperature) and held for 15 to 25 h. This heat treatment produces a soft material that is easily machined, thus saving machining costs. Subsequent heat treatment is applied to regain hardness and strength for service load resistance.
- *Stress Relieving*, wherein the re-heat temperature is kept well below the critical range but high enough to cause residual stresses to be relaxed or eliminated by thermally activated creep.

An enlargement (Fig. A.21) of the *eutectoid* region of the phase diagram of Fig. A.19 illustrates the zones of temperatures appropriate for each specific heat treatment.

Of course, numerous other combinations of exposure temperature and times are possible, and the metallurgist is guided by a documented wealth of knowledge in his specialized field and by his intuition and experience. A special program may be chosen to improve strength, ductility, fracture toughness, or creep characteristics, and we shall note a few in our discussion within the text. It should also be pointed out that the metallurgist has available a number of environments in which to carry out the heat treating process in order to produce most advantageously the effect he seeks. For example, in the rapid quench part of the program, he may use water, oil, air, or even molten salts. If the part is thin, or if he wishes to harden only a surface layer, he is most likely to use water quench because it provides the highest rate of heat removal. But if the part is thick, or if he wishes to harden more volume of the part, he may wish to remove heat more slowly by oil quenching. This may be possible only if the nose of the T-T-T curve, analogous to Fig. A.20, can be missed with the slower cooling rate. Often this is possible only for steel to which alloying elements have been added because these additions tend to move the nose toward the right of the figure. Thus, the choice of alloy steel for a given application may be dictated by the heat treatment that will be required, not by the inherent physical properties of the alloyed steel. For other stages of the heat treatment, the metallurgist may

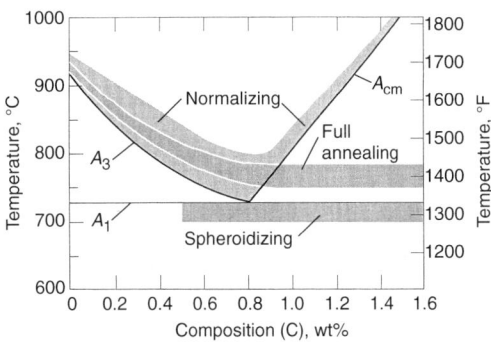

**Fig. A.21** Iron-carbon equilibrium phase diagram in vicinity of the eutectoid showing typical heat treating temperatures for normalizing, annealing, and spheroidizing. Source: Ref A.20

choose special environments, for example, nitrogen or vacuum, if he wishes to exclude air to avoid oxidation and decarburization. Important nuances such as achievable hardness and depth of penetration of hardness are dependent on carbon content as well as other alloying elements. The subject of heat treatment is very arcane and it is not the purpose of this discussion to cover the field broadly. It is discussed mainly to alert the reader that there are many complexities involved and to provide some background as to the terminology sometimes so casually cited.

## Plastic Deformation as a Metallurgical Tool

Plastic deformation is both a tool for materials improvement and a nuisance that has to be dealt with by the metallurgist. It is a nuisance because the forming of a part often inherently involves plastic deformation, which either directly forms cracks or it conditions the material to overbrittleness so that its service performance is impaired. On the other hand, suitably controlled working of a material can impart higher strength, which often cannot be achieved in other ways.

As noted under the discussion of dislocations, the basic mechanism of plastic deformation is the motion of dislocations, and the deformation itself generates additional dislocations. Thus, regions that have been subjected to large deformations will have many dislocation tangles and will be very hard, reducing ductility. For example, if a part is being stamped in a die to form a closed cylindrical container, the corners will undergo very large plastic strains that might exceed the available ductility prior to reaching the extent of shape change required. To improve a worked material before continuing with further stamping, the metallurgist resorts to an annealing process, which changes the condition of the material back to its original high-ductility state, so that it can safely withstand the imposed strains. Annealing also reduces the residual stresses to essentially zero.

**Annealing to Relieve Undesirable Effects of Working.** Annealing consists of exposing the material to a higher temperature for a fixed period of time (commonly one hour or several hours at most) so that adjustment can occur in the microstructure, especially where large plastic deformations have occurred, to improve its metallurgical state. Figure A.22 shows schematically the three regions normally identified in the annealing process of a cold-worked or heavily strained material. In region I (Recovery) the temperature is only moderately increased but is sufficient to allow the internal stresses to relax by virtue of dislocation motion. As seen in the figure, such low-temperature exposure does not greatly affect the microstructure, strength, or ductility, but is beneficial because of residual stress reduction. In region II recrystallization occurs because of the high, unstable internal energy of the large number of dislocations. The recrystallization starts in very local regions, at first in very tiny grains that fill the volume between the existing larger grains in the structure. The temperature at which it starts depends on the melting point of the alloy or pure metal (see Table A.2). At the high temperature of recrystallization, the initiated small grains grow and replace the larger grain structure of the cold-worked material. During this phase, the strength falls substantially and the ductility increases correspondingly. It is somewhere in this phase that the metallurgist seeks to spot an optimum combination of strength and ductility best suited for a specific engineering service. Beyond recrystallization there is phase III wherein the grains continue to grow and coarsen. Little additional benefit occurs with regard to relief of internal stresses and change in physical properties, but the coarser grain size may be disadvantageous

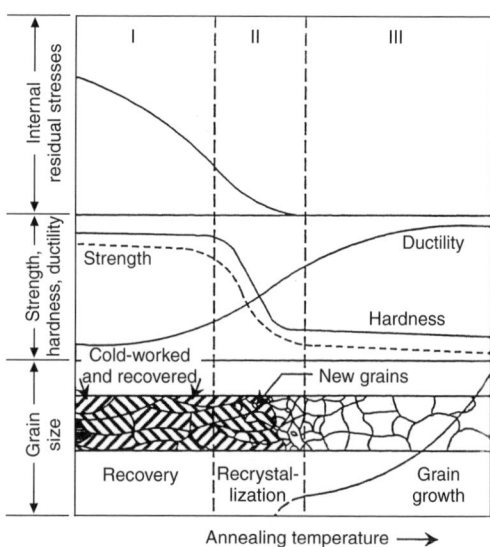

**Fig. A.22** Effect of annealing temperature on cold-worked metals

Table A.2  Physical, thermal, and mechanical properties of selected materials

| Material | Density, lb/in.$^3$ | $E$, 10$^6$ psi | $G$, 10$^6$ psi | $\mu$ | Yield strength $\sigma_{y2}$ or $\sigma_{p1}$, ksi | Tensile strength, $\sigma_u$, ksi | Elongation in 2 in., % | Hardness, HB | Melting point $T_m$ (or range), °F | Coefficient of thermal expansion, $\alpha$ 10$^{-6}$/°F | Thermal conductivity, $k_f$, Btu ft / hr ft$^2$ °F | Specific heat c (Btu/lb/°F or (cal/g)/°C) |
|---|---|---|---|---|---|---|---|---|---|---|---|---|
| **Ferrous metals** | | | | | | | | | | | | |
| Cast iron, gray (ASTM class 25) | 0.260 | 13 | 5.5 | 0.20 | ... | T 24 / C 120(a) | ... | 180 | 2150 | 6.7 | 27 | ... |
| Cast iron, ductile, as cast | 0.26 | 25 | 9.3 | ... | 68 | 90 | 7 | 235 | 2100 | 7.5 | 19 | 0.107 |
| Steel, 0.2% C, hot-rolled | 0.283 | 30 | 12 | 0.27 | 40 | 70 | 35 | 125 | 2760 | 6.7 | 30 | 0.107 |
| Steel, 0.2% C, cold-rolled | 0.283 | 30 | 12 | 0.27 | 65 | 80 | 20 | 160 | 2760 | 6.7 | 30 | 0.107 |
| Steel, 1.0% C, hot-rolled | 0.283 | 30 | 12 | 0.27 | 84 | 143 | 8 | 293 | ... | 7.3 | 30 | 0.107 |
| Steel, 1.0% C. hardened and tempered at 800 °F | 0.283 | 30 | 12 | 0.27 | 138 | 200 | 12 | 388 | ... | ... | 30 | 0.107 |
| Steel, AISI 4640, hardened and tempered at 800 °F | 0.28 | 29 | 11 | 0.30 | 190 | 202 | 15 | 388 | ... | ... | 30 | 0.107 |
| Stainless steel, type 302, cold-rolled | 0.286 | 29 | 10.6 | 0.30 | 100 | 140 | 30 | 280 | 2575 ± 25 | 8.9 | 9.4 | 0.12 |
| **Nonferrous metals** | | | | | | | | | | | | |
| Aluminum, 1100-O (annealed) | 0.098 | 10.0 | 3.75 | 0.33 | 3.5 | 11 | 25 | 23 | 1200 ± 15 | 13.1 | 128 | 0.23 |
| Aluminum alloy, 2024-T4 | 0.100 | 10.6 | 4.0 | 0.33 | 44 | 60 | 12 | ... | 1075 ± 140 | 12.9 | 70 | 0.23 |
| Aluminum alloy, 7075-T6 | 0.101 | 10.4 | 3.9 | 0.33 | 70 | 80 | 10 | 150 | 1035 ± 145 | 13.1 | 75 | 0.23 |
| Brass, yellow, hard | 0.306 | 15 | 5.6 | 0.35 | 60 | 74 | 10 | 180 | 1710 | 10.5 | 69 | 0.09 |
| Phosphor bronze, grade A, hard | 0.320 | 16 | 6 | ... | 65 | 80 | 8 | 160 | 1920 | 9.4 | 40 | 0.09 |
| Magnesium alloy, ZK51A-T5 | 0.066 | 6.5 | 2.4 | 0.35 | 25 | 40 | 8 | 80 | 1175 | 14.5 | 70 | 0.245 |
| Titanium alloy, 6Al-4V | 0.161 | 15.8 | 6.0 | 0.34 | 130 | 140 | 15 | ... | 2881 ± 95 | 4.8 | 3.8 | 0.135 |
| **Metallic elements** | | | | | | | | | | | | |
| Beryllium, extrusions, annealed(b) | 0.066 | 44 | ... | ... | 55 | 90 | 5 | ... | 2340 ± 70 | 6.9 | 92 | 0.52 |
| Cobalt, sintered(b) | 0.32 | 30 | ... | ... | 44 | 100 | 13 | ... | 2723 ± 2 | 6.8 | 40 | 0.099 |
| Columbium, commercial(b) | 0.31 | 22.7 | ... | ... | 24 | 39 | 49 | ... | 4380 | 4.0 | ... | 0.065 |
| Copper, annealed | 0.322 | 17 | 6.4 | 0.33 | 10 | 32 | 45 | 42 | 1980 | 9.3 | 225 | 0.092 |
| Hafnium(b) | 0.412 | 20 | ... | ... | 22 | 59 | 35 | ... | 3800 ± 200 | 3.4 | ... | 0.035 |
| Lead, chemical, rolled | 0.410 | 2 | 0.7 | 0.43 | 2 | 2.5 | 50 | 5 | 621 | 16.4 | 20 | 0.030 |
| Molybdenum, as-rolled | 0.369 | 42 | ... | ... | 75 | 100 | 30 | 250 | 4760 ± 90 | 2.67 | 7.5 | 0.061 |
| Nickel, A, annealed | 0.321 | 30 | 11 | ... | 20 | 70 | 40 | 100 | 2625 ± 10 | 6.6 | 35 | 0.13 |
| Tantalum | 0.600 | 27 | ... | ... | 100 | 110 | 3 | 123 | 5425 ± 90 | 3.6 | 31 | 0.036 |
| Tungsten, hard (sheet)(b) | 0.697 | 50 | ... | ... | ... | 300 | 0 | ... | 6092 | 2.4 | 116 | 0.034 |
| Uranium, annealed(b) | 0.676 | 30 | ... | ... | 25 | 90 | 13 | ... | 2065 | ... | 15 | 0.028 |
| Vanadium, annealed | 0.217 | 21.5 | ... | ... | 55 | 68 | 34 | ... | 3150 ± 90 | 4.3 | ... | 0.120 |
| Zirconium | 0.245 | 12 | ... | ... | 16 | 36 | 31 | 77 | 3380 | 2.9 | ... | 0.118 |

(a) Gray cast iron, a brittle material, has a well-defined compressive ultimate strength. (b) Data courtesy of Crucible Steel Co. (c) Modulus of elasticity measured in tension (T), compression (C), or bending (B). Source: Ref A.7

## Table A.2 Continued

| Material | Specific gravity | Elastic constants $E$(c) ksi | Elastic constants $\mu$ | Tensile strength, $\sigma_{ut}$ (or $\sigma_y$), ksi | Compressive strength, $\sigma_{uc}$ (or $\sigma_y$), ksi | Elongation in 2 in., % | Hardness HR | Max recommended service temp., °F | Coefficient of thermal expansion, $\alpha$ $10^{-6}$/°F | Thermal conductivity, $k$, $\frac{\text{Btu ft}}{\text{hr ft}^2 \,°F}$ | Specific heat $c$ (Btu/lb)/°F or (cal/g)/°C | Dielectric constant (60 cps) |
|---|---|---|---|---|---|---|---|---|---|---|---|---|
| **Thermoplastics** | | | | | | | | | | | | |
| Acrylic plastics, cast, type II | 1.18 | T 400 | 0.4 | 8 | 12 ($\sigma_y$) | 4 | M 100 | 170 | 50 | 0.12 | 0.35 | 4.0 |
| Cellulose acetate, molded, type I | 1.30 | B 250 | ... | 5 | 20 | 32 | R 100 | ... | 70 | 0.15 | 0.36 | 5.5 |
| Polytetrafluoroethylene (Teflon) | 2.2 | T 50; C 80 | 0.4 | 2 | 1 ($\sigma_y$) | 235 | J 85 | 500 | 30 | 0.15 | 0.25 | 2.0 |
| Nylon 66, general-purpose (polymide) | 1.14 | T 410 | 0.4 | 8 | 13 ($\sigma_y$) | 60 | R 118 | 250 | 55 | 0.14 | 0.40 | 3.9 |
| Polystyrene, general-purpose | 1.05 | T 500 | ... | 7 | 14 | 2 | M 80 | 150 | 70 | ... | 0.32 | 2.55 |
| Polyethylene, medium-density | 0.93 | T 60 | 0.45 | 2 | ... | 350 | ... | 180 | 125 | 0.20 | 0.55 | 2.3 |
| **Thermosets** | | | | | | | | | | | | |
| Epoxy, general-purpose, cast | 1.12 | T 650 | ... | 7 | 30 | 4 | M 100 | 176 | 33 | 0.45 | 0.33 | 4.3 |
| Phenolic, molded, type 2, general-purpose | 1.44 | T 1000 | ... | 7.5 | 30 | 0.6 | M 115 | 325 | 21 | 0.13 | 0.38 | 7.0 |
| Silicon, mineral-filled, molded | 1.9 | B 1150 | ... | 4 | 18 | ... | M 88 | 550 | 30 | 0.12 | 0.25 | 4.3 |
| **Rubber, natural, molded** | 0.92 | ... | 0.50 | 3 | ... | 800 | ... | 150 | 90 | ... | ... | 2.4 |
| **Wood** | | | | | | | | | | | | |
| Douglas fir, air dry (parallel to grain) | 0.48 | B 1900 | ... | 8.1 ($\sigma_y$) | 7.4 | ... | ... | ... | ... | 0.77 | ... | ... |
| White oak, air dry (parallel to grain) | 0.67 | B 1620 | ... | 7.9 ($\sigma_y$) | 7.0 | ... | ... | ... | ... | ... | ... | 5.2 |

(a) Gray east iron, a brittle material, has a well-defined compressive ultimate strength. (b) Data courtesy of Crucible Steel Co. (c) Modulus of elasticity measured in tension (T), compression (C), or bending (B). Source: Ref A.7

with regard to fatigue. It is noted, however, that high-temperature creep resistance improves as the grain size increases.

If the material is worked in the recrystallization range, the process is known as hot working. No increase in strength occurs as in cold working because the simultaneous recrystallization nullifies the increase in strength. Of course, the shaping of a part can be accomplished with less mechanical work by hot working, but it is more difficult to maintain the part temperature, oxidation during working becomes a problem, and it is more difficult to maintain fine dimensional tolerances.

**Purposeful Cold Working to Achieve Strength.** While cold working may be a side effect of shaping a part, it may also be a very desirable process to raise the strength of a material that is otherwise too weak. It is especially useful for materials, which cannot be heat treated to a desired strength level. Figure A.23 shows how purposeful cold work can raise the strength of several classes of material. It is not uncommon to be able to double the strength of most of these materials if sufficient cold work is imposed. Of course, imposing such high reductions may become a problem for some materials of initially limited ductility. Forcing the material through a die can be helpful because the die imposes transverse compressive stresses, and the triaxiality of the resulting compressive stress system effectively improves the ductility, which allows more cold work than would be possible by uniaxial tensile loading. Forcing the material through powered rollers can go even further in producing deformation without imposing high tensile drawing loads. Subjecting the material to cyclic loading as, for example, by weaving a sheet alternately over the upper and lower surfaces of successive powered rollers can be helpful in producing a given amount of cold work in a more benign manner (i.e., producing less embrittlement). We discuss in great detail (Chapter 3, "Fatigue Life Relations") how cyclic plastic deformation affects materials properties, especially in relation to low cycle fatigue.

Of course, it should be recognized that accompanying the increases in strength accomplished by cold working, there would generally develop a reduction of ductility. If the metallurgist finds excessive embrittlement as a result of a specified amount cold work, he could follow the cold work by an annealing process. Sometimes he may even benefit more by selective overworking and then annealing to a specified strength and ductility than by achieving the strength or the ductility by monotonic cold working.

**Thermomechanical Processing and Selective Programming of Working and Heat**

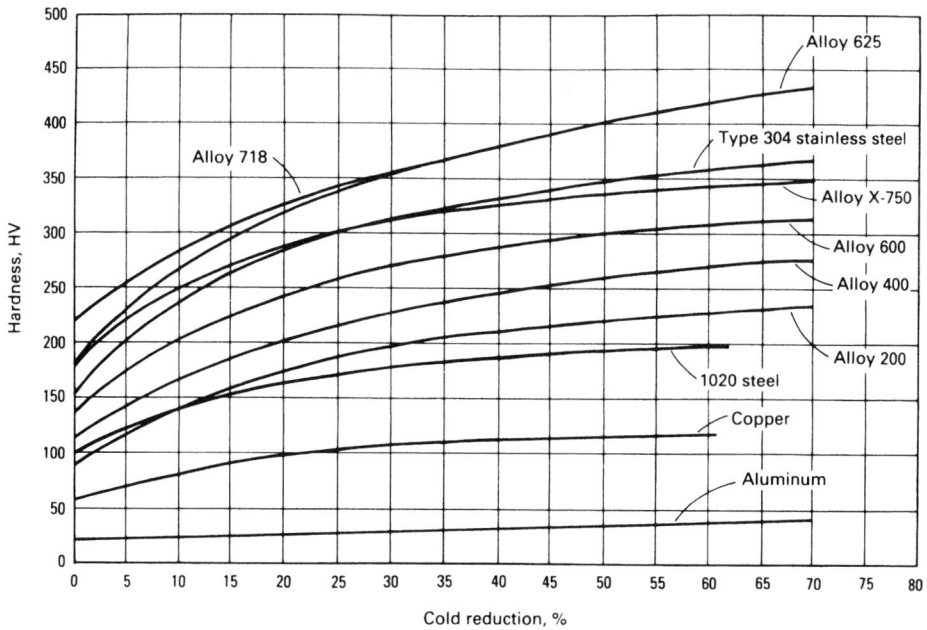

**Fig. A.23** Effect of cold work on hardness for various metals. Source: Ref A.4

**Treating.** In recent years it has become recognized that a carefully selected program of working and heating either simultaneously or successively can produce desirable properties that neither can achieve alone. One such case, for example, is ausforming, introduced by the Ford Motor Company in the 1960s. The process can be explained using Fig. A.20. Suppose, for example, a material is cooled according to path $Q_2$ by quenching into a salt or low-melting alloy, to about 640 °F. If the material is then held at this temperature for about an hour, an appreciable amount of bainite (which is harder than pearlite and tougher than martensite) will form. The process, as shown in Fig. A.20, is called *austempering*. But if the bainite-containing steel is worked while it is being formed the process is labeled as *ausforming*. The working breaks up the bainite into very fine particles, and the alloy then becomes dispersion strengthened by these fine particles. These tough particles can then block dislocation motion, which makes the material very strong. Also, because these particles are extremely fine in a ductile matrix, the material develops exceptionally good fatigue properties associated with both strength and ductility. The properties of ausformed steel are discussed in Chapter 11, "Avoidance, Control, and Repair of Fatigue Damage."

Alloys other than steel can have their properties controlled by appropriate combinations of cold working and heat treating. In Fig. A.24, we use, as an example, the high-temperature cobalt-base superalloy, L-605, studied at NASA (Ref A.32). Samples of the alloy were subjected to various degrees of prior cold reduction of up to 40% followed by aging at 1600 °F for either 200 or 1000 h. As shown in the figure, maximum increases in both strength and ductility (and, hence, toughness) were achieved with 30% prior cold work followed by aging at 1600 °F for 200 h. It was possible to obtain a material with both a maximum in strength as well as maximum in ductility, whereas normally it would be expected that processing for maximum strength would sacrifice ductility and vice versa.

## Review of Important Engineering Materials Systems

To understand the relation of materials to design, it is appropriate first to discuss the common materials. In this section we review the most useful of these materials. It is important to recognize that the choice of material will depend on its availability, cost, processibility to desired properties, physical properties, thermal properties, chemical properties, and how they affect its service performance.

### Steel

Steel is the most common material used in engineering design. It is made from relatively available elements, is inexpensive, and processible to a wide range of desirable properties. Basically, steel is a mixture of iron and carbon. Iron itself is a very ductile material with relatively low strength. It is also very reactive with commonly encountered environments such as air, water (especially if it contains salt), and other chemicals. Because iron is one of the most common and least-expensive elements, it long ago became evident that it could be used as a basic component for material construction. The Iron Age started over 1000 years Before the Common Era (BCE) and has continued until the present time. Over the centuries, many innovations have been introduced in the production of iron, in combining it with carbon to produce basic steel. Alloying it with numerous other elements enhances strength and other desirable characteristics. Processing it by heat treatment and mechanical working produces a truly unique and multipurpose material that has the popularity that steels deserve.

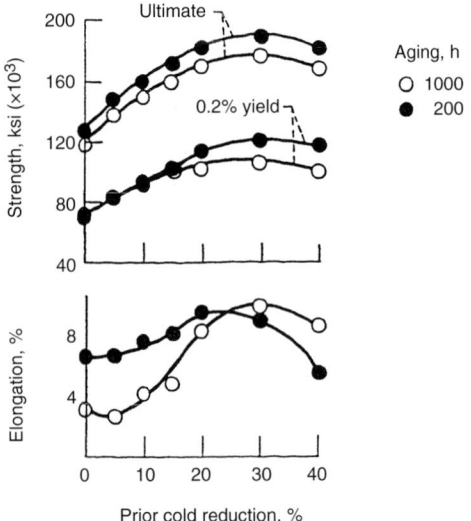

**Fig. A.24** Effect of prior cold reduction on room temperature tensile properties of cobalt-base alloy L-605 after aging either 200 or 1000 h at 870 °C (1600 °F). Source: Ref A.32

It is beyond the scope of this brief review to present a history of the development of steel. However, the following list of facts can be helpful to the nonmetallurgist reader in understanding many of the seeming complexities associated with this product.

**Carbon Content.** The main factor that distinguishes one steel alloy from another is the amount of carbon it contains. Only small amounts of carbon—usually less than 1%—can drastically affect the strength properties of steel. Carbon, either as an element by itself, or as a compound (carbide) in combination with another of the elements, can produce a spectrum of strength (or hardness) that ranges from ultrasoft to ultrahard. Because carbon content is of such importance, it is one of the dominant identifiers describing steel. In designations of steel, one of the numbers used involves the carbon composition. Thus, for example, 1010 steel means that the steel contains $^{10}/_{100}$ or 0.1 wt% (w/o) carbon, while the steel 52100 contains 1.0 w/o carbon (in this case, the last three digits, divided by 100). The first of these two metals is quite soft and weak (but compensates by having a high deformability or ductility) with a tensile strength of 415 MPa (60 ksi), while the second alloy can have a tensile strength of 290 MPa (300 ksi) (and commensurately lower ductility). Each has its own field of use.

**Alloying elements** are added to impart a special property called *hardenability* as well as other unique characteristics. For example, chromium improves hardness and resistance to corrosion; nickel toughens the steel and improves its temperature and corrosion resistance; manganese increases strength and wear resistance; molybdenum increases strength and heat resistance; tungsten improves the high-temperature strength and hardness; and vanadium increases strength and its ability to retain its elasticity. Some elements improve the processing of the steel as well as its properties. Manganese is an active deoxidizer and has fewer tendencies to segregate within the ingot than other common elemental additives. It enables the steel to achieve a fine-grained structure, which improves the toughness and fatigue resistance. Silicon decreases the susceptibility of the steel to decarburization (loss of carbon during processing) and when used together with other elements such as nickel, chromium, and tungsten improves strength and high-temperature oxidation resistance. Copper also increases strength, atmospheric corrosion, and conductivity. Paint also adheres better to steel containing copper than to non-copper-bearing steels. Aluminum, because of its affinity for oxygen atoms, helps to deoxidize steel, and refines its grain size. It is also useful in heat treating steel to higher hardness, especially in case hardening a component when it is desirable to achieve a higher hardness on the surface than in the interior of the part. Boron can be added in very lowconcentration, yet produces a very large effect on the ability to harden even low-carbon steels. It thus makes possible high hardness with very low additions of other alloying elements, conserving the other elements and reducing cost. Lead is sometimes added to steel that is to be subjected to machining operations. The lead reduces friction during machining so that machining speeds can be increased and deeper cuts can be made. Since lead is miscible in the molten steel, upon solidification, the lead is uniformly dispersed and results in easier machining. Hence, such steels are referred to as *free-machining*. The lead is soft and acts as a lubricant as well as a *chip breaker*.

Thus, it can be seen that elemental additives (compositions) are a tool for the metallurgist to achieve desirable physical and structural properties. Not all steels are suitable for all applications, but with care the composition can be optimized for each application. Clever and innovative scientists, engineers, and practitioners have developed the art of alloying over many centuries.

### *Stainless Steel*

The addition of nickel and chromium can render steel corrosion and oxidation resistant. For example, high chromium content (4 wt% minimum, but usually 10 wt% or more) is thought to help the material form a thin passive surface film, probably an oxide that protects against corrosion. This surface film can heal if penetrated and thus continues to offer protection.

Table A.3 shows the four general groups of stainless steels: austenitic (200 and 300 series); ferritic (400 series); martensitic (400 and 500 series); and precipitation hardening (PH series). As noted in the table, the 200 and 300 series are nonmagnetic, whereas the 400, 500, and PH series are magnetic.

Despite their highly desirable corrosion and oxidation resistant attributes (stainless), these steels are costly compared with low-carbon and specialty steels. Production is expensive because of the higher cost of the raw alloying elements,

Table A.3  Characteristics of the families of stainless steels

| Stainless steel series | Crystal structure | Description | Magnetic | Hardenability by Heat treatment | Hardenability by Cold working |
|---|---|---|---|---|---|
| 200 | fcc | Austenitic | No | No | Yes |
| 300 | fcc | Austenitic | No | No | Yes |
| 400 | bcc | Ferritic | Yes | No | Yes |
| 500 and precipitation hardening | bcc | Martensitic | Yes | Yes | No |

fcc, face-centered cubic; bcc, body-centered cubic

which are not as readily available. Stainless steels currently account for only 1.5% of the world's annual steel production. Austenitic stainless steels are more difficult to machine than other steels. Grinding wheels clog readily due to their "gummy" nature. Conventional machine tool cutters tend to dull more quickly and hence suffer undesirable heat buildup in both the workpiece and the tool. This situation is further aggravated by the relatively poor thermal conductivity that is characteristic of austenitic stainless steels. Manufacturing costs associated with stainless steels thus tend to be higher than for conventional carbon steels. Annealed stainless steels exhibit relatively low yield strengths, but work hardens at very high rates. Austenitic stainless steels are noted for their ductility. They exhibit reduction of area (tensile test) values on the order of 30 to 70%, depending on initial degree of work hardening present. Two of the most commonly used structural austenitic stainless steels are types 304 and 316.

## Aluminum Alloys

The second most widely used metal in engineering design is aluminum. Unlike steel, its first use dates back only slightly more than a century ago, when an economically feasible means for smelting bauxite (naturally occurring aluminum ore) was developed by Prof. Charles Hall of Oberlin College*. More expensive to convert into useful ingots than steel, aluminum possesses a number of desirable attributes. It is lightweight, relatively easy to cast, malleable, easy to machine, and can be alloyed and heat treated to strength levels as high as 550 MPa (80 ksi) or more. With care, aluminum alloys are weldable and brazable. Aluminum can be surface anodized with colorfully decorative and durable finishes. All current commercial aircraft utilize weight-saving aluminum extensively in the airframes, spars, and skin. Because of its high conductivity, it is used for electrical wiring and in home cookware. Aluminum has only one-third the density and stiffness of steel. Unalloyed aluminum oxidizes instantaneously in air but quickly forms a reasonably protective oxide surface layer, leading to good weather resistance. Prolonged exposure to saltwater environments, however, catastrophically attacks aluminum.

Alloying elements are added for strength and heat treating purposes. Copper, manganese, silicon, magnesium, and zinc are the dominant alloying elements. A relatively recent alloying addition, lithium, imparts 10% lower density and 10% higher stiffness. A four-digit numerical code clearly identifies the principal alloying elements in accordance with Table A.4. It should be noted that the heat treating process for aluminum alloys is quite different from that for steels. In aluminum alloys, the hardening is a result of precipitation, with time (i.e., aging) of

Table A.4  Wrought aluminum alloy four-digit designation system

| Group | Designation(a) |
|---|---|
| Aluminum, ≥99.00% | 1xxx |
| Aluminum alloys grouped by major alloying element(s) | |
| Copper | 2xxx |
| Manganese | 3xxx |
| Silicon | 4xxx |
| Magnesium | 5xxx |
| Magnesium and silicon | 6xxx |
| Zinc | 7xxx |
| Lithium | 8xxx |
| Unused series | 9xxx |

*According to the *ASM Handbook*, the process for "electrolytic reduction of alumina dissolved in molten cryolite was independently developed by Charles Hall in Ohio and Paul Heroult in France in 1886" (*ASM Handbook*, Vol 2, *Properties and Selection: Nonferrous Alloys and Special-Purpose Materials,* ASM International, 1990, p 3). The technique is commonly referred to as the "Hall-Heroult process."

(a) The first digit identifies the group. The second digit identifies alloy modification; zero is the original alloy. In the previous designation system, modifications were indicated by letters. In the current system, the letter is replaced by the digit corresponding to its position in the alphabet. For example, A17S became 2117. The last two digits identify the aluminum purity or the specific aluminum alloy. For alloys in use prior to the adoption of the four-digit system, the digits are the same as the numbers in the previous designation. For example, alloy 24S is now 2024. Source: The Aluminum Association

hard particles that had been dissolved and held in a state of solid solution at the relatively high solutionizing temperature. As the temperature of the alloy is then lowered and held for a period of time, the hard particles precipitate in a highly dispersed manner throughout the volume and are allowed to grow to an optimum size to achieve maximum hardness and strength. At that point, the temperature is further reduced to stop the thermally activated precipitation and growth process. Varying degrees of hardness, strength, and ductility are achievable by using different heat treating variables.

## Titanium Alloys

It was not until the 1960s and 1970s that titanium alloys began to emerge as viable structural materials, predominantly in the aerospace and chemical industries. Low density with high strength, moderate stiffness, and machinability overshadow the high cost of production for critical applications. Modern, efficient aeronautical gas turbine engines owe their high thrust-to-weight ratio to the introduction of titanium alloys in the cooler-running fan and compressor stages. Tensile strengths higher than 1400 MPa (200 ksi) are attainable with proper alloying and heat treatment. Relatively high ductility is retained in most wrought product forms. Use of titanium alloys as a viable matrix material for metal-matrix composites for moderately high-temperature applications in the aerospace industry has been actively pursued since the 1980s.

The Society of Automotive Engineers Aerospace Materials Specification (AMS) has prepared a numerical designation for commercially available titanium alloys. The first two digits for all titanium alloys is 49, whereas the latter two digits have a fixed definition that codes the chemical composition, heat treatment, and wrought product form. Table A.5 lists detailed information on commercial titanium alloys.

## High-Temperature Alloys (Nickel-Base, Cobalt-Base, and Refractory Alloys)

Only brief mention will be made of some of the metals and alloys designed specifically for high-temperature application. Aeronautical gas turbine engines have required the use of alloys that can operate safely for prolonged periods (10,000 to 20,000 h) in hot gas environments. Turbine hot section gas temperatures frequently exceed the melting temperature of the alloys used. Only because of ceramic oxidation-protective coatings and internal forced cooling is it possible for metals to be used under such harsh conditions. Turbine blade metal temperatures frequently reach 1040 to 1090 °C (1900 to 2000 °F), only a few hundred degrees below the melting point of the alloys. Materials had to be developed concurrently with gas turbine engine development. All commercial gas turbine aircraft engines use some form of nickel- or cobalt-base superalloy that has been intentionally strengthened and alloyed to resist high stresses in a high-temperature oxidizing environment. Resistance to creep deformation (time-dependent, thermally activated deformation mechanism) is also of great concern. For this purpose, a cast material with large grains, rather than wrought with fine grains, has significantly superior resistance to creep deformation and cracking. The smaller the total area of grain boundaries of the large-grained material, the less is the amount of creep that is concentrated in these weak zones of the material. Even single-crystal alloys have been developed to eliminate the notoriously weak (at high temperature) grain boundaries found in polycrystalline alloys.

A series of metals from the periodic table have exceptionally high melting temperatures, high stiffness, and reasonably high-unalloyed strengths. These are the *refractory metals* and include tungsten (melting temperature, $T_m$ = 3400 °C, or 6150 °F), molybdenum ($T_m$ = 2610 °C, or 4730 °F), tantalum ($T_m$ = 3000 °C, or 5430 °F), and niobium (formerly columbium) ($T_m$ = 2470 °C, or 4475 °F). With the exception of niobium, they exhibit low ductility, and their density is quite high, thus severely restricting their use in aeronautical gas turbine engines. In addition, a strong affinity for oxygen renders them virtually useless in an oxidizing atmosphere. A common engineering use for tungsten is filaments for incandescent light bulbs. The high melting temperature permits operation at temperatures in the white hot and highly radiative regime. To prevent oxidation of the filament, bulbs are evacuated of oxygen, backfilled with an inert gas, and sealed.

## Polymers

Polymers are discussed in Chapter 12, "Special Materials," where we assessed the applicability of metallic-based fatigue models to polymer behavior. Herein, we present other basic aspects of polymers.

## Table A.5  Titanium alloys (AMS[a] specifications)

| AMS No. | Mill form | Condition | Alloy(b) | Similar MIL specification |
|---|---|---|---|---|
| 4900 | Plate, sheet, strip | Annealed | Unalloyed; 55 ksi YS | MIL-T-9046 |
| 4901 | Plate, sheet, strip | Annealed | Unalloyed; 70 ksi YS | MIL-T-9046 |
| 4902 | Plate, sheet, strip | Annealed | Unalloyed; 40 ksi YS | MIL-T-9046 |
| 4906 | Sheet, strip; continuously rolled | Annealed | Ti-6Al-4V | ... |
| 4907 | Plate, sheet, strip | Annealed | Ti-6Al-4V-ELI | MIL-T-9046 |
| 4908 | Sheet, strip | Annealed | Ti-8Mn; 110 ksi YS | MIL-T-9046 |
| 4909 | Plate, sheet, strip | Annealed | Ti-5Al-2.5Sn-ELI | MIL-T-9046 |
| 4910 | Plate, sheet, strip | Annealed | Ti-5Al-2.5Sn | MIL-T-9046 |
| 4911 | Plate, sheet, strip | Annealed | Ti-6Al-4V | MIL-T-9046 |
| 4915 | Plate, sheet, strip | Single annealed | Ti-8Al-1Mo-1V | MIL-T-9046 |
| 4916 | Plate, sheet, strip | Duplex annealed | Ti-8Al-1Mo-1V | MIL-T-9046 |
| 4917 | Plate, sheet, strip | Solution treated | Ti-13V-11Cr-3Al | MIL-T-9046 |
| 4918 | Plate, sheet, strip | Annealed | Ti-6Al-6V-2Sn | MIL-T-9046 |
| 4921 | Bar, forgings, rings | Annealed | Unalloyed; 70 ksi YS | MIL-T-9047 |
| 4924 | Bar, forgings, rings | Annealed | Ti-5Al-2.5Sn-ELI; 90 ksi YS | MIL-T-9047 |
| 4926 | Bar, rings | Annealed | Ti-5Al-2.5Sn; 110 ksi YS | MIL-T-9047 |
| 4928 | Bar, forgings | Annealed | Ti-6Al-4V; 120 ksi YS | MIL-T-9047 |
| 4930 | Bar, forgings, rings | Annealed | Ti-6Al-4V-ELI | MIL-T-9047 |
| 4935 | Extrusions | Annealed | Ti-6Al-4V | ... |
| 4936 | Extrusions | | Ti-6Al-6V-2Sn | ... |
| 4941 | Tubing, welded | Annealed | Unalloyed; 40 ksi YS | ... |
| 4942 | Tubing, seamless | Annealed | Unalloyed; 40 ksi YS | ... |
| 4943 | Tubing, seamless | Annealed | Ti-3Al-2.5V | ... |
| 4944 | Tubing, seamless hydraulic | Cold worked and stress relieved | Ti-3Al-2.5V | ... |
| 4951 | Wire, welding | | | ... |
| 4953 | Wire, welding | Annealed | Ti-5Al-2.5Sn | ... |
| 4954 | Wire, welding | | Ti-6Al-4V | ... |
| 4955 | Wire, welding | | Ti-8Al-1Mo-1V | ... |
| 4956 | Wire, welding | | Ti-6Al-4V-ELI | ... |
| 4965 | Bar, forgings, rings | Precipitation heat treated | Ti-6Al-4V | ... |
| 4966 | Forgings | Annealed | Ti-5Al-2.5Sn; 110 ksi YS | MIL-F-83142 |
| 4967 | Bar, forgings | Annealed | Ti-6Al-4V | MIL-T-9047 |
| 4970 | Bar, forgings | Precipitation heat treated | Ti-7Al-4Mo | MIL-T-9047 |
| 4971 | Bar, forgings, rings | Annealed | Ti-6Al-6V-2Sn | MIL-T-9047, MIL-F-83142 |
| 4972 | Bar, rings | Solution treated and stabilized | Ti-8Al-1Mo-1V | ... |
| 4973 | Forgings | Solution treated and stabilized | Ti-8Al-1Mo-1V | ... |
| 4974 | Bar, forgings | Precipitation heat treated | Ti-11Sn-5Zr-2.3Al-1Mo-0.21Si | ... |
| 4975 | Bar, rings | Precipitation heat treated | Ti-6Al-2Sn-4Zr-2Mo | MIL-T-9047 |
| 4976 | Forgings | Precipitation heat treated | Ti-6Al-2Sn-4Zr-2Mo | ... |
| 4977 | Bar, wire | Solution treated | Ti-11.5Mo-6Zr-4.5Sn | MIL-T-9047 |
| 4978 | Bar, forgings, rings | Annealed | Ti-6Al-6V-2Sn; 140 ksi YS | MIL-T-9047, MIL-F-83142 |
| 4979 | Bar, forgings, rings | Precipitation heat treated | Ti-6Al-6V-2Sn | MIL-T-9047, MIL-F-83142 |
| 4980 | Bar, wire | Solution treated at 745 °C (1375 °F) | Ti-11.5Mo-6Zr-4.5Sn | ... |
| 4981 | Bar, forgings | Precipitation heat treated | Ti-6Al-2Sn-4Zr-6Mo | MIL-T-9047 |

(a) Society of Automotive Engineers Aerospace Materials Specification. (b) YS, yield strength. Source: Ref A.33

In contrast with metals, polymers are based on the lighter elements such as hydrogen, carbon, oxygen, fluorine, and sulfur. These elements are abundant, less costly, and of lower density than metals, and therefore are very attractive as engineering materials. Much progress has been made in imparting fine engineering properties to the newer polymeric materials. Today they are used for applications such as gears, automotive bumpers and panels, matrices for fiber-reinforced composites, and put to thousands of other uses that would have been thought unseemingly possible a few decades ago. Polymer research and development remains one of the most active areas of new materials development.

In their simplest form, polymers are made up of long chains of monomers such as illustrated in Fig. A.25. Here the fundamental unit (mer) is ethylene, a bond of carbon and hydrogen atoms, which is a true compound. Each hydrogen atom has one available bond, and carbon has four available bonds that can unite to form a chemical unit. As seen in the figure, two hydrogen atoms use their two single bonds to unite with a carbon

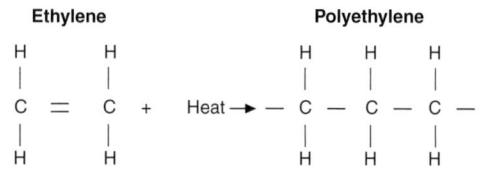

**Fig. A.25** Simple ethylene monomer converted to polyethylene by application of heat

atom. This leaves two of its four available bonds to bond with each other. Each short length of line in the figure represents an atomic bond. The monomer is, however, relatively weak. If heat is supplied to a mass of monomers, the double bond (=) between the carbon atoms can be broken and replaced by a single bond, leaving one bond in each carbon atom free to combine with other atoms. A special case is the conversion to monomeric units of polyethylene, which can then bond with identical units both to the right and left of the carbon atoms. When such bonding occurs, the product is known as a polymer consisting of many monomeric units. A single giant molecule may be made up of several million monomeric units. However, individual molecules in a mixture may have a wide range of sizes. In a sense, each molecule is like a long chain, each monomeric unit consisting of a link in this chain.

**Common Polymers.** A sampling of common engineering polymers is listed in Table A.6. If the hydrogen atoms are replaced by fluorine (also having one bond available) the material becomes Teflon (Dupont, Wilmington, DE) (technically, polytetrafluoroethylene, to describe that the structure is like polyethylene except that the four hydrogen atoms are replaced by fluorine atoms). If only one of the hydrogen atoms is replaced by the compound $CH_3$, the resulting polymer is polypropylene, which has very good fatigue characteristics. If one of the hydrogen atoms is replaced by chlorine, the resulting polymer is polyvinyl chloride. Polymers can also be compounded as copolymers, such as acetate (obtained by polymerization of formaldehyde). More complicated units shown are nylon and polymethylmethacrylate (Plexiglas) and poly carbonates, as seen in the table.

**Cross Linking and Crystallinity.** Without further reinforcement, the polymer chains acting individually do not produce a very rigid structure, and often they lack strength. Two mechanisms have been introduced to improve their structural characteristics. One is cross linking, the other is effective crystallization. Figure A.26 schematically shows the concept of cross linking. As the long molecules cross each other, it is possible to link separate molecules together at their intersections. The linking can be accomplished by heat only, or by the addition of a chemical together with the heating. The result is a more rigid, stronger structure.

Alternately, if the structure can be caused to align the individual molecular chains parallel and close to each other, then these regions may bring the atoms close enough to each other to act almost as if they were crystalline-like, somewhat similar to metals. Polymer "crystalline" regions are caused by long-range ordering but are

**Table A.6  Characteristics of selected engineering polymers**

| Material | | Major characteristics | Applications |
|---|---|---|---|
| Low-density polyethylene | [structure] | Considerable branching; 55–70% crystallinity; excellent insulator; relatively cheap | Film; moldings; cold water plumbing; squeeze bottles |
| Polypropylene | [structure] | Extent of crystallinity depends on stereoregularity; can be highly oriented to form integral hinge with extraordinary fatigue behavior | Hinges; toys; fibers; pipe; sheet; wire covering |
| Acetal copolymer | [structure] | Highly crystalline; thermally stable; excellent fatigue resistance | Speedometer gears; instrument housing; plumbing valves; glands; shower heads |
| Nylon 66 | [structure] | Excellent wear resistance; high strength and good toughness; used as plastic and fiber; highly crystalline; strong affinity for water | Gears and bearings; rollers; wheels; pulleys; power tool housings; light machinery components; fabric |
| Poly(tetrafluoroethylene) (Teflon, Dupont, Wilmington, DE) | [structure] | Extremely high molecular weight; high crystallinity; extraordinary resistance to chemical attack; nonsticking | Coatings for cooking utensils; bearings and gaskets; pipe linings; insulating tape; nonstick; load-bearing pads |
| Poly(vinyl chloride) | [structure] | Primarily amorphous; variable properties through polymeric additions; fire self-extinguishing; fairly brittle when unplasticized; relatively cheap | Floor covering; film; handbags; water pipes; wiring insulation; decorative trim; toys; upholstery |
| Poly(methyl methacrylate) (Plexiglas) | [structure] | Amorphous; brittle; general replacement for glass | Signs; canopies; windows; windshields; sanitary ware |

Source: Ref A.34

not the same as the crystalline nature of metals. Figure A.27 shows schematically how the long-chain molecules can line up, at least in limited regions or domains, to produce local "crystallinity" and hence density. Because of this feature, tensile strength and stiffness increase dramatically (Fig. A.28). Over the range of densities shown, the percent crystallinity increases from about 65 to 96. The crystalline part can also sometimes support higher temperatures.

**Elastomers** are polymers, which use rubber (either material from a tree or chemically synthesized) for their basic structure. At ambient temperatures they are already above their *glass transition temperature,* and would be liquid except for the fact that they are unsaturated, meaning that there are positions along the chain that can still be attached chemically. By using sulfur together with other chemicals, it is possible to link the long chains together at various points, producing a network similar to that already shown in Fig. A.26. This process is known as *vulcanization* (also *curing* or *cross linking*). Because rubber used as the base has a high capacity for elongation (several hundred percent), the resulting product is capable of great deformation but is also strong and wear resistant. Automobile tires are made, for example, by vulcanizing natural or synthetic rubber. Many different raw materials and vulcanization processes compete for the substantial volume of business in this field.

**Thermoplastic, Elastomeric, and Thermosetting Polymers.** Three important categories are recognized in engineering polymers. In the thermoplastics there is little or no cross linking. With increased temperatures the individual long-chain molecules lose their strength and rigidity and they can therefore be molded and formed easily, simply by increasing their temperature. Conventional molds and dies can easily be used to shape a part cheaply. Upon cooling to room temperature, they harden again and become structurally useful. High strength can be achieved at the lower temperatures (mainly room temperature usage), but it is important not to heat these plastics to temperatures where they will again soften. Nylon 6-6, for example, is a material of

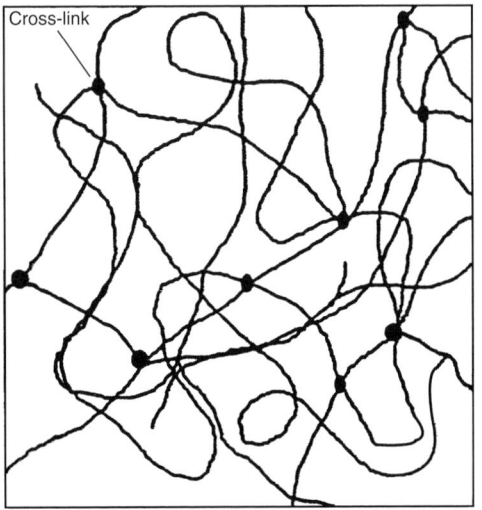

**Fig. A.26** Cross linking in polymers

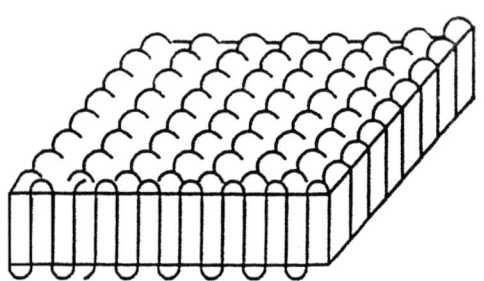

**Fig. A.27** Schematic of crystalline lamellae in polymers. Source: Ref A.34

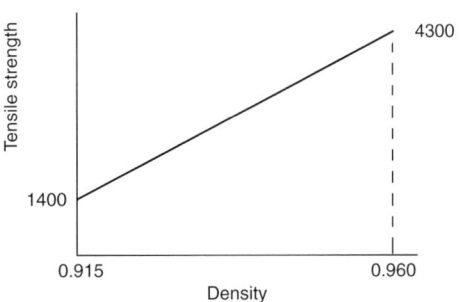

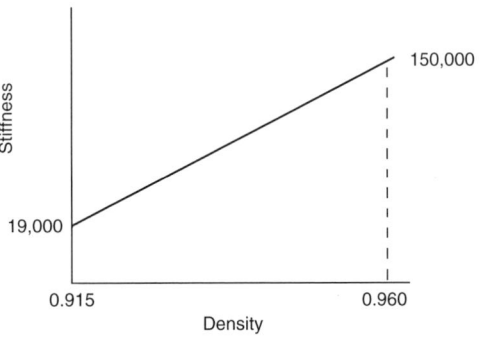

**Fig. A.28** Influence of density on and strength and stiffness of polyethylene. Source: Ref A.35

substantial strength at room temperature but becomes very weak at higher temperatures.

Where it is important to use polymers at high temperatures, thermosetting polymers are a better choice. Upon the initial increase in temperature, they cross link strongly and develop considerable strength, and they retain this strength upon later return to high temperature. Forming and molding such polymers is much more difficult, of course. But by clever control of temperature and processing history, it is usually possible to shape a part just before it sets. The advantage, however, is that the polymer can be used at elevated temperatures.

## Structural Ceramics

Ceramics are discussed in Chapter 12, "Special Materials," where we assessed the applicability of metallic materials-based fatigue life-prediction models to the behavior of ceramics. Other aspects of ceramics are discussed below.

Ceramics are inorganic nonmetallic materials. Since antiquity they have benefited society over a broad range of applications from pottery to art objects. The special characteristics that have made them useful both in art and technology are chemical inertness, hardness, wear resistance, and low electrical and thermal conductivity. It has long been a dream of engineers to adapt the useful properties of ceramics to the modern needs of technology. As required service temperatures increase, engineers have sought to apply high-melting-point ceramics to structures requiring high strength at extremely high-temperature service conditions. While only moderate success has been achieved, the payoff is so enticing that extremely large funding has been allocated in recent years toward this goal. The main problems to be overcome are the improvement of high tensile strength, reliability (i.e., repeatability of performance from one application to the next), mechanical and thermal shock resistance, and resistance to catastrophic crack propagation once a crack starts.

As far back as the 1940s it was recognized that successful application of ceramic rotor blades (see, for example, the work of Freche and Sheflin, Ref A.36) in gas turbine engines could make possible a step increase in the thermal efficiency of these power plants. The development of ceramic gas turbine blades was one of the early projects taken on by the senior author of this book at the newly founded NACA Aircraft Engine Research Laboratory in Cleveland, Ohio. Much effort was expended on developing fastening geometries that could minimize stress concentrations, introduce frictional forces to reduce aerodynamically induced vibrations, and deal with the thermal shocks of starting and stopping the engine (Ref A.37). These did not prove to be easy problems to solve. Among the concepts that emerged was that of combining the advantageous properties of ceramics and metals by mixing them in what we called *ceramals* (later designated as *cermets* at the insistence of the U.S. Air Force, which was a major source of funding). It was thought that the ductility and high conductivity of the metal component would combine with the refractory property of the ceramic component. Unfortunately, it was found, instead, that it was the disadvantageous characteristics of the components that prevailed; the cermets were still brittle, not very strong, and low in melting point.

However, intensive interest has been maintained throughout all the years on various ceramics because of their promise and because of the tangible progress that has been made in improving their underlying properties and in combining them with reinforcement fibers to produce promising composites. Table A.7 (gleaned primarily

Table A.7  Approximate properties of representative engineered ceramics

| Property | $Al_2O_3$ | MgO | $ZrO_2$ | SiC | $Si_3N_4$ | $Si_3N_4$ | BeO | $Al_2TiO_3$ |
|---|---|---|---|---|---|---|---|---|
| Manufacturing method | Hot pressed | . . . | Sintered | Hot pressed | Hot pressed | Reaction bonded | Sintered | . . . |
| Elastic modulus, GPa ($10^6$ psi) | 400 (58.0) | 270 (39.2) | 200 (29.0) | 350 (50.8) | 300 (43.5) | 180 (26.1) | 360 (52.2) | 30 (4.4) |
| Poisson's ratio | 0.22 | 0.17 | 0.25 | 0.2 | 0.28 | 0.23 | 0.25 | 0.2 |
| Hardness, Shore D scale | 2000 | . . . | . . . | 2800 | 1700 | . . . | . . . | . . . |
| Tensile strength, MPa (ksi) | 300 (43.5) | 180 (26.1) | 950 (137.8) | 360 (52.2) | 660 (95.7) | 200 (29.0) | 180 (26.1) | 65 (9.4) |
| Weibull modulus for strength | . . . | . . . | 14 | 10 | 7 | . . . | . . . | 15 |
| Fracture toughness, MPa$\sqrt{m}$ | 4.5 | 3.0 | 10 | 4 | 7 | 2 | 4.8 | . . . |
| Density, g/cm$^3$ | 3.9 | 3.5 | 6.0 | 3.2 | 3.3 | 2.4 | 3.0 | 3.6 |
| Thermal expansion, $10^{-6}$/K | 8 | 12 | 11 | 4 | 3.2 | 2.5 | 8 | 1.8 |
| Thermal conductivity, W/m·K | 30 | 30 | 2.5 | 100 | 35 | 10 | 300 | 2.5 |
| Specific heat, J/g·K | 1.0 | 1.0 | 0.5 | 1.0 | 0.7 | 0.7 | 1.3 | 0.7 |
| Electrical resistivity, Ω·cm | $10^{14}$ | $10^{14}$ | $10^9$–$10^{11}$ | 0.1–100 | $10^7$–$10^{12}$ | $10^7$–$10^{12}$ | $10^{14}$ | $>10^{14}$ |

Source: Ref A.38

from Tables 2.1 to 2.6 and 11.1 of Munz and Fett, Ref A.38) lists some of the representative room temperature properties of ceramics of consideration in current engineering applications. These properties are affected by temperature, and the reader should consult Ref A.38 as well as current handbooks (e.g., Ref A.39) for more up-to-date listings of properties. The reader is cautioned that many of the properties listed are representative values that are subject to considerable variation due to the sensitivity cause by processing and handling. Two of the compounds that have been receiving the most attention for aerospace applications are silicon carbide (SiC) and silicon nitride ($Si_3N_4$). Note the high tensile strength and low thermal expansion coefficient of hot pressed silicon nitride. But silicon carbide has the higher hardness (which was why it had been so popular for use in cutting tool tips). Efforts are being directed at improving strength and other properties by sintering these materials to nearly 100% density because even small porosity reduces strength significantly as can be seen from the empirical relation described by Knudsen (Ref A.40):

$$S_p = (S_o + Kd^{-a})e^{-bP} \quad \text{(Eq A.1)}$$

Here, $S_o$ is the strength of the zero-porosity material; $d$ is the mean grain size; $S_p$ the strength of the ceramic with fractional porosity $P$; and $K$, $a$, and $b$ are material constants. It is clear that strength is decreased with increased porosity and strength is increased with decreased mean grain size.

**Applications of Ceramics.** A brief listing of high-technology applications of ceramics is as follows:

*Aeronautics.* As already mentioned, considerable efforts are in progress to incorporate silicon carbide and silicon nitride into turbine engine components (combustors, guide vanes, blades, disks, exhaust nozzles, etc.). The decomposition temperatures (equivalent to the melting temperature in metals) of these materials are 2500 and 1900 °C (4530 and 3450 °F), respectively. Thermal barrier ceramic coatings are widely used on gas turbine engine hot section metallic components such as combustor liners, guide vanes, and blades. These components utilize backside cooling and the coating acts as an insulating thermal blanket that reduces the heat flux to the substrate metal. Several hundreds of degrees Fahrenheit metal temperature reductions have been achieved. Thus it would be possible to increase gas turbine temperatures substantially, increasing the thermodynamic efficiency. If these materials can be successfully applied, it would also be possible to reduce or eliminate the amount of cooling currently required to operate with metallic materials, thus further increasing efficiency. The problems to be overcome are: suitable fastening methods, thermal shock during starting and stopping, and impact resistance due to foreign objects passing through the engine.

*Space.* Ceramics are also of interest as reentry heat shields, rocket nozzles, and other nonload bearing applications requiring extremely high temperatures. The hypereutectic carbides appear to be most promising for this application. Windows in space vehicles are logical applications of ceramics. Silica was used initially, but with increasing speeds, higher temperatures develop, and more refractory ceramics are needed. Transparent magnesia, alumina, and beryllia have been studied for this purpose. In addition to higher strength than silica, these materials have better resistance to abrasion by micrometeorites.

*Nuclear Applications.* Ceramics are also very important in nuclear applications. They are used both in the fissionable compounds, for example UN, $UO_2$, UC, $UC_2$, and $UBe_{13}$, and as coating material for encasing materials to prevent fissionable material from escaping into the atmosphere. One such application has been uranium carbide particles coated with pyrolytic carbon. They are also used as moderators (controlling the amount of nuclear reaction that takes place by adding non-nuclear material to the mass). For this purpose, BeO is promising because of its thermal shock properties but is a problem because of its toxicity, so special precautions must be exercised.

*Cutting Tools.* Obviously, machining tools are a field for ceramics because of their hardness, wear resistance, chemical stability, and high melting point. Often, only the tip of the tool is replaced by a ceramic, although the entire cutting tool may be ceramic. Alumina is also used as a cutting tool because it is superior to carbide in hardness and chemical stability but is less resistant to mechanical and thermal shock.

*Ball Bearings.* High-speed ball bearings operating at high temperatures can benefit from use of low-density ceramics. This helps significantly to reduce the contact stresses at high rotational speeds. Superior heat resistance of ceramics over metals is also an advantage if slippage occurs. Silicon nitride ball bearings are being used reliably in high-speed turbopumps in the main engines of the U.S. space shuttle as well as in small, high-speed turbopumps for evacuating electron microscopes to ultrahigh vacuum levels.

## Strengthening Mechanisms for Metals and Ceramics

While each material has an inherent strength, there are many ways in which strength can be enhanced. For example, while steel is inherently stronger than aluminum, applying strengthening principles can provide an aluminum alloy that is stronger than many steels. Of course, there is always a price to be paid: added processing, expense, reduction in ductility, and so forth. It is appropriate to be aware of some of the more common ways available to the metallurgist to alter strength, usually increasing it, but sometimes decreasing it in order to achieve some other desirable property. While most of the mechanisms discussed subsequently relate to metals, they may also be applied in some cases to nonmetals. Some of the methods have already been touched on in the previous discussions, but are expanded on here and juxtaposed for completeness. It is important to emphasize that, fundamentally, all the methods are based on one or another role of dislocations, but a complete discussion of the dislocation factors is beyond the scope of this book.

**Solid-Solution Hardening.** We have already pointed out that alloying elements are added to carbon steel in order to impart various desirable characteristics. However, the alloying elements also increase the strength by altering the crystallography. Whether a foreign atom replaces one of the regular lattice sites in the crystal, or whether it ends up as an interstitial, the crystallography is altered by the inequality of the atomic radii and their electrical fields. Figure A.29 (Ref A.22) shows a set of three tensile stress strain curves that dramatize the significant strengthening attributes of increasing additions of a solid solution alloying element. Examples might be very low additions of carbon in iron, silver additions to aluminum, or aluminum additions to copper.

**Strain Hardening.** Subjecting a metal to inelastic strain introduces many new dislocations as discussed previously. The proliferation of dislocations and their entanglement, as illustrated in Fig. A.18, prevents easy glide and raises the stress required for further plastic deformation. Raising the strength in this manner makes the metal harder, hence the term *strain hardening*. Thus, the strain-hardening exponent, as discussed in connection with the tensile test, provides a measure of the increases in stress required for each additional increment of plastic deformation. By providing prior deformation, for example, by passing a material through a pair of rollers, one can precondition the material so that a higher stress is needed to restart new plastic flow, i.e., harden the material. It is clear, however, that the prior deformation may already exhaust some of the ductility of the material so that less is left before fracture occurs in the strain-hardened material. That is, a strain-hardened material, while stronger, may have considerably lower ductility.

**Precipitation Hardening.** In the section "Heat Treatment as a Metallurgical Tool," it was pointed out that by following a prescribed time-temperature history, it is possible for some materials to precipitate a phase that differs in composition and perhaps in crystallography from the matrix. Often these precipitations are very fine particles of what is basically a ceramic or intermetallic, which are very strong and are therefore very resistant to fracture by dislocations moving through the matrix. Instead, they cause new dislocations to be generated in a manner somewhat like the Frank-Read source discussed earlier. By

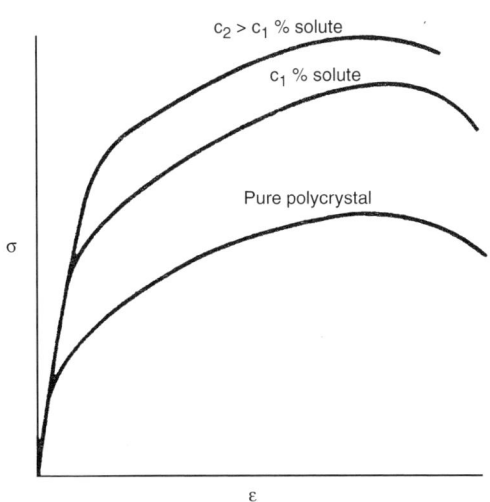

**Fig. A.29** Effect of solute alloy additions on the tensile stress strain curve of metals. Source: Ref A.22

causing many interlinking dislocations their movement is prevented until high stresses are imposed, thereby strengthening the material.

The action is shown schematically in Fig. A.30. As the dislocation line approaches the precipitated particles, and is unable to break them, it bows around them. Eventually, the bowing is so strong that the dislocation becomes unstable and breaks into smaller dislocations around each precipitate, increasing the number, which can then further proliferate.

Precipitation hardening is one of the most common methods of hardening alloys, which have many constituent elements suitable for combining with each other, or with the environment, to form the precipitate. The metallurgist's science is in finding the suitable heat treatment sequence to bring out the optimum particles. It is during the aging process that the particles are precipitated. If the aging is too long, or the temperature too high, the particles may grow too large, or agglomerate into larger sizes, and hence lose their effectiveness. Figure A.31 is interesting in this regard. It shows that strength is critically related to spacing of precipitate particles. Some of the strongest alloys have been achieved by this method.

**Dispersion Strengthening.** The basic hardening principle behind dispersion strengthening is the same as that of precipitation hardening. Fine, hard particles are dispersed into the matrix, and their action is similar to that discussed previously. The distinction is that the dispersoid is artificially introduced into the matrix in the form of fine particles, which the metallurgist is free to choose at random, rather than being required to precipitate from the matrix. Thus, by suitable choice of the dispersed phase it can, for example, be made to be effective at high temperature, whereas a material precipitate may be limited in melting point. For many years the authors' associates at NASA's Lewis Research Center (renamed Glenn Research Center in 1999) did research on dispersion-strengthening materials for high temperature, among them thoria-dispersed (TD)-nickel. For most of the materials studied, the process started with the fine grinding of both matrix and dispersoid-components separately in ball mills for days on end until both were reduced to micron-size particles. Obviously, if the spacing of the dispersoid is to be close, the matrix must also start as a very fine powder. The separate fine powders were then mixed and ball milled together, again for long times until the mixture was homogeneous. The mixture was then consolidated by pressing at a temperature sufficient to diffuse the mixture into a solid. INCO International developed TD-nickel, the first commercial product of this type.

**Grain-Boundary Strengthening.** A very effective way of increasing strength, at least as related to yielding, is to refine the grain size. The reason for this arises from the fact that grain boundaries inhibit slip. Even if grains are oriented so as to start slip on one or more of their slip planes, the slip can progress easily only until the dislocations encounter a grain boundary. From this point there must be enough of a dis-

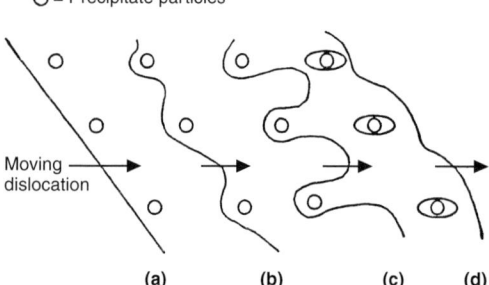

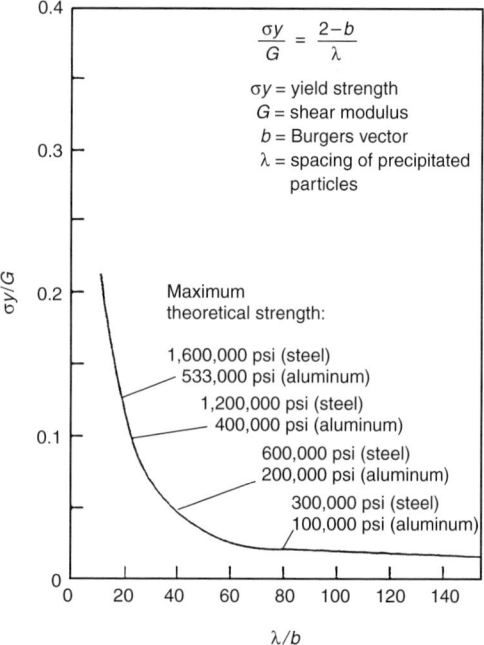

**Fig. A.30** New dislocations generated at precipitated particles as dislocation moves past them

**Fig. A.31** Relationship between theoretical yield strength and spacing of precipitate particles

location buildup to produce a stress concentration high enough to start plastic flow in the adjacent grain, where the slip plane direction may differ appreciably and be less prone to slip. Thus, materials with small grain size, which have many more grain boundaries, generally require a higher stress to cause general yielding.

Figure A.32 shows some results of yield strength as a function of grain size for iron and aluminum. It is seen that yield stress ($\sigma_y$ increases linearly with the inverse of the square root of grain size $d$, leading to the Hall-Petch relation:

$$\sigma_y = \sigma_t + k_y d^{-1/2} \quad \text{(Eq A.2)}$$

Obtaining fine grain size is a problem for the metallurgist. It can be accomplished by adding elements, called grain refiners, to an alloy or it can be accomplished by suitable prior cold work to break up the grains, followed by appropriate heat treatment to produce recrystallization into smaller grains. If large grain size is desired, holding the metal at a temperature within the recrystallization range allows continued grain growth. Dropping the temperature ceases grain growth when the desired size is reached.

It should be added that grain refinement is advantageous for strengthening only at temperatures low enough so that the grain boundaries themselves do not contribute to the strain. At temperatures greater than about half the melting point on an absolute temperature basis, i.e., in the creep range, large grain size may be preferred in order to minimize the number and length of the grain boundaries, particularly those that are perpendicular to the principal loading direction. Thus, directionally solidified materials (grain boundaries oriented parallel with the loading direction) and monocrystal materials have been developed for high-temperature use, as discussed in Chapter 11, "Avoidance, Control, and Repair of Fatigue Damage."

**Diffusionless (Martensitic) Transformations.** As discussed earlier for steels, martensite is a strong phase. Martensitic structures also occur in other alloys, notably, titanium. The concept whereby a nonequilibrium phase can be created is transferable to other alloy systems as well. What is involved here is a transformation of the material in local regions from one crystal structure to another. If the structure to which the material transforms has a higher volume than the structure from which it is derived, high internal residual compressive stresses can develop, which counteract externally applied tension, and effectively strengthen the material. The transformation requires neither time nor the diffusion of atoms within the matrix over large distances because the local composition does not change, only the crystallography. For this reason it is called diffusionless transformation. Of course, it is unstable because reversion of the structure can occur as a result of temperature, strain, or other environmental condition. An important application of the concept is involved in ausformed steels. Here the material is worked in its austenitic range. The martensitic structure that develops can be heat treated not only to higher strength but to equal or higher ductility as well. Figure A.33 shows this characteristic for a series of ausformed steels compared with conventionally heat treated steels. It is seen that the aus-

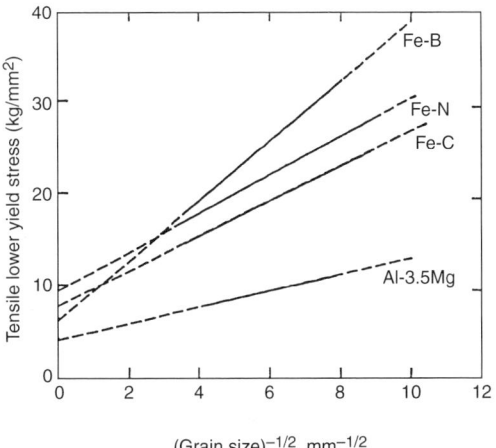

**Fig. A.32** Effect of grain size on yield strength of various alloys. Source: Ref A.41

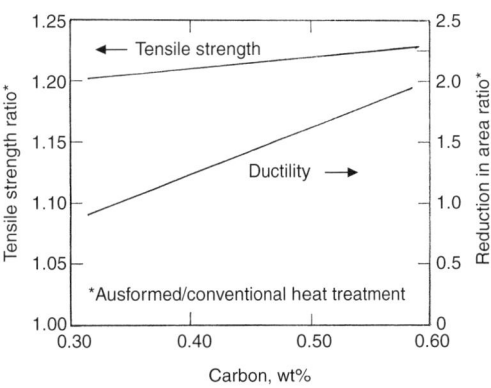

**Fig. A.33** Representation of relative tensile strength and ductility of ausformed steels (highly deformed) compared with the same steels heat treated conventionally

formed steels are more than 20% stronger than conventional steels of the same composition. Of great importance to toughness is the resultant greater ductility at the highest strength levels (highest carbon content) for the ausformed steels. The superior fatigue resistance of ausformed steels is discussed in this book.

The processes involved in diffusionless transformations are very complex and involve many factors that have yet to be fully understood. Research investigators do not agree on the quantitative factors, but the fact of strengthening is not disputed.

**Superlattice Formation.** When different types of atoms occupy two or more lattice sites, a superlattice results. For example, in Fig. A.34(a), a simple cubic structure of one element is surrounded and pierced by a cubic structure of another element. In Fig. A.34(b) the simple cubic structure in one system contains two different elements at some sites, but the superlattice contains only one type of atom. These materials are known as intermetallics because the two types of metals have a very well-defined stoichiometric structure. However, they have very low ductility because the simple cubic system has few slip systems as already discussed. But they are strong and, when incorporated as precipitates in an alloy, can produce strengthening.

**Summary of Some Strengthening Mechanisms in Conventional Structural Materials.** Table A.8 shows a number of conventional structural materials, their approximate yield strengths, and the operative strengthening mechanisms. It is especially noteworthy that most of the strengthening mechanisms are common to all of the metals listed. However, the highest-strength materials within the iron alloy systems involve an additional grain-boundary strengthening mechanism. Grain-boundary strengthening mechanisms are used extensively to improve the high-temperature characteristics of nickel and cobalt-base superalloys.

## Elasticity, Plasticity, and Multiaxiality

Mechanical behavior of materials involves response to stresses and strains. Stress is the force per unit area while strain is the unit change in length of a reference element. Quantitative relations exist between stresses and strains. It is important to distinguish among the most important types of strains that are encountered in engineering applications. The *total* strain is composed of linear elastic (recoverable strain) and nonlinear inelastic (nonrecoverable strain) components. Furthermore, there are two important types of inelastic strains: *plastic* strain due to time-independent dislocation motion and *creep* strain associated with time-dependent, high-temperature, thermally activated diffusion processes. Examples of the engineering stress-strain response of axially loaded cylindrical bars are shown in Fig. A.35 for each of the three conditions mentioned. Each stress-strain response curve shows the initial linear elastic loading portion with the modulus of elasticity $E$ being a proportionality constant. The insert sketches represent a round bar of initial diameter $D_o$, and length $L_o$, with and without an axial tensile engineering stress, $S_A$. After application of the stress, the bar extends by an amount $\Delta L$ to $L_f$, while the diameter $D_0$ decreases by an amount $\Delta D$ to $D_f$. The degree of contraction depends upon the value of Poisson's ratio, which, by def-

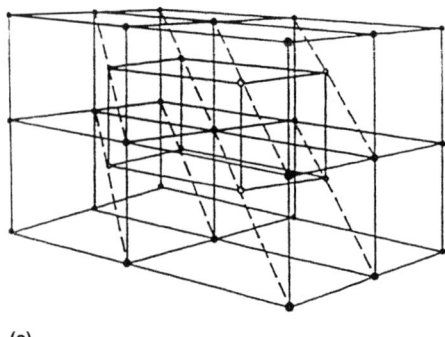

(a)

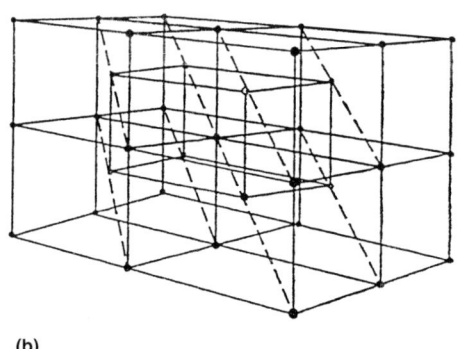

(b)

**Fig. A.34** Strengthening by superlattice formation. (a) CsCl structure: CuZn, AgZn, AgMg, FeAl, CoAl, NiAl. (b) Fe$_3$Al structure: Fe$_3$Al, Cu$_3$Al, Ca$_3$Sb, Fe$_3$Si

## Table A.8 Principal strengthening mechanisms in some common structural materials

| Material and condition | Approximate room temperature yield strength MPa | ksi | Operative strengthening mechanism |
|---|---|---|---|
| **Aluminum alloys** | | | |
| Al-Cu-Mg; heat treated and cold worked (2024-T3) | 340 | 50 | Strain hardening; precipitation strengthening |
| Al-Zn-Mg-Cu; annealed (7075-0) | 100 | 15 | Solution strengthening |
| Al-Zn-Mg-Cu; heat treated (7076-T6) | 500 | 72 | Precipitation strengthening |
| **Copper alloys** | | | |
| Cu (electrolytic); annealed (commercially pure; 0.04% oxygen) | 70 | 10 | Strain hardening (dislocation-dislocation strengthening) |
| Cu (high conductivity); cold worked (oxygen free, high-conductivity; OFHC) | 280 | 40 | Strain hardening |
| Cu-Zn; annealed (brass) | 240 | 35 | Solution strengthened |
| Cu-Be; heat treated (beryllium copper) | 970 | 140 | Precipitation strengthening |
| **Nickel alloys** | | | |
| Ni-Cr-Fe; cold worked (Inconel) | 1030 | 150 | Strain hardening; solution strengthening |
| Ni-Mo-Fe; annealed (Hastelloy B) | 280 | 40 | Solution strengthening |
| Ni-Co-Cr-Mo-Ti-Al; heat treated (known as Udimet 700) | 830(a) | 120(a) | Precipitation strengthening |
| **Iron alloys** | | | |
| Fe; annealed (commercially pure; 0.01% carbon) | 170 | 25 | Solution strengthening; grain-boundary strengthening |
| Fe-C; annealed (1020 steel; 0.2% carbon) | 310 | 45 | Solution strengthening; grain-boundary strengthening |
| Fe-Ni-Mo-Mn-C; heat-treated (4340 steel 0.4% carbon) | 1380 | 200 | Solution strengthening; grain-boundary strengthening |
| Maraging steel 300 (special heat treatment) | 2000 | 290 | Precipitation strengthening; grain-boundary strengthening |
| Cold drawn tungsten wire | 3720 | 540 | Strain hardening; grain-boundary strengthening |

(a) At 650 °C (1200 °F)

inition, is the ratio of the transverse to the axially applied strain. As discussed in detail subsequently, the numerical value of the elastic Poisson's ratio for the elastic strain component $\mu_e$ is on the order of 0.33 for most metals. The plastic Poisson's ratio, $\mu_p$, for the inelastic (plastic or creep) strain component is very close to 0.5. As discussed in greater detail later, the value of 0.5 corresponds to a constant volume condition for inelastic strain, a fact that proves useful in analyzing relations between stresses and strains.

When the axial strain is small it can be defined quite accurately as the fractional increase of an initial length, i.e., *engineering strain e*, where:

$$e = \frac{\Delta L}{L_0} = \frac{(L_f - L_0)}{L_0} \quad \text{(Eq A.3)}$$

where $L_0$ is the initial length and $L_f$ the final length. However, when the strains become large due to inelasticity, engineering strain is no longer as meaningful because it is based on increases in length from the initial length. During any increment of straining, the true strain $\varepsilon$ should be based on the current length (provided, of course, that the straining is uniformly distributed along the length considered). Thus, as length changes from $L_0$ to $L_f$, the increment of axial strain at any intermediate length $L$ can be defined as:

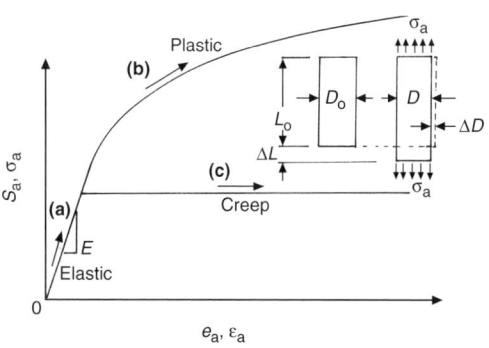

**Fig. A.35** Stress-strain curves featuring (a) linear elastic response, (b) elastic plus plastic response, and (c) elastic plus creep response

$$d\varepsilon = \frac{dL}{L} \quad \text{(Eq A.4)}$$

Thus, the axial true strain is taken as the integral over the interval from $L_0$ to $L_f$ of the instantaneous values of strain increment, i.e.:

$$\varepsilon = \int \frac{dL}{L} \rightarrow \ln\left(\frac{L_f}{L_0}\right) \quad \text{(Eq A.5)}$$

But $L_f$ can also be defined from Eq A.3 in terms of the nominal strain. Thus, $L_f = L_0(1 + e)$, and hence, prior to necking, the true strain can be related to the engineering strain by $\varepsilon = \ln(1 + e)$. Again, it is cautioned that this is true only if the deformation is uniform along the length considered; i.e., there are no regions of localized contractions commonly known as *necking*. Thus, at any point during the straining, if the length $L_0$ is increased to $L_f$, the cross-sectional area $A_0$ ($= \pi(D_0)^2/4$) will decrease uniformly to $A_f$ ($= \pi D_f^2/4$). Assuming constancy of volume for the plastic portion of the deformation, i.e., the plastic Poisson's ratio = 0.5, then $A_0L_0 = A_fL_f$, or $A_0/A_f = L_f/L_0$. We can then also express the relation between the axial true stress $\sigma_a$ and axial nominal stress $S_a$ prior to tensile necking by:

$$S_a = 4P/\pi D_0^2 \quad \sigma = 4P/\pi D_f^2 \quad \text{and}$$
$$D_f = D_0/(1 + e)$$
Thus $\sigma_a = S_a(1 + e)$
$$\quad \text{(Eq A.6)}$$

In the range where both elastic and plastic strains are significant in governing the contraction in diameter, the axial strain relation becomes:

$$\varepsilon_a = (\varepsilon_d/\nu_e) + (1 - 2\nu_e)\sigma_a/E \quad \text{(Eq A.7)}$$

where $\varepsilon = -\ln(D_f/D_o)$ is the diametral (transverse) plastic strain. Because $\ln(D_f/D_o)$ is negative for tensile loading, the first term on the right-hand side of Eq A.7 becomes a positive number.

This relation has been used to control the axial strain according to a desired pattern by monitoring and controlling the more easily measured diametral strain (see, for example, Ref A.42 and A.43).

For purposes of strain visualization, we continue to regard a solid as a system of atoms, which are represented as solid spheres initially touching each other at their periphery, the distance between the centers of the spheres being the atomic spacing (Burgers vector). The actual spacing between the centers of the spheres is a complicated function of the content of the nucleus, the number and location of the electrons, and numerous other factors. For our purposes, we can proceed without having to pursue these complicating factors in detail.

**Hookian Elastic Strain.** When a uniaxial tensile stress is applied to an aggregate of spheres representing atoms, the centers of the spheres separate an amount just sufficient so that the change in the attractive and repulsive forces (see, for example, Ref A.7) between the atoms remain in equilibrium with the externally applied force (stress times area). How much separation, or strain, occurs for a given stress depends on the stress-strain relation of the material. Initially, the strain response is like that of an elastic spring that behaves in the classical linear Hookian sense. When the stress is removed, the atoms return to their original positions as if nothing had happened. No atoms have slipped over one another, nor have any jumped from their original positions. The potential energy of the aggregate of atoms remains exactly the same as before the loading because there is no dissipation of energy due to elastic loading and unloading of the atoms. The axial stress-strain relation in this regime is linear, with the proportionality constant being Young's modulus of elasticity $E$:

$$\sigma_a = E\varepsilon_a \quad \text{(Eq A.8)}$$

where $\sigma_a$ is the axial stress in the direction of the applied elastic strain, $\varepsilon_a$. While $E$ could be calculated by considering the complex interatomic forces, for engineering purposes, it is usually determined directly from accurate measurements of applied stress and resultant strain. Before discussing straining beyond the linear elastic regime, it is appropriate to discuss another very important factor: how much the specimen strains in the direction transverse to the applied strain. The amount of elastic strain in the transverse direction relative to the elastic strain in the axial direction is designated as the elastic Poisson's ratio, $\nu_e$. Thus, for elastic uniaxial loading:

$$\nu_e = -\varepsilon_t/\varepsilon_a = -\varepsilon_t E/\sigma_a \quad \text{(Eq A.9)}$$

While $\nu_e$ is usually directly measured, it is interesting to note that visualizing the atomic aggregate as a group of contacting spheres leads to a value that closely agrees, in most cases, with experimental values. When axially displaced, the atoms are assumed to maintain contact. Figure A.35 shows the model conceived by Shanley (Ref A.44), as described by Juvinall (Ref A.11). In Fig. A.36(a), the solidly stacked contacting spheres (of unit radius, for convenience) are shown when no external load is applied. Under external load, the spheres separate by an amount $2dy$, as shown in Fig. A.36(b) by the continuous circles; the dashed circles show the original unstressed configuration. Because the spheres separate in the axial direction, the laterally adjacent spheres can move inward by an amount $-dx$ in order to continue to make surface contact. By simple geometric consideration it can be shown that $dy/dx = \sqrt{3}$. Because the transverse strain $\varepsilon_t$ $\varepsilon_t$ is $-dx/\sqrt{3}$, and the longitudinal strain $\varepsilon_a$ is $dy$, the elastic Poisson's ratio $\nu_e$ is:

$$\nu_e = -(\varepsilon_t/\varepsilon_a) = -(-dx/\sqrt{3})/(dy)$$
$$= -(-1/\sqrt{3})(1/\sqrt{3}) = 1/3 \quad \text{(Eq A.10)}$$

The calculated value of ⅓ is very close to the observed value of the elastic Poisson's ratio for many materials (Table A.2). Of course, if we know the experimental value, it should be used instead of the value derived from this idealized model. Because the elastic Poisson's ratio is less than 0.5, it can be shown that the atomic lattice dilates (volume increases) during tensile deformation and contracts during compression. This is the classical Lord Kelvin effect: rapid dilatation is accompanied by a small fractional temperature decrease while rapid compression produces a small heating effect. The temperature change is instantaneous with elastic deformation and forms the basis of experimental techniques for pinpointing the location of the peak stresses in a cyclically loaded body. Infrared sensors synchronized with the frequency of loading can detect the coolest (tension) and hottest (compression) regions on the surface of a fatigue loaded body and thereby deduce the durability-critical, highest-stressed regions. Commercial equipment is available that allows for routine measurements in the laboratory or in the field (Ref A.45).

In the past, use has been made of Poisson's ratio in the measurement and control of axial strain during low-cycle fatigue experiments on hourglass-shaped specimens. The transverse measurements permit strain determination at a precise cross section of the specimen rather than over a finite gage length. This technique for conducting fatigue tests is unsuitable for anisotropic materials as well as highly porous cast alloys. Test samples of this configuration are sometimes referred to as *zero gage length* specimens. Diametral strain controlled fatigue has been dropped from ASTM's E 606 standard for fa-

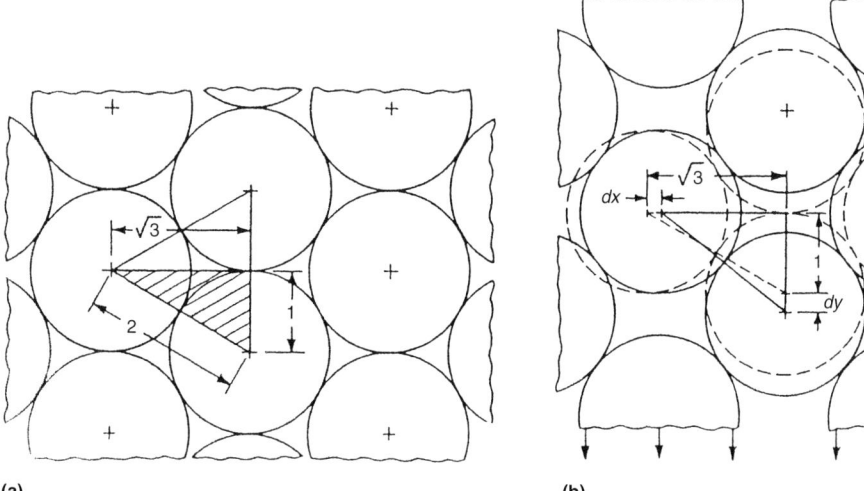

**Fig. A.36** Shanley's model for demonstrating that Poisson's ratio can be estimated by regarding atoms as contacting spheres that displace both axially and transversely under uniaxial load, maintaining surface contact. (a) Model of unstressed atomic crystalline structure. (b) Model of uniaxial stressed atomic crystalline structure. Source: Ref A.11

tigue testing and it does not appear in the new thermomechanical fatigue testing standard.

**Plasticity Relations.** Plasticity, as noted earlier, involves the dislocation gliding of atoms over distances, which may be many hundreds of multiples of the atomic spacing. After each atomic distance moved, the atoms have simply been translated and there is no increase or decrease in volume. Even after displacement over many thousands of atomic distances, the volume change due to the dislocation motion (plastic deformation) per se is still zero. In this discussion, we ignore the miniscule volume changes due to the complex configurations of the many dislocations (and other atomic level defects such as vacancies, subgrain boundaries, etc.) that were created during the plastic deformation. A simple model of hard spheres having sheared over one another is shown schematically in Fig. A.37. It reflects the end result of all the individual atomic motions during dislocation glide. Of course, small volumetric changes do occur due to elastic strains as previously discussed. However, for just the plastic portion of the deformation in a uniaxially loaded specimen, the specimen must contract in area by the same percentage that it increases in length. To accomplish this geometric effect, it turns out that Poisson's ratio, $v_p$, for the plastic portion of the strain must be 0.5. This is demonstrated as follows. For the volume change, $dV = 1 - V_{final}$, of a unit cube to remain zero:

$$dV = 1 - (1 + \varepsilon_p)(1 + v_p\varepsilon_p)$$
$$(1 + v_p\varepsilon_p) = 0 \quad \text{(Eq A.11)}$$

Expanding this expression and neglecting higher powers of $\varepsilon_p$, which itself is assumed to be small, it can readily be shown that $v_p = 0.5$.

The preceding discussion applies directly to uniaxial loading only. It is necessary to generalize the aforementioned relationships to deal with two- and three-dimensional stress and strain states. Generalized multiaxial stress-strain states are presented in the following section.

**Multiaxiality.** While many experiments and applications involve loadings producing predominantly uniaxial stress, there are numerous cases wherein stresses in two or three directions are important in governing fatigue or fracture behavior (See Chapter 5, "Multiaxial Fatigue"). For elastic conditions the usual Hookian equations apply:

$$\varepsilon_x = \frac{\sigma_x}{E} - v_e \frac{\sigma_y + \sigma_z}{E} \quad \text{(Eq A.12)}$$

$$\varepsilon_y = \frac{\sigma_y}{E} - v_e \frac{\sigma_x + \sigma_z}{E} \quad \text{(Eq A.13)}$$

$$\varepsilon_z = \frac{\sigma_y}{E} - v_e \frac{\sigma_x + \sigma_y}{E} \quad \text{(Eq A.14)}$$

These equations are easy to invert (that is, calculate stresses when strains are known, as is often the case).

However, when plasticity occurs, the equations are more complicated, and inverting them is difficult. While the field of plasticity has been developed to a high science, we are not able in this short summary to reflect the state of development. However, it is appropriate to outline some of the fundamental concepts that will be useful to us in the subjects to be covered.

*The Octahedral Planes.* Consider an element in a body where the principal stresses are $\sigma_1$, $\sigma_2$, $\sigma_3$. It turns out that the most important planes to consider from the standpoint of plastic flow are the octahedral planes—the planes that make equal angles with the three principal directions. One such plane is shown as *ABC* in Fig. A.38(a). But there are eight such planes as shown in Fig.

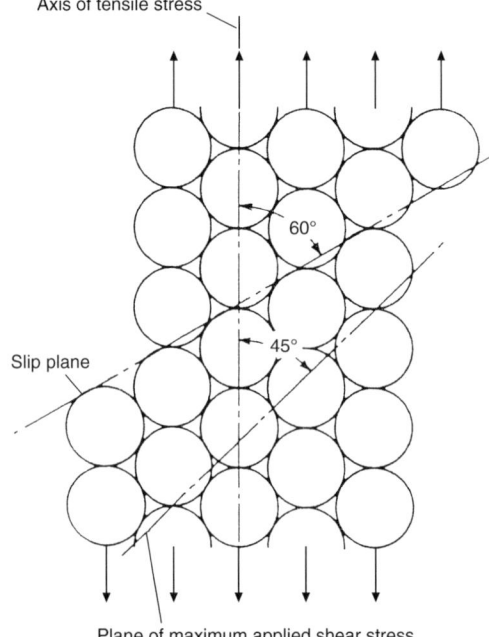

**Fig. A.37** Plastic deformation as atoms slide over each other. Source: Ref A.11

A.38(b). On each of these planes there is a normal octahedral stress $\sigma_{oct}$, and an octahedral shear stress $\tau_{oct}$, which have the values:

$$\sigma_{oct} = \left(\frac{1}{3}\right)(\sigma_1 + \sigma_2 + \sigma_3) \quad \text{(Eq A.15)}$$

$$\tau_{oct} = \left(\frac{1}{3}\right)[(\sigma_1 - \sigma_2)^2 + (\sigma_2 - \sigma_3)^2 + (\sigma_3 - \sigma_1)^2]^{1/2} \quad \text{(Eq A.16)}$$

If 1, 2, and 3 are not the principal directions, but rather any three mutually perpendicular directions along which both normal stresses $\sigma_x$, $\sigma_y$, $\sigma_z$ and shear stress $\tau_{xy}$, $\tau_{yz}$, $\tau_{zx}$ act, the stresses on the octahedral planes are expressed by:

$$\sigma_{oct} = \left(\frac{1}{3}\right)(\sigma_x + \sigma_y + \sigma_z) \quad \text{(Eq A.17)}$$

$$\tau_{oct} = \left(\frac{1}{3}\right)[(\sigma_x - \sigma_y)^2 + (\sigma_y - \sigma_z)^2 + (\sigma_z - \sigma_x)^2 + 6(\tau_{xy}^2 + \tau_{yz}^2 + \tau_{zx}^2)]^{1/2} \quad \text{(Eq A.18)}$$

The octahedral shear stress, or sometimes a simple multiple of it, has been used to determine the relationship between stresses and strains when multiaxial stresses are present. It is remarkable that these equations for octahedral shear and normal stresses became the relevant equations involved in failure theory by at least five independent hypotheses for static failure. The most prominent is the distortion energy criterion of Huber-von Mises-Hencky (see discussion in Ref A.11, p 85–89).

*Relation between Stresses and Strains.* Several theories are available for computing the plastic strains for a given set of stresses. Because we are usually concerned with total strains wherein neither the elastic nor the plastic are negligible relative to each other, we must include both types of strains in the equations. Thus:

$$\varepsilon_1 = \frac{\sigma_1 - \nu(\sigma_2 + \sigma_3)}{E} + \frac{\varepsilon_{eqp}}{\sigma_{eq}}\left[\sigma_1 - \frac{1}{2}(\sigma_2 + \sigma_3)\right] \quad \text{(Eq A.19)}$$

$$\varepsilon_2 = \frac{\sigma_2 - \nu(\sigma_3 + \sigma_1)}{E} + \frac{\varepsilon_{eqp}}{\sigma_{cq}}\left[\sigma_2 - \frac{1}{2}(\sigma_3 + \sigma_1)\right] \quad \text{(Eq A.20)}$$

$$\varepsilon_1 = \frac{\sigma_3 - \nu(\sigma_1 + \sigma_2)}{E} + \frac{\varepsilon_{eqp}}{\sigma_{cq}}\left[\sigma_3 - \frac{1}{2}(\sigma_1 + \sigma_2)\right] \quad \text{(Eq A.21)}$$

where the equivalent stress $\sigma_{eq}$ and equivalent plastic strain $\varepsilon_{eqp}$ are given by:

$$\sigma_{eq} = \frac{\sqrt{2}}{2}[(\sigma_x - \sigma_y)^2 + (\sigma_y - \sigma_z)^2 + (\sigma_z - \sigma_x)^2 + 6(\tau_{xy}^2 + \tau_{yz}^2 + \tau_{zx}^2)]^{1/2} \quad \text{(Eq A.22)}$$

$$\varepsilon_{eq} = \frac{\sqrt{2}}{3}[(\varepsilon_x - \varepsilon_y)^2 + (\varepsilon_y - \varepsilon_z)^2 + (\varepsilon_z - \varepsilon_x)^2 + 6(\gamma_{xy}^2 + \gamma_{yz}^2 + \gamma_{zx}^2)]^{1/2} \quad \text{(Eq A.23)}$$

Here, if $\sigma_x$, $\sigma_y$, $\sigma_z$ are known, it is relatively easy to calculate the strains $\varepsilon_x$, $\varepsilon_y$, $\varepsilon_z$. However, if it is the strains that are known, direct calculation of stresses is difficult. Mendelson and Manson

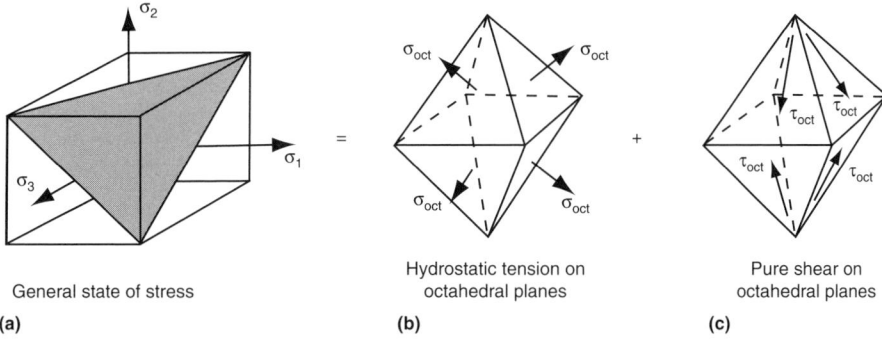

**Fig. A.38** States of stress represented on octahedral planes. (a) General state of stress. (b) Hydrostatic tension on octahedral planes. (c) Pure shear on octahedral planes

(Ref A.46) developed a solution strategy to solve this complex problem.

*Conditions for Development of Multiaxiality.* Even in uniaxial loading the strains are biaxial, as already noted by the strains developed in the transverse direction due to Poisson's ratio. For the elastic part of the axial strain, Poisson's ratio is about 0.33 and for the plastic part of the axial strain, Poisson's ratio is 0.5. When loadings occur in two or more directions, the stresses become multiaxial and, of course, so will the strains. Figure A.39 shows a number of simple types of loading for which multiaxial conditions of stress and strain develop. It is important to recognize that the external loading need not be multiaxial in order for the resultant internal stresses to be multiaxial. An important case is when a crack develops in the specimen under tensile loading as in Fig. A.39(c). At the crack tip nearly equal tensile stresses develop in the two transverse directions to $\sigma_y$. Because of the Poisson's ratio effect, a large transverse contraction occurs, but because highly localized compressive strain is prevented by the adjacent material farther from the crack tip, transverse tensile stresses develop. This stress depends on the thickness of the material wherein the crack exists. If the thickness is high enough, transverse strain may be nearly completely prevented. The transverse stresses can approach, even for elastic conditions, $2/3(\sigma_y)$, and for plastic conditions it can approach $\sigma_y$. Thus, at the crack tip a condition of high triaxial tension can develop even though the external load is uniaxial. Because the triaxial tension prevents local plastic flow (according to the octahedral stress criterion), the root of a crack can act in a very brittle state, even though the material has good ductility in the tensile test wherein the stresses are uniaxial. The problem of brittle fracture at cracks is discussed in Chapter 9, "Crack Mechanics."

## Important Tests for Materials Characterization

Several different types of tests are commonly used in characterizing the mechanical behavior of materials. Often a choice is made of a material for a chosen application based only on the numbers derived from these tests. To determine how realistic these numbers might be to the application, it is important to recognize how the numbers are determined and what their meanings are.

### The Conventional Tension Test

The relation between the strain and the stress produced by a uniaxially applied load is known as the stress-strain curve. It is very important that the specimen be prepared according to strict standards and that the test be conducted in a distinctly specified manner, if the results are to be universally accepted (especially if the matter involves litigation wherein the strength properties of a material are in question).

**Specimen Types.** Two basic types of specimens have been used:

- A gage length of uniform cross-sectional area is provided so that the axial strain along the gage length can be measured as a function of stress. A 50 mm (2 in.) gage length with a diameter of 13 mm (0.5 in.) is common, although other lengths and diameters may be used, especially if material avail-

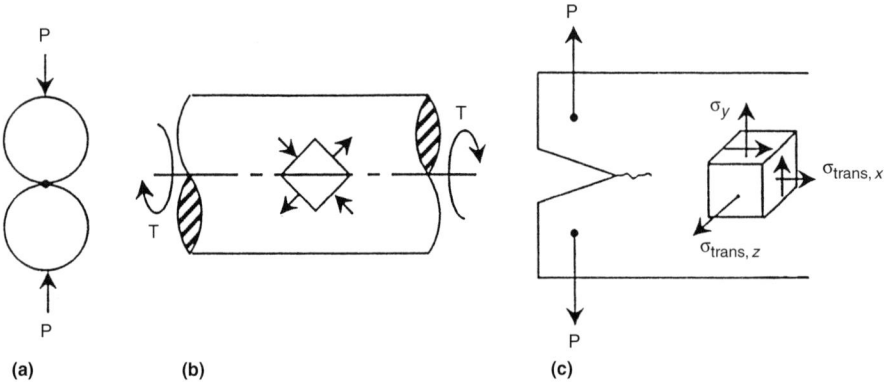

**Fig. A.39** Typical simple types of loading that lead to multiaxial stresses and strains. (a) Bearing. (b) Torque. (c) Cracked body

ability dictates a smaller test section. Thorough specifications for this specimen configuration have been derived by consensus by ASTM International (Ref A.47).
- An hourglass specimen, in which the critical test section has a minimum area, so that the gage length at the minimum section is essentially zero. Strain in the axial direction must then be deduced from measurement at the minimum cross section by a diametral extensometer. Standards no longer exist for this specimen configuration. However, much of the data generated in the past has used this specimen configuration. The specimen has high resistance to compressive buckling under large plastic strains.

Figure A.40 shows schematically the two types of specimens just described. For the case of specimen (b), the radius of curvature of the test section is kept as large as is feasible in order to minimize stress concentration and resultant stress multiaxiality. This specimen also minimizes buckling effects when compression loading is involved, as for example, when the same specimen is used in low-cycle fatigue tests involving large cyclic strains.

Although many variations of geometries may be used, the basic principles are the same, involving either a gage length over a uniform area, or one involving a cross section of minimum area where the transverse strain measurements are made.

**Loading.** With the sophisticated testing machines available today, different control modes of loading can be selected, the most common controls being on force or strain. For tensile testing, the quantity being controlled is increased at a linear rate with time. When the uniform gage length specimen in Fig. A.40(a) is used, it is common to attempt to maintain the axial strain rate constant, while diametral strain rate depends on whether the strain is elastic or plastic. Equation A.7 is useful to control the axial strain rate when the measurement is on the diametral strain. At room temperature the rate of straining does not appreciably affect the stress-strain curve for most engineering alloys. However, when strain rates become very large, or when testing at high temperatures where time-dependent strains develop, special strain rates may be used if they are in the range of rates expected in service applications.

**Common Types of Stress-Strain Curves.** Two generic types of stress-strain curves are commonly encountered. In Fig. A.41, the tests involved are assumed to be strain-rate controlled; for the sake of discussion, let us say that force is applied at whatever rate is required in order to maintain a strain rate of 1% per minute. Figure A.41(a) shows the type of relation commonly observed for soft steels. The initial region $OA$ is linear, the strain being elastic. Dislocations may attempt to move once the higher stresses are achieved, but they are blocked from initial movement, perhaps by the carbon atoms, which are at interstitial sites within the bcc crystal lattice of the steel. At point $A$, the internal stress is sufficient to overcome the resistance to dislocation movement, and a large number of dislocations (often referred to as an *avalanche*) start to move, thus suddenly producing much strain. Because the machine is programmed to apply force that is just sufficient to produce strain at 1% per minute, however, the machine must reduce force to maintain this strain rate. The load falls to point $B$, and is maintained at this level until point $C$, where the originally blocked dislocations exhaust their plastic strain-producing capability. Further plastic strain requires higher loading, along $CD$ to overcome the resistance provided by the newly created dislocations. The stress at point $A$ is called the upper yield strength, that at $B$ the lower yield strength. Accompanying the sudden plastic deformation immediately following the breakaway at the upper yield point, one can readily observe bands of shear on the surface of the specimen. These remarkably distinct bands were noted over a cen-

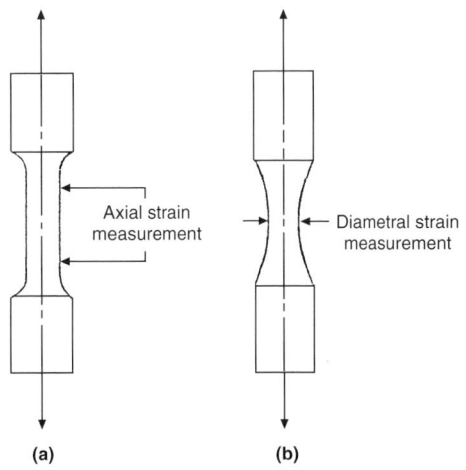

**Fig. A.40** Two types of axial specimens for measuring the stress-strain curve of a material. (a) Uniform gage length. (b) Hourglass configuration

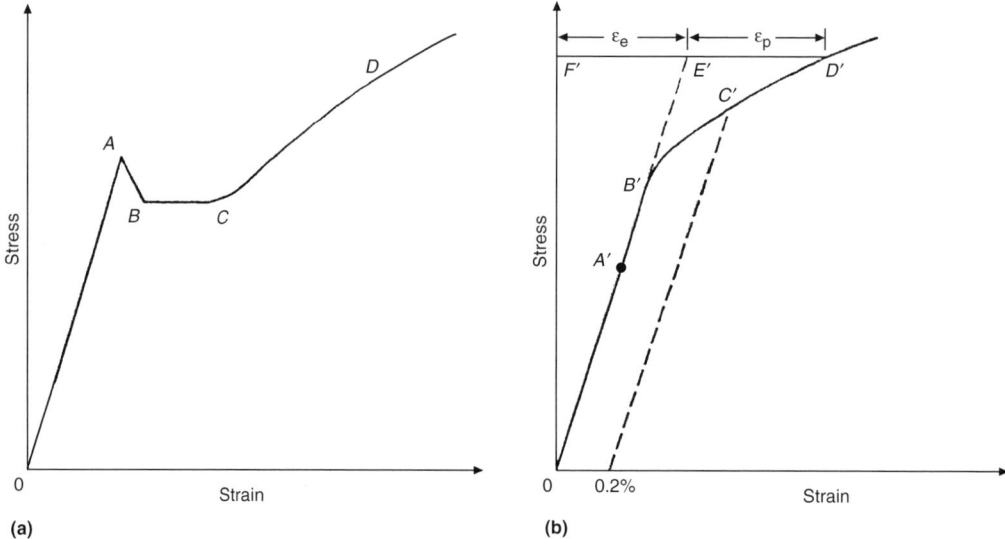

**Fig. A.41** Common types of tensile stress-strain curves showing early regions of plastic flow. (a) Upper and lower yield point. (b) Gradual yielding and strain hardening

tury ago by Lüders, and in his honor are called *Lüders bands*.

While this type of stress-strain curve is very complex, and while it has to be taken into account in the initial yielding of a structure, it is fortunate that we do not have to deal with it extensively in fatigue analysis. As discussed in Chapter 2, "Stress and Strain Cycling," after the first reversal of force, the stress-strain character appears as shown in Fig. A.41(b). In this figure, the onset of dislocation motion with increased load is so gradual that the deviation of the curve $A'B'C'$ from the elastic line is very gradual. Just where it departs (point $B'$) is so difficult to detect, and so depends on the sensitivity and accuracy of the detection equipment, that this point is assigned a lesser engineering significance. It is called the *proportional limit* stress because here the proportionality between stress and strain ceases, but it is rarely used in design. Instead, the point $B$ where a distinct amount of plastic strain develops (commonly 0.2%) is defined as the yield strength, and is the more commonly recognized property for design purposes.

For materials that have a well-behaved stress-strain curve, it has been found that the plastic strain, $\varepsilon_{pl}(E'D')$ bears a power-law relation to the stress $\sigma$:

$$\varepsilon_{pl} = \left(\frac{\sigma}{K}\right)^{1/n} \qquad \text{(Eq A.24)}$$

where $K$ (strength coefficient) and $n$ (strain-hardening exponent) are empirical constants obtained by plotting $\varepsilon_p$ versus $\sigma$ on log-log coordinates. The choice of this form for representing the relation is a convenience for subsequent mathematical manipulations. A novel graphical solution for determining the strain-hardening exponent was developed in 1963 by the junior author (Ref A.48) and is shown in Fig. A.42. A simple derivation reveals that:

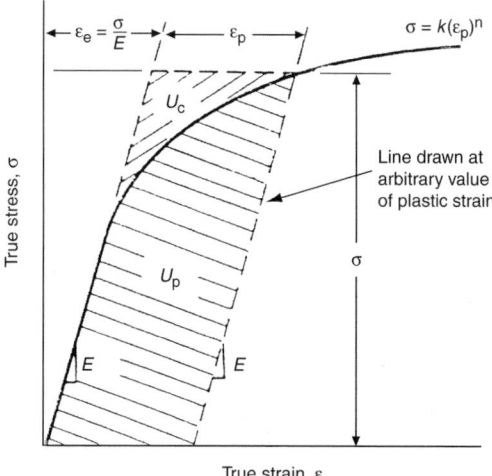

**Fig. A.42** Graphical solution for the strain-hardening exponent ($n = U_c/U_p$) from the conventional stress-strain curve. Source: Ref A.48

$$n = \frac{U_c}{U_p} \quad \text{(Eq A.25)}$$

where $U_c$ is the complementary plastic strain energy (the area above the stress-plastic strain curve) and $U_p$ is the dissipated plastic strain energy (the area below the stress-plastic strain curve) as depicted in Fig. A.41. The graphical solution can be obtained quickly and in the absence of a digital computer or log-log graph paper. Curves that have a sharp breaking point and flow readily with little increase in strength have very low strain-hardening exponents. For example, an elastic-perfectly plastic stress-strain curve (similar to that exhibited by steel during the lower yield point flow regime) has a near zero value for $n$. On the other hand, curves with a large and gradual curvature have a high strain-hardening exponent. The graphical approach allows an experienced observer to glance at a stress-strain curve and closely estimate the value of $n$. The approach also works well for determining the cyclic strain-hardening exponent $n'$ from observation of the shape of the hysteresis loops (Ref A.49). A distinction, however, must be made between the monotonic and cyclic strain-hardening exponents as discussed in Chapter 2, "Stress and Strain Cycling."

The total strain, $\varepsilon$, which is the sum of the elastic and plastic strain $(F'E' + E'D')$ in Fig. A.41 is:

$$\varepsilon = \frac{\sigma}{E} + \left(\frac{\sigma}{K}\right)^{1/n} \quad \text{(Eq A.26)}$$

which is commonly referred to as the Ramberg-Osgood equation (Ref A.50). Although for small stresses and strains, the $\varepsilon$ and $\sigma$ in Eq A.26 may be the engineering values, the deviation from experimental correlation increases for larger plastic strains. Thus, we accept that Eq A.26 applies for the true-stress/true-strain relation, as implied by the use of the $\sigma$ and $\varepsilon$ designation. (Actually, a careful analysis of real stress-strain curves shows some deviations from this characterization; however, we shall not dwell on the deviations, and assume that this formulation is adequate for the purposes we seek in these discussions.)

For stress-strain curves of the type shown in Fig. A.41(a), we do not attempt to fit an equation to the curve in its vicinity of the upper and lower yield point, but such an equation can be fitted for strains large in comparison to the yield strain. In any case, it is not of great concern to us because, as discussed later, the corresponding "cyclic" stress-strain curve sheds the irregular region and is well behaved over the entire strain range, following Eq A.26.

**Brief Analysis of Stress-Strain Curves.** We shall now describe some relations that are useful in estimating fatigue properties from information derived from the tensile test. Consider, for example, a material described by Eq A.26 over the complete stress and strain range up to failure. When the true stress ($\sigma$) is plotted against true total strain ($\varepsilon$), both scales being logarithmic (Fig. A.43) the equation plots as two straight-line segments, with a curved transition segment in the vicinity of their intersection. When plotted on linear engineering stress ($S$) and engineering strain ($e$) scales in Fig. A.44, the curve translates to $OA'B'C'D'E'F'G'$ in accordance with the table inset in the figure. It is not possible to establish the true stress-strain curve beyond the point of maximum load and localized necking from a specimen with an axial extensometer attached over the parallel-sided gage length as in Fig. A.41(a). However, this region of the curve can be determined from diametral strain measurements on an hourglass specimen of Fig. A.41(b) in which the gage length at the minimum section is essentially zero. The true stress is determined as the external divided by the true instantaneous area at the minimum section, and the true strain is determined by converting the nominal diametral strain at the minimum section into nominal axial strain using Eq A.7. The resultant true axial stress-strain curve $OABCDEFG$ is shown in Fig. A.44.

Actually, there is a complication that develops when the true strain becomes large, and the neck

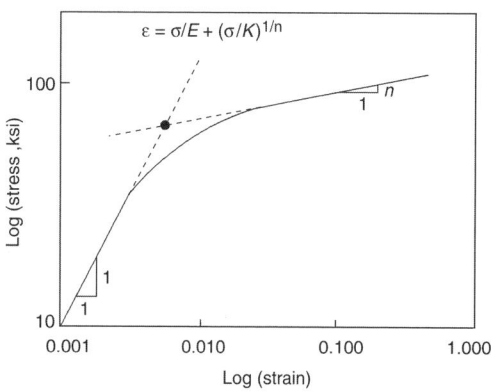

**Fig. A.43** Typical stress-strain curve plotted on log-log coordinates

at the minimum section becomes deep, as indicated in Fig. A.45, so that a substantial triaxial stress state develops. Bridgman (Ref A.51) developed an analysis to correct the axial stress at the minimum section for this triaxiality.

Bridgman considered that at the neck, because of the curvature, a triaxial stress state develops and that at the center of the neck all three principal stresses became equal to σ:

$$\sigma = \frac{(\sigma_x)_{\text{ave}}}{\left(1 + \frac{2R}{a}\right)\left[\ln\left(1 + \frac{a}{2R}\right)\right]} \quad \text{(Eq A.27)}$$

where $(\sigma_x)_{\text{ave}}$ is the average stress as determined by dividing the applied axial force by the minimum cross-sectional area. Thus, to determine the true stress in the neck, the analyst must know the geometric pattern that develops when necking occurs. Experimental observation of this pattern by continuous measurement or by photography can be used to obtain this pattern for a specific material. Aronofsky (Ref A.52) has applied the Bridgman concept to obtain the triaxiality in the neck of a flat tensile specimen.

If a nominal stress-strain curve is determined using an axial extensometer on a straight-sided uniform section specimen such as shown in Fig. A.45(a), the curve may not coincide with $OA'B'C'D'E'F'G'$. In fact, the curve will depend on the gage length of the test specimen. This is explained by the fact that after the maximum load is reached, and the specimen starts to neck, the remainder of the gage length *contracts* as the load decreases. Thus, the elongation will be based on the average strain over the entire gage length, $L_0$ going to $L_f$. But if the necking action takes place over but a small part of the gage length, $L'$, the measured average elongation will be lower than the local elongation. If a plot is made of apparent elongation at each load as the nominal stress (load/initial cross-sectional area), the result will be as shown schematically in Fig. A.44 by $OA'B'C'D'$. Up to

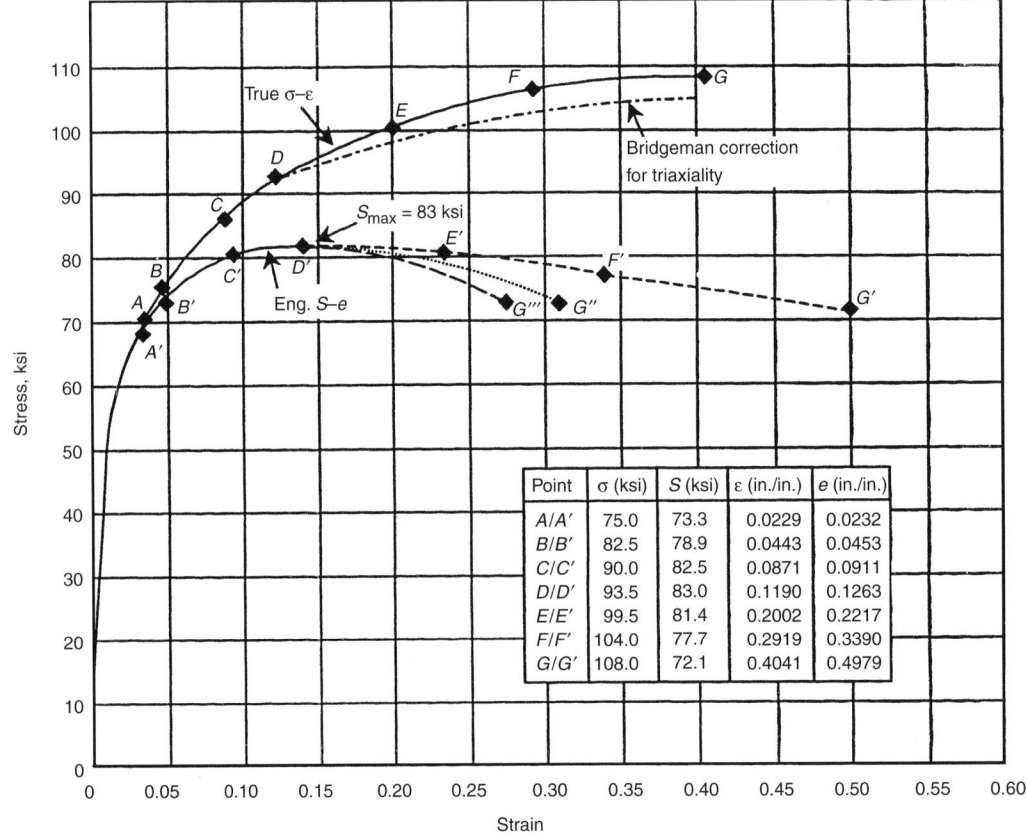

**Fig. A.44** Typical stress-strain curves shown for the two cases where the coordinates are either "true" or nominal values

necking, point $D'$, it may be very close to the engineering curve for a zero gage-length specimen. But beyond the ultimate tensile strength, point $D'$, the curve may fall more rapidly. The apparent ductility (strain at $G''$) will be much lower than the elongation $G'$ for a zero-gage-length specimen. An even longer-gage-length specimen may follow necking with an even lower curve $D'G'''$. Thus, it is clear that elongation at failure must be associated with a specified gage length, and the higher the gage length, the lower is the expected elongation at fracture. But it is also clear that the ultimate tensile strength, which is based on point $D'$, will not depend strongly on the gage length. And, if the fracture stress is based on the cross-sectional area at the point of fracture, it also will not be very sensitive to the gage length.

For fatigue analysis, considerable use is made of two important parameters, true fracture strain and true stress at fracture.

For true fracture strain $\varepsilon_f$:

$$\varepsilon_f = \ln \frac{A_o}{A_f} \qquad \text{(Eq A.28)}$$

where $A_o$ and $A_f$ are the original and final cross-sectional areas, respectively, in the gage length. For true stress at fracture $\sigma_f$:

$$\sigma_f = \frac{P_f}{A_f} \qquad \text{(Eq A.29)}$$

where $P_f$ is the tensile load at the point of fracture.

**Direct Determination of Ultimate Tensile Strength from True Stress-True Strain Parameters.** While a graphical determination of the ultimate tensile strength is useful, a calculation in essentially closed form is possible. Use is made of the fact that at the ultimate tensile stress, the true strain is equal to the strain-hardening exponent, $n$. Thus, the nominal strain can be determined. The true stress can also be calculated for $\varepsilon = n$ by inverting the stress-strain relation, Eq A.26. (The procedure is discussed in the section, "Inversion Techniques".) The nominal stress can then be calculated using Eq A.6. Thus, point $A$ translates to point $A'$, $B$ to $B'$, $C$ to $C'$, and so on. It is now seen that if force (proportional to $S$) were plotted against conventional strain, a maximum develops at $D'$, and then decreases until fracture occurs at $G'$. The nominal stress at the maximum load is called the ultimate tensile strength, $S_u$. Although it has no special physical meaning (an artificial stress obtained by dividing the maximum load that occurs after considerable plastic strain and contraction in area, by the initial cross-sectional area), this ultimate tensile strength is the most often quoted strength property of a material. Experimentally, it is a simple measurement to make: no area determinations are required during straining; just apply a constant strain rate and note when the maximum load is reached, then divide this load by the initial cross-sectional area. Using old-style testing machines the maximum load was observed with a tracking dial gage that moved with the pointer, designating load, but did not retract after the load started to decrease. Fortunately, good correlations were obtained when more meaningful properties were plotted against

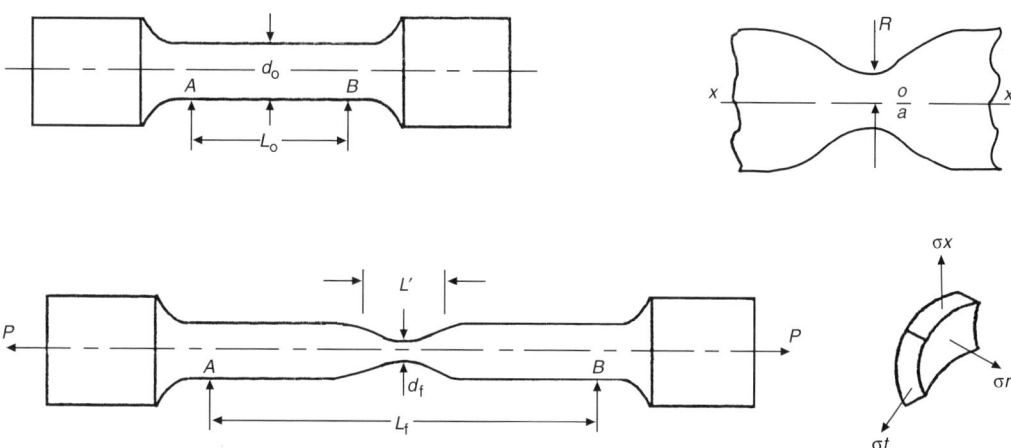

**Fig. A.45** Necking in limited region of a uniform cross-sectional area specimen. (a) Original undeformed specimen. (b) Deformed specimen with necking. (c) Geometry of necked region. (d) Stresses acting at point $O$

$S_u$. We, in fact, use such correlations to advantage.

## The Hardness Test

In estimating fatigue properties it is often advantageous to make use of the hardness of a material because there are remarkably consistent relations between hardness and tensile strength. Many measures of hardness have been developed over the years because it is a property that can be determined on a relatively small volume of material and on finished and semifinished parts. We shall discuss a few measures of hardness that often appear in fatigue literature.

**Brinell Hardness.** In this method, a steel ball is pressed against the surface of the test specimen with a known force. Generally, the ball has a diameter of 10 mm (0.4 in.); for soft metals the load is 500 kg (1100 lb), while for medium and hard materials a 3000 kg (6600 lb) load is used. The Brinell hardness number is determined as the force divided by the surface area of the resulting spherical indentation. Other ball sizes and loadings may be used, and approximately the same hardness number is obtained as long as the ratio of the force to the projected area of the ball, i.e., $(\pi/4)D^2$ is kept constant.

Because the Brinell hardness number relates penetration (a measure of strain) to unit force (a measure of stress) in this highly complex multiaxial stress system, it turns out that a good approximate relation can be determined between the Brinell hardness number and the ultimate tensile strength. We shall discuss this relation in a subsequent section.

**Rockwell Hardness.** Here an indenter is penetrated into the surface of the test piece by a fixed load. In this system, however, the indenter may be a circular cone of diamond material (with a slightly rounded apex) or steel balls of either 1/8 or 1/16 in. diameter. Three combinations are common, each designated by a "scale." For hard materials the "C" scale is used involving a 120° cone with a 150 kg (330 lb) load. The basic measurement made is the amount of penetration produced, but the description of the hardness is according to an arbitrary scale, which for the C scale ranges from 20 to 70 HRC (hardness on a Rockwell C scale) the higher the number, the harder is the material. For the B scale a 1/16 in. steel ball indenter is used with a 100 kg (220 lb) force, and the measurement ranges from 0 to 100 HRB on the B scale. Several Rockwell hardness scales (A, E, K, L, M, and R) have been developed for soft nonmetallic materials. These use a ball indenter of up to 1/2 in. in diameter.

**Vickers Hardness.** In the Vickers hardness test the indenter is a diamond pyramid. Indentation produces a square (approximately), and the diagonals are measured, which together with the force permits the calculation of a "diamond pyramid hardness number." Two important advantages accrue from this test. First, it can be used for materials over a very large hardness range from very soft to very hard metals. Second, the indentation can be very shallow, so that the hardness number refers to a very shallow layer of surface. It is thus particularly useful for determining the surface hardness in cases where the hardness varies appreciably with depth, for example, a surface casting. Also of advantage is that very little surface is needed to conduct the test, and very little of the surface is destroyed. These tests are considered to be nondestructive while providing valuable information on the mechanical characteristics of a material.

**Knoop Hardness.** This type of hardness is very similar to the Vickers test in that a sharp diamond pyramid is used, except that the planes forming the pyramid do not make equal angles with each other. Rather, these are in the ratio of approximately 7:1. The special advantage here is that, to a greater extent even than the Vickers hardness, very small surface areas can be interrogated. For example, the variation of surface hardness due to microconstituent variations can be studied. Similarly, hardness can be determined for very shallow depths and thus have utility for sheet materials.

**Scleroscope Hardness.** This test provides a very easy method for measuring hardness. The tester consists of a hard steel ball that is dropped within a tube onto the test surface. The hardness number is a measure of the rebound of the ball after it impacts the surface. The harder the test material, the higher is the rebound relative to the height from which it was dropped. The advantage here is that no surface measurements are needed; all that is required is to note the amount of rebound on a scale provided within the device.

**Shore Durometer.** Two durometers are used for the hardness testing of soft materials such as plastics and rubber. The hardest of these materials uses the type D durometer, whereas the type A is used for the softest. The two scales cover a broad range. Readings above 90 on the type A scale indicate a switch should be made to the type D scale, and conversely, when readings below 20 are obtained on the type D scale, measurement should be shifted to the type A scale.

**Some Correlations.** Figure A.46 shows some correlations of the various hardness scales for carbon and alloy steels (Ref A.53). As noted in the figure, the values shown are for annealed, normalized, and quenched-and-tempered conditions; they are less accurate for cold-worked conditions and for austenitic steels. An interesting observation that is apparent from the figure is that the Rockwell B scale is applicable for steels of tensile strength lower than about 120 ksi, whereas above this strength the C scale is used. It is also interesting to note that numerically the Knoop, Vickers, and Brinell numbers are very similar to each other and that they are applicable over the entire range of tensile strength from the softest to the hardest steels common in engineering applications. The dotted line shows values for which the hardness numbers on the Vickers, Knoop, and Brinell scales are twice the tensile strength (ksi); leading to the often-used rule of thumb that the hardness numbers for these three scales are approximately twice the tensile strength.

Correlations of hardness for other alloys such as aluminum, copper, and titanium can be found in appropriate handbooks published by ASM International. ASTM International (www.astm.org) has standardized hardness testing procedures for all classes of solid materials ranging from metallics (ASTM E 18-02) to plastics (D 785-98e1) and rubber (D 2240-02).

## The Bend Test

Bend tests are very useful in evaluating brittle materials under static loading and ductile materials in fatigue. In the case of brittle materials (ceramics and very-high-strength metals), attempts to determine strength levels through tensile testing result in lower strengths because of unavoidable bending introduced by small deviations from perfect axial alignment. By using a bending test the maximum stress can be controlled both as to magnitude and location. Two configurations are common, the three-point loading test and the four-point loading test, as illustrated schematically in Fig. A.47.

In the three-point loading test of Fig. A.47(a), the maximum tensile stress occurs under the concentrated central load, at the lowermost fiber. However, there is some distortion of the stress field in the region of the maximum stress due to the central load and the indeterminateness of the precise manner in which the load $P$ is imposed. This complication is usually ignored, and the stress is calculated by the usual bending formula well known in the field of strength of materials. The bending moment $M$ at the critical region of a rectangular cross-section beam is equal to $Pa$:

$$\sigma = \frac{Mc}{I} = Pac \bigg/ \left(\frac{1}{12}\right) bc^4 = \frac{12Pa}{bc^3} \quad \text{(Eq A.30)}$$

For ceramics the stress at fracture is usually designated as the modulus of rupture and is re-

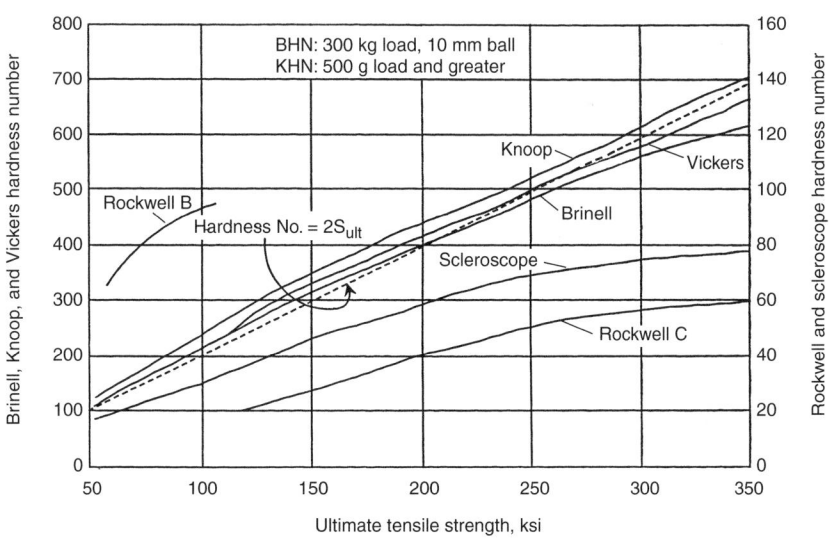

**Fig. A.46** Hardness scale correlations for carbon and alloy steels. Source: Ref A.53

garded as a reasonable representation of strength for comparative purposes among materials. The quantitative accuracy of the value so determined may be questioned for at least two reasons: first, the indeterminacy of the actual stress, and second, the fact that the maximum stress occurs over such a small volume of material. For ceramics, strength depends strongly on volume of material exposed to stress, and usually the Weibull statistical theory of strength is involved in order to obtain more meaningful results. See, for example, the work of Weibull (Ref A.54) for a more complete discussion of this subject.

A more appropriate loading system is that involving four loading points, as shown in Fig. A.47(b). Here, the bending moment is constant over the entire region between the two inner loading stations. The bending moment is still equal to $Pa$, and the outermost fiber stress is still $12Pa/bc^3$, but most of the region is remote from the concentrated applied loads and can be relied on as more accurate, especially if the region between the inner loads is of *slightly* reduced cross section. Furthermore, the volume of material under maximum stress can be controlled to any desired value by suitable spacing of the inner loading points.

The four-point loading system is very common in high-cycle fatigue testing. Rather than cycling the loads, which introduces complications, especially when frequency is high, the specimen is made with a circular cross section and is rotated while the loads and reactions are stationary. In this way, the outermost fibers of the specimen can be successively subjected to tension and compression according to whether they are in the lowermost or uppermost position. Of course, all supports must provide suitable bearings to allow rotation of the test specimen and a drive system for rotating the specimen must be provided. This testing technique is termed *rotating bending*. Further discussion is given in the subsequent section, "Fatigue Life Regimes and Loading Modes."

The formula, Eq A.30, is appropriate as long as the stresses in the specimen remain in the elastic range. For higher loading, which result in plastic deformation, which is of special interest in low-cycle fatigue, it becomes necessary to correct this equation for the resulting plasticity. This subject is discussed in detail in Chapter 7, "Bending of Shafts."

## The Torsion Test

Torsion testing is conducted primarily to simulate the type of loading commonly encountered in torque-transmitting shafts and other types of loading where shear resistance is important. It is also used in fundamental studies because shear is so basic to plastic deformation, and the torsion test is a good way to induce shear without the complication of other types of loading. Usually a torque is applied to a circular shaft and the deformation and fracture properties are observed. Ordinarily the shaft is solid, but on occasion, hollow shafts are used. In the limit a thin circular cylinder permits a reasonably accurate calculation of the average shear stress, $\tau$, across the wall thickness, $t$:

$$\tau = M_T/2\pi tr \qquad \text{(Eq A.31)}$$

where $M_T$ is the torsional moment (torque) and $r$ is the mean radius (from the center axis to the mid-thickness of the wall). Even for fundamental studies the tube thickness generally is maintained at 10% or more of the diameter in order to prevent buckling.

Figure A.48 shows schematically the solid circular cross-section specimen of diameter $D$ used in torsion testing. The maximum shear stress, $\tau_{max}$, is at the surface and for linearly elastic material behavior is given by:

$$\tau_{max} = 16M_T/\pi D^3 \qquad \text{(Eq A.32)}$$

The corresponding shear strain $\gamma$ is:

$$\gamma = \tan\theta \qquad \text{(Eq A.33)}$$

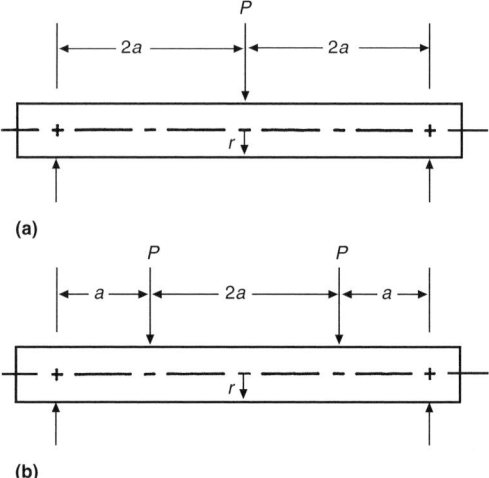

**Fig. A.47** Bending arrangements. (a) Three-point loadings. (b) Four-point loadings

On the surface of the specimen a biaxial state of stress develops, as shown in Fig. A.49. Two principal stresses develop of equal magnitude but opposite sign, and of magnitude equal to the shear stress at the surface. Compare this to a tensile test in which the maximum shear stress is only half the tensile stress. If the shear strength is less than the tensile strength (typical of a ductile material), failure will occur in shear and the true shear strength can be determined in a torsion test. The fracture will be perpendicular to the axis of the specimen. If the tensile strength is less than the shear strength (as is the case for brittle materials), then failure will start normal to the tensile stress $\sigma_1$ that is at 45° to the axis of the specimen. Once the failure starts, the stress distribution can become very complicated and the fracture direction can change. But in general it will appear as a 45° helical fracture.

For thin tubular specimens, ordinarily intended to maintain the shear stress approximately constant through the thickness, the compressive stress $\sigma_3$ can become troublesome. If the specimen is thin this compression can cause buckling and invalidate the quantitative results. It is for this reason that the thickness of a tube is usually kept at least 10% of the diameter.

Table A.9 summarizes the comparison of the stresses and strains in a tension test versus those in a torsion test. In addition to the factors already discussed, it is clear that in a torsion test the axial strain $\varepsilon_2$ is zero. Thus, under torsion the specimen does not change in axial length*. In principle, therefore, the grip ends would not require freedom of axial displacement. In fact, some freedom must be provided in order to avoid axial loading that could develop when large plastic strains are involved. In addition, any temperature rise due either to plastic deformation or external heating causes a sizable expansive thermal strain in the axial direction, which, if not free to move, will induce large compressive stresses in the axial direction.

Torsion is especially interesting in fatigue testing. It is the simplest way of introducing a biaxial stress-strain state for checking some of the theories that relate to multiaxiality effects in fatigue. The tubular specimen also lends itself easily to the introduction of other stress states than that of torsion alone by allowing axial loading and/or internal pressure.

We are especially interested in torsion involving plastic deformation. We refer to a new formulation of fatigue criterion under multiaxiality in Chapter 5, "Multiaxial Fatigue," and we developed a simple formulation for calculating fatigue life in torsion when substantial plasticity occurs.

---

*True only for isotropic materials. Anisotropic materials will exhibit an axial strain in the presence of pure torsional loading.

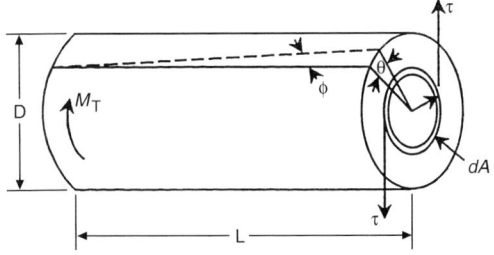

**Fig. A.48** Torsion of a solid bar

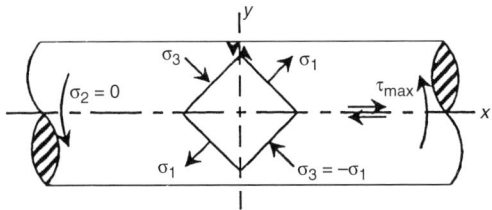

**Fig. A.49** State of stress in torsion

**Table A.9** Comparison of stresses and strains in a tension test with those in a torsion test

| Tension test | Torsion test |
|---|---|
| $\sigma_1 = \sigma_{max}$; $\sigma_2 = \sigma_3 = 0$ | $\sigma_1 = -\sigma_3$; $\sigma_2 = 0$ |
| $\tau_{max} = \dfrac{\sigma_1}{2} = \dfrac{\sigma_{max}}{2}$ | $\tau_{max} = \dfrac{2\sigma_1}{2} = \sigma_{max}$ |
| $\varepsilon_{max} = \varepsilon_1$; $\varepsilon_2 = \varepsilon_3 = -\dfrac{\varepsilon_1}{2}$ | $\varepsilon_{max} = \varepsilon_1 = -\varepsilon_3$; $\varepsilon_2 = 0$ |
| $\gamma_{max} = \dfrac{3\varepsilon_1}{2}$ | $\gamma_{max} = \varepsilon_1 - \varepsilon_3 = 2\varepsilon_1$ |
| $\sigma_{eq} = \dfrac{\sqrt{2}}{2}[(\sigma_1 - \sigma_2)^2 + (\sigma_2 - \sigma_3)^2 + (\sigma_3 - \sigma_1)^2]^{1/2}$ | |
| $\varepsilon_{eq} = \dfrac{\sqrt{2}}{3}[(\varepsilon_1 - \varepsilon_2)^2 + (\varepsilon_2 - \varepsilon_3)^2 + (\varepsilon_3 - \varepsilon_1)^2]^{1/2}$ | |
| $\sigma_{eq} = \sigma_1$ | $\sigma_{eq} = \sqrt{3}\sigma_1$ |
| $\varepsilon_{eq} = \varepsilon_1$ | $\varepsilon_{eq} = \dfrac{2}{\sqrt{3}}\varepsilon_1 = \dfrac{\gamma}{\sqrt{3}}$ |

## Fatigue Testing

### Introduction to Testing under Cyclic Loading

The following sections give an overview introduction to the equipment and procedures used to conduct cyclic stress-strain and fatigue tests. The sections are abbreviated versions of what appear in a recently published ASM International *Handbook* (Ref A.55). Readers requiring a greater depth of coverage should consult this reference. Many other types of mechanical material testing also are covered in this Handbook.

### Fatigue Life Regimes and Loading Modes

Many types of fatigue testing machines have been used to generate fatigue data. Each has unique features designed to accommodate the life regime of interest and the mode of loading: high-cycle fatigue or low-cycle fatigue; force-, deflection-, or strain-control; and whether axial, bending, or torsional loading is required. These distinctions are discussed next.

**High-Cycle Fatigue.** At the high-cycle fatigue (HCF) end of the fatigue spectrum ($10^5$ to $10^8$ cycles to failure and higher), high-frequency cycling is an absolute must. It requires nearly 12 days of cycling at 1000 Hz to reach $10^8$ cycles to failure. Even ultrasonic fatigue testing at 20,000 Hz would require 14 hours to reach this life level. Fortunately, in the HCF regime, the cyclic plasticity is often minuscule so that specimen heating, while ever present, does not restrict testing. Sensitive rise-in-temperature measurements made during high-frequency cycling have, however, been used successfully to help identify HCF strength (Ref A.56). On the macroscopic phenomenological level, specimen behavior is considered to be linearly elastic. Hence, force-controlled testing is adequate, and in practical fact is virtually dictated. Any conventional fatigue extensometer would be of no added value to testing, even if means were found to keep it from being shaken off at higher frequencies due to the high inertia forces during each load reversal.

In the HCF regime, statistical variation in fatigue life is quite large. This dictates multiple test results to provide a sufficient database to establish the fatigue resistance of the material. The exceptionally large amount of total testing time in turn requires multiple fatigue testing machines. Consequently, the cost per machine must be relatively low. The simpler fatigue testing machines based on constant forcing or displacement in plane bending, rotating bending, and direct stress are used almost exclusively for evaluation of very high cycle fatigue resistance of materials. Ultrasonic fatigue testing with frequencies on the order of 20,000 Hz is rarely used because the cyclic strain rates are much higher than found in service. The large difference in strain rate may result in different micromechanisms of straining and hence in different fatigue crack initiation mechanisms.

**Low-Cycle Fatigue.** Conventional low-cycle fatigue (LCF) crack initiation data in the regime of 10 to $10^5$ cycles to failure are best obtained under strain-cycling conditions, leaving little choice but to use a closed-loop, servoelectrohydraulic, direct-stress, axial fatigue testing machine with an extensometer and servostrain control. The cyclic frequency should not be high. Even at 1 Hz, $10^5$ cycles can be reached in just over a day. In fact, higher frequencies would be undesirable at lower life levels where cyclic plasticity could be significant. Energy dissipation due to rapid cyclic plasticity would cause considerable, and highly undesired, specimen heating* (assuming the specimen is not already being heated to an elevated test temperature).

Elevated-temperature fatigue testing, especially low-cycle isothermal creep-fatigue (ICF) and thermomechanical fatigue (TMF), invariably requires servohydraulic, axial strain-controlled cycling capabilities to avoid undesirable strain ratcheting due to plasticity or time-dependent creep. The maximum cyclic rates during TMF cycling are typically quite low (several minutes per cycle) because of the relatively low rate at which temperatures of test samples can be changed in a well-controlled manner. Because of the relatively long testing time per cycle for ICF and TMF tests, and because of the high costs per test and high machine acquisition and maintenance costs, there have been few published investigations** of the statistical nature of ICF and TMF test results. Nuances of high-temperature fatigue testing are discussed in greater detail in a subsequent section.

---

*If cyclically generated heat is not dissipated, the resultant thermal expansion of the specimen test section creates an "apparent" mechanical strain. Under servocontrolled cyclic straining, thermal expansion would be offset by enforced compression and hence compressive mean stress, destroying the original purpose of the test.
**Proprietary databases exist in some large corporations for certain critical materials applications.

The effect of multiaxial stress states on the cyclic stress-strain and fatigue resistance of materials is a perennial question because purely uniaxial stress states are rarely encountered in service loadings. While a wide variety of multiaxial stress states can be generated on laboratory specimens, there is only one specimen that allows direct measurement or control of the multiaxial stress and strain in a convenient manner, and that is torsion. As discussed in a previous section, if the torsion specimen is a thin-walled tube, the shear stresses can be measured or controlled quite accurately. Furthermore, the torsional shear strain can also be measured or controlled with an exceptionally high degree of accuracy. By also applying an axial load to a tubular torsion specimen, a range of multiaxial stress states can be achieved. There have been a few attempts also to include the ability to impose cyclic internal and external pressure to a thin-walled, axially loaded, torsion specimen to achieve an even broader spectrum of multiaxial stress states. Lohr, in 2000, has included an assessment of some such attempts in his summary article (Ref A.58) (overlooked was such a machine developed at the University of Illinois under the guidance of Morrow and Kurath, Ref A.57). These sophisticated multiaxial fatigue-testing machines are quite complex and hence expensive to build and maintain. They have been reserved for crucial evaluation of multiaxial fatigue models and are ill suited for routine database generation.

**Modern Fatigue Testing Equipment.** The state-of-the-art fatigue testing machine in use at the turn of the twenty-first century are highly versatile machines based on closed-loop, servo-controlled, high-pressure hydraulic loading. A schematic of an axially loaded machine is shown in Fig. A.50. The machines are built with a stiff frame and use a hydraulic actuator to apply force and deformation to the centrally located axial specimen. A load cell, mounted in series with the specimen and actuator, senses the axial force carried by the specimen. Typical load cells use resistance strain gages arranged in an electrical four-arm bridge as the sensing element The load cells are generally designed to have low deflection under load so as to keep unnecessary displacement of the hydraulic cylinder to a minimum, thus allowing higher frequency response. The strain across a fixed gage length of the test section of the specimen is sensed by an extensometer. Specimens are clamped securely in place by hydraulically actuated grips that retain the alignment of the axis of the specimen with the loading line. Good axial alignment is nec-

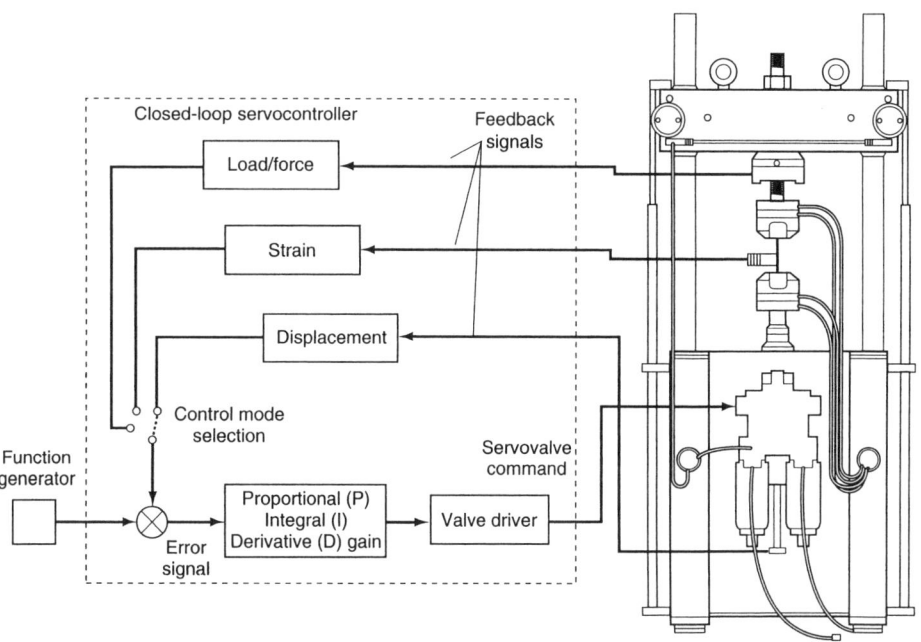

**Fig. A.50** Schematic of modern fatigue testing machine. Source: Ref A.55

essary to avoid undesirable bending moments being applied to the specimen.

The central system of the testing machine is the servovalve that precisely regulates the direction and rate of flow of the hydraulic fluid within the cylinder. The valve driver sends command signals to the servovalve. The valve driver is fed by the error signal, i.e., the algebraic difference between the function generator signal and the sensor signal. The sensor signal may be from the extensometer, load cell, displacement sensor, or any combination of these, or other sensed signals, that are caused to change as the hydraulic cylinder is displaced. Thus, for example, if a strain-controlled test is being performed, the function generator prescribes how the strain in the specimen is to vary with time (for example, sine wave, triangular, etc.). If the strain in the specimen, as measured by the extensometer, exactly matches the programmed strain, the error signal is zero and the test is being perfectly servocontrolled. The function generator and other signals may be analog or digital. Numerous machines in use are controlled by digital computers. Specimen responses from the extensometer and load cell can also be recorded and results stored in computer memory for subsequent playback and for computational data reduction. Not shown in the schematic drawing of Fig. A.50 are analog strip chart (strain vs. time and force vs. time) and $X$-$Y$ (stress vs. strain hysteresis loops) recorders that show instantaneous display and leave a permanent record of the specimen's response behavior. Large testing laboratories with numerous machines can take advantage of a single high-pressure ($\approx$3000 psi) hydraulic supply that distributes fluid as needed to each machine. Hydraulic accumulators are used to isolate any hydraulic pressure fluctuation and interaction between machines. While these fatigue-testing machines were developed for strain-controlled, LCF testing, high-frequency response machines have been developed that can achieve 1000 Hz under closed-loop, servo-load-control. This advance allows such machines to be used advantageously for high-cycle fatigue testing.

## Elevated Temperature Fatigue, Creep-Fatigue, and Thermomechanical Fatigue

The modern fatigue-testing machines just described are particularly well suited for elevated-temperature, LCF testing. This is especially true for closed-loop servo-strain-controlled fatigue testing at temperatures within the creep regime.

Here, thermally activated, time-dependent creep deformation and stress relaxation may occur during cycling or if hold periods at fixed loading conditions are imposed. Typically, cyclic fatigue life is significantly reduced due to creep-fatigue interaction as well as to environmental interactions such as oxidation.

Elevated temperature testing requires modifications and additions to the fatigue testing equipment described previously. A means is needed for heating the specimen and controlling its temperature. Both the grips and the extensometer must be able to function when exposed to the heat applied to the specimen. Depending on the end application of the fatigue results, it may also be necessary to provide an environmental chamber that would surround the specimen. For example, vacuum, or possibly inert gas, chambers are needed to simulate the environment of space. Likewise, a high-pressure hydrogen-rich gas chamber could simulate the environment within the hot-gas path inside a liquid hydrogen/oxygen-fueled rocket engine.

**Specimen Heating.** Induction heating is the most versatile of the various specimen-heating techniques. Secondary high-frequency eddy currents are induced in the outer surface layers of the test sample by coils of copper tubing that are wrapped around the specimen and carry the high primary current. Cooling water is passed through the tubular coils to keep them from getting too hot. A control thermocouple is attached to the specimen test section to provide feedback to a solid-state temperature control module that regulates the current to the primary coil. With multiple thermocouples and the ability to adjust the spacing of the coils, a uniform temperature can be achieved along the specimen gage section. Care must be taken to prevent the thermocouple attachment from being the source of fatigue crack initiation. Induction heating is well suited for thermal cycling fatigue testing because of the low thermal mass of the system. Maximum heating rates are potentially higher than required for most testing. Several alternative forms of heating are available for special test purposes.

*Clamshell furnaces* are equipped with wire resistance or silicon carbide heating elements and are generally used for isothermal testing wherein uniform temperatures are required. A draw back is that high-temperature grips are needed because they tend to be heated to nearly the same temperature as the specimen.

*Parabolic clamshell furnaces* are equipped with radiant quartz lamps. These furnaces are much more thermally responsive than conventional clamshell furnaces. They have been used successfully in thermal-cycling fatigue tests, although they are not as responsive as induction heating systems.

*Direct resistance heating* has been used occasionally (Ref A.59). It offers extraordinarily high heating rates that are nominally uniform across the cross section of the gage length because the heat is generated uniformly throughout the volume of the test section. A very high current of a few hundred amperes is passed through the specimen with a very low voltage drop of generally less than 1 V. The heating technique is well suited for thermomechanical fatigue testing. There are however, two drawbacks: high magnetic fields create electrical interference in the low-voltage sensor and control circuits, and the potential for severe arcing if the high-current circuit is shorted. Burns are possible from the rapid and intense heat of the arcing.

**High-Temperature Specimen Grips.** Modern fatigue testing machine grips are generally complex mechanisms with carefully machined mating surfaces. Because the grips are used repeatedly over a long period of time, they must be free from damage due to wear, surface oxidation, and creep. They are best designed to run at much lower temperatures than the specimens being tested. Grips running at high temperatures eventually suffer damage that disturbs the initial carefully machined alignment and degrades the load-carrying capability. It is good engineering practice to keep the grips outside the heating zone of the furnace. It is also common practice to provide cooling water to the grips used for high-temperature fatigue testing even though the grips are beyond the heated zone of the specimen. A typical hydraulically operated self-aligning grip system that we have is shown in (Fig A.51).

**High-Temperature Extensometry.** High-temperature strain measurement and control requires use of high-temperature probes that are held against the specimen and demarcate the gage length. A high-temperature ceramic is generally the material of choice for the probes. They must be long enough to keep the sensing element of the extensometer far removed from the high-temperature environment surrounding the specimen. Even then, radiation heat shields are typically employed to protect the sensor. Air cooling may be used to keep the sensor cool. Also, the force of the probes pressing against the specimen surface must be kept low enough that damage is not caused at the very small points of

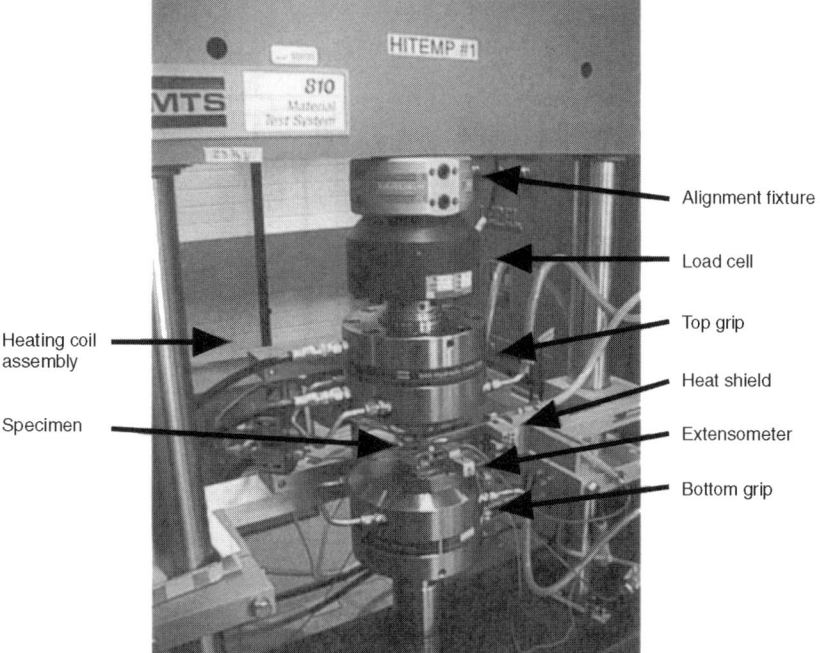

**Fig. A.51** Typical hydraulically operated self-aligning grip system installed on testing machine. Source: Ref A.55

contact where contact stress can be high. Figure A.52 shows a typical high-temperature extensometer set up in conjunction with induction heating coils and hydraulic grips.

**Environmental Chambers for High and Low Temperatures.** Controlled environmental test chambers are necessary to generate fatigue data in simulated service environments. These environments include vacuum, inert gases such as argon or helium, chemically aggressive gases and liquids, and even nuclear radiation.

While environmental degradation of fatigue properties may be most severe at elevated temperatures, there are also needs to control the environment in lower-temperature tests. Cryogenic fatigue testing has been of great concern to the rocket propulsion industry, which uses high-pressure liquid hydrogen ($-253$ °C, or $-423$ °F) as fuel to be mixed with high-pressure liquid oxygen ($-183$ °C, or $-297$ °F) to burn at temperatures of 3300 °C (6000 °F). Test chambers for such extreme conditions are extraordinarily expensive to build and operate in a safe manner. Figure A.53 shows a photograph of a typical vacuum/inert gas test chamber installed at the NASA Glenn Research Center for fatigue testing at temperatures between 25 and about 1090 °C (75 and about 2000 °F). Here, the chamber door is open to expose the specimen, grips, induction heating coils, and the axial extensometer. Chambers such as this and its components are commercially available.

**Variable Temperature Fatigue Testing.** While most high-temperature fatigue testing is isothermal, most service conditions involve periods of operation at high-temperature with other periods of cooling to room temperature prior to the next mission cycle. Variable temperatures are imposed, resulting in thermal gradients and thus stress and strain gradients. As mission usage cycles are repeated, the cyclic temperature and thermal strain variation leads to what is known as thermal fatigue. For many years, high-temperature fatigue laboratory personnel have been developing techniques to measure and study thermal fatigue resistance of high temperature alloys, ceramics, polymers, and composite materials. A chapter in the planned second volume of this book is devoted to describing and understanding these methodologies. In order to simulate thermal fatigue resistance of materials, a testing technique called thermomechanical fatigue (TMF) testing has evolved. The basis of the technique is to utilize conventional high-temperature fatigue specimens and equipment to simulate the thermal fatigue process, but with zero or negligible temperature and strain gradients. This is accomplished by using independent, closed-loop servocontrol of both the mechanical component of strain (i.e., subtracting or otherwise ignoring the free thermal expansion strain) and the temperature*. In this way, one can simulate the temperature and strain phasing of any conceivable thermal fatigue cycle while still being able to measure and control the strains and temperatures. By studying a range of cyclic phases between temperature and strain, the thermal cycling or TMF fatigue resis-

**Fig. A.52** Typical axial extensometer for high-temperature fatigue testing

**Fig. A.53** Environmental chamber for fatigue testing. Source: Ref A.55

---

*Note that rates of temperature variation may be limited by the ability of the test equipment to heat or cool a specimen (while keeping temperature gradients to a minimum) as quickly as actual service thermal fatigue cycles that have large temperature gradients.

tance of a material can be established and modeled such that any service cycle could be addressed analytically and more accurate predictions made of anticipated service lives.

In the extreme, the temperature-strain phasing can be in-phase (wherein the maximum temperature corresponds to the maximum tensile strain), or out-of-phase (with the maximum temperature corresponding to the maximum compressive strain).

From a TMF testing point of view, the biggest challenge is establishing and maintaining the temperature-strain phase relation. If, for example, the temperature doesn't decrease as rapidly as programmed, yet the strain is forced to vary as if the temperature were as programmed, the phasing is upset and the basis of the test is thrown into question. A testing technique developed at NASA Glenn, and referred to as *bithermal fatigue*, completely eliminates such concerns. Simply put, the two extreme thermal fatigue cycles (in-phase or out-of-phase) are characterized by dissecting the cycle into the tensile and compressive halves, each at the isothermal temperature represented by the extreme temperatures of the thermal fatigue cycle. Thus, an in-phase bithermal cycle is conducted just as an isothermal test would be conducted in that all of the mechanical straining is done only in tension isothermally at the maximum temperature. Once the maximum strain has been reached, the force (stress) on the specimen is immediately taken to zero and held under zero load control as the temperature of the specimen is allowed to fall to the minimum thermal cycling temperature. Following a brief thermal stabilization period, the compressive straining portion of the cycle is done isothermally at this minimum temperature. Upon reaching the maximum compressive strain, the load is again forced to zero and controlled there as the temperature of the specimen is raised back to the maximum value. Note that the thermal expansion strain of the specimen ($\alpha\Delta T$) during the temperature excursion up and down takes place with no constraint and hence no thermally induced stress in the axial direction. Once the maximum temperature has been reached and stabilized for a brief period, the entire cycle is repeated. An out-of-phase bithermal cycle is performed in a similar fashion except that the direction of straining is reversed from what it was in the preceding description. The bithermal cycling test avoids many of the problems created by attempting to maintain exacting phase relations between temperature and strain during the more classical TMF cycles.

**Thermal Fatigue Testing of Simulated Hardware.** There are two commonly used techniques for performing controlled thermal cycling tests of components or simulated components. These are mentioned only briefly. Such tests do not yield results that can be considered as independent materials properties. Hence, they lack generality in their application to other geometric components and methods of heating and cooling.

*Simulated Turbine Airfoils in Gas Stream.* Simulated turbine airfoils can be fatigue tested by repeated thermal cycling between hot and cold gas streams (Ref A.60). The thin leading and trailing edges of the airfoil heat and cool much more rapidly than the bulk interior, giving rise to thermal gradients, thermal strain gradients, and thermal stress gradients that cycle between positive and negative values on each cycle. The cyclic range of stresses and strains initiate thermal fatigue cracks that can propagate to failure of the airfoil. Such tests give only a comparative rating of two different airfoil designs, or two different airfoil alloys made to the same geometric design. The differences are due to a few dominant factors: the thermal heat transfer coefficient between the gas and the airfoil (an interface characteristic), the thermal conductivity of the airfoil alloy (a materials property), the coefficient of thermal expansion of the airfoil alloy (materials property), and some inherent cyclic thermal strain resistance of the alloy (materials property).

*Simulated Component Hardware in "Fluidized Bed".* "Fluidized bed" test facilities consist of at least two adjacent containers filled with ceramic particles: one has cold particles, the other, constantly heated particles. In both cases, the particles are kept in constant agitated motion (so that they act like a fluid) by infusing large volumes of air at the base of the container that flow upward through the particles causing them to levitate. The test article is then repeatedly moved between the hot and cold beds. The thermal heat transfer coefficients are quite large for this fluidized media, giving rise to severe thermal gradients that undergo a sign reversal on each transfer from one bed to the other. Highly stylized turbine blades and even integrally bladed automotive turbine disks (blisks) have been tested in a fluidized bed facility (Ref A.61). Figure A.54 shows a stack of such blisks just after leaving a hot fluidized bed.

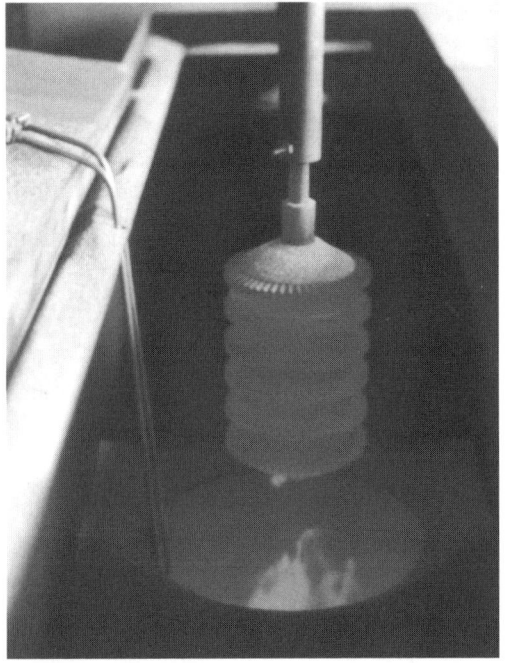

**Fig. A.54** Test components after immersion in hot fluidized bed

## Inversion Techniques

### Inversion of the Double-Term Power-Law Equation: $Y = AX^\alpha + BX^\beta$

Often during fatigue analysis we encounter equations in the form:

$$Y = AX^\alpha + BX^\beta \quad \text{(Eq A.34)}$$

where $A$, $B$, $\alpha$, and $\beta$ are known constants. When $X$ is the known variable it is easy to calculate $Y$; however, when $Y$ is known, it is much more difficult to calculate $X$. For example, a typical relation between strain range $\Delta\varepsilon$ and fatigue life $N_f$ might be, using the Manson-Hirschberg Method of Universal Slopes equation (See Chapter 3, "Fatigue Life Relations"):

$$\Delta\varepsilon = (3.5\sigma_u/E)(N_f)^{-0.12} + (N_f/D)^{-0.60} \quad \text{(Eq A.35)}$$

Thus, if $N_f$ is known, $\Delta\varepsilon$ can readily be calculated. On the other hand, if only $\Delta\varepsilon$ is known, determination of fatigue life is not quite as direct because there is no explicit way to invert this equation. We have, however, developed a formula for inverting this equation by an approximation that is quite accurate, and this approach can be used for other types of equations in this form as well. The basic approach is as follows.

It will be necessary to consider two cases: when $\alpha$ and $\beta$ are both negative as shown in Eq A.35, and when $\alpha$ and $\beta$ are both positive as, for example, we find in a stress-strain curve, Eq A.26.

**Case for Negative Exponents.** Consider first when $\alpha$ and $\beta$ are both negative. Figure A.55 shows the equation plotted graphically. Because each of the power-law terms plot linearly on logarithmic coordinates, we use logarithmic coordinates, resulting in line $AA'$ to represent the first term and line $BB'$ to represent the second term. The curve $LMN$ represents Eq A.34. It is asymptotic to $AA'$ at high values of log $X$ where $AA'$ lies above $BB'$, and asymptotic to $BB'$ at low values of log $X$ where $BB'$ lies above $AA'$. The deviation of the curve from its asymptotes is greatest in the region of point $T$ where two asymptotes intersect, i.e., where:

$$X_T = \left(\frac{B}{A}\right)^{1/(\alpha-\beta)}$$
$$Y_T = \left(\frac{B}{A^{\beta/\alpha}}\right)^{1/(1-\beta/\alpha)} \quad \text{(Eq A.36)}$$

We first attempt a solution in the form:

$$X = X_T[R^{1/\alpha} + R^{1/\beta}]$$

where $R = Y/Y_T$ by analogy to Eq A.37. On a plot such as Fig. A.55, this equation will have the same asymptotes $AA'$ and $BB'$, but its deviation from $LMN$ will be the greatest in the vicinity of $T$. Thus, just making the asymptotes correct is insufficient to ensure a good fit over the entire range of $X$.

We then try a second approximation in the form:

$$X = X_T[R^{z/\alpha} + R^{z/\beta}]^{1/z} \quad \text{(Eq A.37)}$$

This equation also has the same asymptotes $AA'$ and $BB'$ because at low value of log $X$ where $BB'$ lies considerably above $AA'$, the $R^{z/\alpha}$ term can be neglected, and the two terms in $z$ cancel each other. Similarly, of course, at high values of log $X$, the $R^{z/\beta}$ term can be neglected. However, now we have an additional adjustable constant, $z$, which we can use to force the curve of Eq A.37 to conform very closely to the curve in Fig. A.55 over the entire range.

In Ref A.62, where we treated the case for inverting the fatigue life equation, we tried sev-

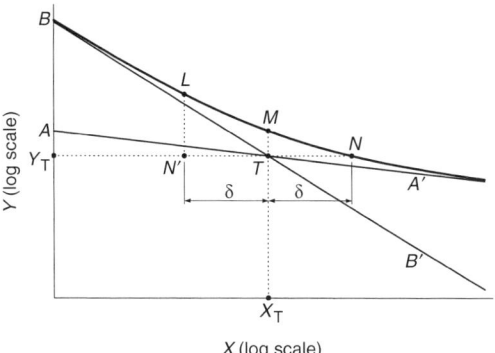

**Fig. A.55** A plot of the equation $Y = AX^\alpha + BX^\beta$ on log-log coordinates

eral approaches for the choice of $z$; the optimum procedure was to force coincidence of the two curves in three additional places:

- Where $X = X_T$ (i.e., point $M$ in Fig. A.55)
- Where $Y = Y_T$ (i.e., point $N$ in Fig. A.55)
- Where $X = X_{N'}$ such that $N'T = \delta = TN$, (i.e., point $L$, Fig. A.55)

With coincidence at these three points, and with both equations having the same asymptotes, we found that the curves of the two equations coincided almost exactly over the entire range of $X$. The equations for $z$ became*:

$$z = S \exp(P \ln^2 R + Q \ln R) \quad \text{(Eq A.38)}$$

where:

$$S = -0.889r$$
$$P = -0.001277r^2 + 0.03893r - 0.0927$$
$$Q = 0.004176r^2 - 0.135r + 0.2309$$
$$r = \beta/\alpha$$

**Case for Positive Exponents.** When the exponents are positive, the curve approaches its asymptotes from below. We could reconstruct the complete analysis for this case; however, complexity can be reduced by introducing a new variable $\phi = 1/X$. Then we arrive at an equation in $\phi$, which is in the same form as Eq A.37 having negative exponents for which the solution of Eq A.38 will apply for the argument $\phi$. Then,

substituting $X = 1/\phi$, we can determine the solutions for $X$. By interesting coincidence, the final solution for the case where $\alpha$ and $\beta$ are positive is identical to that for the case for $\alpha$ and $\beta$ negative, and Eq A.37 and A.38 still apply. However, note that the difference is that in Eq A.38 for $S$, the proper sign for $\beta$ must be accounted for.

Note that while the inverted equation according to this formulation has been found to be very close to the trial and error inversions for all practical cases encountered by the authors (this has been true for $r$ values ranging from 3 to 50), care should be exercised in extending the use of this equation for cases involving too small or too large values of $r$. Furthermore, it should be noted that the aforementioned inversion procedure is applicable only to cases where the constants $A$, $B$, and $r$ are positive.

**Other Applications of the Inversion Equation.** Table A.10 shows other cases for which equations such as Eq A.34 have been encountered in our fatigue studies. We refer to them throughout the book as appropriate. To unify the use of the same equation for all of these applications, the terms $X$, $Y$, $A$, $B$, $\alpha$, and $\beta$ are defined for each case by the terms identified in the table.

An example of the application of the inversion equation is depicted in Fig. A.56. As seen in the figure, the cyclic stress-strain curve for 7075-T6 aluminum at room temperature and the inverted relation given by Eq A.37 are essentially identical. Hence, stress is known directly in terms of applied strain.

## Hysteresis Energy: Framework for Fatigue Analysis

In this book, stress and strain are used as the framework for fatigue analysis. However, numerous attempts also have been made to use hysteresis energy as an alternate framework. The junior author prepared an extensive compilation in 1966 (Ref A.63) of the hysteresis energy expended during fatigue for a broad spectrum of metallic materials. The intent at that time was to use the results to formulate a fatigue life prediction model based on hysteresis energy. But research conducted on metallic materials by many investigators over the ensuing years have made it clear that the stress or strain framework usually was simpler to apply and entirely adequate in most cases. Still, some investigators have used hysteresis energy considerations for ana-

---

*In Ref A.62, the relation was expressed as $z = \exp(P \ln^2 R + Q \ln R + \ln S)$, which is, of course, identical to Eq A.38 because $\exp(\ln S) \equiv S$. The present Eq A.38 is more appropriate for general use because it avoids complications of sign when the equation is generalized to include the case of positive exponents.

**Table A.10** Relations of the form $Y = AX^\alpha + BX^\beta$ commonly encountered in fatigue analysis

| Double power-law relation | X | Y | A | B | α | β |
|---|---|---|---|---|---|---|
| **Strain-life curve:** | | | | | | |
| (i) Actual (obtained from fatigue tests) $\frac{\Delta\varepsilon}{2} = \frac{\sigma'_f}{E}(2N_f)^b + \varepsilon'_f(2N_f)^c$ | $2N_f$ | $\frac{\Delta\varepsilon}{2}$ | $\frac{\sigma'_f}{E}$ | $\varepsilon'_f$ | $b$ | $c$ |
| (ii) Approximate (universal slopes method) $\Delta\varepsilon = 3.5\sigma_u N_f^{-0.12} + D^{0.6} N_f^{-0.6}$ | $N_f$ | $\Delta\varepsilon$ | $3.5\sigma_u$ | $D^{0.6}$ | $-0.12$ | $-0.6$ |
| **Monotonic stress-strain curve:** | | | | | | |
| $\varepsilon = \frac{\sigma}{E} + \left(\frac{\sigma}{K}\right)^{1/n}$ | $\sigma$ | $\varepsilon$ | $\frac{1}{E}$ | $K^{-1/n}$ | 1 | $1/n$ |
| **Cyclic stress-strain curve:** | | | | | | |
| (i) Obtained from relevant test data $\varepsilon = \frac{\sigma}{E} + \left(\frac{\sigma}{K'}\right)^{1/n'}$ | $\sigma$ | $\varepsilon$ | $\frac{1}{E}$ | $(K')^{-1/n'}$ | 1 | $1/n'$ |
| (ii) Derived from strain-life relation $\varepsilon = \frac{\sigma}{E} + \varepsilon'_f\left(\frac{\sigma}{\sigma'_f}\right)^{c/b}$ | $\sigma$ | $\varepsilon$ | $\frac{1}{E}$ | $\varepsilon'_f \sigma'^{-c/b}_f$ | 1 | $c/b$ |
| (iii) "Double-amplitude" (hysteresis) curve $\Delta\varepsilon = \frac{\Delta\sigma}{E} + 2\left(\frac{\Delta\sigma}{2K'}\right)^{1/n'}$ | $\Delta\sigma$ | $\Delta\varepsilon$ | $\frac{1}{E}$ | $2(2K')^{-1/n'}$ | 1 | $1/n'$ |
| **Smith-Watson-Topper parameter:** | | | | | | |
| $\sigma_{max} E \frac{\Delta\varepsilon}{2} = \sigma'^2_f (2N_f)^{2b} + E\varepsilon'_f \sigma'_f (2N_f)^{b+c}$ | $2N_f$ | $\sigma_{max} E \frac{\Delta\varepsilon}{2}$ | $\sigma'^2_f$ | $E\varepsilon'_f \sigma'_f$ | $2b$ | $b+c$ |
| **Neuber notch-strain analysis:** | | | | | | |
| (i) For static loading $K_t^2 \sigma\varepsilon = \frac{\sigma^2_{notch}}{E} + \left(\frac{\sigma^{n+1}_{notch}}{K}\right)^{1/n}$ | $\sigma_{notch}$ | $K_t^2 \sigma\varepsilon$ | $\frac{1}{E}$ | $K^{-1/n}$ | 2 | $\frac{(n+1)}{n}$ |
| (ii) For cyclic loading $K_f^2 \Delta\sigma\Delta\varepsilon = \frac{\Delta\sigma^2_{notch}}{E} + 2\left(\frac{\Delta\sigma^{n+1}_{notch}}{2K'}\right)^{1/n'}$ | $\Delta\sigma_{notch}$ | $K_f^2 \Delta\sigma\Delta\varepsilon$ | $\frac{1}{E}$ | $2(2K')^{-1/n'}$ | 2 | $\frac{(n'+1)}{n'}$ |

lyzing their data, among them the study by Pattin (Ref 12.9) on bone (See Chapter 12, "Special Materials: Polymers, Bone, Composites, and Ceramics"). Other formulations, such as those of Smith-Watson-Topper (Ref A.64) and Walker (Ref A.65) have also been based on hysteresis energy concepts, but the final results are presented in stress-strain terms, so they are not instantly recognized as hysteresis energy based. Both models are more widely known for their ability to model mean stress effects in conjunction with predicting fatigue life of materials and structures.

Because the study of bone (See Chapter 12, "Special Materials: Polymers, Bone, Composites, and Ceramics") has involved hysteresis energy, we present some expectations when the analysis is based on this parameter as derived from metallic behavior. It gives us the opportunity to examine the properties of bone compared with those of metals.

Consider first, an axial specimen of a cyclically stable metal subjected to completely reversed fatigue loading. A stable cyclic stress-strain hysteresis loop is shown in Fig. A.57. The

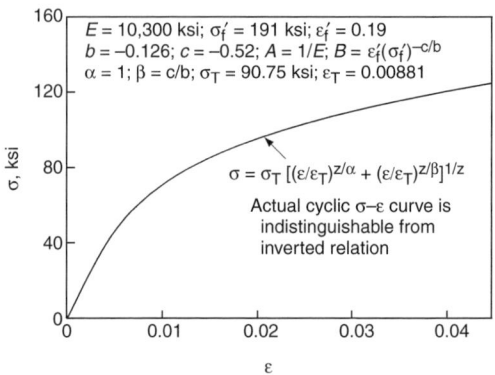

**Fig. A.56** Comparison of the cyclic stress-strain curve and inverted relation for 7075-T6 aluminum

area enclosed by the hysteresis loop represents the mechanical energy per unit volume required to generate the cycle*. The hysteresis energy per cycle can be determined analytically by integrating the area under the stress-strain path over the entire cycle. It is shown in Ref A.63 that the area of the loop is $(1 - n'/1 + n')$ times the area of the parallelogram $\Delta\sigma\Delta\varepsilon_p$ within which the hysteresis loop is contained, i.e.:

$$\Delta W = \left(\frac{1 - n'}{1 + n'}\right)\Delta\sigma\Delta\varepsilon_p \qquad \text{(Eq A.39)}$$

where $n'$ is the cyclic strain-hardening exponent.

Using the strain-life relations expressed in terms of Morrow's notation (see Chapter 3, "Fatigue Life Relations," Ref 3.9):

$$\Delta\sigma = 2\sigma'_f (2N_f)^b \qquad \text{(Eq A.40)}$$

$$\Delta\varepsilon_p = 2\varepsilon'_f (2N_f)^c \qquad \text{(Eq A.41)}$$

where $\sigma'_f$ and $\varepsilon'_f$ are the fatigue strength and fatigue ductility coefficients. Substituting Eq A.40 and A.41 into Eq A.39:

$$\Delta W = 4\left(\frac{1 - n'}{1 + n'}\right)\sigma'_f\varepsilon'_f (2N_f)^{b+c} \qquad \text{(Eq A.42)}$$

A plot of $\Delta W$ versus $N_f$ on log-log coordinates is a straight line of slope $(b + c)$ (Fig. A.58). On average for all metals, $(b + c) = -0.67$ (Ref A.63). Because :

$$W_f = 2\left(\frac{1 - n'}{1 + n'}\right)\sigma'_f\varepsilon'_f (2N_f)^{1+b+c} \qquad \text{(Eq A.43)}$$

For the example of the completely reversed fatigue loading of the metal described by Fig. 12.53, the average value of the slope of a plot of $\log(W_f)$ versus $\log(N_f) = (1 + b + c) \cong 0.33$. This is in contrast to the steeper slope of 0.60 for bone (Fig. 12.43).

Formulating damage in terms of hysteresis energy for metals has basically the same format as formulating damage in terms of elastic and plastic strain ranges. However, there may be advantages to one formulation over the other. For example, it is generally easier to measure and control strain (even plastic strain) than it is to measure and control hysteresis energy. From this standpoint, the more common strain formulation may be more practical to implement than the energy formulation. However, when hysteresis loops lie in the compressive quadrant it may be easier to conceive that damage is being done when the basis is hysteresis energy because energy is a scalar quantity and is always positive. Some correction must still be made for mean stress. It would be a distinct advantage if a hys-

---

*For stable or cyclic strain-hardening metals, the vast majority of this energy is expended as heat, and only a tiny proportion of the energy is stored in the metal in the form of vacancies, dislocations, and other atomic defects (Ref A.66). Cyclic strain-softening metals may even release initially trapped elastic strain energy from its metastable arrangement within the metal. In this case, more heat is generated than mechanical energy expended.

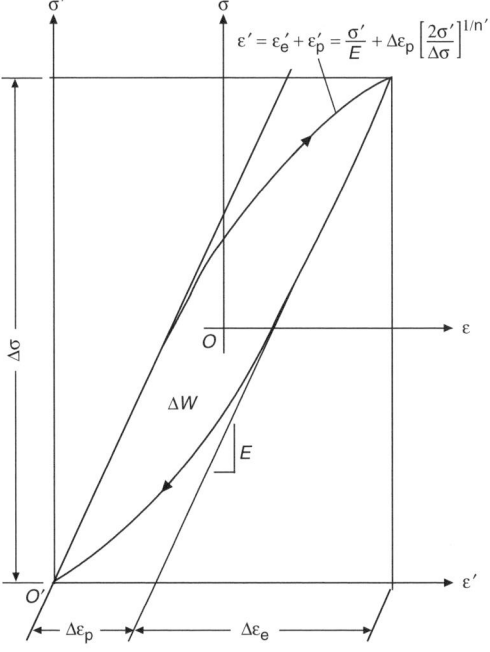

**Fig. A.57** Hysteresis loop of a stable metal in completely reversed cyclic loading. Source: Ref A.63

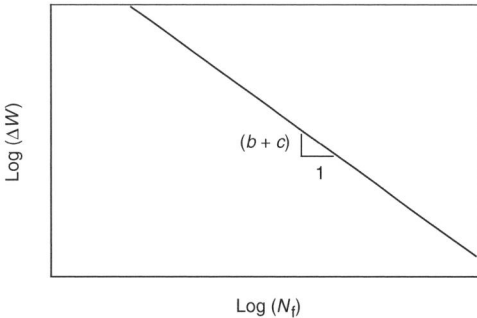

**Fig. A.58** Relationship of hysteresis loop area (energy/unit volume) to fatigue life for metals

teresis energy model automatically accounted for mean stress effects. Unfortunately, the two predominant models (Ref A.64 and A.65) have been formulated only in terms of the tensile hysteresis energy. Significant modification would be required to reformulate these models to also include compressive hysteresis. When tensile stress-strain response is different from compressive, as the material is cycled from tension to compression (as is the case with a portion of Pattin's study, Ref 12.9, and with cast iron), it may be easier to use an energy-based approach. In general, when material behavior differs from that of metals (e.g., bone) it may be more appropriate to use hysteresis energy rather than strain-range considerations.

## Brief Review of the Weibull Concept in Statistical Analysis

Readers are referred to Chapter 7, "Bending of Shafts," and in particular to Ref 7.12 and 7.13 for detailed background information on the basic concept of Weibull. Chapter 12, "Special Materials," also, briefly discusses Weibull statistical theory in relation to the scatter in the behavior of ceramics (see Ref 12.10). The basic hypothesis for all statistical failure theories is that materials contain flaws, and these flaws prevent a material from experimentally achieving its potential strength. Dispersion of fatigue test results, particularly at long lives, is due to dispersion of flaws. Shown in Fig. A.59 is the probability of failure distribution of a group of seemingly identical fatigue coupons versus alternating stress amplitude. Let $S_i$ be the probability that fatigue failure will occur before the specified number of cycles when the material is cycled at stress amplitude $\sigma_i$ for a unit element. Then, the probability that it will survive the stress $\sigma_i$, i.e., exceed the given number of cycles $N$, is $(1 - F_i)$ (this quantity is commonly known as the reliability where $F_i$ is the probability of failure). Similarly, the probability that the second unit element will survive the stress $\sigma_i$ is also $(1 - S_i)$. For the combination of the two elements to survive the stress, both must survive it. Because the probability that two events will both occur is the product of the probabilities that each will occur, it follows that:

$$1 - S_{i,2} = (1 - S_i)(1 - S_i) = (1 - S_i)^2$$
(Eq A.44)

where $S_{i,2}$ is the probability of failure of the double element. If the two elements are in parallel, and the applied load is doubled, the stress will still be $\sigma_i$ and the probability of failure of the double element will still be defined by Eq A.44. Thus, whether the elements are in series or in parallel, the same probability of failure occurs for a double volume subjected to the stress $\sigma_i$. If a volume $V$ contains $v$ such elements with each subjected to stress $\sigma_i$, then whether they are in series or parallel, the probability of failure $S_{i,v}$ is defined by:

$$\begin{aligned}1 - S_{i,v} &= (1 - S_i)_1 \ldots (1 - S_i)_v \\ &= (1 - S_i)^v = (1 - S_i)^V = (1 - S_i)^V\end{aligned}$$
(Eq A.45)

taking logarithms:

$$\log(1 - S_{i,v}) = V \log(1 - S_i)$$
(Eq A.46)

The risk of failure, $R_r$ is defined by Weibull as:

$$R_v = \log(1 - S_{i,v})$$
(Eq A.47)

From Eq A.46:

$$R_v = -V \log(1 - S_i)$$
(Eq A.48)

Consider now a body of arbitrary shape and stress distribution. Each element of volume has a definite risk of rupture that can be determined by writing Eq A.48 for this element:

$$dR_r = -dV \log(1 - S_i)$$
(Eq A.49)

where $dR_r$ is the increment of risk of rupture contributed by the volume $dV$ subjected to stress $\sigma_i$:

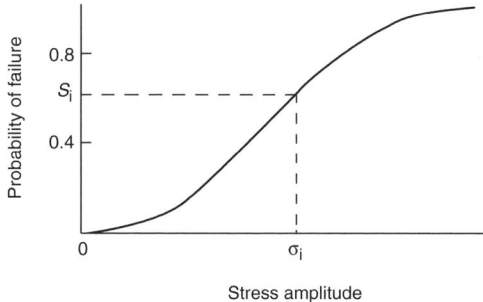

**Fig. A.59** Probability of failure as a function of stress amplitude for a failure to occur before a specified number of cycles

$$dR_r = f(\sigma_i)dV \qquad \text{(Eq A.50)}$$

where:

$$f(\sigma_i) = -\log(1 - S_i) \qquad \text{(Eq A.51)}$$

The risk of rupture for the entire body is obtained by integrating Eq A.50 over the volume of the body:

$$R_r = \int_V f(\sigma_i)dV \qquad \text{(Eq A.52)}$$

A major contribution of Weibull is his suggestion that an analytical form for $f(\sigma_i)$ is:

$$f(\sigma_i) = \left(\frac{\sigma}{\sigma'}\right)^m \qquad \text{(Eq A.53)}$$

where $\sigma'$ is a material constant relating to some inherent strength, $m$ is a constant relating to material homogeneity. The higher the value of $m$, the more homogeneous is the behavior of the material. Weibull has also suggested a more general form of $f(\sigma_i)$:

$$f(\sigma_i) = \left(\frac{\sigma - \sigma''}{\sigma'}\right)^m \qquad \text{(Eq A.54)}$$

where $\sigma''$ is a stress level below which no failures are to be expected within a prescribed lifetime. The use of Eq A.54 permits closer correlation of analytical and experimental results, but the simpler form of Eq A.53 generally is used in the analyses.

## REFERENCES

A.1 A. Nadai, *Theory of Flow and Fracture of Solids,* Vol I, McGraw-Hill, New York, 1950
A.2 C.S. Barrett, *Structure of Metals,* 2nd ed., McGraw-Hill, New York, 1952
A.3 R. Cazaud, *Fatigue of Metals,* A.J. Fenner, Chapman & Hall, Ltd., Trans., London, 1953
A.4 Z.D. Jastrzebski, *Nature and Properties of Engineering Materials,* John Wiley & Sons, New York, 1959
A.5 J.A. Pope, *Metal Fatigue,* Chapman & Hall, Ltd., London, 1959
A.6 G. Sines and J.L. Waisman, *Metal Fatigue,* McGraw-Hill, New York, 1959
A.7 C.W. Richards, *Engineering Materials Science,* Wadsworth Publishing, San Francisco, CA, 1961
A.8 P.G. Forrest, *Fatigue of Metals,* Addison-Wesley, Reading, MA, 1962
A.9 A. Nadai, *Theory of Flow and Fracture of Solids,* Vol II, McGraw-Hill, New York, 1963
A.10 F.A. McClintock and A.S. Argon, *Mechanical Behavior of Materials,* Addison-Wesley, Reading, MA, 1966
A.11 R.C. Juvinall, *Stress, Strain, and Strength,* McGraw-Hill, New York, 1967
A.12 J.Y. Mann, *Fatigue of Materials,* Melbourne University Press, Melbourne, Australia, 1967
A.13 A.F. Madayag, *Metal Fatigue, Theory and Design,* John Wiley & Sons, New York, 1969
A.14 P.J.E. Forsyth, *The Physical Basis of Metal Fatigue,* American Elsevier, New York, 1969
A.15 C.C. Osgood, *Fatigue Design,* John Wiley & Sons, New York, 1970
A.16 C.R. Barrett, W.D. Nix, and A.S. Tetelman, *The Principles of Engineering Materials,* Prentice-Hall, Inc., Englewood Cliffs, New Jersey, 1973
A.17 *ASM Handbook,* Vol 3, *Alloy Phase Diagrams,* ASM International, 1992
A.18 *ASM Handbook,* Vol 19, *Fatigue and Fracture,* ASM International, 1996
A.19 J.A. Collins, *Failure of Materials in Mechanical Design,* John Wiley & Sons, New York, 1981
A.20 W.D. Callister, *Materials Science and Engineering: An Introduction,* 3rd ed., John Wiley & Sons, Inc., New York, 1985
A.21 *Binary Alloy Phase Diagrams,* 2nd ed., T.B. Massalski et al. Ed., ASM International, 1990
A.22 G.E. Dieter, *Mechanical Metallurgy,* 3rd ed., McGraw-Hill, 1986
A.23 N.E. Dowling, *Mechanical Behavior of Materials,* 2nd ed., Prentice Hall, 1999
A.24 *ASM Handbook,* Vol 8, *Mechanical Testing and Evaluation,* ASM International, 2000
A.25 G.I. Taylor, *Proc. R. Soc.* (London), Vol A145, 1934, p 362
A.26 E. Orowan, *Z. Phys.,* Vol 89, 1934, p 605, 614, 634 (in German)
A.27 M. Polanyi, *Z. Phys.,* Vol 89, 1934, p 660 (in German)
A.28 R. de Wit, Theory of Dislocations: An Elementary Introduction, *Mechanical Be-*

havior of Crystalline Solids, NBS Monograph 59, U.S. Dept. of Commerce, March 1963, p 13–34
A.29 D. Hull and D.J. Bacon, *Introduction to Dislocations,* 3rd ed., Pergamon, 1984
A.30 W.T. Read, Jr., *Dislocations in Crystals,* McGraw-Hill, New York, 1953
A.31 A.R. Lang, Studies of Individual Dislocations in Crystals by X-Ray Diffraction Microradiography, *J. Appl. Phys.,* Vol 30, 1959, p 1748–1755
A.32 G.D. Sandrock and L. Leonard, "Cold Reduction as a Means of Reducing Embrittlement of a Cobalt-Base Alloy (L-605)," NASA TN D-3528, National Aeronautics and Space Administration, Washington, DC, 1966
A.33 *Metals Handbook,* 9th ed., Vol 3, *Properties and Selection: Stainless Steels, Tool Materials, and Special-Purpose Metals,* ASM International, 1980, p 353–417
A.34 R.W. Hertzberg, *Deformation and Fracture Mechanics of Engineering Materials,* John Wiley, 1976
A.35 S.S. Schwartz and S.H. Goodman, *Plastic Materials and Processes,* Van Nostrand Reinhold Co., New York, 1982, p 61
A.36 J.C. Freche and R.W. Sheflin, "Investigation of a Gas Turbine with National Bureau of Standards Body 4811 Ceramic Rotor Blades," *NACA Research Memorandum E8G20,* Lewis Flight Propulsion Laboratory, Cleveland, OH, National Advisory Committee for Aeronautics, Oct 28, 1948
A.37 M.P. Hanson, A.J. Meyer, Jr., and S.S. Manson, A Method of Evaluating Loose-Blade Mounting as a Means of Suppressing Turbine and Compressor Blade Vibration, *Proc., Society for Experimental Stress Analysis,* Vol 10 (No. 2), 1952
A.38 D. Munz and T. Fett, *Ceramics: Mechanical Properties, Failure Behavior, Materials Selection,* Springer, Berlin, 1999
A.39 A.F. McLean and D.L. Hartsock, An Overview of the Ceramic Design Process, *Ceramics and Glasses,* Vol 4, Engineered Materials Handbook, ASM International, 1991, p 676–689
A.40 F.P. Knudsen, Dependence of Mechanical Strength of Brittle Polycrystalline Specimens on Porosity and Grain Size, *J. Am. Ceram. Soc.,* Vol 42 (No. 8), 1959, p 376–387
A.41 D. McLean, *Mechanical Properties of Metals,* John Wiley & Sons, New York, 1962, p 199
A.42 M.H. Hirschberg, "A Low Cycle Fatigue Testing Facility," *Manual on Low Cycle Fatigue Testing,* ASTM STP 465, American Society for Testing and Materials, 1969, p 67–86
A.43 J.B. Conway, R.H. Stentz, and J.T. Berling, *Fatigue, Tensile, and Relaxation Behavior of Stainless Steels,* United States Atomic Energy Commission, Oak Ridge, TN, Jan 1975
A.44 F.R. Shanley, Chapter 9, *Strength of Materials,* McGraw-Hill, New York, 1957
A.45 Stress Phototonics, Madison, WI, http://www.stressphotonics.com
A.46 A. Mendelson and S.S. Manson, "Practical Solution of Plastic Deformation Problems in Elastic-Plastic Range," NACA TN 4088, National Advisory Committee for Aeronautics, Washington, DC, Sept 1957
A.47 *Standards of ASTM International,* West Conshohocken, PA, http://www.astm.org
A.48 G.R. Halford, The Strain Hardening Exponent: A New Interpretation and Definition, *Trans., ASM Quarterly,* American Society for Metals, Vol 56 (No. 3), 1963, p 787–788
A.49 G.R. Halford, The Energy Required for Fatigue, *J. Mater.,* American Society of Testing and Materials, Vol 1 (No. 1), 1966, p 3–18
A.50 W. Ramberg and W.R. Osgood, "Description of Stress-Strain Curves by Three Parameters," NACA TN 902, July 1943
A.51 P.W. Bridgman, *Studies in Large Plastic Flow and Fracture with Special Emphasis on the Effects of Hydrostatic Pressure,* 1st ed. (*Metallurgy and Metallurgical Engineering Series*), McGraw-Hill Book Co., New York, 1952
A.52 J. Aronofsky, *J. Appl. Mech.,* Vol 18, 1951, p 75–84
A.53 *Metals Handbook,* 8th ed., Vol 1, *Properties and Selection of Metals,* T. Lyman, Ed., American Society for Metals, 1961
A.54 W. Weibull, A Statistical Distribution of Wide Applicability, *J. Appl. Mech.,* Vol 18, 1951, p 293–297
A.55 G.R. Halford, B.A. Lerch, and M.A. McGaw, Fatigue, Creep-Fatigue, and Thermomechanical Fatigue Life Testing, *Mechanical Testing and Evaluation,* Vol 8, *ASM Handbook,* ASM International, 2000, p 686–716

A.56 W.J. Putnam and J.W. Harsch, "Rise of Temperature" Method of Determining Endurance Limit, *An Investigation of the Fatigue of Metals,* H.F. Moore and J.B. Kommers, Ed., Engineering Experiment Station Bulletin 124, University of Illinois, Oct 1921, p 119–127

A.57 D.L. Morrow and P. Kurath, Proportional Biaxial-Tension Low Cycle Fatigue of Inconel 718, *Biaxial and Multiaxial Fatigue,* EGF 3, *Mechanical Engineering Publications,* University of Illinois, Urbana-Champaign, 1989, p 551–570

A.58 R.D. Lohr, "System Design for Multiaxial High-Strain Fatigue Testing," *Multiaxial Fatigue and Deformation: Testing and Prediction,* ASTM STP 1387, S. Kalluri and P.J. Bonacuse, Ed., American Society for Testing and Materials, 2000, p 355–368

A.59 S.S. Manson, G.R. Halford, and M.H. Hirschberg, Creep-Fatigue Analysis by Strain-Range Partitioning, *Symposium on Design for Elevated Temperature Environment,* S.Y. Zamrik, Ed., American Society of Mechanical Engineers, 1971, p 12–28. See also NASA TM X-67838, 1971.

A.60 D.A. Spera, F.D. Calfo, and P.T. Bizon, "Thermal Fatigue Testing of Simulated Turbine Blades," NASA TM X-67820, 1971

A.61 P.T. Bizon and D.A. Spera, "Thermal-Stress Fatigue Behavior of Twenty-Six Superalloys," *Thermal Fatigue of Materials and Components,* ASTM STP 612, D.A. Spera and D.F. Mowbray, Ed., American Society for Testing and Materials, 1976, p 106–122

A.62 S.S. Manson and U. Muralidharan, A Single Expression Formula for Inverting Strain-Life and Stress-Strain Relationships, *Fatigue and Fracture of Engineering Materials and Structures,* Vol 9 (No. 5), p 343–356. See also NASA CR-165347, Case Western Reserve University, Cleveland, OH, May 1981.

A.63 G.R. Halford, The Energy Required for Fatigue, *J. Mater.,* Vol 1 (No. 1), 1966, p 3–18

A.64 K.N. Smith, P. Watson, and T.H. Topper, A Stress-Strain Function for the Fatigue of Metals, *J. Mater.,* Vol 5 (No. 4a), American Society for Testing and Materials, 1970, p 767–778

A.65 K. Walker, "The Effect of Stress Ratio during Crack Propagation and Fatigue for 2024-T3 and 7075-T6 Aluminum," *Effects of Environment and Complex Load History on Fatigue Life,* ASTM STP 462, American Society for Testing and Materials, 1970, p 1–14

A.66 G.R. Halford, "Stored Energy of Cold Work Changes Induced by Cyclic Deformation," Ph.D. thesis, Department of Theoretical and Applied Mechanics, University of Illinois, Urbana-Champaign, 1966

# Index

## A

**A-ratio,** 81–82
**Absolute life values,** 64
**Aircraft accidents,** 2–4(F,T)
  survey of serious accidents involving fracture fatigue, 5(T)
**Alpha-iron,** 389
**Aluminum alloys**
  bauxite smelting, 400
  heat treating process, 400–401
  wrought aluminum alloy four-digit designation system, 400(T)
**American Institute of Mining, Metallurgical, and Petroleum Engineers (AIME),** 389
**Annealed and hardened steels under cyclic straining, response of,** 13(F)
**Annealing**
  annealing twins, 384
  description of, 393
  relieve undesirable effects of working, 394–397(T)
**Annealing twins,** 384
**Antolovich, Saxena, and Chanani theory,** 261
**Atomic bonding**
  atomic stacking, 381(F)
  covalent bonding, 377
  ionic bonding, 377
  iron cores, 377
  metallic binding, 377(F)
  nonvalance electrons, 377
  valance electrons, 376, 377(F)
**Atomic stacking,** 379–382(F)
  atomic packing factor, 381
  atomic stacking (a) simple cubic crystal. (b) body-centered cubic., 381(F)
  close-packed, 381–382
**Ausformed steels,** 281(F)
**Ausforming,** 392(F)
**Austempering,** 392(F), 398
**Austenite,** 389
**Austenitic steels, typical crack growth equation,** 217
**Autofrettage,** 268, 297

## B

**Bainite,** 390
**Ball and roller bearings (testing methods, multiaxial fatigue),** 107(F)
**Ball bearing races, differential hardness effect,** 282(F)
**Ball bearings, residual stresses,** 308–309(F)
**Battelle Columbus Laboratories,** 1–2(T)
  summary on cost of fracture to economy, 2(T)
**Bend tests**
  four-point loading test, 423, 424(F)
  rotating bending testing technique, 424
  three-point loading test, 423–424(F)
**Bending fracture strength,** 351, 352(F)
**Bending of shafts,** 157–178(F)
  basic approach to treatment of low-cycle fatigue in bending, 157–158(F)
  flexural and rotating bending, 169–172(F)
  flexural bending of a circular cross section, 160–162(F)
  flexural bending of a hollow circular cross section, 166–167(F)
  flexural bending of a hollow rectangular cross section, 162–166(F)
  flexural bending of a solid rectangular cross section, 158–160(F)
  introduction, 157
  summary and illustrations for bending of rectangular and circular shafts, 167–169(F)
  volumetric effects, consideration of, 172–177(F)
**Beneficial initial loading (notch effects),** 189–194(F,T)
  loading sequences for notched specimens, 191(F)
**Biaxial stress,** 6
**Bilinear log (plastic strain range),** 273
**Binary iron-carbon system,** 389
**Bithermal fatigue,** 431
**Body-centered cubic (bcc)**
  atomic packing factor, 381
  body-centered cubic structure, 379(F)
  characteristics of seven crystal systems commonly encountered, 381(T)
  lattice structure, 378–379
**Bone**
  completely reversed fatigue loading, 347–348(F)
  completely reversed fatigue testing, 344–346(F)
  compression-only fatigue testing, 343–344(F)
  introduction, 340
  metallic materials, expected behavior under the types of loading used in tests of bone, 346–349(F)
  modeling of cyclic-softening metal, 348–349
  ratcheting in, 348
  tension-only fatigue testing, 340–343(F)

**Bone** (continued)
 typical study of fatigue behavior, 340–346(F)
 zero-to-compression fatigue loading, 347
 zero-to-tension fatigue loading, 347(F)
**Bravais lattices,** 379, 380(F), 381(T)
**Bridgman concept,** 420–421(F)
**Brinell hardness,** 422
**Brittle fracture**
 brittle manner, 7
 catastrophic brittle fracture of rocket motor during hydrotest, 204(F)
 embrittlement, 203
 examples, 201–204(F)
 Griffith's study on glass, 204–205(F)
 history of, 210
 liquid natural gas tank, 203(F)
 welded World War II ship, 202(F)
**Bubble raft analogy,** 247
 bubble raft with initial 7% vacancies, 248(F)
**Burgers vector,** 385–387(F)

# C

**Cell-size formation,** 249–250
**Cementite,** 390
**Ceramals (cermets),** 405
**Ceramic matrix composites (CMCs),** 350–351(F), 356
**Ceramics**
 advantages, 349
 aeronautics, 406
 applications, examples, 349
 atomic bonding, nature of, 350
 ball bearings, 406–407
 bending versus axial strength, 351, 352(F)
 brief summation regarding fatigue, 355
 brittleness, 350
 CARES, computer code for performing probabilistic design of ceramic components, 351
 ceramals (cermets), 405
 ceramic matrix composites (CMCs), 350–351(F)
 composition and manufacturing of engineered ceramics, 350
 crack bridging, 351
 crack initiation, 353–354(F)
 crack propagation, 354–355
 cutting tools, 406
 disadvantages, 349–350
 fiber reinforcement, 350–351(F)
 fracture strength, 352(F)
 fracture toughness, 352
 introduction, 349
 nuclear applications, 406
 other aspects of fatigue, 355
 overview, 404–406(T)
 space, 406
 strength and durability of, 351–352(F)
 thermal shock, 351–352
**Cermets,** 405
**Circular cross section**
 circular cross-section bar in flexural bending, 160(F)
 flexural bending of, 160–162(F), 163(F), 164(F)
**Circular solid and tubular specimens,** 106–107(F)
**Circular specimens**
 crack initiation for notched circular specimens, 226(F)
 crack initiation, propagation and fracture, 222–233(F,T)
 crack propagation for notched circular specimens, 227(F)
 ductile materials, study of, 222–227(F,T)
 ill-suited for tracking crack growth, 222
 quasi-brittle materials, study of, 227–230(F)
**Clamshell furnaces,** 428
**Coaxing**
 improvement in fatigue life of bearing, 311(F)
 improvement of fatigue life of 1045 steel, 310(F)
 and strain dispersal, 310–312(F)
**Cobalt-base alloy,** 401
**Coffin-Manson.** *see* Manson-Coffin
**Cold rolling,** 298(F), 304
**Compact tension specimen,** 209
 compact tension specimen for fracture testing, 211(F)
**Complex straining histories,** 15–19(F)
 analysis by rainflow counting, 17(F)
 analysis of complex straining pattern, 15(F)
 construction of hysteresis loops for complex straining, corollary of, 19
 double-amplitude template, 15
 illustration of rainflow counting, 17–18(F)
 manual analysis combining rainflow with the double-amplitude stress-strain curve, 18–19(F)
 nonpermissible rule or memory rule, 16
 single-amplitude template, 15
**Composites,** 355–374(F,T)
 applicability of monolithic metal fatigue modeling to composite materials, 372
 basic types and architectures, 357–362(F,T)
 basis for applicability of monolithic mean stress models to MMCs, 365(T), 368–371(F)
 comparison of MMC fatigue results for strain-controlled and force-controlled tests, 371–372(F)
 fatigue modeling of, 360–373(F,T)
 fatigue resistance of constituent components, 360–363
 fiber properties, 364(T)
 fiber-reinforced composites, 357(F), 358
 future trends, 372–373(F)
 inelastic matrix behavior, 370–371
 influence of fibers on matrix properties and behavior, 359(F), 363–364(F,T)
 introduction, 355–357
 linear elastic matrix behavior, 368–370(F)
 matrix properties, 363(T)
 mean stress effects in MMCs, 365–366
 mean stress effects in monolithic metals, 366–367
 mean stress modeling for MMCs, 367–368(F), 369(F)
 mechanical and thermal ply properties, 362(T)
 mechanical effects of matrix on fibers, 364–365
 mechanics of, 360, 361(F), 362(T), 363(T), 364(T)
 nomenclature for equations of Table 12.3, 363(T)
 particulate composites, 357–358
 ply layup architectures, 358–361(F)

# Index / 443

rule of mixtures, 370
*in situ* particulate composites, 357
various forms of fiber-reinforced composites, 357(F)
**Comprehensive model,** 89–102(F)
   basis for comprehensive model for treating stress effects, 90(F)
   basis of the model, 89–91(F)
   characterizing a material, 96–101(F)
   compressive mean stress enhances ductility, 93–94(F)
   compressive mean stress modeling, summary remarks, 101–102
   framework for generalized model for the mean stress effects, 95(F)
   further generalization, 94–95(F)
   illustration using the Morrow notation, 91(F)
   material behavior, some applications to, 95, 96(F)
   six tests to determine constants $Q$ and $P$, 100(T)
   tensile mean stress life reduction, reason for, 92(F)
   tensile mean stress reduces ductility, 92–93(F)
   transition fatigue life, 90
   transition strain range, 90
**Comprehensive model (Manson).** *see* Comprehensive model
**Compressive mean stress**
   comparisons between observed and calculated axial fatigue strengths, 96(F)
   implied effects, 96(F), 97(F)
   modeling, summary remarks, 101–102
**Compressive residual stresses,** 296–297(F)
   commonality of surface treatments for, 299–301(F)
**Compressive stresses,** 296
   compressive residual stresses, 296–297(F)
   stress relaxation, importance of, 296(F)
**Conventional tension test**
   common stress-strain curves, 417–419(F)
   direct determination of ultimate tensile strength from true stress-true strain parameters, 421–422
   loading sequences for notched specimens, 417(F)
   specimen types, 416–417(F)
**Crack arrest,** 221
   tensile overload ratios causing crack arrest, 222(T)
**Crack bridging,** 351
**Crack closure concept**
   application to complex loading, 220–222(F,T)
   crack growth rate retardation pattern after an overload, 221(F)
   illustration of effects of crack closure, 221(F)
   tensile overload ratios causing crack arrest, 222(T)
**Crack effects,** 6–7. *see also* Crack mechanics
**Crack growth**
   crack growth factors, 254–255
   crack-growth striations in fracture surface, 257(F)
   Laird's crack growth model, 255(F)
   realistic model, 255–256(F)
**Crack initiation and propagation, quantitative separation of,** 257–258(F)
**Crack initiation, Manson-Coffin-Basquin equations as a measure of,** 258–259(F)
**Crack initiation (stage I),** 123
   crack initiation and crack propagation phases, distinction between, 256–257(F)

defined, 223(F)
factors controlling: summary, 254
**Crack mechanics,** 201–236(F,T)
   application to fatigue crack growth, 210–212
   brittle fracture, 201–205(F)
   crack initiation, defined, 223
   crack initiation, propagation and fracture in circular specimens, 222–233(F,T)
   discussion, 233–234(F)
   fatigue specimen configuration, 233(F)
   Griffith-Irwin approach for treating, 205–210(F,T)
   introduction, 201
   mean stress-intensity effects, alternate graphical representation of, 214–218(F)
   Paris-Erdogan law, modifications to, 212–218(F,T)
   sequential loading effects, 218–222(F,T)
   tensile overload ratios causing crack arrest, 222(T)
**Crack propagation,** 110, 123
**Crack propagation (stage II)**
   crack initiation and crack propagation phases, 256–257(F)
   loading order effect explained, 123
**Crazes,** 326
**Creep**
   defined, 269
   range, 274
**Crews, John, notched plates, study of,** 187, 188(F)
**Crystal lattice,** 244
**Crystallization theory,** 237–238
**Crystallographic planes,** 237
**Crystallography**
   body-centered cubic (bcc), 378–379(F)
   body-centered cubic structure, 379(F)
   close-packed (crystallographic arrangement), 379
   crystallographic structure, relation to plastic flow, 382–383
   cubic close-packed structure, 379(F)
   face-centered cubic (fcc), 378–379(F)
   face-centered cubic structure, 379(F)
   hexagonal close-packed (hcp), 378–379(F)
   hexagonal close-packed structure, 379(F)
   lattices, 378(F)
   packing of face-centered structure, 382(F)
   packing of hexagonal structure, 382(F)
   point defects, 384–385(F)
   simple cubic structure, 378(F)
   wavy slip, 382
   yield point phenomenon, 382
**Cubic structures,** 378–383(F,T)
   body-centered cubic (bcc), 378–379, 379(F), 381, 381(T)
   Bravais lattices, 379, 380(F), 381(T)
   face-centered cubic (fcc), 379, 379(F), 381(T)
   hexagonal close-packed (hcp), 378, 379, 381, 381(T), 382(F)
   lattices, 378(F), 379(F), 381(T)
**Cumulative damage analysis**
   based on damage curves, 130–131(F)
   counting events for, 123–124(F)
   linear damage rule (LDR), limitations, 125–126(F)
   loading history to be used, 124(F)

**Cumulative fatigue damage,** 123–156(F,T)
   anomalous behavior: synergistic effects, 146–154(F,T)
   concluding remarks, 154–155
   counting events for cumulative damage analysis, 123–124(F)
   damage curves, 130–131(F), 132(F)
   introduction, 123
   linear damage rule (LDR) limitations, 125–126(F)
   loading history to be used in analysis, 124(F)
   loading-order effects, treatment of, 126–130(F)
   mean stress effects, inadvertent, 152–153(F)
   metallurgical instabilities, 153–154
   overview, 6
   potential for double linear rule, 130–146(F)
   summary of linear, nonlinear, and double linear cumulative fatigue damage concepts, 141(T)
**Cyclic hardening or softening, criteria for,** 12–14(F)
**Cyclic mean stress relaxation, overview,** 75–77(F)
**Cyclic plastic deformation,** 237–238, 264
**Cyclic plastic straining,** 13, 46
**Cyclic stress-strain curve**
   blocks of linearly increasing-decreasing straining history for an incremental step test, 21(F)
   comparison of cyclic stress-strain curves, 22(F)
   defined, 10
   incremental step cycling of cold-worked (OFHC) followed by monotonic tension, 22(F)
   methods A—E (based on various features of the curve), 19–22(F)
   typical multi-step straining history, 20(F)

# D

**Damage curve approach (DCA),** 131
**Damage curves**
   application of the double-linear damage curve (DLDC) to illustrative problem, 134–140(F)
   concept of, 131(F)
   construction of double linear damage curve, 133(F)
   cumulative damage analysis, based on, 130–131(F)
   damage curve approach (DCA), 131
   extension of the damage curve concept to double linear damage lines, 132–134(F)
   introduction, 130–131(F)
**Decorating and etching,** 244–245(F)
**Delta-iron,** 389
**Department of Defense (DoD),** 356
**Detrimental initial loading (notch effects),** 189–194(F,T)
   loading sequences for notched specimens, 191(F)
**Diffusionless (martensitic) transformations,** 409–410(F)
**Direct resistance heating,** 429
**Dislocation theory,** 7
**Dislocations**
   agglomeration of, 246–247(F), 248(F), 249(F)
   bubble raft analogy, 247
   Burgers vector, 385–387(F)
   complex dislocations, 387, 388(F)
   edge dislocations, 385–387(F)
   Frank-Read source, 388–389(F)

   multiple dislocations, generation of, 387–389(F)
   precipitation hardening, 408(F)
   screw dislocations, 387(F)
   slip planes, 247
   vacancy formation, 247(F)
**Dispersion strengthening,** 408
**Double-amplitude template,** 10, 15
**Double linear concept predicated on crack initiation and propagation**
   application of the Manson proposal, 126–127(F)
   background (example), 126–130(F)
   difficulties with, 127–130(F)
   potential for double linear rule, 130–146(F)
   two-load-level test results for three engineering alloys, 129(F)
**Double-linear damage curve (DLDC)**
   application to a complex duty cycle, 145(T)
   application to illustrative problem, 135(F)
   applications of, 134–135(F)
   kneepoints, 133
**Double-linear damage rule (DLDR)**
   application to complex history of a compressor disk of a gas-turbine engine, 145–146(T)
   application to several problems, 141–146(F,T)
   applied to block loading involving four loading levels, 144(F)
   applied to block loading involving two loading levels, 143(F)
   multiple-level tests, 143–145(F)
   other comparisons, 145(F)
   resurrection of, by changing the physical basis of the break point, 135–140(F)
   summary of, 140–141(T), 142(F)
   two-level tests, 141–143
**Ductile materials**
   crack initiation, 223(F)
   crack propagation, 223–225(F)
   Manson's hypothesized crack growth relation, 224(F)
   proposal for representing crack initiation lives, 223(F)
   realistic model for crack growth in, 255–256(F)
   study of, 222–227(F,T)
   surface strain, calculation of, 223
   test results, 225–227(F)
**Ductility**
   enhanced by compressive mean stress, 93–94(F)
   reduced by tensile mean stress, 92–93(F)

# E

**Edge dislocations,** 385–387(F)
**Elastic cycling,** 13
**Elastic range**
   classic use of stress concentration data, 181–184(F)
   effect of surface finish on endurance limit of steel, 184(F)
   Heywood's representation of fatigue notch factor, 183
   Juvinall's inclusion of surface finish effects, 183–184(F)
   Kuhn's application of the Neuber equation, 182(F)
   Neuber's fatigue correction, 181–182(F)

notch-sensitivity curves, 183(F)
Peterson's notch analysis, 182–183(F)
stress concentrations factors, determination of, 180–181
stress concentrations in, 179–181(F)
summary for determination of fatigue notch factor, 184
**Elastic strain range, extensions of the equations to include,** 48–49(F)
**Elastomers,** 404(F)
**Elber crack-closure model**
illustration of crack closure, 220(F)
schematic illustration of the results of Elber's closure model, 220(F)
sequential loading effects, 220
**Electric discharge machining (EDM),** 298
**Electron microscopy,** 241–244(F)
scanning reflection electron microscope (SEM), 242–243(F)
scanning transmission system (STEM), 242
transmission electron microscope (TEM), 242–243(F)
**Endurance limit**
anomalous behavior, 51–53(F)
defined, 238
removal of, 54(F)
traditional *S-N* curve, 45
traditional way of representing *S-N* curve of a material with a distinct endurance limit, 54(F)
**Engineering materials systems**
ceramics, 404–406(T)
high-temperature alloys (nickel-base, cobalt-base, refractory alloys), 401
polymers, 401–403(F)
stainless steel, 399–400(T)
steel, 398–399
titanium alloys, 401, 402(T)
**Engineering size crack, concept of,** 223
**Environmental chemical interaction,** 238
**Environmental effects,** 252–253(F)
**Environmental pressure,** 253(F)
**Equilibrium phase diagram,** 389–391(F)
**Equivalent rendering**
effect of fillet geometry on fatigue life of bolt, 288(F)
fabrication and processing, 287
**Eutectic, defined,** 390
**Eutectoid**
defined, 390
eutectoid composition, 391
**Extrusions, defined,** 251

# F

**Face-centered cubic (fcc)**
characteristics of seven crystal systems commonly encountered, 381(T)
face-centered cubic structure, 379(F)
lattice structure, 379
**Fail-safe-designed structures,** 269
**Fatigue**
aircraft accidents, 2–4(F,T)
Amtrak, example of use limitation by fatigue, 4

behavior, important governing factors, 105–107(F)
conventional tests (polymers), 331–337(F,T)
crack initiation, 110
crack propagation, 110
design techniques for improving fatigue resistance, 284–290(F)
development of a fatigue failure, 263–264
examples, 2–4(F,T)
fracture, cost to economy, 1–2(T)
at high temperatures, 269–270
high-versus low-cycle fatigue, 64–66(F)
introduction, 1–8(F,T)
metallic matrix, 360–361
stress and strain cycling, 9–44(F)
summary of failures of 230 components, 1(T)
treatment of as a two-stage process, 261–262
**Fatigue analysis, strain-based approach to.** *see* Strain-based approach to fatigue analysis
**Fatigue crack growth**
application to crack mechanics, 210–212
Manson-Coffin-Basquin equations as a measure of crack initiation, 258–259(F)
**Fatigue cycling,** 245
**Fatigue damage,** 267–324(F,T)
additional mechanisms to be considered at high temperature, 268–270
assembly and use, influence fatigue life, 290–291(F)
avoidance, control, and repair of, 267–324(F,T)
avoidance of strain concentration, 268
choice of materials of inherently favorable fatigue resistance, 268, 270–282(F)
coaxing and strain dispersal, 310–312(F)
cold rolling, 298(F), 304
commonality of surface treatments for compressive residual stresses, 299–304(F)
environment neutralization or exclusion, 291–292(F)
fabrication and processing, 284–290(F)
favorable residual stress, 268
fluid coatings, 293–294(F)
grain size effects, 273–278(F)
high-cycle and low-cycle fatigue, 269
interference fits, 304–305(F)
introduction, 267–270
machining process, 286–287(F)
material cleanliness factor, 272–273(F)
materials compatibility and matching, 282(F)
mechanical and thermomechanical processing, 278–281(F)
mechanical coatings, 294–295(F)
mechanical overload, 305–308(F)
minimization of mechanical damage, 292–293(F)
overview, 7
prestress coatings, 308–309(F)
proof loading, 317–318(F,T)
property conditioning and restoration, 268–269, 310
property restoration by heat treatment, 314–315(F)
residual stresses, overview, 295–297(F)
safe-life and fail-safe design, 269
safe-life and fail-safe design and surveillance, 315–317(F)

**Fatigue damage (continued)**
  shot-peening and similar surface treatments, 297–299(F)
  strain concentration effects, 282–295(F)
  surface protection, 268, 291
  surface removal, 312–313(F)
  thermal treatment, 309–310(F)
  thread rolling, 298(F), 304
**Fatigue life**
  effect of mean stress on fatigue life, 81(F)
  equations, 5
  mean stress effects on, 80–91(F)
  severity of a notch in affecting, 225
**Fatigue life relations,** 45–74(F,T)
  absolute life values, consideration of, 64
  high-versus low-cycle fatigue, 64–66(F)
  introduction, 45–46(F)
  need for new framework, 46(F), 47(F)
  predictive methods, 53–66(F,T)
  quantitative relations involving transition fatigue and strain range, 65–66(F,T)
  S-N curve, traditional, 45–46(F)
  strain-based approach to fatigue analysis, 46–53(F,T)
  strain-controlled fatigue testing, concluding remarks, 66
  transition fatigue life as a criterion for HCF, 64–65(F)
  ultra-high-cycle fatigue, 67–71(F,T)
**Fatigue process**
  illustration of fatigue stages I and II, 256(F)
  stage I and stage II, distinction between, 256–257(F)
**Fatigue property conditioning,** 269
**Fatigue testing**
  environmental test chambers, 430(F)
  fatigue testing equipment, 427–428(F)
  high-cycle fatigue (HCF), 416
  high-temperature extensometry, 429–430(F)
  high-temperature specimen grips, 429(F)
  low-cycle fatigue (LCF), 426–427
  schematic of modern fatigue testing machine, 427(F)
  specimen heating, 428–429
  thermal fatigue testing of simulated hardware, 431, 432(F)
  thermomechanical fatigue (TMF) testing, 430–431
  variable temperature fatigue testing, 430–431
**Ferritic-pearlitic steels, typical crack growth equation,** 217
**Fibers**
  fiber properties, 364(T)
  influence on matrix properties and behavior, 363–364(F)
  mechanical effects of matrix, 364–365
**Flexural and rotating bending**
  difference related to volumetric effects, 172–173(F)
  fatigue lives, comparing, 175–176(F)
  fundamental difference between, 169–172(F)
**Flexural bending**
  circular cross section, 160–162(F), 163(F), 164(F)
  circular cross-section bar in flexural bending, 160(F)
  derivation of equation for δ by integration, 162–165(F)
  hollow circular cross section, 166–167(F)
  hollow rectangular cross section, 162–166(F)
  rectangular cross-section bar in flexural bending, 159(F)
  simple approach to derivation of δ, 165–166(F)
  simple derivation of δ, 166–167(F)
  solid rectangular cross section, 158–160(F)
  verification of the equation δ by numerical integration, 166(F), 167
**Flow lines,** 279–280(F)
  chromatographs showing grain flow for two methods of forming threads, 280(F)
  control of grain flow in bearing races, 279(F)
  flow lines resulting from two methods of forming bolt heads, 279(F)
**Fluid coatings**
  beneficial effects of, 293, 294(F)
  effect of lubricant on fatigue life of threaded steel fastening, 290(F)
**Force versus stress-cycling,** 35–38(F)
  hardening materials, 37–38
  softening materials, 37(F)
  strain-range ratcheting under force cycling of cyclically strain softening 4340 steel, 37(F)
**Forman correction,** 213, 214(F)
**Fracture (due to fatigue), cost to economy,** 1–2(T)
**Fracture mechanics,** 7, 202, 316–317. *see also* Crack mechanics
**Fracture stress,** 351, 352(F)
**Frank-Read source,** 388–389(F)
**Free-machining,** 399
**Fretting**
  fretting corrosion, mechanism of, 292
  fretting corrosion, minimizing, 292–293(F)
  fretting fatigue strength of RC 130B under various conditions, 293(F)
**Functionally graded materials (FGM),** 372(F)

# G

**Geometry degradation,** 238
**Glass transition temperature,** 326, 400–401(T), 404
**Grain size**
  effect of grain size on low-cycle fatigue of aluminum alloy, 275(T)
  effects on fatigue life, 273–278(F,T)
  grain refinement, 409
**Gray cast iron,** 32–35(F)
  advantages/uses of, 32
  combined compound element behavior, 34(F)
  compound spring-slider-gap elements, 33(F)
  computer model behavior under complex straining pattern for gear cast iron, 36(F)
  computer modeled cyclic stress-strain behavior compared with experimental results for gray cast iron, 36(F)
  deformation models in tension and compression for gray iron, 34(F)
  spring-slider-gap model for gray iron with spring constants, 35(F)
**Griffith, A. A.**
  Griffith-Irwin approach for treating crack mechanics, 205–210(F,T)
  Griffith's study on glass, 204–205(F)

**Griffith-Irwin approach**
   American Society for Testing and Materials (ASTM), 209
   equal biaxial stresses at the crack tip, 207
   equations for three basic modes of fracture, 206(T)
   fracture toughness as a critical stress intensity factor, 207–209
   inadequate plain-strain fracture toughness as an explanation for numerous brittle fractures, 209
   overview, 205–206
   plain-strain fracture toughness, measurement of, 209–210(F), 211(F), 212(F,T)
   plane stress and plane strain, distinction between, 207
   stress field ahead of the crack, 206–207(F,T)
   stress-intensity factor, mode 1, 207
   stress-intensity factors for commonly encountered service conditions, 208(F)
   stress-intensity factors, some illustrations of, 207, 208(F)
   stress singularity at the crack tip, 207
   three basic modes of crack surface displacement, 206(F)
**Griffith-Irwin theory.** *see* Griffith-Irwin approach

# H

**Half-life, first known reference,** 49
**Halford-Nachtigall Modified Morrow (HNMM) model,** 32–35(F), 87–88, 367–368(F)
**Hardenability, alloying elements,** 399
**Hardening or softening**
   Cases for which hardening or softening requires special consideration, 41–42(F)
   cases involving use of large factors of safety on fatigue life, 41–42(F)
   excessive or never reaching saturation, 42
**Hardness tests,** 422–423(F)
   Brinell hardness, 422
   correlations, 423(F)
   Knoop hardness, 422
   Rockwell hardness, 422
   Scleroscope hardness, 422
   Shore durometer, 422–423
   Vickers hardness, 422
**Heat treatments**
   annealing, 393
   common, 393–394(F)
   equilibrium phase diagram, 389–391(F)
   as metallurgical tools, 389–394(F)
   normalizing, 393
   quenched and tempered, defined, 393
   quenching as the starting point, 390(F), 391–393(F)
   spheroidizing, 393
   stress relieving, 393
   typical heat treating temperatures, 393(F)
**Hexagonal close-packed (hcp)**
   characteristics of seven crystal systems commonly encountered, 381(T)
   close-packed, 381, 382(F)
   hexagonal close-packed structure, 379(F)
   lattice structure, 378
   packing of hexagonal structure, 382(F)
**High-cycle fatigue (HCF)**
   conditions for, 6
   fatigue testing, 426
   meaning of, 64, 66
   microscopic region (tiny amounts of cyclic plasticity), 269
   reason for distinguishing between LCF and HCF, 64–65(F)
   transition fatigue life as a criterion for HCF, 64–65(F)
**High cyclic life range, volumetric effect in,** 173–176(F)
**High-temperature alloys,** 401
**High-temperature extensometry,** 429–430(F)
**High-temperature specimen grips,** 429(F)
**Hole-expansion (HE),** 268, 297
**Hollow circular cross section**
   simple derivation of $\delta$, 166–167(F)
   verification of the equation $\delta$ by numerical integration, 166(F), 167
**Hollow rectangular cross section**
   derivation of equation for $\delta$ by integration, 162–165(F)
   flexural bending of, 162–166(F)
   simple approach to derivation of $\delta$, 165–166(F)
   simple derivation of $\delta$, 166–167(F)
**Hookian elastic strain,** 412–414(F)
   Lord Kelvin effect, 413
   zero gage length specimens, 413
**Hydrodynamic compression treatment,** 288–289
**Hydrogen bubbles,** 252–253
**Hypereutectoid,** 390
**Hypoeutectoid,** 390
**Hysteresis loops**
   computation of fatigue life, 16–17
   construction, 31–32
   defined, 10
   hysteresis loops for three polymers, 334(F)
   nonpermissible rule or memory rule, 16
   rules used to construct, 16

# I

**In-plane loading,** 358–361(F)
**Inelastic matrix behavior,** 370–371
**Inelastic strains, plastic and creep,** 410(F)
**Interference fits,** 304–305(F)
   Sine-lok fastener, 305, 306(F)
   taper pin interference, 305(F)
   Taperlok fastener, 304(F)
**Intrusions, defined,** 251
**Inversion techniques,** 432–433(F)
   case for negative exponents, 432–433
   case for positive exponents, 433
   inversion of the double-term power-law equation: $Y = AX^{\alpha} + BX^{\beta}$, 432
   other applications of the inversion equation, 433, 434(F,T)
**Irreversible hardening**
   concept of reversible and irreversible hardening, 148–149(F)

**Irreversible hardening (continued)**
example applied to annealed 304 stainless steel, 146–148(F)
overview, 6
simple procedure for treating, 149–152(F,T)
two-level step tests conducted for stainless steel at room temperature, 150(T)
**Irwin, G. R.,** 205. *see also* Griffith-Irwin approach
**Isothermal creep-fatigue (ICF),** 426
**Izod impact energy absorption, defined,** 228

# K

**Kneepoints,** 133–134(F), 143–144(F)
**Knoop hardness,** 422
**Kuhn's application of the Neuber equation,** 182(F)

# L

**Laird's crack growth model**
limitations of and attempts to quantify it, 260–261(F), 262(F)
overview, 255–256(F)
**Laser reflectivity,** 241(F)
**Laser shock peening (LSP),** 268, 297
**Lattices,** 378(F)
bcc (body-centered cubic) structure, 378–379(F)
Bravais lattices, 379, 380(F), 381(T)
cubic structures, 378–383(F,T)
fcc (face-centered cubic) structure, 379(F)
hcp (hexagonal close-packed) structure, 379(F)
**Lead**
adding to steel, 399
chip breaker, as a, 399
**Life reduction, insight into due to tensile mean stress,** 91–93(F)
**Linear damage rule (LDR)**
limitations, 125–126(F)
two-level test results showing inadequacy of LDR, 126(F)
**Linear elastic matrix behavior,** 368–370(F)
**Liquid (liquidus), defined,** 389
**Lord Kelvin effect,** 413
**Low-cycle fatigue (LCF)**
basic approach to treatment of low-cycle fatigue in bending, 157–158(F)
defined, 66
distinguishing between LCF and HCF, reason for, 64–65(F)
fatigue testing, 426–427
high-cycle and low-cycle fatigue, distinction between, 269
macroscopic region (many grain diameters), 269
testing, 426–427
**Low-plasticity burnishing (LPB),** 268, 297
**Lüders bands,** 417–418

# M

**Machining process,** 286–287(F)
equivalent rendering, 287, 288(F)
**Manson-Coffin-Basquin equations, as a measure of crack initiation,** 258–259(F)
**Manson-Coffin-Basquin model,** 49
Morrow's notation for use in the Manson-Coffin-Basquin model, 50(F)
**Manson-Coffin equation (or Coffin-Manson equation)**
defined, 47–48
effect of environment on Manson-Coffin relation of 1113 steel, 54(F)
**Manson-Hirschberg Method of Universal Slopes (MUS),** 332
**Marstressing technique,** 308
**Martensite,** 390
**Martensitic steels, typical crack growth equation,** 217
**Materials**
compatibility and matching, 282(F)
effect of fracture toughness, 272(F)
effect of grain size on low-cycle fatigue of aluminum alloy, 275(T)
flow lines, 279–280(F)
grain size effects, 273–278(F,T)
material cleanliness factor, 272–273(F)
optimum balance among strength, ductility and fracture toughness, 270–271(F), 272(F)
optimum hardness for maximum fatigue strength of steels, 271(F)
**Materials (comprehensive model)**
experimental programs to establish the constants for a new material, 100–101(F,T)
general, 96–97(F)
recharacterizing a material, 97–100(F)
**Matrix**
influence of fibers on properties and behavior, 363–364(F,T)
matrix properties, 363(T)
mechanical effects on fibers, 364–365
metallic matrix, 360–361
**Mean stress,** 75–104(F,T)
A-ratio, 81–82
comprehensive model (Manson), new, 89–102(F)
compressive mean stress, accounting for, 95–96(F)
compressive mean stress modeling, summary remarks on, 101–102
development in complex loading history, 77–79(F)
effect of mean stress on fatigue life, 81(F)
effects on fatigue life, 80–91(F)
force-induced, 75, 76(F)
introduction, 75
mean stresses induced by cycling between fixed strain limits, 76(F)
overview, 6
R-ratio, 82
ratcheting and cyclic mean stress relaxation, 75–77(F)
ratios describing relation of mean to alternating stress, 80–83(F)
strain-induced, 75, 76(F)

stress relaxation pattern in strain-control loading, 79–80(F)
V-ratio, 82–83(F)
**Mean stress effects**
comprehensive model (Manson), new, 89–102(F)
early mean stress theories, 84(F)
early studies of, 83–84(F)
fatigue life, 80–91(F)
framework for generalized model, 95(F)
Halford-Nachtigall modified Morrow model, 87–88
hysteresis loops implied by moving only the elastic due to mean stress, 86(F)
inadvertent, 152–153(F)
Morrow's method, 84–87(F)
under multiaxial loading, 116(F)
recent approaches, 84–89(F)
Smith-Watson-Topper (SWT), stress-strain function of, 88–89(F)
**Mean stress-intensity effects**
alternate graphical representation of, 214–218(F)
detrimental influence of aggressive environment on crack growth, 216(F)
environmental effects, 216(F), 217(F)
threshold stress-intensity factor, 214(F), 216
typical crack growth equations for common steels, 216–218(F)
**Mechanical coatings,** 294–295(F)
**Mechanical damage, minimization of,** 292–293(F)
**Mechanical failures linked to fatigue,** 1(T)
**Mechanical overload**
defined, 305
illustration of crack growth arrest by residual stress due to overload, 307(F)
single overloads, 306–307(F)
**Mechanism of fatigue,** 237–266(F)
bubble raft analogy equipment, 239
bubble raft apparatus for study of "crystal" subjected to reverse shear, 239(F)
crack growth factors, 254–262(F)
crystallization theory, 237–238
decoration and etching, 244–245(F)
early theories of, 237–238
electron microscopy, 241–244(F)
environmental effects, 252–253(F)
foreign particles, role of, 253–254(F)
heat generation detection, 238–239(F)
introduction, 237
laser reflectivity, 241(F)
optical magnification and taper sectioning, 241(F)
photoelasticity as a tool, 240–241(F)
plasticity, role of, 245–250(F)
rolling contact applications, implications regarding, 262–263(F)
slip-band geometry develops at slip band, 251(F)
surface deteriorations, 241(F), 250–252(F)
tools for observation/understanding of, 238–245(F)
treatment as a two-stage process, concluding remarks, 261–262
ultrasonics, 245, 246(F)
x-rays, 244–245(F)

**Metal fatigue**
early recognition as a problem, 5
introduction, 1–8(F,T)
**Metal matrix composites (MMCs),** 356
**Metallic matrix,** 360–361
**Metallurgical tools**
binary iron-carbon system, 389
heat treatments, 389–394(F)
plastic deformation, 394–398(T)
**Microstructural factors**
atomic packing factor, 381
crystalline structures, defects in, 384–385(F)
crystallographic structure, relation to plastic flow, 382–383
crystallography, 378–385(F,T)
line and planar defects, 385
packing atoms and stacking successive planes, 379–382(F)
twinning, alternate form of permanent deformation, 383–384(F)
**Miner's linear damage rule (LDR),** 6, 125. *see also* Linear damage rule (LDR)
**Models, mechanical, strain cycling,** 23–32(F)
**Modulus-of-rupture,** 351, 352(F)
**Moiré plastic-coating method,** 240(F)
**Monotonic loading, tensile stress-strain curve for a ductile polymer,** 326(F)
**Monotonic strain hardening,** 13–14
**Morrow notation**
alternate notation of Morrow, 50(F,T)
illustration using the Morrow notation, 91(F)
Morrow's notation for use in the Manson-Coffin-Basquin model, 50(F)
room temperature fatigue properties for selected materials, 52(T)
**Morrow's method,** 84–87(F)
**Multiaxial fatigue,** 105–122(F,T)
application of multiaxiality concepts to a large range of published information, 116–118(F,T)
body-centered cubic materials for all types of available tests, 118(F)
common theories, 111–116(F)
crack initiation and propagation under different loading, 110(F)
critical planes, 110–111(F)
crystalline structure, 105–106
discussion, 118–119(F)
face-centered cubic materials for all types of available tests, 119(F)
flat-top stress-strain curve, 109(F)
introduction, 105
multiaxiality study, 116–118(F,T)
numerical example for determining stresses when strains are specified, 109(F)
octahedral planes, 110–111(F)
overview, 6
refined cruciform specimen, 106(F)
simple cruciform specimen, 106(F)
stress-based versus strain-based theories, 107–108(F)
stress-strain relationships, 108–109

**Multiaxial fatigue** (continued)
   testing methods, 106–107(F)
   Tresca condition, 112
   triaxial loading conditions, 107
**Multiaxial fatigue theories**
   adding an elastic component to the life relation, 115(F)
   combining multiaxiality with mean-stress effects, 115–116(F)
   extending the life range for applicability of multiaxiality factor, 115(F)
   mean stress effects under multiaxial loading, 116(F)
   methods based on maximum normal stress or on shear stress considerations alone, 111–112
   methods that include both shear and normal stresses, 112–113
   multiaxiality factor, more detailed discussion, 113–115(F)
   multiaxility factor (MF), 112
   Tresca condition, 112
   triaxiality factor (TF), 112
**Multiaxial tension,** 207
**Multiaxiality**
   conditions for development of, 416(F)
   octahedral planes, 414–415(F)
   relation between stresses and strains, 415–416
**Multiaxiality study**
   summary of experimental programs, 116(T)
   types of specimens, 117(T)
**Multiaxility factor (MF),** 112, 113–115(F)

# N

**National Aeronautics and Space Administration (NASA),** 5, 356
**National Bureau of Standards.** *see* National Institute of Standards and Technology (NIST)
**National Institute of Standards and Technology (NIST),** 1
   summary of failures of 230 components, 1(T)
   summary on cost of fracture to economy, 2(T)
**Necking,** 38, 366, 412
**Neuber analysis**
   attempts to generalize, 196(F), 197(F), 197–198(T)
   attempts to improve, 194–196(F)
   corroboratory study, 195(F), 196(F), 197
   discussion of, 194
**Neuber, Heinz,** 180
**Neuber hyperbola,** 185–187(F)
   intersection of Neuber hyperbola with stress-strain curve for monotonic deformation, 186(F)
**Neuber type analysis,** 185(F)
   Neuber condition for small circular hole, 185(F)
**Neuber's fatigue correction,** 181–182(F)
   flank angle, defined, 182(F)
   technical stress concentration factor, 181–182(F)
**Neuber's hypothesis,** 185
**Nickel-base alloy,** 401
**Nominal stress,** 46
**Non-destructive inspection (NDI),** 317(T)

**Nonmetallic inclusions,** 254(F)
   fatigue crack initiation at a surface inclusion, 254(F)
**Normalized Soderberg model,** 368
**Notch, description of,** 179
**Notch effects,** 179–200(F,T)
   classic use of stress concentration data in the elastic range, 181–184(F)
   discussion and concluding remarks, 196(F), 197(T), 198–199(F)
   extension to the plastic range, 184–194(F,T)
   introduction, 179
   loading sequences for notched specimens, 191(F)
   Neuber analysis, attempts to improve, 194–196(F)
   shaft with fillet (a) bending. (b) axial load. (c) torsion., 180(F)
   stress concentrations in the elastic range, 179–181(F)

# O

**Octahedral planes,** 110–111(F)
**Optical magnification,** 241(F)
**Optical microscope,** 241
**Overload (OL),** 221
**Oxygen-free high-conductivity (OFHC),** 12, 22(F), 49(F)

# P

**Parabolic clamshell furnaces,** 428–429
**Paris crack-growth equation**
   small crack behavior, 259–260(F)
   small crack growth behavior doesn't follow long-crack growth behavior, 260(F)
   usefulness and limitations, 259(F)
**Paris-Erdogan law**
   broad spectrum of stress-intensity factor, consideration of, 212–213(F)
   fatigue crack growth beyond the regime of Paris-Erdogan law, 212(F)
   material dependence of exponent $n$, 213(F)
   mean stress-intensity effects, 213–214(F,T)
   modifications to, 212–218(F,T)
   Paris-Erdogan law constants for three structural alloys, 214(T)
   safe-life and fail-safe design and surveillance, 316
**Pearlite,** 390
**Periodic overloads,** 218–219(F)
   crack growth retardation effects of periodic overloads, 218(F)
**Periodic table of elements,** 375–377(F)
   atomic bonds, 376–377(F)
   refractory metals, 401
   valance electrons, 376
**Phase-interference method,** 240–241(F)
**Photoelasticity,** 240–241(F)
   phase-interference method to determine strain distribution at notch tip, 240(F)
   photoelastic-coating method to determine strain distribution at notch tip, 240(F)
**Plain-strain fracture toughness**

Index / 451

compact tension specimen for fracture testing, 211(F)
explanation for brittle fractures, 209
fatigue crack growth beyond the regime of Paris-Erdogan law, 204–205, 212(F)
fracture toughness values for structural engineering alloys, 212(T)
measurement of, 209–210(F,T)
practical fracture toughness specimens, 210(F)
testing, 209

**Plastic deformation**
annealing to relieve undesirable effects of working, 394–397(T)
ausforming, 392(F), 398
austempering, 398
hot working, 397
purposeful cold working to achieve strength, 397(F)
thermomechanical processing and selective programming of working and heat treating, 392(F), 397–398(F)

**Plastic flow**
anomalous behavior, 52
crystallographic structure, relation to, 382–383
fabrication and processing, 285
flexural and rotating bending, 171, 177
flexural bending, 162
high-cycle/low-cycle, distinction between, 269
low cycle fatigue in bending, 157, 158
material cleanliness factor, 271, 273
mechanical and thermomechanical processing, 278, 279
mechanical overload, 305, 307
multiaxial fatigue, 106
Peterson's notch analysis, 183
prestress coatings, 308
residual stresses, 295, 296
rolling (R), 297, 298
surface protection, 291
surface treatments, 301, 304
thermal treatment, 309–310

**Plastic range**
combined loading, beneficial and detrimental effects, 188–194(F,T)
Crews' study of notched plates, 187, 188(F)
elastic loading in region remote from notch, 185–187(F)
Neuber type analysis, 185(F)
notch effects, 184–194(F,T)
stress-strain history in completely reversed loading, 187–188, 189(F)

**Plastic strain,** 5

**Plasticity**
breakup of grains into a substructure, 247–250(F)
cell-size formation, purpose of, 249–250
dislocations, agglomeration of, 246–247(F), 248(F), 249(F)
factors that aggravate the effects of, 250
main effect of, 245
role of in the mechanics of fatigue, 245–250(F)
sketch of dislocations coalescing to form crack nucleus, 246(F)
slip planes, 247

**Plasticity relations,** 414(F)

**Plates (testing methods, multiaxial fatigue), loading to produce biaxial stress-strain states,** 107(F)

**Point defects (crystalline structure)**
impurity (foreign atom), 384–385(F)
interstitial, 384(F)
vacancy, 384(F)

**Polymer-matrix composites (PMCs),** 356

**Polymers**
acrylonitrilebutadiene-styrene (ABS), 330(F)
basic aspects of, 401–403(F)
common engineering polymers, 403(T)
completely reversed strain cycling fatigue, 333(F)
crack growth curves for polycarbonate-resin, 339(F)
crack growth striations, 338(F)
crazing and other phenomena, 326
cross linking and crystallinity, 403–404(F)
cross linking in polymers, 404(F)
cyclic loading, 328–331(F)
cyclic softening, 328–330(F)
cyclic stress-strain curves, 330–331(F)
elastomers, 404(F)
fatigue: conventional tests, 331–337(F,T)
fatigue crack growth, 337–339(F)
fatigue crack growth rates of eight polymers, 339(F)
general comparison with metals, 325–331(F)
hysteresis loops for three polymers, 334(F)
initial hysteresis loop for polycarbonate at room temperature, 328(F)
Manson-Hirschberg Method of Universal Slopes (MUS), 332
methods (3) of analysis to accelerate fatigue property estimation, 334–336(F,T)
microstructure, 325
monotonic loading, 326–328(F)
proximity of test temperature and melting point, 325–326
relationship between tensile and comprehensive yield strength, 326
softening patterns for various polymers, 329(F)
strain cycling, discussion of, 336–337(F)
strain-cycling study and opportunities for accelerated evaluation, 332–336(F,T)
temperature increases associated with testing, 326
thermoplastic, elastomeric, thermosetting, 404–405

**Pratt and Whitney Aircraft (PWA), controlled solidification,** 276(F)

**Precipitation hardening,** 407–408(F)

**Predictive methods,** 53–66(F,T)
based on limited cyclic data, 63–64(F,T)
comparison of predicted and experimental axial fatigue results using universal slopes equation and four-point correlation method, 60(F)
comparison of the two universal slopes equations for high-strength steels, 56(F)
critique of other correlations in the literature, 59–63(F)
four-point correlation method, 59(F)
predictions based only on monotonic tensile properties, 54–59(F,T)
predictive capability of the method of universal slopes equation at cryogenic and ambient temperatures, 57(F)

**Predictive methods (continued)**
 strength and ductility coefficients, 59–60(F)
 strength and ductility exponents, 60–61(F)
 transition fatigue life, 61–62(F)
 universal slopes equation, modified method, 55(T)
 universal slopes equation, original method, 54–55(F)
 universal slopes equation, special uses of, 55–59(F)
 universal slopes equations, comparison between, 55(F)
**Prestress coatings,** 308–309(F)
 marstressing technique, 308
**Prestressing, effects of,** 304
**Proof loading**
 example: heavy vessel, 317
 example: light vessel, 317–318(F,T)
 overview, 317
**Propagation period,** 225
**Property restoration by heat treatment, strain-age hardening,** 314
**Proportional limit stress,** 418

# Q

**Quasi-brittle materials**
 application to notched specimens, 231–232(F)
 application to unnotched specimens, 230–231(F)
 effect of notch toughness on life of smooth specimens, 230(F)
 fracture relation, 229(F)
 illustrative calculations for notched quasi-brittle specimen, 231(F), 232–233
 study of, 227–230(F)
**Quasi-isotropic stacking,** 358–361(F)
**Quenching**
 martensitic transformation, 392–393
 pearlite, 392
 quenched and tempered, defined, 393
 starting point in heat treatment, 390(F), 391–393(F)
 time-temperature-transformation (T-T-T) curve, 391–392(F)

# R

**R-ratio,** 82
**Rainflow counting**
 analysis by, 17(F)
 analysis of a strain history, 18(F)
 combining double-amplitude cyclic stress-strain curve with rainflow concepts to construct hysteresis loops, 17(F)
 illustration of, 17–18(F)
 rainflow cycle-counting procedure, 5
 rules, 17
**Ramberg-Osgood relation,** 38
**Rankine theory,** 111–112
**Rapid-cooling (down-shock),** 352
**Rapid-heating (up-shock),** 352
**Ratcheting and cyclic mean stress relaxation**
 force-control, 77, 78(F)
 mean strain, 77
 schematic patterns of strain ratcheting under force control, 78(F)
 strain-control, 77, 78(F)
**Refined cruciform specimen,** 106(F)
**Refractory alloys,** 401
**Refractory metals,** 401
**Residual stresses**
 ball-bearings inner rings, 308–309(F)
 overview, 295–297(F)
 prestress coatings, 308–309(F)
**Reversible and irreversible hardening, concept of,** 148–149(F)
**Reversible hardening**
 characteristic behavior in two-level test for a reversible hardening material, 149(F)
 concept of reversible and irreversible hardening, 148–149(F)
**Reversible plastic straining,** 5
**Rockwell hardness,** 422
**Rolling contact applications**
 calculation procedure, 262
 fine grain size and material cleanliness, effects on fatigue life, 263
 quantitative study to verify the importance of some of the dominant variables, 263
 subsurface location of critical stress, 262–263(F)
 volumetric effects, 263
**Rolling contact fatigue, volumetric effects,** 263
**Rolling process,** 279–280(F)
**Rolling (R),** 297–298
 effect of rolling threads before/after heat treatment on 220 ksi steel bolts, 298(F)
**Rotating bending**
 determination of bending moment and hysteresis angle in a rotating beam, 170(F)
 flexural and rotating bending, relation between, 169–172(F)
 prediction of rotating-bending fatigue from reversed strain-cycling behavior for four materials, 171(F)
**Rotating shafts, treatment of,** 6

# S

*S-N* **curves**
 anomalous behavior, 51–53, 54(F)
 traditional, 45–46(F)
 traditional way of representing *S-N* curve of a material with a distinct endurance limit, 54(F)
**Safe-life and fail-safe**
 design, 269
 design and surveillance, 315–317(F)
 philosophies, expressed in simple rules, 315
**Safe-life-designed structures,** 269
**Scanning reflection electron microscope (SEM),** 242
**Scanning transmission system (STEM),** 242–243
**Scleroscope hardness,** 422
**Screw dislocations,** 387(F)
**SEM (scanning reflection electron microscope).** *see* Scanning reflection electron microscope (SEM)

Index / 453

**Sequential loading effects**
    application of the crack closure concept to complex loading, 220–222(F,T)
    crack growth rate retardation pattern after an overload, 221(F)
    crack growth retardation effects of periodic overloads, 218(F)
    Elber crack-closure model, 220(F)
    illustration of effects of crack closure, 221(F)
    periodic overloads, effects of, 218–222(F,T)
    tensile overload ratios causing crack arrest, 222(T)
    Wheeler model, 219
    Willenborg model, 219–220
**Shear strain-based transformation,** 51
**Shot peening.** *see* Shot peening (SP)
**Shot peening (SP),** 268, 297, 303, 303(F)
**Simple cruciform specimen,** 106(F)
**Simple shear,** 106–107
**Sine-lok fastener,** 306(F)
**Single-amplitude template,** 10, 15
**Single-amplitude template curve, defined,** 10
**Slip band**
    persistent slip band, 249(F)
    slip-band geometry develops at slip band, 251(F)
**Slip planes,** 247
    crack propagation along slip lines in aluminum, 249(F)
    grain size effects, 273
    persistent slip band, 249(F)
**Smith-Watson-Topper (SWT)**
    analysis applied to fatigue results of 4340 steel, 88(F)
    parameter, 368, 369(F)
    stress-strain function of, 88–89(F)
**Society of Automotive Engineers Aerospace Materials Specification (AMS),** 401
**Solid rectangular cross section, flexural bending of,** 158–160(F)
**Solid-solution hardening,** 407(F)
**Special materials,** 325–374(F,T)
    bone, 340–348(F)
    ceramics, 349–355(F)
    composites, 355–375(F,T)
    polymers, 325–339(F,T)
**Specimen heating,** 428–429
    clamshell furnaces, 428
    direct resistance heating, 429
    parabolic clamshell furnaces, 428–429
**Stable (nonsoftening and nonhardening) metal,** 348
**Stainless steel,** 399–400
    characteristics of the families of stainless steels, 400(T)
**Steel,** 398–399
    alloying elements, 399
    carbon content, 399
    hardenability, 399
    lead, purpose for adding, 399
**STEM (scanning transmission system).** *see* Scanning transmission system (STEM)
**Strain-age hardening,** 314
**Strain aging,** 153–154
**Strain-based approach to fatigue analysis**
    alternate notation of Morrow, 50(F,T)
    anomalous behavior, 50–53(F)
    background, 46–48(F)
    basic equation, 50
    endurance limit, 51–52
    extensions of equations to include elastic strain range, 48–49(F)
    Morrow's notation for use in the Manson-Coffin-Basquin model, 50(F)
    room temperature fatigue properties for selected materials, 52(T)
    strain-based approach, 48(F)
    stress and strain reversals for a material following an idealized stress-strain curve, 47(F)
**Strain concentration effects**
    assembly and use, 290–291(F)
    design aspects, 283–284(F), 285(F), 286
    design techniques for improving fatigue resistance, 284(F)
    environment neutralization or exclusion, 291–292(F)
    fabrication and processing, 284–290(F)
    fluid coatings, 293–294(F)
    mechanical coatings, 294–295(F)
    minimization of mechanical damage, 292–293(F)
    surface protection, 291
    The Welding Institute (TWI), 290
**Strain concentration factors,** 180
**Strain concentrations,** 179, 180
**Strain controlled fatigue testing**
    constraint at notches in structural components, 66
    plastic strain governs fatigue, 66
    thermal fatigue, 66
**Strain cycling**
    alternative criteria for predicting hardening or softening, 38–40(F)
    complex pattern analyzed by mechanical models, 23–32(F)
    constant amplitude cycling of stable materials, 10–11(F)
    conventional mechanical model, 23–25(F), 26(F)
    conventional model involving springs and slider for representing hysteresis, 24(F)
    cyclic hardening materials, 11, 12(F)
    cyclic softening materials, 11–12(F), 13(F)
    cyclic stress-strain curve, 10
    cyclic stress-strain curve and strain history for a straining block, 29(F)
    cyclic stress-strain curve for hardening or softening material, concept of, 10(F), 14–15
    force-deflection diagram for three-element model, 26(F)
    hysteresis loop construction, 31–32
    hysteresis loops, defined, 10
    initial straining, 9–10(F)
    overview, 5
    pattern of response of materials during strain cycling, 12(F)
    patterns of hardening and softening for metals depending on their initial hardness, 11(F)
    reasons for using, 9
    simplified mechanical model using displacement-limited elements, 25–28(F)
    single-amplitude template, 10

**Strain cycling** (continued)
  straining and reverse straining path, shape of, 28(F)
  three-element model, 25(F)
  time-independent stress-strain material behavior, 10
  typical monotonic tensile stress-related curve, 12(F)
  Wetzel's method, 28–31(F)
**Strain hardening,** 238, 389(F), 407
**Strain ratcheting**
  defined, 366–367
  overview, 75–77(F)
  schematic patterns of strain ratcheting under force control, 78(F)
**Strengthening mechanisms**
  diffusionless (martensitic) transformations, 409–410(F)
  dispersion strengthening, 408
  grain-boundary strengthening, 409–410(F)
  precipitation hardening, 407–408(F)
  solid-solution hardening, 407(F)
  strain hardening, 389(F), 407
  summary, 410, 411(T)
  superlattice formation, 410, 411(T)
**Stress and strain cycling,** 9–44(F)
  alternative criteria for predicting hardening or softening, 38–40(F)
  cases for which hardening or softening requires special consideration, 41–42(F)
  complex pattern analyzed by mechanical models, 23–32(F)
  complex straining histories, 15–19(F)
  cyclic stress-strain curve, experimental determination of, 19–23(F)
  force versus stress-cycling, 35–38(F)
  gray cast iron, model analysis, 32–35(F), 36(F)
  introduction, 9
  strain cycling, 9–15(F)
**Stress concentrations**
  caused by abrupt change in cross-sectional area, 179(F)
  data in the elastic range, 181–184(F)
  effects, 6–7
  in the elastic range, 179–181(F)
  factors, determination of, 180–181
  reasons for higher stresses and strains in vicinity of notch, 179–181(F)
**Stress cycling versus force cycling,** 35–38(F)
**Stress effects, basis for comprehensive model for treating stress effects,** 90(F)
**Stress-intensity factor K,** 205
**Stress relaxation pattern in strain-control loading,** 79–80(F)
**Stress reversal,** 46
**Stress-strain curve**
  flat-top stress-strain curve, 109(F)
  ultimate tensile strength, 38
  yield strength, 38
**Stress-strain curves**
  analysis of, 418(F), 419–421(F)
  avalanche, defined, 417
  Bridgman concept, 420–421(F)
  common types, 417–419(F)
  defined, 416

  Lüders bands, 417–418
  proportional limit stress, 418
  two types of axial specimens for measuring Stress-strain curves, 417(F)
**Stress-strain relationships,** 108–109
**Striations (ridge/valley feature)**
  crack-growth striations in fracture surface, 257(F)
  crack initiation and propagation, 257–258
  defined, 255
  fatigue fracture surface of 7075-T6 aluminum showing striations, 256(F)
  grain size effects, 274, 275(T)
  tests in vacuum, 256
**Subgrains (dislocation cells),** 243, 247–249
**Superlattice formation,** 410, 411(T)
**Surface loading,** 303–304(F)
**Surface removal,** 312–313(F)
  engineering-oriented approach to problem of surface removal, 312(F)
  example, 313–314
**Surface treatments**
  autofrettage (AF), 298
  beneficial compressive residual stresses, 268
  commonality of for compressive residual stresses, 299–301(F)
  compensate for fatigue damage due to surface loading, 303–304(F)
  counteracting detrimental surface treatment, 302–303(F)
  general part strengthening and mitigation of stress concentrations, for, 301–302(F)
  hole expansion (HE), 298
  laser shock peening (LSP), 299, 300(F)
  low-plasticity burnishing (LPB), 298, 299(F)
  rolling (R), 297–298(F)
  shot peening (SP), 297

# T

**Taper sectioning,** 241(F), 251
**Taperlok fastener,** 304(F)
**Technical stress concentration factor,** 181–182
**TEM (transmission electron microscope).** *see* Transmission electron microscope (TEM)
**Tensile and compression rheological behavior,** 348
**Tensile mean stress,** 92–93(F)
  insight into cause of life reduction, 91–93(F)
**Tensile prestress,** 304
**Testing methods, multiaxial fatigue**
  ball and roller bearings, 107(F)
  ball or roller bearing on flat surface showing general 3-D state of stress below the surface, 108(F)
  circular solid and tubular specimens, 106–107(F)
  plate specimen, 108(F)
  plates, loading to produce biaxial stress-strain states, 107(F)
  refined cruciform specimen, 106(F)
  simple cruciform specimen, 106(F)
  thin-walled tubular specimen, 107(F)
**Tests for materials characterization,** 416–426(F,T)

bend tests, 423–424(F)
conventional tension test, 416–422(F)
hardness tests, 422–423(F)
torsion test, 424–425(F,T)
**The Welding Institute (TWI),** 290
**Thermal fatigue testing of simulated hardware**
simulated component hardware in "fluidized bed", 431, 432(F)
simulated turbine airfoils in gas stream, 431
**Thermal shock, rapid-heating (up-shock)/rapid-cooling (down-shock),** 352
**Thermal treatment,** 309–310(F)
effect of spot heating on fatigue of simulated weld, 310(F)
**Thermomechanical fatigue (TMF)**
bithermal fatigue, 431
high temperature fatigue testing, 426
testing, basis of, 430–431
**Thermomechanical processing,** 268, 278–281(F)
**Thread rolling,** 298, 304
**Time-independent stress-strain material behavior,** 10
**Titanium alloys,** 401
titanium alloys (AMS[a] specifications), 402(T)
**Torsion test,** 424–425(F,T)
state of stress in torsion, 425(F)
torsion of a solid bar, 425(F)
**Total strain,** 410
**Transition-cycle fatigue,** 66
**Transition fatigue life,** 61–62, 90
graphic relation for transition life vs. Brinell hardness for steels, 62(F)
**Transition point,** 65(F)
**Transition strain range,** 90
**Transmission electron microscope (TEM),** 242–243(F)
**Tresca condition,** 112
**Triaxial loading conditions,** 107
**Triaxial stress,** 6
**Triaxiality factor (TF),** 112
**Twinning,** 383–384(F)
annealing twins, 384
tin cry, 384

## U

**Ultra-high-cycle fatigue,** 67–71(F,T)
extrapolation, general comments on, 70–71
high-cycle-fatigue data for a range of metal and alloys, 69(F)
Langer model, 67(F)
model based on elastic life line of progressively reduced slope, 67–69(F,T)
model containing features of the Langer and the Manson-Coffin-Basquin models, 67(F)
models for treating, 67–69(F,T)
predictions, 69–70(F)
**Ultrasonic machining,** 286
**Uniaxial strain reversal of constant amplitude,** 5
**Universal slopes equations**
comparison of the two universal slopes equations for high-strength steels, 56(F)
comparisons between equations, 55(F)
cryogenic fatigue, 55–58(F)
as developed by Manson and Hirschberg, 54(F)
effects of nuclear radiation on fatigue properties, 58(F)
method of universal slopes applied to irradiated metals, 58
modified method, 55(T)
original method, 54–55(F)
predictive capability of the method of universal slopes equation at cryogenic and ambient temperatures, 57(F)
special uses of, 55–59(F)
test parameters, estimation of, 59
**Upshock/downshock (thermal shock),** 352

## V

**V-ratio,** 82–83(F)
relations between mean stress descriptors, 83(F)
**Vacuum remelting,** 273(F)
**Valance electrons,** 376, 377(F)
**Vickers hardness,** 422
**Volumetric effect**
background, 172–173(F)
extension to consideration of the low-cycle fatigue regime, 177(F)
flexural bending of a circular cross section, 175(F)
flexural bending of a rectangular cross section, 173–174(F)
rotating and flexural bending fatigue lives, comparing, 175–176(F)
rotating bending of a circular cross section, 174–175(F)
schematic for determination of nominal bending stress including the effect of volume, 175(F)
**Vulcanization (also curing or cross linking),** 404(F)

## W

**Walker parameter,** 368, 369(F)
**Wavy slip,** 382
**Weibull's statistical theory,** 6
**Welding,** 287–290(F)
favorable geometry, 288, 289(F)
improvement of the design of the corner of a welded frame, 289(F)
material improvement by heat treatment, 288
mechanical working, 288–289(F)
residual stresses, 289–290
The Welding Institute (TWI), 290
The Welding Research Council (WRC), 290
**Welding Institute (TWI), The,** 290
**Welding Research Council (WRC), The,** 290
**Wetzel's method,** 28–31(F)
**Wheeler model**
relation of crack-tip yield zones in Wheeler crack growth retardation model, 219(F)
sequential loading effects, 219(F)
**Willenborg model, sequential loading effects,** 219–220(F)

**Wöhler, August,** 45–46(F)
    nominal stress, 46(F)
    re-plot of some of Wöhler's data as *S-N* curves for smooth and notched steel specimens, 46(F)
    stress reversal, 46

Wöhler's rotating-cantilever, bending fatigue-testing machine, 45(F)

# Y

**Yield point phenomenon,** 382